FORMULAS FROM GEOMETRY

area A circumference (or perimeter) C volume V curved surface area S altitude h radius r

RIGHT TRIANGLE

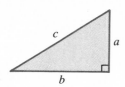

Pythagorean Theorem: $c^2 = a^2 + b^2$

TRIANGLE

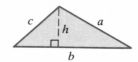

$A = \frac{1}{2}bh$ $C = a + b + c$

EQUILATERAL TRIANGLE

$h = \dfrac{\sqrt{3}}{2}s$ $A = \dfrac{\sqrt{3}}{4}s^2$

RECTANGLE

$A = lw$ $C = 2l + 2w$

PARALLELOGRAM

$A = bh$

TRAPEZOID

$A = \frac{1}{2}(a + b)h$

CIRCLE

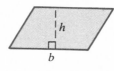

$A = \pi r^2$ $C = 2\pi r$

CIRCULAR SECTOR

$A = \frac{1}{2}r^2\theta$ $s = r\theta$

CIRCULAR RING

$A = \pi(R^2 - r^2)$

RECTANGULAR BOX

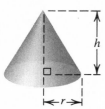

$V = lwh$ $S = 2(hl + lw + hw)$

SPHERE

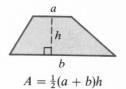

$V = \frac{4}{3}\pi r^3$ $S = 4\pi r^2$

RIGHT CIRCULAR CYLINDER

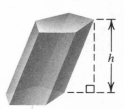

$V = \pi r^2 h$ $S = 2\pi rh$

RIGHT CIRCULAR CONE

$V = \frac{1}{3}\pi r^2 h$ $S = \pi r\sqrt{r^2 + h^2}$

FRUSTUM OF A CONE

$V = \frac{1}{3}\pi h(r^2 + rR + R^2)$

PRISM

$V = Bh$ with B the area of the base

FUNDAMENTALS OF

COLLEGE

ALGEBRA

THE PRINDLE, WEBER & SCHMIDT SERIES IN MATHEMATICS

THE PRINDLE, WEBER & SCHMIDT SERIES IN ADVANCED MATHEMATICS

DEDICATION

This Eighth Edition of *Fundamentals of College Algebra* is dedicated to the memory of Earl W. Swokowski.

I am grateful for having been given the opportunity to work with a great mentor.

A NOTE FROM THE PUBLISHER

Earl W. Swokowski died on June 2, 1992. We at PWS-KENT are grateful to have been associated with one of the premier mathematics authors of the last twenty-five years, and we will miss him.

Earl has left us a rich legacy of extraordinary work and a tradition of excellence. Jeffery A. Cole has worked with Earl over the course of several editions of this precalculus series, and we are confident that he will carry on this tradition of uncompromising quality.

EIGHTH EDITION

FUNDAMENTALS OF

COLLEGE

ALGEBRA

EARL W. SWOKOWSKI

JEFFERY A. COLE
Anoka-Ramsey Community College

PWS PUBLISHING COMPANY
Boston

PWS
Publishing Company

20 Park Plaza
Boston, Massachusetts 02116

PWS Publishing Company is a division of Wadsworth, Inc.

LIBRARY OF CONGRESS CATALOGING-IN-PUBLICATION DATA

Swokowski, Earl William.
    Fundamentals of college algebra / Earl W. Swokowski, Jeffery A.
Cole. — 8th ed.
      p.   cm.
    Includes index.
    ISBN 0-534-93204-5
    1. Algebra.   I. Cole, Jeffery A. (Jeffery Alan).
II. Title.
QA154.2.S96   1992
512.9—dc20   92-28874
                                  CIP

Chapter opening photos appear courtesy of *The Stock Market*; Ch. 1, page 2 © Tom Sanders; Ch. 2, page 52 © David Lawrence; Ch. 3, page 120 © Otto Rogge; Ch. 4, page 224 © Pete Saloutos; Ch. 5, page 304 and Ch. 7, page 446 © Ted Horowitz; Ch. 6, page 360 © Pina Oddo & Livio Sinibaldi.

Printed in the United States of America.
93 94 95 96 97 — 10 9 8 7 6 5 4 3 2

Sponsoring Editor   *Timothy L. Anderson*
Developmental Editor   *Barbara Lovenvirth*
Production Editor   *Elise S. Kaiser*
Manufacturing Coordinator   *Marcia A. Locke*
Production   *Lifland et al., Bookmakers*
Composition   *Syntax International / Monotype Composition Company, Inc.*
Technical Artwork   *Scientific Illustrators*
Interior/Cover Design   *Elise S. Kaiser*
Cover Printer   *New England Book Components*
Text Printer/Binder   *Arcata Graphics / Hawkins*
Cover Image   *Steven Hunt © 1992*

CONTENTS

3 Functions and Graphs 121

4 Polynomial Functions, Rational Functions, and Conic Sections 225

5

Exponential and Logarithmic Functions   305

6

Systems of Equations and Inequalities   361

PREFACE

One of our main goals in preparing the eighth edition of *Fundamentals of College Algebra* was to enhance the clarity of the discussions. We wanted to enable the student to more easily understand the concepts presented, but we did not want to sacrifice the mathematical soundness that has been paramount to the success of this text. By improving the explanations of the more difficult concepts and including step-by-step comments on the solutions of the examples, we have furthered our goal of helping the student while maintaining mathematical rigor.

Treatment of various topics has been updated to include the use of a graphics calculator. Graphics calculator exercises and examples are found throughout the text, wherever they are appropriate to the material being presented.

Suggestions provided by reviewers and instructors for improvements in the arrangement of topics and the writing have resulted in a great deal of reorganization for this edition. The principal changes made to each chapter are highlighted below.

CHANGES FOR THE EIGHTH EDITION

Chapter 1 Discussion of the concept of absolute value has been expanded. The topic of scientific notation has been moved forward to Section 1.1. The binomial theorem, placed in Chapter 1 in the previous edition, is now covered in Chapter 7.

Chapter 2 Application formulas are now included in Section 2.1. Discussion of the use of sign charts for solving inequalities has been expanded and improved.

Chapter 3 Graphics calculator examples and exercises first appear in Section 3.2. Section 3.4 includes a greater emphasis on linear functions. The topic of horizontally stretching or compressing graphs has been added to Section 3.5. Quadratic functions, formerly in Chapter 4, have been moved forward to Section 3.6.

Chapter 4   Added emphasis has been placed on the graphical interpretation of the relationship of a zero of a polynomial and its multiplicity. The discussion of rational functions in Section 4.5 has been extensively rewritten with a more intuitive flavor. The coverage of conic sections, previously in one section, has been expanded to three sections, 4.6, 4.7, and 4.8.

Chapter 5   Two new theorems highlight the fact that exponential and logarithmic functions are one-to-one. Greater emphasis has been placed on changing exponential forms to equivalent logarithmic forms (and vice versa) in the solving of exponential and logarithmic equations.

Chapter 6   Systems of inequalities and linear programming have been moved forward to Sections 6.5 and 6.6, respectively. The test point concept is emphasized in solving systems of inequalities.

Chapter 7   Mathematical induction now appears in Section 7.4, following the discussion of sequences. The topics of mutually exclusive events and independent events have been included in Section 7.8.

FEATURES

Illustrations   Brief demonstrations of the use of definitions, laws, and theorems are provided in the form of illustrations.

Examples   All examples, well structured and graded by difficulty, have been titled for this edition. Many examples contain graphs, charts, or tables to help the student understand procedures and solutions.

Step-by-Step Explanations   In order to help students follow them more easily, step-by-step explanations have been added to the solutions in examples.

Checks   The solutions to some examples are explicitly checked, to remind students to verify that their solutions satisfy the conditions of the problems.

Calculator Examples   Wherever appropriate, examples requiring the use of a graphics calculator have been added to the text. These are designated by a green graphics calculator icon and illustrated with a figure reproduced from a graphics calculator screen.

Calculator Exercises   Exercises specifically designed to be solved with a graphics calculator or algebraic software have been included in appropriate sections. Over 200 of these exercises, designated with a **C**, have been added to this edition.

Exercise Applications   Applications in the exercise sets are identified with titles that show students how the problems are related to practical situations.

Exercise Sets   After beginning with drill problems, exercise sets progress to more challenging problems. Many exercises containing graphs have been added.

Warnings   Interspersed throughout the text are warnings to alert students to common mistakes.

Examples, well structured and graded by difficulty, have been titled for this edition.

Examples requiring the use of a graphics calculator have been added to the text wherever appropriate. These are designated by an icon and illustrated with a figure reproduced from a graphics calculator screen.

Illustrations provide brief demonstrations of the use of definitions, laws, and theorems.

Many examples contain graphs, charts, or tables to help the student understand procedures and solutions.

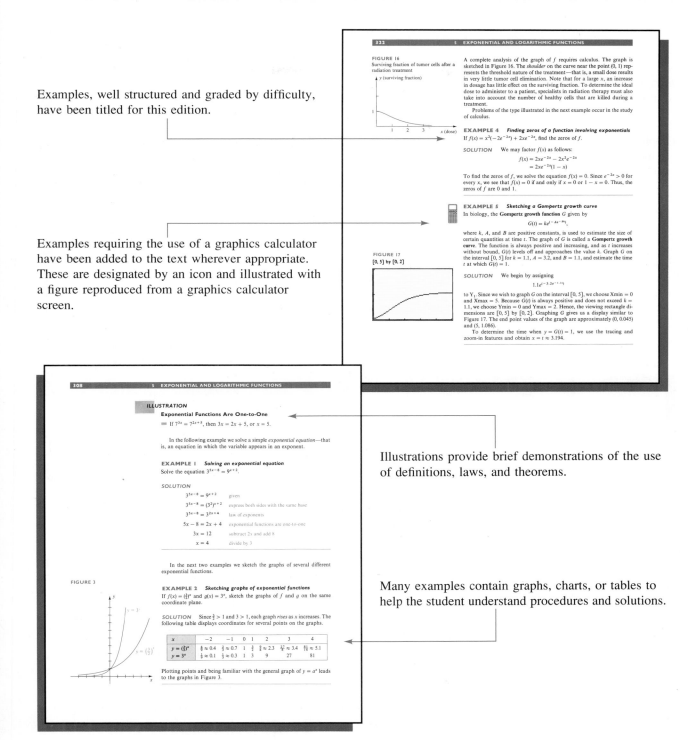

Exercise sets begin with drill problems and progress to more challenging problems.

Graphics calculator exercises, designated by a 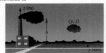, are included in many sections. Over 200 of these exercises have been added to this edition.

Exercise applications are identified to show students how the problems are related to current real-life situations.

5.3 LOGARITHMIC FUNCTIONS   125

Exer. 41–42: Graph f and g on the same coordinate plane, and estimate the solutions of the equation $f(x) = g(x)$.

41 $f(x) = e^{0.5x} - e^{-0.4x}$; $g(x) = x^2 - 2$

42 $f(x) = 0.3e^x$; $g(x) = x^3 - x$

Exer. 43–44: The functions f and g can be used to approximate e^x on the interval $[0, 1]$. Graph f, g, and $y = e^x$ on the same coordinate plane, and compare the accuracy of $f(x)$ and $g(x)$ as an approximation to e^x.

43 $f(x) = x + 1$; $g(x) = 1.72x + 1$

44 $f(x) = \frac{1}{2}x^2 + x + 1$; $g(x) = 0.84x^2 + 0.878x + 1$

Exer. 45–46: Graph f and estimate its zeros.

45 $f(x) = x^2e^x - xe^{x^2} + 0.1$

46 $f(x) = x^3e^x - x^2e^{2x} + 1$

Exer. 47–48: Graph f on the interval $(0, 200]$. Find an approximate equation for the horizontal asymptote.

47 $f(x) = \left(1 + \frac{1}{x}\right)^x$ 48 $f(x) = \left(1 + \frac{2}{x}\right)^x$

Exer. 49–50: Approximate the real root of the equation.

49 $e^{-x} = x$ 50 $e^{3x} = 5 - 2x$

Exer. 51–52: Graph f and determine where f is increasing or is decreasing.

51 $f(x) = xe^x$ 52 $f(x) = x^2e^{-2x}$

53 *Pollution from a smokestack* The concentration C (in units/m^3) of pollution near a ground-level point that is downwind from a smokestack source of height h is sometimes given by

$$C = \frac{Q}{\pi vab}\,e^{-z^2/(2a^2)}[e^{-(x-h)^2/(2b^2)} + e^{-(x+h)^2/(2b^2)}],$$

where Q is the source strength (in units/sec), v is the average wind velocity (in m/sec), z is the height (in meters) above the downwind point, y is the distance from the

downwind point in the direction that is perpendicular to the wind (the cross-wind direction), and a and b are constants that depend on the downwind distance (see the figure).

(a) How does the concentration of pollution change at the ground-level, downwind position ($y = 0$ and $z = 0$) if the height of the smokestack is increased?

(b) How does the concentration of pollution change at ground level ($z = 0$) for a smokestack of fixed height h if a person moves in the cross-wind direction by increasing y?

EXERCISE 53

54 *Pollution concentration* Refer to Exercise 53. If the smokestack height is 100 meters and $b = 12$, use a graph to estimate the height z above the downwind point ($y = 0$) where the maximum pollution concentration occurs. (*Hint:* Let $h = 100$, $b = 12$, and graph the equation $C = e^{-z^2/(2a^2)} + e^{-(z+h)^2/(2b^2)}$.)

55 *Computer chips* For manufacturers of computer chips, it is important to consider the fraction F of chips that will fail after t years of service. This fraction can be sometimes approximated by the formula $F = 1 - e^{-ct}$, where c is a positive constant.

(a) How does the value of c affect the reliability of a chip?

(b) If $c = 0.125$, after how many years will 35% of the chips have failed?

5.3 LOGARITHMIC FUNCTIONS

In Section 5.1 we observed that the exponential function given by $f(x) = a^x$ for $0 < a < 1$ or $a > 1$ is one-to-one. Hence, f has an inverse function f^{-1} (see Section 3.8). This inverse of the exponential function with base a is called the **logarithmic function with base a** and is denoted by $\log_a$. Its

---

348   5 EXPONENTIAL AND LOGARITHMIC FUNCTIONS

and proceed as follows:

$b^w = u$ given

$\log_a b^w = \log_a u$ take $\log_a$ of both sides

$w \log_a b = \log_a u$ law (3) of logarithms

$w = \dfrac{\log_a u}{\log_a b}$ divide by $\log_a b$

Since $w = \log_b u$, we obtain the formula. ■

The following special case of the change of base formula is obtained by letting $u = a$ and using the fact that $\log_a a = 1$:

$$\log_b a = \frac{1}{\log_a b}$$

The change of base formula is sometimes confused with law (2) of logarithms. The following warning could be remembered by the phrase "a quotient of logs is *not* the log of the quotient."

Warning 🖐

$$\frac{\log_a u}{\log_a b} \neq \log_a \frac{u}{b}$$

The most frequently used special cases of the change of base formula are those for $a = 10$ (common logarithms) and $a = e$ (natural logarithms), as stated in the next box.

Special Change of Base Formulas

(1) $\log_b u = \dfrac{\log u}{\log b}$ (2) $\log_b u = \dfrac{\ln u}{\ln b}$

We shall next rework Example 1 using a change of base formula.

EXAMPLE 2 *Using a change of base formula*

Solve the equation $3^x = 21$.

SOLUTION We proceed as follows:

$3^x = 21$ given

$x = \log_3 21$ change to logarithmic form

$= \dfrac{\log 21}{\log 3}$ special change of base formula (1)

Another method is to use special change of base formula (2), obtaining

$$x = \frac{\ln 21}{\ln 3}$$

Warnings are interspersed throughout the text to alert students to common mistakes.

Step-by-step explanations have been added to examples to help students more easily follow the solutions.

Text Art Each figure has been redrawn and relabeled, and graphs have been computer-generated for accuracy, using the latest technology. Additional colors are used to distinguish between different parts of figures. For example, the graph of one function may be shown in blue and that of a second function in red. Labels are the same color as the parts of the figure they identify.

Text Design The text has been completely redesigned to ensure that discussions are easy to follow and important concepts are highlighted. Color is employed pedagogically to clarify complex graphs and to help students visualize applied problems.

Answer Section The answer section at the end of the text provides answers for most of the odd-numbered exercises, as well as answers for all chapter review exercises. Considerable thought and effort were devoted to making this section a learning device for the student. For instance, proofs are given for mathematical induction problems and numerical answers for many exercises are stated in both exact and approximate form. Graphs, proofs, and hints are included wherever appropriate.

Flexibility Syllabi from schools using previous editions attest to the flexibility of this text. Sections and chapters can be arranged in many different ways, depending on objectives and the length of the course.

TEACHING TOOLS FOR THE INSTRUCTOR

Instructor's Solutions Manual Included in the *Instructor's Solutions Manual* are answers to all exercises and reasonably detailed solutions to most exercises. The manual has been thoroughly reviewed for accuracy.

EXPTest This computerized test bank for IBM and compatibles contains over 700 problems, 500 new to this edition. Questions are multiple choice, true-false, and open-ended. Instructors can interact with the program by adding to existing questions and producing individual tests.

EXPTest Sampler with Demonstration Disk The sampler demonstrates the computerized test bank's capabilities and contains sample tests.

ExamBuilder This computerized test bank for the Macintosh has features and questions similar to those of EXPTest, described above. A demo is available.

Test Bank with Chapter Tests All the questions found in EXPTest (and ExamBuilder), along with their answers, appear in print form in the *Test Bank*. Two sample tests are also included for each chapter.

Transparencies Full-color acetates provide enlarged versions of illustrations found in the text. This teaching tool is available to adopters only.

LEARNING TOOLS FOR THE STUDENT

Quick Reference Card Packaged with this edition of the text is a new tool for solving exercises—a formula card. This perforated card, found in the back of the book, will aid students in mastering the key formulas, equations, and graphs in the course. By serving as a quick reference and minimizing the need for page turning, the formula card will reduce the time spent on tedious tasks so that the student can focus on the central concepts and principles of the course.

Student's Solutions Manual (*Cole*) Solutions are given for approximately one-third of the exercises, as well as strategies for solving additional exercises. Many helpful hints and warnings are also included.

INVESTIGATE This text-specific tutorial software package for the Macintosh, IBM, and compatibles helps students review material as needed.

College Algebra: In Simplest Terms A series of videotapes, produced by the Annenberg/CPB Collection, is available to qualified adopters. This lively series shows college algebra at work in everyday situations. Symbols, charts, pictures, and state-of-the-art computer graphics illustrate basic algebraic techniques.

Precalculus in Context: Functioning in the Real World (*Davis/Moran/Murphy*) This lab manual consists of 12 projects that encourage students to explore precalculus concepts. Graphics calculators and computer graphing software are used to solve each experiment and its corresponding exercises.

College Algebra Activities for the TI-81 Graphics Calculator (*Huff/Peterson*) Designed to supplement any college algebra course, this manual offers concise instructions for using the TI-81 graphics calculator, provides examples of how to apply the instructions, and demonstrates techniques for using the calculator as a problem-solving tool. Exercises provide students with additional experience.

TrueBasic for College Algebra (*Kemeny/Kurtz*) This graphing software package is ideal for classroom demonstrations, individual study, and problem-solving.

Student Edition of Theorist This Macintosh-compatible software combines powerful algebra and graphics capabilities with an intuitive, user-friendly interface.

ACKNOWLEDGMENTS Special thanks go to Gary Rockswold, of Mankato State University, for supplying most of the fine assortment of new applied problems and calculator exercises. We also wish to thank Thomas Vanden Eynden, of Thomas More College, for doing a detailed accuracy check of the exercises; Joan Cole, for additional proofreading of the page proofs; and George Morris, of Scientific Illustrators, for creating the mathematically precise art package.

Many changes for this edition are due to the following individuals, who reviewed the manuscript. Their comments about pedagogy and their specific recommendations about the content of precalculus courses helped us to improve the book:

Norma M. Agras
Miami-Dade Community College, South Campus

Daniel D. Anderson
University of Iowa

Duane E. Deal
Ball State University

Marcy Diles
University of Arkansas at Little Rock

David E. Dobbs
University of Tennessee

John P. Edwards, Jr.
Anne Arundel Community College

Norman Eric Ellis
Essex Community College

Lucinda Gallagher
Florida State University

Albert A. Grasser
Oakland Community College

Sarita Gupta
Northern Illinois University

Keith Kuchar
Northern Illinois University

Nancy Long
Trinity Valley Community College

Carl C. Maneri
Wright State University

Eldon L. Miller
University of Mississippi

Roger B. Nelsen
Lewis and Clark College

Bonny J. Peters
St. Petersburg Junior College

John M. Plachy
Metropolitan State College

Richard Quint
Ventura College

Gerald E. Rubin
Marshall University

Jean E. Rubin
Purdue University

Fred Safier
City College of San Francisco

Shirley C. Sorenson
University of Maryland, College Park

Chris R. Siragusa
Cypress College

David Fred Snyder
Southwest Texas State University

Marvel D. Townsend
University of Florida

Jan Vandever
South Dakota State University

Arnold L. Villone
San Diego State University

Bruce Williamson
University of Wisconsin—River Falls

Mary Woods
University of Louisville

We are thankful for the excellent cooperation of the staff of PWS-KENT. Several people in the company deserve special mention. Editors David Geggis and Tim Anderson supervised the project and were continual sources of information and advice. Elise Kaiser coordinated the production of the text and is responsible for the exquisite design. Sally Lifland, Denise Throckmorton, and Gladys Moore, all of Lifland et al., Bookmakers, took exceptional care in seeing that no inconsistencies occurred and offered many helpful suggestions.

In addition to all the persons named here, we express our sincere gratitude to the many students and teachers who have helped shape our views on mathematics education. Please feel free to write to us about any aspect of this text—we value your opinion.

EARL W. SWOKOWSKI
JEFFERY A. COLE

FUNDAMENTALS OF

COLLEGE

ALGEBRA

Real numbers are essential for estimating distances, speeds of objects in motion, and magnitudes of physical quantities.

The word algebra comes from *ilm al-jabr w'al muqabala*, the title of a book written in the ninth century by the Arabian mathematician al-Khworizimi. The title has been translated as the science of restoration and reduction, which means transposing and combining similar terms (of an equation). The Latin transliteration of al-jabr led to the name of the branch of mathematics we now call algebra.

In algebra we use symbols or letters—such as *a*, *b*, *c*, *d*, *x*, *y*—to denote arbitrary numbers. This general nature of algebra is illustrated by the many formulas used in science and industry. As you proceed through this text and go on either to more advanced courses in mathematics or to fields that employ mathematics, you will become more and more aware of the importance and the power of algebraic techniques.

CHAPTER 1

FUNDAMENTAL CONCEPTS OF ALGEBRA

1.1 REAL NUMBERS

Real numbers are used throughout mathematics, and you should be acquainted with symbols that represent them, such as

$$1, \quad 73, \quad -5, \quad \tfrac{49}{12}, \quad \sqrt{2}, \quad 0, \quad \sqrt[3]{-85}, \quad 0.33333\ldots, \quad 596.25,$$

and so on. The **positive integers**, or **natural numbers**, are

$$1, \quad 2, \quad 3, \quad 4, \quad \ldots.$$

The **integers** are often listed as follows:

$$\ldots, \quad -4, \quad -3, \quad -2, \quad -1, \quad 0, \quad 1, \quad 2, \quad 3, \quad 4, \quad \ldots$$

Throughout this chapter lowercase letters $a, b, c, x, y, \ldots$ represent arbitrary real numbers. If a and b denote the same real number, we write $a = b$. An expression of this type is called an **equality**.

If a, b, and c are integers and $c = ab$, then a and b are **factors**, or **divisors**, of c. For example, since

$$6 = 2 \cdot 3 = (-2)(-3) = 1 \cdot 6 = (-1)(-6),$$

we know that $1, -1, 2, -2, 3, -3, 6,$ and -6 are factors of 6.

A positive integer p different from 1 is **prime** if its only positive factors are 1 and p. The first few primes are 2, 3, 5, 7, 11, 13, 17, and 19. The **Fundamental Theorem of Arithmetic** states that every positive integer different from 1 can be expressed as a product of primes in one and only one way (except for order of factors). Some examples are

$$12 = 2 \cdot 2 \cdot 3, \qquad 126 = 2 \cdot 3 \cdot 3 \cdot 7,$$
$$540 = 2 \cdot 2 \cdot 3 \cdot 3 \cdot 3 \cdot 5.$$

A **rational number** is a real number that can be expressed in the form a/b, where a and b are integers and $b \neq 0$. Real numbers that are not rational are **irrational numbers**. One common irrational number is the ratio of the circumference of a circle to its diameter, denoted by π. We sometimes use the notation $\pi \approx 3.1416$ to indicate that π is *approximately equal to* 3.1416.

There is no rational number b such that $b^2 = 2$, where b^2 denotes $b \cdot b$. However, there is an irrational number, denoted by $\sqrt{2}$ (the **square root** of 2), such that $(\sqrt{2})^2 = 2$.

The system of **real numbers** consists of all rational and irrational numbers. Note that every integer a is a rational number, since it can be expressed in the form $a/1$. The diagram in Figure 1 indicates relationships among the types of numbers used in algebra, where a line connecting two rectangles means that the numbers named in the higher rectangle include those in the lower rectangle. The complex numbers, discussed in Chapter 2, contain all real numbers.

FIGURE 1   Types of numbers used in algebra

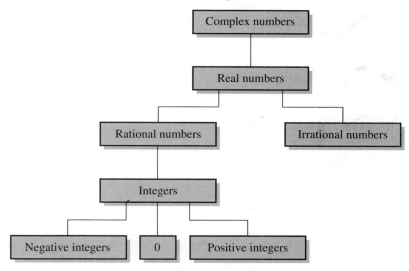

Every real number can be expressed as a decimal. Decimal representations for rational numbers are either terminating or nonterminating and repeating. For example, we can show by using the arithmetic process of long division that

$$\tfrac{5}{4} = 1.25 \qquad \text{and} \qquad \tfrac{177}{55} = 3.2181818\ldots,$$

where the digits 1 and 8 in the representation of $\tfrac{177}{55}$ repeat indefinitely (sometimes written $3.2\overline{18}$). Decimal representations for irrational numbers are always nonterminating and nonrepeating.

The real numbers are **closed relative to the operation of addition** (denoted by $+$); that is, to every pair a, b of real numbers there corresponds exactly one real number $a + b$ called the **sum** of a and b. The real numbers are also **closed relative to multiplication** (denoted by $\cdot$); that is, to every pair a, b of real numbers there corresponds exactly one real number $a \cdot b$ (also denoted by ab) called the **product** of a and b.

Important properties of addition and multiplication are listed in the following chart.

Properties of Real Numbers

Terminology	General case	Meaning
Addition is **commutative**.	$a + b = b + a$	Order is immaterial when adding two numbers.
Addition is **associative**.	$a + (b + c) = (a + b) + c$	Grouping is immaterial when adding three numbers.
0 is the **additive identity**.	$a + 0 = a$	Adding 0 to any number yields the same number.
$-a$ is the **additive inverse**, or negative, of a.	$a + (-a) = 0$	Adding a number and its negative yields 0.
Multiplication is **commutative**.	$ab = ba$	Order is immaterial when multiplying two numbers.
Multiplication is **associative**.	$a(bc) = (ab)c$	Grouping is immaterial when multiplying three numbers.
1 is the **multiplicative identity**.	$a \cdot 1 = a$	Multiplying any number by 1 yields the same number.
If $a \neq 0$, $\dfrac{1}{a}$ is the **multiplicative inverse**, or **reciprocal**, of a.	$a\left(\dfrac{1}{a}\right) = 1$	Multiplying a nonzero number by its reciprocal yields 1.
Multiplication is **distributive** over addition.	$a(b + c) = ab + ac$ and $(a + b)c = ac + bc$	Multiplying a number and a sum of two numbers is equivalent to multiplying each of the two numbers by the number and then adding the products.

Since $a + (b + c)$ and $(a + b) + c$ are always equal, we may use $a + b + c$ to denote this real number. We use abc for either $a(bc)$ or $(ab)c$. Similarly, if four or more real numbers a, b, c, d are added or multiplied, we may write $a + b + c + d$ for their sum and $abcd$ for their product, regardless of how the numbers are grouped or interchanged.

The distributive properties are useful for finding products of many types of expressions involving sums. The next example provides one illustration.

EXAMPLE I   *Using distributive properties*

If p, q, r, and s denote real numbers, show that

$$(p + q)(r + s) = pr + ps + qr + qs.$$

SOLUTION   We use both distributive properties:

$$
\begin{aligned}
(p + q)(r + s) & \\
= p(r + s) + q(r + s) \quad & \text{second distributive property with } c = r + s \\
= (pr + ps) + (qr + qs) \quad & \text{first distributive property} \\
= pr + ps + qr + qs \quad & \text{remove parentheses}
\end{aligned}
$$

The following are basic properties of equality.

Properties of Equality

> If $a = b$ and c is any real number, then
>
> **(1)** $a + c = b + c$
>
> **(2)** $ac = bc$

Properties (1) and (2) state that the same number may be added to both sides of an equality, and both sides of an equality may be multiplied by the same number. We will use these properties extensively in Chapter 2 to help find solutions of equations.

The next result can be proved.

Products Involving Zero

> **(1)** $a \cdot 0 = 0$ for every real number a.
>
> **(2)** If $ab = 0$, then either $a = 0$ or $b = 0$.

When we use the word *or* as we do in (2), we mean that at *least* one of a or b is 0.

Some properties of negatives are listed in the following chart.

Properties of Negatives

Property	Illustration
$-(-a) = a$	$-(-3) = 3$
$(-a)b = -(ab) = a(-b)$	$(-2)3 = -(2 \cdot 3) = 2(-3)$
$(-a)(-b) = ab$	$(-2)(-3) = 2 \cdot 3$
$(-1)a = -a$	$(-1)3 = -3$

The reciprocal $\dfrac{1}{a}$ of a nonzero real number a is often denoted by a^{-1}, as in the next chart.

Notation for Reciprocals

Definition	Illustration
If $a \neq 0$, then $a^{-1} = \dfrac{1}{a}$.	$2^{-1} = \dfrac{1}{2}$ $\left(\dfrac{3}{4}\right)^{-1} = \dfrac{1}{(3/4)} = \dfrac{4}{3}$

Note that if $a \neq 0$, then

$$a \cdot a^{-1} = a\left(\frac{1}{a}\right) = 1.$$

The operations of **subtraction** $(-)$ and **division** $(\div)$ are defined as follows.

Subtraction and Division

Definition	Meaning	Illustration
$a - b = a + (-b)$	To subtract one number from another, add the negative.	$3 - 7 = 3 + (-7)$
$a \div b = a \cdot \left(\dfrac{1}{b}\right)$ $= a \cdot b^{-1}; \; b \neq 0$	To divide one number by a nonzero number, multiply by the reciprocal.	$3 \div 7 = 3 \cdot \left(\dfrac{1}{7}\right)$ $= 3 \cdot 7^{-1}$

We use either a/b or $\dfrac{a}{b}$ for $a \div b$ and refer to a/b as the **quotient of a and b** or the **fraction a over b**. The numbers a and b are the **numerator** and **denominator**, respectively, of a/b. Since 0 has no multiplicative inverse, a/b is not defined if $b = 0$; that is, *division by zero is not defined*. Note that

$$1 \div b = \frac{1}{b} = b^{-1} \quad \text{if } b \neq 0.$$

The following properties of quotients are true, provided all denominators are nonzero real numbers.

Properties of Quotients

Property	Illustration
(1) $\dfrac{a}{b} = \dfrac{c}{d}$ if $ad = bc$	$\dfrac{2}{5} = \dfrac{6}{15}$ because $2 \cdot 15 = 5 \cdot 6$
(2) $\dfrac{ad}{bd} = \dfrac{a}{b}$	$\dfrac{2 \cdot 3}{5 \cdot 3} = \dfrac{2}{5}$
(3) $\dfrac{a}{-b} = \dfrac{-a}{b} = -\dfrac{a}{b}$	$\dfrac{2}{-5} = \dfrac{-2}{5} = -\dfrac{2}{5}$
(4) $\dfrac{a}{b} + \dfrac{c}{b} = \dfrac{a + c}{b}$	$\dfrac{2}{5} + \dfrac{9}{5} = \dfrac{2 + 9}{5} = \dfrac{11}{5}$
(5) $\dfrac{a}{b} + \dfrac{c}{d} = \dfrac{ad + bc}{bd}$	$\dfrac{2}{5} + \dfrac{4}{3} = \dfrac{2 \cdot 3 + 5 \cdot 4}{5 \cdot 3} = \dfrac{26}{15}$
(6) $\dfrac{a}{b} \cdot \dfrac{c}{d} = \dfrac{ac}{bd}$	$\dfrac{2}{5} \cdot \dfrac{7}{3} = \dfrac{2 \cdot 7}{5 \cdot 3} = \dfrac{14}{15}$
(7) $\dfrac{a}{b} \div \dfrac{c}{d} = \dfrac{a}{b} \cdot \dfrac{d}{c} = \dfrac{ad}{bc}$	$\dfrac{2}{5} \div \dfrac{7}{3} = \dfrac{2}{5} \cdot \dfrac{3}{7} = \dfrac{6}{35}$

Real numbers may be represented by points on a line l such that to each real number a there corresponds exactly one point on l and to each point P on l there corresponds one real number. This is called a **one-to-one correspondence**. We first choose an arbitrary point O, called the **origin**,

FIGURE 2

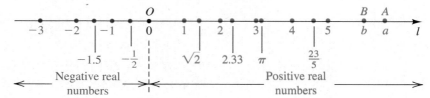

and associate with it the real number 0. Points associated with the integers are then determined by laying off successive line segments of equal length on either side of O, as illustrated in Figure 2. The point corresponding to a rational number, such as $\frac{23}{5}$, is obtained by subdividing these line segments. Points associated with certain irrational numbers, such as $\sqrt{2}$, can be found by construction (see Exercise 41).

The number a that is associated with a point A on l is the **coordinate** of A. We refer to these coordinates as a **coordinate system** and call l a **coordinate line**, or a **real line**. A direction can be assigned to l by taking the **positive direction** to the right and the **negative direction** to the left. The positive direction is noted by placing an arrowhead on l, as shown in Figure 2.

The numbers that correspond to points to the right of O in Figure 2 are **positive real numbers**. Numbers that correspond to points to the left of O are **negative real numbers**. *The real number 0 is neither positive nor negative.*

Note the difference between a negative real number and the *negative of* a real number. In particular, the negative of a real number a can be positive. For example, if a is negative, say $a = -3$, then the negative of a is $-a = -(-3) = 3$, which is positive. In general, we have the following relationships.

Relationships Between
a and $-a$

> **(1)** If a is positive, then $-a$ is negative.
>
> **(2)** If a is negative, then $-a$ is positive.

In the following chart we define the notions of **greater than** and **less than** for real numbers a and b. The symbols $>$ and $<$ are **inequality signs**, and the expressions $a > b$ and $a < b$ are called **inequalities**.

Greater Than or Less Than

Notation	Definition	Terminology
$a > b$	$a - b$ is positive	a is greater than b
$a < b$	$a - b$ is negative	a is less than b

Referring to a coordinate line, if points A and B have coordinates a and b, respectively, then $a > b$ is equivalent to the statement "A is to the right of B," whereas $a < b$ is equivalent to "A is to the left of B."

ILLUSTRATION

Greater Than (>) and Less Than (<)

- $5 > 3$, since $5 - 3 = 2$ is positive.
- $-6 < -2$, since $-6 - (-2) = -6 + 2 = -4$ is negative.
- $\frac{1}{3} > 0.33$, since $\frac{1}{3} - 0.33 = \frac{1}{3} - \frac{33}{100} = \frac{1}{300}$ is positive.
- $7 > 0$, since $7 - 0 = 7$ is positive.
- $-4 < 0$, since $-4 - 0 = -4$ is negative.

The next law enables us to compare, or *order*, any two real numbers.

Trichotomy Law

> If a and b are real numbers, then exactly one of the following is true:
> $$a = b, \qquad a > b, \qquad \text{or} \qquad a < b$$

We refer to the **sign** of a real number as positive or negative if the number is positive or negative, respectively. Two real numbers *have the same sign* if both are positive or both are negative. The numbers have *opposite signs* if one is positive and the other is negative. The following results about the signs of products and quotients of two real numbers a and b can be proved using properties of negatives and quotients.

Laws of Signs

> **(1)** If a and b have the same sign, then ab and $\dfrac{a}{b}$ are positive.
>
> **(2)** If a and b have opposite signs, then ab and $\dfrac{a}{b}$ are negative.

*If a theorem is written in the form "if P, then Q," where P and Q are mathematical statements called the *hypothesis* and *conclusion*, respectively, then the *converse* of the theorem has the form "if Q, then P." If both the theorem and its converse are true, we often write "P if and only if Q."

The **converses*** of the laws of signs are also true. For example, if a quotient is negative, then the numerator and denominator have opposite signs.

The notation $a \geq b$ (*a is greater than or equal to b*) means that either $a > b$ or $a = b$ (but not both). For example, $a^2 \geq 0$ for every real number a. The symbol $a \leq b$ (*a is less than or equal to b*) means that either $a < b$ or $a = b$. The expression $a < b < c$ means that both $a < b$ and $b < c$, and we say that *b is between a and c*. Similarly, the expression $c > b > a$ means that both $c > b$ and $b > a$.

ILLUSTRATION

Ordering Three Real Numbers

- $1 < 5 < \frac{11}{2}$
- $-4 < \frac{2}{3} < \sqrt{2}$
- $3 > -6 > -10$

FIGURE 3

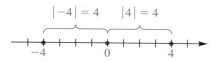

$|-4| = 4 \qquad |4| = 4$

There are other types of inequalities. For example, $a < b \le c$ means both $a < b$ and $b \le c$. Similarly, $a \le b < c$ means both $a \le b$ and $b < c$. Finally, $a \le b \le c$ means both $a \le b$ and $b \le c$.

If a is an integer, then it is the coordinate of some point A on a coordinate line, and the symbol $|a|$ denotes the number of units between A and the origin, without regard to direction. The nonnegative number $|a|$ is called the *absolute value of a.* Referring to Figure 3, we see that for the point with coordinate -4 we have $|-4| = 4$. Similarly, $|4| = 4$. In general, *if a is negative, we change its sign to find $|a|$; if a is nonnegative, then $|a| = a$.* The next definition extends this concept to every real number.

Definition of Absolute Value

The **absolute value** of a real number a, denoted by $|a|$, is defined as follows.

(1) If $a \ge 0$, then $|a| = a$.
(2) If $a < 0$, then $|a| = -a$.

Some special cases of this definition are given in the following illustration.

ILLUSTRATION

The Absolute Value Notation $|a|$

- $|3| = 3$, since $3 > 0$.
- $|-3| = -(-3)$, since $-3 < 0$. Thus, $|-3| = 3$.
- $|2 - \sqrt{2}| = 2 - \sqrt{2}$, since $2 - \sqrt{2} > 0$.
- $|\sqrt{2} - 2| = -(\sqrt{2} - 2)$, since $\sqrt{2} - 2 < 0$. Thus, $|\sqrt{2} - 2| = 2 - \sqrt{2}$.

In the preceding example, $|-3| = |3|$ and $|2 - \sqrt{2}| = |\sqrt{2} - 2|$. In general, we have the following:

$$|a| = |-a| \quad \text{for every real number } a$$

EXAMPLE 2 *Removing an absolute value symbol*

If $x < 1$, rewrite $|x - 1|$ without using the absolute value symbol.

SOLUTION If $x < 1$, then $x - 1$ is negative, and hence, by the definition of absolute value (2),

$$|x - 1| = -(x - 1) = -x + 1 = 1 - x.$$

FIGURE 4

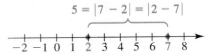

We shall use the concept of absolute value to define the distance between any two points on a coordinate line. First note that the distance between the points with coordinates 2 and 7, shown in Figure 4, equals 5 units. This distance is the difference obtained by subtracting the smaller coordinate from the larger $(7 - 2 = 5)$. If we employ absolute values, then, since $|7 - 2| = |2 - 7|$, it is unnecessary to be concerned about the order of subtraction. This fact motivates the next definition.

Definition of the Distance Between Points on a Coordinate Line

Let a and b be the coordinates of two points A and B, respectively, on a coordinate line. The **distance between A and B**, denoted by $d(A, B)$, is defined by

$$d(A, B) = |b - a|.$$

The number $d(A, B)$ is the length of the line segment AB.

Since $d(B, A) = |a - b|$ and $|b - a| = |a - b|$, we see that

$$d(A, B) = d(B, A).$$

Note that the distance between the origin O and the point A is

$$d(O, A) = |a - 0| = |a|,$$

which agrees with the geometric interpretation of absolute value illustrated in Figure 4. The formula $d(A, B) = |b - a|$ is true regardless of the signs of a and b, as illustrated in the next example.

EXAMPLE 3   *Finding distances between points*

FIGURE 5

Let A, B, C, and D have coordinates -5, -3, 1, and 6, respectively, on a coordinate line, as shown in Figure 5. Find $d(A, B)$, $d(C, B)$, $d(O, A)$, and $d(C, D)$.

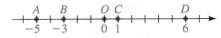

SOLUTION   Using the definition of the distance between points, we obtain the distances:

$$d(A, B) = |-3 - (-5)| = |-3 + 5| = |2| = 2$$
$$d(C, B) = |-3 - 1| = |-4| = 4$$
$$d(O, A) = |-5 - 0| = |-5| = 5$$
$$d(C, D) = |6 - 1| = |5| = 5$$

The concept of absolute value has uses other than finding distances between points; it is employed whenever we are interested in the magnitude or numerical value of a real number without regard to its sign.

In the next section we shall discuss the *exponential notation* a^n, where a is a real number (called the *base*) and n is an integer (called an *exponent*). In particular, for base 10 we have

$$10^0 = 1, \quad 10^1 = 10, \quad 10^2 = 10 \cdot 10 = 100, \quad 10^3 = 10 \cdot 10 \cdot 10 = 1000,$$

and so on. For negative exponents we use reciprocals of the corresponding positive exponent, as follows:

$$10^{-1} = \frac{1}{10^1} = \frac{1}{10}, \quad 10^{-2} = \frac{1}{10^2} = \frac{1}{100}, \quad 10^{-3} = \frac{1}{10^3} = \frac{1}{1000}$$

We can employ this notation to write any finite decimal representation of a real number as a sum of the following type:

$$437.56 = 4(100) + 3(10) + 7(1) + 5(\tfrac{1}{10}) + 6(\tfrac{1}{100})$$
$$= 4(10^2) + 3(10^1) + 7(10^0) + 5(10^{-1}) + 6(10^{-2})$$

In the sciences it is often necessary to work with very large or very small numbers and to compare the relative magnitudes of very large or very small quantities. We usually represent a large or small positive number a in *scientific form*, using $\times$ to denote multiplication.

Scientific Form

$$a = c \times 10^n, \quad \text{where } 1 \le c < 10 \text{ and } n \text{ is an integer}$$

The distance a ray of light travels in one year is approximately 5,900,000,000,000 miles. This number may be written in scientific form as 5.9×10^{12}. The positive exponent 12 indicates that the decimal point should be moved 12 places to the *right*. The notation works equally well for small numbers. To illustrate, the weight of an oxygen molecule is estimated to be

$$0.000\,000\,000\,000\,000\,000\,000\,053 \text{ gram},$$

or, in scientific form, 5.3×10^{-23} gram. The negative exponent indicates that the decimal point should be moved 23 places to the *left*.

ILLUSTRATION

Scientific Form

- $513 = 5.13 \times 10^2$
- $93{,}000{,}000 = 9.3 \times 10^7$
- $0.000\,000\,000\,43 = 4.3 \times 10^{-10}$

- $7.3 = 7.3 \times 10^0$
- $20{,}700 = 2.07 \times 10^4$
- $0.000\,648 = 6.48 \times 10^{-4}$

Many calculators employ scientific form in their display panels: for the number $c \times 10^n$, the 10 is suppressed and the exponent is often shown

FIGURE 6

$$2.025 \;\; 13$$

or

$$2.025 \text{E}13$$

or

$$2.025 \;\; {}^{13}$$

preceded by the letter E. For example, to find $(4,500,000)^2$ on a scientific calculator, we could enter the integer 4,500,000 and press the $\boxed{x^2}$ (or squaring) key, obtaining a display similar to one of those in Figure 6.

We would translate this as 2.025×10^{13}. Thus,

$$(4,500,000)^2 = 20,250,000,000,000.$$

Calculators may also use scientific form in the entry of numbers. The user's manual for your calculator should give specific details.

As a final remark, applied problems often include numbers that are obtained by various types of measurements and, hence, are approximations to exact values. Such answers should be rounded off, since the final result of a calculation cannot be more accurate than the data that have been used. For example, if the length and width of a rectangle are measured to two-decimal-place accuracy, we cannot expect more than two-decimal-place accuracy in the calculated value of the area of the rectangle. For purely *mathematical* work, if values of the length and width of a rectangle are given, we assume that the dimensions are *exact*, and no rounding off is required.

If a number a is written in scientific form as $a = c \times 10^n$ for $1 \leq c < 10$ and if c is rounded off to k decimal places, then we say that a is accurate (or has been rounded off) to $k + 1$ **significant figures**, or **digits**. For example, 37.2638 rounded to 5 significant figures is 3.7264×10^1 or 37.264; to 3 significant figures, 3.73×10^1 or 37.3; and to 1 significant figure, 4×10^1 or 40.

1.1 EXERCISES

Exer. 1–2: If $x < 0$ and $y > 0$, determine the sign of the real number.

1 (a) xy (b) x^2y (c) $\dfrac{x}{y} + x$ (d) $y - x$

2 (a) $\dfrac{x}{y}$ (b) xy^2 (c) $\dfrac{x - y}{xy}$ (d) $y(y - x)$

Exer. 3–6: Replace the symbol □ with either <, >, or = to make the resulting statement true.

3 (a) $-7 \;\square\; -4$ (b) $\dfrac{\pi}{2} \;\square\; 1.57$ (c) $\sqrt{225} \;\square\; 15$

4 (a) $-3 \;\square\; -5$ (b) $\dfrac{\pi}{4} \;\square\; 0.8$ (c) $\sqrt{289} \;\square\; 17$

5 (a) $\frac{1}{11} \;\square\; 0.09$ (b) $\frac{2}{3} \;\square\; 0.6666$ (c) $\frac{22}{7} \;\square\; \pi$

6 (a) $\frac{1}{7} \;\square\; 0.143$ (b) $\frac{5}{6} \;\square\; 0.833$ (c) $\sqrt{2} \;\square\; 1.4$

Exer. 7–8: Express the statement as an inequality.

7 (a) x is negative.

(b) y is nonnegative.

(c) q is less than or equal to π.

(d) d is between 4 and 2.

(e) t is not less than 5.

(f) The negative of z is not greater than 3.

(g) The quotient of p and q is at most 7.

(h) The reciprocal of w is at least 9.

(i) The absolute value of x is greater than 7.

8 (a) b is positive.

(b) s is nonpositive.

(c) w is greater than or equal to -4.

(d) c is between $\frac{1}{5}$ and $\frac{1}{3}$.

(e) p is not greater than -2.

(f) The negative of m is not less than -2.

(g) The quotient of r and s is at least $\frac{1}{5}$.

(h) The reciprocal of f is at most 14.

(i) The absolute value of x is less than 4.

Exer. 9–14: Rewrite the number without using the absolute value symbol, and simplify the result.

9 (a) $|-3-2|$ (b) $|-5|-|2|$ (c) $|7|+|-4|$

10 (a) $|-11+1|$ (b) $|6|-|-3|$ (c) $|8|+|-9|$

11 (a) $(-5)|3-6|$ (b) $|-6|/(-2)$ (c) $|-7|+|4|$

12 (a) $(4)|6-7|$ (b) $5/|-2|$ (c) $|-1|+|-9|$

13 (a) $|4-\pi|$ (b) $|\pi-4|$ (c) $|\sqrt{2}-1.5|$

14 (a) $|\sqrt{3}-1.7|$ (b) $|1.7-\sqrt{3}|$ (c) $|\frac{1}{5}-\frac{1}{3}|$

Exer. 15–18: The given numbers are coordinates of points A, B, and C, respectively, on a coordinate line. Find the distance:

(a) $d(A, B)$ (b) $d(B, C)$

(c) $d(C, B)$ (d) $d(A, C)$

15 $3, 7, -5$ 16 $-6, -2, 4$

17 $-9, 1, 10$ 18 $8, -4, -1$

Exer. 19–24: The two given numbers are coordinates of points A and B, respectively, on a coordinate line. Express the indicated statement as an inequality involving the absolute value symbol.

19 x, 7; $d(A, B)$ is less than 5

20 x, $-\sqrt{2}$; $d(A, B)$ is greater than 1

21 x, -3; $d(A, B)$ is at least 8

22 x, 4; $d(A, B)$ is at most 2

23 4, x; $d(A, B)$ is not greater than 3

24 -2, x; $d(A, B)$ is not less than 2

Exer. 25–32: Rewrite the expression without using the absolute value symbol, and simplify the result.

25 $|3+x|$ if $x<-3$ 26 $|5-x|$ if $x>5$

27 $|2-x|$ if $x<2$ 28 $|7+x|$ if $x\geq-7$

29 $|a-b|$ if $a<b$ 30 $|a-b|$ if $a>b$

31 $|x^2+4|$ 32 $|-x^2-1|$

Exer. 33–40: Replace the symbol $\square$ with either $=$ or $\neq$ to make the resulting statement true for all real numbers a, b, c, and d, whenever the expressions are defined.

33 $\dfrac{ab+ac}{a} \,\square\, b+ac$

34 $\dfrac{ab+ac}{a} \,\square\, b+c$

35 $\dfrac{b+c}{a} \,\square\, \dfrac{b}{a}+\dfrac{c}{a}$

36 $\dfrac{a+c}{b+d} \,\square\, \dfrac{a}{b}+\dfrac{c}{d}$

37 $(a\div b)\div c \,\square\, a\div(b\div c)$

38 $(a-b)-c \,\square\, a-(b-c)$

39 $\dfrac{a-b}{b-a} \,\square\, -1$

40 $-(a+b) \,\square\, -a+b$

41 The point on a coordinate line corresponding to $\sqrt{2}$ may be determined by constructing a right triangle with sides of length 1, as shown in the figure. Determine the points that correspond to $\sqrt{3}$ and $\sqrt{5}$, respectively. (*Hint:* Use the Pythagorean theorem.)

EXERCISE 41

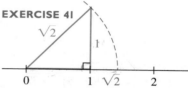

42 A circle of radius 1 rolls along a coordinate line in the positive direction, as shown in the figure. If point P is initially at the origin, find the coordinate of P after

(a) one, (b) two, and (c) ten complete revolutions.

EXERCISE 42

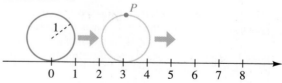

43 Geometric proofs of properties of real numbers were first given by the ancient Greeks. In order to establish the distributive property $a(b+c)=ab+ac$ for positive real numbers a, b, and c, find the area of the rectangle shown in the figure in two ways.

EXERCISE 43

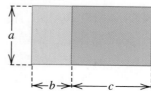

44 Rational approximations to square roots can be found using a formula discovered by the ancient Babylonians. Let x_1 be the first rational approximation for $\sqrt{n}$. If we let

$$x_2 = \frac{1}{2}\left(x_1 + \frac{n}{x_1}\right),$$

then x_2 will be a better approximation for $\sqrt{n}$, and we can repeat the computation with x_2 replacing x_1. Starting with $x_1 = \frac{3}{2}$, find the next two rational approximations for $\sqrt{2}$.

Exer. 45–46: Express the number in scientific form.

45 (a) 427,000   (b) 0.000 000 098   (c) 810,000,000

46 (a) 85,200   (b) 0.000 005 5   (c) 24,900,000

Exer. 47–48: Express the number in decimal form.

47 (a) 8.3×10^5 (b) 2.9×10^{-12} (c) 5.63×10^8

48 (a) 2.3×10^7 (b) 7.01×10^{-9} (c) 1.23×10^{10}

49 *Mass of a hydrogen atom*   The mass of a hydrogen atom is approximately

0.000 000 000 000 000 000 000 001 7 gram.

Express this number in scientific form.

50 *Mass of an electron*   The mass of an electron is approximately 9.1×10^{-31} kilogram. Express this number in decimal form.

51 *Light year*   In astronomy, distances to stars are measured in light years. One light year is the distance a ray of light travels in one year. If the speed of light is approximately 186,000 miles per second, estimate the number of miles in one light year.

52 *Milky Way galaxy*

(a) Astronomers have estimated that the Milky Way galaxy contains 100 billion stars. Express this number in scientific form.

(b) The diameter d of the Milky Way galaxy is estimated as 100,000 light years. Express d in miles. (Refer to Exercise 51.)

53 *Avogadro's number*   The number of hydrogen atoms in a mole is Avogadro's number, 6.02×10^{23}. If one mole of the gas has a mass of 1.01 grams, estimate the mass of a hydrogen atom.

54 *Fish population*   The population dynamics of many fish are characterized by extremely high fertility rates among adults and very low survival rates among the young. A mature halibut may lay as many as 2.5 million eggs, but only 0.00035% of the offspring survive to the age of 3 years. Use scientific form to approximate the number of offspring that live to age 3.

55 *Frames in a movie film*   The longest movie ever made is a 1970 British film that runs for 48 hours. Assuming that the film speed is 24 frames per second, approximate the total number of frames in this film. Express your answer in scientific form.

56 *Large prime numbers*   The number $2^{44,497} - 1$ is prime. At the time that this number was determined to be prime, it took one of the world's fastest computers about 60 days to verify that it was prime. This computer was capable of performing 2×10^{11} calculations per second. Use scientific form to estimate the number of calculations needed to perform this computation. (More recently, in 1992, $2^{756,839} - 1$, a number containing 227,832 digits, was shown to be prime.)

57 *Intelligence quotient*   A person's intelligence quotient (IQ) is determined by multiplying the quotient of his or her mental age and chronological age by 100.

(a) Find the IQ of a 12-year-old child whose mental age is 15.

(b) Find the mental age of a person 15 years old whose IQ is 140.

58 *Earth's surface area*   Water covers 70.8%, or 361×10^6 km^2, of the earth's surface. Approximate the total surface area of the earth.

1.2   EXPONENTS AND RADICALS

If n is a positive integer, the exponential notation a^n, defined in the following chart, represents the product of the real number a with itself n times. We refer to a^n as **a to the nth power** or, simply, *a to the n*. The positive integer n is called the **exponent**, and the real number a is called the **base**.

Exponential Notation

General case (n is any positive integer)	Special cases
$a^n = \underbrace{a \cdot a \cdot a \cdots \cdot a}_{n \text{ factors of } a}$	$a^1 = a$ $a^2 = a \cdot a$ $a^3 = a \cdot a \cdot a$ $a^6 = a \cdot a \cdot a \cdot a \cdot a \cdot a$

The next illustration contains several numerical examples of exponential notation.

ILLUSTRATION

The Exponential Notation a^n

- $5^4 = 5 \cdot 5 \cdot 5 \cdot 5 = 625$
- $(\frac{1}{2})^5 = \frac{1}{2} \cdot \frac{1}{2} \cdot \frac{1}{2} \cdot \frac{1}{2} \cdot \frac{1}{2} = \frac{1}{32}$
- $(-3)^3 = (-3)(-3)(-3) = -27$
- $(-\frac{1}{3})^4 = (-\frac{1}{3})(-\frac{1}{3})(-\frac{1}{3})(-\frac{1}{3}) = (\frac{1}{9})(\frac{1}{9}) = \frac{1}{81}$

It is important to note that if n is a positive integer, then an expression such as $3a^n$ means $3(a^n)$, *not* $(3a)^n$. The real number 3 is the **coefficient** of a^n in the expression $3a^n$. Similarly, $-3a^n$ means $(-3)a^n$, *not* $(-3a)^n$.

ILLUSTRATION

The Notation ca^n

- $5 \cdot 2^3 = 5 \cdot 8 = 40$
- $-5 \cdot 2^3 = -5 \cdot 8 = -40$
- $-2^4 = -(2^4) = -16$
- $3(-2)^3 = 3(-2)(-2)(-2) = 3(-8) = -24$

We next extend the definition of a^n to nonpositive exponents.

Zero and Negative Exponents

Definition ($a \neq 0$)	Illustration
$a^0 = 1$	$3^0 = 1, \; (-\sqrt{2})^0 = 1$
$a^{-n} = \dfrac{1}{a^n}$	$5^{-3} = \dfrac{1}{5^3}, \; 3^{-5} = \dfrac{1}{3^5}$

If m and n are positive integers, then

$$a^m a^n = \underbrace{a \cdot a \cdot a \cdots \cdots a}_{m \text{ factors of } a} \cdot \underbrace{a \cdot a \cdot a \cdots \cdots a}_{n \text{ factors of } a}.$$

Since the total number of factors of a on the right is $m + n$, this expression is equal to a^{m+n}; that is,

$$a^m a^n = a^{m+n}.$$

We can extend this formula to $m \leq 0$ or $n \leq 0$ by using the definitions of zero exponent and negative exponents. This gives us law 1, stated in the next chart.

To prove law 2, we may write, for m and n positive,

$$(a^m)^n = \underbrace{a^m \cdot a^m \cdot a^m \cdots \cdots a^m}_{n \text{ factors of } a^m}$$

and count the number of times a appears as a factor on the right-hand side. Since $a^m = a \cdot a \cdot a \cdots \cdots a$, with a occurring as a factor m times, and since the number of such groups of m factors is n, the total number of factors of a is $m \cdot n$. Thus,

$$(a^m)^n = a^{mn}.$$

The cases $m \leq 0$ or $n \leq 0$ can be proved using the definition of nonpositive exponents. The remaining three laws can be established in similar fashion by counting factors. In laws (4) and (5) we assume that denominators are not 0.

Laws of Exponents for Real Numbers a and b and Integers m and n

Law	Illustration
(1) $a^m a^n = a^{m+n}$	$2^3 \cdot 2^4 = 2^{3+4} = 2^7 = 128$
(2) $(a^m)^n = a^{mn}$	$(2^3)^4 = 2^{3 \cdot 4} = 2^{12} = 4096$
(3) $(ab)^n = a^n b^n$	$(20)^3 = (2 \cdot 10)^3 = 2^3 \cdot 10^3 = 8 \cdot 1000 = 8000$
(4) $\left(\dfrac{a}{b}\right)^n = \dfrac{a^n}{b^n}$	$\left(\dfrac{2}{5}\right)^3 = \dfrac{2^3}{5^3} = \dfrac{8}{125}$
(5) (a) $\dfrac{a^m}{a^n} = a^{m-n}$	$\dfrac{2^5}{2^3} = 2^{5-3} = 2^2 = 4$
(b) $\dfrac{a^m}{a^n} = \dfrac{1}{a^{n-m}}$	$\dfrac{2^3}{2^5} = \dfrac{1}{2^{5-3}} = \dfrac{1}{2^2} = \dfrac{1}{4}$

We usually use 5(a) if $m > n$, and 5(b) if $m < n$.

We can extend laws of exponents to obtain rules such as $(abc)^n = a^n b^n c^n$ and $a^m a^n a^p = a^{m+n+p}$.

Some other examples of the laws of exponents are given in the next illustration.

ILLUSTRATION

Laws of Exponents

- $x^5 x^6 x^2 = x^{5+6+2} = x^{13}$

- $(y^5)^7 = y^{5 \cdot 7} = y^{35}$

- $(3st)^4 = 3^4 s^4 t^4 = 81 s^4 t^4$

- $\left(\dfrac{p}{2}\right)^5 = \dfrac{p^5}{2^5} = \dfrac{p^5}{32}$

- $\dfrac{c^8}{c^3} = c^{8-3} = c^5$

- $\dfrac{u^3}{u^8} = \dfrac{1}{u^{8-3}} = \dfrac{1}{u^5}$

To **simplify** an expression involving powers of real numbers means to change it to an expression in which each real number appears only once and all exponents are positive. *We shall assume that denominators always represent nonzero real numbers.*

EXAMPLE I *Simplifying expressions containing exponents*

Use laws of exponents to simplify the expression:

(a) $(3x^3 y^4)(4xy^5)$ **(b)** $(2a^2 b^3 c)^4$ **(c)** $\left(\dfrac{2r^3}{s}\right)^2 \left(\dfrac{s}{r^3}\right)^3$ **(d)** $(u^{-2} v^3)^{-3}$

SOLUTION

(a)

$$(3x^3 y^4)(4xy^5) = (3)(4)x^3 xy^4 y^5 \qquad \text{rearrange factors}$$
$$= 12x^4 y^9 \qquad \text{law 1}$$

(b)

$$(2a^2 b^3 c)^4 = 2^4 (a^2)^4 (b^3)^4 c^4 \qquad \text{law 3}$$
$$= 16a^8 b^{12} c^4 \qquad \text{law 2}$$

(c)

$$\left(\frac{2r^3}{s}\right)^2 \left(\frac{s}{r^3}\right)^3 = \frac{(2r^3)^2}{s^2} \cdot \frac{s^3}{(r^3)^3} \qquad \text{law 4}$$

$$= \frac{2^2 (r^3)^2}{s^2} \cdot \frac{s^3}{(r^3)^3} \qquad \text{law 3}$$

$$= \left(\frac{4r^6}{s^2}\right)\left(\frac{s^3}{r^9}\right) \qquad \text{law 2}$$

$$= 4\left(\frac{r^6}{r^9}\right)\left(\frac{s^3}{s^2}\right) \qquad \text{rearrange factors}$$

$$= 4\left(\frac{1}{r^3}\right)(s) \qquad \text{laws 5(b) and 5(a)}$$

$$= \frac{4s}{r^3} \qquad \text{rearrange factors}$$

(d)

$$(u^{-2} v^3)^{-3} = (u^{-2})^{-3}(v^3)^{-3} \qquad \text{law 3}$$
$$= u^6 v^{-9} \qquad \text{law 2}$$
$$= \frac{u^6}{v^9} \qquad \text{definition of } a^{-n}$$

The following theorem is useful for problems that involve negative exponents.

Theorem on Negative Exponents

(1) $\dfrac{a^{-m}}{b^{-n}} = \dfrac{b^n}{a^m}$ **(2)** $\left(\dfrac{a}{b}\right)^{-n} = \left(\dfrac{b}{a}\right)^n$

PROOF   Using properties of negative exponents and quotients, we obtain

(1) $\dfrac{a^{-m}}{b^{-n}} = \dfrac{1/a^m}{1/b^n} = \dfrac{1}{a^m} \cdot \dfrac{b^n}{1} = \dfrac{b^n}{a^m}$

(2) $\left(\dfrac{a}{b}\right)^{-n} = \dfrac{a^{-n}}{b^{-n}} = \dfrac{b^n}{a^n} = \left(\dfrac{b}{a}\right)^n$ ∎

EXAMPLE 2   *Simplifying expressions containing negative exponents*

Simplify:

(a) $\dfrac{8x^3 y^{-5}}{4x^{-1} y^2}$ **(b)** $\left(\dfrac{u^2}{2v}\right)^{-3}$

SOLUTION   We apply the preceding theorem and laws of exponents.

(a) $\dfrac{8x^3 y^{-5}}{4x^{-1} y^2} = \dfrac{8x^3}{4y^2} \cdot \dfrac{y^{-5}}{x^{-1}}$ rearrange quotients so that negative exponents are in one fraction

$= \dfrac{8x^3}{4y^2} \cdot \dfrac{x^1}{y^5}$ theorem on negative exponents (1)

$= \dfrac{2x^4}{y^7}$ law 1 of exponents

(b) $\left(\dfrac{u^2}{2v}\right)^{-3} = \left(\dfrac{2v}{u^2}\right)^3$ theorem on negative exponents (2)

$= \dfrac{2^3 v^3}{(u^2)^3}$ laws 4 and 3 of exponents

$= \dfrac{8v^3}{u^6}$ law 2 of exponents

We next define the **principal nth root** $\sqrt[n]{a}$ of a real number a.

Definition of $\sqrt[n]{a}$

Let n be a positive integer greater than 1 and let a be a real number.

(1) If $a = 0$, then $\sqrt[n]{a} = 0$.

(2) If $a > 0$, then $\sqrt[n]{a}$ is the *positive* real number b such that $b^n = a$.

(3) (a) If $a < 0$ and n is odd, then $\sqrt[n]{a}$ is the *negative* real number b such that $b^n = a$.

(b) If $a < 0$ and n is even, then $\sqrt[n]{a}$ is not a real number.

Complex numbers, discussed in Section 2.4, are needed to define $\sqrt[n]{a}$ if $a < 0$ and n is an *even* positive integer, since for all real numbers b, $b^n \geq 0$ whenever n is even.

If $n = 2$, we write $\sqrt{a}$ instead of $\sqrt[2]{a}$ and call $\sqrt{a}$ the **principal square root** of a or, simply, the **square root** of a. The number $\sqrt[3]{a}$ is the (principal) **cube root** of a.

ILLUSTRATION

The Principal nth Root $\sqrt[n]{a}$

- $\sqrt{16} = 4$, since $4^2 = 16$.
- $\sqrt[5]{\frac{1}{32}} = \frac{1}{2}$, since $\left(\frac{1}{2}\right)^5 = \frac{1}{32}$.
- $\sqrt[3]{-8} = -2$, since $(-2)^3 = -8$.
- $\sqrt[4]{-16}$ is not a real number.

Note that $\sqrt{16} \neq \pm 4$, since, by definition, roots of positive real numbers are positive.

To complete our terminology, the expression $\sqrt[n]{a}$ is a **radical**, the number a is the **radicand**, and n is the **index** of the radical. The symbol $\sqrt{}$ is called a **radical sign**.

If $\sqrt{a} = b$, then $b^2 = a$; that is, $(\sqrt{a})^2 = a$. If $\sqrt[3]{a} = b$, then $b^3 = a$, or $(\sqrt[3]{a})^3 = a$. Generalizing this pattern gives us property 1 in the next chart.

Properties of $\sqrt[n]{}$
(n **is a positive integer**)

Property	Illustration
(1) $(\sqrt[n]{a})^n = a$, if $\sqrt[n]{a}$ exists	$(\sqrt{5})^2 = 5$, $(\sqrt[3]{-8})^3 = -8$
(2) $\sqrt[n]{a^n} = a$, if $a \geq 0$	$\sqrt{5^2} = 5$, $\sqrt[3]{2^3} = 2$
(3) $\sqrt[n]{a^n} = a$, if $a < 0$ and n is odd	$\sqrt[3]{(-2)^3} = -2$, $\sqrt[5]{(-2)^5} = -2$
(4) $\sqrt[n]{a^n} = \lvert a \rvert$, if $a < 0$ and n is even	$\sqrt{(-3)^2} = \lvert -3 \rvert = 3$, $\sqrt[4]{(-2)^4} = \lvert -2 \rvert = 2$

If $a \geq 0$, then property 4 reduces to property 2. We also see from property 4 that

$$\sqrt{x^2} = |x|$$

for every real number x. In particular, if $x \geq 0$, then $\sqrt{x^2} = x$; however, if $x < 0$, then $\sqrt{x^2} = -x$, which is positive.

The three laws listed in the next chart are true for positive integers m and n, *provided the indicated roots exist.*

Laws of Radicals

Law	Illustration
(1) $\sqrt[n]{ab} = \sqrt[n]{a}\,\sqrt[n]{b}$	$\sqrt{50} = \sqrt{25 \cdot 2} = \sqrt{25}\,\sqrt{2} = 5\sqrt{2}$
	$\sqrt[3]{-108} = \sqrt[3]{(-27)(4)} = \sqrt[3]{-27}\,\sqrt[3]{4} = -3\sqrt[3]{4}$
(2) $\sqrt[n]{\dfrac{a}{b}} = \dfrac{\sqrt[n]{a}}{\sqrt[n]{b}}$	$\sqrt[3]{\dfrac{5}{8}} = \dfrac{\sqrt[3]{5}}{\sqrt[3]{8}} = \dfrac{\sqrt[3]{5}}{2}$
(3) $\sqrt[m]{\sqrt[n]{a}} = \sqrt[mn]{a}$	$\sqrt{\sqrt[3]{64}} = \sqrt[3(2)]{64} = \sqrt[6]{2^6} = 2$

The radicands in laws 1 and 2 involve products and quotients. Care must be taken if sums or differences occur in the radicand. The following chart contains two particular warnings concerning mistakes that frequently occur in working with radicals containing sums.

Warnings

If $a \neq 0$ and $b \neq 0$	Illustration
$\sqrt{a^2 + b^2} \neq a + b$	$\sqrt{3^2 + 4^2} = \sqrt{25} = 5 \neq 3 + 4 = 7$
$\sqrt{a + b} \neq \sqrt{a} + \sqrt{b}$	$\sqrt{4 + 9} = \sqrt{13} \neq \sqrt{4} + \sqrt{9} = 5$

If c is a real number and c^n occurs as a factor in a radical of index n, then we can remove c from the radicand if the sign of c is taken into account. For example, if $c > 0$, or if $c < 0$ and n is *odd*, then

$$\sqrt[n]{c^n d} = \sqrt[n]{c^n}\,\sqrt[n]{d} = c\sqrt[n]{d},$$

provided $\sqrt[n]{d}$ exists. If $c < 0$ and n is *even*, then

$$\sqrt[n]{c^n d} = \sqrt[n]{c^n}\,\sqrt[n]{d} = |c|\sqrt[n]{d},$$

provided $\sqrt[n]{d}$ exists.

ILLUSTRATION

Removing nth Powers from $\sqrt[n]{}$

- $\sqrt[5]{x^7} = \sqrt[5]{x^5 \cdot x^2} = \sqrt[5]{x^5}\,\sqrt[5]{x^2} = x\sqrt[5]{x^2}$
- $\sqrt[3]{x^7} = \sqrt[3]{x^6 \cdot x} = \sqrt[3]{(x^2)^3 x} = \sqrt[3]{(x^2)^3}\,\sqrt[3]{x} = x^2\sqrt[3]{x}$
- $\sqrt{x^2 y} = \sqrt{x^2}\,\sqrt{y} = |x|\sqrt{y}$
- $\sqrt{x^6} = \sqrt{(x^3)^2} = |x^3|$
- $\sqrt[4]{x^6 y^3} = \sqrt[4]{x^4 \cdot x^2 y^3} = \sqrt[4]{x^4}\,\sqrt[4]{x^2 y^3} = |x|\sqrt[4]{x^2 y^3}$

Note: To avoid considering absolute values, *in examples and exercises involving radicals in this chapter, we shall assume that all letters—a, b, c, d, x, y, and so on—that appear in radicands represent positive real numbers, unless otherwise specified.*

As illustrated in the preceding illustration and in the following examples, if the index of a radical is n, then we rearrange the radicand, isolating a factor of the form p^n, where p may consist of several letters. We then remove $\sqrt[n]{p^n} = p$ from the radical as previously indicated. Thus, in Example 3(b) the index of the radical is 3 and we rearrange the radicand into *cubes*, obtaining a factor p^3, with $p = 2xy^2z$. In part (c) the index of the radical is 2 and we rearrange the radicand into *squares*, obtaining a factor p^2, with $p = 3a^3b^2$.

To *simplify a radical* means to remove factors from the radical until no factor in the radicand has an exponent greater than or equal to the index of the radical and the index is as low as possible.

EXAMPLE 3   *Removing factors from radicals*

Simplify the radical (all letters denote positive real numbers):

(a) $\sqrt[3]{320}$ **(b)** $\sqrt[3]{16x^3y^8z^4}$ **(c)** $\sqrt{3a^2b^3}\sqrt{6a^5b}$

SOLUTION

(a)
$$\sqrt[3]{320} = \sqrt[3]{64 \cdot 5} \qquad \text{factor out the largest cube in 320}$$
$$= \sqrt[3]{4^3}\sqrt[3]{5} \qquad \text{law 1 of radicals}$$
$$= 4\sqrt[3]{5} \qquad \text{property 2 of } \sqrt[n]{}$$

(b)
$$\sqrt[3]{16x^3y^8z^4} = \sqrt[3]{(2^3x^3y^6z^3)(2y^2z)} \qquad \text{rearrange radicand into cubes}$$
$$= \sqrt[3]{(2xy^2z)^3(2y^2z)} \qquad \text{laws 2 and 3 of exponents}$$
$$= \sqrt[3]{(2xy^2z)^3}\sqrt[3]{2y^2z} \qquad \text{law 1 of radicals}$$
$$= 2xy^2z\sqrt[3]{2y^2z} \qquad \text{property 2 of } \sqrt[n]{}$$

(c)
$$\sqrt{3a^2b^3}\sqrt{6a^5b} = \sqrt{3a^2b^3 \cdot 2 \cdot 3a^5b} \qquad \text{law 1 of radicals}$$
$$= \sqrt{(3^2a^6b^4)(2a)} \qquad \text{rearrange radicand into squares}$$
$$= \sqrt{(3a^3b^2)^2(2a)} \qquad \text{laws 2 and 3 of exponents}$$
$$= \sqrt{(3a^3b^2)^2}\sqrt{2a} \qquad \text{law 1 of radicals}$$
$$= 3a^3b^2\sqrt{2a} \qquad \text{property 2 of } \sqrt[n]{}$$

If the denominator of a quotient contains a factor of the form $\sqrt[n]{a^k}$, with $k < n$ and $a > 0$, then multiplying numerator and denominator by $\sqrt[n]{a^{n-k}}$ will eliminate the radical from the denominator, since

$$\sqrt[n]{a^k}\sqrt[n]{a^{n-k}} = \sqrt[n]{a^{k+n-k}} = \sqrt[n]{a^n} = a.$$

This process is called **rationalizing a denominator**. Some special cases are listed in the following chart.

**Rationalizing Denominators
of Quotients** ($a > 0$)

Factor in denominator	Multiply numerator and denominator by	Resulting factor
$\sqrt{a}$	$\sqrt{a}$	$\sqrt{a}\sqrt{a} = \sqrt{a^2} = a$
$\sqrt[3]{a}$	$\sqrt[3]{a^2}$	$\sqrt[3]{a}\sqrt[3]{a^2} = \sqrt[3]{a^3} = a$
$\sqrt[7]{a^3}$	$\sqrt[7]{a^4}$	$\sqrt[7]{a^3}\sqrt[7]{a^4} = \sqrt[7]{a^7} = a$

The next example illustrates this technique.

EXAMPLE 4   *Rationalizing denominators*

Rationalize the denominator:

(a) $\dfrac{1}{\sqrt{5}}$ **(b)** $\dfrac{1}{\sqrt[3]{x}}$ **(c)** $\sqrt{\dfrac{2}{3}}$ **(d)** $\sqrt[5]{\dfrac{x}{y^2}}$

SOLUTION

(a)
$$\frac{1}{\sqrt{5}} = \frac{1}{\sqrt{5}}\frac{\sqrt{5}}{\sqrt{5}} = \frac{\sqrt{5}}{\sqrt{5^2}} = \frac{\sqrt{5}}{5}$$

(b)
$$\frac{1}{\sqrt[3]{x}} = \frac{1}{\sqrt[3]{x}}\frac{\sqrt[3]{x^2}}{\sqrt[3]{x^2}} = \frac{\sqrt[3]{x^2}}{\sqrt[3]{x^3}} = \frac{\sqrt[3]{x^2}}{x}$$

(c)
$$\sqrt{\frac{2}{3}} = \frac{\sqrt{2}}{\sqrt{3}} = \frac{\sqrt{2}}{\sqrt{3}}\frac{\sqrt{3}}{\sqrt{3}} = \frac{\sqrt{2\cdot 3}}{\sqrt{3^2}} = \frac{\sqrt{6}}{3}$$

(d)
$$\sqrt[5]{\frac{x}{y^2}} = \frac{\sqrt[5]{x}}{\sqrt[5]{y^2}} = \frac{\sqrt[5]{x}}{\sqrt[5]{y^2}}\frac{\sqrt[5]{y^3}}{\sqrt[5]{y^3}} = \frac{\sqrt[5]{xy^3}}{\sqrt[5]{y^5}} = \frac{\sqrt[5]{xy^3}}{y}$$

If we use a calculator to find decimal approximations of radicals, there is no advantage in rationalizing denominators, such as $1/\sqrt{5} = \sqrt{5}/5$ or $\sqrt{2/3} = \sqrt{6}/3$, as we did in Example 4(a) and (c). However, for *algebraic* simplifications, changing expressions to such forms is sometimes desirable. Similarly, in advanced mathematics courses such as calculus, changing $1/\sqrt[3]{x}$ to $\sqrt[3]{x^2}/x$, as in Example 4(b), could make a problem *more* complicated. In such courses it is simpler to work with the expression $1/\sqrt[3]{x}$ than with its rationalized form.

We next use radicals to define *rational exponents*.

**Definition of
Rational Exponents**

Let m/n be a rational number, where n is a positive integer greater than 1. If a is a real number such that $\sqrt[n]{a}$ exists, then

(1) $a^{1/n} = \sqrt[n]{a}$

(2) $a^{m/n} = (\sqrt[n]{a})^m = \sqrt[n]{a^m}$

(3) $a^{m/n} = (a^{1/n})^m = (a^m)^{1/n}$

When evaluating $a^{m/n}$ in (2), we usually use $(\sqrt[n]{a})^m$; that is, we take the nth root of a first and then raise that result to the mth power, as shown in the following illustration.

ILLUSTRATION

The Exponential Notation $a^{m/n}$

- $x^{1/3} = \sqrt[3]{x}$
- $x^{3/5} = (\sqrt[5]{x})^3 = \sqrt[5]{x^3}$
- $125^{2/3} = (\sqrt[3]{125})^2 = (\sqrt[3]{5^3})^2 = 5^2 = 25$
- $(\frac{32}{243})^{3/5} = (\sqrt[5]{\frac{32}{243}})^3 = (\sqrt[5]{(\frac{2}{3})^5})^3 = (\frac{2}{3})^3 = \frac{8}{27}$

The laws of exponents are true for rational exponents and also for *irrational* exponents, such as $3^{\sqrt{2}}$ or 5^π, considered in Chapter 5.

To simplify an expression involving rational powers of letters that represent real numbers, we change it to an expression in which each letter appears only once and all exponents are positive. As we did with radicals, we shall assume that all letters represent positive real numbers unless otherwise specified.

EXAMPLE 5   *Simplifying rational powers*

Simplify:

(a) $(-27)^{2/3}(4)^{-5/2}$ **(b)** $(r^2 s^6)^{1/3}$ **(c)** $\left(\dfrac{2x^{2/3}}{y^{1/2}}\right)^2\left(\dfrac{3x^{-5/6}}{y^{1/3}}\right)$

SOLUTION

(a) $(-27)^{2/3}(4)^{-5/2} = (\sqrt[3]{-27})^2(\sqrt{4})^{-5}$ definition of rational exponents

$= (-3)^2(2)^{-5}$ take roots

$= \dfrac{(-3)^2}{2^5}$ definition of negative exponent

$= \dfrac{9}{32}$ take powers

(b) $(r^2 s^6)^{1/3} = (r^2)^{1/3}(s^6)^{1/3}$ law 3 of exponents

$= r^{2/3}s^2$ law 2 of exponents

(c) $\left(\dfrac{2x^{2/3}}{y^{1/2}}\right)^2\left(\dfrac{3x^{-5/6}}{y^{1/3}}\right) = \left(\dfrac{4x^{4/3}}{y}\right)\left(\dfrac{3x^{-5/6}}{y^{1/3}}\right)$ laws of exponents

$= \dfrac{(4 \cdot 3)x^{4/3 - 5/6}}{y^{1 + (1/3)}}$ law 1 of exponents

$= \dfrac{12x^{8/6 - 5/6}}{y^{4/3}}$ common denominator

$= \dfrac{12x^{1/2}}{y^{4/3}}$ simplify

Rational exponents are useful for problems involving radicals that do not have the same index, as illustrated in the next example.

EXAMPLE 6 *Combining radicals*

Change to an expression containing one radical, $\sqrt[n]{a^m}$:

(a) $\sqrt[3]{a}\sqrt{a}$ (b) $\dfrac{\sqrt[4]{a}}{\sqrt[3]{a^2}}$

SOLUTION   Introducing rational exponents, we obtain

(a) $\sqrt[3]{a}\sqrt{a} = a^{1/3}a^{1/2} = a^{(1/3)+(1/2)} = a^{5/6} = \sqrt[6]{a^5}$

(b) $\dfrac{\sqrt[4]{a}}{\sqrt[3]{a^2}} = \dfrac{a^{1/4}}{a^{2/3}} = a^{(1/4)-(2/3)} = a^{-5/12} = \dfrac{1}{a^{5/12}} = \dfrac{1}{\sqrt[12]{a^5}}$

In Exercises 1.2, whenever an index of a radical is even (or a rational exponent m/n with n even is employed), assume that the letters that appear in the radicand denote positive real numbers unless otherwise specified.

1.2 EXERCISES

Exer. 1–10: Express the number in the form a/b, where a and b are integers.

1 $\left(-\frac{2}{3}\right)^4$

2 $(-3)^3$

3 $\dfrac{2^{-3}}{3^{-2}}$

4 $\dfrac{2^0 + 0^2}{2 + 0}$

5 $-2^4 + 3^{-1}$

6 $\left(-\frac{3}{2}\right)^4 - 2^{-4}$

7 $16^{-3/4}$

8 $9^{5/2}$

9 $(-0.008)^{2/3}$

10 $(0.008)^{-2/3}$

Exer. 11–46: Simplify.

11 $\left(\frac{1}{2}x^4\right)(16x^5)$

12 $(-3x^{-2})(4x^4)$

13 $\dfrac{(2x^3)(3x^2)}{(x^2)^3}$

14 $\dfrac{(2x^2)^3}{4x^4}$

15 $\left(\frac{1}{6}a^5\right)(-3a^2)(4a^7)$

16 $(-4b^3)\left(\frac{1}{6}b^2\right)(-9b^4)$

17 $\dfrac{(6x^3)^2}{(2x^2)^3}$

18 $\dfrac{(3y^3)(2y^2)^2}{(y^4)^3}$

19 $(3u^7v^3)(4u^4v^{-5})$

20 $(x^2yz^3)(-2xz^2)(x^3y^{-2})$

21 $(8x^4y^{-3})\left(\frac{1}{2}x^{-5}y^2\right)$

22 $\left(\dfrac{4a^2b}{a^3b^2}\right)\left(\dfrac{5a^2b}{2b^4}\right)$

23 $\left(\frac{1}{3}x^4y^{-3}\right)^{-2}$

24 $(-2xy^2)^5\left(\dfrac{x^7}{8y^3}\right)$

25 $(3y^3)^4(4y^2)^{-3}$

26 $(-3a^2b^{-5})^3$

27 $(-2r^4s^{-3})^{-2}$

28 $(2x^2y^{-5})(6x^{-3}y)\left(\frac{1}{3}x^{-1}y^3\right)$

29 $(5x^2y^{-3})(4x^{-5}y^4)$

30 $(-2r^2s)^5(3r^{-1}s^3)^2$

31 $\left(\dfrac{3x^5y^4}{x^0y^{-3}}\right)^2$

32 $(4a^2b)^4\left(\dfrac{-a^3}{2b}\right)^2$

33 $(4a^{3/2})(2a^{1/2})$

34 $(-6x^{7/5})(2x^{8/5})$

35 $(3x^{5/6})(8x^{2/3})$

36 $(8r)^{1/3}(2r^{1/2})$

37 $(27a^6)^{-2/3}$

38 $(25z^4)^{-3/2}$

39 $(8x^{-2/3})x^{1/6}$

40 $(3x^{1/2})(-2x^{5/2})$

41 $\left(\dfrac{-8x^3}{y^{-6}}\right)^{2/3}$

42 $\left(\dfrac{-y^{3/2}}{y^{-1/3}}\right)^3$

43 $\left(\dfrac{x^6}{9y^{-4}}\right)^{-1/2}$

44 $\left(\dfrac{c^{-4}}{16d^8}\right)^{3/4}$

45 $\dfrac{(x^6y^3)^{-1/3}}{(x^4y^2)^{-1/2}}$

46 $a^{4/3}a^{-3/2}a^{1/6}$

Exer. 47–52: Rewrite the expression using rational exponents.

47 $\sqrt[4]{x^3}$

48 $\sqrt[3]{x^5}$

49 $\sqrt[3]{(a+b)^2}$

50 $\sqrt{a+\sqrt{b}}$

51 $\sqrt{x^2+y^2}$

52 $\sqrt[3]{r^3-s^3}$

Exer. 53–56: Rewrite the expression using a radical.

53 (a) $4x^{3/2}$ (b) $(4x)^{3/2}$

54 (a) $4 + x^{3/2}$ (b) $(4+x)^{3/2}$

55 (a) $8 - y^{1/3}$ (b) $(8-y)^{1/3}$

56 (a) $8y^{1/3}$ (b) $(8y)^{1/3}$

Exer. 57–80: Simplify the expression and rationalize the denominator when appropriate.

57 $\sqrt{81}$

58 $\sqrt[3]{-125}$

59 $\sqrt[5]{-64}$

60 $\sqrt[4]{256}$

61 $\dfrac{1}{\sqrt[3]{2}}$

62 $\sqrt{\dfrac{1}{7}}$

63 $\sqrt{9x^{-4}y^6}$

64 $\sqrt{16a^8b^{-2}}$

65 $\sqrt[3]{8a^6b^{-3}}$

66 $\sqrt[4]{81r^5s^8}$

67 $\sqrt{\dfrac{3x}{2y^3}}$

68 $\sqrt{\dfrac{1}{3x^3y}}$

69 $\sqrt[3]{\dfrac{2x^4y^4}{9x}}$

70 $\sqrt[3]{\dfrac{3x^2y^5}{4x}}$

71 $\sqrt[4]{\dfrac{5x^8y^3}{27x^2}}$

72 $\sqrt[4]{\dfrac{x^7y^{12}}{125x}}$

73 $\sqrt[5]{\dfrac{5x^7y^2}{8x^3}}$

74 $\sqrt[5]{\dfrac{3x^{11}y^3}{9x^2}}$

75 $\sqrt[4]{(3x^5y^{-2})^4}$

76 $\sqrt[6]{(2u^{-3}v^4)^6}$

77 $\sqrt[5]{\dfrac{8x^3}{y^4}}\sqrt[5]{\dfrac{4x^4}{y^2}}$

78 $\sqrt{5xy^7}\sqrt{10x^3y^3}$

79 $\sqrt[3]{3t^4v^2}\sqrt[3]{-9t^{-1}v^4}$

80 $\sqrt[3]{(2r-s)^3}$

Exer. 81–84: Simplify the expression, assuming x and y may be negative.

81 $\sqrt{x^6y^4}$

82 $\sqrt{x^4y^{10}}$

83 $\sqrt[4]{x^5(y-1)^6}$

84 $\sqrt[4]{(x+2)^{12}y^7}$

Exer. 85–90: Replace the symbol □ with either = or ≠ to make the resulting statement true, whenever the expression has meaning. Give a reason for your answer.

85 $(a^r)^2 \ \square \ a^{(r2)}$

86 $(a^2+1)^{1/2} \ \square \ a+1$

87 $a^xb^y \ \square \ (ab)^{xy}$

88 $\sqrt{a^r} \ \square \ (\sqrt{a})^r$

89 $\sqrt[n]{\dfrac{1}{c}} \ \square \ \dfrac{1}{\sqrt[n]{c}}$

90 $a^{1/k} \ \square \ \dfrac{1}{a^k}$

91 *Savings account*   One of the oldest banks in the United States is the Bank of America, founded in 1812. If \$200 had been deposited at that time into an account that paid 4% annual interest, then 180 years later the amount would have grown to $200(1.04)^{180}$ dollars. Approximate this amount to the nearest cent.

92 *Viewing distance*   On a clear day, the distance d (in miles) that can be seen from the top of a tall building of height h (in feet) can be approximated by $d = 1.2\sqrt{h}$. Approximate the distance that can be seen from the top of the Chicago Sears Tower, which is 1454 ft tall.

93 *Length of a halibut*   The length-weight relationship for Pacific halibut can be approximated by $L = 0.46\sqrt[3]{W}$, where W is in kilograms and L is in meters. The largest documented halibut weighed 230 kg. Estimate its length.

94 *Weight of a whale*   The length-weight relationship for the sei whale can be approximated by $W = 0.0016L^{2.43}$, where W is in tons and L is in feet. Estimate the weight of a whale that is 25 ft long.

95 *Weight lifters' handicaps*   O'Carroll's formula is used to handicap weight lifters. If a lifter who weighs b kg lifts w kg of weight, then the handicapped weight W is given by

$$W = \dfrac{w}{\sqrt[3]{b-35}}.$$

Suppose two lifters weighing 75 kg and 120 kg lift weights of 180 kg and 250 kg, respectively. Use O'Carroll's formula to determine the superior weight lifter.

96 *Body surface area*   A person's body surface area S (in square feet) can be approximated by

$$S = (0.1091)w^{0.425}h^{0.725},$$

where height h is in inches and weight w is in pounds.

(a) Estimate S for a person 6 ft tall and weighing 175 lb.

(b) If a person is 5 ft 6 in. tall, what effect does a 10% increase in weight have on S?

1.3 ALGEBRAIC EXPRESSIONS

We sometimes use the notation and terminology of sets to describe mathematical relationships. A **set** is a collection of objects of some type, and the objects are called **elements** of the set. Capital letters $R, S, T, \ldots$ are often used to denote sets and lowercase letters $a, b, x, y, \ldots$ usually represent elements of sets. Throughout this book, $\mathbb{R}$ denotes the set of real numbers and $\mathbb{Z}$ denotes the set of integers.

Two sets S and T are **equal**, denoted by $S = T$, if S and T contain exactly the same elements. We write $S \neq T$ if S and T are not equal. Additional notation and terminology are listed in the following chart.

Notation or terminology	Meaning	Illustration
$a \in S$	a is an element of S.	$3 \in \mathbb{Z}$
$a \notin S$	a is not an element of S.	$\frac{3}{5} \notin \mathbb{Z}$
S is a **subset** of T	Every element of S is an element of T.	$\mathbb{Z}$ is a subset of $\mathbb{R}$
Constant	A letter or symbol that represents a *specific* element of a set.	$5, -\sqrt{2}, \pi$
Variable	A letter or symbol that represents *any* element of a set	Let x denote any real number.

We usually use letters near the end of the alphabet such as x, y, and z for variables, and letters near the beginning of the alphabet such as a, b, and c for constants. Throughout this text, unless otherwise specified, variables represent real numbers.

If the elements of a set S have a certain property, we sometimes write $S = \{x: \ \}$ and state the property describing the variable x in the space after the colon. The expression involving the braces and colon is read "the set of all x such that," where we complete the phrase by stating the desired property. For example, $\{x: x > 3\}$ is read "the set of all x such that x is greater than 3."

For finite sets we sometimes list all the elements of the set within braces. Thus, if the set T consists of the first five positive integers, we may write $T = \{1, 2, 3, 4, 5\}$. When we describe sets in this way, the order used in listing the elements is irrelevant, so we could also write $T = \{1, 3, 2, 4, 5\}$, $T = \{4, 3, 2, 5, 1\}$, and so on.

If we begin with any collection of variables and real numbers, then an **algebraic expression** is the result obtained by applying additions, subtractions, multiplications, divisions, powers, or the taking of roots to this collection. If specific numbers are substituted for the variables in an algebraic expression, the resulting number is called the **value** of the expression for these numbers. The **domain** of an algebraic expression consists of all real numbers that are represented by the variables. *Unless otherwise specified, we assume that the domain consists of the values of the variables that do not make the expression meaningless, in the sense that denominators cannot equal zero and roots always exist.* Some illustrations are given in the following chart.

Algebraic Expressions

Illustration	Domain	Typical value
$x^3 - 5x + \dfrac{6}{\sqrt{x}}$	all $x > 0$	At $x = 4$: $$4^3 - 5(4) + \frac{6}{\sqrt{4}} = 64 - 20 + 3$$ $$= 47$$
$\dfrac{2xy + (3/x^2)}{\sqrt[3]{y^2 - 1}}$	all $x \neq 0$ and all $y \neq \pm 1$	At $x = 1$ and $y = 3$: $$\frac{2(1)(3) + (3/1^2)}{\sqrt[3]{3^2 - 1}} = \frac{6 + 3}{\sqrt[3]{8}} = \frac{9}{2}$$

If x is a variable, then a **monomial** in x is an expression of the form ax^n, where a is a real number and n is a nonnegative integer. A **binomial** is a sum of two monomials, and a **trinomial** is a sum of three monomials. A *polynomial in x* is a sum of any number of monomials in x. Another way of stating this is as follows.

Definition of Polynomial

A **polynomial in x** is a sum of the form
$$a_n x^n + a_{n-1} x^{n-1} + \cdots + a_1 x + a_0,$$
where n is a nonnegative integer and each coefficient a_k is a real number. If $a_n \neq 0$, then the polynomial is said to have **degree n.**

Each expression $a_k x^k$ in the sum is a **term** of the polynomial. If a coefficient a_k is zero, we usually delete the term $a_k x^k$. The coefficient a_k of the highest power of x is called the **leading coefficient** of the polynomial.

The next chart contains specific illustrations of polynomials.

Polynomials

Example	Leading coefficient	Degree
$3x^4 + 5x^3 + (-7)x + 4$	3	4
$x^8 + 9x^2 + (-2)x$	1	8
$-5x^2 + 1$	-5	2
$7x + 2$	7	1
8	8	0

By definition, two polynomials are **equal** if and only if they have the same degree and the coefficients of like powers of x are equal. If all the coefficients of a polynomial are zero, it is called the **zero polynomial** and is denoted by 0. However, by convention, the degree of the zero polynomial is *not* zero but, instead, is undefined. If c is a *nonzero real number*, then c is a polynomial of degree 0. Such polynomials (together with the zero polynomial) are **constant polynomials**.

If a coefficient of a polynomial is negative, we usually use a minus sign between appropriate terms. To illustrate,

$$3x^2 + (-5)x + (-7) = 3x^2 - 5x - 7.$$

We may also consider polynomials in variables other than x. For example, $\frac{2}{5}z^2 - 3z^7 + 8 - \sqrt{5}z^4$ is a polynomial in z of degree 7. We often arrange the terms of a polynomial in order of decreasing powers of the variable; thus, we write

$$\frac{2}{5}z^2 - 3z^7 + 8 - \sqrt{5}z^4 = -3z^7 - \sqrt{5}z^4 + \frac{2}{5}z^2 + 8.$$

We may regard a polynomial in x as an algebraic expression obtained by employing a finite number of additions, subtractions, and multiplications involving x. If an algebraic expression contains divisions or roots involving a variable x, then it is not a polynomial in x.

ILLUSTRATION

Nonpolynomials

- $\dfrac{1}{x} + 3x$ - $\dfrac{x-5}{x^2+2}$ - $3x^2 + \sqrt{x} - 2$

Since polynomials represent real numbers, we may use the properties described in Section 1.1. In particular, if additions, subtractions, and multiplications are carried out with polynomials, we may simplify the results by using properties of real numbers.

EXAMPLE I *Adding and subtracting polynomials*

(a) Find the sum $(x^3 + 2x^2 - 5x + 7) + (4x^3 - 5x^2 + 3)$.

(b) Find the difference $(x^3 + 2x^2 - 5x + 7) - (4x^3 - 5x^2 + 3)$.

SOLUTION

(a) To obtain the sum of any two polynomials in x, we may add coefficients of like powers of x.

$$
\begin{aligned}
&(x^3 + 2x^2 - 5x + 7) + (4x^3 - 5x^2 + 3) \\
&= (1 + 4)x^3 + (2 - 5)x^2 - 5x + (7 + 3) \qquad \text{add coefficients of like powers of } x \\
&= 5x^3 - 3x^2 - 5x + 10 \qquad\qquad\qquad\quad \text{simplify}
\end{aligned}
$$

The grouping in the first step was shown for completeness. You may omit this step after you become proficient with such manipulations.

(b) When subtracting polynomials, we first remove parentheses, noting that the minus sign preceding the second pair of parentheses changes the sign of *each* term of that polynomial.

$$(x^3 + 2x^2 - 5x + 7) - (4x^3 - 5x^2 + 3)$$

$$= x^3 + 2x^2 - 5x + 7 - 4x^3 + 5x^2 - 3 \qquad \text{remove parentheses}$$

$$= (1 - 4)x^3 + (2 + 5)x^2 - 5x + (7 - 3) \qquad \begin{array}{l}\text{add coefficients of}\\ \text{like powers of } x\end{array}$$

$$= -3x^3 + 7x^2 - 5x + 4 \qquad \text{simplify}$$

EXAMPLE 2 *Multiplying binomials*

Find the product $(4x + 5)(3x - 2)$.

SOLUTION Since $3x - 2 = 3x + (-2)$, we may proceed as in Example 1 of Section 1.1:

$$(4x + 5)(3x - 2)$$

$$= (4x)(3x) + (4x)(-2) + (5)(3x) + (5)(-2) \qquad \text{distributive properties}$$

$$= 12x^2 - 8x + 15x - 10 \qquad \text{multiply}$$

$$= 12x^2 + 7x - 10 \qquad \text{simplify}$$

After becoming proficient working problems of the type in Example 2, you may wish to perform the first two steps mentally and proceed directly to the final form.

In the next example we illustrate different methods for finding the product of two polynomials.

EXAMPLE 3 *Multiplying polynomials*

Find the product $(x^2 + 5x - 4)(2x^3 + 3x - 1)$.

SOLUTION

Method I We begin by using a distributive property, treating the polynomial $2x^3 + 3x - 1$ as a single real number:

$$(x^2 + 5x - 4)(2x^3 + 3x - 1)$$

$$= x^2(2x^3 + 3x - 1) + 5x(2x^3 + 3x - 1) - 4(2x^3 + 3x - 1)$$

We next use another distributive property three times and simplify the result, obtaining

$$(x^2 + 5x - 4)(2x^3 + 3x - 1)$$

$$= 2x^5 + 3x^3 - x^2 + 10x^4 + 15x^2 - 5x - 8x^3 - 12x + 4$$

$$= 2x^5 + 10x^4 - 5x^3 + 14x^2 - 17x + 4.$$

(continued)

Note that the three monomials in the first polynomial were multiplied by each of the three monomials in the second polynomial, giving us a total of nine terms.

Method 2 We list the polynomials vertically and multiply, leaving spaces for powers of x that have zero coefficients, as follows:

$$
\begin{array}{l}
\quad 2x^3 \quad\; +3x \quad\; -1 \\
\quad\;\, x^2 \quad\; +5x \quad\; -4 \\
\hline
2x^5 \qquad\qquad +3x^3 \;-\; x^2 \qquad\qquad\qquad = x^2 \; (2x^3 + 3x - 1) \\
\qquad\quad 10x^4 \qquad\qquad\quad +15x^2 \;-\; 5x \qquad\quad = 5x \; (2x^3 + 3x - 1) \\
\qquad\qquad\qquad -8x^3 \qquad\qquad\quad -12x \;+4 = -4 \, (2x^3 + 3x - 1) \\
\hline
2x^5 \;+10x^4 \;-5x^3 \;+14x^2 \;-17x \;+4 = \text{sum of the above}
\end{array}
$$

In practice, we would omit the reasons (equalities) listed on the right in the last four lines.

We may consider polynomials in more than one variable. For example, a polynomial in *two* variables, x and y, is a finite sum of terms, each of the form $ax^m y^k$ for some real number a and nonnegative integers m and k. An example is

$$3x^4 y + 2x^3 y^5 + 7x^2 - 4xy + 8y - 5.$$

Other polynomials may involve three variables—such as x, y, z—or, for that matter, *any* number of variables. Addition, subtraction, and multiplication are performed using properties of real numbers, just as for polynomials in one variable.

The next example illustrates division of a polynomial by a monomial.

EXAMPLE 4   *Dividing a polynomial by a monomial*

Express as a polynomial in x and y:

$$\frac{6x^2 y^3 + 4x^3 y^2 - 10xy}{2xy}$$

SOLUTION

$$
\begin{aligned}
\frac{6x^2 y^3 + 4x^3 y^2 - 10xy}{2xy} &= \frac{6x^2 y^3}{2xy} + \frac{4x^3 y^2}{2xy} - \frac{10xy}{2xy} \qquad \text{divide each term by } 2xy \\
&= 3xy^2 + 2x^2 y - 5 \qquad\qquad\quad \text{simplify}
\end{aligned}
$$

The products listed in the next chart occur so frequently that they deserve special attention. You can check the validity of each formula by multiplication. In (2) and (3) the symbol $\pm$ is read "plus or minus." When employing these formulas, we use either the top sign on both sides or the bottom sign on both sides. Thus, (2) is actually *two* formulas:

$$(x + y)^2 = x^2 + 2xy + y^2 \quad \text{and} \quad (x - y)^2 = x^2 - 2xy + y^2$$

Similarly, (3) represents two formulas.

Product Formulas

Formula	Illustration
(1) $(x + y)(x - y) = x^2 - y^2$	$(2a + 3)(2a - 3) = (2a)^2 - 3^2$ $= 4a^2 - 9$
(2) $(x \pm y)^2 = x^2 \pm 2xy + y^2$	$(2a - 3)^2 = (2a)^2 - 2(2a)(3) + (3)^2$ $= 4a^2 - 12a + 9$
(3) $(x \pm y)^3 = x^3 \pm 3x^2y$ $+ 3xy^2 \pm y^3$	$(2a + 3)^3 = (2a)^3 + 3(2a)^2(3)$ $+ 3(2a)(3)^2 + (3)^3$ $= 8a^3 + 36a^2 + 54a + 27$

Several other illustrations of the product formulas are given in the next example.

EXAMPLE 5   *Using product formulas*

Find the product:

(a) $(2r^2 - \sqrt{s})(2r^2 + \sqrt{s})$ **(b)** $\left(\sqrt{c} + \dfrac{1}{\sqrt{c}}\right)^2$ **(c)** $(2a - 5b)^3$

SOLUTION

(a) We use product formula 1 with $x = 2r^2$ and $y = \sqrt{s}$:

$$(2r^2 - \sqrt{s})(2r^2 + \sqrt{s}) = (2r^2)^2 - (\sqrt{s})^2$$
$$= 4r^4 - s$$

(b) We use product formula 2 with $x = \sqrt{c}$ and $y = \dfrac{1}{\sqrt{c}}$:

$$\left(\sqrt{c} + \frac{1}{\sqrt{c}}\right)^2 = (\sqrt{c})^2 + 2\sqrt{c} \cdot \frac{1}{\sqrt{c}} + \left(\frac{1}{\sqrt{c}}\right)^2$$

$$= c + 2 + \frac{1}{c}$$

Note that the last expression is *not* a polynomial.

(c) We use product formula 3 with $x = 2a$ and $y = 5b$:

$$(2a - 5b)^3 = (2a)^3 - 3(2a)^2(5b) + 3(2a)(5b)^2 - (5b)^3$$
$$= 8a^3 - 60a^2b + 150ab^2 - 125b^3$$

If a polynomial is a product of other polynomials, then each polynomial in the product is a **factor** of the original polynomial. **Factoring** is the process of expressing a sum of terms as a product. For example, since $x^2 - 9 = (x + 3)(x - 3)$, the polynomials $x + 3$ and $x - 3$ are factors of $x^2 - 9$.

Factoring is an important process in mathematics, since it may be used to reduce the study of a complicated expression to the study of several simpler expressions. For example, properties of the polynomial $x^2 - 9$ can be determined by examining the factors $x + 3$ and $x - 3$. As we shall see in Chapter 2, another important use for factoring is in finding solutions of equations.

We shall be interested primarily in **nontrivial factors** of polynomials—that is, factors that contain polynomials of positive degree. However, if the coefficients are restricted to *integers*, then we usually remove a common integral factor from each term of the polynomial. For example,

$$4x^2y + 8z^3 = 4(x^2y + 2z^3).$$

A polynomial with coefficients in some set S of numbers is **prime**, or **irreducible** over S, if it cannot be written as a product of two polynomials of positive degree with coefficients in S. A polynomial may be irreducible over one set S but not over another. For example, $x^2 - 2$ is irreducible over the rational numbers, since it cannot be expressed as a product of two polynomials of positive degree that have *rational* coefficients. However, $x^2 - 2$ is *not* irreducible over the real numbers, since we can write

$$x^2 - 2 = (x + \sqrt{2})(x - \sqrt{2}).$$

Similarly, $x^2 + 1$ is irreducible over the real numbers, but, as we shall see in Section 2.4, not over the complex numbers.

Every polynomial $ax + b$ of degree 1 is irreducible.

Before we factor a polynomial, we must specify the number system (or set) from which the coefficients of the factors are to be chosen. In this chapter we shall use the rule that *if a polynomial has integral coefficients, then the factors should be polynomials with integral coefficients.* To **factor a polynomial** means to express it as a product of irreducible polynomials.

ILLUSTRATION

Factored Polynomials

■ $x^2 + x - 6 = (x + 3)(x - 2)$ ■ $4x^2 - 9y^2 = (2x - 3y)(2x + 3y)$

It is usually difficult to factor polynomials of degree greater than 2. In simple cases the following factoring formulas may be useful. Each formula can be verified by multiplying the factors on the right of the equals sign. It can be shown that the factors $x^2 + xy + y^2$ and $x^2 - xy + y^2$ in the difference and sum of two cubes, respectively, are irreducible over the real numbers.

Factoring Formulas

Formula	Illustration
(1) Difference of two squares: $x^2 - y^2 = (x + y)(x - y)$	$9a^2 - 16 = (3a)^2 - (4)^2 = (3a + 4)(3a - 4)$
(2) Difference of two cubes: $x^3 - y^3 = (x - y)(x^2 + xy + y^2)$	$\begin{aligned} 8a^3 - 27 &= (2a)^3 - (3)^3 \\ &= (2a - 3)[(2a)^2 + (2a)(3) + (3)^2] \\ &= (2a - 3)(4a^2 + 6a + 9) \end{aligned}$
(3) Sum of two cubes: $x^3 + y^3 = (x + y)(x^2 - xy + y^2)$	$\begin{aligned} 125a^3 + 1 &= (5a)^3 + (1)^3 \\ &= (5a + 1)[(5a)^2 - (5a)(1) + (1)^2] \\ &= (5a + 1)(25a^2 - 5a + 1) \end{aligned}$

Several other illustrations of the use of factoring formulas are given in the next two examples.

EXAMPLE 6 *Difference of two squares*

Factor the polynomial:

(a) $25r^2 - 49s^2$ **(b)** $81x^4 - y^4$ **(c)** $16x^4 - (y - 2z)^2$

SOLUTION

(a) We apply the difference of two squares formula, with $x = 5r$ and $y = 7s$:

$$25r^2 - 49s^2 = (5r)^2 - (7s)^2 = (5r + 7s)(5r - 7s)$$

(b) We write $81x^4 = (9x^2)^2$ and $y^4 = (y^2)^2$ and apply the difference of two squares formula twice:

$$\begin{aligned} 81x^4 - y^4 &= (9x^2)^2 - (y^2)^2 \\ &= (9x^2 + y^2)(9x^2 - y^2) \\ &= (9x^2 + y^2)[(3x)^2 - (y)^2] \\ &= (9x^2 + y^2)(3x + y)(3x - y) \end{aligned}$$

(c) We write $16x^4 = (4x^2)^2$ and apply the difference of two squares formula:

$$\begin{aligned} 16x^4 - (y - 2z)^2 &= (4x^2)^2 - (y - 2z)^2 \\ &= [(4x^2) + (y - 2z)][(4x^2) - (y - 2z)] \\ &= (4x^2 + y - 2z)(4x^2 - y + 2z) \end{aligned}$$

EXAMPLE 7 *Sum and difference of two cubes*

Factor the polynomial:

(a) $a^3 + 64b^3$ **(b)** $8c^6 - 27d^9$

SOLUTION

(a) We apply the sum of two cubes formula, with $x = a$ and $y = 4b$:

$$a^3 + 64b^3 = a^3 + (4b)^3$$
$$= (a + 4b)[a^2 - a(4b) + (4b)^2]$$
$$= (a + 4b)(a^2 - 4ab + 16b^2)$$

(b) We apply the difference of two cubes formula, with $x = 2c^2$ and $y = 3d^3$:

$$8c^6 - 27d^9 = (2c^2)^3 - (3d^3)^3$$
$$= (2c^2 - 3d^3)[(2c^2)^2 + (2c^2)(3d^3) + (3d^3)^2]$$
$$= (2c^2 - 3d^3)(4c^4 + 6c^2d^3 + 9d^6)$$

A factorization of a trinomial $px^2 + qx + r$, where p, q, and r are integers, must be of the form

$$px^2 + qx + r = (ax + b)(cx + d),$$

where a, b, c, and d are integers. It follows that

$$ac = p, \qquad bd = r, \qquad \text{and} \qquad ad + bc = q.$$

Only a limited number of choices for a, b, c, and d satisfy these conditions. If none of the choices work, then $px^2 + qx + r$ is irreducible. Trying the various possibilities, as depicted in the next example, is called the **method of trial and error**. This method is also applicable to trinomials of the form $px^2 + qxy + ry^2$, in which case the factorization must be of the form $(ax + by)(cx + dy)$.

EXAMPLE 8   *Factoring a trinomial by trial and error*

Factor $6x^2 - 7x - 3$.

SOLUTION   If we write

$$6x^2 - 7x - 3 = (ax + b)(cx + d),$$

then the following relationships must be true:

$$ac = 6, \qquad bd = -3, \qquad \text{and} \qquad ad + bc = -7$$

If we assume that a and c are both positive, then all possible values are given in the following table:

a	1	6	2	3
c	6	1	3	2

Thus, if $6x^2 - 7x - 3$ is factorable, then one of the following is true:

$$6x^2 - 7x - 3 = (x + b)(6x + d)$$
$$6x^2 - 7x - 3 = (6x + b)(x + d)$$
$$6x^2 - 7x - 3 = (2x + b)(3x + d)$$
$$6x^2 - 7x - 3 = (3x + b)(2x + d)$$

We next consider all possible values for b and d. Since $bd = -3$, these are as follows:

b	1	-1	3	-3
d	-3	3	-1	1

Trying various (possibly all) values, we arrive at $b = -3$ and $d = 1$; that is,

$$6x^2 - 7x - 3 = (2x - 3)(3x + 1).$$

As a check, you should multiply the final factorization to see whether the given polynomial is obtained.

The method of trial and error illustrated in Example 8 can be long and tedious if the coefficients of the polynomial are large and have many prime factors. For simple cases it is often possible to arrive at the correct choice rapidly.

EXAMPLE 9 *Factoring polynomials*

Factor:

(a) $12x^2 - 36xy + 27y^2$ **(b)** $4x^4y - 11x^3y^2 + 6x^2y^3$

SOLUTION

(a) Since each term has 3 as a factor, we begin by writing

$$12x^2 - 36xy + 27y^2 = 3(4x^2 - 12xy + 9y^2).$$

A factorization of $4x^2 - 12xy + 9y^2$ as a product of two first-degree polynomials must be of the form

$$4x^2 - 12xy + 9y^2 = (ax + by)(cx + dy)$$

with

$$ac = 4, \qquad bd = 9, \qquad \text{and} \qquad ad + bc = -12.$$

Using the method of trial and error, as in Example 8, we obtain

$$4x^2 - 12xy + 9y^2 = (2x - 3y)(2x - 3y) = (2x - 3y)^2.$$

(continued)

Thus,

$$12x^2 - 36xy + 27y^2 = 3(4x^2 - 12xy + 9y^2) = 3(2x - 3y)^2.$$

(b) Since each term has x^2y as a factor, we begin by writing

$$4x^4y - 11x^3y^2 + 6x^2y^3 = x^2y(4x^2 - 11xy + 6y^2).$$

By trial and error, we obtain the factorization

$$4x^4y - 11x^3y^2 + 6x^2y^3 = x^2y(4x - 3y)(x - 2y).$$

If a sum contains four or more terms, it may be possible to group the terms in a suitable manner and then find a factorization by using distributive properties. This technique, called **factoring by grouping**, is illustrated in the next example.

EXAMPLE 10   *Factoring by grouping*

Factor:

(a) $4ac + 2bc - 2ad - bd$ **(b)** $3x^3 + 2x^2 - 12x - 8$

(c) $x^2 - 16y^2 + 10x + 25$

SOLUTION

(a) We group the first two terms and the last two terms and then proceed as follows:

$$4ac + 2bc - 2ad - bd = (4ac + 2bc) - (2ad + bd)$$
$$= 2c(2a + b) - d(2a + b)$$

At this stage we have not factored the given expression because the right-hand side has the form

$$2ck - dk \quad \text{with } k = 2a + b.$$

However, if we factor out k, then

$$2ck - dk = (2c - d)k = (2c - d)(2a + b).$$

Hence,

$$4ac + 2bc - 2ad - bd = 2c(2a + b) - d(2a + b)$$
$$= (2c - d)(2a + b).$$

(b) We group the first two terms and the last two terms and then proceed as follows:

$$3x^3 + 2x^2 - 12x - 8 = (3x^3 + 2x^2) - (12x + 8)$$
$$= x^2(3x + 2) - 4(3x + 2)$$
$$= (x^2 - 4)(3x + 2)$$

Finally, using the difference of two squares formula for $x^2 - 4$, we obtain the factorization:

$$3x^3 + 2x^2 - 12x - 8 = (x + 2)(x - 2)(3x + 2)$$

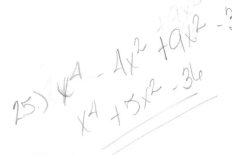

(c) First we rearrange and group terms, and then we apply the difference of two squares formula, as follows:

$$x^2 - 16y^2 + 10x + 25 = (x^2 + 10x + 25) - 16y^2$$
$$= (x + 5)^2 - (4y)^2$$
$$= [(x + 5) + 4y][(x + 5) - 4y]$$
$$= (x + 4y + 5)(x - 4y + 5)$$

1.3 EXERCISES

Exer. 1–44: Express as a polynomial.

1 $(3x^3 + 4x^2 - 7x + 1) + (9x^3 - 4x^2 - 6x)$

2 $(7x^3 + 2x^2 - 11x) + (-3x^3 - 2x^2 + 5x - 3)$

3 $(4x^3 + 5x - 3) - (3x^3 + 2x^2 + 5x - 7)$

4 $(6x^3 - 2x^2 + x - 2) - (8x^2 + x + 2)$

5 $(2x + 5)(3x - 7)$ 6 $(3x - 4)(2x + 9)$

7 $(5x + 7y)(3x + 2y)$ 8 $(4x - 3y)(x - 5y)$

9 $(2u + 3)(u - 4) + 4u(u - 2)$

10 $(3u - 1)(u + 2) + 7u(u + 1)$

11 $(3x + 5)(2x^2 + 9x - 5)$

12 $(7x - 4)(x^3 - x^2 + 6)$

13 $(t^2 + 2t - 5)(3t^2 - t + 2)$

14 $(r^2 - 8r - 2)(-r^2 + 3r - 1)$

15 $(x + 1)(2x^2 - 2)(x^3 + 5)$

16 $(2x - 1)(x^2 - 5)(x^3 - 1)$

17 $\dfrac{8x^2y^3 - 10x^3y}{2x^2y}$

18 $\dfrac{6a^3b^3 - 9a^2b^2 + 3ab^4}{3ab^2}$

19 $\dfrac{3u^3v^4 - 2u^5v^2 + (u^2v^2)^2}{u^3v^2}$

20 $\dfrac{6x^2yz^3 - xy^2z}{xyz}$

21 $(2x + 3y)(2x - 3y)$

22 $(5x + 4y)(5x - 4y)$ 23 $(x^2 + 2y)(x^2 - 2y)$

24 $(3x + y^3)(3x - y^3)$ 25 $(x^2 + 9)(x^2 - 4)$

26 $(x^2 + 1)(x^2 - 16)$ 27 $(3x + 2y)^2$

28 $(5x - 4y)^2$ 29 $(x^2 - 3y^2)^2$

30 $(2x^2 + 5y^2)^2$ 31 $(x + 2)^2(x - 2)^2$

32 $(x + y)^2(x - y)^2$ 33 $(\sqrt{x} + \sqrt{y})(\sqrt{x} - \sqrt{y})$

34 $(\sqrt{x} + \sqrt{y})^2(\sqrt{x} - \sqrt{y})^2$

35 $(x^{1/3} - y^{1/3})(x^{2/3} + x^{1/3}y^{1/3} + y^{2/3})$

36 $(x^{1/3} + y^{1/3})(x^{2/3} - x^{1/3}y^{1/3} + y^{2/3})$

37 $(x - 2y)^3$ 38 $(x + 3y)^3$

39 $(2x + 3y)^3$ 40 $(3x - 4y)^3$

41 $(a + b - c)^2$ 42 $(x^2 + x + 1)^2$

43 $(2x + y - 3z)^2$ 44 $(x - 2y + 3z)^2$

Exer. 45–100: Factor the polynomial.

45 $rs + 4st$ 46 $4u^2 - 2uv$

47 $3a^2b^2 - 6a^2b$ 48 $10xy + 15xy^2$

49 $3x^2y^3 - 9x^3y^2$ 50 $16x^5y^2 + 8x^3y^3$

51 $15x^3y^5 - 25x^4y^2 + 10x^6y^4$

52 $121r^3s^4 + 77r^2s^4 - 55r^4s^3$

53 $8x^2 - 53x - 21$ 54 $7x^2 + 10x - 8$

55 $x^2 + 3x + 4$ 56 $3x^2 - 4x + 2$

57 $6x^2 + 7x - 20$ 58 $12x^2 - x - 6$

59 $12x^2 - 29x + 15$ 60 $21x^2 + 41x + 10$

61 $4x^2 - 20x + 25$ 62 $9x^2 + 24x + 16$

63 $25z^2 + 30z + 9$ 64 $16z^2 - 56z + 49$

65 $45x^2 + 38xy + 8y^2$ 66 $50x^2 + 45xy - 18y^2$

67 $36r^2 - 25t^2$ 68 $81r^2 - 16t^2$

69 $z^4 - 64w^2$ 70 $9y^4 - 121x^2$

71 $x^4 - 4x^2$ 72 $x^3 - 25x$

73 $x^2 + 25$ 74 $4x^2 + 9$

$(4x+3)(16x^2 - 12x + 9)$

75 $75x^2 - 48y^2$

76 $64x^2 - 36y^2$

77 $64x^3 + 27$

78 $125x^3 - 8$

79 $64x^3 - y^6$

80 $216x^9 + 125y^3$

81 $343x^3 + y^9$

82 $x^6 - 27y^3$

83 $2ax - 6bx + ay - 3by$

84 $2ay^2 - axy + 6xy - 3x^2$

✳ 85 $3x^3 + 3x^2 - 27x - 27$

86 $5x^3 + 10x^2 - 20x - 40$

87 $x^4 + 2x^3 - x - 2$

88 $x^4 - 3x^3 + 8x - 24$

89 $a^3 - a^2b + ab^2 - b^3$

90 $6w^8 + 17w^4 + 12$

91 $a^6 - b^6$

92 $x^8 - 16$

93 $x^2 + 4x + 4 - 9y^2$

94 $x^2 - 4y^2 - 6x + 9$

95 $y^2 - x^2 + 8y + 16$

96 $y^2 + 9 - 6y - 4x^2$

97 $y^6 + 7y^3 - 8$

98 $8c^6 + 19c^3 - 27$

99 $x^{16} - 1$

100 $4x^3 + 4x^2 + x$

Exer. 101–102: The ancient Greeks gave geometric proofs of the factoring formulas for the difference of two squares and the difference of two cubes. Establish the formula for the special case described.

101 Find the areas of regions I and II in the figure to establish the difference of two squares formula for the special case $x > y$.

EXERCISE 101

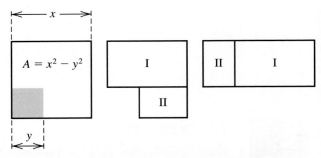

102 Find the volumes of boxes I, II, and III in the figure to establish the difference of two cubes formula for the special case $x > y$.

EXERCISE 102

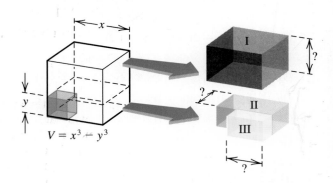

1.4 FRACTIONAL EXPRESSIONS

A **fractional expression** is a quotient of two algebraic expressions. As a special case, a **rational expression** is a quotient p/q of two *polynomials* p and q. Since division by zero is not allowed, the domain of p/q consists of all real numbers except those that make the denominator zero. Two illustrations are given in the next chart.

Rational Expressions

Example	Domain
$\dfrac{6x^2 - 5x + 4}{x^2 - 9}$	All $x \neq \pm 3$
$\dfrac{x^3 - 3x^2y + 4y^2}{y - x^3}$	All x and y such that $y \neq x^3$

In most of our work we will be concerned with rational expressions in which both numerator and denominator are polynomials in only one variable.

Since the variables in a rational expression represent real numbers, we may use the properties of quotients in Section 1.1, replacing the letters a, b, c, and d with polynomials. The following property is of particular importance, where $bd \neq 0$:

$$\frac{ad}{bd} = \frac{a}{b} \cdot \frac{d}{d} = \frac{a}{b} \cdot 1 = \frac{a}{b}$$

We sometimes describe this simplification process by saying that *a common nonzero factor in the numerator and denominator of a quotient may be canceled*. In practice, we usually show this cancellation by means of a slash through the common factor, as in the following illustration, where all denominators are assumed to be nonzero.

ILLUSTRATION

Canceled Common Factors

■ $\dfrac{a\cancel{d}}{b\cancel{d}} = \dfrac{a}{b}$ ■ $\dfrac{m\cancel{n}}{\cancel{n}pq} = \dfrac{m}{pq}$ ■ $\dfrac{\cancel{p}q\cancel{r}}{\cancel{r}p v} = \dfrac{q}{v}$

A rational expression is *simplified*, or *reduced to lowest terms*, if the numerator and the denominator have no common polynomial factors of positive degree and no common integral factors greater than 1. To simplify a rational expression, we factor both the numerator and the denominator into prime factors and then, assuming the factors in the denominator are not zero, cancel common factors, as in the following illustration.

ILLUSTRATION

Simplified Rational Expressions

■ $\dfrac{3x^2 - 5x - 2}{x^2 - 4} = \dfrac{(3x + 1)(x - 2)}{(x + 2)(x - 2)} \overset{\text{if } x \neq 2}{=} \dfrac{3x + 1}{x + 2}$

■ $\dfrac{2 - x - 3x^2}{6x^2 - x - 2} = \dfrac{-(3x^2 + x - 2)}{6x^2 - x - 2} = -\dfrac{(3x - 2)(x + 1)}{(3x - 2)(2x + 1)} \overset{\text{if } x \neq 2}{=} -\dfrac{x + 1}{2x + 1}$

■ $\dfrac{(x^2 + 8x + 16)(x - 5)}{(x^2 - 5x)(x^2 - 16)} = \dfrac{(x + 4)^2(x - 5)}{x(x - 5)(x + 4)(x - 4)} \overset{\text{if } x \neq 5, \, x \neq -4}{=} \dfrac{x + 4}{x(x - 4)}$

As shown in the next example, when simplifying a product or quotient of rational expressions, we often use properties of quotients to obtain one rational expression. Then we factor the numerator and denominator and cancel common factors, as we did in the preceding illustration.

EXAMPLE I   *Products and quotients of rational expressions*

Perform the indicated operation and simplify:

(a) $\dfrac{x^2 - 6x + 9}{x^2 - 1} \cdot \dfrac{2x - 2}{x - 3}$ (b) $\dfrac{x + 2}{2x - 3} \div \dfrac{x^2 - 4}{2x^2 - 3x}$

SOLUTION

(a) $\dfrac{x^2 - 6x + 9}{x^2 - 1} \cdot \dfrac{2x - 2}{x - 3} = \dfrac{(x^2 - 6x + 9)(2x - 2)}{(x^2 - 1)(x - 3)}$ property of quotients

$= \dfrac{(x - 3)^2 \cdot 2(x - 1)}{(x + 1)(x - 1)(x - 3)}$ factor all polynomials

if $x \neq 3, x \neq 1$

$= \dfrac{2(x - 3)}{x + 1}$ cancel common factors

(b) $\dfrac{x + 2}{2x - 3} \div \dfrac{x^2 - 4}{2x^2 - 3x} = \dfrac{x + 2}{2x - 3} \cdot \dfrac{2x^2 - 3x}{x^2 - 4}$ property of quotients

$= \dfrac{(x + 2)x(2x - 3)}{(2x - 3)(x + 2)(x - 2)}$ property of quotients factor all polynomials

if $x \neq -2, x \neq \frac{3}{2}$

$= \dfrac{x}{x - 2}$ cancel common factors

To add or subtract two rational expressions, we usually find a common denominator and use the following properties of quotients:

$$\frac{a}{d} + \frac{c}{d} = \frac{a + c}{d}; \qquad \frac{a}{d} - \frac{c}{d} = \frac{a - c}{d}$$

If the denominators of the expressions are not the same, we may obtain a common denominator by multiplying the numerator and denominator of each fraction by a suitable expression. We usually use the **least common denominator (lcd)** of the two quotients. To find the lcd we factor each denominator into primes and then form the product of the different prime factors, using the *largest* exponent that appears with each prime factor. Let us begin with a numerical example of this technique.

EXAMPLE 2   *Adding fractions using the lcd*

Express as a simplified rational number:

$$\frac{7}{24} + \frac{5}{18}$$

SOLUTION The prime factorizations of the denominators 24 and 18 are $24 = 2^3 \cdot 3$ and $18 = 2 \cdot 3^2$. To find the lcd we form the product of the different prime factors, using the largest exponent associated with each factor. This gives us $2^3 \cdot 3^2$. We now change each fraction to an equivalent fraction with denominator $2^3 \cdot 3^2$ and add:

$$\frac{7}{24} + \frac{5}{18} = \frac{7}{2^3 \cdot 3} + \frac{5}{2 \cdot 3^2}$$

$$= \frac{7}{2^3 \cdot 3} \cdot \frac{3}{3} + \frac{5}{2 \cdot 3^2} \cdot \frac{2^2}{2^2}$$

$$= \frac{21}{2^3 \cdot 3^2} + \frac{20}{2^3 \cdot 3^2}$$

$$= \frac{41}{2^3 \cdot 3^2}$$

$$= \frac{41}{72}$$

The method for finding the lcd for rational expressions is analogous to the process illustrated in Example 2. The only difference is that we use factorizations of polynomials instead of integers.

EXAMPLE 3 *Sums and differences of rational expressions*

Perform the operations and simplify:

$$\frac{6}{x(3x - 2)} + \frac{5}{3x - 2} - \frac{2}{x^2}$$

SOLUTION The denominators are already in factored form. The lcd is $x^2(3x - 2)$. To obtain three quotients having the denominator $x^2(3x - 2)$, we multiply the numerator and denominator of the first quotient by x, those of the second by x^2, and those of the third by $3x - 2$, which gives us

$$\frac{6}{x(3x - 2)} + \frac{5}{3x - 2} - \frac{2}{x^2} = \frac{6}{x(3x - 2)} \cdot \frac{x}{x} + \frac{5}{3x - 2} \cdot \frac{x^2}{x^2} - \frac{2}{x^2} \cdot \frac{3x - 2}{3x - 2}$$

$$= \frac{6x}{x^2(3x - 2)} + \frac{5x^2}{x^2(3x - 2)} - \frac{2(3x - 2)}{x^2(3x - 2)}$$

$$= \frac{6x + 5x^2 - 2(3x - 2)}{x^2(3x - 2)}$$

$$= \frac{5x^2 + 4}{x^2(3x - 2)}.$$

EXAMPLE 4   *Simplifying sums of rational expressions*

Perform the operations and simplify:

$$\frac{2x+5}{x^2+6x+9}+\frac{x}{x^2-9}+\frac{1}{x-3}$$

SOLUTION   We begin by factoring denominators:

$$\frac{2x+5}{x^2+6x+9}+\frac{x}{x^2-9}+\frac{1}{x-3}=\frac{2x+5}{(x+3)^2}+\frac{x}{(x+3)(x-3)}+\frac{1}{x-3}$$

Since the lcd is $(x+3)^2(x-3)$, we multiply the numerator and denominator of the first quotient by $x-3$, those of the second by $x+3$, and those of the third by $(x+3)^2$ and then add:

$$\frac{(2x+5)(x-3)}{(x+3)^2(x-3)}+\frac{x(x+3)}{(x+3)^2(x-3)}+\frac{(x+3)^2}{(x+3)^2(x-3)}$$

$$=\frac{(2x^2-x-15)+(x^2+3x)+(x^2+6x+9)}{(x+3)^2(x-3)}$$

$$=\frac{4x^2+8x-6}{(x+3)^2(x-3)}=\frac{2(2x^2+4x-3)}{(x+3)^2(x-3)}$$

It is sometimes necessary to simplify a **complex fraction**—that is, a quotient in which the numerator and/or denominator is a fractional expression—as illustrated in the following examples.

EXAMPLE 5   *Simplifying complex fractions*

Simplify

$$\frac{1-\dfrac{2}{x+1}}{x-\dfrac{1}{x}}.$$

SOLUTION   We change the numerator and denominator of the given expression into single quotients and then use the rule for dividing quotients.

$$\frac{1-\dfrac{2}{x+1}}{x-\dfrac{1}{x}}=\frac{\dfrac{(x+1)-2}{x+1}}{\dfrac{x^2-1}{x}}$$ combine expressions in numerator and denominator

$$=\frac{\dfrac{x-1}{x+1}}{\dfrac{x^2-1}{x}}$$ simplify

$$= \frac{x-1}{x+1} \cdot \frac{x}{x^2-1} \qquad \text{property of quotients}$$

$$= \frac{(x-1)x}{(x+1)(x+1)(x-1)} \qquad \begin{array}{l}\text{property of quotients;}\\ \text{factor } x^2-1\end{array}$$

$$\text{if } x \neq 1$$
$$\downarrow$$
$$= \frac{x}{(x+1)^2} \qquad \text{cancel } x-1$$

An alternative method is to multiply the numerator and denominator of the given expression by $x(x+1)$, the lcd of $2/(x+1)$ and $1/x$, and then simplify the result.

Some quotients that are not rational expressions contain denominators of the form $a + \sqrt{b}$ or $\sqrt{a} + \sqrt{b}$. As in the next example, these quotients can be simplified by multiplying the numerator and denominator by the **conjugate** $a - \sqrt{b}$ or $\sqrt{a} - \sqrt{b}$, respectively. Of course, if $a - \sqrt{b}$ appears, multiply by $a + \sqrt{b}$ instead.

EXAMPLE 6 *Rationalizing a denominator*

Rationalize the denominator of

$$\frac{1}{\sqrt{x} + \sqrt{y}}.$$

SOLUTION

$$\frac{1}{\sqrt{x}+\sqrt{y}} = \frac{1}{\sqrt{x}+\sqrt{y}} \cdot \frac{\sqrt{x}-\sqrt{y}}{\sqrt{x}-\sqrt{y}} \qquad \begin{array}{l}\text{multiply numerator and denominator}\\ \text{by the conjugate of } \sqrt{x}+\sqrt{y}\end{array}$$

$$= \frac{\sqrt{x}-\sqrt{y}}{(\sqrt{x})^2-(\sqrt{y})^2} \qquad \begin{array}{l}\text{property of quotients and difference}\\ \text{of squares}\end{array}$$

$$= \frac{\sqrt{x}-\sqrt{y}}{x-y} \qquad \text{law of radicals}$$

In calculus it is sometimes necessary to rationalize the *numerator* of a quotient, as shown in the following example.

EXAMPLE 7 *Rationalizing a numerator*

If $h \neq 0$, rationalize the numerator of

$$\frac{\sqrt{x+h} - \sqrt{x}}{h}.$$

SOLUTION

$$\frac{\sqrt{x + h} - \sqrt{x}}{h} = \frac{\sqrt{x + h} - \sqrt{x}}{h} \cdot \frac{\sqrt{x + h} + \sqrt{x}}{\sqrt{x + h} + \sqrt{x}}$$

multiply numerator and denominator by the conjugate of $\sqrt{x + h} - \sqrt{x}$

$$= \frac{(\sqrt{x + h})^2 - (\sqrt{x})^2}{h(\sqrt{x + h} + \sqrt{x})}$$

property of quotients and difference of two squares

$$= \frac{(x + h) - x}{h(\sqrt{x + h} + \sqrt{x})}$$

law of radicals

$$= \frac{\cancel{h}}{\cancel{h}(\sqrt{x + h} + \sqrt{x})}$$

simplify

$$= \frac{1}{\sqrt{x + h} + \sqrt{x}}$$

cancel $h \neq 0$

It may seem as though we have accomplished very little, since radicals occur in the denominator. In calculus, however, it is of interest to determine what is true if h is very close to zero. Note that if we use the *given* expression we obtain the following:

If $h \approx 0$, then $\dfrac{\sqrt{x + h} - \sqrt{x}}{h} \approx \dfrac{\sqrt{x + 0} - \sqrt{x}}{0} = \dfrac{0}{0}$,

a meaningless expression. If we use the *rationalized* form, however, we obtain the following information:

If $h \approx 0$, then $\dfrac{\sqrt{x + h} - \sqrt{x}}{h} = \dfrac{1}{\sqrt{x + h} + \sqrt{x}} \approx \dfrac{1}{\sqrt{x} + \sqrt{x}} = \dfrac{1}{2\sqrt{x}}$.

Certain problems in calculus require simplifying expressions of the type given in the next example.

EXAMPLE 8 *Simplifying a fractional expression*

Simplify, if $h \neq 0$:

$$\frac{\dfrac{1}{(x + h)^2} - \dfrac{1}{x^2}}{h}$$

SOLUTION

$$\frac{\dfrac{1}{(x+h)^2} - \dfrac{1}{x^2}}{h} = \frac{\dfrac{x^2 - (x+h)^2}{(x+h)^2 x^2}}{h} \qquad \text{combine quotients in numerator}$$

$$= \frac{x^2 - (x^2 + 2xh + h^2)}{(x+h)^2 x^2} \cdot \frac{1}{h} \qquad \begin{array}{l}\text{square } x+h;\text{ use a property}\\ \text{of quotients}\end{array}$$

$$= \frac{x^2 - x^2 - 2xh - h^2}{(x+h)^2 x^2 h} \qquad \text{remove parentheses}$$

$$= \frac{-\cancel{h}(2x + h)}{(x+h)^2 x^2 \cancel{h}} \qquad \text{simplify; factor out } -h$$

$$= -\frac{2x + h}{(x+h)^2 x^2} \qquad \text{cancel } h \neq 0$$

Problems of the type given in the next example also occur in calculus.

EXAMPLE 9 *Simplifying a fractional expression*

Simplify

$$\frac{3x^2(2x+5)^{1/2} - x^3(\tfrac{1}{2})(2x+5)^{-1/2}(2)}{[(2x+5)^{1/2}]^2}.$$

SOLUTION One way to simplify the expression is as follows:

$$\frac{3x^2(2x+5)^{1/2} - x^3(\tfrac{1}{2})(2x+5)^{-1/2}(2)}{[(2x+5)^{1/2}]^2}$$

$$= \frac{3x^2(2x+5)^{1/2} - \dfrac{x^3}{(2x+5)^{1/2}}}{2x+5} \qquad \text{definition of negative exponent}$$

$$= \frac{\dfrac{3x^2(2x+5) - x^3}{(2x+5)^{1/2}}}{2x+5} \qquad \text{combine terms in numerator}$$

$$= \frac{6x^3 + 15x^2 - x^3}{(2x+5)^{1/2}} \cdot \frac{1}{2x+5} \qquad \text{property of quotients}$$

$$= \frac{5x^3 + 15x^2}{(2x+5)^{3/2}} \qquad \text{simplify}$$

$$= \frac{5x^2(x+3)}{(2x+5)^{3/2}} \qquad \text{factor numerator}$$

(continued)

An alternative simplification is to eliminate the negative power $(2x + 5)^{-1/2}$ in the given expression, as follows:

$$\frac{3x^2(2x + 5)^{1/2} - x^3(\tfrac{1}{2})(2x + 5)^{-1/2}(2)}{[(2x + 5)^{1/2}]^2} \cdot \frac{(2x + 5)^{1/2}}{(2x + 5)^{1/2}}$$ multiply numerator and denominator by $(2x + 5)^{1/2}$

$$= \frac{3x^2(2x + 5) - x^3}{(2x + 5)(2x + 5)^{1/2}}$$ property of quotients and law of exponents

The remainder of the simplification is similar.

A third method of simplification is to first factor out the *greatest common factor* (*gcf*) of the numerator. The gcf is the product of the common factors, each raised to the smallest nonzero exponent. In this case, the common factors are x and $(2x + 5)$, and the smallest exponents are 2 and $-\tfrac{1}{2}$, respectively. Thus, the gcf is $x^2(2x + 5)^{-1/2}$, and we factor the numerator and simplify as follows:

$$\frac{x^2(2x + 5)^{-1/2}[3(2x + 5)^1 - x]}{(2x + 5)^1} = \frac{x^2(5x + 15)}{(2x + 5)^{3/2}} = \frac{5x^2(x + 3)}{(2x + 5)^{3/2}}$$

1.4 EXERCISES

Exer. 1–4: Write the expression as a simplified rational number.

1 $\dfrac{3}{50} + \dfrac{7}{30}$

2 $\dfrac{4}{63} + \dfrac{5}{42}$

3 $\dfrac{5}{24} - \dfrac{3}{20}$

4 $\dfrac{11}{54} - \dfrac{7}{72}$

Exer. 5–44: Simplify the expression.

5 $\dfrac{2x^2 + 7x + 3}{2x^2 - 7x - 4}$

6 $\dfrac{2x^2 + 9x - 5}{3x^2 + 17x + 10}$

7 $\dfrac{y^2 - 25}{y^3 - 125}$

8 $\dfrac{y^2 - 9}{y^3 + 27}$

9 $\dfrac{12 + r - r^2}{r^3 + 3r^2}$

10 $\dfrac{10 + 3r - r^2}{r^4 + 2r^3}$

11 $\dfrac{9x^2 - 4}{3x^2 - 5x + 2} \cdot \dfrac{9x^4 - 6x^3 + 4x^2}{27x^4 + 8x}$

12 $\dfrac{4x^2 - 9}{2x^2 + 7x + 6} \cdot \dfrac{4x^4 + 6x^3 + 9x^2}{8x^7 - 27x^4}$

13 $\dfrac{5a^2 + 12a + 4}{a^4 - 16} \div \dfrac{25a^2 + 20a + 4}{a^2 - 2a}$

14 $\dfrac{a^3 - 8}{a^2 - 4} \div \dfrac{a}{a^3 + 8}$

15 $\dfrac{6}{x^2 - 4} - \dfrac{3x}{x^2 - 4}$

16 $\dfrac{15}{x^2 - 9} - \dfrac{5x}{x^2 - 9}$

17 $\dfrac{2}{3s + 1} - \dfrac{9}{(3s + 1)^2}$

18 $\dfrac{4}{(5s - 2)^2} + \dfrac{s}{5s - 2}$

19 $\dfrac{2}{x} + \dfrac{3x + 1}{x^2} - \dfrac{x - 2}{x^3}$

20 $\dfrac{5}{x} - \dfrac{2x - 1}{x^2} + \dfrac{x + 5}{x^3}$

21 $\dfrac{3t}{t + 2} + \dfrac{5t}{t - 2} - \dfrac{40}{t^2 - 4}$

22 $\dfrac{t}{t + 3} + \dfrac{4t}{t - 3} - \dfrac{18}{t^2 - 9}$

23 $\dfrac{4x}{3x - 4} + \dfrac{8}{3x^2 - 4x} + \dfrac{2}{x}$

24 $\dfrac{12x}{2x + 1} - \dfrac{3}{2x^2 + x} + \dfrac{5}{x}$

25 $\dfrac{2x}{x + 2} - \dfrac{8}{x^2 + 2x} + \dfrac{3}{x}$

26 $\dfrac{5x}{2x + 3} - \dfrac{6}{2x^2 + 3x} + \dfrac{2}{x}$

27 $\dfrac{p^4 + 3p^3 - 8p - 24}{p^3 - 2p^2 - 9p + 18}$

28 $\dfrac{2ac + bc - 6ad - 3bd}{6ac + 2ad + 3bc + bd}$

29 $3 + \dfrac{5}{u} + \dfrac{2u}{3u + 1}$

30 $4 + \dfrac{2}{u} - \dfrac{3u}{u + 5}$

31 $\dfrac{2x+1}{x^2+4x+4} - \dfrac{6x}{x^2-4} + \dfrac{3}{x-2}$

32 $\dfrac{2x+6}{x^2+6x+9} + \dfrac{5x}{x^2-9} + \dfrac{7}{x-3}$

33 $\dfrac{\dfrac{b}{a} - \dfrac{a}{b}}{\dfrac{1}{a} - \dfrac{1}{b}}$

34 $\dfrac{\dfrac{1}{x+2} - 3}{\dfrac{4}{x} - x}$

35 $\dfrac{\dfrac{x}{y^2} - \dfrac{y}{x^2}}{\dfrac{1}{y^2} - \dfrac{1}{x^2}}$

36 $\dfrac{\dfrac{r}{s} + \dfrac{s}{r}}{\dfrac{r^2}{s^2} - \dfrac{s^2}{r^2}}$

37 $\dfrac{\dfrac{5}{x+1} + \dfrac{2x}{x+3}}{\dfrac{x}{x+1} + \dfrac{7}{x+3}}$

38 $\dfrac{\dfrac{3}{w} - \dfrac{6}{2w+1}}{\dfrac{5}{w} + \dfrac{8}{2w+1}}$

39 $\dfrac{(x+h)^2 - 3(x+h) - (x^2 - 3x)}{h}$

40 $\dfrac{(x+h)^3 + 5(x+h) - (x^3 + 5x)}{h}$

41 $\dfrac{\dfrac{1}{(x+h)^3} - \dfrac{1}{x^3}}{h}$

42 $\dfrac{\dfrac{1}{x+h} - \dfrac{1}{x}}{h}$

43 $\dfrac{\dfrac{4}{3x+3h-1} - \dfrac{4}{3x-1}}{h}$

44 $\dfrac{\dfrac{5}{2x+2h+3} - \dfrac{5}{2x+3}}{h}$

Exer. 45–50: Rationalize the denominator.

45 $\dfrac{\sqrt{t}+5}{\sqrt{t}-5}$

46 $\dfrac{\sqrt{t}-4}{\sqrt{t}+4}$

47 $\dfrac{81x^2 - 16y^2}{3\sqrt{x} - 2\sqrt{y}}$

48 $\dfrac{16x^2 - y^2}{2\sqrt{x} - \sqrt{y}}$

49 $\dfrac{1}{\sqrt[3]{a} - \sqrt[3]{b}}$ (*Hint:* Multiply numerator and denominator by $\sqrt[3]{a^2} + \sqrt[3]{ab} + \sqrt[3]{b^2}$.)

50 $\dfrac{1}{\sqrt[3]{x} + \sqrt[3]{y}}$

Exer. 51–56: Rationalize the numerator.

51 $\dfrac{\sqrt{a} - \sqrt{b}}{a^2 - b^2}$

52 $\dfrac{\sqrt{b} + \sqrt{c}}{b^2 - c^2}$

53 $\dfrac{\sqrt{2(x+h)+1} - \sqrt{2x+1}}{h}$

54 $\dfrac{\sqrt{x} - \sqrt{x+h}}{h\sqrt{x}\sqrt{x+h}}$

55 $\dfrac{\sqrt{1-x-h} - \sqrt{1-x}}{h}$

56 $\dfrac{\sqrt[3]{x+h} - \sqrt[3]{x}}{h}$ (*Hint:* Compare with Exercise 49.)

Exer. 57–60: Express as a sum of terms of the form ax^r, where r is a rational number.

57 $\dfrac{4x^2 - x + 5}{x^{2/3}}$

58 $\dfrac{x^2 + 4x - 6}{\sqrt{x}}$

59 $\dfrac{(x^2 + 2)^2}{x^5}$

60 $\dfrac{(\sqrt{x} - 3)^2}{x^3}$

Exer. 61–64: Express as a quotient.

61 $x^{-3} + x^2$

62 $x^{-4} - x$

63 $x^{-1/2} - x^{3/2}$

64 $x^{-2/3} + x^{7/3}$

Exer. 65–76: Simplify the expression.

65 $(2x^2 - 3x + 1)(4)(3x+2)^3(3) + (3x+2)^4(4x-3)$

66 $(6x-5)^3(2)(x^2+4)(2x) + (x^2+4)^2(3)(6x-5)^2(6)$

67 $(x^2-4)^{1/2}(3)(2x+1)^2(2) + (2x+1)^3(\tfrac{1}{2})(x^2-4)^{-1/2}(2x)$

68 $(3x+2)^{1/3}(2)(4x-5)(4) + (4x-5)^2(\tfrac{1}{3})(3x+2)^{-2/3}(3)$

69 $(3x+1)^6(\tfrac{1}{2})(2x-5)^{-1/2}(2) + (2x-5)^{1/2}(6)(3x+1)^5(3)$

70 $(x^2+9)^4(-\tfrac{1}{3})(x+6)^{-4/3} + (x+6)^{-1/3}(4)(x^2+9)^3(2x)$

71 $\dfrac{(6x+1)^3(27x^2+2) - (9x^3+2x)(3)(6x+1)^2(6)}{(6x+1)^6}$

72 $\dfrac{(x^2-1)^4(2x) - x^2(4)(x^2-1)^3(2x)}{(x^2-1)^8}$

73 $\dfrac{(x^2+4)^{1/3}(3) - (3x)(\tfrac{1}{3})(x^2+4)^{-2/3}(2x)}{[(x^2+4)^{1/3}]^2}$

74 $\dfrac{(1-x^2)^{1/2}(2x) - x^2(\tfrac{1}{2})(1-x^2)^{-1/2}(-2x)}{[(1-x^2)^{1/2}]^2}$

75 $\dfrac{(4x^2+9)^{1/2}(2) - (2x+3)(\tfrac{1}{2})(4x^2+9)^{-1/2}(8x)}{[(4x^2+9)^{1/2}]^2}$

76 $\dfrac{(3x+2)^{1/2}(\tfrac{1}{3})(2x+3)^{-2/3}(2) - (2x+3)^{1/3}(\tfrac{1}{2})(3x+2)^{-1/2}(3)}{[(3x+2)^{1/2}]^2}$

(*continued*)

CHAPTER I REVIEW EXERCISES

1 Express as a simplified rational number:

  (a) $(\frac{2}{3})(-\frac{5}{8})$ (b) $\frac{3}{4} + \frac{6}{5}$ (c) $\frac{5}{8} - \frac{6}{7}$ (d) $\frac{3}{4} \div \frac{6}{5}$

2 Replace the symbol □ with either <, >, or = to make the resulting statement true.

  (a) -0.1 □ -0.001 (b) $\sqrt{9}$ □ -3 (c) $\frac{1}{6}$ □ 0.166

3 Express as an inequality:

  (a) x is negative.

  (b) a is between $\frac{1}{2}$ and $\frac{1}{3}$.

  (c) The absolute value of x is not greater than 4.

4 Rewrite without using the absolute value symbol, and simplify:

  (a) $|-7|$ (b) $\dfrac{|-5|}{-5}$ (c) $|3^{-1} - 2^{-1}|$

5 If points A, B, and C on a coordinate line have coordinates -8, 4, and -3, respectively, find the distance:

  (a) $d(A, C)$ (b) $d(C, A)$ (c) $d(B, C)$

6 Determine whether the expression is true for all values of the variables, whenever the expression is defined.

  (a) $(x + y)^2 = x^2 + y^2$ (b) $\dfrac{1}{\sqrt{x + y}} = \dfrac{1}{\sqrt{x}} + \dfrac{1}{\sqrt{y}}$

  (c) $\dfrac{1}{\sqrt{c} - \sqrt{d}} = \dfrac{\sqrt{c} + \sqrt{d}}{c - d}$

7 Express the number in scientific form.

  (a) 93,700,000,000   (b) 0.000 004 02

8 Express the number in decimal form.

  (a) 6.8×10^7 (b) 7.3×10^{-4}

Exer. 9–10: Rewrite the expression without using the absolute value symbol, and simplify the result.

9 $|x + 3|$ if $x \leq -3$

10 $|(x - 2)(x - 3)|$ if $2 < x < 3$

Exer. 11–12: Express the number in the form a/b, where a and b are integers.

11 $-3^2 + 2^0 + 27^{-2/3}$ 12 $(\frac{1}{2})^0 - 1^2 + 16^{-3/4}$

Exer. 13–38: Simplify the expression and rationalize the denominator when appropriate.

13 $(3a^2b)^2(2ab^3)$ 14 $\dfrac{6r^3y^2}{2r^5y}$

15 $\dfrac{(3x^2y^{-3})^{-2}}{x^{-5}y}$ 16 $\left(\dfrac{a^{2/3}b^{3/2}}{a^2b}\right)^6$

17 $(-2p^2q)^3\left(\dfrac{p}{4q^2}\right)^2$ 18 $c^{-4/3}c^{3/2}c^{1/6}$

19 $\left(\dfrac{xy^{-1}}{\sqrt{z}}\right)^4 \div \left(\dfrac{x^{1/3}y^2}{z}\right)^3$ 20 $\left(\dfrac{-64x^3}{z^6y^9}\right)^{2/3}$

21 $[(a^{2/3}b^{-2})^3]^{-1}$ 22 $\dfrac{(3u^2v^5w^{-4})^3}{(2uv^{-3}w^2)^4}$

23 $\dfrac{r^{-1} + s^{-1}}{(rs)^{-1}}$ 24 $(u + v)^3(u + v)^{-2}$

25 $s^{5/2}s^{-4/3}s^{-1/6}$ 26 $x^{-2} - y^{-1}$

27 $\sqrt[3]{(x^4y^{-1})^6}$ 28 $\sqrt[3]{8x^5y^3z^4}$

29 $\dfrac{1}{\sqrt[3]{4}}$ 30 $\sqrt{\dfrac{a^2b^3}{c}}$

31 $\sqrt[3]{4x^2y}\sqrt[3]{2x^5y^2}$ 32 $\sqrt[4]{(-4a^3b^2c)^2}$

33 $\dfrac{1}{\sqrt{t}}\left(\dfrac{1}{\sqrt{t}} - 1\right)$ 34 $\sqrt{\sqrt[3]{(c^3d^6)^4}}$

35 $\dfrac{\sqrt{12x^4y}}{\sqrt{3x^2y^5}}$ 36 $\sqrt[3]{(a + 2b)^3}$

37 $\sqrt[3]{\dfrac{1}{2\pi^2}}$ 38 $\sqrt[3]{\dfrac{x^2}{9y}}$

Exer. 39–42: Rationalize the denominator.

39 $\dfrac{1 - \sqrt{x}}{1 + \sqrt{x}}$ 40 $\dfrac{1}{\sqrt{a} + \sqrt{a - 2}}$

41 $\dfrac{81x^2 - y^2}{3\sqrt{x} + \sqrt{y}}$ 42 $\dfrac{3 + \sqrt{x}}{3 - \sqrt{x}}$

Exer. 43–58: Express as a polynomial.

43 $(3x^3 - 4x^2 + x - 7) + (x^4 - 2x^3 + 3x^2 + 5)$

44 $(4z^4 - 3z^2 + 1) - z(z^3 + 4z^2 - 4)$

45 $(x + 4)(x + 3) - (2x - 1)(x - 5)$

46 $(4x - 5)(2x^2 + 3x - 7)$

47 $(3y^3 - 2y^2 + y + 4)(y^2 - 3)$

48 $(3x + 2)(x - 5)(5x + 4)$

49 $(a - b)(a^3 + a^2b + ab^2 + b^3)$

50 $\dfrac{9p^4q^3 - 6p^2q^4 + 5p^3q^2}{3p^2q^2}$

51 $(3a - 5b)(2a + 7b)$

52 $(4r^2 - 3s)^2$

53 $(13a^2 + 4b)(13a^2 - 4b)$

54 $(a^3 - a^2)^2$

55 $(2a + b)^3$

56 $(c^2 - d^2)^3$

57 $(3x + 2y)^2(3x - 2y)^2$

58 $(a + b + c + d)^2$

Exer. 59–74: Factor the polynomial.

59 $60xw + 70w$

60 $2r^4s^3 - 8r^2s^5$

61 $28x^2 + 4x - 9$

62 $16a^4 + 24a^2b^2 + 9b^4$

63 $2wy + 3yx - 8wz - 12zx$

64 $2c^3 - 12c^2 + 3c - 18$

65 $8x^3 + 64y^3$

66 $u^3v^4 - u^6v$

67 $p^8 - q^8$

68 $x^4 - 8x^3 + 16x^2$

69 $w^6 + 1$

70 $3x + 6$

71 $x^2 + 25$

72 $x^2 - 49y^2 - 14x + 49$

73 $x^5 - 4x^3 + 8x^2 - 32$

74 $4x^4 + 12x^3 + 20x^2$

Exer. 75–86: Simplify the expression.

75 $\dfrac{6x^2 - 7x - 5}{4x^2 + 4x + 1}$

76 $\dfrac{r^3 - t^3}{r^2 - t^2}$

77 $\dfrac{6x^2 - 5x - 6}{x^2 - 4} \div \dfrac{2x^2 - 3x}{x + 2}$

78 $\dfrac{2}{4x - 5} - \dfrac{5}{10x + 1}$

79 $\dfrac{7}{x + 2} + \dfrac{3x}{(x + 2)^2} - \dfrac{5}{x}$

80 $\dfrac{x + x^{-2}}{1 + x^{-2}}$

81 $\dfrac{1}{x} - \dfrac{2}{x^2 + x} - \dfrac{3}{x + 3}$

82 $(a^{-1} + b^{-1})^{-1}$

83 $\dfrac{x + 2 - \dfrac{3}{x + 4}}{\dfrac{x}{x + 4} + \dfrac{1}{x + 4}}$

84 $\dfrac{\dfrac{x}{x + 2} - \dfrac{4}{x + 2}}{x - 3 - \dfrac{6}{x + 2}}$

85 $(x^2 + 1)^{3/2}(4)(x + 5)^3 + (x + 5)^4(\tfrac{3}{2})(x^2 + 1)^{1/2}(2x)$

86 $\dfrac{(4 - x^2)(\tfrac{1}{3})(6x + 1)^{-2/3}(6) - (6x + 1)^{1/3}(-2x)}{(4 - x^2)^2}$

Exer. 87–88: Some calculators use an algorithm similar to the following to approximate $\sqrt{N}$ for a positive real number N: Let $x_1 = N/2$ and find successive approximations $x_2, x_3, \ldots$ by using

$$x_2 = \frac{1}{2}\left(x_1 + \frac{N}{x_1}\right), \quad x_3 = \frac{1}{2}\left(x_2 + \frac{N}{x_2}\right), \quad \ldots$$

until the desired accuracy is obtained. Use this method to approximate the radical to 6-decimal-place accuracy.

87 $\sqrt{5}$

88 $\sqrt{18}$

89 *Red blood cells in a body* The body of an average person contains 5.5 liters of blood and about 5 million red blood cells per cubic millimeter of blood. Given that $1 \text{ L} = 10^6 \text{ mm}^3$, estimate the number of red blood cells in an average person's body.

90 *Heartbeats in a lifetime* A healthy heart beats 70 to 90 times per minute. Estimate the number of heartbeats in the lifetime of an individual who lives to age 80.

91 *Body surface area* At age 2 years, a typical boy is 86 cm tall and weighs 13 kg. Use the DuBois and DuBois formula, $S = (0.007184)w^{0.425}h^{0.725}$, where w is weight and h is height, to find the body surface area S (in square meters).

92 *Adiabatic expansion* A gas is said to expand *adiabatically* if there is no loss or gain of heat. The formula for the adiabatic expansion of air is $pv^{1.4} = c$, where p is the pressure, v is the volume, and c is a constant. If, at a certain instant, the pressure is 40 dyne/cm^2 and the volume is 60 cm^3, find the value of c (a *dyne* is the unit of force in the cgs system).

Equations may be used to specify the distance between objects that are traveling at different speeds and in different directions.

CHAPTER 2

EQUATIONS

AND

INEQUALITIES

Methods for solving equations date back to the Babylonians (2000 B.C.), who described equations in words instead of the variables—x, y, and so on—that we use today. Major advances in finding solutions of equations then took place in Italy in the sixteenth century and continued throughout the world well into the nineteenth century. In modern times, computers are used to approximate solutions of very complicated equations.

Inequalities that involve variables have now attained the same level of importance as equations, and they are used extensively in applications of mathematics. In this chapter we shall discuss several methods for solving basic equations and inequalities. ■

2.1 EQUATIONS

An **equation** is a statement that two quantities or expressions are equal. Equations are employed in every field that uses real numbers. As an illustration, the equation

$$d = rt, \quad \text{or} \quad \text{distance} = (\text{rate})(\text{time}),$$

is used in solving problems involving an object moving at a constant rate of speed. If the rate r is 45 mi/hr (miles per hour), then the distance d (in miles) traveled after time t (in hours) is given by

$$d = 45t.$$

For example, if $t = 2$ hr, then $d = 45 \cdot 2 = 90$ mi. If we wish to find how long it takes the object to travel 75 miles, we let $d = 75$ and *solve* the equation

$$75 = 45t \quad \text{or, equivalently,} \quad 45t = 75.$$

Dividing both sides of the last equation by 45, we obtain

$$t = \tfrac{75}{45} = \tfrac{5}{3}.$$

Thus, if $r = 45$ mi/hr, then the time required to travel 75 miles is $1\tfrac{2}{3}$ hours, or 1 hour and 40 minutes.

Note that the equation $d = rt$ contains three variables: d, r, and t. In much of our work in this chapter we shall consider equations that contain only one variable. The following chart applies to a variable x, but any other variable may be considered. The abbreviations LS and RS in the second illustration stand for the equation's left side and right side.

Terminology	Definition	Illustration
Equation in x	A statement of equality involving one variable x	$x^2 - 5 = 4x$
Solution, or **root**, of an equation in x	A number a that yields a true statement when substituted for x	5 is a solution of $x^2 - 5 = 4x$, since substitution gives us LS: $5^2 - 5 = 25 - 5 = 20$ and RS: $4 \cdot 5 = 20$, and $20 = 20$ is a true statement.
A number a **satisfies** an equation in x	a is a solution of the equation	5 satisfies $x^2 - 5 = 4x$.
Equivalent equations	Equations that have exactly the same solutions	$2x + 1 = 7$ $2x = 7 - 1$ $2x = 6$ $x = 3$
Solve an equation in x	Find all solutions of the equation	To solve $(x + 3)(x - 5) = 0$, set each factor equal to 0: $x + 3 = 0, \quad x - 5 = 0$, obtaining the solutions -3 and 5.

An **algebraic equation** in x contains only algebraic expressions such as polynomials, rational expressions, radicals, and so on. An equation of this type is called a **conditional equation** if there are numbers in the domains of the expressions that are not solutions. For example, the equation $x^2 = 9$ is conditional, since the number $x = 4$ (and others) is not a solution. If *every* number in the domains of the expressions in an algebraic equation is a solution, the equation is called an **identity**.

Sometimes it is difficult to determine whether an equation is conditional or an identity. An identity will often be indicated when, after properties of real numbers are applied, an equation of the form $p = p$ is obtained, where p is some expression. To illustrate, if we multiply both sides of the equation

$$\frac{x}{x^2 - 4} = \frac{x}{(x + 2)(x - 2)}$$

by $x^2 - 4$, we obtain $x = x$. This alerts us to the fact that we may have an identity on our hands; it does not, however, prove anything. A standard method for verifying that an equation is an identity is to show, using properties of real numbers, that the expression which appears on one side of the given equation can be transformed into the expression which appears on the other side of the given equation. That is easy to do in the preceding illustration, since we know that $x^2 - 4 = (x + 2)(x - 2)$. Of course, to show that an equation is not an identity, we need only find one real number in the domain of the variable that fails to satisfy the original equation.

The most basic equation in algebra is the *linear equation*, defined in the next chart, where a and b denote real numbers.

Terminology	Definition	Illustration
Linear equation in x	An equation that can be written in the form $ax + b = 0$, where $a \neq 0$	$4x + 5 = 0$ $4x = -5$ $x = -\frac{5}{4}$

The illustration in the preceding chart indicates a typical method of solving a linear equation. Following the same procedure, we see that

$$\text{if} \quad ax + b = 0, \quad \text{then} \quad x = -\frac{b}{a},$$

provided $a \neq 0$. Thus, a linear equation has exactly one solution.

We sometimes solve an equation by making a list of equivalent equations, each in some sense simpler than the preceding one, ending the list with an equation from which the solutions can be easily obtained. We often simplify an equation by adding the same expression to both sides or by subtracting from both sides. We can also multiply or divide both sides of an equation by an expression that represents a *nonzero* real number.

In the following examples, the phrases in color indicate how an equivalent equation was obtained from the preceding equation. To shorten these phrases we have, as in Example 1, used "add 7" instead of the more accurate but lengthy *add 7 to both sides*. Similarly, "subtract $2x$" is used for *subtract $2x$ from both sides*, and "divide by 4" means *divide both sides by* 4.

EXAMPLE I   *Solving a linear equation*

Solve the equation $6x - 7 = 2x + 5$.

SOLUTION   The equations in the following list are equivalent:

$$6x - 7 = 2x + 5 \qquad \text{given}$$
$$(6x - 7) + 7 = (2x + 5) + 7 \qquad \text{add 7}$$
$$6x = 2x + 12 \qquad \text{simplify}$$
$$6x - 2x = (2x + 12) - 2x \qquad \text{subtract } 2x$$
$$4x = 12 \qquad \text{simplify}$$
$$\frac{4x}{4} = \frac{12}{4} \qquad \text{divide by 4}$$
$$x = 3 \qquad \text{simplify}$$

$\checkmark$ **Check $x = 3$**   LS: $6(3) - 7 = 18 - 7 = 11$

RS: $2(3) + 5 = 6 + 5 = 11$

Since $11 = 11$ is a true statement, $x = 3$ checks as a solution.

As indicated in the preceding example, we often check a solution by substituting it into the given equation. Such checks may detect errors introduced through incorrect manipulations or mistakes in arithmetic.

We say that the equation given in Example 1 *has the solution $x = 3$*. Similarly, we would say that the equation $x^2 = 4$ *has solutions $x = 2$ and $x = -2$*.

The next example illustrates that a seemingly complicated equation may simplify to a linear equation.

EXAMPLE 2   *Solving an equation*

Solve the equation $(8x - 2)(3x + 4) = (4x + 3)(6x - 1)$.

SOLUTION   The equations in the following list are equivalent:

$$(8x - 2)(3x + 4) = (4x + 3)(6x - 1) \qquad \text{given}$$
$$24x^2 + 26x - 8 = 24x^2 + 14x - 3 \qquad \text{multiply factors}$$

$$26x - 8 = 14x - 3 \qquad \text{subtract } 24x^2$$
$$12x - 8 = -3 \qquad \text{subtract } 14x$$
$$12x = 5 \qquad \text{add } 8$$
$$x = \tfrac{5}{12} \qquad \text{divide by } 12$$

Hence, the solution of the given equation is $\tfrac{5}{12}$.

We did not check the preceding solution because each step yields an equivalent equation; however, when you are working exercises or taking a test, it is always a good idea to check answers to guard against arithmetic errors.

If an equation contains rational expressions, we often eliminate denominators by multiplying both sides by the lcd of these expressions. If we multiply both sides by an expression that equals zero for some value of x, then the resulting equation may not be equivalent to the original equation, as illustrated in the following example.

EXAMPLE 3   *An equation with no solutions*

Solve $\dfrac{3x}{x-2} = 1 + \dfrac{6}{x-2}$.

SOLUTION

$$\frac{3x}{x-2} = 1 + \frac{6}{x-2} \qquad \text{given}$$

$$\left(\frac{3x}{x-2}\right)(x-2) = (1)(x-2) + \left(\frac{6}{x-2}\right)(x-2) \qquad \text{multiply by } x-2$$

$$3x = (x-2) + 6 \qquad \text{simplify}$$

$$3x = x + 4 \qquad \text{simplify}$$

$$2x = 4 \qquad \text{subtract } x$$

$$x = 2 \qquad \text{divide by } 2$$

✓ **Check $x = 2$**   LS: $\dfrac{3(2)}{(2)-2} = \dfrac{6}{0}$

Since division by 0 is not permissible, $x = 2$ is not a solution. Hence, *the given equation has no solutions*.

In the process of solving an equation, we may obtain, as a *possible* solution, a number that is *not* a solution of the given equation. Such a number is called an **extraneous solution**, or **extraneous root**, of the given equation. In Example 3, $x = 2$ is an extraneous solution (root) of the given equation.

The following guidelines may also be used to solve the equation in Example 3. In this case, observing guideline 2 would make it unnecessary to check the extraneous solution $x = 2$.

Guidelines for Solving an Equation Containing Rational Expressions

1 Determine the lcd of the rational expressions.

2 Find the values of the variable that make the lcd zero. These are *not* solutions, because they yield at least one zero denominator when substituted into the given equation.

3 Multiply each term of the equation by the lcd and simplify, thereby eliminating all of the denominators.

4 Solve the equation obtained in guideline 3.

5 The solutions of the given equation are the solutions found in guideline 4, with the exclusion of the values found in guideline 2.

We shall follow these guidelines in the next example.

EXAMPLE 4   *An equation containing rational expressions*

Solve $\dfrac{3}{2x - 4} - \dfrac{5}{x + 3} = \dfrac{2}{x - 2}$.

SOLUTION

Guideline 1   Rewriting the denominator $2x - 4$ as $2(x - 2)$, we see that the lcd of the three rational expressions is $2(x - 2)(x + 3)$.

Guideline 2   The values of x that make the lcd $2(x - 2)(x + 3)$ zero are 2 and -3, so these numbers cannot be solutions of the equation.

Guideline 3   Multiplying each term of the equation by the lcd and simplifying gives us the following:

$$\frac{3}{2(x - 2)} \, 2(x - 2)(x + 3) - \frac{5}{x + 3} \, 2(x - 2)(x + 3)$$

$$= \frac{2}{x - 2} \, 2(x - 2)(x + 3)$$

$$3(x + 3) - 10(x - 2) = 4(x + 3) \quad \text{cancel like factors}$$

$$3x + 9 - 10x + 20 = 4x + 12 \quad \text{multiply factors}$$

Guideline 4 We solve the last equation obtained in guideline 3.

$$3x - 10x - 4x = 12 - 9 - 20 \quad \text{subtract } 4x, 9, \text{ and } 20$$

$$-11x = -17 \quad \text{combine like terms}$$

$$x = \tfrac{17}{11} \quad \text{divide by } -11$$

Guideline 5 Since $\tfrac{17}{11}$ is not included among the values (2 and -3) that make the lcd zero (guideline 2), we see that $x = \tfrac{17}{11}$ is a solution of the given equation.

We shall not check the solution $x = \tfrac{17}{11}$ by substitution, because the arithmetic involved is complicated. It is simpler to carefully check the algebraic manipulations used in each step.

Formulas involving several variables occur in many applications of mathematics. Sometimes it is necessary to solve for a specific variable in terms of the remaining variables that appear in the formula, as the next two examples illustrate.

EXAMPLE 5 *Relationship between temperature scales*

The Celsius and Fahrenheit temperature scales are shown on the thermometer in Figure 1. The relationship between the temperature readings C and F is given by $C = \tfrac{5}{9}(F - 32)$. Solve for F.

SOLUTION To *solve for F* we must obtain a formula that has F by itself on one side of the equals sign and does not have F on the other side. We may do this as follows:

$$C = \tfrac{5}{9}(F - 32) \quad \text{given}$$

$$\tfrac{9}{5}C = F - 32 \quad \text{multiply by } \tfrac{9}{5}$$

$$\tfrac{9}{5}C + 32 = F \quad \text{add 32}$$

$$F = \tfrac{9}{5}C + 32 \quad \text{equivalent equation}$$

EXAMPLE 6 *Resistors connected in parallel*

In electrical theory, the formula

$$\frac{1}{R} = \frac{1}{R_1} + \frac{1}{R_2}$$

is used to find the total resistance R when two resistors R_1 and R_2 are connected in parallel, as illustrated in Figure 2. Solve for R_1.

FIGURE I

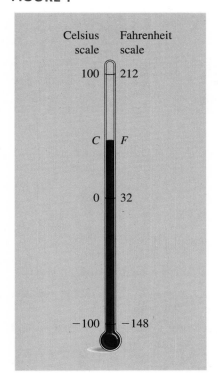

FIGURE 2

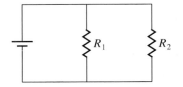

SOLUTION   We first multiply both sides of the given equation by the lcd of the three fractions and then solve for R_1, as follows:

$$\frac{1}{R} = \frac{1}{R_1} + \frac{1}{R_2} \qquad \text{given}$$

$$\frac{1}{R} \cdot RR_1R_2 = \frac{1}{R_1} \cdot RR_1R_2 + \frac{1}{R_2} \cdot RR_1R_2 \qquad \text{multiply by the lcd, } RR_1R_2$$

$$R_1R_2 = RR_2 + RR_1 \qquad \text{cancel common factors}$$

$$R_1R_2 - RR_1 = RR_2 \qquad \text{collect } R_1 \text{ terms on one side}$$

$$R_1(R_2 - R) = RR_2 \qquad \text{factor out } R_1$$

$$R_1 = \frac{RR_2}{R_2 - R} \qquad \text{divide by } R_2 - R$$

An alternative method of solution is to first solve for $\frac{1}{R_1}$:

$$\frac{1}{R} = \frac{1}{R_1} + \frac{1}{R_2} \qquad \text{given}$$

$$\frac{1}{R_1} + \frac{1}{R_2} = \frac{1}{R} \qquad \text{equivalent equation}$$

$$\frac{1}{R_1} = \frac{1}{R} - \frac{1}{R_2} \qquad \text{subtract } \frac{1}{R_2}$$

$$\frac{1}{R_1} = \frac{R_2 - R}{RR_2} \qquad \text{combine fractions}$$

If two nonzero numbers are equal, then so are their reciprocals. Hence,

$$R_1 = \frac{RR_2}{R_2 - R}.$$

2.1 EXERCISES

Exer. 1–44: Solve the equation.

1 $-3x + 4 = -1$

2 $2x - 2 = -9$

3 $4x - 3 = -5x + 6$

4 $5x - 4 = 2(x - 2)$

5 $4(2y + 5) = 3(5y - 2)$

6 $6(2y + 3) - 3(y - 5) = 0$

7 $\frac{1}{5}x + 2 = 3 - \frac{2}{7}x$

8 $\frac{5}{3}x - 1 = 4 + \frac{2}{3}x$

9 $0.3(3 + 2x) + 1.2x = 3.2$

10 $1.5x - 0.7 = 0.4(3 - 5x)$

11 $\frac{3 + 5x}{5} = \frac{4 - x}{7}$

12 $\frac{2x - 9}{4} = 2 + \frac{x}{12}$

13 $\frac{13 + 2x}{4x + 1} = \frac{3}{4}$

14 $\frac{3}{7x - 2} = \frac{9}{3x + 1}$

15 $8 - \frac{5}{x} = 2 + \frac{3}{x}$

16 $\frac{3}{y} + \frac{6}{y} - \frac{1}{y} = 11$

17 $(3x - 2)^2 = (x - 5)(9x + 4)$

18 $(x + 5)^2 + 3 = (x - 2)^2$

19 $(5x - 7)(2x + 1) - 10x(x - 4) = 0$

20 $(2x + 9)(4x - 3) = 8x^2 - 12$

21 $\dfrac{3x + 1}{6x - 2} = \dfrac{2x + 5}{4x - 13}$ 22 $\dfrac{5x + 2}{10x - 3} = \dfrac{x - 8}{2x + 3}$

23 $\dfrac{2}{5} + \dfrac{4}{10x + 5} = \dfrac{7}{2x + 1}$ 24 $\dfrac{-5}{3x - 9} + \dfrac{4}{x - 3} = \dfrac{5}{6}$

25 $\dfrac{3}{2x - 4} - \dfrac{5}{3x - 6} = \dfrac{3}{5}$ 26 $\dfrac{9}{2x + 6} - \dfrac{7}{5x + 15} = \dfrac{2}{3}$

27 $2 - \dfrac{5}{3x - 7} = 2$ 28 $\dfrac{6}{2x + 11} + 5 = 5$

29 $\dfrac{1}{2x - 1} = \dfrac{4}{8x - 4}$ 30 $\dfrac{4}{5x + 2} - \dfrac{12}{15x + 6} = 0$

31 $\dfrac{7}{y^2 - 4} - \dfrac{4}{y + 2} = \dfrac{5}{y - 2}$

32 $\dfrac{4}{2u - 3} + \dfrac{10}{4u^2 - 9} = \dfrac{1}{2u + 3}$

33 $(x + 3)^3 - (3x - 1)^2 = x^3 + 4$

34 $(x - 1)^3 = (x + 1)^3 - 6x^2$

35 $\dfrac{9x}{3x - 1} = 2 + \dfrac{3}{3x - 1}$ 36 $\dfrac{2x}{2x + 3} + \dfrac{6}{4x + 6} = 5$

37 $\dfrac{1}{x + 4} + \dfrac{3}{x - 4} = \dfrac{3x + 8}{x^2 - 16}$

38 $\dfrac{2}{2x + 3} + \dfrac{4}{2x - 3} = \dfrac{5x + 6}{4x^2 - 9}$

39 $\dfrac{4}{x + 2} + \dfrac{1}{x - 2} = \dfrac{5x - 6}{x^2 - 4}$

40 $\dfrac{2}{2x + 5} + \dfrac{3}{2x - 5} = \dfrac{10x + 5}{4x^2 - 25}$

41 $\dfrac{2}{2x + 1} - \dfrac{3}{2x - 1} = \dfrac{-2x + 7}{4x^2 - 1}$

42 $\dfrac{3}{2x + 5} + \dfrac{4}{2x - 5} = \dfrac{14x + 3}{4x^2 - 25}$

43 $\dfrac{5}{2x + 3} + \dfrac{4}{2x - 3} = \dfrac{14x + 3}{4x^2 - 9}$

44 $\dfrac{-3}{x + 4} + \dfrac{7}{x - 4} = \dfrac{-5x + 4}{x^2 - 16}$

Exer. 45–50: Show that the equation is an identity.

45 $(4x - 3)^2 - 16x^2 = 9 - 24x$

46 $(3x - 4)(2x + 1) + 5x = 6x^2 - 4$

47 $\dfrac{x^2 - 9}{x + 3} = x - 3$ 48 $\dfrac{x^3 + 8}{x + 2} = x^2 - 2x + 4$

49 $\dfrac{3x^2 + 8}{x} = \dfrac{8}{x} + 3x$ 50 $\dfrac{49x^2 - 25}{7x - 5} = 7x + 5$

Exer. 51–52: For what value of c is the number a a solution of the equation?

51 $4x + 1 + 2c = 5c - 3x + 6;\quad a = -2$

52 $3x - 2 + 6c = 2c - 5x + 1;\quad a = 4$

Exer. 53–54: Determine whether the two equations are equivalent.

53 (a) $\dfrac{7x}{x - 5} = \dfrac{42}{x - 5},$ $x = 6$

   (b) $\dfrac{7x}{x - 5} = \dfrac{35}{x - 5},$ $x = 5$

54 (a) $\dfrac{8x}{x - 7} = \dfrac{72}{x - 7},$ $x = 9$

   (b) $\dfrac{8x}{x - 7} = \dfrac{56}{x - 7},$ $x = 7$

Exer. 55–56: Determine values for a and b such that $\frac{5}{3}$ is a solution of the equation.

55 $ax + b = 0$ 56 $ax^2 + bx = 0$

Exer. 57–58: Determine which equation is not equivalent to the equation preceding it.

57    $x^2 - x - 2 = x^2 - 4$
   $(x + 1)(x - 2) = (x + 2)(x - 2)$
         $x + 1 = x + 2$
              $1 = 2$

58       $5x + 6 = 4x + 3$
     $x^2 + 5x + 6 = x^2 + 4x + 3$
   $(x + 2)(x + 3) = (x + 1)(x + 3)$
         $x + 2 = x + 1$
              $2 = 1$

Exer. 59–72: The formula occurs in the indicated application. Solve for the specified variable.

59 $I = Prt$ for P (simple interest)

60 $C = 2\pi r$ for r (circumference of a circle)

61   $A = \frac{1}{2}bh$ for h (area of a triangle)

62   $V = \frac{1}{3}\pi r^2 h$ for h (volume of a cone)

63   $F = g\dfrac{mM}{d^2}$ for m (Newton's law of gravitation)

64   $R = \dfrac{V}{I}$ for I (Ohm's law in electrical theory)

65   $P = 2l + 2w$ for w (perimeter of a rectangle)

66   $A = P + Prt$ for r (principal plus interest)

67   $A = \frac{1}{2}(b_1 + b_2)h$ for b_1 (area of a trapezoid)

68   $s = \frac{1}{2}gt^2 + v_0 t$ for v_0 (distance an object falls)

69   $S = \dfrac{p}{q + p(1 - q)}$ for q (Amdahl's law for supercomputers)

70   $S = 2(lw + hw + hl)$ for h (surface area of a rectangular box)

71   $\dfrac{1}{f} = \dfrac{1}{p} + \dfrac{1}{q}$ for q (lens equation)

72   $\dfrac{1}{R} = \dfrac{1}{R_1} + \dfrac{1}{R_2} + \dfrac{1}{R_3}$ for R_2 (three resistors connected in parallel)

2.2   APPLIED PROBLEMS

Equations are often used to solve *applied problems*—that is, problems that involve applications of mathematics to other fields. Because of the unlimited variety of applied problems, it is difficult to state specific rules for finding solutions. The following guidelines may be helpful, provided the problem can be formulated in terms of an equation in one variable.

Guidelines for Solving Applied Problems

1   If the problem is stated in writing, read it carefully several times and think about the given facts, together with the unknown quantity that is to be found.

2   Introduce a letter to denote the unknown quantity. This is one of the most crucial steps in the solution. Phrases containing words such as *what*, *find*, *how much*, *how far*, or *when* should alert you to the unknown quantity.

3   If appropriate, draw a picture and label it.

4   List the known facts, together with any relationships that involve the unknown quantity. A relationship may be described by an equation in which written statements, instead of letters or numbers, appear on one or both sides of the equals sign.

5   After analyzing the list in guideline 4, formulate an equation that describes precisely what is stated in words.

6   Solve the equation formulated in guideline 5.

7   Check the solutions obtained in guideline 6 by referring to the original statement of the problem. Verify that the solution agrees with the stated conditions.

The use of these guidelines is illustrated in the next example.

EXAMPLE I   *Test average*

A student in an algebra course has test scores of 64 and 78. What score on a third test will give the student an average of 80?

SOLUTION

Guideline 1   Read the problem at least one more time.

Guideline 2   The unknown quantity is the score on the third test, so we let

$$x = \text{score on the third test.}$$

Guideline 3   A picture or diagram is unnecessary for this problem.

Guideline 4   (a) Known facts are scores of 64 and 78 on the first two tests.
(b) A relationship that involves x is the average score of 64, 78, and x. Thus,

$$\text{average score} = \frac{64 + 78 + x}{3}.$$

Guideline 5   Since the average score in guideline 4 is to be 80, we consider the equation

$$\frac{64 + 78 + x}{3} = 80.$$

Guideline 6   We solve the equation formulated in guideline 5:

$$64 + 78 + x = 80 \cdot 3 \qquad \text{multiply by 3}$$
$$142 + x = 240 \qquad \text{simplify}$$
$$x = 98 \qquad \text{subtract 142}$$

Guideline 7   Check: If the three test scores are 64, 78, and 98, then the average is

$$\frac{64 + 78 + 98}{3} = \frac{240}{3} = 80,$$

as desired.

In the remaining examples, try to identify the explicit guidelines that are used in solutions.

EXAMPLE 2   *Calculating a presale price*

A clothing store holding a clearance sale advertises that all prices have been discounted 20%. If a shirt is on sale for $28, what was its presale price?

SOLUTION   Since the unknown quantity is the presale price, we let

$$x = \text{presale price}.$$

We next note the following facts:

$$0.20x = \text{discount of 20\% on presale price}$$
$$28 = \text{sale price}$$

The sale price is determined as follows:

$$(\text{presale price}) - (\text{discount}) = \text{sale price}$$

Translating the last equation into symbols and then solving gives us

$x - 0.20x = 28$	formulate an equation
$0.80x = 28$	subtract $0.20x$ from $1x$
$x = \dfrac{28}{0.80} = 35.$	divide by 0.80

The presale price was $35.

✓ **Check**   If a $35 shirt is discounted 20%, then the discount (in dollars) is $(0.20)(35) = 7$ and the sale price is $35 - 7$, or $28.

Banks and other financial institutions pay interest on investments. Usually this interest is *compounded* (as described in Section 5.1); however, if money is invested or loaned for a short period of time, *simple interest* may be paid using the following formula.

Simple Interest Formula

> If a sum of money P (the **principal**) is invested at a simple interest rate r (expressed as a decimal), then the **simple interest** I at the end of t years is
>
> $$I = Prt.$$

The following table illustrates simple interest for three cases.

Principal P	Interest rate r	Number of years t	Interest $I = Prt$
$1000	$8\% = 0.08$	1	$1000(0.08)(1) = \$80$
$2000	$6\% = 0.06$	$1\frac{1}{2}$	$2000(0.06)(1.5) = \$180$
$3200	$5\frac{1}{2}\% = 0.055$	2	$3200(0.055)(2) = \$352$

EXAMPLE 3   *Investing money in two stocks*

An investment firm has $100,000 to invest for a client and decides to invest it in two stocks A and B. The expected annual rate of return, or simple interest, for stock A is 15%, but there is some risk involved, and the client does not wish to invest more than $50,000 in this stock. The annual rate of return on the more stable stock B is anticipated to be 10%. Determine whether there is a way of investing the money so that the annual interest is

(a) $12,000   **(b)** $13,000

SOLUTION   The annual interest is given by $I = Pr$, which comes from the simple interest formula $I = Prt$ with $t = 1$. If we let x denote the amount invested in stock A, then $100,000 - x$ will be invested in stock B. This leads to the following equalities:

$$x = \text{amount invested in stock A at } 15\%$$

$$100,000 - x = \text{amount invested in stock B at } 10\%$$

$$0.15x = \text{annual interest from stock A}$$

$$0.10(100,000 - x) = \text{annual interest from stock B}$$

Adding the interest from both stocks, we obtain

$$\text{total annual interest} = 0.15x + 0.10(100,000 - x).$$

Simplifying the right side gives us

$$(*) \qquad \text{total annual interest} = 10,000 + 0.05x.$$

(a) The total annual interest is $12,000 if

$$
\begin{array}{lll}
10,000 + 0.05x = 12,000 & \text{from } (*) \\
0.05x = 2000 & \text{subtract } 10,000 \\
x = \dfrac{2000}{0.05} = 40,000 & \text{divide by } 0.05
\end{array}
$$

Thus, $40,000 should be invested in stock A, and the remaining $60,000 should be invested in stock B. Since the amount invested in stock A is not more than $50,000, this manner of investing the money meets the requirement of the client.

✓   *Check*   If $40,000 is invested in stock A and $60,000 in stock B, then the total annual interest is

$$40,000(0.15) + 60,000(0.10) = 6000 + 6000 = 12,000.$$

(continued)

(b) The total annual interest is $13,000 if

$$10,000 + 0.05x = 13,000 \qquad \text{from (∗)}$$

$$0.05x = 3000 \qquad \text{subtract 10,000}$$

$$x = \frac{3000}{0.05} = 60,000 \qquad \text{divide by 0.05}$$

Thus, $60,000 should be invested in stock A and the remaining $40,000 in stock B. This plan does *not* meet the requirement that not more than $50,000 is to be invested in stock A. Hence the firm cannot invest the client's money in stocks A and B such that the total annual interest is $13,000.

In certain applications, it is necessary to combine two substances to obtain a prescribed mixture, as illustrated in the next two examples.

EXAMPLE 4   *Mixing chemicals*

A chemist has 10 milliliters of a solution that contains a 30% concentration of acid. How many milliliters of pure acid must be added in order to increase the concentration to 50%?

SOLUTION   Since the unknown quantity is the amount of pure acid to add, we let

$$x = \text{the number of mL of pure acid to be added.}$$

To help visualize the problem, let us draw a picture, as in Figure 3, and attach appropriate labels.

FIGURE 3

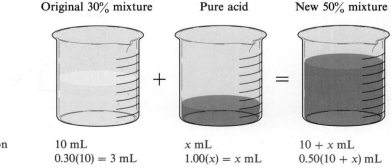

	Original 30% mixture	Pure acid	New 50% mixture
Total amount of solution	10 mL	x mL	$10 + x$ mL
Amount of pure acid	$0.30(10) = 3$ mL	$1.00(x) = x$ mL	$0.50(10 + x)$ mL

Since we can express the amount of pure acid in the final solution as either $3 + x$ (from the first two beakers) or $0.50(10 + x)$, we obtain the equation

$$3 + x = 0.50(10 + x).$$

We now solve for x:

$$3 + x = 5 + 0.5x \qquad \text{multiply factors}$$

$$0.5x = 2 \qquad \text{subtract } 0.5x \text{ and } 3$$

$$x = \frac{2}{0.5} = 4 \qquad \text{divide by } 0.05$$

Hence, 4 milliliters of the acid should be added to the original solution.

✓ **Check**   If 4 milliliters of acid is added to the given solution, then the new solution contains 14 milliliters, 7 milliliters of which is pure acid. This is the desired 50% concentration.

EXAMPLE 5   *Replacing antifreeze*

A radiator contains 8 quarts of a mixture of water and antifreeze. If 40% of the mixture is antifreeze, how much of the mixture should be drained and replaced by pure antifreeze so that the resultant mixture will contain 60% antifreeze?

SOLUTION   Let

$$x = \text{the number of qt of mixture to be drained.}$$

Since there were 8 quarts in the original 40% mixture, we may depict the problem as in Figure 4.

FIGURE 4

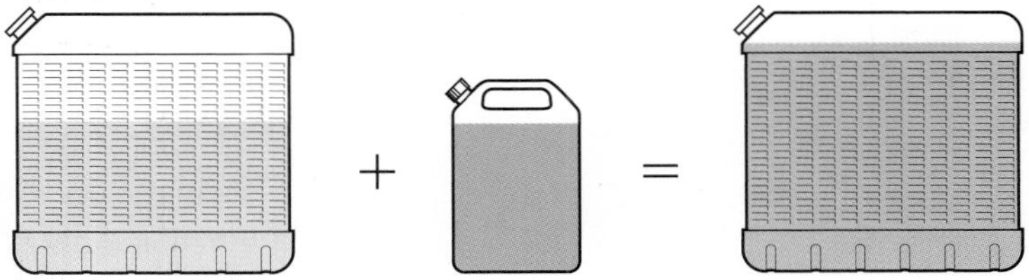

Original 40% mixture,
less amount drained   Pure antifreeze   New 60% mixture

	Original 40% mixture, less amount drained	Pure antifreeze	New 60% mixture
Total amount	$(8 - x)$ qt	x qt	8 qt
Amount of pure antifreeze	$0.40(8 - x)$ qt	$1.00(x) = x$ qt	$0.60(8) = 4.8$ qt

Since the number of quarts of pure antifreeze in the final mixture can be expressed as either $0.40(8 - x) + x$ or 4.8, we obtain the equation

$$0.40(8 - x) + x = 4.8.$$

(continued)

We now solve for x:

$$3.2 - 0.4x + x = 4.8 \qquad \text{multiply factors}$$

$$0.6x = 1.6 \qquad \text{combine } x \text{ terms and subtract 3.2}$$

$$x = \frac{1.6}{0.6} = \frac{16}{6} = \frac{8}{3} \qquad \text{divide by 0.6}$$

Thus, $\frac{8}{3}$ quarts should be drained from the original mixture.

✓ **Check**   Let us first note that the amount of antifreeze in the original 8-quart mixture was 0.4(8), or 3.2 quarts. In draining $\frac{8}{3}$ quarts of the original 40% mixture, we lose $0.4(\frac{8}{3})$ quarts of antifreeze, and hence $3.2 - 0.4(\frac{8}{3})$ quarts of antifreeze remain after draining. If we then add $\frac{8}{3}$ quarts of pure antifreeze, the amount of antifreeze in the final mixture is

$$3.2 - 0.4(\tfrac{8}{3}) + \tfrac{8}{3} = 4.8 \text{ qt.}$$

This number, 4.8, is 60% of 8.

EXAMPLE 6   *Comparing times traveled by cars*

Two cities are connected by means of a highway. A car leaves city B at 1:00 P.M. and travels at a constant rate of 40 mi/hr toward city C. Thirty minutes later, another car leaves B and travels toward C at a constant rate of 55 mi/hr. If the lengths of the cars are disregarded, at what time will the second car reach the first car?

SOLUTION   Let t denote the number of hours after 1:00 P.M. traveled by the first car. Since the second car leaves B at 1:30 P.M., it has traveled $\frac{1}{2}$ hour less than the first. This leads to the following table.

Car	Rate (mi/hr)	Hours traveled	Miles traveled
First car	40	t	$40t$
Second car	55	$t - \frac{1}{2}$	$55(t - \frac{1}{2})$

The schematic drawing in Figure 5 illustrates possible positions of the cars t hours after 1:00 P.M. The second car reaches the first car when the number of miles traveled by the two cars is equal—that is, when

$$55(t - \tfrac{1}{2}) = 40t.$$

We now solve for t:

$$55t - \tfrac{55}{2} = 40t \qquad \text{multiply factors}$$

$$15t = \tfrac{55}{2} \qquad \text{subtract } 40t \text{ and add } \tfrac{55}{2}$$

$$t = \tfrac{55}{30} = \tfrac{11}{6} \qquad \text{divide by 15}$$

FIGURE 5

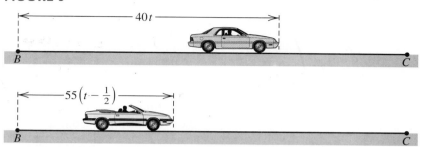

Thus, t is $1\frac{5}{6}$ hours or, equivalently, 1 hour 50 minutes after 1:00 P.M. Consequently, the second car reaches the first at 2:50 P.M.

✓ *Check* At 2:50 P.M. the first car has traveled for $1\frac{5}{6}$ hours, and its distance from B is $40(\frac{11}{6}) = \frac{220}{3}$ mi. At 2:50 P.M. the second car has traveled for $1\frac{1}{3}$ hours and is $55(\frac{4}{3}) = \frac{220}{3}$ mi from B. Hence, they are together at 2:50 P.M.

EXAMPLE 7 *Constructing a grain-elevator hopper*

FIGURE 6

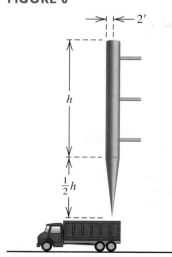

A grain-elevator hopper is to be constructed as shown in Figure 6, with a right circular cylinder of radius 2 feet and altitude h feet on top of a right circular cone whose altitude is one-half that of the cylinder. What value of h will make the total volume V of the hopper 500 cubic feet?

SOLUTION If V_{cylinder} and V_{cone} denote the volumes (in ft^3) of the cylinder and cone, respectively, then, using the formulas for volume stated on the end papers at the front of the text, we obtain the following:

$$V_{\text{cylinder}} = \pi r^2 h = \pi(2)^2 h = 4\pi h$$

$$V_{\text{cone}} = \tfrac{1}{3}\pi r^2 h = \tfrac{1}{3}\pi(2)^2(\tfrac{1}{2}h) = \tfrac{2}{3}\pi h$$

Since the total volume V of the hopper is to be 500 ft^3, we must have

$4\pi h + \tfrac{2}{3}\pi h = 500$	$V_{\text{cylinder}} + V_{\text{cone}} = V$
$12\pi h + 2\pi h = 1500$	multiply by 3
$14\pi h = 1500$	combine terms
$h = \dfrac{1500}{14\pi} \approx 34.1$ ft	divide by 14π

EXAMPLE 8   *Time required to do a job*

Two pumps are available for filling a gasoline storage tank. Pump A, used alone, can fill the tank in 3 hours, and pump B, used alone, can fill it in 4 hours. If both pumps are used simultaneously, how long will it take to fill the tank?

SOLUTION   Let t denote the number of hours needed for A and B to fill the tank if used simultaneously. It is convenient to introduce the *part* of the tank filled in 1 hour as follows:

$$\tfrac{1}{3} = \text{part of the tank filled by A in 1 hr}$$

$$\tfrac{1}{4} = \text{part of the tank filled by B in 1 hr}$$

$$\frac{1}{t} = \text{part of the tank filled by A } and \text{ B in 1 hr}$$

Using the fact that

$$\left(\begin{array}{c}\text{part filled by}\\ \text{A in 1 hr}\end{array}\right) + \left(\begin{array}{c}\text{part filled by}\\ \text{B in 1 hr}\end{array}\right) = \left(\begin{array}{c}\text{part filled by}\\ \text{A } and \text{ B in 1 hr}\end{array}\right),$$

we obtain

$$\frac{1}{3} + \frac{1}{4} = \frac{1}{t}, \quad \text{or} \quad \frac{7}{12} = \frac{1}{t}.$$

Taking the reciprocal of each side of the last equation gives us $t = \tfrac{12}{7}$. Thus, if pumps A and B are used simultaneously, the tank will be filled in $1\tfrac{5}{7}$ hours, or approximately 1 hour 43 minutes.

2.2 EXERCISES

1 **Test scores** A student in an algebra course has test scores of 75, 82, 71, and 84. What score on the next test will raise the student's average to 80?

2 **Final class average** Before the final exam, a student has test scores of 72, 80, 65, 78, and 60. If the final exam counts as one-third of the final grade, what score must the student receive in order to have a final average of 76?

3 **Gross pay** A worker's take-home pay is $492, after deductions totaling 40% of the gross pay have been subtracted. What is the gross pay?

4 **Cost of dining out** A couple does not wish to spend more than $70 for dinner at a restaurant. If a sales tax of 6% is added to the bill and they plan to tip 15% after the tax has been added, what is the most they can spend for the meal?

5 **Cost of insulation** The cost of installing insulation in a particular two-bedroom home is $1080. Present monthly heating costs average $60, but the insulation is expected to reduce heating costs by 10%. How many months will it take to recover the cost of the insulation?

6 Overtime pay A workman's basic hourly wage is $10, but he receives one and a half times his hourly rate for any hours worked in excess of 40 per week. If his paycheck for the week is $595, how many hours of overtime did he work?

7 Savings accounts An algebra student has won $100,000 in a lottery and wishes to deposit it in savings accounts in two financial institutions. One account pays 8% simple interest, but deposits are insured only to $50,000. The second account pays 6.4% simple interest, and deposits are insured up to $100,000. Determine whether the money can be deposited so that it is fully insured and earns annual interest of $7,500.

8 Municipal funding A city government has approved the construction of a $50 million sports arena. Up to $30 million will be raised by selling bonds that pay simple interest at a rate of 12% annually. The remaining amount (up to $40 million) will be obtained by borrowing money from an insurance company at a simple interest rate of 10%. Determine whether the arena can be financed so that the annual interest is $5.2 million.

9 Movie attendance Six hundred people attended the premiere of a motion picture. Adult tickets cost $5, and children were admitted for $2. If box office receipts totaled $2400, how many children attended the premiere?

10 Hourly pay A consulting engineer's time is billed at $60 per hour, and her assistant's is billed at $20 per hour. A customer received a bill for $580 for a certain job. If the assistant worked 5 hours less than the engineer, how much time did each bill on the job?

11 Preparing a glucose solution In a certain medical test designed to measure carbohydrate tolerance, an adult drinks 7 ounces of a 30% glucose solution. When the test is administered to a child, the glucose concentration must be decreased to 20%. How much 30% glucose solution and how much water should be used to prepare 7 ounces of 20% glucose solution?

12 Preparing eye drops A pharmacist is to prepare 15 milliliters of special eye drops for a glaucoma patient. The eye-drop solution must have a 2% active ingredient, but the pharmacist only has 10% solution and 1% solution in stock. How much of each type of solution should be used to fill the prescription?

13 Preparing an alloy British sterling silver is a copper-silver alloy that is 7.5% copper by weight. How many grams of pure copper and how many grams of British sterling silver should be used to prepare 200 grams of a copper-silver alloy that is 10% copper by weight?

14 Drug concentration Theophylline, an asthma medicine, is to be prepared from an elixir with a drug concentration of 5 mg/mL and a cherry-flavored syrup that is to be added to hide the taste of the drug. How much of each must be used to prepare 100 milliliters of solution with a drug concentration of 2 mg/mL?

15 Walking rates Two children, who are 224 meters apart, start walking toward each other at the same instant at rates of 1.5 m/sec and 2 m/sec, respectively (see the figure).

(a) When will they meet?

(b) How far will each have walked?

EXERCISE 15

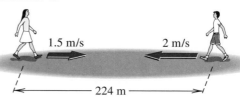

16 Running rates A runner starts at the beginning of a runners' path and runs at a constant rate of 6 mi/hr. Five minutes later a second runner begins at the same point, running at a rate of 8 mi/hr and following the same course. How long will it take the second runner to reach the first?

17 Snowplow speed At 6 A.M. a snowplow, traveling at a constant speed, begins to clear a highway leading out of town. At 8 A.M. an automobile begins traveling the highway at a speed of 30 mi/hr and reaches the plow 30 minutes later. Find the speed of the snowplow.

18 Two-way radio range Two children own two-way radios that have a maximum range of 2 miles. One leaves a certain point at 1:00 P.M., walking due north at a rate of 4 mi/hr. The other leaves the same point at 1:15 P.M., traveling due south at 6 mi/hr. When will they be unable to communicate with one another?

19 Rowing rate A boy can row a boat at a constant rate of 5 mi/hr in still water, as indicated in the figure shown on page 72. He rows upstream for 15 minutes and then rows downstream, returning to his starting point in another 12 minutes.

(a) Find the rate of the current.

(b) Find the total distance traveled.

EXERCISE 19

5 mi/hr   Upstream net speed = $5 - x$ mi/hr

x mi/hr

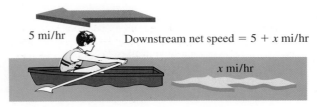

5 mi/hr   Downstream net speed = $5 + x$ mi/hr

x mi/hr

20 *Gas mileage*   A salesperson purchased an automobile that was advertised as averaging 25 mi/gal in the city and 40 mi/gal on the highway. A recent sales trip that covered 1800 miles required 51 gallons of gasoline. Assuming that the advertised mileage estimates were correct, how many miles were driven in the city?

21 *Distance to a target*   A bullet is fired horizontally at a target, and the sound of its impact is heard 1.5 seconds later. If the speed of the bullet is 3300 ft/sec and the speed of sound is 1100 ft/sec, how far away is the target?

22 *Jogging rates*   A woman begins jogging at 3:00 P.M., running due north at a 6-minute-mile pace. Later, she reverses direction and runs due south at a 7-minute-mile pace. If she returns to her starting point at 3:45 P.M., find the total number of miles run.

23 *Fencing a region*   A farmer plans to use 180 feet of fencing to enclose a rectangular region, using part of a straight river bank instead of fencing as one side of the rectangle, as shown in the figure. Find the area of the region if the length of the side parallel to the river bank is

(a) twice the length of an adjacent side.

(b) one-half the length of an adjacent side.

(c) the same as the length of an adjacent side.

EXERCISE 23

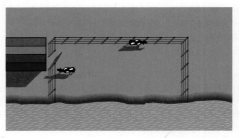

24 *House dimensions*   Shown in the figure is a cross section of a design for a two-story home. The center height h of the second story has not yet been determined. Find h such that the second story will have the same cross-sectional area as the first story.

EXERCISE 24

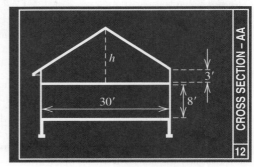

25 *Window dimensions*   A stained-glass window is being designed in the shape of a rectangle surmounted by a

EXERCISE 25

semicircle, as shown in the figure on page 72. The width of the window is to be 3 feet, but the height h is yet to be determined. If 24 ft² of glass is to be used, find the height h.

26 *Drainage ditch dimensions* Every cross section of a drainage ditch is an isosceles trapezoid with a small base of 3 feet and a height of 1 foot, as shown in the figure. Determine the width of the larger base that would give the ditch a cross-sectional area of 5 ft².

EXERCISE 26

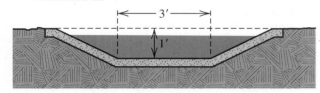

27 *Constructing a silo* A large grain silo is to be constructed in the shape of a circular cylinder with a hemisphere attached to the top (see the figure). The diameter of the silo is to be 30 feet, but the height is yet to be determined. Find the height h of the silo that will result in a capacity of $11,250\pi$ ft³.

EXERCISE 27

28 *Dimensions of a cone* The wafer cone shown in the figure is to hold 8 in.³ of ice cream when filled to the bottom. The diameter of the cone is 2 inches, and the top of the ice cream has the shape of a hemisphere. Find the height h of the cone.

EXERCISE 28

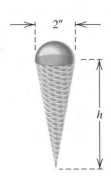

29 *Lawn mowing rates* It takes a boy 90 minutes to mow the lawn, but his sister can mow it in 60 minutes. How long would it take them to mow the lawn if they worked together, using two lawn mowers?

30 *Filling a swimming pool* With water from one hose, a swimming pool can be filled in 8 hours. A second, larger hose used alone can fill the pool in 5 hours. How long would it take to fill the pool if both hoses were used simultaneously?

31 *Delivering newspapers* It takes a girl 45 minutes to deliver the newspapers on her route; however, if her brother helps, it takes them only 20 minutes. How long would it take her brother to deliver the newspapers by himself?

32 *Emptying a tank* A water tank can be emptied by using one pump for 5 hours. A second, smaller pump can empty the tank in 8 hours. If the larger pump is started at 1:00 P.M., at what time should the smaller pump be started so that the tank will be emptied at 5:00 P.M.?

33 *Baseball stats* After playing 100 games, a major-league baseball team has a record of 0.650; that is, the team has won 65% of its games. If the team wins only 50% of its games for the remainder of the season, after how many additional games will its record be 0.600?

34 *Ohm's law* In electrical theory, Ohm's law states that $I = V/R$, where I is the current in amperes, V is the electromotive force in volts, and R is the resistance in ohms. In a certain circuit $V = 110$ and $R = 50$. If V and R are to be changed by the same numerical amount, what change in them will cause I to double?

2.3   QUADRATIC EQUATIONS

FIGURE 7

A toy rocket is launched vertically upward from level ground, as illustrated in Figure 7. If the initial speed is 120 ft/sec and the only force acting is gravity, then the rocket's height h (in feet) above the ground after t seconds is given by

$$h = -16t^2 + 120t.$$

Some values of h for the first 7 seconds of flight are listed in the following table.

t (sec)	0	1	2	3	4	5	6	7
h (ft)	0	104	176	216	224	200	144	56

We see from the table that, as it ascended, the rocket was 180 feet above the ground at some time between $t = 2$ and $t = 3$. As it descended, the rocket was 180 feet above the ground at some time between $t = 5$ and $t = 6$. To find the exact values of t for which $h = 180$ ft, we must solve the equation

$$180 = -16t^2 + 120t,$$

or $\qquad\qquad 16t^2 - 120t + 180 = 0.$

As indicated in the next chart, an equation of this type is called a *quadratic equation* in t. After developing a formula for solving such equations, we will return to this problem in Example 11 and find the exact times at which the rocket was 180 feet above the ground.

Terminology	Definition	Illustrations
Quadratic equation in x	An equation that can be written in the form $ax^2 + bx + c = 0$, where $a \neq 0$	$4x^2 = 8 - 11x$ $x(3 + x) = 5$ $4x = x^2$

One method for solving a quadratic equation uses the fact that if p and q are expressions in a variable x, then $pq = 0$ if and only if $p = 0$ or $q = 0$. It follows that if $ax^2 + bx + c$ can be written as a product of two first-degree polynomials, then solutions can be found by setting each factor equal to 0, as illustrated in the next two examples. This technique is called the **method of factoring**.

EXAMPLE 1   *Solving an equation by factoring*

Solve $3x^2 = 10 - x$.

SOLUTION   To use the method of factoring, *it is essential that only the number 0 appear on one side of the equation.* Thus, we proceed as follows:

$$3x^2 = 10 - x \quad \text{given}$$

$$3x^2 + x - 10 = 0 \qquad \text{add } x - 10$$

$$(3x - 5)(x + 2) = 0 \qquad \text{factor}$$

$$3x - 5 = 0, \quad x + 2 = 0 \qquad \text{set each factor equal to 0}$$

$$x = \tfrac{5}{3}, \qquad x = -2 \qquad \text{solve for } x$$

Hence, the solutions of the given equation are $\tfrac{5}{3}$ and -2.

EXAMPLE 2   *Solving an equation by factoring*

Solve $x^2 + 16 = 8x$.

SOLUTION   We proceed as in Example 1:

$$x^2 + 16 = 8x \quad \text{given}$$

$$x^2 - 8x + 16 = 0 \qquad \text{subtract } 8x$$

$$(x - 4)(x - 4) = 0 \qquad \text{factor}$$

$$x - 4 = 0, \quad x - 4 = 0 \qquad \text{set each factor equal to 0}$$

$$x = 4, \qquad x = 4 \qquad \text{solve for } x$$

Thus, the given quadratic equation has one solution, 4.

Since $x - 4$ appears as a factor twice in the previous solution, we call 4 a **double root**, or **root of multiplicity 2**, of the equation $x^2 + 16 = 8x$.

If a quadratic equation has the form $x^2 = d$, for some $d > 0$, then $x^2 - d = 0$ or, equivalently,

$$(x + \sqrt{d})(x - \sqrt{d}) = 0.$$

Setting each factor equal to zero gives us the solutions $-\sqrt{d}$ and $\sqrt{d}$. We frequently use the symbol $\pm\sqrt{d}$ (*plus or minus* $\sqrt{d}$) to represent both $\sqrt{d}$ and $-\sqrt{d}$. Thus, for $d > 0$, we have proved the following result. (The case $d < 0$ requires the system of complex numbers discussed in Section 2.4.)

A Special Quadratic Equation

> If $x^2 = d$,   then $x = \pm\sqrt{d}$.

The process of solving $x^2 = d$ as indicated in the preceding box is referred to as *taking the square root of both sides of the equation.* Note that if $d > 0$ we obtain both a positive square root and a negative square root, not just the principal square root defined in Section 1.2.

EXAMPLE 3   *Solving equations of the form $x^2 = d$*

Solve

(a) $x^2 = 5$ **(b)** $(x + 3)^2 = 5$

SOLUTION

(a)

$$x^2 = 5 \qquad \text{given}$$

$$x = \pm\sqrt{5} \qquad \text{take the square root}$$

Thus, the solutions are $\sqrt{5}$ and $-\sqrt{5}$.

(b)

$$(x + 3)^2 = 5 \qquad \text{given}$$

$$x + 3 = \pm\sqrt{5} \qquad \text{take the square root}$$

$$x = -3 \pm \sqrt{5} \qquad \text{subtract 3}$$

Thus, the solutions are $-3 + \sqrt{5}$ and $-3 - \sqrt{5}$.

In the work to follow we will replace an expression of the form $x^2 + kx$ by $(x + d)^2$, where k and d are real numbers. This procedure, called **completing the square** for $x^2 + kx$, calls for adding $(k/2)^2$, as described in the next box. (The same procedure is used for $x^2 - kx$.)

Completing the Square

> To complete the square for $x^2 + kx$ or $x^2 - kx$, add $\left(\dfrac{k}{2}\right)^2$; that is,
> *add the square of half the coefficient of x.*
>
> **(1)** $x^2 + kx + \left(\dfrac{k}{2}\right)^2 = \left(x + \dfrac{k}{2}\right)^2$
>
> **(2)** $x^2 - kx + \left(\dfrac{k}{2}\right)^2 = \left(x - \dfrac{k}{2}\right)^2$

EXAMPLE 4   *Completing the square*

Complete the square for

(a) $x^2 + 3x$ **(b)** $x^2 - 3x$

SOLUTION   The square of half the coefficient of x is $(\frac{3}{2})^2$ (note that we may disregard the sign of the coefficient of x). Thus,

(a) $x^2 + 3x + (\frac{3}{2})^2 = (x + \frac{3}{2})^2$

(b) $x^2 - 3x + (\frac{3}{2})^2 = (x - \frac{3}{2})^2$

In the next example we solve a quadratic equation by completing a square.

EXAMPLE 5 *Solving a quadratic equation by completing the square*

Solve $x^2 - 5x + 3 = 0$.

SOLUTION   It is convenient to first rewrite the equation so that only terms involving x are on the left, as follows:

$$x^2 - 5x + 3 = 0 \qquad \text{given}$$

$$x^2 - 5x = -3 \qquad \text{subtract 3}$$

$$x^2 - 5x + \left(\tfrac{5}{2}\right)^2 = -3 + \left(\tfrac{5}{2}\right)^2 \qquad \text{complete the square,}$$
adding $\left(\tfrac{5}{2}\right)^2$ to *both* sides

$$\left(x - \tfrac{5}{2}\right)^2 = \tfrac{13}{4} \qquad \text{equivalent equation}$$

$$x - \tfrac{5}{2} = \pm\sqrt{\tfrac{13}{4}} \qquad \text{take the square root}$$

$$x = \frac{5}{2} \pm \frac{\sqrt{13}}{2} = \frac{5 \pm \sqrt{13}}{2} \qquad \text{add } \frac{5}{2}$$

Thus, the solutions of the equation are $(5 + \sqrt{13})/2 \approx 4.3$ and $(5 - \sqrt{13})/2 \approx 0.7$.

We may solve a quadratic equation $ax^2 + bx + c = 0$ with $a \neq 1$ by adding a step to the procedure used in the preceding example. After rewriting the equation so that only terms involving x are on the left,

$$ax^2 + bx = -c,$$

we divide both sides by a, obtaining

$$x^2 + \frac{b}{a}x = -\frac{c}{a}.$$

We then complete the square by adding $\left(\dfrac{b}{2a}\right)^2$ to both sides. This technique is used in the proof of the following important formula.

Quadratic Formula

If $a \neq 0$, the roots of $ax^2 + bx + c = 0$ are given by

$$x = \frac{-b \pm \sqrt{b^2 - 4ac}}{2a}.$$

PROOF   We shall assume that $b^2 - 4ac \geq 0$ so that $\sqrt{b^2 - 4ac}$ is a real number. (The case in which $b^2 - 4ac < 0$ will be discussed in the next section.) Let us proceed as follows:

$$ax^2 + bx + c = 0 \qquad \text{given}$$

$$ax^2 + bx = -c \qquad \text{subtract } c$$

$$x^2 + \frac{b}{a}x = -\frac{c}{a} \qquad \text{divide by } a$$

$$x^2 + \frac{b}{a}x + \left(\frac{b}{2a}\right)^2 = \left(\frac{b}{2a}\right)^2 - \frac{c}{a} \qquad \text{complete the square}$$

$$\left(x + \frac{b}{2a}\right)^2 = \frac{b^2 - 4ac}{4a^2} \qquad \text{equivalent equation}$$

$$x + \frac{b}{2a} = \pm\sqrt{\frac{b^2 - 4ac}{4a^2}} \qquad \text{take the square root}$$

$$x = -\frac{b}{2a} \pm \sqrt{\frac{b^2 - 4ac}{4a^2}} \qquad \text{subtract } \frac{b}{2a}$$

We may write the radical in the last equation as

$$\pm\sqrt{\frac{b^2 - 4ac}{4a^2}} = \pm\frac{\sqrt{b^2 - 4ac}}{\sqrt{(2a)^2}} = \pm\frac{\sqrt{b^2 - 4ac}}{|2a|}.$$

Since $|2a| = 2a$ if $a > 0$, or $|2a| = -2a$ if $a < 0$, we see that in all cases

$$x = -\frac{b}{2a} \pm \frac{\sqrt{b^2 - 4ac}}{2a} = \frac{-b \pm \sqrt{b^2 - 4ac}}{2a}. \quad \blacksquare$$

Note that if the quadratic formula is executed properly, it is unnecessary to check the solutions.

The number $b^2 - 4ac$ under the radical sign in the quadratic formula is called the **discriminant** of the quadratic equation. The discriminant can be used to determine the nature of the roots of the equation, as in the following chart.

Value of the discriminant $b^2 - 4ac$	Nature of the roots of $ax^2 + bx + c = 0$
Positive value	Two real and unequal roots
0	One root of multiplicity 2
Negative value	No real root

The discriminant in the next two examples is positive. In Example 8 the discriminant is 0.

EXAMPLE 6   *Using the quadratic formula*

Solve $4x^2 + x - 3 = 0$.

SOLUTION   Let $a = 4$, $b = 1$, and $c = -3$ in the quadratic formula:

$$x = \frac{-1 \pm \sqrt{(1)^2 - 4(4)(-3)}}{2(4)}$$

$$= \frac{-1 \pm \sqrt{49}}{8}$$

$$= \frac{-1 \pm 7}{8}$$

Hence, the solutions are

$$x = \frac{-1 + 7}{8} = \frac{3}{4} \quad \text{and} \quad x = \frac{-1 - 7}{8} = -1.$$

Example 6 can also be solved by factoring. Writing $(4x - 3)(x + 1) = 0$ and setting each factor equal to zero gives us $x = \frac{3}{4}$ and $x = -1$.

EXAMPLE 7   *Using the quadratic formula*

Solve $2x(3 - x) = 3$.

SOLUTION   To use the quadratic formula we must write the equation in the form $ax^2 + bx + c = 0$. The following equations are equivalent:

$$2x(3 - x) = 3 \quad \text{given}$$

$$6x - 2x^2 = 3 \quad \text{multiply factors}$$

$$-2x^2 + 6x - 3 = 0 \quad \text{subtract 3}$$

$$2x^2 - 6x + 3 = 0 \quad \text{multiply by } -1$$

We now let $a = 2$, $b = -6$, and $c = 3$ in the quadratic formula, obtaining

$$x = \frac{-(-6) \pm \sqrt{(-6)^2 - 4(2)(3)}}{2(2)}$$

$$= \frac{6 \pm \sqrt{12}}{4} = \frac{6 \pm 2\sqrt{3}}{4}.$$

Since 2 is a factor of the numerator and denominator, we can simplify the last fraction as follows:

$$\frac{2(3 \pm \sqrt{3})}{2 \cdot 2} = \frac{3 \pm \sqrt{3}}{2}$$

Hence, the solutions are $\dfrac{3 + \sqrt{3}}{2} \approx 2.37$ and $\dfrac{3 - \sqrt{3}}{2} \approx 0.63$.

The following example illustrates the case of a double root.

EXAMPLE 8   *Using the quadratic formula*

Solve $9x^2 - 30x + 25 = 0$.

SOLUTION   Let $a = 9, b = -30$, and $c = 25$ in the quadratic formula:

$$x = \frac{-(-30) \pm \sqrt{(-30)^2 - 4(9)(25)}}{2(9)}$$

$$= \frac{30 \pm \sqrt{900 - 900}}{18}$$

$$= \frac{30 \pm 0}{18} = \frac{5}{3}$$

Consequently, the equation has one (double) root, $\frac{5}{3}$.

EXAMPLE 9   *Clearing an equation of fractions*

Solve $\dfrac{2x}{x - 3} + \dfrac{5}{x + 3} = \dfrac{36}{x^2 - 9}$.

SOLUTION   Using the guidelines stated in Section 2.1 for solving an equation containing rational expressions, we multiply by the lcd, $(x + 3)(x - 3)$, remembering that, by guideline 2, the numbers $(-3$ and $3)$ that make the lcd zero cannot be solutions. Thus we proceed as follows:

$$\frac{2x}{x - 3} + \frac{5}{x + 3} = \frac{36}{x^2 - 9} \qquad \text{given}$$

$$2x(x + 3) + 5(x - 3) = 36 \qquad \text{multiply by the lcd, } (x + 3)(x - 3)$$

$$2x^2 + 6x + 5x - 15 - 36 = 0 \qquad \text{multiply factors and subtract 36}$$

$$2x^2 + 11x - 51 = 0 \qquad \text{simplify}$$

$$(2x + 17)(x - 3) = 0 \qquad \text{factor}$$

$$2x + 17 = 0, \qquad x - 3 = 0 \qquad \text{set each factor equal to 0}$$

$$x = -\tfrac{17}{2}, \qquad x = 3 \qquad \text{solve for } x$$

Since $x = 3$ cannot be a solution, we see that $x = -\frac{17}{2}$ is the only solution of the given equation.

Many applied problems lead to quadratic equations. One is illustrated in the following example.

EXAMPLE 10 *Constructing a rectangular box*

A box with a square base and no top is to be made from a square piece of tin by cutting out a 3-inch square from each corner and folding up the sides. If the box is to hold 48 in.³, what size piece of tin should be used?

FIGURE 8

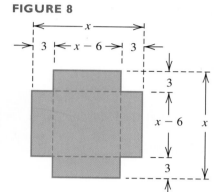

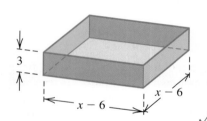

SOLUTION   We begin by drawing the picture in Figure 8, letting x denote the unknown length of the side of the piece of tin. Subsequently, each side of the base of the box will have length $x - 3 - 3 = x - 6$.

Since the area of the base of the box is $(x - 6)^2$ and the height is 3, we obtain

$$\text{volume of box} = 3(x - 6)^2.$$

Since the box is to hold 48 in.³,

$$3(x - 6)^2 = 48.$$

We now solve for x:

$$(x - 6)^2 = 16 \qquad \text{divide by 3}$$
$$x - 6 = \pm 4 \qquad \text{take the square root}$$
$$x = 6 \pm 4 \qquad \text{add 6}$$

Consequently,

$$x = 10 \qquad \text{or} \qquad x = 2.$$

✓ *Check*   Referring to Figure 8, we see that $x = 2$ is unacceptable, since no box is possible in this case. However, if we begin with a 10-inch square of tin, cut out 3-inch corners, and fold, we obtain a box having dimensions 4 inches, 4 inches, and 3 inches. The box has the desired volume of 48 in.³. Thus, 10 inches is the answer to the problem.

As illustrated in Example 10, even though an equation is formulated correctly, it is possible to arrive at meaningless solutions because of the physical nature of a given problem. Such solutions should be discarded. For example, we would not accept the answer -7 years for the age of an individual or $\sqrt{50}$ for the number of automobiles in a parking lot.

In the next example we solve the applied problem discussed at the beginning of this section.

EXAMPLE 11 *Finding the height of a toy rocket*

The height above ground h (in feet) of a toy rocket, t seconds after it is launched, is given by $h = -16t^2 + 120t$. When will the rocket be 180 feet above the ground?

SOLUTION   Using $h = -16t^2 + 120t$, we obtain

$$180 = -16t^2 + 120t \quad \text{let } h = 180$$

$$16t^2 - 120t + 180 = 0 \quad \text{add } 16t^2 - 120t$$

$$4t^2 - 30t + 45 = 0. \quad \text{divide by 4}$$

Applying the quadratic formula with $a = 4$, $b = -30$, and $c = 45$ gives us

$$t = \frac{-(-30) \pm \sqrt{(-30)^2 - 4(4)(45)}}{2(4)} = \frac{30 \pm \sqrt{180}}{8}$$

$$= \frac{30 \pm 6\sqrt{5}}{8} = \frac{15 \pm 3\sqrt{5}}{4}.$$

Hence, the rocket is 180 feet above the ground at the following times:

$$t = \frac{15 - 3\sqrt{5}}{4} \approx 2.07 \text{ sec}$$

$$t = \frac{15 + 3\sqrt{5}}{4} \approx 5.43 \text{ sec}$$

2.3 EXERCISES

Exer. 1–14: Solve the equation by factoring.

1 $6x^2 + x - 12 = 0$

2 $4x^2 + x - 14 = 0$

3 $15x^2 - 12 = -8x$

4 $15x^2 - 14 = 29x$

5 $2x(4x + 15) = 27$

6 $x(3x + 10) = 77$

7 $75x^2 + 35x - 10 = 0$

8 $48x^2 + 12x - 90 = 0$

9 $12x^2 + 60x + 75 = 0$

10 $4x^2 - 72x + 324 = 0$

11 $\dfrac{2x}{x + 3} + \dfrac{5}{x} - 4 = \dfrac{18}{x^2 + 3x}$

12 $\dfrac{5x}{x - 2} + \dfrac{3}{x} + 2 = \dfrac{-6}{x^2 - 2x}$

13 $\dfrac{5x}{x - 3} + \dfrac{4}{x + 3} = \dfrac{90}{x^2 - 9}$

14 $\dfrac{3x}{x - 2} + \dfrac{1}{x + 2} = \dfrac{-4}{x^2 - 4}$

Exer. 15–16: Determine whether the two equations are equivalent.

15 (a) $x^2 = 16$, $x = 4$ (b) $x = \sqrt{9}$, $x = 3$

16 (a) $x^2 = 25$, $x = 5$ (b) $x = \sqrt{64}$, $x = 8$

Exer. 17–24: Solve the equation.

17 $x^2 = 169$

18 $x^2 = 361$

19 $25x^2 = 9$

20 $16x^2 = 49$

21 $(x - 3)^2 = 17$

22 $(x + 4)^2 = 31$

23 $4(x + 2)^2 = 11$

24 $9(x - 1)^2 = 7$

Exer. 25–26: Determine the values of d that complete the square for the expression.

25 (a) $x^2 + 9x + d$ (b) $x^2 - 8x + d$

(c) $x^2 + dx + 36$ (d) $x^2 + dx + \frac{49}{4}$

26 (a) $x^2 + 13x + d$ (b) $x^2 - 6x + d$

(c) $x^2 + dx + 25$ (d) $x^2 + dx + \frac{81}{4}$

Exer. 27–30: Solve by completing the square.

27 $x^2 + 6x + 7 = 0$

28 $x^2 - 8x + 11 = 0$

29 $4x^2 - 12x - 11 = 0$

30 $4x^2 + 20x + 13 = 0$

Exer. 31–44: Solve by using the quadratic formula.

31 $6x^2 - x = 2$ 32 $5x^2 + 13x = 6$

33 $x^2 + 4x + 2 = 0$ 34 $x^2 - 6x - 3 = 0$

35 $2x^2 - 3x - 4 = 0$ 36 $3x^2 + 5x + 1 = 0$

37 $\frac{3}{2}z^2 - 4z - 1 = 0$ 38 $\frac{5}{3}s^2 + 3s + 1 = 0$

39 $\dfrac{5}{w^2} - \dfrac{10}{w} + 2 = 0$ 40 $\dfrac{x+1}{3x+2} = \dfrac{x-2}{2x-3}$

41 $4x^2 + 81 = 36x$ 42 $24x + 9 = -16x^2$

43 $\dfrac{5x}{x^2+9} = -1$ 44 $\frac{1}{7}x^2 + 1 = \frac{4}{7}x$

Exer. 45–46: Use the quadratic formula to solve the equation for (a) x **in terms of** y **and (b)** y **in terms of** x.

45 $4x^2 - 4xy + 1 - y^2 = 0$ 46 $2x^2 - xy = 3y^2 + 1$

Exer. 47–50: Solve for the specified variable.

47 $K = \frac{1}{2}mv^2$ for v (kinetic energy)

48 $F = g\dfrac{mM}{d^2}$ for d (Newton's law of gravitation)

49 $A = 2\pi r(r + h)$ for r (surface area of a closed cylinder)

50 $s = \frac{1}{2}gt^2 + v_0 t$ for t (distance an object falls)

51 *Velocity of a gas* When a hot gas exits a cylindrical smokestack, its velocity varies throughout a circular cross section of the smokestack, with the gas near the center of the cross section having a greater velocity than the gas near the perimeter. This phenomenon can be described by the formula

$$V = V_{\max}\left[1 - \left(\frac{r}{r_0}\right)^2\right],$$

where $V_{\max}$ is the maximum velocity of the gas, r_0 is the radius of the smokestack, and V is the velocity of the gas at a distance r from the center of the circular cross section. Solve this formula for r.

52 *Density of the atmosphere* For altitudes h up to 10,000 meters, the density D of the earth's atmosphere (in kg/m³) can be approximated by the formula

$$D = 1.225 - (1.12 \times 10^{-4})h + (3.24 \times 10^{-9})h^2.$$

Approximate the altitude if the density of the atmosphere is 0.74 kg/m³.

53 *Dimensions of a tin can* A manufacturer of tin cans wishes to construct a right circular cylindrical can of height 20 centimeters and capacity 3000 cm³ (see the figure). Find the inner radius r of the can.

EXERCISE 53

54 *Constructing a rectangular box* Refer to Example 10. A box with an open top is to be constructed by cutting 3-inch squares from the corners of a rectangular sheet of tin whose length is twice its width. What size sheet will produce a box having a volume of 60 in.³?

55 *Baseball toss* A baseball is thrown straight upward with an initial speed of 64 ft/sec. The number of feet s above the ground after t seconds is given by the equation $s = -16t^2 + 64t$.

   (a) When will the baseball be 48 feet above the ground?

   (b) When will it hit the ground?

56 *Braking distance* The distance that a car travels between the time the driver makes the decision to hit the brakes and the time the car actually stops is called the braking distance. For a certain car traveling v mi/hr, the braking distance d (in feet) is given by $d = v + (v^2/20)$.

   (a) Find the braking distance when v is 55 mi/hr.

   (b) If a driver decides to brake 120 feet from a stop sign, how fast can the car be going and still stop by the time it reaches the sign?

57 *Temperature of boiling water* The temperature T (in °C) at which water boils is related to the elevation h (in meters above sea level) by the formula

$$h = 1000(100 - T) + 580(100 - T)^2$$

for $95 \le T \le 100$.

   (a) At what elevation does water boil at a temperature of 98 °C?

   (b) The elevation of Mt. Everest is approximately 8840 meters. Estimate the temperature at which water boils at the top of this mountain. (*Hint:* Use the quadratic formula with $x = 100 - T$.)

58 *Coulomb's law* A particle of charge -1 is located on a coordinate line at $x = -2$, and a particle of charge -2 is located at $x = 2$, as shown in the figure. If a particle of charge $+1$ is located at a position x between -2 and 2, Coulomb's law in electrical theory asserts that the net force F acting on this particle is given by

$$F = \frac{-k}{(x + 2)^2} + \frac{2k}{(2 - x)^2}$$

for some constant $k > 0$. Determine the position at which the net force is zero.

EXERCISE 58

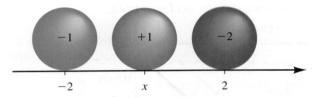

59 *Dimensions of a sidewalk* A rectangular plot of ground having dimensions 26 feet by 30 feet is surrounded by a walk of uniform width. If the area of the walk is 240 ft^2, what is its width?

60 *Designing a poster* A 24-by-36-inch sheet of paper is to be used for a poster, with the shorter side at the bottom. The margins at the sides and top are to have the same width, and the bottom margin is to be twice as wide as the other margins. Find the width of the margins if the printed area is to be 661.5 in.^{2}.

61 *Fencing a garden* A square vegetable garden is to be enclosed with a fence. If the fence costs \$1 per foot and the cost of preparing the soil is \$0.50 per ft^2, determine the size of the garden that can be enclosed for \$120.

62 *Fencing a region* A farmer plans to enclose a rectangular region, using part of his barn for one side and fencing for the other three sides. If the side parallel to the barn is to be twice the length of an adjacent side, and the area of the region is to be 128 ft^2, how many feet of fencing should be purchased?

63 *Planning a freeway* The boundary of a city is a circle of diameter 5 miles. As shown in the figure, a straight highway runs through the center of the city from A to B. The highway department is planning to build a 6-mile-long freeway from A to a point P on the outskirts and then to B. Find the distance from A to P. (*Hint: APB* is a right triangle.)

EXERCISE 63

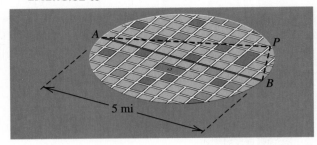

5 mi

64 *City expansion* The boundary of a city is a circle of diameter 10 miles. Within the last decade, the city has grown in area by approximately 16π mi^2 (about 50 mi^2). Assuming the city was always circular in shape, find the corresponding change in distance from the center of the city to the boundary.

65 *Distance between airplanes* An airplane flying north at 200 mi/hr passed over a point on the ground at 2:00 P.M. Another airplane at the same altitude passed over the point at 2:30 P.M., flying east at 400 mi/hr (see the figure).

(a) If t denotes the time in hours after 2:30 P.M., express the distance d between the airplanes in terms of t.

(b) At what time after 2:30 P.M. were the airplanes 500 miles apart?

EXERCISE 65

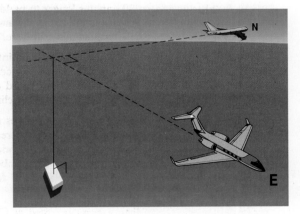

66 *Two-way radio range* Two surveyors with two-way radios leave the same point at 9:00 A.M., one walking due south at 4 mi/hr and the other due west at 3 mi/hr. How long can they communicate with one another if each radio has a maximum range of 2 miles?

67 *Constructing a pizza box* A pizza box with a square base is to be made from a rectangular sheet of cardboard by cutting six 1-inch squares from the corners and the middle sections and folding up the sides (see the figure). If the area of the base is to be 144 in.^{2}, what size piece of cardboard should be used?

EXERCISE 67

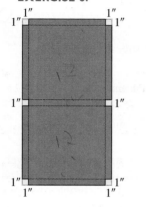

68 *Constructing wire frames* Two square wire frames are to be constructed from a piece of wire 100 inches long. If the area enclosed by one frame is to be one-half the area enclosed by the other, find the dimensions of each frame. (Disregard the thickness of the wire.)

69 *Canoeing rate* The speed of the current in a stream is 5 mi/hr. It takes a canoeist 30 minutes longer to paddle 1.2 miles upstream than to paddle the same distance downstream. What is the canoeist's rate in still water?

70 *Height of a cliff* If a rock is dropped from a cliff into an ocean, it travels approximately $16t^2$ feet in t seconds. If the splash is heard 4 seconds later and the speed of sound is 1100 ft/sec, approximate the height of the cliff.

71 *Quantity discount* A company sells running shoes to dealers for $40 per pair if less than 50 pairs are ordered. If 50 or more pairs are ordered (up to 600), the price per pair is reduced at a rate of $0.04 times the number ordered. How many pairs can a dealer purchase for $8400?

72 *Price of a CD player* When a popular brand of CD player is priced at $300 per unit, a store sells 15 units per week. Each time the price is reduced by $10, however, the sales increase by 2 per week. What selling price will result in weekly revenues of $7000?

73 *Dimensions of an oil drum* A closed cylindrical oil drum of height 4 feet is to be constructed so that the total surface area is 10π ft^2. Find the diameter of the drum.

74 *Dimensions of a vitamin tablet* The rate at which a tablet of vitamin C begins to dissolve depends on the surface area of the tablet. One brand of tablet is 2 centimeters long and is in the shape of a cylinder with hemispheres of diameter 0.5 centimeter attached to both ends, as shown in the figure. A second brand of tablet is to be manufactured in the shape of a right circular cylinder of altitude 0.5 centimeter.

(a) Find the diameter of the second tablet so that its surface area is equal to that of the first tablet.

(b) Find the volume of each tablet.

EXERCISE 74

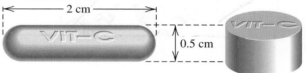

Exer. 75–76: During a nuclear explosion, a fireball will be produced having a maximum volume V_0. For temperatures below 2000 K and a given explosive force, the volume V of the fireball t seconds after the explosion can be estimated using the given formula. (Note that the kelvin is abbreviated as K, not °K.) Approximate t when V is 95% of V_0.

75 $V/V_0 = 0.8197 + 0.007752t + 0.0000281t^2$
(20-kiloton explosion)

76 $V/V_0 = 0.831 + 0.00598t + 0.0000919t^2$
(10-megaton explosion)

Exer. 77–78: When computations are carried out on a calculator, the quadratic formula will not always give accurate results if b^2 is large in comparison to ac, because one of the roots will be close to zero and difficult to approximate.

(a) Use the quadratic formula to approximate the roots of the given equation.

(b) To obtain a better approximation for the root near zero, rationalize the numerator to change

$$x = \frac{-b \pm \sqrt{b^2 - 4ac}}{2a} \quad \text{to} \quad x = \frac{2c}{-b \mp \sqrt{b^2 - 4ac}},$$

and use the second formula.

77 $x^2 + 4,500,000x - 0.96 = 0$

78 $x^2 - 73,000,000x + 2.01 = 0$

2.4 COMPLEX NUMBERS

Complex numbers are needed to find solutions of equations that cannot be solved using only the set $\mathbb{R}$ of real numbers. The following chart illustrates several simple quadratic equations and the types of numbers required for solutions.

Equation	Solutions	Type of numbers required
$x^2 = 9$	$3, -3$	Integers
$x^2 = \frac{9}{4}$	$\frac{3}{2}, -\frac{3}{2}$	Rational numbers
$x^2 = 5$	$\sqrt{5}, -\sqrt{5}$	Irrational numbers
$x^2 = -5$	?	Complex numbers

The solutions of the first three equations in the chart are in $\mathbb{R}$; however, since squares of real numbers are never negative, $\mathbb{R}$ does not contain the solutions of $x^2 = -5$. To solve this equation we need the **complex number system** $\mathbb{C}$, which contains both $\mathbb{R}$ and numbers whose squares are negative.

We begin by introducing the **imaginary unit**, denoted by i, which has the following properties.

Properties of i

$$i = \sqrt{-1}, \qquad i^2 = -1$$

Because its square is negative, the letter i does not represent a real number. It is a new mathematical entity that will enable us to obtain $\mathbb{C}$. Since i, together with $\mathbb{R}$, is to be contained in $\mathbb{C}$, we must consider products of the form bi for a real number b and also expressions of the form $a + bi$ for real numbers a and b. The next chart provides definitions we shall use.

Terminology	Definition
Complex number	$a + bi$, where a and b are real numbers and $i^2 = -1$
Equality	$a + bi = c + di$ if and only if $a = c$ and $b = d$
Sum	$(a + bi) + (c + di) = (a + c) + (b + d)i$
Product	$(a + bi)(c + di) = (ac - bd) + (ad + bc)i$

It is unnecessary to memorize the definitions of addition and multiplication of complex numbers given in the preceding chart. Instead, *we may treat all symbols as having properties of real numbers, with exactly one* *exception: we replace i^2 by -1.* Thus, for the product $(a + bi)(c + di)$ we simply use the distributive laws and the fact that

$$(bi)(di) = bdi^2 = bd(-1) = -bd.$$

EXAMPLE 1   *Addition and multiplication of complex numbers*

Express in the form $a + bi$, where a and b are real numbers:

(a) $(3 + 4i) + (2 + 5i)$ **(b)** $(3 + 4i)(2 + 5i)$

SOLUTION

(a) $(3 + 4i) + (2 + 5i) = (3 + 2) + (4 + 5)i = 5 + 9i$

(b) $(3 + 4i)(2 + 5i) = (3 + 4i)2 + (3 + 4i)(5i)$

$$= 6 + 8i + 15i + 20i^2$$
$$= 6 + 23i + 20(-1)$$
$$= -14 + 23i$$

The set $\mathbb{R}$ of real numbers may be identified with the set of complex numbers of the form $a + 0i$. It is also convenient to denote the complex number $0 + bi$ by bi. Thus,

$$(a + 0i) + (0 + bi) = (a + 0) + (0 + b)i = a + bi.$$

Hence we may regard $a + bi$ as the sum of two complex numbers a and bi (that is, $a + 0i$ and $0 + bi$). We sometimes call a the **real part** and b the **imaginary part** of the complex number $a + bi$.

We can now solve an equation such as $x^2 = -5$. Specifically, since

$$(\sqrt{5}i)(\sqrt{5}i) = (\sqrt{5})^2 i^2 = 5(-1) = -5,$$

we see that one solution is $\sqrt{5}i$ and another is $-\sqrt{5}i$.

In the next chart we define the difference of complex numbers and multiplication of a complex number by a real number.

Terminology	Definition
Difference	$(a + bi) - (c + di) = (a - c) + (b - d)i$
Multiplication by a real number k	$k(a + bi) = ka + (kb)i$

If we are asked to write an expression in the form $a + bi$, we shall also accept the form $a - di$, since $a - di = a + (-d)i$.

EXAMPLE 2   *Operations with complex numbers*

Express in the form $a + bi$, where a and b are real numbers:

(a) $4(2 + 5i) - (3 - 4i)$ **(b)** $(4 - 3i)(2 + i)$ **(c)** $i(3 - 2i)^2$ **(d)** i^{51}

SOLUTION

(a) $4(2 + 5i) - (3 - 4i) = 8 + 20i - 3 + 4i = 5 + 24i$

(b) $(4 - 3i)(2 + i) = 8 - 6i + 4i - 3i^2 = 11 - 2i$

(c) $i(3 - 2i)^2 = i(9 - 12i + 4i^2) = i(5 - 12i) = 5i - 12i^2 = 12 + 5i$

(d) Taking successive powers of i, we obtain

$$i^1 = i, \quad i^2 = -1, \quad i^3 = -i, \quad i^4 = 1,$$

and then the cycle starts over:

$$i^5 = i, \quad i^6 = i^2 = -1, \quad \text{and so on.}$$

In particular,

$$i^{51} = i^{48}i^3 = (i^4)^{12}i^3 = (1)^{12}i^3 = i^3 = -i.$$

The following concept has important uses in working with complex numbers.

Definition of the Conjugate of a Complex Number

The **conjugate** of $a + bi$ is $a - bi$.

Since $a - bi = a + (-bi)$, it follows that the conjugate of $a - bi$ is

$$a - (-bi) = a + bi.$$

Therefore, $a + bi$ and $a - bi$ are conjugates of one another.

ILLUSTRATION

Conjugates

Complex number	Conjugate
$5 + 7i$	$5 - 7i$
$5 - 7i$	$5 + 7i$
$4i$	$-4i$
3	3

The following two properties are consequences of the definitions of sum and product of complex numbers.

Properties of conjugates	Illustration
$(a + bi) + (a - bi) = 2a$	$(4 + 3i) + (4 - 3i) = 4 + 4 = 2 \cdot 4$
$(a + bi)(a - bi) = a^2 + b^2$	$(4 + 3i)(4 - 3i) = 4^2 - (3i)^2 = 4^2 - 3^2 i^2 = 4^2 + 3^2$

Note that *the sum and the product of a complex number and its conjugate are real numbers*. Conjugates are useful for finding the **multiplicative inverse** $\dfrac{1}{a + bi}$ of $a + bi$ or for simplifying the quotient $\dfrac{a + bi}{c + di}$ of two complex numbers, as illustrated in the next example.

EXAMPLE 3 *Quotients of complex numbers*

Express in the form $a + bi$, where a and b are real numbers:

(a) $\dfrac{1}{9 + 2i}$ **(b)** $\dfrac{7 - i}{3 - 5i}$

SOLUTION

(a) $\dfrac{1}{9 + 2i} = \dfrac{1}{9 + 2i} \cdot \dfrac{9 - 2i}{9 - 2i} = \dfrac{9 - 2i}{81 + 4} = \dfrac{9}{85} - \dfrac{2}{85} i$

(b) $\dfrac{7 - i}{3 - 5i} = \dfrac{7 - i}{3 - 5i} \cdot \dfrac{3 + 5i}{3 + 5i} = \dfrac{21 + 35i - 3i - 5i^2}{9 + 25}$

$\qquad\qquad = \dfrac{26 + 32i}{34} = \dfrac{13}{17} + \dfrac{16}{17} i$

If p is a positive real number, then the equation $x^2 = -p$ has solutions in $\mathbb{C}$. One solution is $\sqrt{p}\,i$, since

$$(\sqrt{p}\,i)^2 = (\sqrt{p})^2 i^2 = p(-1) = -p.$$

Similarly, $-\sqrt{p}\,i$ is also a solution.

The definition of $\sqrt{-r}$ in the next chart is motivated by $(\sqrt{r}\,i)^2 = -r$ for $r > 0$. When using this definition, take care *not* to write $\sqrt{ri}$ when $\sqrt{r}\,i$ is intended.

Terminology	Definition	Illustrations
Principal square root $\sqrt{-r}$ for $r > 0$	$\sqrt{-r} = \sqrt{r}\,i$	$\sqrt{-9} = \sqrt{9}\,i = 3i$ $\sqrt{-5} = \sqrt{5}\,i$ $\sqrt{-1} = \sqrt{1}\,i = i$

The radical sign must be used with caution when the radicand is negative. For example, the formula $\sqrt{a}\sqrt{b} = \sqrt{ab}$, which holds for positive real numbers, is not true when a and b are both negative, as shown below:

$$\sqrt{-3}\sqrt{-3} = (\sqrt{3}\,i)(\sqrt{3}\,i) = (\sqrt{3})^2 i^2 = 3(-1) = -3,$$

but

$$\sqrt{(-3)(-3)} = \sqrt{9} = 3.$$

Hence,

$$\sqrt{-3}\sqrt{-3} \ne \sqrt{(-3)(-3)}.$$

If only *one* of a or b is negative, then $\sqrt{a}\sqrt{b} = \sqrt{ab}$. In general, we shall not apply laws of radicals if radicands are negative. Instead, we shall change the form of radicals before performing any operations, as illustrated in the next example.

EXAMPLE 4   *Working with square roots of negative numbers*

Express in the form $a + bi$, where a and b are real numbers:

$$(5 - \sqrt{-9})(-1 + \sqrt{-4})$$

SOLUTION   First we use the definition $\sqrt{-r} = \sqrt{r}\,i$, and then we simplify:

$$\begin{aligned}
(5 - \sqrt{-9})(-1 + \sqrt{-4}) &= (5 - \sqrt{9}\,i)(-1 + \sqrt{4}\,i) \\
&= (5 - 3i)(-1 + 2i) \\
&= -5 + 10i + 3i - 6i^2 \\
&= -5 + 13i + 6 = 1 + 13i
\end{aligned}$$

In the proof of the quadratic formula in the previous section we saw that if a, b, and c are real numbers such that $b^2 - 4ac \ge 0$, and if $a \ne 0$, then the solutions of the quadratic equation $ax^2 + bx + c = 0$ are given by

$$x = \frac{-b \pm \sqrt{b^2 - 4ac}}{2a}.$$

We may now extend this fact to include $b^2 - 4ac < 0$. The same manipulations used to obtain the quadratic formula, together with the developments in this section, show that if $b^2 - 4ac < 0$, then the solutions of $ax^2 + bx + c = 0$ are the two *complex* numbers given above. Notice that the solutions are conjugates of each other.

EXAMPLE 5   *A quadratic equation with complex solutions*

Solve $5x^2 + 2x + 1 = 0$.

SOLUTION   Applying the quadratic formula with $a = 5$, $b = 2$, and $c = 1$, we see that

$$x = \frac{-2 \pm \sqrt{4 - 20}}{10} = \frac{-2 \pm \sqrt{-16}}{10} = \frac{-2 \pm 4i}{10} = \frac{-1 \pm 2i}{5}.$$

Thus, the solutions of the equation are $-\frac{1}{5} + \frac{2}{5}i$ and $-\frac{1}{5} - \frac{2}{5}i$.

EXAMPLE 6   *An equation with complex solutions*

Solve $x^3 - 1 = 0$.

SOLUTION   Using the difference of two cubes factoring formula in Section 1.3, we write $x^3 - 1 = 0$ as

$$(x - 1)(x^2 + x + 1) = 0.$$

Setting each factor equal to zero and solving the resulting equations, we obtain the solutions

$$1, \quad \frac{-1 \pm \sqrt{1 - 4}}{2} = \frac{-1 \pm \sqrt{3}\,i}{2}$$

or, equivalently,

$$1, \quad -\frac{1}{2} + \frac{\sqrt{3}}{2}\,i, \quad -\frac{1}{2} - \frac{\sqrt{3}}{2}\,i.$$

The three solutions of $x^3 - 1 = 0$ are called the **cube roots of unity**.

2.4 EXERCISES

Exer. 1–34: Write the expression in the form $a + bi$, where a and b are real numbers.

1 $(5 - 2i) + (-3 + 6i)$

2 $(-5 + 7i) + (4 + 9i)$

3 $(7 - 6i) - (-11 - 3i)$

4 $(-3 + 8i) - (2 + 3i)$

5 $(3 + 5i)(2 - 7i)$

6 $(-2 + 6i)(8 - i)$

7 $(1 - 3i)(2 + 5i)$

8 $(8 + 2i)(7 - 3i)$

9 $(5 - 2i)^2$

10 $(6 + 7i)^2$

11 $i(3 + 4i)^2$

12 $i(2 - 7i)^2$

13 $(3 + 4i)(3 - 4i)$

14 $(4 + 9i)(4 - 9i)$

15 i^{43}

16 i^{92}

17 i^{73}

18 i^{66}

19 $\dfrac{3}{2 + 4i}$

20 $\dfrac{5}{2 - 7i}$

21 $\dfrac{1 - 7i}{6 - 2i}$

22 $\dfrac{2 + 9i}{-3 - i}$

23 $\dfrac{-4 + 6i}{2 + 7i}$

24 $\dfrac{-3 - 2i}{5 + 2i}$

25 $\dfrac{4 - 2i}{-5i}$

26 $\dfrac{-2 + 6i}{3i}$

27 $(2 + 5i)^3$

28 $(3 - 2i)^3$

29 $(2 - \sqrt{-4})(3 - \sqrt{-16})$

30 $(-3 + \sqrt{-25})(8 - \sqrt{-36})$

31 $\dfrac{4 + \sqrt{-81}}{7 - \sqrt{-64}}$

32 $\dfrac{5 - \sqrt{-121}}{1 + \sqrt{-25}}$

33 $\dfrac{\sqrt{-36}\sqrt{-49}}{\sqrt{-16}}$

34 $\dfrac{\sqrt{-25}}{\sqrt{-16}\sqrt{-81}}$

Exer. 35–38: Find the values of x and y.

35 $8 + (3x + y)i = 2x - 4i$

36 $(x - y) + 3i = 7 + yi$

37 $(3x + 2y) - y^3 i = 9 - 27i$

38 $x^3 + (2x - y)i = -8 - 3i$

Exer. 39–54: Find the solutions of the equation.

39 $x^2 - 6x + 13 = 0$

40 $x^2 - 2x + 26 = 0$

41 $x^2 + 4x + 13 = 0$

42 $x^2 + 8x + 17 = 0$

43 $x^2 - 5x + 20 = 0$

44 $x^2 + 3x + 6 = 0$

45 $4x^2 + x + 3 = 0$

46 $-3x^2 + x - 5 = 0$

47 $x^3 + 125 = 0$

48 $x^3 - 27 = 0$

49 $x^4 = 256$

50 $x^4 = 81$

51 $4x^4 + 25x^2 + 36 = 0$

52 $27x^4 + 21x^2 + 4 = 0$

53 $x^3 + 3x^2 + 4x = 0$

54 $8x^3 - 12x^2 + 2x - 3 = 0$

Exer. 55–60: If $z = a + bi$ is a complex number, its conjugate is often denoted by $\bar{z}$; that is, $\bar{z} = a - bi$. Verify the property.

55 $\overline{z + w} = \bar{z} + \bar{w}$

56 $\overline{z - w} = \bar{z} - \bar{w}$

57 $\overline{z \cdot w} = \bar{z} \cdot \bar{w}$

58 $\overline{z/w} = \bar{z}/\bar{w}$

59 $\bar{z} = z$ if and only if z is real.

60 $\overline{z^2} = (\bar{z})^2$

2.5 OTHER TYPES OF EQUATIONS

The equations considered in previous sections are inadequate for many problems. For example, in applications it is often necessary to consider powers x^k with $k > 2$. Some equations involve absolute values or radicals. In this section we give examples of equations of these types that can be solved using elementary methods.

EXAMPLE I   *Solving an equation containing an absolute value*

Solve the equation $|x - 5| = 3$.

SOLUTION   If a and b are real numbers with $b > 0$, then $|a| = b$ if and only if $a = b$ or $a = -b$. Hence if $|x - 5| = 3$, then either

$$x - 5 = 3 \quad \text{or} \quad x - 5 = -3.$$

Solving for x gives us

$$x = 5 + 3 = 8 \quad \text{or} \quad x = 5 - 3 = 2.$$

Thus, the given equation has two solutions, 8 and 2.

For an equation such as

$$2|x - 5| + 3 = 11,$$

we first isolate the absolute value expression by subtracting 3 and dividing by 2, to obtain

$$|x - 5| = \frac{11 - 3}{2} = 4,$$

and then proceed as in Example 1.

If an equation is in factored form *with zero on one side*, then we may often obtain solutions by setting each factor equal to zero. For example, if p, q, and r are expressions in x and if $pqr = 0$, then either $p = 0$, $q = 0$, or $r = 0$. In the next example we factor by grouping terms.

EXAMPLE 2 *Solving an equation using grouping*

Solve $x^3 + 2x^2 - x - 2 = 0$.

SOLUTION

$$x^3 + 2x^2 - x - 2 = 0 \qquad \text{given}$$

$$x^2(x + 2) - 1(x + 2) = 0 \qquad \text{group terms}$$

$$(x^2 - 1)(x + 2) = 0 \qquad \text{factor out } x + 2$$

$$(x + 1)(x - 1)(x + 2) = 0 \qquad \text{factor } x^2 - 1$$

$$x + 1 = 0, \quad x - 1 = 0, \quad x + 2 = 0 \qquad \text{set each factor equal to 0}$$

$$x = -1, \quad x = 1, \quad x = -2 \qquad \text{solve for } x$$

EXAMPLE 3 *An equation containing rational exponents*

Solve $x^{3/2} = x^{1/2}$.

SOLUTION

$$x^{3/2} = x^{1/2} \qquad \text{given}$$

$$x^{3/2} - x^{1/2} = 0 \qquad \text{subtract } x^{1/2}$$

$$x^{1/2}(x - 1) = 0 \qquad \text{factor out } x^{1/2}$$

$$x^{1/2} = 0, \quad x - 1 = 0 \qquad \text{set each factor equal to 0}$$

$$x = 0, \quad x = 1 \qquad \text{solve for } x$$

In Example 3 it would have been incorrect to divide both sides of the equation $x^{3/2} = x^{1/2}$ by $x^{1/2}$, obtaining $x = 1$, since the solution $x = 0$ would be lost. In general, *avoid dividing both sides of an equation by an expression that contains variables*; always *factor* instead.

If an equation involves radicals or fractional exponents, we often raise both sides to a positive power. The solutions of the new equation always contain the solutions of the given equation. For example, the solutions of

$$2x - 3 = \sqrt{x + 6}$$

are also solutions of

$$(2x - 3)^2 = (\sqrt{x + 6})^2.$$

In some cases the new equation has *more* solutions than the given equation. To illustrate, if we are given the equation $x = 3$ and we square both sides, we obtain $x^2 = 9$. Note that the given equation $x = 3$ has only one solution, 3, but the new equation $x^2 = 9$ has two solutions, 3 and -3. Any solution of the new equation that is not a solution of the given equation is an extraneous solution. Since extraneous solutions may occur, *it is absolutely essential to check all solutions obtained after raising both sides of an equation to an even power.* Such checks are unnecessary if both sides are raised to an *odd* power, because in this case extraneous solutions are not introduced.

EXAMPLE 4   *Solving an equation containing a radical*

Solve $\sqrt[3]{x^2 - 1} = 2$.

SOLUTION

$$\sqrt[3]{x^2 - 1} = 2 \qquad \text{given}$$

$$(\sqrt[3]{x^2 - 1})^3 = 2^3 \qquad \text{cube both sides}$$

$$x^2 - 1 = 8 \qquad \text{property of } \sqrt[n]{\;}$$

$$x^2 = 9 \qquad \text{add 1}$$

$$x = \pm 3 \qquad \text{take the square root}$$

Thus, the given equation has two solutions, 3 and -3. Except to detect algebraic errors, a check is unnecessary, since we raised both sides to an odd power.

In the last solution, we used the phrase *cube both sides* of $\sqrt[3]{x^2 - 1} = 2$. In general, for the equation $x^{m/n} = a$, we raise both sides to the power n/m (the reciprocal of m/n) to solve for x. If m is odd, we obtain $x = a^{n/m}$, but if m is even, we have $x = \pm a^{n/m}$. If n is even, extraneous solutions may occur.

In the next two examples, before we raise both sides of the equation to a power, we *isolate a radical*—that is, we consider an equivalent equation in which only the radical appears on one side.

EXAMPLE 5   *Solving an equation containing a radical*

Solve $3 + \sqrt{3x + 1} = x$.

SOLUTION

$$3 + \sqrt{3x + 1} = x \qquad \text{given}$$

$$\sqrt{3x + 1} = x - 3 \qquad \text{isolate the radical}$$

$$(\sqrt{3x + 1})^2 = (x - 3)^2 \qquad \text{square both sides}$$

$$3x + 1 = x^2 - 6x + 9 \qquad \text{simplify}$$

$$x^2 - 9x + 8 = 0 \qquad \text{subtract } 3x + 1$$

$$(x - 1)(x - 8) = 0 \qquad \text{factor}$$

$$x - 1 = 0, \quad x - 8 = 0 \qquad \text{set each factor equal to } 0$$

$$x = 1, \qquad x = 8 \qquad \text{solve for } x$$

We raised both sides to an even power, so checks are required.

✓ **Check $x = 1$**   LS: $3 + \sqrt{3(1) + 1} = 3 + \sqrt{4} = 3 + 2 = 5$

RS: 1

Since $5 \neq 1$, $x = 1$ is not a solution.

✓ **Check $x = 8$**   LS: $3 + \sqrt{3(8) + 1} = 3 + \sqrt{25} = 3 + 5 = 8$

RS: 8

Since $8 = 8$ is a true statement, $x = 8$ is a solution.
Hence, the given equation has one solution, $x = 8$.

In order to solve an equation involving several radicals, it may be necessary to raise both sides to powers several times, as in the next example.

EXAMPLE 6   *Solving an equation containing radicals*
Solve $\sqrt{2x - 3} - \sqrt{x + 7} + 2 = 0$.

SOLUTION

$$\sqrt{2x - 3} - \sqrt{x + 7} + 2 = 0 \qquad \text{given}$$

$$\sqrt{2x - 3} = \sqrt{x + 7} - 2 \qquad \text{isolate } \sqrt{2x - 3}$$

$$2x - 3 = (x + 7) - 4\sqrt{x + 7} + 4 \qquad \text{square both sides}$$

$$x - 14 = -4\sqrt{x + 7} \qquad \text{isolate the radical term}$$

$$x^2 - 28x + 196 = 16(x + 7) \qquad \text{square both sides}$$

$$x^2 - 28x + 196 = 16x + 112 \qquad \text{multiply factors}$$

$$x^2 - 44x + 84 = 0 \qquad \text{subtract } 16x + 112$$

$$(x - 42)(x - 2) = 0 \qquad \text{factor}$$

$$x - 42 = 0, \quad x - 2 = 0 \qquad \text{set each factor equal to } 0$$

$$x = 42, \qquad x = 2 \qquad \text{solve for } x$$

(continued)

A check is required, since both sides were raised to an even power.

✓ *Check* $x = 42$ LS: $\sqrt{84-3} - \sqrt{42+7} + 2 = 9 - 7 + 2 = 4$

RS: 0

Since $4 \neq 0$, $x = 42$ is not a solution.

✓ *Check* $x = 2$ LS: $\sqrt{4-3} - \sqrt{2+7} + 2 = 1 - 3 + 2 = 0$

RS: 0

Since $0 = 0$ is a true statement, $x = 2$ is a solution.

Hence, the given equation has one solution, $x = 2$.

An equation is of **quadratic type** if it can be written in the form

$$au^2 + bu + c = 0,$$

where $a \neq 0$ and u is an expression in some variable. If we find the solutions in terms of u, then the solutions of the given equation can be obtained by referring to the specific form of u.

EXAMPLE 7   *Solving an equation of quadratic type*

Solve $x^{2/3} + x^{1/3} - 6 = 0$.

SOLUTION   Since $x^{2/3} = (x^{1/3})^2$, the form of the equation suggests that we let $u = x^{1/3}$, as in the second line below:

$$
\begin{aligned}
x^{2/3} + x^{1/3} - 6 &= 0 && \text{given} \\
u^2 + u - 6 &= 0 && \text{let } u = x^{1/3} \\
(u + 3)(u - 2) &= 0 && \text{factor} \\
u + 3 = 0, \quad u - 2 &= 0 && \text{set each factor equal to 0} \\
u = -3, \quad u &= 2 && \text{solve for } u \\
x^{1/3} = -3, \quad x^{1/3} &= 2 && u = x^{1/3} \\
x = -27, \quad x &= 8 && \text{cube both sides}
\end{aligned}
$$

A check is unnecessary, since we did not raise both sides to an even power. Hence, the given equation has two solutions, -27 and 8.

An alternative method is to factor the left side of the given equation as follows:

$$x^{2/3} + x^{1/3} - 6 = (x^{1/3} + 3)(x^{1/3} - 2)$$

By setting each factor equal to 0, we obtain the solutions.

EXAMPLE 8 *Solving an equation of quadratic type*

Solve $x^4 - 3x^2 + 1 = 0$.

SOLUTION Since $x^4 = (x^2)^2$, the form of the equation suggests that we let $u = x^2$, as in line two below:

$$x^4 - 3x^2 + 1 = 0 \qquad \text{given}$$

$$u^2 - 3u + 1 = 0 \qquad \text{let } u = x^2$$

$$u = \frac{3 \pm \sqrt{9 - 4}}{2} = \frac{3 \pm \sqrt{5}}{2} \qquad \text{quadratic formula}$$

$$x^2 = \frac{3 \pm \sqrt{5}}{2} \qquad u = x^2$$

$$x = \pm\sqrt{\frac{3 \pm \sqrt{5}}{2}} \qquad \text{take the square root}$$

Thus, there are four solutions:

$$\sqrt{\frac{3 + \sqrt{5}}{2}}, \quad -\sqrt{\frac{3 + \sqrt{5}}{2}}, \quad \sqrt{\frac{3 - \sqrt{5}}{2}}, \quad -\sqrt{\frac{3 - \sqrt{5}}{2}}$$

Using a calculator, we obtain the approximations ± 1.62 and ± 0.62. A check is unnecessary because we did not raise both sides of an equation to an even power.

EXAMPLE 9 *Determining the route of a ferry*

A passenger ferry makes trips from a town to an island community that is 7 miles downshore from the town and 3 miles off a straight shoreline. As shown in Figure 9, the ferry travels along the shoreline to some point and then proceeds directly to the island. If the ferry travels 12 mi/hr along the shoreline and 10 mi/hr as it moves out to sea, determine the routes that have a travel time of 45 minutes.

SOLUTION Let x denote the distance traveled along the shoreline. This leads to the sketch in Figure 10, where d is the distance from a point on the shoreline to the island. Refer to the indicated right triangle:

$$d^2 = (7 - x)^2 + 3^2 \qquad \text{Pythagorean theorem}$$

$$= 49 - 14x + x^2 + 9 \qquad \text{square terms}$$

$$= x^2 - 14x + 58 \qquad \text{simplify}$$

Taking the square root of both sides and noting that $d > 0$, we obtain

$$d = \sqrt{x^2 - 14x + 58}.$$

(continued)

FIGURE 9

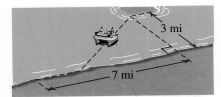

FIGURE 10

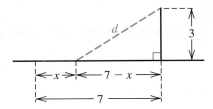

Using distance = (rate)(time) or, equivalently, time = (distance)/(rate) gives us the following table.

	Along the shoreline	Away from shore
Distance (mi)	x	$\sqrt{x^2 - 14x + 58}$
Rate (mi/hr)	12	10
Time (hr)	$\dfrac{x}{12}$	$\dfrac{\sqrt{x^2 - 14x + 58}}{10}$

The time for the complete trip is the sum of the two expressions in the last row of the table. Since the rate is in mi/hr, we must, for consistency, express this time (45 minutes) as $\frac{3}{4}$ hour. Thus, we have the following:

$$\frac{x}{12} + \frac{\sqrt{x^2 - 14x + 58}}{10} = \frac{3}{4} \qquad \text{total time for trip}$$

$$\frac{\sqrt{x^2 - 14x + 58}}{10} = \frac{3}{4} - \frac{x}{12} \qquad \text{subtract } \frac{x}{12}$$

$$6\sqrt{x^2 - 14x + 58} = 45 - 5x \qquad \text{multiply by the lcd, 60}$$

$$6\sqrt{x^2 - 14x + 58} = 5(9 - x) \qquad \text{factor}$$

$$36(x^2 - 14x + 58) = 25(9 - x)^2 \qquad \text{square both sides}$$

$$11x^2 - 54x + 63 = 0 \qquad \text{simplify}$$

$$(x - 3)(11x - 21) = 0 \qquad \text{factor}$$

$$x - 3 = 0, \quad 11x - 21 = 0 \qquad \text{set each factor equal to 0}$$

$$x = 3, \qquad x = \frac{21}{11} \qquad \text{solve for } x$$

A check verifies that these numbers are also solutions of the original equation. Hence, there are two possible routes with a travel time of 45 minutes: the ferry may travel along the shoreline either 3 miles or $\frac{21}{11} \approx 1.9$ miles before proceeding to the island.

2.5 EXERCISES

Exer. 1–50: Solve the equation.

1 $|x + 4| = 11$

2 $|x - 5| = 2$

3 $|3x - 2| + 3 = 7$

4 $2|5x + 2| - 1 = 5$

5 $3|x + 1| - 2 = -11$

6 $|x - 2| + 5 = 5$

7 $9x^3 - 18x^2 - 4x + 8 = 0$

8 $3x^3 - 4x^2 - 27x + 36 = 0$

9 $4x^4 + 10x^3 = 6x^2 + 15x$

10 $15x^5 - 20x^4 = 6x^3 - 8x^2$

11 $y^{3/2} = 5y$ **12** $y^{4/3} = -3y$

13 $\sqrt{7 - 5x} = 8$ **14** $\sqrt{2x - 9} = \frac{1}{3}$

15 $2 + \sqrt[3]{1 - 5t} = 0$ **16** $\sqrt[3]{6 - s^2} + 5 = 0$

17 $\sqrt[5]{2x^2 + 1} - 2 = 0$ **18** $\sqrt[4]{2x^2 - 1} = x$

19 $\sqrt{7 - x} = x - 5$ **20** $\sqrt{3 - x} - x = 3$

21 $3\sqrt{2x - 3} + 2\sqrt{7 - x} = 11$

22 $\sqrt{2x + 15} - 2 = \sqrt{6x + 1}$

23 $x = 4 + \sqrt{4x - 19}$ **24** $x = 3 + \sqrt{5x - 9}$

25 $x + \sqrt{5x + 19} = -1$ **26** $x - \sqrt{-7x - 24} = -2$

27 $\sqrt{7 - 2x} - \sqrt{5 + x} = \sqrt{4 + 3x}$

28 $4\sqrt{1 + 3x} + \sqrt{6x + 3} = \sqrt{-6x - 1}$

29 $\sqrt{11 + 8x} + 1 = \sqrt{9 + 4x}$

30 $2\sqrt{x} - \sqrt{x - 3} = \sqrt{5 + x}$

31 $\sqrt{2\sqrt{x + 1}} = \sqrt{3x - 5}$ **32** $\sqrt{5\sqrt{x}} = \sqrt{2x - 3}$

33 $\sqrt{1 + 4\sqrt{x}} = \sqrt{x} + 1$ **34** $\sqrt{x + 1} = \sqrt{x - 1}$

35 $x^4 - 25x^2 + 144 = 0$ **36** $2x^4 - 10x^2 + 8 = 0$

37 $5y^4 - 7y^2 + 1 = 0$ **38** $3y^4 - 5y^2 + 1 = 0$

39 $36x^{-4} - 13x^{-2} + 1 = 0$ **40** $x^{-2} - 2x^{-1} - 35 = 0$

41 $3x^{2/3} + 4x^{1/3} - 4 = 0$ **42** $2y^{1/3} - 3y^{1/6} + 1 = 0$

43 $6w - 23w^{1/2} + 20 = 0$

44 $2x^{-2/3} - 7x^{-1/3} - 15 = 0$

45 $\left(\dfrac{t}{t + 1}\right)^2 - \dfrac{2t}{t + 1} - 8 = 0$

46 $6u^{-1/2} - 13u^{-1/4} + 6 = 0$

47 $27x^3 = (x + 5)^3$

48 $16x^4 = (x - 4)^4$

49 $\sqrt[3]{x} = 2\sqrt[4]{x}$ (*Hint:* Raise both sides to the least common multiple of 3 and 4.)

50 $\sqrt{x + 3} = \sqrt[4]{2x + 6}$

Exer. 51–52: Find the real solutions of the equation.

51 (a) $x^{5/3} = 32$ (b) $x^{4/3} = 16$

    (c) $x^{2/3} = -36$ (d) $x^{3/4} = 125$

    (e) $x^{3/2} = -27$

52 (a) $x^{3/5} = -27$ (b) $x^{2/3} = 25$

    (c) $x^{4/3} = -49$ (d) $x^{3/2} = 27$

    (e) $x^{3/4} = -8$

Exer. 53–55: Solve for the specified variable.

53 $T = 2\pi\sqrt{\dfrac{l}{g}}$ for l (period of a pendulum)

54 $d = \frac{1}{2}\sqrt{4R^2 - C^2}$ for C (segments of circles)

55 $S = \pi r\sqrt{r^2 + h^2}$ for h (surface area of a cone)

56 *Nuclear experiments* Nuclear experiments performed in the ocean vaporize large quantities of salt water. Salt boils and turns into vapor at 1738 K. After being vaporized by a 10-megaton force, the salt takes at least 8–10 seconds to cool enough to crystallize. The amount of salt A that has crystallized t seconds after an experiment is sometimes calculated using the formula $A = k\sqrt{t/T}$, where k and T are constants. Solve this equation for t.

57 *Windmill power* The power P (in watts) generated by a windmill that has efficiency E is given by the formula $P = 0.31ED^2V^3$, where D is the diameter (in feet) of the windmill blades and V is the wind velocity (in ft/sec). Approximate the wind velocity necessary to generate 10,000 watts if $E = 42\%$ and $D = 10$.

58 *Withdrawal resistance of nails* The *withdrawal resistance* of a nail indicates its holding strength in wood. A formula that is used for bright common nails is $P = 15{,}700S^{5/2}RD$, where P is the maximum withdrawal resistance (in pounds), S is the specific gravity of the wood at 12% moisture content, R is the radius of the nail (in inches), and D is the depth (in inches) that the nail has penetrated the wood. A 6d (six-penny) bright, common nail of length 2 inches and diameter 0.113 inch is driven completely into a piece of Douglas fir. If it requires a maximum force of 380 pounds to remove the nail, approximate the specific gravity of Douglas fir.

59 *The effect of price on demand* The demand for a commodity usually depends on its price. If other factors do not affect the demand, then the quantity Q purchased at price P (in cents) is given by $Q = kP^{-c}$, where k and c are positive constants. If $k = 10^5$ and $c = \frac{1}{2}$, find the price that will result in the purchase of 5000 items.

60 *The urban heat island* Urban areas have higher average air temperatures than rural areas, as a result of the presence of buildings, asphalt, and concrete. This phenomenon has become known as the *urban heat island*.

The temperature difference T (in °C) between urban and rural areas near Montreal, with a population P between 1000 and 1,000,000, can be described by the formula $T = 0.25P^{1/4}/\sqrt{v}$, where v is the average wind speed (in mi/hr) and $v \geq 1$. If $T = 3$ and $v = 5$, find P.

61 *Dimensions of a sand pile*   As sand leaks out of a certain container, it forms a pile that has the shape of a right circular cone whose altitude is always one-half the diameter d of the base. What is d at the instant at which 144 cm³ of sand has leaked out?

EXERCISE 61

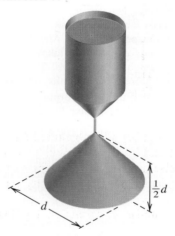

d $\frac{1}{2}d$

62 *Inflating a weather balloon*   The volume of a spherical weather balloon is $10\frac{2}{3}$ ft³. In order to lift a transmitter and meteorological equipment, the balloon is inflated with an additional $25\frac{1}{3}$ ft³ of helium. How much does its diameter increase?

63 *The cube rule in political science*   The cube rule in political science is an empirical formula that is said to predict the percentage y of seats in the U.S. House of Representatives that will be won by a political party from the popular vote for the party's presidential candidate. If x denotes the percentage of the popular vote for a party's presidential candidate, then the cube rule states that

$$y = \frac{x^3}{x^3 + (1 - x)^3}.$$

What percentage of the popular vote will the presidential candidate need in order for the candidate's party to win 60% of the House seats?

64 *Dimensions of a conical cup*   A conical paper cup is to have a height of 3 inches. Find the radius of the cone that will result in a surface area of 6π in.²

65 *Installing a power line*   A power line is to be installed across a river that is 1 mile wide to a town that is 5 miles downstream (see the figure). It costs $7500 per mile to lay the cable underwater and $6000 per mile to lay it overland. Determine how the cable should be installed if $35,000 has been allocated for this project.

EXERCISE 65

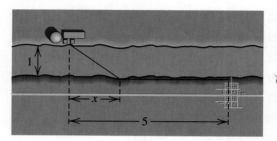

1   x 5

66 *Calculating human growth*   Adolphe Quetelet (1796–1874), the director of the Brussels Observatory from 1832 to 1874, was the first person to attempt to fit a mathematical expression to human growth data. If h denotes height in meters and t denotes age in years, Quetelet's formula for males in Brussels can be expressed as

$$h + \frac{h}{h_M - h} = at + \frac{h_0 + t}{1 + \frac{4}{3}t},$$

with $h_0 = 0.5$, the height at birth; $h_M = 1.684$, the final adult male height; and $a = 0.545$.

(a) Find the expected height of a 12-year-old male.

(b) At what age should 50% of the adult height be reached?

67 The equation $\frac{1}{3}\sqrt[3]{x} - x + 2 = 0$ has a root near 2. To approximate this root, rewrite the equation as $x = \frac{1}{3}\sqrt[3]{x} + 2$. Let $x_1 = 2$ and find successive approximations $x_2, x_3, \ldots$ by using the formulas

$$x_2 = \frac{1}{3}\sqrt[3]{x_1} + 2, \qquad x_3 = \frac{1}{3}\sqrt[3]{x_2} + 2, \qquad \ldots$$

until four-decimal-place accuracy is obtained.

68 The equation $2x + \frac{1}{x^4 + x + 2} = 0$ has a root near 0. Use a procedure similar to that in Exercise 67 to approximate this root to four-decimal-place accuracy.

2.6 INEQUALITIES

Consider the inequality

$$2x + 3 > 11,$$

where x is a variable. As illustrated in the following table, certain numbers yield true statements when substituted for x, and others yield false statements.

x	$2x + 3 > 11$	Conclusion
3	$9 > 11$	False statement
4	$11 > 11$	False statement
5	$13 > 11$	True statement
6	$15 > 11$	True statement

If a true statement is obtained when a number a is substituted for x, then a is a **solution** of the inequality. Thus, 5 and 6 are solutions of $2x + 3 > 11$, but 3 and 4 are not solutions. To **solve** an inequality means to find *all* solutions. Two inequalities are **equivalent** if they have exactly the same solutions.

Most inequalities have an infinite number of solutions. To illustrate, the solutions of the inequality

$$2 < x < 5$$

consist of *every* real number x between 2 and 5. We call this set of numbers an **open interval** and denote it by $(2, 5)$. The **graph** of the open interval $(2, 5)$ is the set of all points on a coordinate line that lie between (but do not include) the points corresponding to $x = 2$ and $x = 5$. The graph is represented by shading an appropriate part of the axis, as shown in Figure 11. We refer to this process as **sketching the graph** of the interval. The numbers 2 and 5 are called the **end points** of the interval $(2, 5)$. The parentheses in the notation $(2, 5)$ and in Figure 11 are used to indicate that the end points of the interval are not included.

If we wish to include an end point, we use a bracket instead of a parenthesis. For example, the solutions of the inequality $2 \leq x \leq 5$ are denoted by $[2, 5]$ and referred to as a **closed interval**. The graph of $[2, 5]$ is sketched in Figure 12, where brackets indicate that end points are included. We shall also consider **half-open intervals** $[a, b)$ and $(a, b]$ and **infinite intervals**, as described in the following chart. The symbol ∞ (read *infinity*) used for infinite intervals is merely a notational device and does *not* represent a real number.

FIGURE 11

FIGURE 12

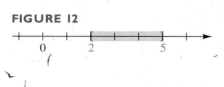

Intervals

Notation	Inequality	Graph
(a, b)	$a < x < b$	
$[a, b]$	$a \le x \le b$	
$[a, b)$	$a \le x < b$	
$(a, b]$	$a < x \le b$	
(a, ∞)	$x > a$	
$[a, \infty)$	$x \ge a$	
$(-\infty, b)$	$x < b$	
$(-\infty, b]$	$x \le b$	
$(-\infty, \infty)$	$-\infty < x < \infty$	

Methods for solving inequalities in x are similar to those used for solving equations. In particular, we often use properties of inequalities to replace a given inequality with a list of equivalent inequalities, ending with an inequality from which solutions are easily obtained. The properties in the following chart can be proved for real numbers a, b, c, and d.

Properties of Inequalities

Property	Illustration
(1) If $a < b$ and $b < c$, then $a < c$.	$2 < 5$ and $5 < 9$, so $2 < 9$.
(2) If $a < b$, then $a + c < b + c$ and $a - c < b - c$.	$2 < 7$, so $2 + 3 < 7 + 3$ and $2 - 3 < 7 - 3$.
(3) If $a < b$ and $c > 0$, then $ac < bc$ and $\dfrac{a}{c} < \dfrac{b}{c}$.	$2 < 5$ and $3 > 0$, so $2 \cdot 3 < 5 \cdot 3$ and $\dfrac{2}{3} < \dfrac{5}{3}$.
(4) If $a < b$ and $c < 0$, then $ac > bc$ and $\dfrac{a}{c} > \dfrac{b}{c}$.	$2 < 5$ and $-3 < 0$, so $2(-3) > 5(-3)$ and $\dfrac{2}{-3} > \dfrac{5}{-3}$.

It is important to remember that multiplying or dividing both sides of an inequality by a negative real number *reverses* the inequality sign (see property 4). Properties similar to those above are true for other inequalities and for $\le$ or $\ge$. Thus, if $a > b$, then $a + c > b + c$; if $a \ge b$ and $c < 0$, then $ac \le bc$; and so on.

If x represents a real number, then, by property 2, adding or subtracting the same expression containing x on both sides of an inequality yields an equivalent inequality. By property 3, we may multiply or divide both sides of an inequality by an expression containing x if we are certain that the expression is positive for all values of x under consideration. To illustrate, multiplication or division by $x^4 + 3x^2 + 5$ would be permissible, since this expression is always positive. If we multiply or divide both sides of an inequality by an expression that is always negative, such as $-7 - x^2$, then, by property 4, the inequality sign is reversed.

In examples we shall describe solutions of inequalities by means of intervals and also represent them graphically.

EXAMPLE 1 *Solving an inequality*

Solve the inequality $-3x + 4 < 11$.

SOLUTION

$$
\begin{aligned}
-3x + 4 &< 11 && \text{given} \\
(-3x + 4) - 4 &< 11 - 4 && \text{subtract 4} \\
-3x &< 7 && \text{simplify} \\
\frac{-3x}{-3} &> \frac{7}{-3} && \text{divide by } -3; \\
&&& \text{reverse the inequality sign} \\
x &> -\tfrac{7}{3} && \text{simplify}
\end{aligned}
$$

FIGURE 13

$-\frac{7}{3}$ 0

Thus, the solutions of $-3x + 4 < 11$ consist of all real numbers x such that $x > -\frac{7}{3}$. This is the interval $(-\frac{7}{3}, \infty)$ sketched in Figure 13.

EXAMPLE 2 *Solving an inequality*

Solve $4x - 3 < 2x + 5$.

SOLUTION

$$
\begin{aligned}
4x - 3 &< 2x + 5 && \text{given} \\
(4x - 3) + 3 &< (2x + 5) + 3 && \text{add 3} \\
4x &< 2x + 8 && \text{simplify} \\
4x - 2x &< (2x + 8) - 2x && \text{subtract } 2x \\
2x &< 8 && \text{simplify} \\
\frac{2x}{2} &< \frac{8}{2} && \text{divide by 2} \\
x &< 4 && \text{simplify}
\end{aligned}
$$

FIGURE 14

0 4

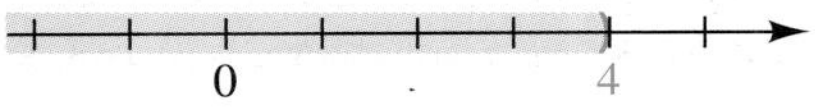

Hence, the solutions of the given inequality consist of all real numbers x such that $x < 4$. This is the interval $(-\infty, 4)$ sketched in Figure 14.

EXAMPLE 3   *Solving an inequality*

Solve $-6 < 2x - 4 < 2$.

SOLUTION   A real number x is a solution of the given inequality if and only if it is a solution of *both* of the inequalities

$$-6 < 2x - 4 \quad \text{and} \quad 2x - 4 < 2.$$

The first inequality is solved as follows:

$$-6 < 2x - 4 \qquad \text{given}$$
$$-6 + 4 < (2x - 4) + 4 \quad \text{add 4}$$
$$-2 < 2x \qquad \text{simplify}$$
$$\frac{-2}{2} < \frac{2x}{2} \qquad \text{divide by 2}$$
$$-1 < x \qquad \text{simplify}$$
$$x > -1 \qquad \text{equivalent inequality}$$

The second inequality is then solved:

$$2x - 4 < 2 \qquad \text{given}$$
$$2x < 6 \qquad \text{add 4}$$
$$x < 3 \qquad \text{divide by 2}$$

Thus, x is a solution of the given inequality if and only if *both*

$$x > -1 \quad \text{and} \quad x < 3;$$

that is, $-1 < x < 3$.

Hence, the solutions are all numbers in the open interval $(-1, 3)$ sketched in Figure 15.

An alternative (and shorter) method is to solve both inequalities simultaneously:

$$-6 < 2x - 4 < 2 \qquad \text{given}$$
$$-6 + 4 < 2x < 2 + 4 \qquad \text{add 4}$$
$$-2 < 2x < 6 \qquad \text{simplify}$$
$$-1 < x < 3 \qquad \text{divide by 2}$$

FIGURE 15

$-1 \quad 0 \qquad 3$

EXAMPLE 4   *Solving an inequality*

Solve $-5 \le \dfrac{4 - 3x}{2} < 1$.

SOLUTION   A number x is a solution of the given inequality if and only if

$$-5 \le \frac{4 - 3x}{2} \quad \text{and} \quad \frac{4 - 3x}{2} < 1.$$

We can either work with each inequality separately or solve both simultaneously, as follows:

$$-5 \le \frac{4 - 3x}{2} < 1 \qquad \text{given}$$

$$-10 \le 4 - 3x < 2 \qquad \text{multiply by 2}$$

$$-10 - 4 \le -3x < 2 - 4 \qquad \text{subtract 4}$$

$$-14 \le -3x < -2 \qquad \text{simplify}$$

$$\frac{-14}{-3} \ge \frac{-3x}{-3} > \frac{-2}{-3} \qquad \text{divide by } -3$$

$$\tfrac{14}{3} \ge x > \tfrac{2}{3} \qquad \text{simplify}$$

$$\tfrac{2}{3} < x \le \tfrac{14}{3} \qquad \text{equivalent inequality}$$

FIGURE 16

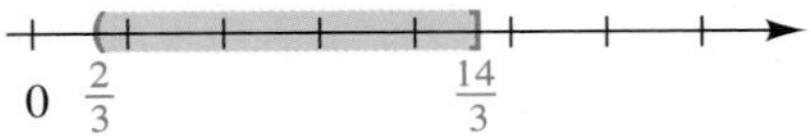

Thus, the solutions of the inequality are all numbers in the half-open interval $(\tfrac{2}{3}, \tfrac{14}{3}]$ sketched in Figure 16.

EXAMPLE 5   *Solving a rational inequality*

Solve $\dfrac{1}{x - 2} > 0$.

FIGURE 17

SOLUTION   Since the numerator is positive, the fraction is positive if and only if $x - 2 > 0$ or, equivalently, $x > 2$. Thus, the solutions are all numbers in the infinite interval $(2, \infty)$ sketched in Figure 17.

EXAMPLE 6   *Using a lens formula*

FIGURE 18

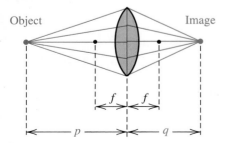

Object   Image

As illustrated in Figure 18, if a convex lens has focal length f centimeters and if an object is placed a distance p centimeters from the lens with $p > f$, then the distance q from the lens to the image is related to p and f by the formula

$$\frac{1}{p} + \frac{1}{q} = \frac{1}{f}.$$

If $f = 5$ cm, how close must the object be to the lens for the image to be more than 12 centimeters from the lens?

SOLUTION   Since $f = 5$, the given formula may be written as

$$\frac{1}{p} + \frac{1}{q} = \frac{1}{5}.$$

We wish to determine the values of q such that $q > 12$. Let us first solve for q:

$$5q + 5p = pq \qquad \text{multiply by the lcd, } 5pq$$

$$q(5 - p) = -5p \qquad \text{collect } q \text{ terms on one side and factor}$$

$$q = -\frac{5p}{5 - p} = \frac{5p}{p - 5} \qquad \text{divide by } 5 - p$$

To solve the inequality $q > 12$, we proceed as follows:

$$\frac{5p}{p - 5} > 12 \qquad\qquad q = \frac{5p}{p - 5}$$

$$5p > 12(p - 5) \qquad \text{allowable, since } p > f \text{ implies } p - 5 > 0$$

$$-7p > -60 \qquad \text{multiply factors and collect } p \text{ terms on one side}$$

$$p < \tfrac{60}{7} \qquad \text{divide by } -7$$

Since $p > 5$, we obtain

$$5 < p < \tfrac{60}{7}.$$

FIGURE 19

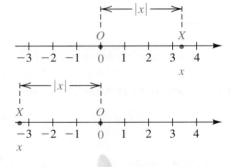

If a point X on a coordinate line has coordinate x, as shown in Figure 19, then X is to the right of the origin O if $x > 0$ and to the left of O if $x < 0$. From Section 1.1, the distance $d(O, X)$ between O and X is the *nonnegative* real number given by

$$d(O, X) = |x - 0| = |x|.$$

It follows that the solutions of an inequality such as $|x| < 3$ consist of the coordinates of all points whose distance from O is less than 3. This is the open interval $(-3, 3)$ sketched in Figure 20. Thus,

$$|x| < 3 \quad \text{is equivalent to} \quad -3 < x < 3.$$

Similarly, for $|x| > 3$, the distance between O and a point with coordinate x is greater than 3; that is,

$$|x| > 3 \quad \text{is equivalent to} \quad x < -3 \text{ or } x > 3.$$

FIGURE 20

The graph of the solutions to $|x| > 3$ is sketched in Figure 21. We often use the **union symbol** $\cup$ and write

$$(-\infty, -3) \cup (3, \infty)$$

FIGURE 21

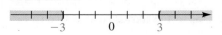

to denote all real numbers that are in either $(-\infty, -3)$ or $(3, \infty)$.

The **intersection symbol** $\cap$ is used to denote the elements that are *common* to two sets. For example,

$$(-\infty, 3) \cap (-3, \infty) = (-3, 3),$$

since the intersection of $(-\infty, 3)$ and $(-3, \infty)$ consists of all real numbers x such that both $x < 3$ *and* $x > -3$.

The preceding discussion may be generalized to obtain the following.

Properties of Absolute Values $(b > 0)$

> **(1)** $|a| < b$ is equivalent to $-b < a < b$.
>
> **(2)** $|a| > b$ is equivalent to $a < -b$ or $a > b$.

In the next example we use property 1 with $a = x - 3$ and $b = 0.5$.

EXAMPLE 7 *Solving an inequality containing an absolute value*

Solve $|x - 3| < 0.5$.

SOLUTION

$$|x - 3| < 0.5 \qquad \text{given}$$
$$-0.5 < x - 3 < 0.5 \qquad \text{property 1}$$
$$-0.5 + 3 < (x - 3) + 3 < 0.5 + 3 \quad \text{isolate } x$$
$$2.5 < x < 3.5 \qquad \text{simplify}$$

FIGURE 22

Thus, the solutions are the real numbers in the open interval $(2.5, 3.5)$. The graph is sketched in Figure 22.

In the next example we use property 2 with $a = 2x + 3$ and $b = 9$.

EXAMPLE 8 *Solving an inequality containing an absolute value*

Solve $|2x + 3| > 9$.

SOLUTION

$$|2x + 3| > 9 \quad \text{given}$$
$$2x + 3 < -9 \quad \text{or} \quad 2x + 3 > 9 \quad \text{property 2}$$
$$2x < -12 \quad \text{or} \quad 2x > 6 \quad \text{subtract 3}$$
$$x < -6 \quad \text{or} \quad x > 3 \quad \text{divide by 2}$$

FIGURE 23

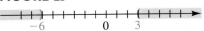

Consequently, the solutions of the inequality $|2x + 3| > 9$ consist of the numbers in $(-\infty, -6) \cup (3, \infty)$. The graph is sketched in Figure 23.

The trichotomy law in Section 1.1 states that for any real numbers a and b, exactly one of the following is true:

$$a > b, \quad a < b, \quad \text{or} \quad a = b$$

Thus, after solving $|2x + 3| > 9$ in Example 8, we readily obtain the solutions for $|2x + 3| < 9$ and $|2x + 3| = 9$—namely, $(-6, 3)$ and $\{-6, 3\}$, respectively. Note that the union of these three sets of solutions is necessarily the set $\mathbb{R}$ of real numbers.

When using the notation $a < x < b$, we must have $a < b$. Thus, *it is incorrect to write the solution $x < -6$ or $x > 3$ in Example 8 as $3 < x < -6$.* Another misuse of inequality notation is to write $a < x > b$, since when several inequality symbols are used in one expression, *they must point in the same direction.*

2.6 EXERCISES

1 Given $-7 < -3$, determine the inequality obtained if

  (a) 5 is added to both sides

  (b) 4 is subtracted from both sides

  (c) both sides are multiplied by $\frac{1}{3}$

  (d) both sides are multiplied by $-\frac{1}{3}$

2 Given $4 > -5$, determine the inequality obtained if

  (a) 7 is added to both sides

  (b) -5 is subtracted from both sides

  (c) both sides are divided by 6

  (d) both sides are divided by -6

Exer. 3–12: Express the inequality as an interval, and sketch its graph.

3 $x < -2$ **4** $x \leq 5$

5 $x \geq 4$ **6** $x > -3$

7 $-2 < x \leq 4$ **8** $-3 \leq x < 5$

9 $3 \leq x \leq 7$ **10** $-3 < x < -1$

11 $5 > x \geq -2$ **12** $-3 \geq x > -5$

Exer. 13–20: Express the interval as an inequality in the variable x.

13 $(-5, 8]$ **14** $[0, 4)$

15 $[-4, -1]$ **16** $(3, 7)$

17 $[4, \infty)$ **18** $(-3, \infty)$

19 $(-\infty, -5)$ **20** $(-\infty, 2]$

Exer. 21–68: Solve the inequality, and express the solutions in terms of intervals whenever possible.

21 $3x - 2 > 14$ **22** $2x + 5 \leq 7$

23 $-2 - 3x \geq 2$ **24** $3 - 5x < 11$

25 $2x + 5 < 3x - 7$ **26** $x - 8 > 5x + 3$

27 $9 + \frac{1}{3}x \geq 4 - \frac{1}{2}x$ **28** $\frac{1}{4}x + 7 \leq \frac{1}{3}x - 2$

29 $-3 < 2x - 5 < 7$ **30** $4 \geq 3x + 5 > -1$

31 $3 \leq \dfrac{2x - 3}{5} < 7$ **32** $-2 < \dfrac{4x + 1}{3} \leq 0$

33 $4 > \dfrac{2 - 3x}{7} \geq -2$ **34** $5 \geq \dfrac{6 - 5x}{3} > 2$

35 $0 \leq 4 - \frac{1}{3}x < 2$ **36** $-2 < 3 + \frac{1}{4}x \leq 5$

37 $(2x - 3)(4x + 5) \leq (8x + 1)(x - 7)$

38 $(x - 3)(x + 3) \geq (x + 5)^2$

39 $(x - 4)^2 > x(x + 12)$

40 $2x(6x + 5) < (3x - 2)(4x + 1)$

41 $\dfrac{4}{3x + 2} \geq 0$ **42** $\dfrac{3}{2x + 5} \leq 0$

43 $\dfrac{-2}{4 - 3x} > 0$ **44** $\dfrac{-3}{2 - x} < 0$

45 $\dfrac{2}{(1-x)^2} > 0$

46 $\dfrac{4}{x^2+4} < 0$

47 $|x| < 3$

48 $|x| \le 7$

49 $|x| \ge 5$

50 $|-x| > 2$

51 $|x+3| < 0.01$

52 $|x-4| \le 0.03$

53 $|x+2| + 0.1 \ge 0.2$

54 $|x-3| - 0.3 > 0.1$

55 $|2x+5| < 4$

56 $|3x-7| \ge 5$

57 $-\frac{1}{3}|6-5x| + 2 \ge 1$

58 $2|-11-7x| - 2 > 10$

59 $|7x+2| > -2$

60 $|6x-5| \le -2$

61 $|3x-9| > 0$

62 $|5x+2| \le 0$

63 $\left|\dfrac{2-3x}{5}\right| \ge 2$

64 $\left|\dfrac{2x+5}{3}\right| < 1$

65 $\dfrac{3}{|5-2x|} < 2$

66 $\dfrac{2}{|2x+3|} \ge 5$

67 $1 < |x-2| < 4$

68 $2 < |2x-1| < 3$

Exer. 69–70: Solve part (a) and use that answer to determine the answers to parts (b) and (c).

69 (a) $|x+5| = 3$ (b) $|x+5| < 3$ (c) $|x+5| > 3$

70 (a) $|x-3| < 2$ (b) $|x-3| = 2$ (c) $|x-3| > 2$

Exer. 71–74: Express the statement in terms of an inequality involving an absolute value.

71 The weight w of a wrestler must be within 2 pounds of 148 pounds.

72 The radius r of a ball bearing must be within 0.01 centimeter of 1 centimeter.

73 The difference of two temperatures T_1 and T_2 within a chemical mixture must be between 5 °C and 10 °C.

74 The arrival time t of train B must be at least 5 minutes different from the 4:00 P.M. arrival time of train A.

75 *Temperature scales* Temperature readings on the Fahrenheit and Celsius scales are related by the formula $C = \frac{5}{9}(F - 32)$. What values of F correspond to the values of C such that $30 \le C \le 40$?

76 *Hooke's law* According to Hooke's law, the force F (in pounds) required to stretch a certain spring x inches beyond its natural length is given by $F = (4.5)x$ (see the figure). If $10 \le F \le 18$, what are the corresponding values for x?

EXERCISE 76

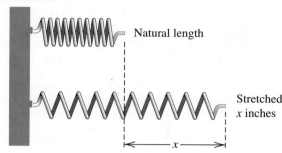

Natural length

Stretched x inches

x

77 *Ohm's law* Ohm's law in electrical theory states that if R denotes the resistance of an object (in ohms), V the potential difference across the object (in volts), and I the current that flows through it (in amperes), then $R = V/I$. If the voltage is 110, what values of the resistance will result in a current that does not exceed 10 amperes?

78 *Electrical resistance* If two resistors R_1 and R_2 are connected in parallel in an electrical circuit, the net resistance R is given by

$$\frac{1}{R} = \frac{1}{R_1} + \frac{1}{R_2}.$$

If $R_1 = 10$ ohms, what values of R_2 will result in a net resistance of less than 5 ohms?

79 *Linear magnification* Shown in the figure is a simple magnifier consisting of a convex lens. The object to be magnified is positioned so that the distance p from the lens is less than the focal length f. The linear magnification M is the ratio of the image size to the object size. It is shown in physics that $M = f/(f - p)$. If $f = 6$ cm, how far should the object be placed from the lens so that its image appears at least three times as large? (Compare with Example 6.)

EXERCISE 79

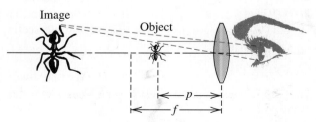

Image

Object

p

f

80 *Drug concentration* To treat arrhythmia (irregular heartbeat), a drug is fed intravenously into the bloodstream. Suppose that the concentration c of the drug after t hours is given by $c = 3.5t/(t + 1)$ mg/L. If the minimum therapeutic level is 1.5 mg/L, determine when this level is exceeded.

81 *Business expenditure* A construction firm is trying to decide which of two models of a crane to purchase. Model A costs $50,000 and requires $4000 per year to maintain. Model B has an initial cost of $40,000 and a maintenance cost of $5500 per year. For how many years must model A be used before it becomes more economical than B?

82 *Buying a car* A consumer is trying to decide whether to purchase car A or car B. Car A costs $10,000 and has an mpg rating of 30, and insurance is $550 per year. Car B costs $12,000 and has an mpg rating of 50, and insurance is $600 per year. Assume that the consumer drives 15,000 miles per year and that the price of gas remains constant at $1.25 per gallon. Based only on these facts, determine how long it will take for the total cost of car B to become less than that of car A.

2.7 MORE ON INEQUALITIES

To solve an inequality involving polynomials of degree greater than 1, we shall express each polynomial as a product of linear factors $ax + b$ and/or irreducible quadratic factors $ax^2 + bx + c$. If any such factor is not zero in an interval, then it is either positive throughout the interval or negative throughout the interval. Hence, if we choose any k in the interval and if the factor is positive (or negative) for $x = k$, then it is positive (or negative) throughout the interval. The value of the factor at $x = k$ is called a **test value** at k. This concept is exhibited in the following example.

EXAMPLE I   *Solving a quadratic inequality*

Solve $2x^2 - x < 3$.

SOLUTION   To use test values *it is essential to have* 0 *on one side* of the inequality sign. Thus, we proceed as follows:

$$2x^2 - x < 3 \quad \text{given}$$

$$2x^2 - x - 3 < 0 \quad \text{make one side 0}$$

$$(x + 1)(2x - 3) < 0 \quad \text{factor}$$

The factors $x + 1$ and $2x - 3$ are zero at -1 and $\frac{3}{2}$, respectively. The corresponding points on a coordinate line (see Figure 24) determine the nonintersecting intervals

$$(-\infty, -1), \quad (-1, \tfrac{3}{2}), \quad \text{and} \quad (\tfrac{3}{2}, \infty).$$

FIGURE 24

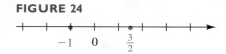

$-1 \quad 0 \quad \frac{3}{2}$

We may find the signs of $x + 1$ and $2x - 3$ in each interval by using a test value. To illustrate, if we choose $k = -10$ in $(-\infty, -1)$, the values of both $x + 1$ and $2x - 3$ are negative, and hence they are negative throughout $(-\infty, -1)$. A similar procedure for the remaining two intervals gives us the following *sign chart*, where the terminology *resulting sign* in the last row refers to the sign obtained by applying laws of signs to the product of the factors. Note that the resulting sign is positive or negative according to whether the number of negative signs of factors is even or odd, respectively.

Interval	$(-\infty, -1)$	$(-1, \frac{3}{2})$	$(\frac{3}{2}, \infty)$
Sign of $x + 1$	$-$	$+$	$+$
Sign of $2x - 3$	$-$	$-$	$+$
Resulting sign	$+$	$-$	$+$

Sometimes it is convenient to represent the signs of $x + 1$ and $2x - 3$ by using a coordinate line and a *sign diagram*, of the type illustrated in Figure 25. The vertical lines indicate where the factors are zero, and signs of factors are shown above the coordinate line. The resulting signs are shown in red.

FIGURE 25

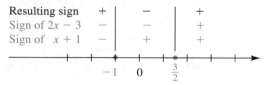

The solutions of $(x + 1)(2x - 3) < 0$ are the values of x for which the product of the factors is *negative*—that is, where the resulting sign is negative. This corresponds to the open interval $(-1, \frac{3}{2})$.

In future examples we will use either a sign chart or a sign diagram, but not both. When working exercises, you should choose the method of solution with which you feel most comfortable.

EXAMPLE 2 *Solving a quadratic inequality*

Solve $x^2 > 7x - 10$.

SOLUTION

$$x^2 > 7x - 10 \qquad \text{given}$$
$$x^2 - 7x + 10 > 0 \qquad \text{make one side 0}$$
$$(x - 2)(x - 5) > 0 \qquad \text{factor}$$

The factors are zero at 2 and 5. The corresponding points on a coordinate line (see Figure 26) determine the nonintersecting intervals

$$(-\infty, 2), \quad (2, 5), \quad \text{and} \quad (5, \infty).$$

(continued)

FIGURE 26

As in Example 1, we may use test values to obtain the following sign chart.

Interval	$(-\infty, 2)$	$(2, 5)$	$(5, \infty)$
Sign of $x - 2$	$-$	$+$	$+$
Sign of $x - 5$	$-$	$-$	$+$
Resulting sign	$+$	$-$	$+$

The solutions of $(x - 2)(x - 5) > 0$ are the values of x for which the resulting sign is *positive*. Thus, the solution of the given inequality is the union $(-\infty, 2) \cup (5, \infty)$.

EXAMPLE 3   *Using a sign diagram to solve an inequality*

Solve $\dfrac{(x + 2)(3 - x)}{(x + 1)(x^2 + 1)} \le 0$.

SOLUTION   Since 0 is already on the right side of the inequality and the left side is factored, we may proceed directly to the sign diagram in Figure 27, where the vertical lines indicate the zeros $(-2, -1, \text{and } 3)$ of the factors. Since the quadratic factor $x^2 + 1$ is always positive, it has no effect on the sign of the quotient and hence may be omitted from the diagram.

FIGURE 27

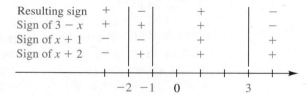

The various signs of the factors can be found using test values. Alternatively, we need only remember that as x increases, the sign of a linear factor $ax + b$ changes from negative to positive if the coefficient a of x is positive, and the sign changes from positive to negative if a is negative.

To determine where the quotient is less than or equal to 0, we first note from the sign diagram that it is *negative* for numbers in $(-2, -1) \cup (3, \infty)$. Since the quotient is 0 at $x = -2$ and $x = 3$, but is *undefined* at $x = -1$, the numbers -2 and 3 are also solutions. Thus, the solution of the given inequality is $[-2, -1) \cup [3, \infty)$.

EXAMPLE 4   *Using a sign diagram to solve an inequality*

Solve $\dfrac{(2x + 1)^2(x - 1)}{x(x^2 - 1)} \ge 0$.

SOLUTION   Rewriting the inequality as

$$\frac{(2x + 1)^2(x - 1)}{x(x + 1)(x - 1)} \geq 0,$$

we see that $x - 1$ is a factor of both the numerator and the denominator. Thus, *assuming that $x - 1 \neq 0$ (that is, $x \neq 1$)*, we may cancel this factor and reduce our search for solutions to the case

$$\frac{(2x + 1)^2}{x(x + 1)} \geq 0 \quad \text{and} \quad x \neq 1.$$

We next observe that this quotient is 0 if $2x + 1 = 0$ (that is, if $x = -\frac{1}{2}$). Hence, $-\frac{1}{2}$ is a solution. To find the remaining solutions, we construct the sign diagram in Figure 28. We do not include $(2x + 1)^2$ in the sign diagram, since this expression is always positive if $x \neq -\frac{1}{2}$ and so has no effect on the sign of the quotient. Referring to the resulting sign and remembering that $-\frac{1}{2}$ is a solution but 1 is *not* a solution, we see that the solutions of the given inequality are given by

$$(-\infty, -1) \cup \{-\tfrac{1}{2}\} \cup (0, 1) \cup (1, \infty).$$

FIGURE 28

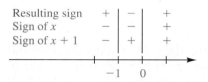

Resulting sign	+	−	+
Sign of x	−	−	+
Sign of $x + 1$	−	+	+

$-1 \quad 0$

EXAMPLE 5   *Using a sign diagram to solve an inequality*

Solve $\dfrac{x + 1}{x + 3} \leq 2$.

SOLUTION   A common mistake in solving such an inequality is to first multiply both sides by $x + 3$. If we did so, we would have to consider several different cases, since $x + 3$ may be positive, negative, or zero. A simpler method is to first obtain an equivalent inequality that has 0 on the right side and proceed from there:

$$\frac{x + 1}{x + 3} \leq 2 \qquad \text{given}$$

$$\frac{x + 1}{x + 3} - 2 \leq 0 \qquad \text{make one side 0}$$

$$\frac{x + 1 - 2(x + 3)}{x + 3} \leq 0 \qquad \text{combine into one fraction}$$

$$\frac{-x - 5}{x + 3} \leq 0 \qquad \text{simplify}$$

$$\frac{x + 5}{x + 3} \geq 0 \qquad \text{multiply by } -1$$

(continued)

Note that the direction of the inequality is changed in the last step, since we multiplied by a negative number. This multiplication was performed for convenience, so that all factors would have positive coefficients of x.

The factors $x + 5$ and $x + 3$ are 0 at $x = -5$ and $x = -3$, respectively. This leads to the sign diagram in Figure 29, where the signs are determined as in previous examples. We see from the diagram that the resulting sign, and hence the sign of the quotient, is positive in $(-\infty, -5) \cup (-3, \infty)$. The quotient is 0 at $x = -5$ and undefined at $x = -3$. Therefore, the solution of $(x + 5)/(x + 3) \geq 0$ is $(-\infty, -5] \cup (-3, \infty)$.

FIGURE 29

An alternative method is to begin by multiplying both sides of the given inequality by $(x + 3)^2$, *assuming that* $x \neq -3$. In this case, $(x + 3)^2 > 0$, and the multiplication is permissible; however, after the resulting inequality is solved, the value $x = -3$ must be excluded.

EXAMPLE 6   *Determining minimum therapeutic levels*

For a drug to have a beneficial effect, its concentration in the bloodstream must exceed a certain value, which is called the *minimum therapeutic level*. Suppose that the concentration c (in mg/L) of a particular drug t hours after it is taken orally is given by

$$c = \frac{20t}{t^2 + 4}.$$

If the minimum therapeutic level is 4 mg/L, determine when this level is exceeded.

SOLUTION   The minimum therapeutic level, 4 mg/L, is exceeded if $c > 4$. Thus, we must solve the inequality

$$\frac{20t}{t^2 + 4} > 4.$$

Since $t^2 + 4 > 0$ for every t, we may proceed as follows:

$$20t > 4t^2 + 16 \qquad \text{multiply by } t^2 + 4$$

$$-4t^2 + 20t - 16 > 0 \qquad \text{make one side 0}$$

$$t^2 - 5t + 4 < 0 \qquad \text{divide by the common factor } -4$$

$$(t - 1)(t - 4) < 0 \qquad \text{factor}$$

The factors in the last inequality are 0 when $t = 1$ and $t = 4$. These are the times at which c is *equal* to 4. As in previous examples, we may use a sign chart or sign diagram (with $t \geq 0$) to show that $(t - 1)(t - 4) < 0$ for every t in the interval $(1, 4)$. Hence, the minimum therapeutic level is exceeded if $1 < t < 4$.

Because graphs in a coordinate plane are introduced in the next chapter, it would be premature to demonstrate here the use of a graphics calculator or computer software to solve inequalities in x. Such methods will be considered later in the text.

Some basic properties of inequalities were stated at the beginning of the last section. The following additional properties are sometimes helpful for solving certain inequalities. Proofs of the properties are given after the chart.

Additional Properties of Inequalities

Property	Illustration
(1) If $0 < a < b$, then $\dfrac{1}{a} > \dfrac{1}{b}$.	If $0 < \dfrac{1}{x} < 4$, then $\dfrac{1}{1/x} > \dfrac{1}{4}$, or $x > \dfrac{1}{4}$.
(2) If $0 < a < b$, then $0 < a^2 < b^2$.	If $0 < \sqrt{x} < 4$, then $0 < (\sqrt{x})^2 < 4^2$, or $0 < x < 16$.
(3) If $0 < a < b$, then $0 < \sqrt{a} < \sqrt{b}$.	If $0 < x^2 < 4$, then $0 < \sqrt{x^2} < \sqrt{4}$, or $0 < |x| < 2$.

PROOFS

(1) If $0 < a < b$, then $a \cdot \dfrac{1}{ab} < b \cdot \dfrac{1}{ab}$, or $\dfrac{1}{b} < \dfrac{1}{a}$; that is, $\dfrac{1}{a} > \dfrac{1}{b}$.

(2) If $0 < a < b$, then $a \cdot a < a \cdot b$ and $b \cdot a < b \cdot b$, so $a^2 < b^2$.

(3) If $0 < a < b$, then $b - a > 0$ or, equivalently,

$$(\sqrt{b} + \sqrt{a})(\sqrt{b} - \sqrt{a}) > 0.$$

Dividing both sides of the last inequality by $\sqrt{b} + \sqrt{a}$, we obtain $\sqrt{b} - \sqrt{a} > 0$; that is, $\sqrt{b} > \sqrt{a}$. ■

2.7 EXERCISES

Exer. 1–40: Solve the inequality, and express the solutions in terms of intervals whenever possible.

1 $(3x + 1)(5 - 10x) > 0$

2 $(2 - 3x)(4x - 7) \geq 0$

3 $(x + 2)(x - 1)(4 - x) \leq 0$

4 $(x - 5)(x + 3)(-2 - x) < 0$

5 $x^2 - x - 6 < 0$

6 $x^2 + 4x + 3 \geq 0$

7 $x^2 - 2x - 5 > 3$

8 $x^2 - 4x - 17 \leq 4$

9 $x(2x + 3) \geq 5$

10 $x(3x - 1) \leq 4$

11 $6x - 8 > x^2$

12 $x + 12 \leq x^2$

13 $x^2 < 16$

14 $x^2 > 9$

15 $25x^2 - 9 < 0$

16 $25x^2 - 9x < 0$

17 $16x^2 \geq 9x$

18 $16x^2 > 9$

19 $x^4 + 5x^2 \geq 36$

20 $x^4 + 15x^2 < 16$

21 $x^3 + 2x^2 - 4x - 8 \geq 0$

22 $2x^3 - 3x^2 - 2x + 3 \leq 0$

23 $\dfrac{x^2(x + 2)}{(x + 2)(x + 1)} \leq 0$

24 $\dfrac{(x^2 + 1)(x - 3)}{x^2 - 9} \geq 0$

25 $\dfrac{x^2 - x}{x^2 + 2x} \leq 0$

26 $\dfrac{(x + 3)^2(2 - x)}{(x + 4)(x^2 - 4)} \leq 0$

27 $\dfrac{x - 2}{x^2 - 3x - 10} \geq 0$

28 $\dfrac{x + 5}{x^2 - 7x + 12} \leq 0$

29 $\dfrac{-3x}{x^2 - 9} > 0$

30 $\dfrac{2x}{16 - x^2} < 0$

31 $\dfrac{x + 1}{2x - 3} > 2$

32 $\dfrac{x - 2}{3x + 5} \leq 4$

33 $\dfrac{1}{x - 2} \geq \dfrac{3}{x + 1}$

34 $\dfrac{2}{2x + 3} \leq \dfrac{2}{x - 5}$

35 $\dfrac{4}{3x - 2} \leq \dfrac{2}{x + 1}$

36 $\dfrac{3}{5x + 1} \geq \dfrac{1}{x - 3}$

37 $\dfrac{x}{3x - 5} \leq \dfrac{2}{x - 1}$

38 $\dfrac{x}{2x - 1} \geq \dfrac{3}{x + 2}$

39 $x^3 > x$

40 $x^4 \geq x^2$

Exer. 41–42: As a particle moves along a straight path, its speed v (in cm/sec) at time t (in seconds) is given by the equation. For what subintervals of the given time interval $[a, b]$ will its speed be at least k cm/sec?

41 $v = t^3 - 3t^2 - 4t + 20$; $[0, 5]$; $k = 8$

42 $v = t^4 - 4t^2 + 10$; $[1, 6]$; $k = 10$

43 *Vertical leap record* *Guinness Book of World Records* reports that German shepherds can make vertical leaps of over 10 feet when scaling walls. If the distance s (in feet) off the ground after t seconds is given by the equation $s = -16t^2 + 24t + 1$, for how many seconds is the dog more than 9 feet off the ground?

44 *Height of a projected object* If an object is projected vertically upward from ground level with an initial velocity of 320 ft/sec, then its distance s above the ground after t seconds is given by $s = -16t^2 + 320t$. For what values of t will the object be more than 1536 feet above the ground?

45 *Braking distance* The braking distance d (in feet) of a certain car traveling v mi/hr is given by $d = v + (v^2/20)$. Determine the velocities that result in braking distances of less than 75 feet.

46 *Gas mileage* The number of miles M that a certain compact car can travel on 1 gallon of gasoline is related to its speed v (in mi/hr) by

$$M = -\tfrac{1}{30}v^2 + \tfrac{5}{2}v \quad \text{for } 0 < v < 70.$$

For what speeds will M be at least 45?

47 *Salmon propagation* For a particular salmon population, the relationship between the number S of spawners and the number R of offspring that survive to maturity is given by the formula $R = 4500S/(S + 500)$. Under what conditions is $R > S$?

48 *Population density* The population density D (in people/mi^2) in a large city is related to the distance x from the center of the city by $D = 5000x/(x^2 + 36)$. In what areas of the city does the population density exceed 400 people/mi^2?

49 *Weight in space* After an astronaut is launched into space, the astronaut's weight decreases until a state of weightlessness is achieved. The weight of a 125-pound astronaut at an altitude of x km above sea level is given by

$$W = 125\left(\frac{6400}{6400 + x}\right)^2.$$

At what altitudes is the astronaut's weight less than 5 pounds?

50 *Lorentz contraction formula* The Lorentz contraction formula in relativity theory relates the length L of an object moving at a velocity of v mi/sec with respect to an observer to its length L_0 at rest. If c is the speed of light, then

$$L^2 = L_0^2\left(1 - \frac{v^2}{c^2}\right).$$

For what velocities will L be less than $\tfrac{1}{2}L_0$? State the answer in terms of c.

CHAPTER 2 REVIEW EXERCISES

Exer. 1–24: Solve the equation.

1 $\dfrac{3x+1}{5x+7} = \dfrac{6x+11}{10x-3}$

2 $2 - \dfrac{1}{x} = 1 + \dfrac{4}{x}$

3 $\dfrac{2}{x+5} - \dfrac{3}{2x+1} = \dfrac{5}{6x+3}$

4 $\dfrac{7}{x-2} - \dfrac{6}{x^2-4} = \dfrac{3}{2x+4}$

5 $\dfrac{1}{\sqrt{x}} - 2 = \dfrac{1-2\sqrt{x}}{\sqrt{x}}$

6 $2x^2 + 5x - 12 = 0$

7 $x(3x+4) = 5$

8 $\dfrac{x}{3x+1} = \dfrac{x-1}{2x+3}$

9 $(x-2)(x+1) = 3$

10 $4x^4 - 33x^2 + 50 = 0$

11 $x^{2/3} - 2x^{1/3} - 15 = 0$

12 $20x^3 + 8x^2 - 35x - 14 = 0$

13 $5x^2 = 2x - 3$

14 $x^2 + \tfrac{1}{3}x + 2 = 0$

15 $6x^4 + 29x^2 + 28 = 0$

16 $x^4 - 3x^2 + 1 = 0$

17 $|4x-1| = 7$

18 $2|2x+1| + 1 = 19$

19 $\dfrac{1}{x} + 6 = \dfrac{5}{\sqrt{x}}$

20 $\sqrt[3]{4x-5} - 2 = 0$

21 $\sqrt{7x+2} + x = 6$

22 $\sqrt{x+4} = \sqrt[4]{6x+19}$

23 $\sqrt{3x+1} - \sqrt{x+4} = 1$

24 $x^{4/3} = 16$

Exer. 25–26: Solve the equation by completing the square.

25 $3x^2 - 12x + 3 = 0$

26 $x^2 + 10x + 38 = 0$

Exer. 27–44: Solve the inequality, and express the solutions in terms of intervals whenever possible.

27 $(x-3)^2 \le 0$

28 $10 - 7x < 4 + 2x$

29 $-\dfrac{1}{2} < \dfrac{2x+3}{5} < \dfrac{3}{2}$

30 $(3x-1)(10x+4) \ge (6x-5)(5x-7)$

31 $\dfrac{6}{10x+3} < 0$

32 $|4x+7| < 21$

33 $2|3-x| + 1 > 5$

34 $-2|x-3| + 1 \ge -5$

35 $|16-3x| \ge 5$

36 $2 < |x-6| < 4$

37 $10x^2 + 11x > 6$

38 $x(x-3) \le 10$

39 $\dfrac{x^2(3-x)}{x+2} \le 0$

40 $\dfrac{x^2-x-2}{x^2+4x+3} \le 0$

41 $\dfrac{3}{2x+3} < \dfrac{1}{x-2}$

42 $\dfrac{x+1}{x^2-25} \le 0$

43 $x^3 > x^2$

44 $(x^2-x)(x^2-5x+6) < 0$

Exer. 45–48: Solve for the specified variable.

45 $V = \tfrac{4}{3}\pi r^3$ for r (volume of a sphere)

46 $F = \dfrac{\pi P R^4}{8VL}$ for R (Poiseuille's law for fluids)

47 $c = \sqrt{4h(2R-h)}$ for h (base of a circular segment)

48 $V = \tfrac{1}{3}\pi h(r^2 + R^2 + rR)$ for r (volume of a frustum of a cone)

Exer. 49–54: Express in the form $a + bi$, where a and b are real numbers.

49 $(7+5i) - (-8+3i)$

50 $(4+2i)(-5+4i)$

51 $(3+8i)^2$

52 $\dfrac{1}{9 - \sqrt{-4}}$

53 $\dfrac{6-3i}{2+7i}$

54 $\dfrac{20-8i}{4i}$

55 *Electrical resistance* When two resistors R_1 and R_2 are connected in parallel, the net resistance R is given by $1/R = (1/R_1) + (1/R_2)$. If $R_1 = 5$ ohms, what value of R_2 will make the net resistance 2 ohms?

56 *Investment income* An investor has a choice of two investments: a bond fund and a stock fund. The bond fund yields 7.186% interest annually, which is nontaxable at both the federal and state levels. Suppose the investor pays federal income tax at a rate of 28% and state income tax at a rate of 7%. Determine what the annual yield must be on the taxable stock fund so that the two funds pay the same amount of net interest income to the investor.

57 *Gold and silver mixture*   A ring that weighs 80 grams is made of gold and silver. By measuring the displacement of the ring in water, it has been determined that the ring has a volume of 5 cm^3. Gold weighs 19.3 g/cm^3, and silver weighs 10.5 g/cm^3. How many grams of gold does the ring contain?

58 *Preparing hospital food*   A hospital dietitian wishes to prepare a 10-ounce meat-vegetable dish that will provide 7 grams of protein. If an ounce of the vegetable portion supplies $\frac{1}{2}$ gram of protein and an ounce of meat supplies 1 gram of protein, how much of each should be used?

59 *Preparing a bactericide*   A solution of ethyl alcohol that is 75% alcohol by weight is to be used as a bactericide. The solution is to be made by adding water to a 95% ethyl alcohol solution. How many grams of each should be used to prepare 400 grams of the bactericide?

60 *Solar heating*   A large solar heating panel requires 120 gallons of a fluid that is 30% antifreeze. The fluid comes in either a 50% solution or a 20% solution. How many gallons of each should be used to prepare the 120-gallon solution?

61 *Fuel consumption*   A boat has a 10-gallon gasoline tank and travels at 20 mi/hr with a fuel consumption of 16 mi/gal when operated at full throttle in still water. The boat is moving upstream into a 5-mi/hr current. How far upstream can the boat travel and return on 10 gallons of gasoline if it is operated at full throttle during the entire trip?

62 *Train travel*   A high-speed train makes a 400-mile non-stop run between two major cities in $5\frac{1}{2}$ hours. The train travels 100 mi/hr in the country, but safety regulations require that it travel only 25 mi/hr when passing through smaller, intermediate cities. How many hours are spent traveling through the smaller cities?

63 *Windspeed*   An airplane flew with the wind for 30 minutes and returned the same distance in 45 minutes. If the cruising speed of the airplane was 320 mi/hr, what was the speed of the wind?

64 *Passing speed*   An automobile 20 feet long overtakes a truck 40 feet long that is traveling at 50 mi/hr (see the figure). At what constant speed must the automobile travel in order to pass the truck in 5 seconds?

EXERCISE 64

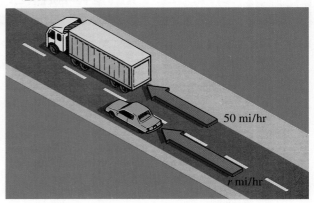

50 mi/hr

r mi/hr

65 *Filling a bin*   An extruder can fill an empty bin in 2 hours, and a packaging crew can empty a full bin in 5 hours. If a bin is half full when an extruder begins to fill it and a crew begins to empty it, how long will it take to fill the bin?

66 *Gasoline mileage*   A sales representative for a company estimates that automobile gasoline consumption averages 28 mpg on the highway and 22 mpg in the city. A recent trip covered 627 miles, and 24 gallons of gasoline was used. How much of the trip was spent driving in the city?

67 *City expansion*   The longest drive to the center of a square city from the outskirts is 10 miles. Within the last decade the city has expanded in area by 50 mi^2. Assuming the city has always been square in shape, find the corresponding change in the longest drive to the center of the city.

68 *Dimensions of a cell membrane*   The membrane of a cell is a sphere of radius 6 microns. What change in the radius will increase the surface area of the membrane by 25%?

69 *Highway travel*   A north-south highway intersects an east-west highway at a point P. An automobile crosses P at 10:00 A.M., traveling east at a constant rate of 20 mi/hr. At that same instant another automobile is 2 miles north of P, traveling south at 50 mi/hr.

   (a) Find a formula for the distance d between the automobiles t hours after 10:00 A.M.

   (b) At approximately what time will the automobiles be 104 miles apart?

70 *Fencing a kennel*   A kennel owner has 270 feet of fencing material to be used to divide a rectangular area into 10 equal pens, as shown in the figure. Find dimensions that would allow 100 ft^2 for each pen.

EXERCISE 70

71 *Dimensions of an aquarium*   An open-topped aquarium is to be constructed with 6-foot-long sides and square ends, as shown in the figure.

(a) Find the height of the aquarium if the volume is to be 48 ft^3.

(b) Find the height if 44 ft^2 of glass is to be used.

EXERCISE 71

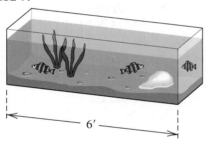

72 *Dimensions of a pool*   The length of a rectangular pool is to be four times its width, and a sidewalk of width 6 feet will surround the pool. If a total area of 1440 ft^2 has been set aside for construction, what are the dimensions of the pool?

73 *Dimensions of a bath*   A contractor wishes to design a rectangular sunken bath with 40 ft^2 of bathing area. A 1-foot-wide tile strip is to surround the bathing area. The total length of the tiled area is to be twice the width. Find the dimensions of the bathing area.

74 *Population growth*   The population P (in thousands) of a small town is expected to increase according to the formula

$$P = 15 + \sqrt{3t + 2},$$

where t is time in years. When will the population be 20,000?

75 *Boyle's law*   Boyle's law for a certain gas states that if the temperature is constant, then $pv = 200$, where p is the pressure (in lb/in.^{2}) and v is the volume (in in.^{3}). If $25 \leq v \leq 50$, what is the corresponding range for p?

76 *Sales commission*   A recent college graduate has job offers for a sales position in two computer firms. Job A pays $25,000 per year plus a 5% commission. Job B pays only $20,000 per year, but the commission rate is 10%. How much yearly business must the salesman do for the second job to be more lucrative?

77 *Speed of sound*   The speed of sound in air at 0 °C (or 273 K) is 1087 ft/sec, but this speed increases as the temperature rises. The speed v of sound at temperature T in K is given by $v = 1087\sqrt{T/273}$. At what temperatures does the speed of sound exceed 1100 ft/sec?

78 *Period of a pendulum*   If the length of the pendulum in a grandfather clock is l centimeters, then its period T (in seconds) is given by $T = 2\pi\sqrt{l/g}$, where g is a gravitational constant. If, under certain conditions, $g = 980$ and $98 \leq l \leq 100$, what is the corresponding range for T?

79 *Orbit of a satellite*   For a satellite to maintain an orbit of altitude h kilometers, its velocity (in km/sec) must equal $626.4/\sqrt{h + R}$, where $R = 6372$ km is the radius of the earth. What velocities will result in orbits with an altitude of more than 100 kilometers from the earth's surface?

80 *Fencing a region*   There is 100 feet of fencing available to enclose a rectangular region. For what widths will the fenced region contain at least 600 ft^2?

81 *Planting an apple orchard*   The owner of an apple orchard estimates that if 24 trees are planted per acre, then each mature tree will yield 600 apples per year. For each additional tree planted per acre, the number of apples produced by each tree decreases by 12 per year. How many trees should be planted per acre to obtain at least 16,416 apples per year?

82 *Apartment rentals*   A real estate company owns 180 efficiency apartments, which are fully occupied when the rent is $300 per month. The company estimates that for each $10 increase in rent, 5 apartments will become unoccupied. What rent should be charged in order to pay the monthly bills, which total $54,400?

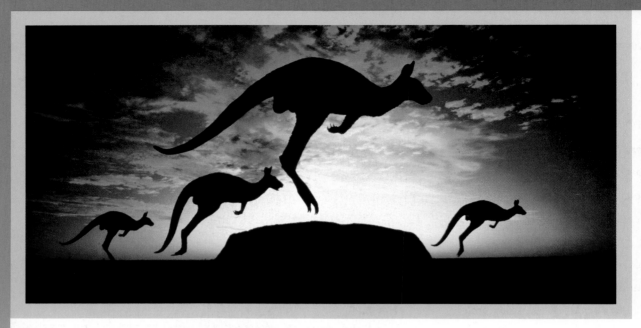

Flights of leaping animals typically have parabolic paths.

The mathematical term *function* (or its Latin equivalent) dates back to the late seventeenth century, when calculus was in the early stages of development. This important concept is now the backbone of advanced courses in mathematics and is indispensable in every field of science.

In this chapter we study properties of functions using algebraic and graphical methods that include plotting points, determining symmetries, and making horizontal and vertical shifts. These techniques are adequate for obtaining rough sketches of graphs that help us understand properties of functions; modern-day methods, however, employ sophisticated computer software and advanced mathematics to generate extremely accurate graphical representations of functions. ■

CHAPTER 3

FUNCTIONS

AND

GRAPHS

3.1   RECTANGULAR COORDINATE SYSTEMS

*The term *Cartesian* is used in honor of the French mathematician and philosopher René Descartes (1596–1650), who was one of the first to employ such coordinate systems.

In Section 1.1 we discussed how to assign a real number (coordinate) to each point on a line. We shall now show how to assign an **ordered pair** (a, b) of real numbers to each point in a plane. Although we have also used the notation (a, b) to denote an open interval, there is little chance for confusion, since it should always be clear from our discussion whether (a, b) represents a point or an interval.

We introduce a **rectangular,** or **Cartesian,*** **coordinate system** in a plane by means of two perpendicular coordinate lines, called **coordinate axes**, that intersect at the **origin** O, as shown in Figure 1. We often refer to the horizontal line as the **x-axis** and the vertical line as the **y-axis** and label them x and y, respectively. The plane is then a **coordinate plane**, or an **xy-plane**. The coordinate axes divide the plane into four parts called the **first, second, third**, and **fourth quadrants**, labeled I, II, III, and IV, respectively (see Figure 1). Points on the axes do not belong to any quadrant.

Each point P in an xy-plane may be assigned an ordered pair (a, b), as shown in Figure 1. We call a the **x-coordinate** (or **abscissa**) of P, and b the **y-coordinate** (or **ordinate**). We say that P has *coordinates* (a, b) and refer to the *point* (a, b), or the *point* $P(a, b)$. Conversely, every ordered pair (a, b) determines a point P with coordinates a and b. We **plot a point** by using a dot, as illustrated in Figure 2.

FIGURE 1   **FIGURE 2**

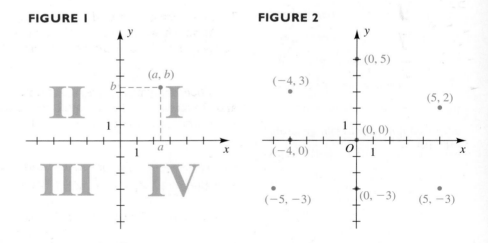

We may use the following formula to find the distance between two points in a coordinate plane.

Distance Formula

> The distance $d(P_1, P_2)$ between any two points $P_1(x_1, y_1)$ and $P_2(x_2, y_2)$ in a coordinate plane is
> $$d(P_1, P_2) = \sqrt{(x_2 - x_1)^2 + (y_2 - y_1)^2}.$$

FIGURE 3

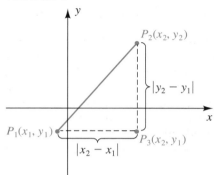

PROOF If $x_1 \neq x_2$ and $y_1 \neq y_2$, then, as illustrated in Figure 3, the points P_1, P_2, and $P_3(x_2, y_1)$ are vertices of a right triangle. By the Pythagorean theorem,

$$[d(P_1, P_2)]^2 = [d(P_1, P_3)]^2 + [d(P_3, P_2)]^2.$$

From the figure we see that

$$d(P_1, P_3) = |x_2 - x_1| \qquad \text{and} \qquad d(P_3, P_2) = |y_2 - y_1|.$$

Since $|a|^2 = a^2$ for every real number a, we may write

$$[d(P_1, P_2)]^2 = (x_2 - x_1)^2 + (y_2 - y_1)^2.$$

Taking the square root of each side of the last equation and using the fact that $d(P_1, P_2) \geq 0$ gives us the distance formula.

If $y_1 = y_2$, the points P_1 and P_2 lie on the same horizontal line, and

$$d(P_1, P_2) = |x_2 - x_1| = \sqrt{(x_2 - x_1)^2}.$$

Similarly, if $x_1 = x_2$, the points are on the same vertical line, and

$$d(P_1, P_2) = |y_2 - y_1| = \sqrt{(y_2 - y_1)^2}.$$

These are special cases of the distance formula.

Although we referred to the points shown in Figure 3, our proof is independent of the positions of P_1 and P_2. ▄

When applying the distance formula, note that $d(P_1, P_2) = d(P_2, P_1)$ and, hence, the order in which we subtract the x-coordinates and the y-coordinates of the points is immaterial.

FIGURE 4

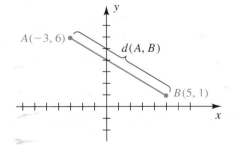

EXAMPLE I   *Finding the distance between points*

Plot the points $A(-3, 6)$ and $B(5, 1)$, and find the distance $d(A, B)$.

SOLUTION   The points are plotted in Figure 4. By the distance formula,

$$d(A, B) = \sqrt{[5 - (-3)]^2 + (1 - 6)^2}$$
$$= \sqrt{8^2 + (-5)^2}$$
$$= \sqrt{64 + 25} = \sqrt{89} \approx 9.43.$$

EXAMPLE 2   *Showing that a triangle is a right triangle*

(a) Plot $A(-1, -3)$, $B(6, 1)$, and $C(2, -5)$, and show that triangle ABC is a right triangle.

(b) Find the area of triangle ABC.

SOLUTION

FIGURE 5

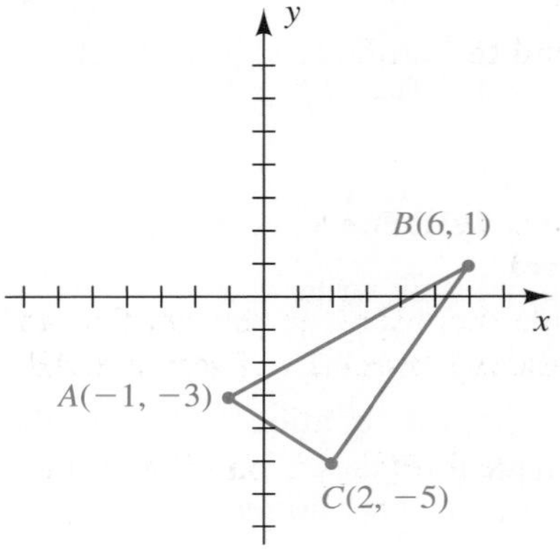

(a) The points are plotted in Figure 5. From geometry, triangle ABC is a right triangle if the sum of the squares of two of its sides is equal to the square of the remaining side. By the distance formula,

$$d(A, B) = \sqrt{(6+1)^2 + (1+3)^2} = \sqrt{49 + 16} = \sqrt{65}$$

$$d(B, C) = \sqrt{(2-6)^2 + (-5-1)^2} = \sqrt{16 + 36} = \sqrt{52}$$

$$d(A, C) = \sqrt{(2+1)^2 + (-5+3)^2} = \sqrt{9 + 4} = \sqrt{13}.$$

Since $d(A, B) = \sqrt{65}$ is the largest of the three values, the condition to be satisfied is

$$[d(A, B)]^2 = [d(B, C)]^2 + [d(A, C)]^2.$$

Substituting the values found using the distance formula, we obtain

$$[d(A, B)]^2 = (\sqrt{65})^2 = 65$$

and

$$[d(B, C)]^2 + [d(A, C)]^2 = (\sqrt{52})^2 + (\sqrt{13})^2 = 52 + 13 = 65.$$

Thus, the triangle is a right triangle with hypotenuse AB.

(b) The area of a triangle with base b and altitude h is $\frac{1}{2}bh$. Referring to Figure 5, we let

$$b = d(B, C) = \sqrt{52} \qquad \text{and} \qquad h = d(A, C) = \sqrt{13}.$$

Hence, the area of triangle ABC is

$$\tfrac{1}{2}bh = \tfrac{1}{2}\sqrt{52}\sqrt{13} = \tfrac{1}{2} \cdot 2\sqrt{13}\sqrt{13} = 13.$$

FIGURE 6

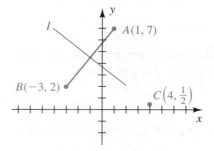

EXAMPLE 3   *Applying the distance formula*

Given $A(1, 7)$, $B(-3, 2)$, and $C(4, \frac{1}{2})$, prove that C is on the perpendicular bisector of segment AB.

SOLUTION   The points A, B, C and the perpendicular bisector l are illustrated in Figure 6. From plane geometry, l can be characterized by either of the following conditions:

(1) l is the line perpendicular to segment AB at its midpoint.

(2) l is the set of all points equidistant from the end points of segment AB.

We shall use condition (2) to show that C is on l by verifying that

$$d(A, C) = d(B, C).$$

We apply the distance formula:

$$d(A, C) = \sqrt{(4 - 1)^2 + (\tfrac{1}{2} - 7)^2} = \sqrt{3^2 + (-\tfrac{13}{2})^2} = \sqrt{9 + \tfrac{169}{4}} = \sqrt{\tfrac{205}{4}}$$

$$d(B, C) = \sqrt{[4 - (-3)]^2 + (\tfrac{1}{2} - 2)^2} = \sqrt{7^2 + (-\tfrac{3}{2})^2} = \sqrt{49 + \tfrac{9}{4}} = \sqrt{\tfrac{205}{4}}$$

Thus, C is equidistant from A and B, and the verification is complete.

EXAMPLE 4   *Finding a formula that describes a perpendicular bisector*

Given $A(1, 7)$ and $B(-3, 2)$, find a formula that expresses the fact that an arbitrary point $P(x, y)$ is on the perpendicular bisector l of segment AB.

SOLUTION   By condition (2) of Example 3, $P(x, y)$ is on l if and only if $d(A, P) = d(B, P)$; that is,

$$\sqrt{(x - 1)^2 + (y - 7)^2} = \sqrt{[x - (-3)]^2 + (y - 2)^2}.$$

To obtain a simpler formula, let us square both sides and simplify terms of the resulting equation, as follows:

$$(x - 1)^2 + (y - 7)^2 = [x - (-3)]^2 + (y - 2)^2$$
$$x^2 - 2x + 1 + y^2 - 14y + 49 = x^2 + 6x + 9 + y^2 - 4y + 4$$
$$-2x + 1 - 14y + 49 = 6x + 9 - 4y + 4$$
$$-8x - 10y = -37$$
$$8x + 10y = 37$$

Note that, in particular, the last formula is true for the coordinates of the point $C(4, \tfrac{1}{2})$ in Example 3, since if $x = 4$ and $y = \tfrac{1}{2}$, then substitution in $8x + 10y$ gives us

$$8 \cdot 4 + 10 \cdot \tfrac{1}{2} = 37.$$

In Example 9 of Section 3.3, we will find a formula for the perpendicular bisector of a segment using condition (1) of Example 3.

We can find the midpoint of a line segment by using the following formula.

Midpoint Formula

> The midpoint M of the line segment from $P_1(x_1, y_1)$ to $P_2(x_2, y_2)$ is
>
> $$\left(\frac{x_1 + x_2}{2}, \frac{y_1 + y_2}{2}\right).$$

FIGURE 7

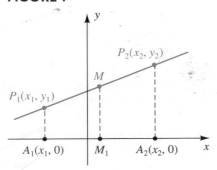

The lines through P_1 and P_2 parallel to the y-axis intersect the x-axis at $A_1(x_1, 0)$ and $A_2(x_2, 0)$. From plane geometry, the line through the midpoint M parallel to the y-axis bisects the segment A_1A_2 at point M_1 (see Figure 7). If $x_1 < x_2$, then $x_2 - x_1 > 0$, and hence $d(A_1, A_2) = x_2 - x_1$. Since M_1 is halfway from A_1 to A_2, the x-coordinate of M_1 is equal to the x-coordinate of A_1 plus one-half the distance from A_1 to A_2, that is,

$$x\text{-coordinate of } M_1 = x_1 + \tfrac{1}{2}(x_2 - x_1).$$

The expression on the right side of the last equation simplifies to

$$\frac{x_1 + x_2}{2}.$$

This is the *average* of the numbers x_1 and x_2. It follows that the x-coordinate of M is also $(x_1 + x_2)/2$. Similarly, the y-coordinate of M is $(y_1 + y_2)/2$. These formulas hold for all positions of P_1 and P_2.

EXAMPLE 5 *Finding a midpoint*

Find the midpoint M of the line segment from $P_1(-2, 3)$ to $P_2(4, -2)$, and verify that $d(P_1, M) = d(P_2, M)$.

FIGURE 8

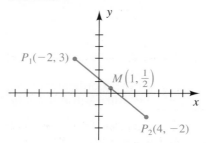

SOLUTION By the midpoint formula, the coordinates of M are

$$\left(\frac{-2 + 4}{2}, \frac{3 + (-2)}{2}\right), \quad \text{or} \quad \left(1, \frac{1}{2}\right).$$

The three points P_1, P_2, and M are plotted in Figure 8. By the distance formula,

$$d(P_1, M) = \sqrt{(1 + 2)^2 + (\tfrac{1}{2} - 3)^2} = \sqrt{9 + \tfrac{25}{4}}$$

$$d(P_2, M) = \sqrt{(1 - 4)^2 + (\tfrac{1}{2} + 2)^2} = \sqrt{9 + \tfrac{25}{4}}.$$

Hence, $d(P_1, M) = d(P_2, M)$.

3.1 EXERCISES

1 Plot the points $A(5, -2)$, $B(-5, -2)$, $C(5, 2)$, $D(-5, 2)$, $E(3, 0)$, and $F(0, 3)$ on a coordinate plane.

2 Plot the points $A(-3, 1)$, $B(3, 1)$, $C(-2, -3)$, $D(0, 3)$, and $E(2, -3)$ on a coordinate plane. Draw the line segments AB, BC, CD, DE, and EA.

3 Plot the points $A(0, 0)$, $B(1, 1)$, $C(3, 3)$, $D(-1, -1)$, and $E(-2, -2)$. Describe the set of all points of the form (a, a), where a is a real number.

4 Plot the points $A(0, 0)$, $B(1, -1)$, $C(3, -3)$, $D(-1, 1)$, and $E(-3, 3)$. Describe the set of all points of the form $(a, -a)$, where a is a real number.

Exer. 5–6: Find the coordinates of the points A–F.

5

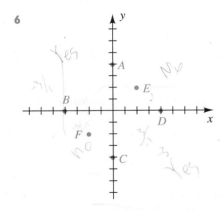

6

11 $A(-5, 0)$, $B(-2, -2)$ **12** $A(6, 2)$, $B(6, -2)$

13 $A(7, -3)$, $B(3, -3)$ **14** $A(-4, 7)$, $B(0, -8)$

Exer. 15–16: Show that the triangle with vertices A, B, and C is a right triangle, and find its area.

15

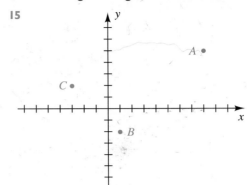

16

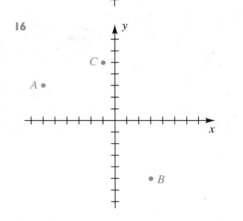

17 Show that $A(-4, 2)$, $B(1, 4)$, $C(3, -1)$, and $D(-2, -3)$ are vertices of a square.

18 Show that $A(-4, -1)$, $B(0, -2)$, $C(6, 1)$, and $D(2, 2)$ are vertices of a parallelogram.

19 Given $A(-3, 8)$, find the coordinates of the point B such that $C(5, -10)$ is the midpoint of segment AB.

20 Given $A(5, -8)$ and $B(-6, 2)$, find the point on segment AB that is three-fourths of the way from A to B.

Exer. 7–8: Describe the set of all points $P(x, y)$ in a coordinate plane that satisfy the given condition.

7 (a) $x = -2$ (b) $y = 3$ (c) $x \geq 0$

  (d) $xy > 0$ (e) $y < 0$ (f) $x = 0$

8 (a) $y = -2$ (b) $x = -4$ (c) $x/y < 0$

  (d) $xy = 0$ (e) $y > 1$ (f) $y = 0$

Exer. 9–14: (a) Find the distance $d(A, B)$ between A and B. (b) Find the midpoint of the segment AB.

9 $A(4, -3)$, $B(6, 2)$ **10** $A(-2, -5)$, $B(4, 6)$

Exer. 21–22: Prove that C is on the perpendicular bisector of segment AB.

21 $A(-4, -3)$, $B(6, 1)$, $C(5, -11)$

22 $A(-3, 2)$, $B(5, -4)$, $C(7, 7)$

Exer. 23–24: Find a formula that expresses the fact that an arbitrary point $P(x, y)$ is on the perpendicular bisector l of segment AB.

23 $A(-4, -3)$, $B(6, 1)$ **24** $A(-3, 2)$, $B(5, -4)$

25 Find a formula that expresses the fact that $P(x, y)$ is a distance 5 from the origin. Describe the set of all such points.

26 Find a formula that states that $P(x, y)$ is a distance $r > 0$ from a fixed point $C(h, k)$. Describe the set of all such points.

27 Find all points on the y-axis that are a distance 6 from $P(5, 3)$.

28 Find all points on the x-axis that are a distance 5 from $P(-2, 4)$.

29 Find the point with coordinates of the form $(2a, a)$ that is in the third quadrant and is a distance 5 from $P(1, 3)$.

30 Find all points with coordinates of the form (a, a) that are a distance 3 from $P(-2, 1)$.

31 For what values of a is the distance between $P(a, 3)$ and $Q(5, 2a)$ greater than $\sqrt{26}$?

32 Given $A(-2, 0)$ and $B(2, 0)$, find a formula not containing radicals that expresses the fact that the sum of the distances from $P(x, y)$ to A and to B, respectively, is 5.

33 Prove that the midpoint of the hypotenuse of any right triangle is equidistant from the vertices. (*Hint:* Label the vertices of the triangle $O(0, 0)$, $A(a, 0)$, and $B(0, b)$.)

34 Prove that the diagonals of any parallelogram bisect each other. (*Hint:* Label three of the vertices of the parallelogram $O(0, 0)$, $A(a, b)$, and $C(0, c)$.)

3.2 GRAPHS OF EQUATIONS

Graphs are often used to illustrate changes in quantities. A graph in the business section of a newspaper may show the fluctuation of the Dow-Jones average during a given month; a meteorologist might use a graph to indicate how the air temperature varied throughout a day; a cardiologist employs graphs (electrocardiograms) to analyze heart irregularities; an engineer or physicist may turn to a graph to illustrate the manner in which the pressure of a confined gas increases as the gas is heated. Such visual aids usually reveal the behavior of quantities more readily than a long table of numerical values.

Two quantities are sometimes related by means of an equation or formula that involves two variables. In this section we discuss how to represent such an equation geometrically, by a graph in a coordinate plane. The graph may then be used to discover properties of the quantities that are not evident from the equation alone. The following chart introduces the basic concept of *the graph of an equation* in two variables x and y. Of course, other letters can also be used for the variables.

Terminology	Definition	Illustration
Solution of an equation in x and y	An ordered pair (a, b) that yields a true statement if $x = a$ and $y = b$	$(2, 3)$ is a solution of $y^2 = 5x - 1$, since substituting $x = 2$ and $y = 3$ gives us LS: $3^2 = 9$ RS: $5(2) - 1 = 10 - 1 = 9$

For each solution (a, b) of an equation in x and y there is a point $P(a, b)$ in a coordinate plane. The set of all such points is called the **graph** of the equation. To *sketch the graph of an equation*, we illustrate the significant features of the graph in a coordinate plane. In simple cases, a graph can be sketched by plotting few, if any, points. For a complicated equation, plotting points may give very little information about the graph. In such cases, methods of calculus or computer graphics are often employed. Let us begin with a simple example.

EXAMPLE 1 *Sketching a simple graph by plotting points*

Sketch the graph of the equation $y = 2x - 1$.

SOLUTION   We wish to find the points (x, y) in a coordinate plane that correspond to the solutions of the equation. It is convenient to list coordinates of several such points in a table, where for each x we obtain the value for y from $y = 2x - 1$:

FIGURE 9

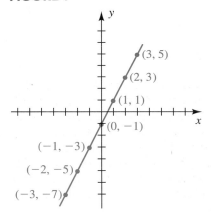

x	-3	-2	-1	0	1	2	3
y	-7	-5	-3	-1	1	3	5

The points with these coordinates appear to lie on a line, and we can sketch the graph in Figure 9. Ordinarily, the few points we have plotted would not be enough to illustrate the graph of an equation; however, in this elementary case we can be reasonably sure that the graph is a line. In the next section we will establish this fact.

It is impossible to sketch the entire graph in Example 1, because we can assign values to x that are numerically as large as desired. Nevertheless, we call the drawing in Figure 9 *the graph of the equation* or *a sketch of the graph*. In general, the sketch of a graph should illustrate its essential features so that the remaining (unsketched) parts are self-evident. If a graph terminates at some point (as would be the case for a half-line or line segment), we place a dot at the appropriate *end point* of the graph. As a final general remark, *if tics on the coordinate axes are not labeled* (as in Figure 9), *then each tic represents one unit*. We shall label tics only when different units are used on the axes. For *arbitrary* graphs, where units of measurement are irrelevant, we omit tics completely (see, for example, Figures 13 and 14).

EXAMPLE 2   *Sketching the graph of an equation*

Sketch the graph of $y = x^2 - 3$.

SOLUTION   Substituting values for x and finding the corresponding values of y using $y = x^2 - 3$, we obtain a table of coordinates for several points on the graph:

x	-3	-2	-1	0	1	2	3
y	6	1	-2	-3	-2	1	6

Larger values of $|x|$ produce larger values of y. For example, the points $(4, 13), (5, 22)$, and $(6, 33)$ are on the graph, as are $(-4, 13), (-5, 22)$, and $(-6, 33)$. Plotting the points given by the table and drawing a smooth curve through these points (in the order of increasing values of x) gives us the sketch in Figure 10.

FIGURE 10

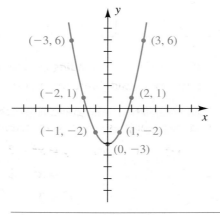

The graph in Figure 10 is a **parabola**, and the y-axis is the **axis of the parabola**. The lowest point $(0, -3)$ is the **vertex** of the parabola, and we say that the parabola **opens upward**. If we invert the graph, then the parabola **opens downward** and the vertex is the highest point on the graph. In general, the graph of *any* equation of the form $y = ax^2 + c$ with $a \neq 0$ is a parabola with vertex $(0, c)$, opening upward if $a > 0$ or downward if $a < 0$. If $c = 0$, the equation reduces to $y = ax^2$ and the vertex is at the origin $(0, 0)$. Parabolas may also open to the right or to the left (see Example 4) or in other directions.

We shall use the following terminology to describe where the graph of an equation in x and y intersects the x-axis or the y-axis.

Intercepts of the Graph of an Equation in x and y

Terminology	Definition	Graphical interpretation	How to find
x-intercepts	The x-coordinates of points where the graph intersects the x-axis		Let $y = 0$ and solve for x.
y-intercepts	The y-coordinates of points where the graph intersects the y-axis		Let $x = 0$ and solve for y.

EXAMPLE 3   *Finding x-intercepts and y-intercepts*

Find the x- and y-intercepts of the graph of $y = x^2 - 3$.

SOLUTION   The graph is sketched in Figure 10 (Example 2). We find the intercepts as stated in the preceding chart.

(1) *x-intercepts:*
$$y = x^2 - 3 \qquad \text{given}$$
$$0 = x^2 - 3 \qquad \text{let } y = 0$$
$$x^2 = 3 \qquad \text{equivalent equation}$$
$$x = \pm\sqrt{3} \approx \pm 1.732 \quad \text{take the square root}$$

Thus, the x-intercepts are $-\sqrt{3}$ and $\sqrt{3}$. The points at which the graph crosses the x-axis are $(-\sqrt{3}, 0)$ and $(\sqrt{3}, 0)$.

(2) *y-intercepts:*
$$y = x^2 - 3 \qquad \text{given}$$
$$y = 0 - 3 = -3 \quad \text{let } x = 0$$

Thus, the y-intercept is -3, and the point at which the graph crosses the y-axis is $(0, -3)$.

If the coordinate plane in Figure 10 is folded along the y-axis, the graph that lies in the left half of the plane coincides with that in the right half. We say that **the graph is symmetric with respect to the y-axis**. A graph is symmetric with respect to the y-axis provided that the point $(-x, y)$ is on the graph whenever (x, y) is on the graph. The graph of $y = x^2 - 3$ in Example 2 has this property, since substitution of $-x$ for x yields the same equation:

$$y = (-x)^2 - 3 = x^2 - 3$$

This substitution is an application of symmetry test (1) in the following chart. Two other types of symmetry and the appropriate tests are also listed. The graphs of $x = y^2$ and $4y = x^3$ in the illustration column are discussed in Examples 4 and 5.

Symmetries of Graphs of Equations in x and y

Terminology	Graphical interpretation	Test for symmetry	Illustration
The graph is symmetric with respect to the y-axis	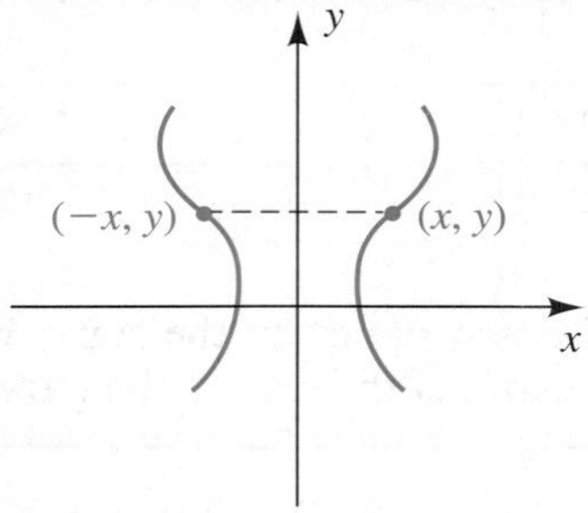	(1) Substitution of $-x$ for x leads to the same equation.	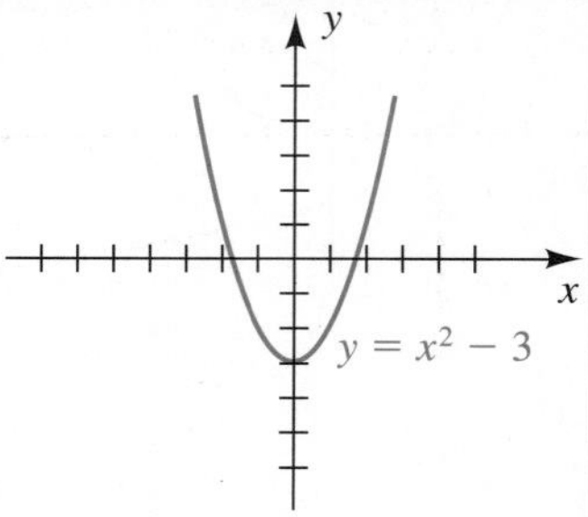
The graph is symmetric with respect to the x-axis.	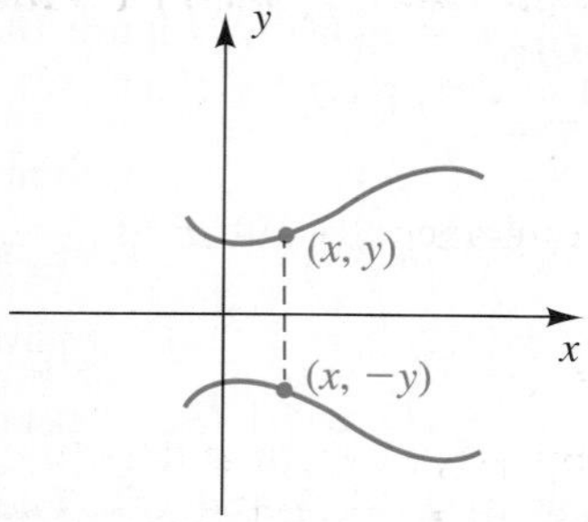	(2) Substitution of $-y$ for y leads to the same equation.	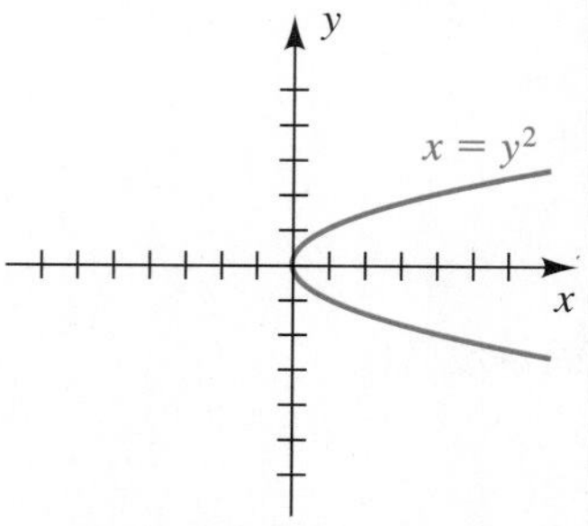
The graph is symmetric with respect to the origin.	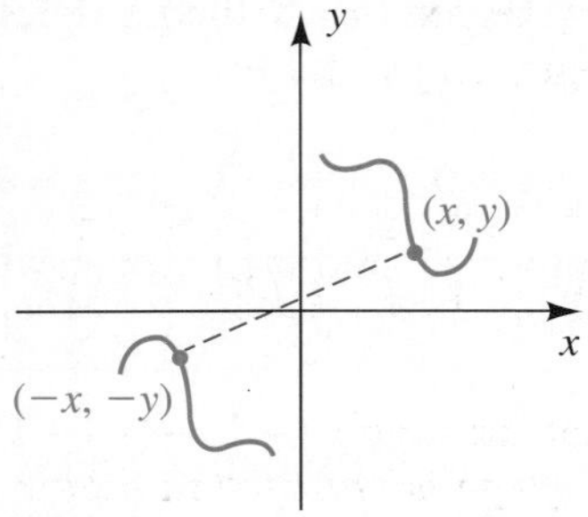	(3) Simultaneous substitution of $-x$ for x and $-y$ for y leads to the same equation.	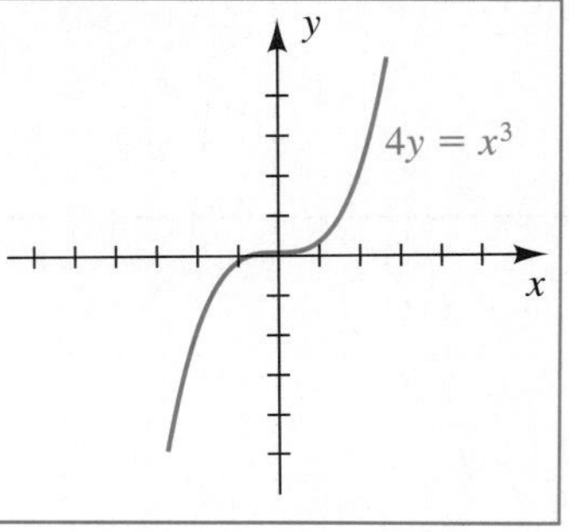

If a graph is symmetric with respect to an axis, it is sufficient to determine the graph in half of the coordinate plane, since we can sketch the remainder by taking a *mirror image*, or *reflection*, through the appropriate axis.

EXAMPLE 4   *A graph that is symmetric with respect to the x-axis*

Sketch the graph of the equation $y^2 = x$.

FIGURE 11

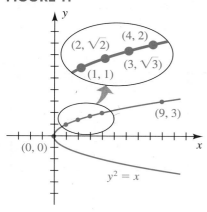

SOLUTION   Since substitution of $-y$ for y does not change the equation, the graph is symmetric with respect to the x-axis (see symmetry test (2)). Thus, it is sufficient to find points with nonnegative y-coordinates and then reflect through the x-axis. Since $y^2 = x$, the y-coordinates of points above the x-axis are given by $y = \sqrt{x}$. Coordinates of some points on the graph are listed below. The graph is sketched in Figure 11.

x	0	1	2	3	4	9
y	0	1	$\sqrt{2} \approx 1.4$	$\sqrt{3} \approx 1.7$	2	3

The graph is a parabola that opens to the right, with its vertex at the origin. In this case the x-axis is the axis of the parabola.

EXAMPLE 5   *A graph that is symmetric with respect to the origin*

Sketch the graph of $4y = x^3$.

FIGURE 12

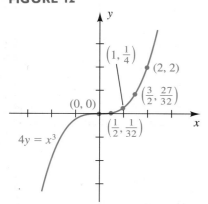

SOLUTION   If we simultaneously substitute $-x$ for x and $-y$ for y, then

$$4(-y) = (-x)^3, \qquad \text{or} \qquad -4y = -x^3.$$

Multiplying both sides by -1, we see that the last equation has the same solutions as the equation $4y = x^3$. Hence, from symmetry test (3), the graph is symmetric with respect to the origin. The following table lists coordinates of some points on the graph.

x	0	$\frac{1}{2}$	1	$\frac{3}{2}$	2	$\frac{5}{2}$
y	0	$\frac{1}{32}$	$\frac{1}{4}$	$\frac{27}{32}$	2	$\frac{125}{32}$

By symmetry we see that the points $(-1, -\frac{1}{4})$, $(-2, -2)$, and so on, are also on the graph. The graph is sketched in Figure 12.

FIGURE 13

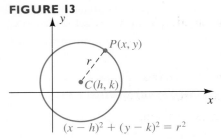

Standard Equation of a Circle with Center (h, k), Radius r

If $C(h, k)$ is a point in a coordinate plane, then a circle with center C and radius $r > 0$ consists of all points in the plane that are r units from C. As shown in Figure 13, a point $P(x, y)$ is on the circle provided $d(C, P) = r$ or, by the distance formula,

$$\sqrt{(x - h)^2 + (y - k)^2} = r.$$

The above equation is equivalent to the following equation.

$$\boxed{(x - h)^2 + (y - k)^2 = r^2}$$

If $h = 0$ and $k = 0$, this equation reduces to $x^2 + y^2 = r^2$, which is an equation of a circle of radius r with center at the origin (see Figure 14). If $r = 1$, we call the graph a **unit circle**.

FIGURE 14

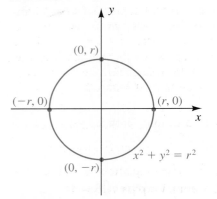

EXAMPLE 6   *Finding an equation of a circle*

Find an equation of the circle that has center $C(-2, 3)$ and contains the point $D(4, 5)$.

SOLUTION   The circle is illustrated in Figure 15. Since D is on the circle, the radius r is $d(C, D)$. By the distance formula,

$$r = \sqrt{(4 + 2)^2 + (5 - 3)^2} = \sqrt{36 + 4} = \sqrt{40}.$$

Using the standard equation of a circle with $h = -2$, $k = 3$, and $r = \sqrt{40}$, we obtain

$$(x + 2)^2 + (y - 3)^2 = 40.$$

By squaring terms and simplifying the last equation, we may write it as

$$x^2 + y^2 + 4x - 6y - 27 = 0.$$

FIGURE 15

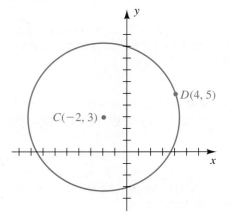

As in the solution to Example 6, squaring terms of an equation of the form $(x - h)^2 + (y - k)^2 = r^2$ and simplifying leads to an equation of the form

$$x^2 + y^2 + ax + by + c = 0,$$

where a, b, and c are real numbers. Conversely, if we begin with this equation, it is always possible, by *completing squares*, to obtain an equation of the form

$$(x - h)^2 + (y - k)^2 = d.$$

This method will be illustrated in Example 7. If $d > 0$, the graph is a circle with center (h, k) and radius $r = \sqrt{d}$. If $d = 0$, the graph consists of only one point (h, k). Finally, if $d < 0$, the equation has no real solutions, and hence there is no graph.

EXAMPLE 7 *Finding the center and radius of a circle*

Find the center and radius of the circle with equation

$$x^2 + y^2 - 4x + 6y - 3 = 0.$$

SOLUTION   We begin by rewriting the equation as follows:

$$(x^2 - 4x + \underline{}) + (y^2 + 6y + \underline{}) = 3 + \underline{} + \underline{},$$

where the underscored spaces represent numbers to be determined. Next we complete the squares for the expressions within parentheses, taking care to add the appropriate numbers to *both* sides of the equation. To complete the square for an expression of the form $x^2 + ax$, we add the square of half the coefficient of x (that is, $(a/2)^2$) to both sides of the equation (see page 76). Similarly, for $y^2 + by$, we add $(b/2)^2$ to both sides. In this example, $a = -4$, $b = 6$, $(a/2)^2 = (-2)^2 = 4$, and $(b/2)^2 = 3^2 = 9$. These additions lead to

$$(x^2 - 4x + \underline{4}) + (y^2 + 6y + \underline{9}) = 3 + \underline{4} + \underline{9} \qquad \text{completing the squares}$$

$$(x - 2)^2 + (y + 3)^2 = 16 \qquad \text{equivalent equation}$$

Comparing the last equation with the standard equation of a circle, we conclude that the circle has center $(2, -3)$ and radius $\sqrt{16} = 4$.

In some applications it is necessary to work with only one-half of a circle—that is, a **semicircle**. The next example indicates how to find equations for certain semicircles with centers at the origin.

EXAMPLE 8 *Finding equations of semicircles*

Find equations for the upper half, lower half, right half, and left half of the circle $x^2 + y^2 = 81$.

SOLUTION   The graph of $x^2 + y^2 = 81$ is a circle of radius 9 with center at the origin (compare with Figure 14). To find equations for the upper and lower halves, we solve for y in terms of x:

$$x^2 + y^2 = 81 \qquad \text{given}$$

$$y^2 = 81 - x^2 \qquad \text{subtract } x^2$$

$$y = \pm\sqrt{81 - x^2} \qquad \text{take the square root}$$

(continued)

Since $\sqrt{81 - x^2} \geq 0$, it follows that the upper half of the circle has the equation $y = \sqrt{81 - x^2}$ and the lower half is given by $y = -\sqrt{81 - x^2}$, as illustrated in Figure 16(a) and (b).

FIGURE 16

(a) $y = \sqrt{81 - x^2}$

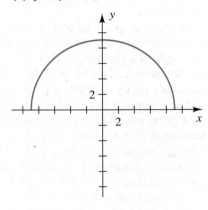

(b) $y = -\sqrt{81 - x^2}$

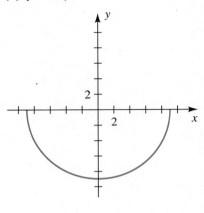

(c) $x = \sqrt{81 - y^2}$

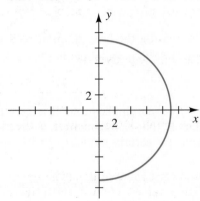

(d) $x = -\sqrt{81 - y^2}$

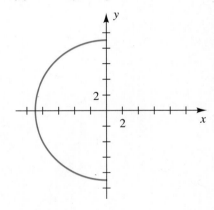

Similarly, to find equations for the right and left halves, we solve $x^2 + y^2 = 81$ for x in terms of y, obtaining

$$x = \pm\sqrt{81 - y^2}.$$

Since $\sqrt{81 - y^2} \geq 0$, it follows that the right half of the circle has the equation $x = \sqrt{81 - y^2}$, and the left half is given by $x = -\sqrt{81 - y^2}$, as illustrated in Figure 16(c) and (d).

Engineers, scientists, and mathematicians often use graphics calculators or computer software to obtain sketches of graphs. These devices have the capability to *zoom in* (or *zoom out*) on any part of a graph, allowing the *user* to approximate x-intercepts, y-intercepts, high points, low points, and other important aspects of the graph. Because of the great variety of graphics calculators and computer software, we shall not discuss methods of using any specific type. User's manuals usually give adequate instructions.

The term **graphing utility** refers to either a graphics calculator or a computer equipped with appropriate software packages. The **viewing rectangle** of a graphing utility is simply the portion of the xy-plane shown on the screen. The boundaries (sides) of the viewing rectangle can be manually set by assigning a minimum x value (Xmin), a maximum x value (Xmax), a minimum y value (Ymin), and a maximum y value (Ymax). In examples, we often use the standard (or default) values for the viewing rectangle. If we want a different view of the graph, we use the phrase "using [Xmin, Xmax] by [Ymin, Ymax]" to indicate the change in the viewing rectangle. The standard values of the viewing rectangle depend on the dimensions (in pixels) of the graphing utility screen.

In many applications it is essential to find the points at which the graphs of two equations in x and y intersect. To approximate such points of intersection with a graphing utility, it is usually necessary to solve each equation for y in terms of x. For example, suppose one equation is

$$4x^2 - 3x + 2y + 6 = 0.$$

Solving for y gives us

$$y = \frac{-4x^2 + 3x - 6}{2} = -2x^2 + \frac{3}{2}x - 3.$$

The graph of the equation is then found by *making the assignment*

$$Y_1 = -2x^2 + \tfrac{3}{2}x - 3$$

in the graphing utility. (The symbol Y_1 indicates the *first* equation, or the first y value.) We also solve the second equation for y in terms of x and make the assignment

$$Y_2 = \text{an expression in } x.$$

Pressing appropriate keys gives us sketches of the graphs, which we refer to as the graphs of Y_1 and Y_2. We then use the zoom-in and *tracing* features of the graphing utility to estimate the coordinates of the points of intersection. When estimating coordinates in examples, we usually find one-decimal-place approximations unless otherwise specified.

In the next example we demonstrate this technique for the graphs discussed previously in Examples 1 and 2.

EXAMPLE 9   *Estimating points of intersection of graphs*

Use a graphing utility to estimate the points of intersection of the graphs of $y = 2x - 1$ and $y = x^2 - 3$.

SOLUTION   We first make the assignments

$$Y_1 = 2x - 1 \qquad \text{and} \qquad Y_2 = x^2 - 3.$$

FIGURE 17 $[-15, 15]$ by $[-10, 10]$

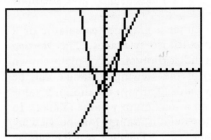

Using a standard viewing rectangle, $[-15, 15]$ by $[-10, 10]$, we see from the graphs of Y_1 and Y_2 in Figure 17 that there are two points of intersection: P_1 in quadrant I and P_2 in quadrant III.

We next either manually move the cursor or use a tracing feature (consult your user's manual for specific directions) to get close to P_1. Using the zoom-in feature, we estimate the coordinates of P_1 as (2.7, 4.5).

We can now either zoom out or redraw the original graphs to view P_2. Using the tracing and zoom-in features, we obtain $(-0.7, -2.5)$ as approximate coordinates of P_2.

EXAMPLE 10   *Estimating points of intersection of graphs*

Use a graphing utility to estimate the points of intersection of the circles $x^2 + y^2 = 25$ and $x^2 + y^2 - 4y = 12$.

SOLUTION   We solve the first equation for y in terms of x, as follows:

$$x^2 + y^2 = 25 \qquad \text{given}$$
$$y^2 = 25 - x^2 \qquad \text{subtract } x^2$$
$$y = \pm\sqrt{25 - x^2} \qquad \text{take the square root}$$

Note that the last equation represents *two* sets of y values: $y = \sqrt{25 - x^2}$ (the upper half of the circle) and $y = -\sqrt{25 - x^2}$ (the lower half of the circle). To display the entire circle on the graphing utility, we make the following assignments:

$$Y_1 = \sqrt{25 - x^2} \qquad \text{and} \qquad Y_2 = -Y_1$$

(We often assign Y_2 in terms of Y_1 to avoid repetitive key stroking, thereby speeding up the graphing process and possibly eliminating errors.)

At this point you should graph Y_1 and Y_2 using the standard viewing rectangle. If the circle has an oval shape, consult your user's manual to find out which dimensions for the viewing rectangle will yield a more circular shape.

We may regard the equation of the second circle as a quadratic $ay^2 + by + c = 0$ in y by rearranging terms as follows:

$$y^2 - 4y + (x^2 - 12) = 0$$

Applying the quadratic formula with $a = 1$, $b = -4$, and $c = x^2 - 12$ ($x^2 - 12$ is considered to be the constant term, since it does not contain the variable y) gives us

$$y = \frac{-(-4) \pm \sqrt{(-4)^2 - 4(1)(x^2 - 12)}}{2(1)}$$

$$= \frac{4 \pm \sqrt{16 - 4(x^2 - 12)}}{2}$$

$$= \frac{4 \pm 2\sqrt{4 - (x^2 - 12)}}{2} = 2 \pm \sqrt{16 - x^2}.$$

(It is unnecessary to simplify the equation as much as we have, but the simplified form is easier to enter in a graphing utility.)

We now make the assignments

$$Y_3 = \sqrt{16 - x^2}, \qquad Y_4 = 2 + Y_3, \qquad \text{and} \qquad Y_5 = 2 - Y_3.$$

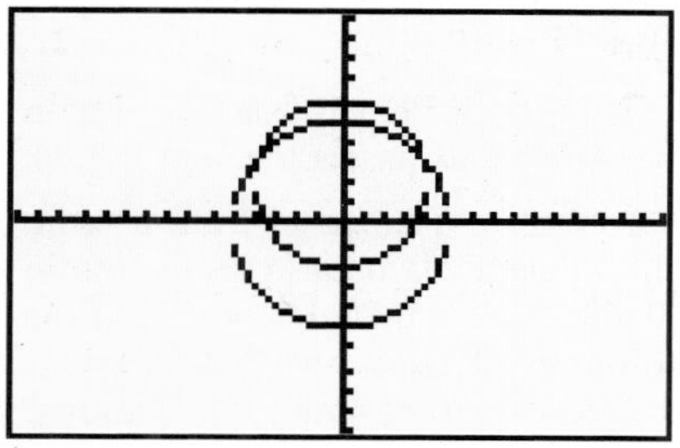

(If a Y_5 is not available on your graphing utility, consult the user's manual for information about displaying additional graphs on the screen.) We then select Y_1, Y_2, Y_4, and Y_5 to be graphed, obtaining a display similar to Figure 18. There are two points of intersection. Zooming in on the point in the first quadrant, we estimate its coordinates as (3.8, 3.25). Since both circles are symmetric with respect to the y-axis, the other point of intersection is approximately $(-3.8, 3.25)$.

FIGURE 18 $[-15, 15]$ by $[-10, 10]$

It should be noted that the approximate solutions found in Examples 9 and 10 do not satisfy the given equations because of the inaccuracy of the estimates made from the graph. In Chapter 6 we will discuss how to find the *exact* values for the points of intersection.

3.2 EXERCISES

Exer. 1–20: Sketch the graph of the equation, and label the x- and y-intercepts.

1 $y = 2x - 3$

2 $y = 3x + 2$

3 $y = -x + 1$

4 $y = -2x - 3$

5 $y = -4x^2$

6 $y = \frac{1}{3}x^2$

7 $y = 2x^2 - 1$

8 $y = -x^2 + 2$

9 $x = \frac{1}{4}y^2$

10 $x = -2y^2$

11 $x = -y^2 + 3$

12 $x = 2y^2 - 4$

13 $y = -\frac{1}{2}x^3$

14 $y = \frac{1}{2}x^3$

15 $y = x^3 - 8$

16 $y = -x^3 + 1$

17 $y = \sqrt{x}$

18 $y = \sqrt{-x}$

19 $y = \sqrt{x} - 4$

20 $y = \sqrt{x - 4}$

Exer. 21–22: Use tests for symmetry to determine which graphs in the indicated exercises are symmetric with respect to (a) the y-axis, (b) the x-axis, and (c) the origin.

21 The odd-numbered exercises in 1–20

22 The even-numbered exercises in 1–20

Exer. 23–34: Sketch the graph of the circle or semicircle.

23 $x^2 + y^2 = 11$

24 $x^2 + y^2 = 7$

25 $(x + 3)^2 + (y - 2)^2 = 9$

26 $(x - 4)^2 + (y + 2)^2 = 4$

27 $(x + 3)^2 + y^2 = 16$

28 $x^2 + (y - 2)^2 = 25$

29 $4x^2 + 4y^2 = 25$

30 $9x^2 + 9y^2 = 1$

31 $y = -\sqrt{16 - x^2}$

32 $y = \sqrt{4 - x^2}$

33 $x = \sqrt{9 - y^2}$

34 $x = -\sqrt{25 - y^2}$

Exer. 35–46: Find an equation of the circle that satisfies the stated conditions.

35 Center $C(2, -3)$, radius 5

36 Center $C(-4, 1)$, radius 3

37 Center $C(\frac{1}{4}, 0)$, radius $\sqrt{5}$

38 Center $C(\frac{3}{4}, -\frac{2}{3})$, radius $3\sqrt{2}$

39 Center $C(-4, 6)$, passing through $P(1, 2)$

40 Center at the origin, passing through $P(4, -7)$

41 Center $C(-3, 6)$, tangent to the y-axis

42 Center $C(4, -1)$, tangent to the x-axis

43 Tangent to both axes, center in the second quadrant, radius 4

44 Tangent to both axes, center in the fourth quadrant, radius 3

45 End points of a diameter $A(4, -3)$ and $B(-2, 7)$

46 End points of a diameter $A(-5, 2)$ and $B(3, 6)$

Exer. 47–56: Find the center and radius of the circle with the given equation.

47 $x^2 + y^2 - 4x + 6y - 36 = 0$

48 $x^2 + y^2 + 8x - 10y + 37 = 0$

49 $x^2 + y^2 + 4y - 117 = 0$

50 $x^2 + y^2 - 10x + 18 = 0$

51 $2x^2 + 2y^2 - 12x + 4y - 15 = 0$

52 $9x^2 + 9y^2 + 12x - 6y + 4 = 0$

53 $x^2 + y^2 + 4x - 2y + 5 = 0$

54 $x^2 + y^2 - 6x + 4y + 13 = 0$

55 $x^2 + y^2 - 2x - 8y + 19 = 0$

56 $x^2 + y^2 + 4x + 6y + 16 = 0$

Exer. 57–60: Find equations for the upper half, lower half, right half, and left half of the circle.

57 $x^2 + y^2 = 36$

58 $(x + 3)^2 + y^2 = 64$

59 $(x - 2)^2 + (y + 1)^2 = 49$

60 $(x - 3)^2 + (y - 5)^2 = 4$

Exer. 61–62: Determine whether the point P is inside, outside, or on the circle with center C and radius r.

61 (a) $P(2, 3)$,　　$C(4, 6)$,　　$r = 4$

　　(b) $P(4, 2)$,　　$C(1, -2)$,　　$r = 5$

　　(c) $P(-3, 5)$,　$C(2, 1)$,　　$r = 6$

62 (a) $P(3, 8)$,　　$C(-2, -4)$,　$r = 13$

　　(b) $P(-2, 5)$,　$C(3, 7)$,　　$r = 6$

　　(c) $P(1, -2)$,　$C(6, -7)$,　$r = 7$

Exer. 63–64: For the given circle, find (a) the x-intercepts and (b) the y-intercepts.

63 $x^2 + y^2 - 4x - 6y + 4 = 0$

64 $x^2 + y^2 - 10x + 4y + 13 = 0$

65 Find an equation of the circle that is concentric with $x^2 + y^2 + 4x - 6y + 4 = 0$ and passes through $P(2, 6)$.

66 *Radio broadcasting ranges* The signal from a radio station has a circular range of 50 miles. A second radio station, located 100 miles east and 80 miles north of the first station, has a range of 80 miles. Are there locations where signals can be received from both radio stations? Explain your answer.

Exer. 67–70: The figure represents the screen of a graphing utility, where Y_1 and Y_2 are y-value assignments for two equations in x. Express, in interval form, the x-values such that $Y_1 < Y_2$ for the indicated viewing rectangle. Assume that the x- and y-values of each point of intersection are integers.

67

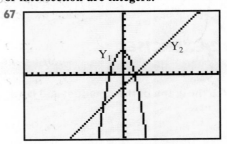

68

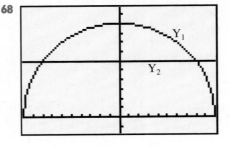

69

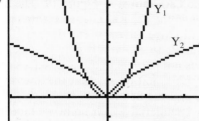

70

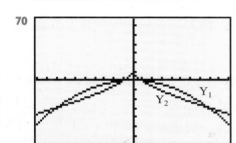

Exer. 71–72: Graph the equation, and estimate the x-intercepts.

71 $y = x^3 - \frac{9}{10}x^2 - \frac{43}{25}x + \frac{24}{25}$

72 $y = x^4 + 0.85x^3 - 2.46x^2 - 1.07x + 0.51$

Exer. 73–76: Graph the two equations on the same coordinate plane, and estimate the coordinates of their points of intersection.

73 $y = x^3 + x;$ $\qquad\qquad x^2 + y^2 = 1$

74 $y = 3x^4 - \frac{3}{2};$ $\qquad\qquad x^2 + y^2 = 1$

75 $x^2 + (y - 1)^2 = 1;$ $\qquad (x - \frac{5}{4})^2 + y^2 = 1$

76 $(x + 1)^2 + (y - 1)^2 = \frac{1}{4};$ $(x + \frac{1}{2})^2 + (y - \frac{1}{2})^2 = 1$

3.3 LINES

One of the basic concepts in geometry is that of a *line*. In this section we will restrict our discussion to lines that lie in a coordinate plane. This will allow us to use algebraic methods to study their properties. Two of our principal objectives may be stated as follows:

1. Given a line l in a coordinate plane, find an equation whose graph corresponds to l.

2. Given an equation of a line l in a coordinate plane, sketch the graph of the equation.

The following concept is fundamental to the study of lines.

Definition of Slope of a Line

Let l be a line that is not parallel to the y-axis, and let $P_1(x_1, y_1)$ and $P_2(x_2, y_2)$ be distinct points on l. The **slope m** of l is

$$m = \frac{y_2 - y_1}{x_2 - x_1}.$$

If l is parallel to the y-axis, then the slope of l is not defined.

FIGURE 19

(a) Positive slope (line rises)

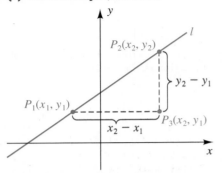

(b) Negative slope (line falls)

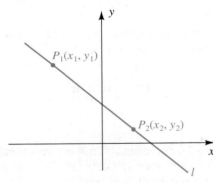

Typical points P_1 and P_2 on a line l are shown in Figure 19. The numerator $y_2 - y_1$ in the formula for m is the vertical change in direction from P_1 to P_2 and may be positive, negative, or zero. The denominator $x_2 - x_1$ is the horizontal change from P_1 to P_2, and it may be positive or negative, but never zero, because l is not parallel to the y-axis if a slope exists. In Figure 19(a) the slope is positive, and we say that the line *rises*. In Figure 19(b) the slope is negative, and the line *falls*.

In finding the slope of a line it is immaterial which point we label as P_1 and which as P_2, since

$$\frac{y_2 - y_1}{x_2 - x_1} = \frac{y_2 - y_1}{x_2 - x_1} \cdot \frac{(-1)}{(-1)} = \frac{y_1 - y_2}{x_1 - x_2}.$$

If the points are labeled so that $x_1 < x_2$, as in Figure 19, then $x_2 - x_1 > 0$, and hence the slope is positive, negative, or zero, depending on whether $y_2 > y_1$, $y_2 < y_1$, or $y_2 = y_1$.

The definition of slope is independent of the two points that are chosen on l. If other points $P_1'(x_1', y_1')$ and $P_2'(x_2', y_2')$ are used, then, as in Figure 20, the triangle with vertices P_1', P_2', and $P_3'(x_2', y_1')$ is similar to the triangle with vertices P_1, P_2, and $P_3(x_2, y_1)$. Since the ratios of corresponding sides of similar triangles are equal,

$$\frac{y_2 - y_1}{x_2 - x_1} = \frac{y_2' - y_1'}{x_2' - x_1'}.$$

FIGURE 20

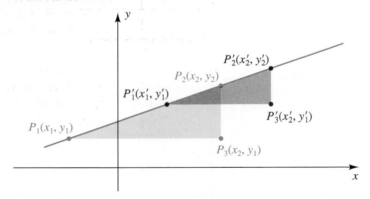

EXAMPLE 1   *Finding slopes*

Sketch the line through each pair of points, and find its slope:

(a) $A(-1, 4)$ and $B(3, 2)$ **(b)** $A(2, 5)$ and $B(-2, -1)$

(c) $A(4, 3)$ and $B(-2, 3)$ **(d)** $A(4, -1)$ and $B(4, 4)$

SOLUTION   The lines are sketched in Figure 21. We use the definition of slope to find the slope of each line.

FIGURE 21

(a) $m = -\frac{1}{2}$

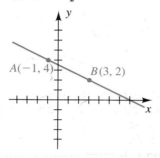

$A(-1, 4)$ $B(3, 2)$

(b) $m = \frac{3}{2}$

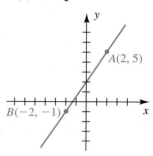

$A(2, 5)$

$B(-2, -1)$

(c) $m = 0$

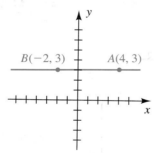

$B(-2, 3)$ $A(4, 3)$

(d) m undefined

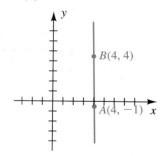

$B(4, 4)$

$A(4, -1)$

(a)
$$m = \frac{2 - 4}{3 - (-1)} = \frac{-2}{4} = -\frac{1}{2}$$

(b)
$$m = \frac{5 - (-1)}{2 - (-2)} = \frac{6}{4} = \frac{3}{2}$$

(c)
$$m = \frac{3 - 3}{-2 - 4} = \frac{0}{-6} = 0$$

(d) The slope is undefined because the line is parallel to the y-axis. Note that if the formula for m is used, the denominator is zero.

FIGURE 22

(a) $m = \frac{5}{3}$

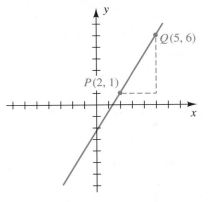

(b) $m = -\frac{5}{3}$

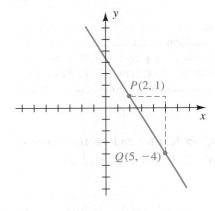

EXAMPLE 2 *Sketching a line with a given slope*

Sketch a line through $P(2, 1)$ that has

(a) slope $\frac{5}{3}$ **(b)** slope $-\frac{5}{3}$

SOLUTION   If the slope of a line is a/b and b is positive, then for every change of b units in the horizontal direction, the line rises or falls $|a|$ units, depending on whether a is positive or negative.

(a) If $P(2, 1)$ is on the line and $m = \frac{5}{3}$, we can obtain another point on the line by starting at P and moving 3 units to the right and 5 units upward. This gives us the point $Q(5, 6)$, and the line is determined as in Figure 22(a).

(b) If $P(2, 1)$ is on the line and $m = -\frac{5}{3}$, we move 3 units to the right and 5 units downward, obtaining the line through $Q(5, -4)$, as in Figure 22(b).

The diagram in Figure 23 indicates the slopes of several lines through the origin. The line that lies on the x-axis has slope $m = 0$. If this line is rotated about O in the counterclockwise direction (as indicated by the blue arrow), the slope is positive and increases, reaching the value 1 when the line bisects the first quadrant and continuing to increase as the line gets closer to the y-axis. If we rotate the line of slope $m = 0$ in the *clockwise* direction (as indicated by the red arrow), the slope is negative, reaching the value -1 when the line bisects the second quadrant and becoming large and negative as the line gets closer to the y-axis.

FIGURE 23

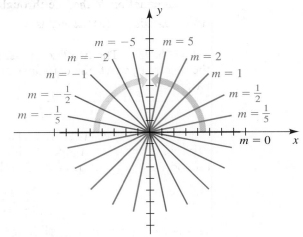

Lines that are horizontal or vertical have simple equations, as indicated in the following chart.

Terminology	Definition	Graph	Equation
Horizontal line	A line parallel to the x-axis	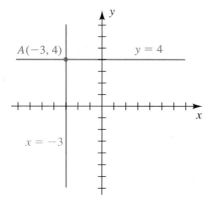	$y = b$
Vertical line	A line parallel to the y-axis		$x = a$

A common error is to regard the graph of $y = b$ as consisting of only the one point $(0, b)$. If we express the equation in the form $0 \cdot x + y = b$, we see that the value of x is immaterial; thus, the graph of $y = b$ consists of the points (x, b) for every x and hence is a horizontal line. Similarly, the graph of $x = a$ is the vertical line consisting of all points (a, y), where y is a real number.

FIGURE 24

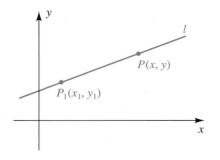

EXAMPLE 3 *Finding equations of horizontal and vertical lines*

Find an equation of the line through $A(-3, 4)$ that is parallel to

(a) the x-axis **(b)** the y-axis

SOLUTION The two lines are sketched in Figure 24. As indicated in the preceding chart, the equations are $y = 4$ for part (a) and $x = -3$ for part (b).

FIGURE 25

Let us next find an equation of a line l through a point $P_1(x_1, y_1)$ with slope m. If $P(x, y)$ is any point with $x \neq x_1$ (see Figure 25), then P is on l if and only if the slope of the line through P_1 and P is m—that is, if

$$\frac{y - y_1}{x - x_1} = m.$$

This equation may be written in the form

$$y - y_1 = m(x - x_1).$$

Note that (x_1, y_1) is a solution of the last equation, and hence the points on l are precisely the points that correspond to the solutions. This equation for l is referred to as the **point-slope form**.

**Point-Slope Form for
the Equation of a Line**

An equation for the line through the point (x_1, y_1) with slope m is

$$y - y_1 = m(x - x_1).$$

The point-slope form is only one possibility for an equation of a line. There are an unlimited number of equivalent equations. We sometimes simplify the equation obtained using the point-slope form to either

$$ax + by = c \quad \text{or} \quad ax + by + d = 0,$$

where a, b, and c are integers with no common factor, $a > 0$, and $d = -c$.

EXAMPLE 4   *Finding an equation of a line through two points*
Find an equation of the line through $A(1, 7)$ and $B(-3, 2)$.

FIGURE 26

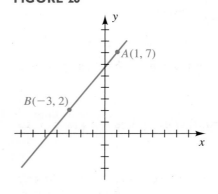

SOLUTION   The line is sketched in Figure 26. The formula for the slope m gives us

$$m = \frac{7 - 2}{1 - (-3)} = \frac{5}{4}.$$

We may use the coordinates of either A or B for (x_1, y_1) in the point-slope form. Using $A(1, 7)$ gives us

$$y - 7 = \tfrac{5}{4}(x - 1) \quad \text{point-slope form}$$
$$4(y - 7) = 5(x - 1) \quad \text{multiply by 4}$$
$$4y - 28 = 5x - 5 \quad \text{multiply factors}$$
$$-5x + 4y = 23 \quad \text{subtract } 5x \text{ and add 28}$$
$$5x - 4y = -23 \quad \text{multiply by } -1$$

The last equation is one of the desired forms for an equation of a line. Another is $5x - 4y + 23 = 0$.

FIGURE 27

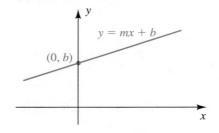

The point-slope form for the equation of a line may be rewritten as $y = mx - mx_1 + y_1$, which is of the form

$$y = mx + b$$

with $b = -mx_1 + y_1$. The real number b is the y-intercept of the graph, as indicated in Figure 27. Since the equation $y = mx + b$ displays the slope m and y-intercept b of l, it is called the **slope-intercept form** for the equation of a line. Conversely, if we start with $y = mx + b$, we may write

$$y - b = m(x - 0).$$

Comparing this equation with the point-slope form, we see that the graph is a line with slope m and passing through the point $(0, b)$. This gives us the next result.

Slope-Intercept Form for the Equation of a Line

The graph of $y = mx + b$ is a line having slope m and y-intercept b.

EXAMPLE 5 *Expressing an equation in slope-intercept form*

Express the equation $2x - 5y = 8$ in slope-intercept form.

SOLUTION Our goal is to solve the given equation for y to obtain the form $y = mx + b$. We may proceed as follows:

$$2x - 5y = 8 \qquad \text{given}$$
$$-5y = -2x + 8 \qquad \text{subtract } 2x$$
$$y = \left(\frac{-2}{-5}\right)x + \left(\frac{8}{-5}\right) \qquad \text{divide by } -5$$
$$y = \tfrac{2}{5}x + (-\tfrac{8}{5}) \qquad \text{equivalent equation}$$

The last equation is the slope-intercept form $y = mx + b$ with slope $m = \tfrac{2}{5}$ and y-intercept $b = -\tfrac{8}{5}$.

It follows from the point-slope form that every line is a graph of an equation

$$ax + by = c,$$

where a, b, and c are real numbers and a and b are not both zero. We call such an equation a **linear equation** in x and y. Let us show, conversely, that the graph of $ax + by = c$, with a and b not both zero, is always a line. If $b \neq 0$, we may solve for y, obtaining

$$y = \left(-\frac{a}{b}\right)x + \frac{c}{b},$$

which, by the slope-intercept form, is an equation of a line with slope $-a/b$ and y-intercept c/b. If $b = 0$ but $a \neq 0$, we may solve for x, obtaining $x = c/a$, which is the equation of a vertical line with x-intercept c/a. This discussion establishes the following result.

General Form for the Equation of a Line

The graph of a linear equation $ax + by = c$ is a line, and conversely, every line is the graph of a linear equation.

For simplicity, we use the terminology *the line $ax + by = c$* rather than *the line with equation $ax + by = c$*.

EXAMPLE 6   *Sketching the graph of a linear equation*

Sketch the graph of $2x - 5y = 8$.

FIGURE 28

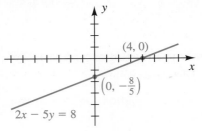

SOLUTION   We know from the preceding discussion that the graph is a line, so it is sufficient to find two points on the graph. Let us find the *x*- and *y*-intercepts by substituting $y = 0$ and $x = 0$, respectively, in $2x - 5y = 8$.

> *x-intercept:*   If $y = 0$, then $2x = 8$, or $x = 4$.
>
> *y-intercept:*   If $x = 0$, then $-5y = 8$, or $y = -\frac{8}{5}$.

Plotting the intercepts $(4, 0)$ and $(0, -\frac{8}{5})$ and drawing a line through them gives us the graph in Figure 28.

The following theorem specifies the relationship between parallel lines (lines in a plane that do not intersect) and slope.

Theorem on Slopes of Parallel Lines

> Two nonvertical lines are parallel if and only if they have the same slope.

FIGURE 29

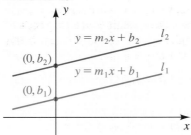

PROOF   Let l_1 and l_2 be distinct lines of slopes m_1 and m_2, respectively. If the *y*-intercepts are b_1 and b_2 (see Figure 29), then, by the slope-intercept form, the lines have equations

$$y = m_1 x + b_1 \quad \text{and} \quad y = m_2 x + b_2.$$

The lines intersect at some point (x, y) if and only if the values of y are equal for some x—that is, if

$$m_1 x + b_1 = m_2 x + b_2,$$

or

$$(m_1 - m_2)x = b_2 - b_1.$$

The last equation can be solved for x if and only if $m_1 - m_2 \neq 0$. We have shown that the lines l_1 and l_2 intersect if and only if $m_1 \neq m_2$. Hence they do *not* intersect (are parallel) if and only if $m_1 = m_2$. ■

EXAMPLE 7   *Finding an equation of a line parallel to a given line*

Find an equation of the line through $P(5, -7)$ that is parallel to the line $6x + 3y = 4$.

SOLUTION   We first express the given equation in slope-intercept form:

$$6x + 3y = 4 \qquad \text{given}$$

$$3y = -6x + 4 \qquad \text{subtract } 6x$$

$$y = -2x + \tfrac{4}{3} \qquad \text{divide by 3}$$

FIGURE 30

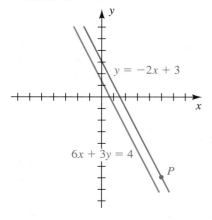

$y = -2x + 3$

$6x + 3y = 4$

P

The last equation is in slope-intercept form $y = mx + b$ with slope $m = -2$ and y-intercept $\frac{4}{3}$. Since parallel lines have the same slope, the required line also has slope -2. Using the point $P(5, -7)$ gives us the following:

$$y - (-7) = -2(x - 5) \quad \text{point-slope form}$$
$$y + 7 = -2x + 10 \quad \text{simplify}$$
$$y = -2x + 3 \quad \text{subtract 7}$$

The last equation is in slope-intercept form and shows that the parallel line we have found has y-intercept 3. This line and the given line are sketched in Figure 30.

The next theorem gives us information about **perpendicular lines** (lines that intersect at a right angle).

Theorem on Slopes of Perpendicular Lines

Two lines with slope m_1 and m_2 are perpendicular if and only if
$$m_1 m_2 = -1.$$

FIGURE 31

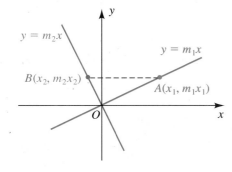

$y = m_2 x$

$y = m_1 x$

$B(x_2, m_2 x_2)$

$A(x_1, m_1 x_1)$

O

PROOF For simplicity, let us consider the special case of two lines that intersect at the origin O, as illustrated in Figure 31. Equations of these lines are $y = m_1 x$ and $y = m_2 x$. If, as in the figure, we choose points $A(x_1, m_1 x_1)$ and $B(x_2, m_2 x_2)$ different from O on the lines, then the lines are perpendicular if and only if angle AOB is a right angle. Applying the Pythagorean theorem, we know that angle AOB is a right angle if and only if

$$[d(A, B)]^2 = [d(O, B)]^2 + [d(O, A)]^2$$

or, by the distance formula,

$$(x_2 - x_1)^2 + (m_2 x_2 - m_1 x_1)^2 = x_2^2 + (m_2 x_2)^2 + x_1^2 + (m_1 x_1)^2.$$

Squaring terms, simplifying, and factoring gives us

$$-2m_1 m_2 x_1 x_2 - 2x_1 x_2 = 0$$
$$-2x_1 x_2 (m_1 m_2 + 1) = 0.$$

Since both x_1 and x_2 are not zero, we may divide both sides by $-2x_1 x_2$, obtaining $m_1 m_2 + 1 = 0$. Thus, the lines are perpendicular if and only if $m_1 m_2 = -1$.

The same type of proof may be given if the lines intersect at *any* point (a, b). ∎

A convenient way to remember the conditions on slopes of perpendicular lines is to note that m_1 and m_2 must be *negative reciprocals* of one another—that is, $m_1 = -1/m_2$ and $m_2 = -1/m_1$.

FIGURE 32

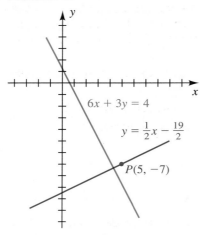

$6x + 3y = 4$

$y = \frac{1}{2}x - \frac{19}{2}$

$P(5, -7)$

EXAMPLE 8 *Finding an equation of a line perpendicular to a given line*

Find the slope-intercept form for the line through $P(5, -7)$ that is perpendicular to the line $6x + 3y = 4$.

SOLUTION   We considered the line $6x + 3y = 4$ in Example 7 and found that its slope is -2. Hence, the slope of the required line is the negative reciprocal $-[1/(-2)]$, or $\frac{1}{2}$. Using $P(5, -7)$ gives us the following:

$$y - (-7) = \tfrac{1}{2}(x - 5) \quad \text{point-slope form}$$

$$y + 7 = \tfrac{1}{2}x - \tfrac{5}{2} \quad \text{simplify}$$

$$y = \tfrac{1}{2}x - \tfrac{19}{2} \quad \text{subtract 7}$$

The last equation is in slope-intercept form and shows that the perpendicular line has y-intercept $-\frac{19}{2}$. This line and the given line are sketched in Figure 32.

EXAMPLE 9 *Finding an equation of a perpendicular bisector*

Given $A(-3, 1)$ and $B(5, 4)$, find an equation of the perpendicular bisector of the line segment AB.

SOLUTION   The line segment AB and its perpendicular bisector l are shown in Figure 33, where M is the midpoint of AB. We calculate the following:

FIGURE 33

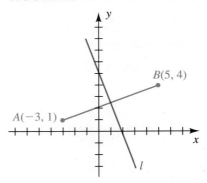

$B(5, 4)$

$A(-3, 1)$

l

Coordinates of M: $\left(\dfrac{-3 + 5}{2}, \dfrac{1 + 4}{2}\right) = \left(1, \dfrac{5}{2}\right)$ midpoint formula

Slope of AB: $\dfrac{4 - 1}{5 - (-3)} = \dfrac{3}{8}$ slope formula

Slope of l: $-\dfrac{1}{\frac{3}{8}} = -\dfrac{8}{3}$ negative reciprocal of $\dfrac{3}{8}$

Using the point $M(1, \frac{5}{2})$ and slope $-\frac{8}{3}$ gives us the following equivalent equations for l:

$$y - \tfrac{5}{2} = -\tfrac{8}{3}(x - 1) \quad \text{point-slope form}$$

$$6y - 15 = -16(x - 1) \quad \text{multiply by the lcd, 6}$$

$$6y - 15 = -16x + 16 \quad \text{multiply}$$

$$16x + 6y = 31 \quad \text{add } 16x \text{ and } 15$$

Two variables x and y are **linearly related** if $y = ax + b$, where a and b are real numbers and $a \neq 0$. Linear relationships between variables occur frequently in applied problems. The following example gives one illustration.

EXAMPLE 10 *Relating air temperature to altitude*

The relationship between the air temperature T (in °F) and the altitude h (in feet above sea level) is approximately linear for $0 \leq h \leq 20{,}000$. If the temperature at sea level is 60°, an increase of 5000 feet in altitude lowers the air temperature about 18°.

(a) Express T in terms of h, and sketch the graph on an hT-coordinate system.

(b) Approximate the air temperature at an altitude of 15,000 feet.

(c) Approximate the altitude at which the temperature is 0°.

SOLUTION

(a) If T is linearly related to h, then

$$T = ah + b$$

for some constants a and b. Referring to the given data, we note that $T = 60°$ when $h = 0$ ft (sea level). Hence, the T-intercept is 60, and the temperature T at any altitude h is

$$T = ah + 60.$$

Again referring to the given data, we note that when $h = 5000$ ft, the temperature $T = 60° - 18° = 42°$. Hence

$$42 = a(5000) + 60 \qquad \text{let } T = 42 \text{ and } h = 5000$$

$$-18 = 5000a \qquad \text{subtract 60}$$

$$a = -\tfrac{18}{5000} = -\tfrac{9}{2500} \qquad \text{divide by 5000}$$

Substituting for a in $T = ah + 60$ gives us the following formula for T:

$$T = -\tfrac{9}{2500}h + 60$$

The graph is sketched in Figure 34, with different scales on the axes.

(b) Using the last formula for T obtained in part (a), we find that the temperature (in °F) when $h = 15{,}000$ is

$$T = -\tfrac{9}{2500}(15{,}000) + 60 = -54 + 60 = 6.$$

(c) To find the altitude h that corresponds to $T = 0°$, we proceed as follows:

$$T = -\tfrac{9}{2500}h + 60 \qquad \text{from part (a)}$$

$$0 = -\tfrac{9}{2500}h + 60 \qquad \text{let } T = 0$$

$$\tfrac{9}{2500}h = 60 \qquad \text{add } \tfrac{9}{2500}h$$

$$h = 60 \cdot \tfrac{2500}{9} \qquad \text{multiply by } \tfrac{2500}{9}$$

$$h = \frac{50{,}000}{3} \approx 16{,}667 \text{ ft} \qquad \text{simplify and approximate}$$

FIGURE 34

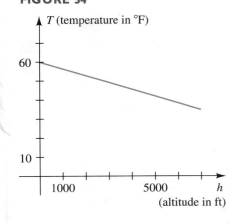

In Chapter 6 we will discuss algebraic methods for finding the point of intersection of two nonparallel lines. In some applications, numbers that occur in equations of lines are irrational or involve approximate data. In such cases it may be sufficient to estimate the coordinates of the point using a graphing utility, as in the next example.

EXAMPLE 11   *Estimating the point of intersection of two lines*

Suppose the following lines were obtained using approximate data:

$$1.018x + 0.230y = 0.447$$
$$1.847x + 4.538y = 1.414$$

Use a graphing utility to estimate the coordinates of the point of intersection to two decimal places.

FIGURE 35 $[-15, 15]$ by $[-10, 10]$

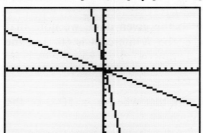

SOLUTION   Solving each equation for y, we obtain

$$y = (0.447 - 1.018x)/0.230,$$
$$y = (1.414 - 1.847x)/4.538.$$

We next assign the preceding expressions for y to Y_1 and Y_2, respectively. A display of the graph (similar to Figure 35) shows that there is a point of intersection near the origin. Using the tracing and zoom-in features yields the approximate solution $(0.41, 0.15)$.

Problems similar to this one are often encountered in statistical analysis when the *method of least squares* is employed to find a *linear regression line*.

3.3 EXERCISES

Exer. 1–6: Sketch the line through A and B, and find its slope m.

1 $A(-3, 2)$, $B(5, -4)$

2 $A(4, -1)$, $B(-6, -3)$

3 $A(2, 5)$, $B(-7, 5)$

4 $A(5, -1)$, $B(5, 6)$

5 $A(-3, 2)$, $B(-3, 5)$

6 $A(4, -2)$, $B(-3, -2)$

Exer. 7–10: Use slopes to show that the points are vertices of the specified polygon.

7 $A(-3, 1)$, $B(5, 3)$, $C(3, 0)$, $D(-5, -2)$;   parallelogram

8 $A(2, 3)$, $B(5, -1)$, $C(0, -6)$, $D(-6, 2)$;   trapezoid

9 $A(6, 15)$, $B(11, 12)$, $C(-1, -8)$, $D(-6, -5)$;   rectangle

10 $A(1, 4)$, $B(6, -4)$, $C(-15, -6)$;   right triangle

11 If three consecutive vertices of a parallelogram are $A(-1, -3)$, $B(4, 2)$, and $C(-7, 5)$, find the fourth vertex.

12 Let $A(x_1, y_1)$, $B(x_2, y_2)$, $C(x_3, y_3)$, and $D(x_4, y_4)$ denote the vertices of an arbitrary quadrilateral. Show that the line segments joining midpoints of adjacent sides form a parallelogram.

Exer. 13–14: Sketch the graph of $y = mx$ for the given values of m.

13 $m = 3, -2, \frac{2}{3}, -\frac{1}{4}$ 14 $m = 5, -3, \frac{1}{2}, -\frac{1}{3}$

Exer. 15–16: Sketch the graph of the line through P for each value of m.

15 $P(3, 1)$; $m = \frac{1}{2}, -1, -\frac{1}{5}$

16 $P(-2, 4)$; $m = 1, -2, -\frac{1}{2}$

Exer. 17–18: Sketch the graphs of the lines on the same coordinate plane.

17 $y = x + 3$, $y = x + 1$, $y = -x + 1$

18 $y = -2x - 1$, $y = -2x + 3$, $y = \frac{1}{2}x + 3$

Exer. 19–30: Find a general form of an equation of the line through the point A that satisfies the given condition.

19 $A(5, -2)$

  (a) parallel to the y-axis
  (b) perpendicular to the y-axis

20 $A(-4, 2)$

  (a) parallel to the x-axis
  (b) perpendicular to the x-axis

21 $A(5, -3)$;   slope -4

22 $A(-1, 4)$;   slope $\frac{2}{3}$

23 $A(4, 0)$;   slope -3

24 $A(0, -2)$;   slope 5

25 $A(4, -5)$;   through $B(-3, 6)$

26 $A(-1, 6)$; x-intercept 5

27 $A(2, -4)$;   parallel to the line $5x - 2y = 4$

28 $A(-3, 5)$;   parallel to the line $x + 3y = 1$

29 $A(7, -3)$;   perpendicular to the line $2x - 5y = 8$

30 $A(4, 5)$;   perpendicular to the line $3x + 2y = 7$

Exer. 31–34: Find the slope-intercept form of the line that satisfies the given conditions.

31 x-intercept 4, y-intercept -3

32 x-intercept -5, y-intercept -1

33 Through $A(5, 2)$ and $B(-1, 4)$

34 Through $A(-2, 1)$ and $B(3, 7)$

Exer. 35–36: Find a general form of an equation for the perpendicular bisector of the segment AB.

35 $A(3, -1)$, $B(-2, 6)$ 36 $A(4, 2)$, $B(-2, 10)$

Exer. 37–38: Find an equation for the line that bisects the given quadrants.

37 II and IV   38 I and III

Exer. 39–42: Use the slope-intercept form to find the slope and y-intercept of the given line, and sketch its graph.

39 $2x = 15 - 3y$ 40 $7x = -4y - 8$

41 $4x - 3y = 9$ 42 $x - 5y = -15$

Exer. 43–44: Find an equation of the line shown in the figure.

43 (a)

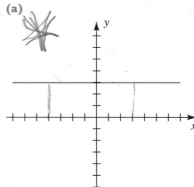

(b)

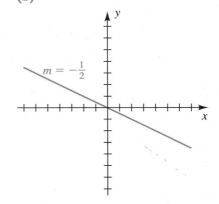

$m = -\frac{1}{2}$

(c)

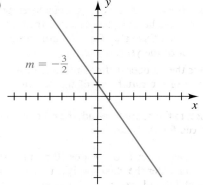

$m = -\dfrac{3}{2}$

(b)

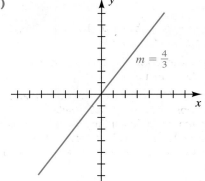

$m = \dfrac{4}{3}$

(d)

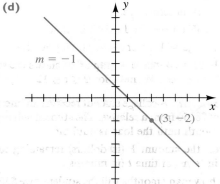

$m = -1$

$(3, -2)$

(c)

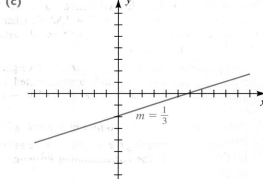

$m = \dfrac{1}{3}$

(d)

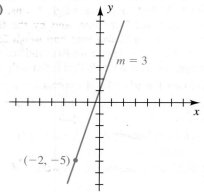

$m = 3$

$(-2, -5)$

44 (a)

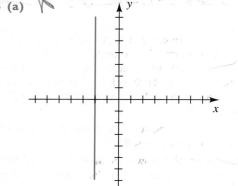

Exer. 45–46: If a line l has nonzero x- and y-intercepts a and b, respectively, then its *intercept form* is

$$\frac{x}{a} + \frac{y}{b} = 1.$$

Find the intercept form for the given line.

45 $4x - 2y = 6$ **46** $x - 3y = -2$

47 Find an equation of the circle that has center $C(3, -2)$ and is tangent to the line $y = 5$.

48 Find an equation of the line that is tangent to the circle $x^2 + y^2 = 25$ at the point $P(3, 4)$.

49 *Fetal growth*   The growth of a fetus more than 12 weeks old can be approximated by the formula $L = 1.53t - 6.7$, where L is the length in centimeters and t is the age in weeks. Prenatal length can be determined by ultrasound. Approximate the age of a fetus whose length is 28 centimeters.

50 *Estimating salinity*   Salinity of the ocean refers to the amount of dissolved material found in a sample of seawater. Salinity S can be estimated from the amount C of chlorine in seawater using $S = 0.03 + 1.805C$, where S and C are measured by weight in parts per thousand. Approximate C if S is 0.35.

51 *Weight of a humpback whale*   The expected weight W (in tons) of a humpback whale can be approximated from its length L (in feet) by the formula $W = 1.70L - 42.8$ for $30 \leq L \leq 50$.

  (a) Estimate the weight of a 40-foot humpback whale.

  (b) If the error in estimating the length could be as large as 2 feet, what is the corresponding error for the weight estimate?

52 *Growth of a blue whale*   Newborn blue whales are approximately 24 feet long and weigh 3 tons. Young whales are nursed for 7 months, and by the time of weaning they are often 53 feet long and weigh 23 tons. Let L and W denote the length (in feet) and the weight (in tons), respectively, of a whale that is t months of age.

  (a) If L and t are linearly related, express L in terms of t.

  (b) What is the daily increase in the length of a young whale? (Use 1 month = 30 days.)

  (c) If W and t are linearly related, express W in terms of t.

  (d) What is the daily increase in the weight of a young whale?

53 *Baseball stats*   Suppose a major league baseball player has hit 5 home runs in the first 14 games, and he keeps up this pace throughout the 162-game season.

  (a) Express the number y of home runs in terms of the number x of games played.

  (b) How many home runs will the player hit for the season?

54 *Cheese production*   A cheese manufacturer produces 18,000 pounds of cheese from January 1 through March 24. Suppose that this rate of production continues for the remainder of the year.

  (a) Express the number y of pounds of cheese produced in terms of the number x of the day in a 365-day year.

  (b) Predict, to the nearest pound, the number of pounds produced for the year.

55 *Childhood weight*   A baby weighs 10 pounds at birth, and three years later the child's weight is 30 pounds. Assume that childhood weight W (in pounds) is linearly related to age t (in years).

  (a) Express W in terms of t.

  (b) What is W on the child's sixth birthday?

  (c) At what age will the child weigh 70 pounds?

  (d) Sketch, on a tW-plane, a graph that shows the relationship between W and t for $0 \leq t \leq 12$.

56 *Loan repayment*   A college student receives an interest-free loan of $8250 from a relative. The student will repay $125 per month until the loan is paid off.

  (a) Express the amount P (in dollars) remaining to be paid in terms of time t (in months).

  (b) After how many months will the student owe $5000?

  (c) Sketch, on a tP-plane, a graph that shows the relationship between P and t for the duration of the loan.

57 *Vaporizing water*   The amount of heat H (in joules) required to convert one gram of water into vapor is linearly related to the temperature T (in °C) of the atmosphere. At 10 °C this conversion requires 2480 joules, and each increase in temperature of 15 °C lowers the amount of heat needed by 40 joules. Express H in terms of T.

58 *Aerobic power*   In exercise physiology, aerobic power P is defined in terms of maximum oxygen intake. For altitudes up to 1800 meters, aerobic power is optimal—that is, 100%. Beyond 1800 meters, P decreases linearly from the maximum of 100% to a value near 40% at 5000 meters.

  (a) Express aerobic power P in terms of altitude h (in meters) for $1800 \leq h \leq 5000$.

  (b) Estimate aerobic power in Mexico City (altitude: 2400 meters), the site of the 1968 Summer Olympic Games.

59 *Urban heat island* The urban heat island phenomenon has been observed in Tokyo. The average temperature was 13.5 °C in 1915, and since then has risen 0.032 °C per year.

(a) Assuming that temperature T (in °C) is linearly related to time t (in years) and that $t = 0$ corresponds to 1915, express T in terms of t.

(b) Predict the average temperature in the year 2000.

60 *Rising ground temperature* In 1870 the average ground temperature in Paris was 11.8 °C. Since then it has risen at a nearly constant rate, reaching 13.5 °C in 1969.

(a) Express the temperature T (in °C) in terms of time t (in years), where $t = 0$ corresponds to 1870 and $0 \le t \le 100$.

(b) During what year was the average ground temperature 12.5 °C?

61 *Business expenses* The owner of an ice cream franchise must pay the parent company $1000 per month plus 5% of the monthly revenue R. Operating cost of the franchise includes a fixed cost of $2600 per month for items such as utilities and labor. The cost of ice cream and supplies is 50% of the revenue.

(a) Express the owner's monthly expense E in terms of R.

(b) Express the monthly profit P in terms of R.

(c) Determine the monthly revenue needed to break even.

62 *Drug dosage* Pharmarcological products must specify recommended dosages for adults and children. Two formulas for modification of adult dosage levels for young children are

$$\text{Cowling's rule:} \quad y = \tfrac{1}{24}(t + 1)a$$

and $\quad$ Friend's rule: $\quad y = \tfrac{2}{25}ta$,

where a denotes adult dose (in milligrams) and t denotes the age of the child (in years).

(a) If $a = 100$, graph the two linear equations on the same coordinate plane for $0 \le t \le 12$.

(b) For what age do the two formulas specify the same dosage?

63 *Video game* In the video game shown in the figure, an airplane flies from left to right along the path given by $y = 1 + (1/x)$ and shoots bullets in the tangent direction at creatures placed along the x-axis at $x = 1, 2, 3, 4, 5$.

From calculus, the slope of the tangent line to the path at $P(1, 2)$ is $m = -1$ and at $Q(\tfrac{3}{2}, \tfrac{5}{3})$ is $m = -\tfrac{4}{9}$. Determine whether a creature will be hit if bullets are shot when the airplane is at

(a) P $\qquad\qquad\qquad$ (b) Q

EXERCISE 63

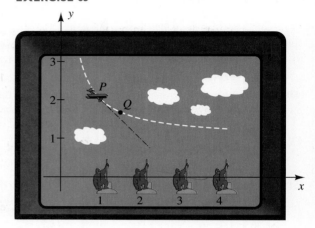

64 *Temperature scales* The relationship between the temperature reading F on the Fahrenheit scale and the temperature reading C on the Celsius scale is given by $C = \tfrac{5}{9}(F - 32)$.

(a) Find the temperature at which the reading is the same on both scales.

(b) When is the Fahrenheit reading twice the Celsius reading?

65 *Vertical wind shear* Vertical wind shear occurs when wind speed varies at different heights above the ground. Wind shear is of great importance to pilots during takeoffs and landings. If the wind speed is v_1 at height h_1 and v_2 at height h_2, then the average wind shear s is given by the slope formula

$$s = \frac{v_2 - v_1}{h_2 - h_1}.$$

If the wind speed at ground level is 22 mi/hr and s has been determined to be 0.07, find the wind speed 185 feet above the ground.

66 *Vertical wind shear* In the study of vertical wind shear, the formula

$$\frac{v_1}{v_2} = \left(\frac{h_1}{h_2}\right)^P$$

is sometimes used, where P is a variable that depends on the terrain and structures near ground level. In Montreal, average daytime values for P with north winds over 29 mi/hr were determined to be 0.13. If a 32 mi/hr north wind is measured 20 feet above the ground, approximate the average wind shear (see Exercise 65) between 20 feet and 200 feet.

 Exer. 67–68: Graph the lines on the same coordinate plane and estimate the coordinates of the points of intersection. Identify the polygon determined by the lines.

67 $2x - y = -1$; $\quad x + 2y = -2$; $\quad 3x + y = 11$

68 $10x - 42y = -7.14$; $\quad 8.4x + 2y = -3.8$;
$\quad 0.5x - 2.1y = 2.73$; $\quad 16.8x + 4y = 14$

Exer. 69–70: The given points were found using empirical methods. Determine whether they lie on the same line $y = ax + b$, and if so, find the values of a and b.

69 $A(-1.3, -1.3598)$, $\quad B(-0.55, -1.11905)$,
$\quad C(1.2, -0.5573)$, $\quad D(3.25, 0.10075)$

70 $A(-0.22, 1.6968)$, $\quad B(-0.12, 1.6528)$,
$\quad C(1.3, 1.028)$, $\quad D(1.45, 0.862)$

3.4 DEFINITION OF FUNCTION

The notion of **correspondence** occurs frequently in everyday life. Some examples are given in the following illustration.

ILLUSTRATION

Correspondence

- To each book in a library there corresponds the number of pages in the book.
- To each human being there corresponds a birth date.
- If the temperature of the air is recorded throughout a day, then to each instant of time there corresponds a temperature.

Each correspondence in the previous illustration involves two sets, D and E. In the first illustration, D denotes the set of books in a library and E the set of positive integers. To each book x in D there corresponds a positive integer y in E—namely, the number of pages in the book.

We sometimes depict correspondences by diagrams of the type shown in Figure 36, where the sets D and E are represented by points within regions in a plane. The curved arrow indicates that the element y of E corresponds to the element x of D. The two sets may have elements in common. As a matter of fact, we often have $D = E$. It is important to note that *to each x in D there corresponds exactly one y in E.* However, the same element of E may correspond to different elements of D. For example, two books may have the same number of pages, two people may have the same birthday, and the temperature may be the same at different times.

FIGURE 36

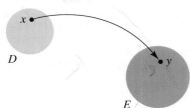

In most of our work, D and E will be sets of numbers. To illustrate, let both D and E denote the set $\mathbb{R}$ of real numbers, and to each real number x let us assign its square x^2. This gives us a correspondence from $\mathbb{R}$ to $\mathbb{R}$.

Each of our illustrations of a correspondence is a *function*, which we define as follows.

Definition of Function

> A **function** f from a set D to a set E is a correspondence that assigns to each element x of D exactly one element y of E.

The element y of E is the **value** of f at x and is denoted by $f(x)$, read "f of x." The set D is the **domain** of the function. The **range** of f is the subset R of E consisting of all possible values $f(x)$ for x in D.

Consider the diagram in Figure 37. The curved arrows indicate that the elements $f(w)$, $f(z)$, $f(x)$, and $f(a)$ of E correspond to the elements w, z, x, and a of D. *To each element in D there is assigned exactly one function value in E*; however, different elements of D, such as w and z in Figure 37, may have the same value in E.

FIGURE 37

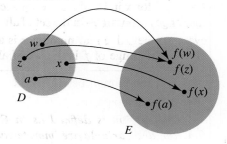

The symbols

$$D \xrightarrow{f} E, \quad f : D \to E, \quad \text{and}$$

signify that f is a function from D to E. Initially the notations f and $f(x)$ may be confusing. Remember that f is used to represent the function. It is neither in D nor in E. However, $f(x)$ is an element of the range E—the element that the function f assigns to the element x, which is in the domain D.

Two functions f and g from D to E are **equal**, and we write

$$f = g \quad \text{provided} \quad f(x) = g(x) \quad \text{for every } x \text{ in } D.$$

For example, if $g(x) = \frac{1}{2}(2x^2 - 6) + 3$ and $f(x) = x^2$ for every x in $\mathbb{R}$, then $g = f$.

EXAMPLE 1   *Finding function values*

Let f be the function with domain $\mathbb{R}$ such that $f(x) = x^2$ for every x in $\mathbb{R}$.

(a) Find $f(-6)$, $f(\sqrt{3})$, $f(a + b)$, and $f(a) + f(b)$, where a and b are real numbers.

(b) What is the range of f?

SOLUTION

(a) We find values of f by substituting for x in the equation $f(x) = x^2$:

$$f(-6) = (-6)^2 = 36$$
$$f(\sqrt{3}) = (\sqrt{3})^2 = 3$$
$$f(a + b) = (a + b)^2 = a^2 + 2ab + b^2$$
$$f(a) + f(b) = a^2 + b^2$$

(b) By definition, the range of f consists of all numbers of the form $f(x) = x^2$ for x in $\mathbb{R}$. Since the square of every real number is nonnegative, the range is contained in the set of all nonnegative real numbers. Moreover, every nonnegative real number c is a value of f, since $f(\sqrt{c}) = (\sqrt{c})^2 = c$. Hence, the range of f is the set of all nonnegative real numbers.

If a function is defined as in Example 1, the symbols used for the function and variable are immaterial; that is, expressions such as $f(x) = x^2$, $f(s) = s^2$, $g(t) = t^2$, and $k(r) = r^2$ all define the same function. This is true because if a is any number in the domain, then the same value a^2 is obtained regardless of which expression is employed.

In the remainder of our work, the phrase f *is a function* will mean that the domain and range are sets of real numbers. If a function is defined by means of an expression, as in Example 1, and the domain D is not stated, then we will consider D to be the totality of real numbers x such that $f(x)$ is real. This is sometimes called the **implied domain** of f. To illustrate, if $f(x) = \sqrt{x - 2}$, then the implied domain is the set of real numbers x such that $\sqrt{x - 2}$ is real—that is, $x - 2 \geq 0$, or $x \geq 2$. Thus, the domain is the infinite interval $[2, \infty)$. If x is in the domain, we say that f *is defined at* x or that $f(x)$ *exists*. If a set S is contained in the domain, f *is defined on* S. The terminology f *is undefined at* x means that x is not in the domain of f.

EXAMPLE 2   Finding function values

Let $g(x) = \dfrac{\sqrt{4 + x}}{1 - x}$.

(a) Find the domain of g.

(b) Find $g(5)$, $g(-2)$, $g(-a)$, and $-g(a)$.

SOLUTION

(a) The expression $\sqrt{4 + x}/(1 - x)$ is a real number if and only if the radicand $4 + x$ is nonnegative and the denominator $1 - x$ is not equal to 0. Thus, $g(x)$ exists if and only if

$$4 + x \ge 0 \qquad \text{and} \qquad 1 - x \ne 0$$

or, equivalently, $x \ge -4$ and $x \ne 1$.

We may express the domain in terms of intervals as $[-4, 1) \cup (1, \infty)$.

(b) To find values of g, we substitute for x:

$$g(5) = \frac{\sqrt{4 + 5}}{1 - 5} = \frac{\sqrt{9}}{-4} = -\frac{3}{4}$$

$$g(-2) = \frac{\sqrt{4 + (-2)}}{1 - (-2)} = \frac{\sqrt{2}}{3}$$

$$g(-a) = \frac{\sqrt{4 + (-a)}}{1 - (-a)} = \frac{\sqrt{4 - a}}{1 + a}$$

$$-g(a) = -\frac{\sqrt{4 + a}}{1 - a} = \frac{\sqrt{4 + a}}{a - 1}$$

FIGURE 38

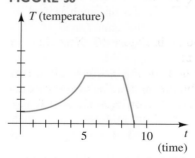

Graphs are often used to describe the variation of physical quantities. For example, a scientist may use the graph in Figure 38 to indicate the temperature T of a certain solution at various times t during an experiment. The sketch shows that the temperature increased gradually from time $t = 0$ to time $t = 5$, did not change between $t = 5$ and $t = 8$, and then decreased rapidly from $t = 8$ to $t = 9$.

Similarly, if f is a function, we may use a graph to indicate the change in $f(x)$ as x varies through the domain of f. Specifically, we have the following definition.

Definition of Graph of a Function

The **graph of a function** f is the graph of the equation $y = f(x)$ for x in the domain of f.

FIGURE 39

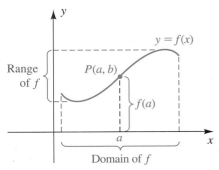

We often attach the label $y = f(x)$ to a sketch of the graph. If $P(a, b)$ is a point on the graph, then the y-coordinate b is the function value $f(a)$, as illustrated in Figure 39. The figure displays the domain of f (the set of possible values of x) and the range of f (the corresponding values of y). Although we have pictured the domain and range as closed intervals, they may be infinite intervals or other sets of real numbers.

Since there is exactly one value $f(a)$ for each a in the domain of f, only *one* point on the graph of f has x-coordinate a. Thus, *every vertical line intersects the graph of a function in at most one point.* Consequently, the graph of a function cannot be a figure such as a circle, in which a vertical line may intersect the graph in more than one point.

The x-intercepts of the graph of a function f are the solutions of the equation $f(x) = 0$. These numbers are called the **zeros** of the function. The y-intercept of the graph is $f(0)$, if it exists.

EXAMPLE 3 *Sketching the graph of a function*

Let $f(x) = \sqrt{x - 1}$.

(a) Sketch the graph of f.
(b) Find the domain and range of f.

SOLUTION

(a) By definition, the graph of f is the graph of the equation $y = \sqrt{x - 1}$. The following table lists coordinates of several points on the graph.

FIGURE 40

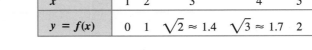

x	1	2	3	4	5	6
$y = f(x)$	0	1	$\sqrt{2} \approx 1.4$	$\sqrt{3} \approx 1.7$	2	$\sqrt{5} \approx 2.2$

Plotting points, we obtain the sketch shown in Figure 40. Note that the x-intercept is 1 and there is no y-intercept.
(b) Referring to Figure 40, note that the domain of f consists of all real numbers x such that $x \geq 1$ or, equivalently, the interval $[1, \infty)$. The range of f is the set of all real numbers y such that $y \geq 0$ or, equivalently, $[0, \infty)$.

In Example 3, as x increases, the function value $f(x)$ also increases, and we say that the graph of f *rises* (see Figure 40). A function of this type is said to be *increasing*. For certain functions, $f(x)$ decreases as x increases. In this case the graph *falls*, and f is a *decreasing* function. In general, we shall consider functions that increase or decrease on an interval I, as described in the following chart, where x_1 and x_2 denote numbers in I.

Terminology	Definition	Graphical interpretation
f is **increasing** on an interval I	$f(x_1) < f(x_2)$ whenever $x_1 < x_2$	
f is **decreasing** on an interval I	$f(x_1) > f(x_2)$ whenever $x_1 < x_2$	
f is **constant** on an interval I	$f(x_1) = f(x_2)$ for every x_1 and x_2	

If $f(x) = c$ for every real number x, then f is called a **constant function**. We shall use the phrases *f is increasing* and *f(x) is increasing* interchangeably. We shall do the same with the terms *decreasing* and *constant*.

EXAMPLE 4 *Using a graph to find domain, range, and where a function increases or decreases*

Let $f(x) = \sqrt{9 - x^2}$.

(a) Sketch the graph of f.

(b) Find the domain and range of f.

(c) Find the intervals on which f is increasing or is decreasing.

SOLUTION

(a) By definition, the graph of f is the graph of the equation $y = \sqrt{9 - x^2}$. We know from our work with circles in Section 3.2 that the graph of $x^2 + y^2 = 9$ is a circle of radius 3 with center at the origin. Solving the equation $x^2 + y^2 = 9$ for y gives us $y = \pm\sqrt{9 - x^2}$. It follows that the graph of f is the *upper half* of the circle, as illustrated in Figure 41.

(b) Referring to Figure 41, we see that the domain of f is the closed interval $[-3, 3]$, and the range of f is the interval $[0, 3]$.

(c) The graph rises as x increases from -3 to 0, so f is increasing on the closed interval $[-3, 0]$. Thus, as shown in the preceding chart, if $x_1 < x_2$ in $[-3, 0]$, then $f(x_1) < f(x_2)$ (note that *possibly* $x_1 = -3$ or $x_2 = 0$).

The graph falls as x increases from 0 to 3, so f is decreasing on the closed interval $[0, 3]$. In this case, the chart indicates that if $x_1 < x_2$ in $[0, 3]$, then $f(x_1) > f(x_2)$ (note that *possibly* $x_1 = 0$ or $x_2 = 3$).

FIGURE 41

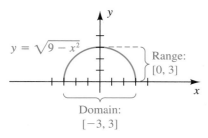

$y = \sqrt{9 - x^2}$

Range: [0, 3]

Domain: [−3, 3]

The following type of function is the most basic in algebra.

Definition of Linear Function

A function f is a **linear function** if

$$f(x) = ax + b,$$

where x is any real number and a and b are constants.

The graph of f in the preceding definition is the graph of $y = ax + b$, which, by the slope-intercept form, is a line with slope a and y-intercept b. Thus, *the graph of a linear function is a line*. Since $f(x)$ exists for every x, the domain of f is $\mathbb{R}$. As illustrated in the next example, if $a \neq 0$, then the range of f is also $\mathbb{R}$.

EXAMPLE 5 *Sketching the graph of a linear function*

Let $f(x) = 2x + 3$.

(a) Sketch the graph of f.

(b) Find the domain and range of f.

(c) Determine where f is increasing or is decreasing.

FIGURE 42

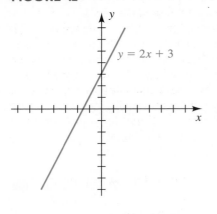

$y = 2x + 3$

SOLUTION

(a) Since $f(x)$ has the form $ax + b$, with $a = 2$ and $b = 3$, f is a linear function. The graph of $y = 2x + 3$ is the line with slope 2 and y-intercept 3, illustrated in Figure 42.

(b) We see from the graph that x and y may be any real numbers, so both the domain and the range of f are $\mathbb{R}$.

(c) Since the slope a is positive, the graph of f rises as x increases; that is, $f(x_1) < f(x_2)$ whenever $x_1 < x_2$. Thus, f is increasing throughout its domain.

In applications it is sometimes necessary to determine a specific linear function from given data, as in the next example.

EXAMPLE 6   *Finding a linear function*

If f is a linear function such that $f(-2) = 5$ and $f(6) = 3$, find $f(x)$, where x is any real number.

FIGURE 43

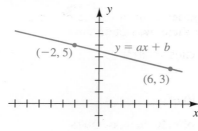

$(-2, 5)$

$y = ax + b$

$(6, 3)$

SOLUTION   By the definition of linear function, $f(x) = ax + b$, where a and b are constants. Moreover, the given function values tell us that the points $(-2, 5)$ and $(6, 3)$ are on the graph of f—that is, on the line $y = ax + b$ illustrated in Figure 43. The slope a of this line is

$$a = \frac{5 - 3}{-2 - 6} = \frac{2}{-8} = -\frac{1}{4},$$

and hence $f(x)$ has the form

$$f(x) = -\tfrac{1}{4}x + b.$$

To find the value of b, we may use the fact that $f(6) = 3$, as follows:

$$f(6) = -\tfrac{1}{4}(6) + b \quad \text{let } x = 6 \text{ in } f(x) = -\tfrac{1}{4}x + b$$
$$3 = -\tfrac{3}{2} + b \quad\quad f(6) = 3$$
$$b = 3 + \tfrac{3}{2} = \tfrac{9}{2} \quad\quad \text{solve for } b$$

Thus, the linear function satisfying $f(-2) = 5$ and $f(6) = 3$ is

$$f(x) = -\tfrac{1}{4}x + \tfrac{9}{2}.$$

Many formulas that occur in mathematics and the sciences determine functions. For instance, the formula $A = \pi r^2$ for the area A of a circle of radius r assigns to each positive real number r exactly one value of A. This determines a function f such that $f(r) = \pi r^2$, and we may write $A = f(r)$.

The letter r, which represents an arbitrary number from the domain of f, is called an **independent variable**. The letter A, which represents a number from the range of f, is a **dependent variable**, since its value depends on the number assigned to r. If two variables r and A are related in this manner, we say that *A is a function of r*. In applications, the independent variable and dependent variable are sometimes referred to as the **input variable** and **output variable**, respectively. As another example, if an automobile travels at a uniform rate of 50 mi/hr, then the distance d (miles) traveled in time t (hours) is given by $d = 50t$, and hence *the distance d is a function of time t.*

EXAMPLE 7   *Expressing the volume of a tank as a function of its radius*

A steel storage tank for propane gas is to be constructed in the shape of a right circular cylinder of altitude 10 feet with a hemisphere attached to each end. The radius r is yet to be determined. Express the volume V (in ft^3) of the tank as a function of r (in ft).

FIGURE 44

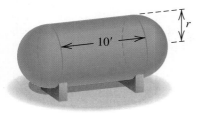

SOLUTION   The tank is illustrated in Figure 44. We may find the volume of the cylindrical part of the tank by multiplying the altitude 10 by the area πr^2 of the base of the cylinder. This gives us

$$\text{volume of cylinder} = 10(\pi r^2) = 10\pi r^2.$$

The two hemispherical ends, taken together, form a sphere of radius r. Using the formula for the volume of a sphere, we obtain

$$\text{volume of the two ends} = \tfrac{4}{3}\pi r^3.$$

Thus, the volume V of the tank is

$$V = \tfrac{4}{3}\pi r^3 + 10\pi r^2.$$

This formula expresses V as a function of r. In factored form,

$$V = \tfrac{1}{3}\pi r^2(4r + 30) = \tfrac{2}{3}\pi r^2(2r + 15).$$

FIGURE 45

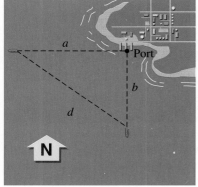

EXAMPLE 8   *Expressing a distance as a function of time*

Two ships leave port at the same time, one sailing west at a rate of 17 mi/hr and the other sailing south at 12 mi/hr. If t is the time (in hours) after their departure, express the distance d between the ships as a function of t.

SOLUTION   To help visualize the problem, we begin by drawing a picture and labeling it, as in Figure 45. By the Pythagorean theorem,

$$d^2 = a^2 + b^2, \qquad \text{or} \qquad d = \sqrt{a^2 + b^2}.$$

(continued)

Since distance = (rate)(time) and the rates are 17 and 12, respectively,

$$a = 17t \quad \text{and} \quad b = 12t.$$

Substitution in $d = \sqrt{a^2 + b^2}$ gives us

$$d = \sqrt{(17t)^2 + (12t)^2} = \sqrt{289t^2 + 144t^2} = \sqrt{433t^2} \approx (20.8)t.$$

Ordered pairs can be used to obtain an alternative approach to functions. We first observe that a function f from D to E determines the following set W of ordered pairs:

$$W = \{(x, f(x)): x \text{ is in } D\}.$$

Thus, W consists of all ordered pairs such that the first number x is in D and the second number is the function value $f(x)$. In Example 1, where $f(x) = x^2$, W is the set of all ordered pairs of the form (x, x^2). It is important to note that, *for each x, there is exactly one ordered pair (x, y) in W having x in the first position.*

Conversely, if we begin with a set W of ordered pairs such that each x in D appears exactly once in the first position of an ordered pair, then W determines a function. Specifically, for each x in D there is exactly one pair (x, y) in W, and by letting y correspond to x, we obtain a function with domain D. The range consists of all real numbers y that appear in the second position of the ordered pairs.

It follows from the preceding discussion that the next statement could also be used as a definition of function.

Alternative Definition of Function

A **function** with domain D is a set W of ordered pairs such that, for each x in D, there is exactly one ordered pair (x, y) in W having x in the first position.

In terms of the preceding definition, the ordered pairs $(x, \sqrt{x - 1})$ determine the function of Example 3 given by $f(x) = \sqrt{x - 1}$. Note, however, that if

$$W = \{(x, y): x^2 = y^2\},$$

then W is *not* a function, since for a given x, there may be more than one pair in W with x in the first position. For example, if $x = 2$, then both $(2, 2)$ and $(2, -2)$ are in W.

In the next example we illustrate how some of the concepts presented in this section may be studied with the aid of a graphing utility. Hereafter, when making assignments on a graphing utility, we will frequently refer to variables such as Y_1 and Y_2 as the *functions* Y_1 and Y_2.

EXAMPLE 9   *Analyzing the graph of a function*

Let $f(x) = x^{2/3} - 3$.

(a) Find $f(-2)$.

(b) Sketch the graph of f.

(c) State the domain and range of f.

(d) State the intervals on which f is increasing or is decreasing.

(e) Estimate the x-intercepts of the graph to one-decimal-place accuracy.

SOLUTION

(a) A representation of f on a computational device may have the form

$$X^{\wedge}(2/3) - 3 \quad \text{or} \quad (X^{\wedge}(1/3))^{\wedge}2 - 3 \quad \text{or} \quad (X^{\wedge}2)^{\wedge}(1/3) - 3.$$

We assign one of these expressions to the function Y_1. To find the value of Y_1 at $x = -2$, we first assign the value -2 to a memory location identified as X. This is usually done with a "store" or "assign" operation on a computational device. We next determine the value of Y_1 by requesting the computational device to indicate the contents of the memory location that contains the value of Y_1. We often refer to this process of finding a value of Y_1 as "querying Y_1." Upon querying Y_1, we see that its value is approximately -1.41 (that is, $f(-2) \approx -1.41$). It should be noted that not all computational devices process rational exponents in the same fashion. If you did not obtain this answer for $f(-2)$, change your representation of Y_1 before proceeding.

(b) Using the viewing rectangle $[-15, 15]$ by $[-10, 10]$ to graph Y_1 gives us a display similar to that of Figure 46.

(c) The domain of f is $\mathbb{R}$, since we may input any value for x. The figure indicates that $y \geq -3$, so we conclude that the range of f is $[-3, \infty)$.

(d) From the figure, we see that f is decreasing on $(-\infty, 0]$ and is increasing on $[0, \infty)$.

(e) Using the tracing and zoom-in features, we find that the positive x-intercept in Figure 46 is approximately 5.2. Since f is symmetric with respect to the y-axis, the negative x-intercept is about -5.2.

FIGURE 46 $[-15, 15]$ by $[-10, 10]$

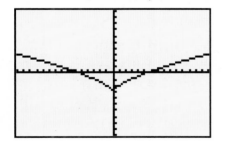

3.4 EXERCISES

1 If $f(x) = -x^2 - x - 4$, find $f(-2)$, $f(0)$, and $f(4)$.

2 If $f(x) = -x^3 - x^2 + 3$, find $f(-3)$, $f(0)$, and $f(2)$.

3 If $f(x) = \sqrt{x - 4} - 3x$, find $f(4)$, $f(8)$, and $f(13)$.

4 If $f(x) = \dfrac{x}{x - 3}$, find $f(-2)$, $f(0)$, and $f(3)$.

Exer. 5–8: If a and h are real numbers, find

(a) $f(a)$ (b) $f(-a)$ (c) $-f(a)$ (d) $f(a + h)$

(e) $f(a) + f(h)$ (f) $\dfrac{f(a + h) - f(a)}{h}$, if $h \neq 0$

5 $f(x) = 5x - 2$ **6** $f(x) = 3 - 4x$

7 $f(x) = x^2 - x + 3$

8 $f(x) = 2x^2 + 3x - 7$

Exer. 9–12: If a is a positive real number, find

(a) $g\left(\dfrac{1}{a}\right)$ **(b)** $\dfrac{1}{g(a)}$ **(c)** $g(\sqrt{a})$ **(d)** $\sqrt{g(a)}$

9 $g(x) = 4x^2$

10 $g(x) = 2x - 5$

11 $g(x) = \dfrac{2x}{x^2 + 1}$

12 $g(x) = \dfrac{x^2}{x + 1}$

Exer. 13–14: For the graph of the function f sketched in the figure, determine

(a) the domain   **(b)** the range   **(c)** $f(1)$
(d) all x such that $f(x) = 1$ **(e)** all x such that $f(x) > 1$

13

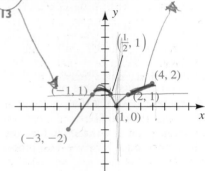

14

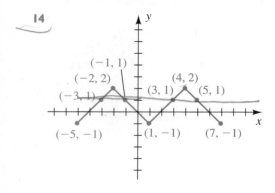

Exer. 15–26: Find the domain of f.

15 $f(x) = \sqrt{2x + 7}$

16 $f(x) = \sqrt{8 - 3x}$

17 $f(x) = \sqrt{9 - x^2}$

18 $f(x) = \sqrt{x^2 - 25}$

19 $f(x) = \dfrac{x + 1}{x^3 - 4x}$

20 $f(x) = \dfrac{4x}{6x^2 + 13x - 5}$

21 $f(x) = \dfrac{\sqrt{2x - 3}}{x^2 - 5x + 4}$

22 $f(x) = \dfrac{\sqrt{4x - 3}}{x^2 - 4}$

23 $f(x) = \dfrac{x - 4}{\sqrt{x - 2}}$

24 $f(x) = \dfrac{1}{(x - 3)\sqrt{x + 3}}$

25 $f(x) = \sqrt{x + 2} + \sqrt{2 - x}$

26 $f(x) = \sqrt{(x - 2)(x - 6)}$

Exer. 27–36: (a) Sketch the graph of f. **(b)** Find the domain D and range R of f. **(c)** Find the intervals on which f is increasing, is decreasing, or is constant.

27 $f(x) = 3x - 2$

28 $f(x) = -2x + 3$

29 $f(x) = 4 - x^2$

30 $f(x) = x^2 - 1$

31 $f(x) = \sqrt{x + 4}$

32 $f(x) = \sqrt{4 - x}$

33 $f(x) = -2$

34 $f(x) = 3$

35 $f(x) = -\sqrt{36 - x^2}$

36 $f(x) = \sqrt{16 - x^2}$

Exer. 37–38: If a linear function f satisfies the given conditions, find $f(x)$.

37 $f(-3) = 1$ and $f(3) = 2$

38 $f(-2) = 7$ and $f(4) = -2$

Exer. 39–48: Determine whether the set W of ordered pairs is a function in the sense of the alternative definition of function on page 166.

39 $W = \{(x, y): 2y = x^2 + 5\}$

40 $W = \{(x, y): x = 3y + 2\}$

41 $W = \{(x, y): x^2 + y^2 = 4\}$

42 $W = \{(x, y): y^2 - x^2 = 1\}$

43 $W = \{(x, y): y = 3\}$ **44** $W = \{(x, y): x = 3\}$

45 $W = \{(x, y): xy = 0\}$ **46** $W = \{(x, y): x + y = 0\}$

47 $W = \{(x, y): |y| = |x|\}$ **48** $W = \{(x, y): y < x\}$

49 Constructing a box From a rectangular piece of cardboard having dimensions 20 inches × 30 inches, an open

EXERCISE 49

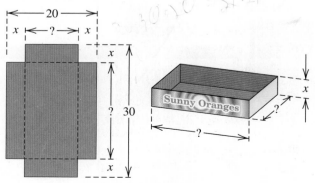

box is to be made by cutting out identical squares of area x^2 from each corner and turning up the sides (see the figure on page 168). Express the volume V of the box as a function of x.

50 *Constructing a storage tank* Refer to Example 7. A steel storage tank for propane gas is to be constructed in the shape of a right circular cylinder of altitude 10 feet with a hemisphere attached to each end. The radius r is yet to be determined. Express the surface area S of the tank as a function of r.

51 *Dimensions of a building* A small office unit is to contain 500 ft^2 of floor space. A simplified model is shown in the figure.

(a) Express the length y of the building as a function of the width x.

(b) If the walls cost \$100 per running foot, express the cost C of the walls as a function of the width x. (Disregard the wall space above the doors and the thickness of the walls.)

EXERCISE 51

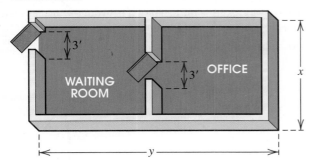

52 *Dimensions of an aquarium* An aquarium of height 1.5 feet is to have a volume of 6 ft^3. Let x denote the length of the base and y the width (see the figure).

EXERCISE 52

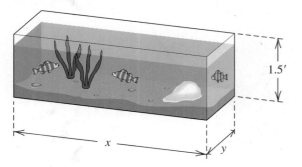

(a) Express y as a function of x.

(b) Express the total number S of square feet of glass needed as a function of x.

53 *Childhood growth* For children between ages 6 and 10, height y (in inches) is frequently a linear function of age t (in years). The height of a certain child is 48 inches at age 6 and 50.5 inches at age 7.

(a) Express y as a function of t.

(b) Sketch the line in part (a), and interpret the slope.

(c) Predict the height of the child at age 10.

54 *Radioactive contamination* It has been estimated that 1000 curies of a radioactive substance introduced at a point on the surface of the open sea would spread over an area of 40,000 km^2 in 40 days. Assuming that the area covered by the radioactive substance is a linear function of time t and is always circular in shape, express the radius r of the contamination as a function of t.

55 *Distance to a hot-air balloon* A hot-air balloon is released at 1:00 P.M. and rises vertically at a rate of 2 m/sec. An observation point is situated 100 meters from a point on the ground directly below the balloon (see the figure). If t denotes the time (in seconds) after 1:00 P.M., express the distance d between the balloon and the observation point as a function of t.

EXERCISE 55

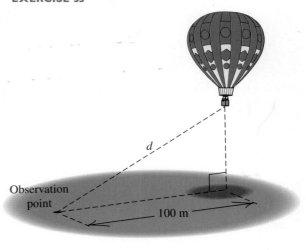

56 Triangle ABC is inscribed in a semicircle of diameter 15 (see the figure on the following page).

(a) If x denotes the length of side AC, express the length y of side BC as a function of x. (*Hint:* Angle ACB is a right angle.)

(b) Express the area $\mathscr{A}$ of triangle ABC as a function of x, and state the domain of this function.

EXERCISE 56

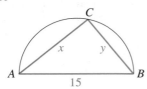

57 *Distance to the earth* From an exterior point P that is h units from a circle of radius r, a tangent line is drawn to the circle (see the figure). Let y denote the distance from the point P to the point of tangency T.

(a) Express y as a function of h. (*Hint:* If C is the center of the circle, then PT is perpendicular to CT.)

(b) If r is the radius of the earth and h is the altitude of a space shuttle, then y is the maximum distance to the earth that an astronaut can see from the shuttle. In particular, if $h = 200$ mi and $r \approx 4000$ mi, approximate y.

EXERCISE 57

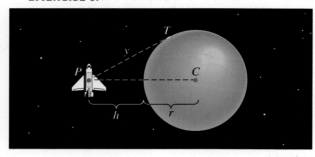

58 *Length of a tightrope* The figure illustrates the apparatus for a tightrope walker. Two poles are set 50 feet apart, but the point of attachment P for the rope is yet to be determined.

(a) Express the length L of the rope as a function of the distance x from P to the ground.

(b) If the total walk is to be 75 feet, determine the distance from P to the ground.

EXERCISE 58

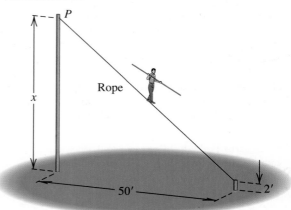

59 *Airport runway* The relative positions of an aircraft runway and a 20-foot-tall control tower are shown in the figure. The beginning of the runway is at a perpendicular distance of 300 feet from the base of the tower. If x denotes the distance an airplane has moved down the runway, express the distance d between the airplane and the control booth as a function of x.

EXERCISE 59

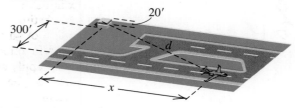

60 *Destination time* A man in a rowboat that is 2 miles from the nearest point A on a straight shoreline wishes to reach a house located at a point B that is 6 miles

EXERCISE 60

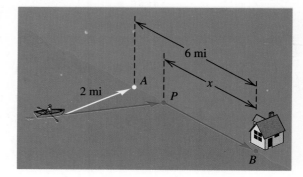

farther down the shoreline (see the figure). He plans to row to a point P that is between A and B and is x miles from the house, and then he will walk the remainder of the distance. Suppose he can row at a rate of 3 mi/hr and can walk at a rate of 5 mi/hr. If T is the total time required to reach the house, express T as a function of x.

c **Exer. 61–64:** (a) **Sketch the graph of f on the given interval $[a, b]$.** (b) **Estimate the range of f on $[a, b]$.** (c) **Estimate the intervals on which f is increasing or is decreasing.**

61 $f(x) = \dfrac{x^{1/3}}{1 + x^4}$; $[-2, 2]$

62 $f(x) = x^4 - 0.4x^3 - 0.8x^2 + 0.2x + 0.1$; $[-1, 1]$

63 $f(x) = x^5 - 3x^2 + 1$; $[-0.7, 1.4]$

64 $f(x) = \dfrac{1 - x^3}{1 + x^4}$; $[-4, 4]$

c **Exer. 65–66:** In Exercises 51–52 of Section 2.5, algebraic methods were used to find solutions to each of the following equations. Now solve the equation graphically by assigning the expression on the left side to Y_1 and the number on the right side to Y_2 and then finding the x-coordinates of all points of intersection of the two graphs.

65 (a) $x^{5/3} = 32$ (b) $x^{4/3} = 16$ (c) $x^{2/3} = -36$
   (d) $x^{3/4} = 125$ (e) $x^{3/2} = -27$

66 (a) $x^{3/5} = -27$ (b) $x^{2/3} = 25$ (c) $x^{4/3} = -49$
   (d) $x^{3/2} = 27$ (e) $x^{3/4} = -8$

3.5 GRAPHS OF FUNCTIONS

In this section we discuss aids for sketching graphs of certain types of functions. In particular, a function f is called **even** if $f(-x) = f(x)$ for every x in its domain. In this case, the equation $y = f(x)$ is not changed if $-x$ is substituted for x, and hence, from symmetry test (1) of Section 3.2, the graph of an even function is symmetric with respect to the y-axis.

A function f is called **odd** if $f(-x) = -f(x)$ for every x in its domain. If we apply symmetry test (3) of Section 3.2 to the equation $y = f(x)$, we see that the graph of an odd function is symmetric with respect to the origin.

These facts are summarized in the first two columns of the next chart.

Terminology	Definition	Illustration	Symmetry of graph
f is an **even function**.	$f(-x) = f(x)$ for every x in the domain.	$y = f(x) = x^2$	y-axis
f is an **odd function**.	$f(-x) = -f(x)$ for every x in the domain.	$y = f(x) = x^3$	the origin

EXAMPLE I *Determining whether a function is even or odd*

Determine whether f is even, odd, or neither even nor odd.

(a) $f(x) = 3x^4 - 2x^2 + 5$ (b) $f(x) = 2x^5 - 7x^3 + 4x$

(c) $f(x) = x^3 + x^2$

SOLUTION   In each case the domain of f is $\mathbb{R}$. To determine whether f is even or odd, we begin by examining $f(-x)$, where x is any real number.

(a) $f(-x) = 3(-x)^4 - 2(-x)^2 + 5$ substitute $-x$ for x in $f(x)$

$\qquad\quad = 3x^4 - 2x^2 + 5$ simplify

$\qquad\quad = f(x)$ definition of f

Since $f(-x) = f(x)$, f is an even function.

(b) $f(-x) = 2(-x)^5 - 7(-x)^3 + 4(-x)$ substitute $-x$ for x in $f(x)$

$\qquad\quad = -2x^5 + 7x^3 - 4x$ simplify

$\qquad\quad = -(2x^5 - 7x^3 + 4x)$ factor out -1

$\qquad\quad = -f(x)$ definition of f

Since $f(-x) = -f(x)$, f is an odd function.

(c) $f(-x) = (-x)^3 + (-x)^2$ substitute $-x$ for x in $f(x)$

$\qquad\quad = -x^3 + x^2$ simplify

Since $f(-x) \neq f(x)$ and $f(-x) \neq -f(x)$ (note that $-f(x) = -x^3 - x^2$), the function f is neither even nor odd.

In the next example, we consider the **absolute value function** f, defined by $f(x) = |x|$.

EXAMPLE 2   *Sketching the graph of the absolute value function*

Let $f(x) = |x|$.

(a) Determine whether f is even or odd.

(b) Sketch the graph of f.

(c) Find the intervals on which f is increasing or is decreasing.

SOLUTION

(a) The domain of f is $\mathbb{R}$, because the absolute value of x exists for every real number x. If x is in $\mathbb{R}$, then

$$f(-x) = |-x| = |x| = f(x).$$

Thus, f is an even function, since $f(-x) = f(x)$.

(b) Since f is even, its graph is symmetric with respect to the y-axis. If $x \geq 0$, then $|x| = x$, and therefore the first quadrant part of the graph coincides with the line $y = x$. Sketching this half-line and using symmetry gives us Figure 47.

(c) Referring to the graph, we see that f is decreasing on $(-\infty, 0]$ and is increasing on $[0, \infty)$.

FIGURE 47

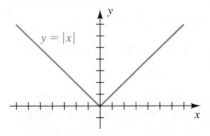

$y = |x|$

If we know the graph of $y = f(x)$, it is easy to sketch the graphs of

$$y = f(x) + c \quad \text{and} \quad y = f(x) - c$$

for any positive real number c. As in the next chart, for $y = f(x) + c$, we add c to the y-coordinate of each point on the graph of $y = f(x)$. This *shifts* the graph of f *upward* a distance c. For $y = f(x) - c$ with $c > 0$, we subtract c from each y-coordinate, thereby shifting the graph of f a distance c *downward*. These are called **vertical shifts** of graphs.

Vertically Shifting the Graph of $y = f(x)$

Equation	Effect on graph	Graphical interpretation
$y = f(x) + c$ with $c > 0$	The graph of f is shifted vertically upward a distance c.	
$y = f(x) - c$ with $c > 0$	The graph of f is shifted vertically downward a distance c.	

FIGURE 48

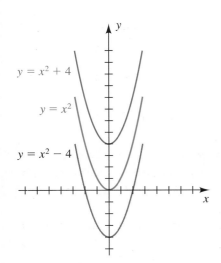

EXAMPLE 3 *Vertically shifting a graph*

Sketch the graph of f:

(a) $f(x) = x^2$ **(b)** $f(x) = x^2 + 4$ **(c)** $f(x) = x^2 - 4$

SOLUTION We shall sketch all graphs on the same coordinate plane.

(a) Since $f(-x) = (-x)^2 = x^2 = f(x),$

the function f is even, and hence its graph is symmetric with respect to the y-axis. Several points on the graph of $y = x^2$ are (0, 0), (1, 1), (2, 4), and (3, 9). Drawing a smooth curve through these points and reflecting through the y-axis gives us the sketch in Figure 48. The graph is a parabola with vertex at the origin and opening upward.

(continued)

(b) To sketch the graph of $y = x^2 + 4$, we add 4 to the y-coordinate of each point on the graph of $y = x^2$; that is, we shift the graph in part (a) upward 4 units, as shown in the figure.

(c) To sketch the graph of $y = x^2 - 4$, we decrease y-coordinates of $y = x^2$ by 4; that is, we shift the graph in part (a) downward 4 units.

We can also consider **horizontal shifts** of graphs. Specifically, if $c > 0$, consider the graphs of $y = f(x)$ and $y = f(x - c)$ sketched on the same coordinate plane, as illustrated in the next chart. Since

$$f(a) = f([a + c] - c),$$

we see that the point with x-coordinate a on the graph of $y = f(x)$ has the same y-coordinate as the point with x-coordinate $a + c$ on the graph of $y = f(x - c)$. This implies that the graph of $y = f(x - c)$ can be obtained by shifting the graph of $y = f(x)$ *to the right* a distance c. Similarly, the graph of $y = f(x + c)$ can be obtained by shifting the graph of f *to the left* a distance c, as shown in the chart.

Horizontally Shifting the Graph of $y = f(x)$

Equation	Effect on graph	Graphical interpretation
$y = f(x - c)$ with $c > 0$	The graph of f is shifted horizontally to the right a distance c.	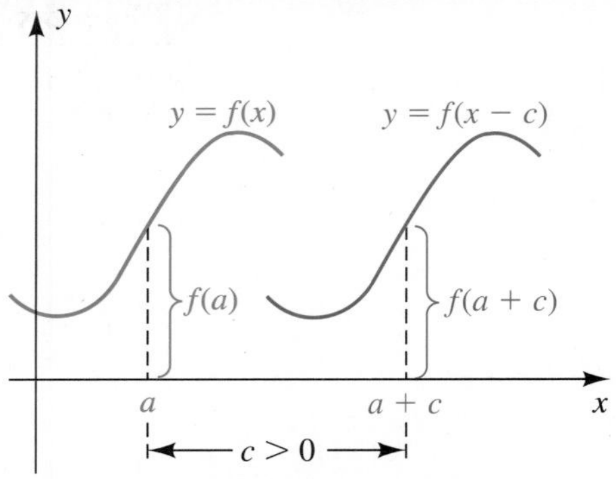
$y = f(x + c)$ with $c > 0$	The graph of f is shifted horizontally to the left a distance c.	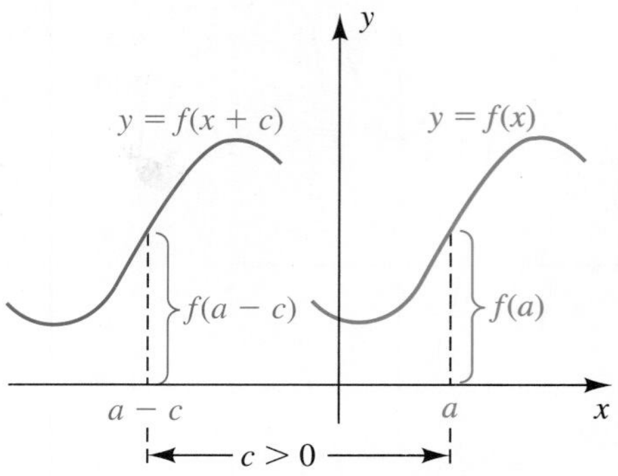

FIGURE 49

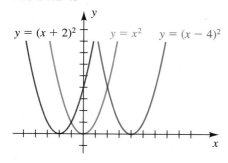

Horizontal and vertical shifts are also referred to as *translations*.

EXAMPLE 4   *Horizontally shifting a graph*

Sketch the graph of f:

(a) $f(x) = (x - 4)^2$ **(b)** $f(x) = (x + 2)^2$

SOLUTION   The graph of $y = x^2$ is sketched in Figure 49.

(a) Shifting the graph of $y = x^2$ to the right 4 units gives us the graph of $y = (x - 4)^2$, shown in the figure.

(b) Shifting the graph of $y = x^2$ to the left 2 units leads to the graph of $y = (x + 2)^2$, shown in the figure.

To obtain the graph of $y = cf(x)$ for some real number c, we may *multiply* the y-coordinates of points on the graph of $y = f(x)$ by c. For example, if $y = 2f(x)$, we double y-coordinates; or if $y = \frac{1}{2}f(x)$, we multiply each y-coordinate by $\frac{1}{2}$. This procedure is referred to as **vertically stretching** the graph of f (if $c > 1$) or **vertically compressing** the graph (if $0 < c < 1$) and is summarized in the following chart.

Vertically Stretching or Compressing the Graph of $y = f(x)$

Equation	Effect on graph	Graphical interpretation
$y = cf(x)$ with $c > 1$	The graph of f is stretched vertically by a factor c.	$y = cf(x)$ with $c > 1$ $\quad$ $y = f(x)$
$y = cf(x)$ with $0 < c < 1$	The graph of f is compressed vertically by a factor $1/c$.	$y = cf(x)$ with $0 < c < 1$ $\quad$ $y = f(x)$

FIGURE 50

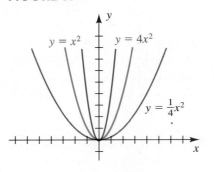

EXAMPLE 5   *Vertically stretching or compressing a graph*

Sketch the graph of the equation:

(a) $y = 4x^2$ **(b)** $y = \frac{1}{4}x^2$

SOLUTION

(a) To sketch the graph of $y = 4x^2$, we may refer to the graph of $y = x^2$ in Figure 50 and multiply the y-coordinate of each point by 4. This stretches the graph of $y = x^2$ vertically by a factor 4 and gives us a narrower parabola that is sharper at the vertex, as illustrated in the figure.

(b) The graph of $y = \frac{1}{4}x^2$ may be sketched by multiplying y-coordinates of points on the graph of $y = x^2$ by $\frac{1}{4}$. This compresses the graph of $y = x^2$ vertically by a factor $1/\frac{1}{4} = 4$ and gives us a wider parabola that is flatter at the vertex, as shown in Figure 50.

FIGURE 51

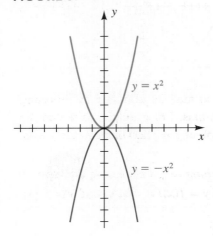

We may obtain the graph of $y = -f(x)$ by multiplying the y-coordinate of each point on the graph of $y = f(x)$ by -1. Thus, every point (a, b) on the graph of $y = f(x)$ that lies above the x-axis determines a point $(a, -b)$ on the graph of $y = -f(x)$ that lies below the x-axis. Similarly, if (c, d) lies below the x-axis (that is, $d < 0$), then $(c, -d)$ lies above the x-axis. The graph of $y = -f(x)$ is a **reflection** of the graph of $y = f(x)$ through the x-axis.

EXAMPLE 6   *Reflecting a graph through the x-axis*

Sketch the graph of $y = -x^2$.

SOLUTION   The graph may be found by plotting points; however, since the graph of $y = x^2$ is familiar to us, we sketch it as in Figure 51, and then multiply y-coordinates of points by -1. This procedure gives us the reflection through the x-axis indicated in the figure.

Sometimes it is useful to compare the graphs of $y = f(x)$ and $y = f(cx)$ if $c \neq 0$. In this case the function values $f(x)$ for

$$a \leq x \leq b$$

are the same as the function values $f(cx)$ for

$$a \leq cx \leq b \quad \text{or, equivalently,} \quad \frac{a}{c} \leq x \leq \frac{b}{c}.$$

This implies that the graph of f is **horizontally compressed** (if $c > 1$) or **horizontally stretched** (if $0 < c < 1$), as summarized in the following chart.

Horizontally Compressing or Stretching the Graph of $y = f(x)$

Equation	Effect on graph	Graphical interpretation
$y = f(cx)$ with $c > 1$	The graph of f is compressed horizontally by a factor c.	$y = f(x)$, $y = f(cx)$ with $c > 1$
$y = f(cx)$ with $0 < c < 1$	The graph of f is stretched horizontally by a factor $1/c$.	$y = f(x)$, $y = f(cx)$ with $0 < c < 1$

If $c < 0$, then the graph of $y = f(cx)$ may be obtained by reflecting the graph of $y = f(|c|x)$ through the y-axis. For example, to sketch the graph of $y = f(-2x)$, we reflect the graph of $y = f(2x)$ through the y-axis.

EXAMPLE 7 *Horizontally stretching or compressing a graph*

If $f(x) = x^3 - 4x^2$, sketch the graphs of $y = f(x)$, $y = f(2x)$, and $y = f(\frac{1}{2}x)$.

SOLUTION We have the following:

$$y = f(x) = x^3 - 4x^2 = x^2(x - 4)$$
$$y = f(2x) = (2x)^3 - 4(2x)^2 = 8x^3 - 16x^2 = 8x^2(x - 2)$$
$$y = f(\tfrac{1}{2}x) = (\tfrac{1}{2}x)^3 - 4(\tfrac{1}{2}x)^2 = \tfrac{1}{8}x^3 - x^2 = \tfrac{1}{8}x^2(x - 8)$$

Note that the x-intercepts of the graph of $y = f(2x)$ are 0 and 2, which are $\frac{1}{2}$ the x-intercepts of 0 and 4 for $y = f(x)$. This indicates a horizontal compression by a factor 2.

The x-intercepts of the graph of $y = f(\frac{1}{2}x)$ are 0 and 8, which are 2 times the x-intercepts for $y = f(x)$. This indicates a horizontal stretching by a factor $1/\frac{1}{2} = 2$.

The graphs, obtained by using a graphing utility with viewing rectangle $[-6, 15]$ by $[-10, 4]$, are shown in Figure 52.

FIGURE 52 $[-6, 15]$ by $[-10, 4]$

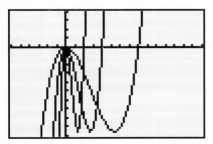

Functions are sometimes described by more than one expression, as in the next examples. We call such functions **piecewise-defined functions**.

EXAMPLE 8   *Sketching the graph of a piecewise-defined function*

Sketch the graph of the function f if

$$f(x) = \begin{cases} 2x + 3 & \text{if } x < 0 \\ x^2 & \text{if } 0 \le x < 2 \\ 1 & \text{if } x \ge 2 \end{cases}$$

FIGURE 53

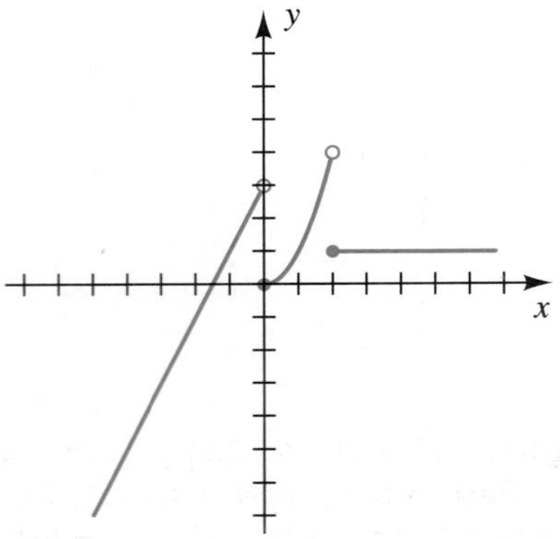

SOLUTION   If $x < 0$, then $f(x) = 2x + 3$ and the graph of f coincides with the line $y = 2x + 3$. This gives that portion of the graph to the left of the y-axis, sketched in Figure 53. The small circle indicates that the point $(0, 3)$ is *not* on the graph.

If $0 \le x < 2$, we use x^2 to find values of f, and therefore this part of the graph of f coincides with the parabola $y = x^2$, as indicated in the figure. Note that the point $(2, 4)$ is not on the graph.

Finally, if $x \ge 2$, the values of f are always 1. Thus, the graph of f for $x \ge 2$ is the horizontal half-line illustrated in Figure 53.

If x is a real number, we define the symbol $[\![x]\!]$ as follows:

$$[\![x]\!] = n, \quad \text{where } n \text{ is the greatest integer such that } n \le x$$

If we identify $\mathbb{R}$ with points on a coordinate line, then n is the first integer to the *left* of (or *equal* to) x.

ILLUSTRATION

The Symbol $[\![x]\!]$

- $[\![0.5]\!] = 0$
- $[\![1.8]\!] = 1$
- $[\![\sqrt{5}]\!] = 2$
- $[\![3]\!] = 3$
- $[\![-3]\!] = -3$
- $[\![-2.7]\!] = -3$
- $[\![-\sqrt{3}]\!] = -2$
- $[\![-0.5]\!] = -1$

The **greatest integer function** f is defined by $f(x) = [\![x]\!]$.

EXAMPLE 9 *Sketching the graph of the greatest integer function*

Sketch the graph of the greatest integer function.

SOLUTION The *x*- and *y*-coordinates of some points on the graph may be listed as follows:

FIGURE 54

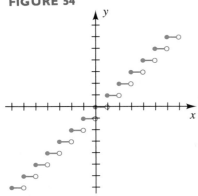

Values of x	$f(x) = [\![x]\!]$
$\vdots$	$\vdots$
$-2 \le x < -1$	-2
$-1 \le x < 0$	-1
$0 \le x < 1$	0
$1 \le x < 2$	1
$2 \le x < 3$	2
$\vdots$	$\vdots$

Whenever *x* is between successive integers, the corresponding part of the graph is a segment of a horizontal line. Part of the graph is sketched in Figure 54. The graph continues indefinitely to the right and to the left.

The next example involves absolute values.

FIGURE 55

(a)

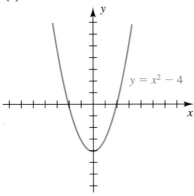

$y = x^2 - 4$

EXAMPLE 10 *Sketching the graph of an equation containing an absolute value*

Sketch the graph of $y = |x^2 - 4|$.

SOLUTION The graph of $y = x^2 - 4$ was sketched in Figure 48 and is resketched in Figure 55(a). We note the following facts:

(1) If $x \le -2$ or $x \ge 2$, then $x^2 - 4 \ge 0$, and hence $|x^2 - 4| = x^2 - 4$.

(2) If $-2 < x < 2$, then $x^2 - 4 < 0$, and hence $|x^2 - 4| = -(x^2 - 4)$.

It follows from (1) that the graphs of $y = |x^2 - 4|$ and $y = x^2 - 4$ coincide for $|x| \ge 2$. We see from (2) that if $|x| < 2$, then the graph of $y = |x^2 - 4|$ is the reflection of the graph of $y = x^2 - 4$ through the *x*-axis. This gives us the sketch in Figure 55(b).

(b)

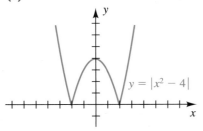

$y = |x^2 - 4|$

In Chapter 2 we used algebraic methods to solve inequalities involving absolute values of polynomials of degree 1, such as

$$|2x - 5| < 7 \quad \text{and} \quad |5x + 2| \geq 3.$$

Much more complicated inequalities can be investigated using a graphing utility, as illustrated in the next example.

EXAMPLE 11   *Solving an absolute value inequality graphically*

Estimate the solutions of

$$|0.14x^2 - 13.72| > |0.58x| + 11.$$

SOLUTION   To solve the inequality, we make the assignments

$$Y_1 = \text{ABS}(0.14x^2 - 13.72) \quad \text{and} \quad Y_2 = \text{ABS}(0.58x) + 11$$

FIGURE 56 [−30, 30] by [0, 40]

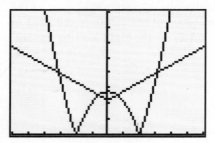

and estimate the values of x for which the graph of Y_1 is *above* the graph of Y_2. After perhaps several trials, we choose the viewing rectangle [−30, 30] by [0, 40], obtaining graphs similar to those in Figure 56, where each tic represents 5 units. Since there is symmetry with respect to the y-axis, it is sufficient to find the x-coordinates of the points of intersection of the graphs for $x > 0$. Using tracing and zoom-in features, we obtain $x \approx 2.80$ and $x \approx 15.52$. Referring to Figure 56, we obtain the (approximate) solution

$$(-\infty, -15.52) \cup (-2.80, 2.80) \cup (15.52, \infty).$$

3.5 EXERCISES

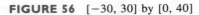

Exer. 1–10: Determine whether f is even, odd, or neither even nor odd.

1 $f(x) = 5x^3 + 2x$

2 $f(x) = |x| - 3$

3 $f(x) = 3x^4 + 2x^2 - 5$

4 $f(x) = 7x^5 - 4x^3$

5 $f(x) = 8x^3 - 3x^2$

6 $f(x) = 12$

7 $f(x) = \sqrt{x^2 + 4}$

8 $f(x) = 3x^2 - 5x + 1$

9 $f(x) = \sqrt[3]{x^3 - x}$

10 $f(x) = x^3 - \dfrac{1}{x}$

Exer. 11–24: Sketch, on the same coordinate plane, the graphs of f for the given values of c. (Make use of symmetry, shifting, stretching, compressing, or reflecting.)

11 $f(x) = |x| + c; \quad c = -3, 1, 3$

12 $f(x) = |x - c|; \quad c = -3, 1, 3$

13 $f(x) = -x^2 + c; \quad c = -4, 2, 4$

14 $f(x) = 2x^2 - c; \quad c = -4, 2, 4$

15 $f(x) = 2\sqrt{x} + c; \quad c = -3, 0, 2$

16 $f(x) = \sqrt{9 - x^2} + c$; $c = -3, 0, 2$

17 $f(x) = \frac{1}{2}\sqrt{x - c}$; $c = -2, 0, 3$

18 $f(x) = -\frac{1}{2}(x - c)^2$; $c = -2, 0, 3$

19 $f(x) = c\sqrt{4 - x^2}$; $c = -2, 1, 3$

20 $f(x) = (x + c)^3$; $c = -2, 1, 2$

21 $f(x) = cx^3$; $c = -\frac{1}{3}, 1, 2$

22 $f(x) = (cx)^3 + 1$; $c = -1, 1, 4$

23 $f(x) = \sqrt{cx} - 1$; $c = -1, \frac{1}{9}, 4$

24 $f(x) = -\sqrt{16 - (cx)^2}$; $c = 1, \frac{1}{2}, 4$

Exer. 25–26: The graph of a function f with domain [0, 4] is shown in the figure. Sketch the graph of the given equation.

25

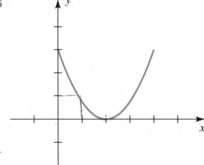

(a) $y = f(x + 3)$ (b) $y = f(x - 3)$

(c) $y = f(x) + 3$ (d) $y = f(x) - 3$

(e) $y = -3f(x)$ (f) $y = -\frac{1}{3}f(x)$

(g) $y = f(-\frac{1}{2}x)$ (h) $y = f(2x)$

(i) $y = -f(x + 2) - 3$ (j) $y = f(x - 2) + 3$

26

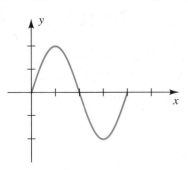

(a) $y = f(x - 2)$ (b) $y = f(x + 2)$

(c) $y = f(x) - 2$ (d) $y = f(x) + 2$

(e) $y = -2f(x)$ (f) $y = -\frac{1}{2}f(x)$

(g) $y = f(-2x)$ (h) $y = f(\frac{1}{2}x)$

(i) $y = -f(x + 4) - 2$ (j) $y = f(x - 4) + 2$

Exer. 27–30: The graph of a function f is shown, together with graphs of three other functions (a), (b), and (c). Use properties of symmetry, shifts, and reflecting to find equations for graphs (a), (b), and (c) in terms of f.

27

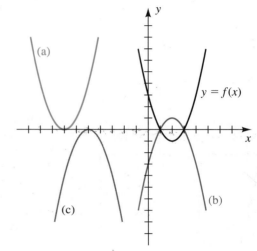

28

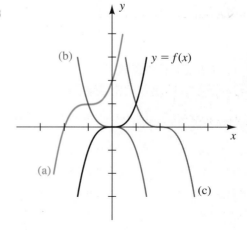

29

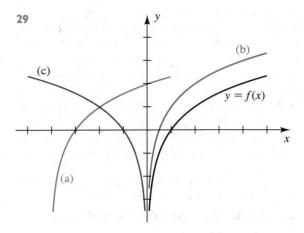

30

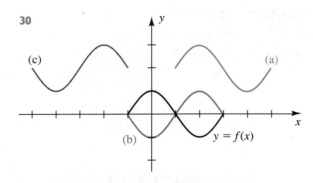

Exer. 31–36: Sketch the graph of f.

31 $f(x) = \begin{cases} 3 & \text{if } x \le -1 \\ -2 & \text{if } x > -1 \end{cases}$

32 $f(x) = \begin{cases} -1 & \text{if } x \text{ is an integer} \\ -2 & \text{if } x \text{ is not an integer} \end{cases}$

33 $f(x) = \begin{cases} 3 & \text{if } x < -2 \\ -x + 1 & \text{if } |x| \le 2 \\ -3 & \text{if } x > 2 \end{cases}$

34 $f(x) = \begin{cases} -2x & \text{if } x < -1 \\ x^2 & \text{if } -1 \le x < 1 \\ -2 & \text{if } x \ge 1 \end{cases}$

35 $f(x) = \begin{cases} x + 2 & \text{if } x \le -1 \\ x^3 & \text{if } |x| < 1 \\ -x + 3 & \text{if } x \ge 1 \end{cases}$

36 $f(x) = \begin{cases} x - 3 & \text{if } x \le -2 \\ -x^2 & \text{if } -2 < x < 1 \\ -x + 4 & \text{if } x \ge 1 \end{cases}$

Exer. 37–38: The symbol $[\![x]\!]$ denotes values of the greatest integer function. Sketch the graph of f.

37 (a) $f(x) = [\![x - 3]\!]$ **(b)** $f(x) = [\![x]\!] - 3$

   (c) $f(x) = 2[\![x]\!]$ **(d)** $f(x) = [\![2x]\!]$

   (e) $f(x) = [\![-x]\!]$

38 (a) $f(x) = [\![x + 2]\!]$ **(b)** $f(x) = [\![x]\!] + 2$

   (c) $f(x) = \frac{1}{2}[\![x]\!]$ **(d)** $f(x) = [\![\frac{1}{2}x]\!]$

   (e) $f(x) = -[\![-x]\!]$

Exer. 39–40: Explain why the graph of the equation is not the graph of a function.

39 $x = y^2$ **40** $x = -|y|$

41 If f is an odd function and g is an even function, is fg even, odd, or neither even nor odd?

42 There is one function with domain $\mathbb{R}$ that is both even and odd. Find that function.

Exer. 43–46: Sketch the graph of the equation.

43 $y = |9 - x^2|$ **44** $y = |x^3 - 1|$

45 $y = |\sqrt{x} - 1|$ **46** $y = ||x| - 1|$

Exer. 47–48: For the graph of $y = f(x)$ shown in the figure, sketch the graph of $y = |f(x)|$.

47

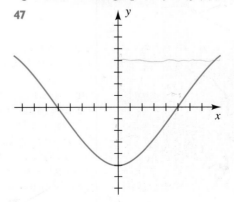

48

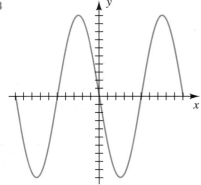

49 *Income tax rates* A certain country taxes the first $20,000 of an individual's income at a rate of 15%, and all income over $20,000 is taxed at 20%. Find a piecewise-defined function T that specifies the total tax on an income of x dollars.

50 *Telephone rates* A telephone company charges 25 cents for a long-distance call that does not exceed one minute; for longer calls it charges 15 cents for each additional minute. Find a piecewise-defined function C that specifies the total cost of a long-distance call of x minutes.

51 *Royalty rates* A certain paperback sells for $12. The author is paid royalties of 10% on the first 10,000 copies sold, 12.5% on the next 5000 copies, and 15% on any additional copies. Find a piecewise-defined function R that specifies the total royalties if x copies are sold.

52 *Electricity rates* An electric company charges its customers $0.0577 per kilowatt-hour (kWh) for the first 1000 kWh used, $0.0532 for the next 4000 kWh, and $0.0511 for any kWh over 5000. Find a piecewise-defined function C for a customer's bill of x kWh.

c Exer. 53–56: Estimate the solutions of the inequality.

53 $|1.3x + 2.8| < 1.2x + 5$

54 $|0.3x| - 2 > 2.2 - 0.63x^2$

55 $|1.2x^2 - 10.8| > 1.36x + 4.08$

56 $|\sqrt{16 - x^2} - 3| < 0.12x^2 - 0.3$

3.6 QUADRATIC FUNCTIONS

If $a \neq 0$, then the graph of $y = ax^2$ is a parabola with vertex at the origin $(0, 0)$, opening upward if $a > 0$ or downward if $a < 0$ (see, for example, Figures 50 and 51). In this section we show that the graph of an equation of the form

$$y = ax^2 + bx + c$$

FIGURE 57

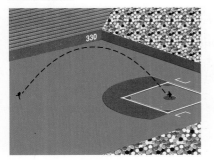

can be obtained by vertical and/or horizontal shifts of the graph of $y = ax^2$ and hence is also a parabola. An important application of such equations is to describe the trajectory, or path, of an object near the surface of the earth when the only force acting on the object is gravitational attraction. To illustrate, if an outfielder on a baseball team throws a ball into the infield, as illustrated in Figure 57, and if air resistance and other outside forces are negligible, then the path of the ball is a parabola. If suitable coordinate axes are introduced, then the path coincides with the graph of $y = ax^2 + bx + c$ for some a, b, and c. We call the function determined by this equation a *quadratic function*.

Definition of Quadratic Function

A function f is a **quadratic function** if

$$f(x) = ax^2 + bx + c,$$

where a, b, and c are real numbers with $a \neq 0$.

If $b = c = 0$ in the preceding definition, then $f(x) = ax^2$, and the graph is a parabola with vertex at the origin. If $b = 0$ and $c \neq 0$, then

$$f(x) = ax^2 + c,$$

and, from our discussion of vertical shifts in Section 3.5, the graph is a parabola with vertex at the point $(0, c)$ on the y-axis. The following example contains additional illustrations.

FIGURE 58

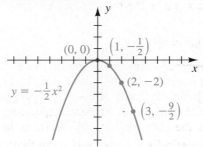

EXAMPLE 1   *Sketching the graph of a quadratic function*

Sketch the graph of f if

(a) $f(x) = -\frac{1}{2}x^2$ **(b)** $f(x) = -\frac{1}{2}x^2 + 4$

SOLUTION

(a) Since f is even, the graph of f (that is, of $y = -\frac{1}{2}x^2$) is symmetric with respect to the y-axis. It is similar in shape to but wider than the parabola $y = -x^2$, sketched in Figure 51 of Section 3.5. Several points on the graph are $(0, 0)$, $(1, -\frac{1}{2})$, $(2, -2)$, and $(3, -\frac{9}{2})$. Plotting and using symmetry, we obtain the sketch in Figure 58.

(b) To find the graph of $y = -\frac{1}{2}x^2 + 4$, we shift the graph of $y = -\frac{1}{2}x^2$ upward a distance 4, obtaining the sketch in Figure 59.

FIGURE 59

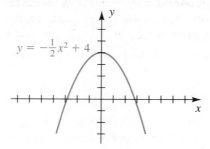

If $f(x) = ax^2 + bx + c$ and $b \neq 0$, then, by completing the square, we can change the form to

$$f(x) = a(x - h)^2 + k$$

for some real numbers h and k. This technique is illustrated in the next example.

EXAMPLE 2   *Expressing a quadratic function as*
$$f(x) = a(x - h)^2 + k$$

If $f(x) = 3x^2 + 24x + 50$, express $f(x)$ in the form $a(x - h)^2 + k$.

SOLUTION   Before completing the square, *it is essential that we factor out the coefficient of x^2 from the first two terms of $f(x)$*, as follows:

$$f(x) = 3x^2 + 24x + 50 \qquad \text{given}$$
$$= 3(x^2 + 8x + \quad) + 50 \qquad \text{factor out 3 from } 3x^2 + 24x$$

We now complete the square for the expression $x^2 + 8x$ within the parentheses by adding the square of half the coefficient of x—that is, $(\frac{8}{2})^2$, or 16. However, if we add 16 to the expression within parentheses, then, because of the factor 3, we are actually adding 48 to $f(x)$. Hence, we must compensate by subtracting 48:

$$
\begin{array}{ll}
f(x) = 3(x^2 + 8x + \quad) + 50 & \text{given} \\
\quad = 3(x^2 + 8x + 16) + (50 - 48) & \text{complete the square for } x^2 + 8x \\
\quad = 3(x + 4)^2 + 2 & \text{equivalent equation}
\end{array}
$$

The last expression has the form $a(x - h)^2 + k$ with $a = 3$, $h = -4$, and $k = 2$.

If $f(x) = ax^2 + bx + c$, then, by completing the square as in Example 2, we see that the graph of f is the same as the graph of an equation of the form

$$ y = a(x - h)^2 + k. $$

The graph of this equation can be obtained from the graph of $y = ax^2$ shown in Figure 60(a) by means of a horizontal and a vertical shift, as follows. First, as in Figure 60(b), we obtain the graph of $y = a(x - h)^2$ by shifting the graph of $y = ax^2$ either to the left or to the right, depending on the sign of h (the figure illustrates the case with $h > 0$). Next, as in Figure 60(c), we shift the graph in (b) vertically a distance $|k|$ (the figure illustrates the case with $k > 0$). It follows that *the graph of a quadratic function is a parabola with a vertical axis.*

The sketch in Figure 60(c) illustrates one possible graph of the equation $y = ax^2 + bx + c$. If $a > 0$, the point (h, k) is the lowest point on the parabola, and the function f has a **minimum value** $f(h) = k$. If $a < 0$, the

FIGURE 60

(a)

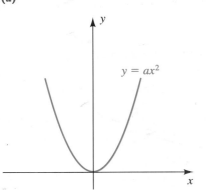

(b)

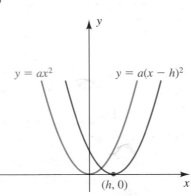

(c)

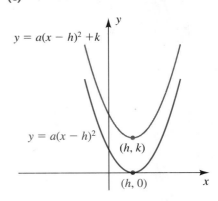

parabola opens downward, and the point (h, k) is the highest point on the parabola. In this case, the function f has a **maximum value** $f(h) = k$.

We have obtained the following result.

Standard Equation of a Parabola with Vertical Axis

> The graph of the equation
>
> $$y = a(x - h)^2 + k$$
>
> for $a \neq 0$ is a parabola that has vertex $V(h, k)$ and a vertical axis. The parabola opens upward if $a > 0$ or downward if $a < 0$.

For convenience, we often refer to the *parabola $y = ax^2 + bx + c$* when considering the graph of this equation.

EXAMPLE 3   *Finding a standard equation of a parabola*

Express $y = 2x^2 - 6x + 4$ as a standard equation of a parabola with a vertical axis. Find the vertex and sketch the graph.

FIGURE 61

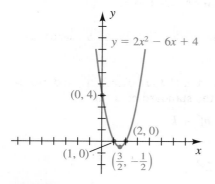

SOLUTION

$$
\begin{aligned}
y &= 2x^2 - 6x + 4 && \text{given} \\
&= 2(x^2 - 3x + \phantom{\tfrac{9}{4}}) + 4 && \text{factor out 2 from } 2x^2 - 6x \\
&= 2(x^2 - 3x + \tfrac{9}{4}) + (4 - \tfrac{9}{2}) && \text{complete the square for } x^2 - 3x \\
&= 2(x - \tfrac{3}{2})^2 - \tfrac{1}{2} && \text{equivalent equation}
\end{aligned}
$$

The last equation has the form of the standard equation of a parabola with $a = 2$, $h = \frac{3}{2}$, and $k = -\frac{1}{2}$. Hence, the vertex $V(h, k)$ of the parabola is $V(\frac{3}{2}, -\frac{1}{2})$. Since $a = 2 > 0$, the parabola opens upward.

To find the y-intercept of the graph of $y = 2x^2 - 6x + 4$, we let $x = 0$, obtaining $y = 4$. To find the x-intercepts, we let $y = 0$ and solve the equation $2x^2 - 6x + 4 = 0$ or the equivalent equation $2(x - 1)(x - 2) = 0$, obtaining $x = 1$ and $x = 2$. Plotting the vertex and using the x- and y-intercepts provides enough points for a reasonably accurate sketch (see Figure 61).

FIGURE 62

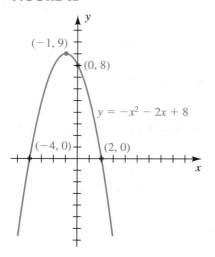

EXAMPLE 4   *Finding a standard equation of a parabola*

Express $y = -x^2 - 2x + 8$ as a standard equation of a parabola with a vertical axis. Find the vertex and sketch the graph.

SOLUTION

$$
\begin{aligned}
y &= -x^2 - 2x + 8 && \text{given} \\
&= -(x^2 + 2x +) + 8 && \text{factor out } -1 \text{ from } -x^2 - 2x \\
&= -(x^2 + 2x + 1) + (8 + 1) && \text{complete the square for } x^2 + 2x \\
&= -(x + 1)^2 + 9 && \text{equivalent equation}
\end{aligned}
$$

This is the standard equation of a parabola with $h = -1, k = 9$, and hence the vertex is $(-1, 9)$. Since $a = -1 < 0$, the parabola opens downward.

The y-intercept of the graph of $y = -x^2 - 2x + 8$ is the constant term, 8. To find the x-intercepts, we solve $-x^2 - 2x + 8 = 0$ or, equivalently, $x^2 + 2x - 8 = 0$. Factoring gives us $(x + 4)(x - 2) = 0$, and hence the intercepts are $x = -4$ and $x = 2$. Using this information gives us the sketch in Figure 62.

FIGURE 63

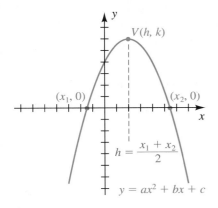

If a parabola $y = ax^2 + bx + c$ has x-intercepts x_1 and x_2, as illustrated in Figure 63 for the case $a < 0$, then the axis of the parabola is the vertical line $x = (x_1 + x_2)/2$ through the midpoint of $(x_1, 0)$ and $(x_2, 0)$. Therefore, the x-coordinate h of the vertex (h, k) is $h = (x_1 + x_2)/2$. Some special cases are illustrated in Figures 61 and 62.

In the following example we find an equation of a parabola from given data.

EXAMPLE 5   *Finding an equation of a parabola with a given vertex*

Find an equation of a parabola that has vertex $V(2, 3)$ and a vertical axis and passes through the point $(5, 1)$.

FIGURE 64

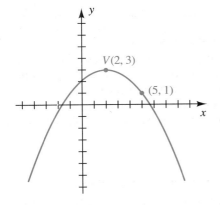

SOLUTION   Figure 64 shows the vertex V, the point $(5, 1)$, and a possible position of the parabola. Using the standard equation

$$y = a(x - h)^2 + k$$

with $h = 2$ and $k = 3$ gives us

$$y = a(x - 2)^2 + 3.$$

To find a, we use the fact that $(5, 1)$ is on the parabola and so is a solution of the last equation. Thus,

$$1 = a(5 - 2)^2 + 3, \quad \text{or} \quad a = -\tfrac{2}{9}.$$

Hence, an equation for the parabola is

$$y = -\tfrac{2}{9}(x - 2)^2 + 3.$$

The next theorem gives us a simple formula for locating the vertex of a parabola.

Theorem for Locating the Vertex of a Parabola

The vertex of the parabola $y = ax^2 + bx + c$ has x-coordinate

$$-\frac{b}{2a}.$$

PROOF   Let us begin by writing $y = ax^2 + bx + c$ as

$$y = a\left(x^2 + \frac{b}{a}x + \right) + c.$$

Next we complete the square by adding $\left(\frac{1}{2}\frac{b}{a}\right)^2$ to the expression within parentheses:

$$y = a\left(x^2 + \frac{b}{a}x + \frac{b^2}{4a^2}\right) + \left(c - \frac{b^2}{4a}\right)$$

Note that if $b^2/(4a^2)$ is added *inside* the parentheses, then, because of the factor a on the *outside*, we have actually added $b^2/(4a)$ to y. Therefore, we must compensate by subtracting $b^2/(4a)$. The last equation may be written

$$y = a\left(x + \frac{b}{2a}\right)^2 + \left(c - \frac{b^2}{4a}\right).$$

This is the equation of a parabola that has vertex (h, k) with $h = -b/(2a)$ and $k = c - b^2/(4a)$. ∎

It is unnecessary to remember the formula for the y-coordinate of the vertex of the parabola in the preceding result. Once the x-coordinate has been found, we can calculate the y-coordinate by substituting $-b/(2a)$ for x in the equation of the parabola.

EXAMPLE 6   *Finding the vertex of a parabola*

Find the vertex of the parabola $y = 2x^2 - 6x + 4$.

SOLUTION   We considered this parabola in Example 3 and found the vertex by completing the square. We shall now use the vertex formula with $a = 2$ and $b = -6$, obtaining the x-coordinate

$$\frac{-b}{2a} = \frac{-(-6)}{2(2)} = \frac{6}{4} = \frac{3}{2}.$$

We next find the y-coordinate by substituting $\frac{3}{2}$ for x in the given equation:

$$y = 2(\tfrac{3}{2})^2 - 6(\tfrac{3}{2}) + 4 = -\tfrac{1}{2}$$

Thus, the vertex is $(\tfrac{3}{2}, -\tfrac{1}{2})$ (see Figure 61).

Since the graph of $f(x) = ax^2 + bx + c$ for $a \neq 0$ is a parabola, we can use the vertex formula to help find the maximum or minimum value of a quadratic function. Specifically, since the x-coordinate of the vertex V is $-b/(2a)$, the y-coordinate of V is the function value $f(-b/(2a))$. Moreover, since the parabola opens downward if $a < 0$ and upward if $a > 0$, this function value is the maximum or minimum value, respectively, of f. We may summarize these facts as follows.

Theorem on the Maximum or Minimum Value of a Quadratic Function

If $f(x) = ax^2 + bx + c$, where $a \neq 0$, then $f\left(-\dfrac{b}{2a}\right)$ is

(1) the maximum value of f if $a < 0$

(2) the minimum value of f if $a > 0$

We shall use this theorem in the next example.

EXAMPLE 7 *Finding the maximum value of a quadratic function*

A long rectangular sheet of metal, 12 inches wide, is to be made into a rain gutter by turning up two sides so that they are perpendicular to the sheet. How many inches should be turned up to give the gutter its greatest capacity?

FIGURE 65

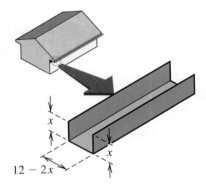

$12 - 2x$

SOLUTION The gutter is illustrated in Figure 65. If x denotes the number of inches turned up on each side, the width of the base of the gutter is $12 - 2x$ inches. The capacity will be greatest when the cross-sectional area of the rectangle with sides of lengths x and $12 - 2x$ has its greatest value. Letting $f(x)$ denote this area, we have

$$f(x) = x(12 - 2x)$$
$$= 12x - 2x^2$$
$$= -2x^2 + 12x,$$

which has the form $f(x) = ax^2 + bx + c$ with $a = -2$, $b = 12$, and $c = 0$. Since f is a quadratic function and $a = -2 < 0$, it follows from the preceding theorem that the maximum value of f occurs at

$$x = -\frac{b}{2a} = -\frac{12}{2(-2)} = 3.$$

Thus, 3 inches should be turned up on each side to achieve maximum capacity.

As an alternative solution, we may note that the graph of the function $f(x) = x(12 - 2x)$ has x-intercepts at $x = 0$ and $x = 6$. Hence, the average of the intercepts,

$$x = \frac{0 + 6}{2} = 3,$$

is the x-coordinate of the vertex of the parabola and the value that yields the maximum capacity.

In Chapter 2 we solved quadratic equations and inequalities algebraically. The next example indicates how they can be solved with the aid of a graphing utility.

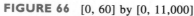

EXAMPLE 8   *Analyzing the flight of a projectile*

A projectile is fired vertically upward from a height of 600 feet above the ground. Its height $h(t)$ in feet, after t seconds, is given by

$$h(t) = -16t^2 + 803t + 600.$$

(a) Determine a reasonable viewing rectangle that includes all pertinent features of the graph of h.

(b) Use a graphing utility to estimate when the height is 5000 feet.

(c) Determine when the projectile will be more than 5000 feet above the ground.

(d) How long will the projectile be in flight?

SOLUTION

(a) The graph of h is a parabola that opens downward. To estimate Ymax (note that we use x and y interchangeably with t and h), let us approximate the maximum value of h. Using

$$t = -\frac{b}{2a} = -\frac{803}{2(-16)} \approx 25.1,$$

we see that the maximum height is approximately $h(25) = 10{,}675$.

The projectile rises for approximately the first 25 seconds, and because its height at $t = 0$, 600 feet, is small in comparison to 10,675, it will take only slightly more than an additional 25 seconds to fall to the ground. Since h and t are positive, a reasonable viewing rectangle is

$$[0, 60] \quad \text{by} \quad [0, 11{,}000].$$

(b) We wish to estimate where the graph of h intersects the horizontal line $h(t) = 5000$, so we make the assignments

$$Y_1 = -16x^2 + 803x + 600$$

and

$$Y_2 = 5000$$

and obtain a display similar to Figure 66. It is important to remember that the graph of Y_1 shows only the height at time t—it is *not* the path of the projectile, which is vertical. Using the tracing and zoom-in features, we find that the smallest value of t for which $h(t) = 5000$ is about 6.3 seconds.

Since the vertex is on the axis of the parabola, the other time at which $h(t)$ is 5000 is approximately $25.1 - 6.3$, or 18.8, seconds *after* $t = 25.1$—that is, at $t \approx 25.1 + 18.8 = 43.9$ sec.

FIGURE 66   [0, 60] by [0, 11,000]

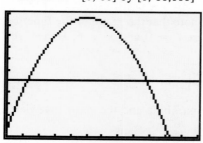

(c) The projectile is more than 5000 feet above the ground when the graph of the parabola in Figure 66 is above the horizontal line—that is, when

$$6.3 < t < 43.9.$$

(d) The projectile will be in flight until $h(t) = 0$. This corresponds to the x-intercept in Figure 66. Using the tracing and zoom-in features, we obtain $t \approx 50.9$ sec. (Note that since the y-intercept is not zero, it is incorrect to merely double the t value of the vertex to find the total time of the flight; however, this *would* be acceptable for problems in which $h(0) = 0$.)

We will discuss parabolas further in Section 4.6.

3.6 EXERCISES

Exer. 1–4: Find the standard equation of any parabola that has vertex V.

1 $V(-3, 1)$ 2 $V(4, -2)$

3 $V(0, -3)$ 4 $V(-2, 0)$

Exer. 5–12: Express $f(x)$ in the form $a(x - h)^2 + k$.

5 $f(x) = -x^2 - 4x - 8$ 6 $f(x) = x^2 - 6x + 11$

7 $f(x) = 2x^2 - 12x + 22$ 8 $f(x) = 5x^2 + 20x + 17$

9 $f(x) = -3x^2 - 6x - 5$

10 $f(x) = -4x^2 + 16x - 13$

11 $f(x) = -\frac{3}{4}x^2 + 9x - 34$

12 $f(x) = \frac{2}{5}x^2 - \frac{12}{5}x + \frac{23}{5}$

Exer. 13–22: (a) Find the maximum or minimum value of $f(x)$. (b) Use the quadratic formula to find the zeros of f. (c) Sketch the graph of f.

13 $f(x) = x^2 - 4x$ 14 $f(x) = -x^2 - 6x$

15 $f(x) = -12x^2 + 11x + 15$

16 $f(x) = 6x^2 + 7x - 24$

17 $f(x) = 9x^2 + 24x + 16$ 18 $f(x) = -4x^2 + 4x - 1$

19 $f(x) = x^2 + 4x + 9$ 20 $f(x) = -3x^2 - 6x - 6$

21 $f(x) = -2x^2 + 20x - 43$

22 $f(x) = 2x^2 - 4x - 11$

Exer. 23–26: Find the standard equation of the parabola shown in the figure.

23

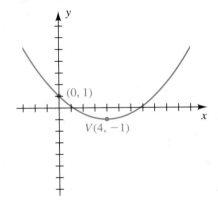

24

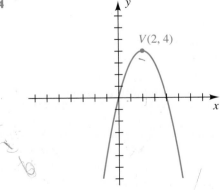

25

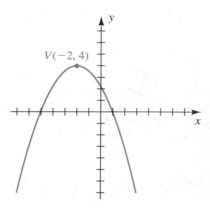

26

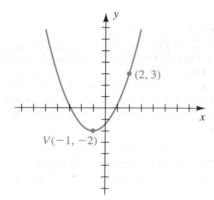

Exer. 27–32: Find the standard equation of a parabola that has a vertical axis and satisfies the given conditions.

27 Vertex $(0, -2)$, passing through $(3, 25)$

28 Vertex $(0, 5)$, passing through $(2, -3)$

29 Vertex $(3, 5)$, x-intercept 0

30 Vertex $(4, -7)$, x-intercept -4

31 x-intercepts -3 and 5, highest point has y-coordinate 4

32 x-intercepts 8 and 0, lowest point has y-coordinate -48

Exer. 33–34: Ozone occurs at all levels of the earth's atmosphere. The density of ozone varies both seasonally and latitudinally. At Edmonton, Canada, the density $D(h)$ of ozone (in 10^{-3} cm/km) for altitudes h between 20 kilometers and 35 kilometers was determined experimentally. For each $D(h)$ and season, approximate the altitude at which the density of ozone is greatest.

33 $D(h) = -0.058h^2 + 2.867h - 24.239$ (autumn)

34 $D(h) = -0.078h^2 + 3.811h - 32.433$ (spring)

35 *Infant growth rate*   The growth rate y (in pounds per month) of an infant is related to present weight x (in pounds) by the formula $y = cx(21 - x)$, where c is a positive constant and $0 < x < 21$. At what weight does the maximum growth rate occur?

36 *Gasoline mileage*   The number of miles M that a certain automobile can travel on one gallon of gasoline at a speed of v mi/hr is given by

$$M = -\tfrac{1}{30}v^2 + \tfrac{5}{2}v \quad \text{for } 0 < v < 70.$$

(a) Find the most economical speed for a trip.

(b) Find the largest value of M.

37 *Height of a projectile*   An object is projected vertically upward from the top of a building with an initial velocity of 144 ft/sec. Its distance $s(t)$ in feet above the ground after t seconds is given by $s(t) = -16t^2 + 144t + 100$.

(a) Find its maximum distance above the ground.

(b) Find the height of the building.

38 *Flight of a projectile*   An object is projected vertically upward with an initial velocity of v_0 ft/sec, and its distance $s(t)$ in feet above the ground after t seconds is given by $s(t) = -16t^2 + v_0t$.

(a) If the object hits the ground after 12 seconds, find its initial velocity v_0.

(b) Find its maximum distance above the ground.

39 Find two positive real numbers whose sum is 40 and whose product is a maximum.

40 Find two real numbers whose difference is 40 and whose product is a minimum.

41 *Constructing cages*   One thousand feet of chain-link fence is to be used to construct six animal cages, as shown in the figure.

EXERCISE 41

(a) Express the width y as a function of the length x.

(b) Express the total enclosed area A of the exhibit as a function of x.

(c) Find the dimensions that maximize the enclosed area.

42 Fencing a field A farmer wishes to put a fence around a rectangular field and then divide the field into three rectangular plots by placing two fences parallel to one of the sides. If the farmer can afford only 1000 yards of fencing, what dimensions will give the maximum rectangular area?

43 Leaping animals Flights of leaping animals typically have parabolic paths. The figure illustrates a frog jump superimposed on a coordinate plane. The length of the leap is 9 feet, and the maximum height off the ground is 3 feet. Find a standard equation for the path of the frog.

EXERCISE 43

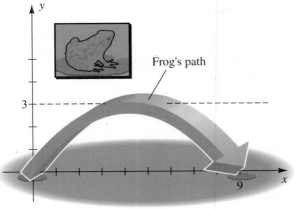

44 The human cannonball In the 1940s, the human cannonball stunt was performed regularly by Emmanuel Zacchini for The Ringling Brothers and Barnum & Bailey Circus. The tip of the cannon rose 15 feet off the ground, and the total horizontal distance traveled was 175 feet. When the cannon is aimed at an angle of 45°, an equation of the parabolic flight (see the figure) has the form $y = ax^2 + x + c$.

(a) Use the given information to find an equation of the flight.

(b) Find the maximum height attained by the human cannonball.

EXERCISE 44

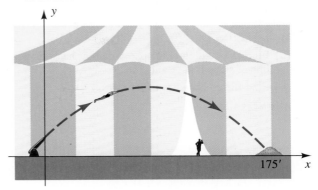

45 Shape of a suspension bridge One section of a suspension bridge has its weight uniformly distributed between twin towers that are 400 feet apart and rise 90 feet above the horizontal roadway (see the figure). A cable strung between the tops of the towers has the shape of a parabola, and its center point is 10 feet above the roadway. Suppose coordinate axes are introduced, as shown in the figure.

(a) Find an equation for the parabola.

(b) Nine equally spaced vertical cables are used to support the bridge (see the figure). Find the total length of these supports.

EXERCISE 45

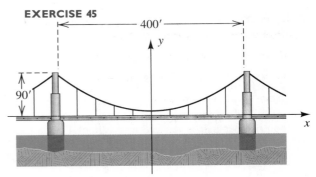

46 Designing a highway Traffic engineers are designing a stretch of highway that will connect a horizontal highway with one having a 20% grade (that is, slope $\frac{1}{5}$), as illustrated in the figure. The smooth transition is to take place over a horizontal distance of 800 feet, with a parabolic piece of highway used to connect points A and B.

If the equation of the parabolic segment is of the form $y = ax^2 + bx + c$, it can be shown that the slope of the tangent line at the point $P(x, y)$ on the parabola is given by $m = 2ax + b$.

(a) Find an equation of the parabola that has a tangent line of slope 0 at A and $\frac{1}{5}$ at B.

(b) Find the coordinates of B.

EXERCISE 46

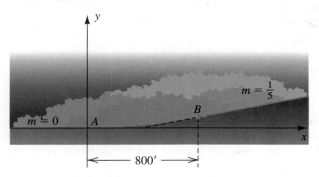

47 **Parabolic doorway** A doorway has the shape of a parabolic arch and is 9 feet high at the center and 6 feet wide at the base. If a rectangular box 8 feet high must fit through the doorway, what is the maximum width the box can have?

48 **Wire rectangle** A piece of wire 24 inches long is bent into the shape of a rectangle having width x and length y.

(a) Express y as a function of x.

(b) Express the area A of the rectangle as a function of x.

(c) Show that the area A is greatest if the rectangle is a square.

49 **Quantity discount** A company sells running shoes to dealers at a rate of $40 per pair if less than 50 pairs are ordered. If a dealer orders 50 or more pairs (up to 600), the price per pair is reduced at a rate of 4 cents times the number ordered. What size order will produce the maximum amount of money for the company?

50 **Group discount** A travel agency offers group tours at a rate of $60 per person for the first 30 participants. For larger groups—up to 90—each person receives a $0.50 discount for every participant in excess of 30. For example, if 31 people participate, then the cost per person is $59.50. Determine the size of the group that will produce the maximum amount of money for the agency.

51 **Cable TV fee** A cable television firm presently serves 5000 households and charges $20 per month. A marketing survey indicates that each decrease of $1 in the monthly charge will result in 500 new customers. Let $R(x)$ denote the total monthly revenue when the monthly charge is x dollars.

(a) Determine the revenue function R.

(b) Sketch the graph of R and find the value of x that results in maximum monthly revenue.

52 **Apartment rentals** A real estate company owns 180 efficiency apartments, which are fully occupied when the rent is $300 per month. The company estimates that for each $10 increase in rent, 5 apartments will become unoccupied. What rent should be charged so that the company will receive the maximum monthly income?

c Exer. 53–54: Graph $y = x^3 - x^{1/3}$ and f on the same coordinate plane, and estimate the points of intersection.

53 $f(x) = x^2 - x - \frac{1}{4}$

54 $f(x) = -x^2 + 0.5x + 0.4$

c 55 Graph, on the same coordinate plane, $y = ax^2 + x + 1$ for $a = \frac{1}{4}, \frac{1}{2}, 1, 2$, and 4, and describe how the value of a affects the graph.

c 56 Graph, on the same coordinate plane, $y = x^2 + bx + 1$ for $b = 0, \pm 1, \pm 2$, and ± 3, and describe how the value of b affects the graph.

3.7 OPERATIONS ON FUNCTIONS

Functions are often defined using sums, differences, products, and quotients of various expressions. For example, if

$$h(x) = x^2 + \sqrt{5x + 1},$$

we may regard $h(x)$ as a sum of values of the functions f and g given by

$$f(x) = x^2 \quad \text{and} \quad g(x) = \sqrt{5x + 1}.$$

We call h the *sum* of f and g and denote it by $f + g$. Thus,

$$h(x) = (f + g)(x) = x^2 + \sqrt{5x + 1}.$$

In general, if f and g are *any* functions, we use the terminology and notation given in the following chart.

Terminology	Function value
sum $f + g$	$(f + g)(x) = f(x) + g(x)$
difference $f - g$	$(f - g)(x) = f(x) - g(x)$
product fg	$(fg)(x) = f(x)g(x)$
quotient $\dfrac{f}{g}$	$\left(\dfrac{f}{g}\right)(x) = \dfrac{f(x)}{g(x)}$

The domains of $f + g$, $f - g$, and fg are the intersection I of the domains of f and g—that is, the numbers that are *common* to both domains. The domain of f/g is the subset of I consisting of all x in I such that $g(x) \neq 0$.

EXAMPLE I   *Finding function values of $f + g$, $f - g$, fg, and f/g*

If $f(x) = 3x - 2$ and $g(x) = x^3$, find $(f + g)(2)$, $(f - g)(2)$, $(fg)(2)$, and $(f/g)(2)$.

SOLUTION   Since $f(2) = 3(2) - 2 = 4$ and $g(2) = 2^3 = 8$, we have

$$(f + g)(2) = f(2) + g(2) = 4 + 8 = 12$$

$$(f - g)(2) = f(2) - g(2) = 4 - 8 = -4$$

$$(fg)(2) = f(2)g(2) = (4)(8) = 32$$

$$\left(\frac{f}{g}\right)(2) = \frac{f(2)}{g(2)} = \frac{4}{8} = \frac{1}{2}$$

EXAMPLE 2   *Finding $(f + g)(x)$, $(f - g)(x)$, $(fg)(x)$, and $(f/g)(x)$*

If $f(x) = \sqrt{4 - x^2}$ and $g(x) = 3x + 1$, find $(f + g)(x)$, $(f - g)(x)$, $(fg)(x)$, and $(f/g)(x)$, and state the domains of the respective functions.

SOLUTION   The domain of f is the closed interval $[-2, 2]$, and the domain of g is $\mathbb{R}$. The intersection of these domains is $[-2, 2]$, which is the domain of $f + g$, $f - g$, and fg. For the domain of f/g we exclude each number x in $[-2, 2]$ such that $g(x) = 3x + 1 = 0$ (namely, $x = -\frac{1}{3}$).

(continued)

Thus, we have the following:

$$(f + g)(x) = \sqrt{4 - x^2} + (3x + 1), \qquad -2 \leq x \leq 2$$

$$(f - g)(x) = \sqrt{4 - x^2} - (3x + 1), \qquad -2 \leq x \leq 2$$

$$(fg)(x) = \sqrt{4 - x^2}(3x + 1), \qquad -2 \leq x \leq 2$$

$$\left(\frac{f}{g}\right)(x) = \frac{\sqrt{4 - x^2}}{3x + 1}, \qquad -2 \leq x \leq 2 \text{ and } x \neq -\frac{1}{3}$$

A function f is a **polynomial function** if $f(x)$ is a polynomial—that is, if

$$f(x) = a_n x^n + a_{n-1} x^{n-1} + \cdots + a_1 x + a_0,$$

where the coefficients $a_0, a_1, \ldots, a_n$ are real numbers and the exponents are nonnegative integers. A polynomial function may be regarded as a sum of functions whose values are of the form cx^k, where c is a real number and k is a nonnegative integer. Note that the quadratic functions considered in the previous section are polynomial functions.

An **algebraic function** is a function that can be expressed in terms of finite sums, differences, products, quotients, or roots of polynomial functions.

ILLUSTRATION

Algebraic Function

■ $f(x) = 5x^4 - 2\sqrt[3]{x} + \dfrac{x(x^2 + 5)}{\sqrt{x^3 + \sqrt{x}}}$

Functions that are not algebraic are **transcendental**. The exponential and logarithmic functions considered in Chapter 5 are examples of transcendental functions.

In the remainder of this section we shall discuss how two functions f and g may be used to obtain the *composite functions* $f \circ g$ and $g \circ f$ (read "f circle g" and "g circle f," respectively). Functions of this type are very important in calculus. The function $f \circ g$ is defined as follows.

Definition of Composite Function

The **composite function** $f \circ g$ of two functions f and g is defined by

$$(f \circ g)(x) = f(g(x)).$$

The domain of $f \circ g$ is the set of all x in the domain of g such that $g(x)$ is in the domain of f.

Figure 67 is a schematic diagram that illustrates relationships among f, g, and $f \circ g$. Note that for x in the domain of g, *first we find $g(x)$* (which must be in the domain of f) and then, *second, we find $f(g(x))$*.

FIGURE 67

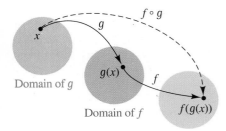

Domain of g

$g(x)$

f

Domain of f

$f(g(x))$

$f \circ g$

g

x

For the composite function $g \circ f$, we reverse this order, first finding $f(x)$ and second finding $g(f(x))$. The domain of $g \circ f$ is the set of all x in the domain of f such that $f(x)$ is in the domain of g.

Since the notation $g(x)$ is read "g of x," we sometimes say that *g is a function of x*. For the composite function $f \circ g$, the notation $f(g(x))$ is read "*f* of *g* of *x*," and we could regard f as a function of $g(x)$. In this sense, *a composite function is a function of a function* or, more precisely, a function of another function's values.

EXAMPLE 3 *Finding composite functions*

Let $f(x) = x^2 - 1$ and $g(x) = 3x + 5$.

(a) Find $(f \circ g)(x)$ and the domain of $f \circ g$.

(b) Find $(g \circ f)(x)$ and the domain of $g \circ f$.

(c) Find $f(g(2))$ in two different ways: first using the functions f and g separately and second using the composite function $f \circ g$.

SOLUTION

(a)

$$(f \circ g)(x) = f(g(x)) \qquad \text{definition of } f \circ g$$
$$= f(3x + 5) \qquad \text{definition of } g$$
$$= (3x + 5)^2 - 1 \qquad \text{definition of } f$$
$$= 9x^2 + 30x + 24 \qquad \text{simplify}$$

The domain of both f and g is $\mathbb{R}$. Since for each x in $\mathbb{R}$ (the domain of g), the function value $g(x)$ is in $\mathbb{R}$ (the domain of f), the domain of $f \circ g$ is also $\mathbb{R}$.

(b)

$$(g \circ f)(x) = g(f(x)) \qquad \text{definition of } g \circ f$$
$$= g(x^2 - 1) \qquad \text{definition of } f$$
$$= 3(x^2 - 1) + 5 \qquad \text{definition of } g$$
$$= 3x^2 + 2 \qquad \text{simplify}$$

Since for each x in $\mathbb{R}$ (the domain of f), the function value $f(x)$ is in $\mathbb{R}$ (the domain of g), the domain of $g \circ f$ is $\mathbb{R}$.

(c) To find $f(g(2))$ using $f(x) = x^2 - 1$ and $g(x) = 3x + 5$ separately, we may proceed as follows:

$$g(2) = 3(2) + 5 = 11$$

$$f(g(2)) = f(11) = 11^2 - 1 = 120$$

To find $f(g(2))$ using $f \circ g$, we refer to part (a), where we found

$$(f \circ g)(x) = f(g(x)) = 9x^2 + 30x + 24.$$

Hence,

$$f(g(2)) = 9(2)^2 + 30(2) + 24$$
$$= 36 + 60 + 24 = 120.$$

Note that in Example 3, $f(g(x))$ and $g(f(x))$ are not always the same; that is, $f \circ g \neq g \circ f$.

If two functions f and g both have domain $\mathbb{R}$, then the domain of $f \circ g$ and $g \circ f$ is also $\mathbb{R}$. This was illustrated in Example 3. The next example shows that the domain of a composite function may differ from those of the two given functions.

EXAMPLE 4   *Finding composite functions*

Let $f(x) = x^2 - 16$ and $g(x) = \sqrt{x}$.

(a) Find $(f \circ g)(x)$ and the domain of $f \circ g$.

(b) Find $(g \circ f)(x)$ and the domain of $g \circ f$.

SOLUTION   We first note that the domain of f is $\mathbb{R}$ and the domain of g is the set of all nonnegative real numbers—that is, the interval $[0, \infty)$. We may proceed as follows.

(a)
$$
\begin{aligned}
(f \circ g)(x) &= f(g(x)) &&\text{definition of } f \circ g \\
&= f(\sqrt{x}) &&\text{definition of } g \\
&= (\sqrt{x})^2 - 16 &&\text{definition of } f \\
&= x - 16 &&\text{simplify}
\end{aligned}
$$

If we consider only the final expression $x - 16$, we might be led to believe that the domain of $f \circ g$ is $\mathbb{R}$, since $x - 16$ is defined for every real number x. However, this is not the case. By definition, the domain of $f \circ g$ is the set of all x in $[0, \infty)$ (the domain of g) such that $g(x)$ is in $\mathbb{R}$ (the domain of f). Since $g(x) = \sqrt{x}$ is in $\mathbb{R}$ for every x in $[0, \infty)$, it follows that the domain of $f \circ g$ is $[0, \infty)$.

(b)
$$
\begin{aligned}
(g \circ f)(x) &= g(f(x)) &&\text{definition of } g \circ f \\
&= g(x^2 - 16) &&\text{definition of } f \\
&= \sqrt{x^2 - 16} &&\text{definition of } g
\end{aligned}
$$

By definition, the domain of $g \circ f$ is the set of all x in $\mathbb{R}$ (the domain of f) such that $f(x) = x^2 - 16$ is in $[0, \infty)$ (the domain of g). The statement "$x^2 - 16$ is in $[0, \infty)$" is equivalent to each of the inequalities

$$ x^2 - 16 \geq 0, \qquad x^2 \geq 16, \qquad |x| \geq 4. $$

Thus, the domain of $g \circ f$ is the union $(-\infty, -4] \cup [4, \infty)$. Note that this domain is different from the domains of both f and g.

The next example illustrates how special values of composite functions may sometimes be obtained from tables.

EXAMPLE 5 *Finding composite function values from tables*

Several values of two functions f and g are listed in the following tables.

x	1	2	3	4
$f(x)$	3	4	2	1

x	1	2	3	4
$g(x)$	4	1	3	2

Find $(f \circ g)(2)$, $(g \circ f)(2)$, $(f \circ f)(2)$, and $(g \circ g)(2)$.

SOLUTION Using the definition of composite function and referring to the tables above, we obtain

$$(f \circ g)(2) = f(g(2)) = f(1) = 3$$
$$(g \circ f)(2) = g(f(2)) = g(4) = 2$$
$$(f \circ f)(2) = f(f(2)) = f(4) = 1$$
$$(g \circ g)(2) = g(g(2)) = g(1) = 4.$$

In some applied problems it is necessary to express a quantity y as a function of time t. The following example illustrates that it is often easier to introduce a third variable x, express x as a function of t (that is, $x = g(t)$), express y as a function of x (that is, $y = f(x)$), and finally form the composite function given by $y = f(x) = f(g(t))$.

EXAMPLE 6 *Using a composite function to find the volume of a balloon*

A meteorologist is inflating a spherical balloon with helium gas. If the radius of the balloon is changing at a rate of 1.5 cm/sec, express the volume V of the balloon as a function of time t (in seconds).

SOLUTION Let x denote the radius of the balloon. If we assume that the radius is 0 initially, then after t seconds

$$x = 1.5t. \quad \text{radius of balloon after } t \text{ seconds}$$

To illustrate, after 1 second, the radius is 1.5 centimeters; after 2 seconds, it is 3.0 centimeters; after 3 seconds, it is 4.5 centimeters; and so on.

Next we write

$$V = \tfrac{4}{3}\pi x^3. \quad \text{volume of a sphere of radius } x$$

This gives us a composite function relationship in which V is a function of x and x is a function of t. By substitution, we obtain

$$V = \tfrac{4}{3}\pi x^3 = \tfrac{4}{3}\pi(1.5t)^3 = \tfrac{4}{3}\pi(\tfrac{3}{2}t)^3 = \tfrac{4}{3}\pi(\tfrac{27}{8}t^3).$$

Simplifying, we obtain the following formula for V as a function of t:

$$V = \tfrac{9}{2}\pi t^3$$

If f and g are functions such that

$$y = f(u) \qquad \text{and} \qquad u = g(x),$$

then substituting for u in $y = f(u)$ yields

$$y = f(g(x)).$$

For certain problems in calculus we *reverse* this procedure; that is, given $y = h(x)$ for some function h, we find a *composite function form* $y = f(u)$ and $u = g(x)$ such that $h(x) = f(g(x))$.

EXAMPLE 7   *Finding a composite function form*

Express $y = (2x + 5)^8$ as a composite function form.

SOLUTION   Suppose, for a real number x, we wanted to evaluate $(2x + 5)^8$ by using a calculator. We would first calculate $2x + 5$ and then raise the result to the eighth power. This suggests that we let

$$u = 2x + 5 \qquad \text{and} \qquad y = u^8,$$

which is a composite function form for $y = (2x + 5)^8$.

The method used in the preceding example can be extended to other functions. In general, suppose we are given $y = h(x)$. To choose the *inside* expression $u = g(x)$ in a composite function form, ask the following question: If a calculator were being used, which part of the expression $h(x)$ would be evaluated first? This often leads to a suitable choice for $u = g(x)$. After choosing u, refer to $h(x)$ to determine $y = f(u)$. The following illustration contains typical problems.

ILLUSTRATION

Composite Function Forms

Function value	Choice for $u = g(x)$	Choice for $y = f(u)$
$y = (x^3 - 5x + 1)^4$	$u = x^3 - 5x + 1$	$y = u^4$
$y = \sqrt{x^2 - 4}$	$u = x^2 - 4$	$y = \sqrt{u}$
$y = \dfrac{2}{3x + 7}$	$u = 3x + 7$	$y = \dfrac{2}{u}$

The composite function form is never unique. For example, consider the first expression in the preceding illustration:

$$y = (x^3 - 5x + 1)^4$$

If n is any nonzero integer, we could choose

$$u = (x^3 - 5x + 1)^n \qquad \text{and} \qquad y = u^{4/n}.$$

Thus, there are an *unlimited* number of composite function forms. Generally, our goal is to choose a form such that the expression for y is simple, as we did in the illustration.

The next example illustrates how a graphing utility can help determine the domain of a composite function. We use the same functions that appeared in Example 4.

EXAMPLE 8 *Graphically analyzing a composite function*

Let $f(x) = x^2 - 16$ and $g(x) = \sqrt{x}$.

(a) Find $f(g(3))$.

(b) Sketch $y = (f \circ g)(x)$ and use the graph to find the domain of $f \circ g$.

SOLUTION

(a) We begin by making the assignments

$$Y_1 = \sqrt{x} \quad \text{and} \quad Y_2 = (Y_1)^2 - 16.$$

Note that we have substituted Y_1 for x in $f(x)$ and assigned this expression to Y_2, much the same way as we substituted $g(x)$ for x in Example 4.

Next we store the value 3 in the memory location for x and then query the value of Y_2. We see that the value of Y_2 at 3 is -13; that is, $f(g(3)) = -13$.

(b) To determine a viewing rectangle for the graph of $f \circ g$, we first note that $f(x) \geq -16$ for all x and therefore choose Ymin *less* than -16; say, Ymin $= -20$. If we want the rectangle to have a vertical dimension of 40, we must choose Ymax $= 20$.

If your screen is in 1:1 proportion (horizontal:vertical), then a reasonable choice for [Xmin, Xmax] would be [$-10, 30$], a horizontal dimension of 40. If your screen is in 3:2 proportion, choose [Xmin, Xmax] to be [$-10, 50$], a horizontal dimension of 60.

Selecting Y_2 and then displaying the graph of Y_2 using the viewing rectangle [$-10, 50$] by [$-20, 20$] gives us a graph similar to Figure 68. We see that the graph is a half-line with end point $(0, -16)$. Thus, the domain of Y_2 is all $x \geq 0$.

FIGURE 68   [$-10, 50$] by [$-20, 20$]

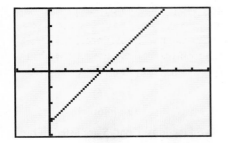

3.7 EXERCISES

Exer. 1–2: Find

(a) $(f + g)(3)$

(b) $(f - g)(3)$

(c) $(fg)(3)$

(d) $(f/g)(3)$

1 $f(x) = x + 3, \quad g(x) = x^2$

2 $f(x) = -x^2, \quad g(x) = 2x - 1$

Exer. 3–8: Find

(a) $(f + g)(x), (f - g)(x), (fg)(x),$ and $(f/g)(x)$

(b) the domain of $f + g, f - g,$ and fg

(c) the domain of f/g

3 $f(x) = x^2 + 2, \quad g(x) = 2x^2 - 1$

4 $f(x) = x^2 + x,$ $g(x) = x^2 - 3$

5 $f(x) = \sqrt{x + 5},$ $g(x) = \sqrt{x + 5}$

6 $f(x) = \sqrt{3 - 2x},$ $g(x) = \sqrt{x + 4}$

7 $f(x) = \dfrac{2x}{x - 4},$ $g(x) = \dfrac{x}{x + 5}$

8 $f(x) = \dfrac{x}{x - 2},$ $g(x) = \dfrac{3x}{x + 4}$

Exer. 9–10: Find

(a) $(f \circ g)(x)$ (b) $(g \circ f)(x)$

(c) $(f \circ f)(x)$ (d) $(g \circ g)(x)$

9 $f(x) = 2x - 1,$ $g(x) = -x^2$

10 $f(x) = 3x^2,$ $g(x) = x - 1$

Exer. 11–20: Find

(a) $(f \circ g)(x)$ (b) $(g \circ f)(x)$

(c) $f(g(-2))$ (d) $g(f(3))$

11 $f(x) = 2x - 5,$ $g(x) = 3x + 7$

12 $f(x) = 5x + 2,$ $g(x) = 6x - 1$

13 $f(x) = 3x^2 + 4,$ $g(x) = 5x$

14 $f(x) = 3x - 1,$ $g(x) = 4x^2$

15 $f(x) = 2x^2 + 3x - 4,$ $g(x) = 2x - 1$

16 $f(x) = 5x - 7,$ $g(x) = 3x^2 - x + 2$

17 $f(x) = 4x,$ $g(x) = 2x^3 - 5x$

18 $f(x) = x^3 + 2x^2,$ $g(x) = 3x$

19 $f(x) = |x|,$ $g(x) = -7$

20 $f(x) = 5,$ $g(x) = x^2$

Exer. 21–34: Find (a) $(f \circ g)(x)$ and the domain of $f \circ g$ and (b) $(g \circ f)(x)$ and the domain of $g \circ f$.

21 $f(x) = x^2 - 3x,$ $g(x) = \sqrt{x + 2}$

22 $f(x) = \sqrt{x - 15},$ $g(x) = x^2 + 2x$

23 $f(x) = x^2 - 4,$ $g(x) = \sqrt{3x}$

24 $f(x) = -x^2 + 1,$ $g(x) = \sqrt{x}$

25 $f(x) = \sqrt{x - 2},$ $g(x) = \sqrt{x + 5}$

26 $f(x) = \sqrt{3 - x},$ $g(x) = \sqrt{x + 2}$

27 $f(x) = \sqrt{3 - x},$ $g(x) = \sqrt{x^2 - 16}$

28 $f(x) = x^3 + 5,$ $g(x) = \sqrt[3]{x - 5}$

29 $f(x) = \dfrac{3x + 5}{2},$ $g(x) = \dfrac{2x - 5}{3}$

30 $f(x) = \dfrac{1}{x - 1},$ $g(x) = x - 1$

31 $f(x) = x^2,$ $g(x) = \dfrac{1}{x^3}$

32 $f(x) = \dfrac{x}{x - 2},$ $g(x) = \dfrac{3}{x}$

33 $f(x) = \dfrac{x - 1}{x - 2},$ $g(x) = \dfrac{x - 3}{x - 4}$

34 $f(x) = \dfrac{x + 2}{x - 1},$ $g(x) = \dfrac{x - 5}{x + 4}$

Exer. 35–36: Solve the equation $(f \circ g)(x) = 0$.

35 $f(x) = x^2 - 2,$ $g(x) = x + 3$

36 $f(x) = x^2 - x - 2,$ $g(x) = 2x - 1$

37 Several values of two functions f and g are listed in the following tables:

x	5	6	7	8
$f(x)$	8	7	6	5

x	5	6	7	8
$g(x)$	7	8	6	5

Find

(a) $(f \circ g)(6)$ (b) $(g \circ f)(6)$

(c) $(f \circ f)(6)$ (d) $(g \circ g)(6)$

38 Several values of two functions T and S are listed in the following tables:

t	0	1	2	3
$T(t)$	2	3	1	0

x	0	1	2	3
$S(x)$	1	0	3	2

Find

(a) $(T \circ S)(1)$ (b) $(S \circ T)(1)$

(c) $(T \circ T)(1)$ (d) $(S \circ S)(1)$

39 If $D(t) = \sqrt{400 + t^2}$ and $R(x) = 20x$, find $(D \circ R)(x)$.

40 If $S(r) = 4\pi r^2$ and $D(t) = 2t + 5$, find $(S \circ D)(t)$.

41 *Spreading fire*   A fire has started in a dry open field and is spreading in the form of a circle. If the radius of this

circle increases at the rate of 6 ft/min, express the total fire area A as a function of time t (in minutes).

42 Dimensions of a balloon A spherical balloon is being inflated at a rate of $\frac{9}{2}\pi$ ft^3/min. Express its radius r as a function of time t (in minutes), assuming that $r = 0$ when $t = 0$.

43 Dimensions of a sand pile The volume of a conical pile of sand is increasing at a rate of 243π ft^3/min, and the height of the pile always equals the radius r of the base. Express r as a function of time t (in minutes), assuming that $r = 0$ when $t = 0$.

44 Diameter of a cube The diameter d of a cube is the distance between two opposite vertices. Express d as a function of the edge x of the cube. (*Hint:* First express the diagonal y of a face as a function of x.)

45 Altitude of a balloon A hot-air balloon rises vertically from ground level as a rope attached to the base of the balloon is released at the rate of 5 ft/sec (see the figure). The pulley that releases the rope is 20 feet from a platform where passengers board the balloon. Express the altitude h of the balloon as a function of time t.

EXERCISE 45

46 Tightrope walker Refer to Exercise 58 of Section 3.4. Starting at the lowest point, the tightrope walker moves up the rope at a steady rate of 2 ft/sec. If the rope is at-

tached 30 feet up the pole, express the height h of the walker above the ground as a function of time t. (*Hint:* Let d denote the total distance traveled along the wire. First express d as a function of t, and then h as a function of d.)

47 Airplane take-off Refer to Exercise 59 of Section 3.4. When the airplane is 500 feet down the runway, it has reached a speed of 150 ft/sec (or about 102 mi/hr), which it will maintain until take-off. Express the distance d of the plane from the control tower as a function of time t (in seconds). (*Hint:* In the figure, first write x as a function of t.)

48 Cable corrosion A 100-foot-long cable of diameter 4 inches is submerged in seawater. Because of corrosion, the surface area of the cable decreases at the rate of 750 in.^{2} per year. Express the diameter d of the cable as a function of time t (in years). (Disregard corrosion at the ends of the cable.)

Exer. 49–56: Find a composite function form for y.

49 $y = (x^2 + 3x)^{1/3}$

50 $y = \sqrt[4]{x^4 - 16}$

51 $y = \dfrac{1}{(x-3)^4}$

52 $y = 4 + \sqrt{x^2 + 1}$

53 $y = (x^4 - 2x^2 + 5)^5$

54 $y = \dfrac{1}{(x^2 + 3x - 5)^3}$

55 $y = \dfrac{\sqrt{x+4} - 2}{\sqrt{x+4} + 2}$

56 $y = \dfrac{\sqrt[3]{x}}{1 + \sqrt[3]{x}}$

57 If $f(x) = \sqrt{x} - 1$ and $g(x) = x^3 + 1$, approximate $(f \circ g)(0.0001)$. In order to avoid calculating a zero value for $(f \circ g)(0.0001)$, rewrite the formula for $f \circ g$ as

$$\frac{x^3}{\sqrt{x^3 + 1} + 1}.$$

58 If $f(x) = \dfrac{x^3}{x^2 + x + 2}$ and $g(x) = (\sqrt{3x} - x^3)^{3/2}$, approximate

$$\frac{(f + g)(1.12) - (f/g)(1.12)}{[(f \circ f)(5.2)]^2}.$$

3.8 INVERSE FUNCTIONS

A function f may have the same value for different numbers in its domain. For example, if $f(x) = x^2$, then $f(2) = 4$ and $f(-2) = 4$, but $2 \neq -2$. For *the inverse of a function* to be defined, it is essential that different numbers in the domain *always* give different values of f. Such functions are called *one-to-one functions*.

Definition of One-to-One Function

A function f with domain D and range R is a **one-to-one function** if either of the following equivalent conditions is satisfied:

(1) Whenever $a \neq b$ in D, then $f(a) \neq f(b)$ in R.

(2) Whenever $f(a) = f(b)$ in R, then $a = b$ in D.

FIGURE 69

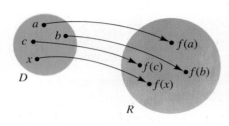

The arrow diagram in Figure 69 illustrates a one-to-one function. Note that each function value in the range R corresponds to *exactly one* element in the domain D. The function illustrated in Figure 37 of Section 3.4 is not one-to-one, since $f(w) = f(z)$, but $w \neq z$.

EXAMPLE 1   Determining whether a function is one-to-one

(a) If $f(x) = 3x + 2$, prove that f is one-to-one.

(b) If $g(x) = x^4 + 2x^2$, prove that g is not one-to-one.

SOLUTION

(a) We shall use condition (2) of the preceding definition. Thus, suppose that $f(a) = f(b)$ for some numbers a and b in the domain of f. This gives us

$$3a + 2 = 3b + 2 \qquad \text{definition of } f(x)$$
$$3a = 3b \qquad \text{subtract 2}$$
$$a = b \qquad \text{divide by 3}$$

Hence, f is one-to-one.

(b) Showing that a function *is* one-to-one requires a *general* proof, as in part (a). To show that g is *not* one-to-one we need only find two distinct real numbers in the domain that produce the same function value. For example, $-1 \neq 1$, but $g(-1) = g(1)$. In fact, since g is an even function, $f(-a) = f(a)$ for every real number a.

FIGURE 70

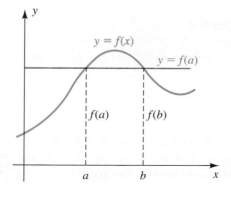

If we know the graph of a function f, it is easy to determine whether f is one-to-one. For example, the function whose graph is sketched in Figure 70 is not one-to-one, since $a \neq b$, but $f(a) = f(b)$. Note that the horizontal line $y = f(a)$ (or $y = f(b)$) intersects the graph in more than one point. In general, we may use the following graphical test to determine whether a function is one-to-one.

Horizontal Line Test

> A function f is one-to-one if and only if every horizontal line intersects the graph of f in at most one point.

Since every increasing function or decreasing function passes the horizontal line test, we obtain the following result.

Theorem: Increasing or Decreasing Functions Are One-to-One

> **(1)** A function that is increasing throughout its domain is one-to-one.
>
> **(2)** A function that is decreasing throughout its domain is one-to-one.

Let f be a one-to-one function with domain D and range R. Thus, for each number y in R, there is *exactly one* number x in D such that $y = f(x)$, as illustrated by the arrow in Figure 71(a). We may, therefore, define a function g from R to D by means of the following rule:

$$x = g(y)$$

As in Figure 71(b), *g reverses the correspondence given by* f. We call g the *inverse function* of f, as in the next definition.

FIGURE 71

(a) $y = f(x)$ **(b)** $x = g(y)$

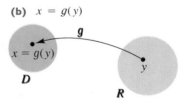

Definition of Inverse Function

> Let f be a one-to-one function with domain D and range R. A function g with domain R and range D is the **inverse function** of f, provided the following condition is true for every x in D and every y in R:
>
> $$y = f(x) \quad \text{if and only if} \quad x = g(y)$$

Remember that for the inverse of a function f to be defined, *it is absolutely essential that f be one-to-one*. The following theorem, stated without proof, is useful to verify that a function g is the inverse of f.

Theorem on Inverse Functions

> Let f be a one-to-one function with domain D and range R. If g is a function with domain R and range D, then g is the inverse function of f if and only if both of the following conditions are true:
>
> **(1)** $g(f(x)) = x$ for every x in D.
>
> **(2)** $f(g(y)) = y$ for every y in R.

FIGURE 72

(a) First f, then g

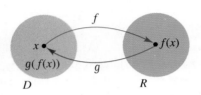

(b)

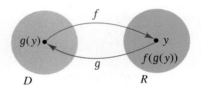

Conditions (1) and (2) of the preceding theorem are illustrated in Figure 72(a) and (b), respectively, where the blue arrow indicates that f is a function from D to R and the red arrow indicates that g is a function from R to D.

Note that in Figure 72(a) we first apply f to the number x in D, obtaining the function value $f(x)$ in R, and then apply g to $f(x)$, obtaining the number $g(f(x))$ in D. Condition (1) of the theorem states that $g(f(x)) = x$ for every x; that is, g *reverses* the correspondence given by f.

In Figure 72(b) we use the opposite order for the functions. We first apply g to the number y in R, obtaining the function value $g(y)$ in D, and then apply f to $g(y)$, obtaining the number $f(g(y))$ in R. Condition (2) of the theorem states that $f(g(y)) = y$ for every y; that is, f *reverses* the correspondence given by g.

If a function f has an inverse function g, we often denote g by f^{-1}. The -1 used in this notation should not be mistaken for an exponent; that is, $f^{-1}(y)$ *does not mean* $1/[f(y)]$. The reciprocal $1/[f(y)]$ may be denoted by $[f(y)]^{-1}$. It is important to remember the following facts about the domain and range of f and f^{-1}.

Domain and Range of f and f^{-1}

> domain of f^{-1} = range of f
>
> range of f^{-1} = domain of f

When we discuss functions, we often let x denote an arbitrary number in the domain. Thus, for the inverse function f^{-1}, we may wish to consider $f^{-1}(x)$, *where x is in the domain R of f^{-1}*. In this event, the two conditions in the theorem on inverse functions are written as follows:

(1) $f^{-1}(f(x)) = x$ for every x in the domain of f

(2) $f(f^{-1}(x)) = x$ for every x in the domain of f^{-1}

Figure 72 contains a hint for finding the inverse of a one-to-one function in certain cases: If possible, *we solve the equation $y = f(x)$ for x in terms of y*, obtaining an equation of the form $x = g(y)$. If the two conditions $g(f(x)) = x$ and $f(g(x)) = x$ are true for every x in the domains of f and g, respectively, then g is the required inverse function f^{-1}. The following guidelines summarize this procedure; in guideline 2, in anticipation of finding f^{-1}, we write $x = f^{-1}(y)$ instead of $x = g(y)$.

Guidelines for Finding f^{-1} in Simple Cases

> 1 Verify that f is a one-to-one function throughout its domain.
> 2 Solve the equation $y = f(x)$ for x in terms of y, obtaining an equation of the form $x = f^{-1}(y)$.
> 3 Verify the following two conditions:
>   a $f^{-1}(f(x)) = x$ for every x in the domain of f
>   b $f(f^{-1}(x)) = x$ for every x in the domain of f^{-1}

The success of this method depends on the nature of the equation $y = f(x)$, since we must be able to solve for x in terms of y. For this reason, we include the phrase *in simple cases* in the title of the guidelines. We shall follow these guidelines in the next three examples.

EXAMPLE 2 *Finding the inverse of a function*

Let $f(x) = 3x - 5$. Find the inverse function of f.

SOLUTION

Guideline 1 The graph of the linear function f is a line of slope 3, and hence f is increasing throughout $\mathbb{R}$. Thus, f is one-to-one and the inverse function f^{-1} exists. Moreover, since the domain and range of f is $\mathbb{R}$, the same is true for f^{-1}.

Guideline 2 Solve the equation $y = f(x)$ for x:

$$y = 3x - 5 \quad \text{let } y = f(x)$$

$$x = \frac{y + 5}{3} \quad \text{solve for } x \text{ in terms of } y$$

We now formally let $x = f^{-1}(y)$; that is,

$$f^{-1}(y) = \frac{y + 5}{3}.$$

Since the symbol used for the variable is immaterial, we may also write

$$f^{-1}(x) = \frac{x + 5}{3},$$

where x is in the domain of f^{-1}.

Guideline 3 Since the domain and range of both f and f^{-1} is $\mathbb{R}$, we must verify conditions (a) and (b) for every real number x. We proceed as follows:

(a)
$$f^{-1}(f(x)) = f^{-1}(3x - 5) \quad \text{definition of } f$$

$$= \frac{(3x - 5) + 5}{3} \quad \text{definition of } f^{-1}$$

$$= x \quad \text{simplify}$$

(continued)

(b)
$$f(f^{-1}(x)) = f\left(\frac{x+5}{3}\right) \qquad \text{definition of } f^{-1}$$

$$= 3\left(\frac{x+5}{3}\right) - 5 \qquad \text{definition of } f$$

$$= x \qquad \text{simplify}$$

These verifications prove that the inverse function of f is given by

$$f^{-1}(x) = \frac{x+5}{3}.$$

EXAMPLE 3   *Finding the inverse of a function*

Let $f(x) = x^2 - 3$ for $x \geq 0$. Find the inverse function of f.

SOLUTION

FIGURE 73

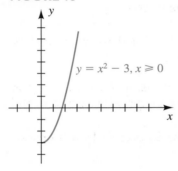

Guideline 1 The graph of f is sketched in Figure 73. The domain of f is $[0, \infty)$, and the range is $[-3, \infty)$. Since f is increasing, it is one-to-one and hence has an inverse function f^{-1} with domain $[-3, \infty)$ and range $[0, \infty)$.

Guideline 2 We consider the equation

$$y = x^2 - 3$$

and solve for x, obtaining

$$x = \pm\sqrt{y+3}.$$

Since x is nonnegative, we reject $x = -\sqrt{y+3}$ and let

$$f^{-1}(y) = \sqrt{y+3} \qquad \text{or, equivalently,} \qquad f^{-1}(x) = \sqrt{x+3}.$$

(Note that if the function f had domain $x \leq 0$, we would choose the function $f^{-1}(x) = -\sqrt{x+3}$.)

Guideline 3 We verify conditions (a) and (b) for x in the domains of f and f^{-1}, respectively.

(a) $f^{-1}(f(x)) = f^{-1}(x^2 - 3) = \sqrt{(x^2 - 3) + 3} = \sqrt{x^2} = x$ for $x \geq 0$

(b) $f(f^{-1}(x)) = f(\sqrt{x+3})$
$$= (\sqrt{x+3})^2 - 3 = (x+3) - 3 = x \text{ for } x \geq -3$$

Thus, the inverse function is given by

$$f^{-1}(x) = \sqrt{x+3} \quad \text{for } x \geq -3.$$

There is an interesting relationship between the graph of a function f and the graph of its inverse function f^{-1}. We first note that $b = f(a)$ is

FIGURE 74

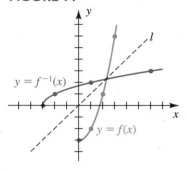

FIGURE 75

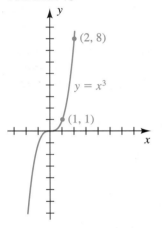

FIGURE 76

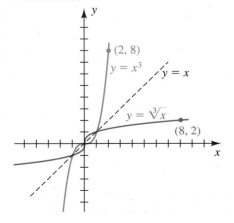

equivalent to $a = f^{-1}(b)$. These equations imply that *the point (a, b) is on the graph of f if and only if the point (b, a) is on the graph of f^{-1}.*

As an illustration, in Example 3 we found that the functions f and f^{-1} given by

$$f(x) = x^2 - 3 \qquad \text{and} \qquad f^{-1}(x) = \sqrt{x + 3}$$

are inverse functions of each other, provided that x is suitably restricted. Some points on the graph of f are $(0, -3)$, $(1, -2)$, $(2, 1)$, and $(3, 6)$. Corresponding points on the graph of f^{-1} are $(-3, 0)$, $(-2, 1)$, $(1, 2)$, and $(6, 3)$. The graphs of f and f^{-1} are sketched on the same coordinate plane in Figure 74. If the page is folded along the line $y = x$ that bisects quadrants I and III (as indicated by the dashes in the figure), then the graphs of f and f^{-1} coincide. The two graphs are *reflections* of each other through the line $y = x$, or are *symmetric* with respect to this line. This is typical of the graph of every function f that has an inverse function f^{-1} (see Exercise 34).

EXAMPLE 4 **The relationship between the graphs of f and f^{-1}**

Let $f(x) = x^3$. Find the inverse function f^{-1} of f, and sketch the graphs of f and f^{-1} on the same coordinate plane.

SOLUTION The graph of f is sketched in Figure 75. Note that f is an odd function, and hence the graph is symmetric with respect to the origin.

Guideline 1 Since f is increasing throughout its domain $\mathbb{R}$, it is one-to-one and hence has an inverse function f^{-1}.

Guideline 2 We consider the equation

$$y = x^3$$

and solve for x by taking the cube root of each side, obtaining

$$x = y^{1/3} = \sqrt[3]{y}.$$

We now let

$$f^{-1}(y) = \sqrt[3]{y} \qquad \text{or, equivalently,} \qquad f^{-1}(x) = \sqrt[3]{x}.$$

Guideline 3 We verify conditions (a) and (b):

(a) $f^{-1}(f(x)) = f^{-1}(x^3) = \sqrt[3]{x^3} = x$ for every x in $\mathbb{R}$

(b) $f(f^{-1}(x)) = f(\sqrt[3]{x}) = (\sqrt[3]{x})^3 = x$ for every x in $\mathbb{R}$

The graph of f^{-1} (that is, the graph of the equation $y = \sqrt[3]{x}$) may be obtained by reflecting the graph in Figure 75 through the line $y = x$, as shown in Figure 76. Three points on the graph of f^{-1} are $(0, 0)$, $(1, 1)$, and $(8, 2)$.

In the next example we illustrate how some of the concepts presented in this section are helpful in determining the sketch of a graph with the aid of a graphing utility.

EXAMPLE 5   *Sketching the graph of the inverse of a function*

(a) Sketch the graph of $f(x) = \dfrac{x}{\sqrt{x^2 + 1}}$.

(b) Explain why f is one-to-one, and sketch the graph of f^{-1}.

SOLUTION

FIGURE 77 $[-15, 15]$ by $[-10, 10]$

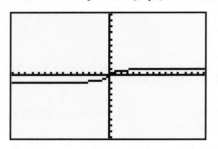

(a) We assign $x/\sqrt{x^2 + 1}$ to Y_1 and use a standard viewing rectangle to obtain a display similar to Figure 77. The graph is close to 1 if x is large positive and close to -1 if x is large negative. To improve the graph, we change the viewing rectangle to $[-15, 15]$ by $[-1, 1]$, obtaining a display similar to Figure 78.

(b) By the horizontal line test, the function f appears to be one-to-one, with domain $\mathbb{R}$ and range $(-1, 1)$. (These facts can be proved algebraically; however, we shall not do so here.) Hence, f has an inverse function f^{-1} that has domain $(-1, 1)$ and range $\mathbb{R}$.

To find f^{-1}, we solve the equation $y = f(x)$ for x in terms of y, as follows:

FIGURE 78 $[-15, 15]$ by $[-1, 1]$

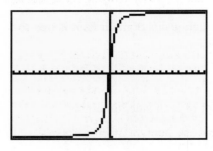

$$y = \frac{x}{\sqrt{x^2 + 1}} \qquad \text{given}$$

$$y\sqrt{x^2 + 1} = x \qquad \text{multiply by } \sqrt{x^2 + 1}$$

$$y^2(x^2 + 1) = x^2 \qquad \text{square both sides}$$

$$y^2x^2 + y^2 = x^2 \qquad \text{multiply}$$

$$(y^2 - 1)x^2 = -y^2 \qquad \text{combine } x^2 \text{ terms}$$

$$x^2 = \frac{y^2}{1 - y^2} \qquad \text{divide by } 1 - y^2 \text{ and change signs}$$

FIGURE 79 $[-1, 1]$ by $[-15, 15]$

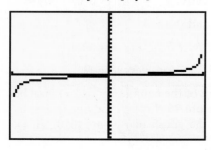

$$x = \pm\sqrt{\frac{y^2}{1 - y^2}} = \pm\frac{\sqrt{y^2}}{\sqrt{1 - y^2}} \qquad \text{take the square root}$$

$$x = \pm\frac{|y|}{\sqrt{1 - y^2}} = \pm\frac{y}{\sqrt{1 - y^2}} \qquad \sqrt{y^2} = |y|$$

Referring to Figure 78, we see that x and y are *always both positive or both negative*, so the last equation can be simplified to

$$x = \frac{y}{\sqrt{1 - y^2}}.$$

Hence,

$$f^{-1}(y) = \frac{y}{\sqrt{1 - y^2}} \qquad \text{or, equivalently,} \qquad f^{-1}(x) = \frac{x}{\sqrt{1 - x^2}}.$$

Using the viewing rectangle $[-1, 1]$ by $[-15, 15]$ gives us a display similar to Figure 79.

3.8 EXERCISES

Exer. 1–12: Determine whether the function f is one-to-one.

1 $f(x) = 3x - 7$

2 $f(x) = \dfrac{1}{x - 2}$

3 $f(x) = x^2 - 9$

4 $f(x) = x^2 + 4$

5 $f(x) = \sqrt{x}$

6 $f(x) = \sqrt[3]{x}$

7 $f(x) = |x|$

8 $f(x) = 3$

9 $f(x) = \sqrt{4 - x^2}$

10 $f(x) = 2x^3 - 4$

11 $f(x) = \dfrac{1}{x}$

12 $f(x) = \dfrac{1}{x^2}$

Exer. 13–16: Use the theorem on inverse functions to prove that f and g are inverse functions of each other, and sketch the graphs of f and g on the same coordinate plane.

13 $f(x) = 3x - 2;$ $\qquad$ $g(x) = \dfrac{x + 2}{3}$

14 $f(x) = x^2 + 5, x \leq 0;$ $\quad$ $g(x) = -\sqrt{x - 5}, x \geq 5$

15 $f(x) = -x^2 + 3, x \geq 0;$ $\quad$ $g(x) = \sqrt{3 - x}, x \leq 3$

16 $f(x) = x^3 - 4;$ $\qquad$ $g(x) = \sqrt[3]{x + 4}$

Exer. 17–32: Find the inverse function of f.

17 $f(x) = 3x + 5$

18 $f(x) = 7 - 2x$

19 $f(x) = \dfrac{1}{3x - 2}$

20 $f(x) = \dfrac{1}{x + 3}$

21 $f(x) = \dfrac{3x + 2}{2x - 5}$

22 $f(x) = \dfrac{4x}{x - 2}$

23 $f(x) = 2 - 3x^2, x \leq 0$

24 $f(x) = 5x^2 + 2, x \geq 0$

25 $f(x) = 2x^3 - 5$

26 $f(x) = -x^3 + 2$

27 $f(x) = \sqrt{3 - x}$

28 $f(x) = \sqrt{4 - x^2}, 0 \leq x \leq 2$

29 $f(x) = \sqrt[3]{x} + 1$

30 $f(x) = (x^3 + 1)^5$

31 $f(x) = x$

32 $f(x) = -x$

33 **(a)** Prove that the function defined by $f(x) = ax + b$ (a linear function) for $a \neq 0$ has an inverse function, and find $f^{-1}(x)$.

 (b) Does a constant function have an inverse? Explain.

34 Show that the graph of f^{-1} is the reflection of the graph of f through the line $y = x$ by verifying the following conditions:

(1) If $P(a, b)$ is on the graph of f, then $Q(b, a)$ is on the graph of f^{-1}.

(2) The midpoint of line segment PQ is on the line $y = x$.

(3) The line PQ is perpendicular to the line $y = x$.

Exer. 35–38: The graph of a one-to-one function f is shown. **(a)** Use the reflection property to sketch the graph of f^{-1}. **(b)** Find the domain D and range R of the function f. **(c)** Find the domain D_1 and range R_1 of the inverse function f^{-1}.

35

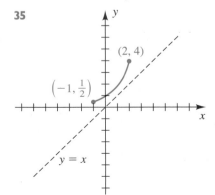

36

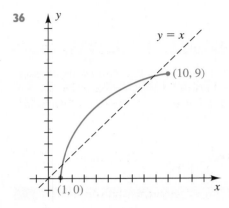

37

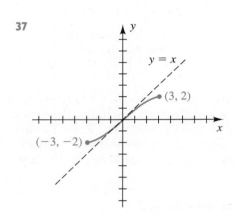

38

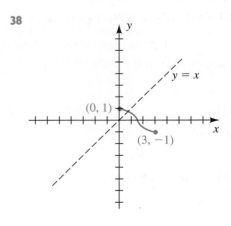

39 Verify that $f(x) = f^{-1}(x)$ if

(a) $f(x) = -x + b$ (b) $f(x) = \dfrac{ax + b}{cx - a}$ for $c \neq 0$

(c) $f(x)$ has the following graph:

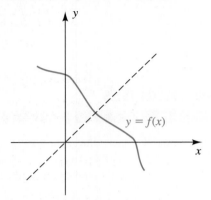

40 Let n be any positive integer. Find the inverse function of f if

(a) $f(x) = x^n$ for $x \geq 0$

(b) $f(x) = x^{m/n}$ for $x \geq 0$ and m any positive integer

c **Exer. 41–42: Use the graph of f to determine whether f is one-to-one.**

41 $f(x) = 0.4x^5 - 0.4x^4 + 1.2x^3 - 1.2x^2 + 0.8x - 0.8$

42 $f(x) = \dfrac{x - 8}{x^{2/3} + 4}$

c **Exer. 43–44: Graph f on the given interval. (a) Estimate the largest interval $[a, b]$ with $a < 0 < b$ on which f is one-to-one. (b) If g is the function with domain $[a, b]$ such that $g(x) = f(x)$ for $a \leq x \leq b$, estimate the domain and range of g^{-1}.**

43 $f(x) = 2.1x^3 - 2.98x^2 - 2.11x + 3;$ $[-1, 2]$

44 $f(x) = 0.05x^4 - 0.24x^3 - 0.15x^2 + 1.18x + 0.24;$
$[-2, 2]$

c **45** *Density of the ozone layer* The density D (in 10^{-3} cm/km) of the ozone layer at altitudes x between 3 and 15 kilometers during winter at Edmonton, Canada, was determined experimentally to be

$$D = 0.0833x^2 - 0.4996x + 3.5491.$$

Express x as a function of D.

3.9 VARIATION

In some scientific investigations, the terminology of *variation* or *proportion* is used to describe relationships between variable quantities. In the following chart, k is a nonzero real number called a **constant of variation** or a **constant of proportionality**.

Terminology	General formula	Illustration
y **varies directly** as x, or y is **directly proportional** to x	$y = kx$	$C = 2\pi r$, where C is the circumference of a circle, r is the radius, and $k = 2\pi$
y **varies inversely** as x, or y is **inversely proportional** to x	$y = \dfrac{k}{x}$	$I = \dfrac{110}{R}$, where I is the current in an electrical circuit, R is the resistance, and $k = 110$ is the voltage

The variable x in the chart can also represent a power. For example, the formula $A = \pi r^2$ states that the area A of a circle varies directly as the *square* of the radius r, where π is the constant of variation. Similarly, the formula $V = \frac{4}{3}\pi r^3$ states that the volume V of a sphere is directly proportional to the *cube* of the radius. In this case the constant of proportionality is $\frac{4}{3}\pi$.

EXAMPLE I *Directly proportional variables*

Suppose a variable q is directly proportional to a variable z.

(a) If $q = 12$ when $z = 5$, determine the constant of proportionality.

(b) Find the value of q when $z = 7$.

SOLUTION   Since q is directly proportional to z,

$$q = kz,$$

where k is a constant of proportionality.

(a) Substituting $q = 12$ and $z = 5$ gives us

$$12 = k \cdot 5, \quad \text{or} \quad k = \tfrac{12}{5}.$$

(b) Since $k = \frac{12}{5}$, the formula $q = kz$ has the specific form

$$q = \tfrac{12}{5}z.$$

Thus, when $z = 7$,

$$q = \tfrac{12}{5} \cdot 7 = \tfrac{84}{5} = 16.8.$$

The following guidelines may be used to solve applied problems that involve variation or proportion.

Guidelines for Solving Variation Problems

> **1** Write a *general* formula that involves the variables and a constant of variation (or proportion) k.
>
> **2** Find the value of k in guideline 1 by using the initial data given in the statement of the problem.
>
> **3** Substitute the value of k found in guideline 2 into the formula of guideline 1, obtaining a *specific* formula that involves the variables.
>
> **4** Use the new data to solve the problem.

We shall follow these guidelines in the solution of the next example.

EXAMPLE 2   *Pressure and volume as inversely proportional quantities*

If the temperature remains constant, the pressure of an enclosed gas is inversely proportional to the volume. The pressure of a certain gas within a spherical balloon of radius 9 inches is 20 lb/in.². If the radius of the balloon increases to 12 inches, approximate the new pressure of the gas.

SOLUTION

Guideline 1 If we denote the pressure by P (in lb/in.²) and the volume by V (in in.³), then since P is inversely proportional to V,

$$P = \frac{k}{V}$$

for some constant of proportionality k.

Guideline 2 We find the constant of proportionality k in guideline 1. Since the volume V of a sphere of radius r is $V = \frac{4}{3}\pi r^3$, the initial volume of the balloon is $V = \frac{4}{3}\pi(9)^3 = 972\pi$ in.³. This leads to the following:

$$20 = \frac{k}{972\pi} \qquad P = 20 \text{ when } V = 972\pi$$

$$k = 20(972\pi) = 19,440\pi \quad \text{solve for } k$$

Guideline 3 Substituting $k = 19,440\pi$ into $P = k/V$, we find that the pressure corresponding to any volume V is given by

$$P = \frac{19,440\pi}{V}.$$

Guideline 4 If the new radius of the balloon is 12 inches, then

$$V = \frac{4}{3}\pi(12)^3 = 2304\pi \text{ in.}^3.$$

Substituting this number for V in the formula obtained in guideline 3 gives us

$$P = \frac{19{,}440\pi}{2304\pi} = \frac{135}{16} = 8.4375.$$

Thus, the pressure decreases to approximately 8.4 lb/in.^{2} when the radius increases to 12 inches.

There are other types of variation. If x, y, and z are variables and $y = kxz$ for some real number k, we say that y *varies directly as the product of x and z* or y **varies jointly as x and z**. If $y = k(x/z)$, then y *varies directly as x and inversely as z*. As a final illustration, if a variable w varies directly as the product of x and the cube of y and inversely as the square of z, then

$$w = k\frac{xy^3}{z^2},$$

where k is a constant of proportionality.

EXAMPLE 3   *Combining several types of variation*

A variable w varies directly as the product of u and v and inversely as the square of s.

(a) If $w = 20$ when $u = 3$, $v = 5$, and $s = 2$, find the constant of variation.

(b) Find the value of w when $u = 7$, $v = 4$, and $s = 3$.

SOLUTION   A general formula for w is

$$w = k\frac{uv}{s^2},$$

where k is a constant of variation.

(a) Substituting $w = 20$, $u = 3$, $v = 5$, and $s = 2$ gives us

$$20 = k\frac{3 \cdot 5}{2^2}, \quad \text{or} \quad k = \frac{80}{15} = \frac{16}{3}.$$

(b) Since $k = \frac{16}{3}$, the specific formula for w is

$$w = \frac{16}{3}\frac{uv}{s^2}.$$

Thus, when $u = 7$, $v = 4$, and $s = 3$,

$$w = \frac{16}{3}\frac{7 \cdot 4}{3^2} = \frac{448}{27} \approx 16.6.$$

In the next example we again follow the guidelines stated in this section.

EXAMPLE 4   *Finding the support load of a rectangular beam*

The weight that can be safely supported by a beam with a rectangular cross section varies directly as the product of the width and square of the depth of the cross section and inversely as the length of the beam. If a 2-inch by 4-inch beam that is 8 feet long safely supports a load of 500 pounds, what weight can be safely supported by a 2-inch by 8-inch beam that is 10 feet long? (Assume that the width is the *shorter* dimension of the cross section.)

SOLUTION

Guideline 1   If the width, depth, length, and weight are denoted by w, d, l, and W, respectively, then a general formula for W is

$$W = k \frac{wd^2}{l},$$

where k is a constant of variation.

Guideline 2   To find the value of k in guideline 1, we see from the given data that

$$500 = k \frac{2(4^2)}{8}, \quad \text{or} \quad k = 125.$$

Guideline 3   Substituting $k = 125$ into the formula of guideline 1 gives us the specific formula

$$W = 125 \frac{wd^2}{l}.$$

Guideline 4   To answer the question, we substitute $w = 2$, $d = 8$, and $l = 10$ into the formula found in guideline 3, obtaining

$$W = 125 \cdot \frac{2 \cdot 8^2}{10} = 1600 \text{ lb.}$$

3.9 EXERCISES

Exer. 1–12: Express the statement as a formula that involves the given variables and a constant of proportionality k, and then determine the value of k from the given conditions.

1   u is directly proportional to v. If $v = 30$, then $u = 12$.

2   s varies directly as t. If $t = 10$, then $s = 18$.

3   r varies directly as s and inversely as t. If $s = -2$ and $t = 4$, then $r = 7$.

4   w varies directly as z and inversely as the square root of u. If $z = 2$ and $u = 9$, then $w = 6$.

5   y is directly proportional to the square of x and inversely proportional to the cube of z. If $x = 5$ and $z = 3$, then $y = 25$.

6 q is inversely proportional to the sum of x and y. If $x = 0.5$ and $y = 0.7$, then $q = 1.4$.

7 z is directly proportional to the product of the square of x and the cube of y. If $x = 7$ and $y = -2$, then $z = 16$.

8 r is directly proportional to the product of s and v and inversely proportional to the cube of p. If $s = 2$, $v = 3$, and $p = 5$, then $r = 40$.

9 y is directly proportional to x and inversely proportional to the square of z. If $x = 4$ and $z = 3$, then $y = 16$.

10 y is directly proportional to x and inversely proportional to the sum of r and s. If $x = 3$, $r = 5$, and $s = 7$, then $y = 2$.

11 y is directly proportional to the square root of x and inversely proportional to the cube of z. If $x = 9$ and $z = 2$, then $y = 5$.

12 y is directly proportional to the square of x and inversely proportional to the square root of z. If $x = 5$ and $z = 16$, then $y = 10$.

13 *Liquid pressure* The pressure P acting at a point in a liquid is directly proportional to the distance d from the surface of the liquid to the point.

(a) Express P as a function of d by means of a formula that involves a constant of proportionality k.

(b) In a certain oil tank, the pressure at a depth of 2 feet is 118 lb/ft^3. Find the value of k in part (a).

(c) Find the pressure at a depth of 5 feet for the oil tank in part (b).

14 *Hooke's law* Hooke's law states that the force F required to stretch a spring x units beyond its natural length is directly proportional to x.

(a) Express F as a function of x by means of a formula that involves a constant of proportionality k.

(b) A weight of 4 pounds stretches a certain spring from its natural length of 10 inches to a length of 10.3 inches. Find the value of k in part (b).

(c) What weight will stretch the spring in part (b) to a length of 11.5 inches?

15 *Electrical resistance* The electrical resistance R of a wire varies directly as its length l and inversely as the square of its diameter d.

(a) Express R in terms of l, d, and a constant of variation k.

(b) A wire 100 feet long of diameter 0.01 inch has a resistance of 25 ohms. Find the value of k in part (a).

(c) Find the resistance of a wire made of the same material that has a diameter of 0.015 inch and is 50 feet long.

16 *Intensity of illumination* The intensity of illumination I from a source of light varies inversely as the square of the distance d from the source.

(a) Express I in terms of d and a constant of variation k.

(b) A searchlight has an intensity of 1,000,000 candlepower at a distance of 50 feet. Find the value of k in part (a).

(c) Approximate the intensity of the searchlight in (b) at a distance of 1 mile.

17 *Period of a pendulum* The period P of a simple pendulum—that is, the time required for one complete oscillation—is directly proportional to the square root of its length l.

(a) Express P in terms of l and a constant of proportionality k.

(b) If a pendulum 2 feet long has a period of 1.5 seconds, find the value of k in part (a).

(c) Find the period of a pendulum 6 feet long.

18 *Dimensions of a human limb* A circular cylinder is sometimes used in physiology as a simple representation of a human limb.

(a) Express the volume V of a cylinder in terms of its length L and the square of its circumference C.

(b) The formula obtained in part (a) can be used to approximate the volume of a limb from length and circumference measurements. Suppose the (average) circumference of a human forearm is 22 centimeters and the average length is 27 centimeters. Approximate the volume of the forearm to the nearest cm^3.

19 *Period of a planet* Kepler's third law states that the period T of a planet (the time needed to make one complete revolution about the sun) is directly proportional to the $\frac{3}{2}$ power of its average distance d from the sun.

(a) Express T as a function of d by means of a formula that involves a constant of proportionality k.

(b) For the planet Earth, $T = 365$ days and $d = 93$ million miles. Find the value of k in part (a).

(c) Estimate the period of Venus if its average distance from the sun is 67 million miles.

20 *Range of a projectile* It is known from physics that the range R of a projectile is directly proportional to the square of its velocity v.

(a) Express R as a function of v by means of a formula that involves a constant of proportionality k.

(b) A motorcycle daredevil has made a jump of 150 feet. If the speed coming off the ramp was 70 mi/hr, find the value of k in part (a).

(c) If the daredevil can reach a speed of 80 mi/hr coming off the ramp and maintain proper balance, estimate the possible length of the jump.

21 *Automobile skid marks*   The speed V at which an automobile was traveling before the brakes were applied can sometimes be estimated from the length L of the skid marks. Assume that V is directly proportional to the square root of L.

(a) Express V as a function of L by means of a formula that involves a constant of proportionality k.

(b) For a certain automobile on a dry surface, $L = 50$ ft when $V = 35$ mi/hr. Find the value of k in part (a).

(c) Estimate the initial speed of the automobile in part (b) if the skid marks are 150 feet long.

22 *Coulomb's law*   Coulomb's law in electrical theory states that the force F of attraction between two oppositely charged particles varies directly as the product of the magnitudes Q_1 and Q_2 of the charges and inversely as the square of the distance d between the particles.

(a) Find a formula for F in terms of Q_1, Q_2, d, and a constant of variation k.

(b) What is the effect of reducing the distance between the particles by a factor of one-fourth?

23 *Threshold weight*   Threshold weight W is defined to be that weight beyond which risk of death increases significantly. For middle-aged males, W is directly proportional to the third power of the height h.

(a) Express W as a function of h by means of a formula that involves a constant of proportionality k.

(b) For a 6-foot male, W is about 200 pounds. Find the value of k in part (a).

(c) Estimate, to the nearest pound, the threshold weight for an individual who is 5 feet 6 inches tall.

24 *The ideal gas law*   The ideal gas law states that the volume V that a gas occupies is directly proportional to the product of the number n of moles of gas and the temperature T (in °K) and is inversely proportional to the pressure P (in atmospheres).

(a) Express V in terms of n, T, P, and a constant of proportionality k.

(b) What is the effect on the volume if the number of moles is doubled and both the temperature and the pressure are reduced by a factor of one-half?

25 *Poiseuille's law*   Poiseuille's law states that the blood flow rate F (in L/min) through a major artery is directly proportional to the product of the fourth power of the radius r and the blood pressure P.

(a) Express F in terms of P, r, and a constant of proportionality k.

(b) During heavy exercise, normal blood flow rates sometimes triple. If the radius of a major artery increases by 10%, approximately how much harder must the heart pump?

26 *Trout population*   Suppose 200 trout are caught, tagged, and released in a lake's general population. Let T denote the number of tagged fish that are recaptured when a sample of n trout are caught at a later date. The validity of the mark-recapture method for estimating the lake's total trout population is based on the assumption that T is directly proportional to n. If 10 tagged trout are recovered from a sample of 300, estimate the total trout population of the lake.

27 *Radioactive decay of radon gas*   When uranium disintegrates into lead, one step in the process is the radioactive decay of radium into radon gas. Radon enters through the soil into home basements, where it presents a health hazard if inhaled. In the simplest case of radon detection, a sample of air with volume V is taken. After equilibrium has been established, the radioactive decay D of the radon gas is counted with efficiency E over time t. The radon concentration C present in the sample of air varies directly as the product of D and E and inversely as the product of V and t. For a fixed radon concentration C and time t, find the change in the radioactive decay count D if V is doubled and E is reduced by 20%.

28 *Radon concentration*   Refer to Exercise 27. Find the change in the radon concentration C if D increases by 30%, t increases by 60%, V decreases by 10%, and E remains constant.

Exer. 29–32: Examine the expression for the given set of data points of the form (x, y). Find the constant of variation and a formula that describes how y varies with respect to x.

29 y/x;   {(0.6, 0.72), (1.2, 1.44), (4.2, 5.04), (7.1, 8.52), (9.3, 11.16)}

30 xy; $\{(0.2, -26.5), (0.4, -13.25), (0.8, -6.625),$
$(1.6, -3.3125), (3.2, -1.65625)\}$

31 x^2y; $\{(0.16, -394.53125), (0.8, -15.78125),$
$(1.6, -3.9453125), (3.2, -0.986328125)\}$

32 y/x^3; $\{(0.11, 0.00355377), (0.56, 0.46889472),$
$(1.2, 4.61376), (2.4, 36.91008)\}$

Test 2-19
17-54

CHAPTER 3 REVIEW EXERCISES

1 Describe the set of all points (x, y) in a coordinate plane such that $y/x < 0$.

2 Show that the triangle with vertices $A(3, 1)$, $B(-5, -3)$, and $C(4, -1)$ is a right triangle, and find its area.

3 Given $P(-5, 9)$ and $Q(-8, -7)$, find

　(a) the distance $d(P, Q)$

　(b) the midpoint of the segment PQ

　(c) a point R such that Q is the midpoint of PR

4 Find all points on the y-axis that are a distance 13 from $P(12, 6)$.

5 For what values of a is the distance between $P(a, 1)$ and $Q(-2, a)$ less than 3?

6 Find an equation of the circle that has center $C(7, -4)$ and passes through $P(-3, 3)$.

7 Find an equation of the circle that has end points of a diameter $A(8, 10)$ and $B(-2, -14)$.

8 Find an equation for the left half of the circle $(x + 2)^2 + y^2 = 9$.

9 Find the slope of the line through $C(11, -5)$ and $D(-8, 6)$.

10 Show that $A(-3, 1)$, $B(1, -1)$, $C(4, 1)$, and $D(3, 5)$ are vertices of a trapezoid.

11 Find an equation of the line through $A(\frac{1}{2}, -\frac{1}{3})$ that is

　(a) parallel to the line $6x + 2y + 5 = 0$

　(b) perpendicular to the line $6x + 2y + 5 = 0$

12 Express $8x + 3y - 24 = 0$ in slope-intercept form.

13 Find an equation of the circle that has center $C(-5, -1)$ and is tangent to the line $x = 4$.

14 Find an equation of the line that has x-intercept -3 and passes through the center of the circle $x^2 + y^2 - 4x + 10y + 26 = 0$.

15 Find a general form of an equation of the line through $P(4, -3)$ with slope 5.

16 Given $A(-1, 2)$ and $B(3, -4)$, find a general form of an equation for the perpendicular bisector of segment AB.

Exer. 17–18: Find the center and radius of the circle with the given equation.

17 $x^2 + y^2 - 12y + 31 = 0$

18 $4x^2 + 4y^2 + 24x - 16y + 39 = 0$

19 If $f(x) = \dfrac{x}{\sqrt{x + 3}}$, find

　(a) $f(1)$　　(b) $f(-1)$　　(c) $f(0)$　　(d) $f(-x)$

　(e) $-f(x)$　　(f) $f(x^2)$　　(g) $[f(x)]^2$

20 Find the domain and range of f if

　(a) $f(x) = \sqrt{3x - 4}$　　(b) $f(x) = \dfrac{1}{(x + 3)^2}$

Exer. 21–22: Find $\dfrac{f(a + h) - f(a)}{h}$ if $h \neq 0$.

21 $f(x) = -x^2 + x + 5$

22 $f(x) = \dfrac{1}{x + 2}$

23 Find a linear function f such that $f(1) = 2$ and $f(3) = 7$.

24 Determine whether f is even, odd, or neither even nor odd.

　(a) $f(x) = \sqrt[3]{x^3 + 4x}$　　(b) $f(x) = \sqrt[3]{3x^2 - x^3}$

　(c) $f(x) = \sqrt[3]{x^4 + 3x^2 + 5}$

Exer. 25–38: Sketch the graph of the equation and label the x- and y-intercepts.

25 $x + 5 = 0$

26 $2y - 7 = 0$

27 $2y + 5x - 8 = 0$

28 $x = 3y + 4$

29 $9y + 2x^2 = 0$

30 $3x - 7y^2 = 0$

31 $y = \sqrt{1 - x}$ **32** $y = (x - 1)^3$

33 $y^2 = 16 - x^2$

34 $x^2 + y^2 + 4x - 16y + 64 = 0$

35 $x^2 + y^2 - 8x = 0$ **36** $x = -\sqrt{9 - y^2}$

37 $y = (x - 3)^2 - 2$ **38** $y = -x^2 - 2x + 3$

Exer. 39–48: **(a)** Sketch the graph of f. **(b)** Find the domain D and range R of f. **(c)** Find the intervals on which f is increasing, is decreasing, or is constant.

39 $f(x) = \dfrac{1 - 3x}{2}$ **40** $f(x) = 1000$

41 $f(x) = |x + 3|$ **42** $f(x) = -\sqrt{10 - x^2}$

43 $f(x) = 1 - \sqrt{x + 1}$ **44** $f(x) = \sqrt{2 - x}$

45 $f(x) = 9 - x^2$ **46** $f(x) = x^2 + 6x + 16$

47 $f(x) = \begin{cases} x^2 & \text{if } x < 0 \\ 3x & \text{if } 0 \le x < 2 \\ 6 & \text{if } x \ge 2 \end{cases}$ **48** $f(x) = 1 + 2[\![x]\!]$

49 Sketch the graphs of the following equations, making use of shifting, stretching, or reflecting:

(a) $y = \sqrt{x}$ **(b)** $y = \sqrt{x + 4}$ **(c)** $y = \sqrt{x} + 4$

(d) $y = 4\sqrt{x}$ **(e)** $y = \frac{1}{4}\sqrt{x}$ **(f)** $y = -\sqrt{x}$

50 The graph of a function f with domain $[-3, 3]$ is shown in the figure. Sketch the graph of the given equation.

(a) $y = f(x - 2)$ **(b)** $y = f(x) - 2$

(c) $y = f(-x)$ **(d)** $y = f(2x)$

(e) $y = f(\frac{1}{2}x)$ **(f)** $y = f^{-1}(x)$

EXERCISE 50

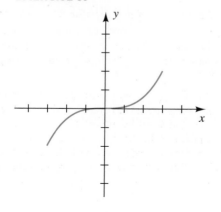

Exer. 51–54: Find an equation for the graph shown in the figure.

51

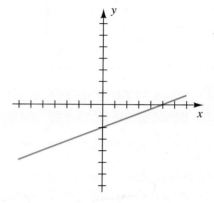

52

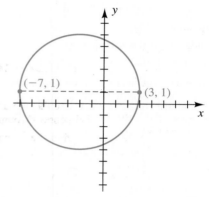

53

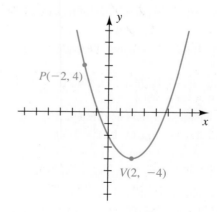

54

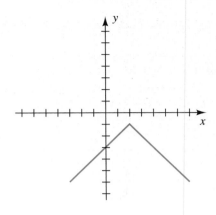

Exer. 55–56: Find the maximum or minimum value of $f(x)$.

55 $f(x) = 5x^2 + 30x + 49$

56 $f(x) = -3x^2 + 30x - 82$

57 Express the function $f(x) = -2x^2 + 12x - 14$ in the form $a(x - h)^2 + k$.

58 Find the standard equation of a parabola with a vertical axis that has vertex $V(3, -2)$ and passes through $(5, 4)$.

59 If $f(x) = \sqrt{4 - x^2}$ and $g(x) = \sqrt{x}$, find the domain of

  (a) fg **(b)** f/g

60 If $f(x) = 8x - 1$ and $g(x) = \sqrt{x - 2}$, find

  (a) $(f \circ g)(2)$ **(b)** $(g \circ f)(2)$

Exer. 61–62: Find (a) $(f \circ g)(x)$ **and (b)** $(g \circ f)(x)$.

61 $f(x) = 2x^2 - 5x + 1$, $g(x) = 3x + 2$

62 $f(x) = \sqrt{3x + 2}$, $g(x) = 1/x^2$

Exer. 63–64: Find (a) $(f \circ g)(x)$ **and the domain of** $f \circ g$ **and (b)** $(g \circ f)(x)$ **and the domain of** $g \circ f$.

63 $f(x) = \sqrt{25 - x^2}$, $g(x) = \sqrt{x - 3}$

64 $f(x) = \dfrac{x}{3x + 2}$, $g(x) = \dfrac{2}{x}$

65 Find a composite function form for $y = \sqrt[3]{x^2 - 5x}$.

66 Is $f(x) = 2x^3 - 5$ a one-to-one function?

Exer. 67–68: (a) Find $f^{-1}(x)$. **(b) Sketch the graphs of** f **and** f^{-1} **on the same coordinate plane.**

67 $f(x) = 10 - 15x$ **68** $f(x) = 9 - 2x^2$, $x \le 0$

69 Suppose y is directly proportional to the cube root of x and inversely proportional to the square of z. Find the constant of proportionality if $y = 6$ when $x = 8$ and $z = 3$.

70 *Discus throw* Based on Olympic records, the winning distance for the discus throw can be approximated by $d = 175 + 1.75t$, where d is in feet and $t = 0$ corresponds to the year 1948.

  (a) Predict the winning distance for the 1996 Summer Olympics in Atlanta.

  (b) Estimate the year in which the winning distance will be 270 feet.

71 *House appreciation* Six years ago a house was purchased for $89,000. This year it was appraised at $125,000. Assume that the value V of the house after its purchase is a linear function of time t (in years).

  (a) Express V in terms of t.

  (b) How many years after the purchase date was the house worth $103,000?

72 *Temperature scales* The freezing point of water is $0\,°C$, or $32\,°F$, and the boiling point is $100\,°C$, or $212\,°F$.

  (a) Express the Fahrenheit temperature F as a linear function of the Celsius temperature C.

  (b) What temperature increase in $°F$ corresponds to an increase in temperature of $1\,°C$?

73 *Gasoline mileage* Suppose the cost of driving an automobile is a linear function of the number x of miles driven and that gasoline costs $1.25 per gallon. A certain automobile presently gets 20 mi/gal, and a tune-up that will improve gasoline mileage by 10% costs $50.

  (a) Express the cost C_1 of driving without a tune-up in terms of x.

  (b) Express the cost C_2 of driving with a tune-up in terms of x.

  (c) How many miles must the automobile be driven after a tune-up to make the cost of the tune-up worthwhile?

74 *Constructing a storage shelter* An open rectangular storage shelter consisting of two vertical sides, 4 feet wide, and a flat roof is to be attached to an existing structure, as illustrated in the figure on the following page. The flat roof is made of tin and costs $5 per square foot, and the two sides are made of plywood costing $2 per square foot.

(a) If \$400 is available for construction, express the length y as a function of the height x.

(b) Express the volume V inside the shelter as a function of x.

EXERCISE 74

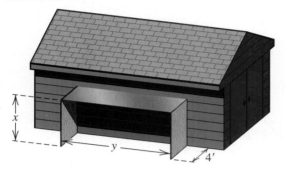

75 Constructing a cylindrical container A company plans to manufacture a container having the shape of a right circular cylinder, open at the top, and having a capacity of 24π in.³ If the cost of the material for the bottom is 30 cents per in.² and that for the curved sides is 10 cents per in.², express the total cost C of the material as a function of the radius r of the base of the container.

76 Filling a pool A cross section of a rectangular pool of dimensions 80 feet by 40 feet is shown in the figure. The pool is being filled with water at a rate of 10 ft³ per minute.

(a) Express the volume V of the water in the pool as a function of time t.

(b) Express V as a function of the depth h at the deep end for $0 \le h \le 6$ and then for $6 < h \le 9$.

(c) Express h as a function of t for $0 \le h \le 6$ and then for $6 < h \le 9$.

EXERCISE 76

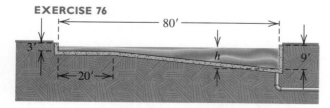

77 Filtering water Suppose 5 in.³ of water is poured into a conical filter and subsequently drips into a cup, as shown in the figure. Let x denote the height of the water

in the filter, and let y denote the height of the water in the cup.

(a) Express the radius r shown in the figure as a function of x. (*Hint:* Use similar triangles.)

(b) Express the height y of the water in the cup as a function of x. (*Hint:* What is the sum of the two volumes shown in the figure?)

EXERCISE 77

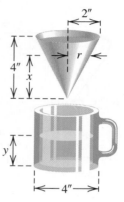

78 Frustum of a cone The shape of the first spacecraft in the Apollo program was a frustum of a right circular cone—a solid formed by truncating a cone by a plane parallel to its base. For the frustum shown in the figure, the radii a and b have already been determined.

(a) Use similar triangles to express y as a function of h.

(b) Derive a formula for the volume of the frustum as a function of h.

(c) If $a = 6$ ft and $b = 3$ ft, for what value of h is the volume of the frustum 600 ft³?

EXERCISE 78

79 *Distance between ships* At 1:00 P.M. ship A is 30 miles due south of ship B and is sailing north at a rate of 15 mi/hr. If ship B is sailing west at a rate of 10 mi/hr, find the time at which the distance *d* between the ships is minimal (see the figure).

EXERCISE 79

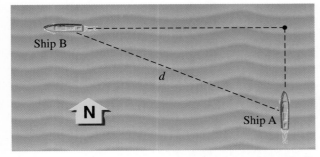

80 *Dimensions of a race track* The interior of a half-mile race track consists of a rectangle with semicircles at two opposite ends. Find the dimensions that will maximize the area of the rectangle.

81 *Vertical leaps* When a particular basketball player leaps straight up for a dunk, the player's distance $f(t)$ (in feet) off the floor after t seconds is given by the formula $f(t) = -\frac{1}{2}gt^2 + 16t$, where g is a gravitational constant.

(a) If $g = 32$, find the player's hang time—that is, the total number of seconds that the player is in the air.

(b) Find the player's vertical leap—that is, the maximum distance of the player's feet from the floor.

(c) On the moon, $g = \frac{32}{6}$. Rework parts (a) and (b) for the player on the moon.

82 *Trajectory of a rocket* A rocket is fired up a hillside, following a path given by $y = -0.016x^2 + 1.6x$. The hillside has slope $\frac{1}{5}$, as illustrated in the figure.

(a) Where does the rocket land?

(b) Find the maximum height of the rocket *above the ground*.

EXERCISE 82

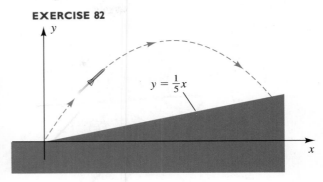

$$y = \frac{1}{5}x$$

83 *Telephone calls* In a certain county, the average number of telephone calls per day between any two cities is directly proportional to the product of their populations and inversely proportional to the square of the distance between them. Cities A and B are 25 miles apart and have populations of 10,000 and 5000, respectively. Telephone records indicate an average of 2000 calls per day between the two cities. Estimate the average number of calls per day between city A and another city of 15,000 people that is 100 miles from A.

84 *Power of a wind rotor* The power P generated by a wind rotor is directly proportional to the product of the square of the area A swept out by the blades and the third power of the wind velocity v. Suppose the diameter of the circular area swept out by the blades is 10 feet, and $P = 3000$ watts when $v = 20$ mi/hr. Find the power generated when the wind velocity is 30 mi/hr.

Many geometric figures that occur in everyday life can be represented by graphs of polynomial and rational functions.

Polynomial functions are the most basic functions in mathematics, because they are defined only in terms of addition, subtraction, and multiplication. In applications it is often necessary to sketch their graphs and to find (or approximate) their zeros. In the first part of this chapter we discuss results that are useful in obtaining this information. We then turn our attention to quotients of polynomial functions—that is, rational functions.

In the last three sections we consider the conic sections: parabolas, ellipses, and hyperbolas. A remarkable fact about conic sections is that although they were studied thousands of years ago by the ancient Greeks, they are far from obsolete. In particular, they are important for present-day investigations in outer space and for the study of the behavior of atomic particles. ■

CHAPTER 4

POLYNOMIAL FUNCTIONS, RATIONAL FUNCTIONS, AND CONIC SECTIONS

4.1   GRAPHS OF POLYNOMIAL FUNCTIONS OF DEGREE GREATER THAN 2

If f is a polynomial function with real coefficients of degree n, then

$$f(x) = a_n x^n + a_{n-1} x^{n-1} + \cdots + a_1 x + a_0,$$

with $a_n \neq 0$. The special cases listed in the following chart were discussed in Chapter 3.

Degree of f	Form of $f(x)$	Graph of f
0	$f(x) = a_0$	A horizontal line with y-intercept a_0
1	$f(x) = a_1 x + a_0$	A line of slope a_1 and y-intercept a_0
2	$f(x) = a_2 x^2 + a_1 x + a_0$	A parabola with a vertical axis

FIGURE 1

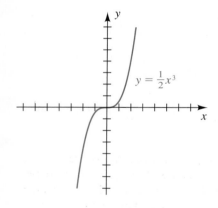

In this section we shall discuss graphs of polynomial functions of degree greater than 2.

If f has degree n *and all the coefficients except a_n are zero*, then

$$f(x) = ax^n \quad \text{for some } a = a_n \neq 0.$$

In this case, if $n = 1$, the graph of f is a line through the origin. If $n = 2$, the graph is a parabola with vertex at the origin. Two illustrations with $n = 3$ are given in the next example.

EXAMPLE 1   *Sketching graphs of $y = ax^3$*

Sketch the graph of f if

(a) $f(x) = \frac{1}{2}x^3$ **(b)** $f(x) = -\frac{1}{2}x^3$

SOLUTION

FIGURE 2

(a) The following table lists several points on the graph of $y = \frac{1}{2}x^3$.

x	0	$\frac{1}{2}$	1	$\frac{3}{2}$	2	$\frac{5}{2}$
y	0	$\frac{1}{16} \approx 0.06$	$\frac{1}{2}$	$\frac{27}{16} \approx 1.7$	4	$\frac{125}{16} \approx 7.8$

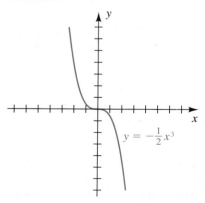

Since f is an odd function, the graph of f is symmetric with respect to the origin, and hence points such as $(-\frac{1}{2}, -\frac{1}{16})$ and $(-1, -\frac{1}{2})$ are also on the graph. The graph is sketched in Figure 1.

(b) If $y = -\frac{1}{2}x^3$, the graph can be obtained from that in part (a) by multiplying all y-coordinates by -1 (that is, by reflecting the graph in part (a) through the x-axis). This gives us the sketch in Figure 2.

If $f(x) = ax^n$ and n is an *odd* positive integer, then f is an odd function and the graph of f is symmetric with respect to the origin, as illustrated in Figures 1 and 2. For $a > 0$, the graph is similar in shape to that in Figure 1; however, as either n or a increases, the graph rises more rapidly for $x > 1$. If $a < 0$, we reflect the graph through the x-axis, as in Figure 2.

If $f(x) = ax^n$ and n is an *even* positive integer, then f is an even function and the graph of f is symmetric with respect to the y-axis, as illustrated in Figure 3 for the case $a = 1$. Note that as the exponent increases, the graph becomes flatter at the origin. It also rises more rapidly for $x > 1$. If $a < 0$, the graph lies below the x-axis.

FIGURE 3

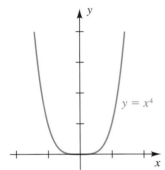
$y = x^4$

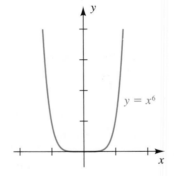
$y = x^6$

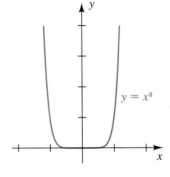
$y = x^8$

FIGURE 4

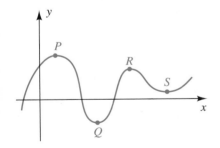

A complete analysis of graphs of polynomial functions of degree greater than 2 requires methods that are used in calculus. As the degree increases, the graphs usually become more complicated. They always have a smooth appearance, however, with a number of high points and low points, such as P, Q, R, and S in Figure 4. Such points are sometimes called **turning points** for the graph. Each function value (y-coordinate) corresponding to a high or low point is called an **extremum** of the function f. At an extremum, f changes from an increasing function to a decreasing function, or vice versa.

The intermediate value theorem specifies another important property of polynomial functions. The proof is omitted, since it requires advanced mathematics.

Intermediate Value Theorem for Polynomial Functions

If f is a polynomial function and $f(a) \neq f(b)$ for $a < b$, then f takes on every value between $f(a)$ and $f(b)$ in the interval $[a, b]$.

The intermediate value theorem states that if w is any number between $f(a)$ and $f(b)$, there is at least one number c between a and b such that $f(c) = w$. If we regard the graph of f as extending continuously from the

FIGURE 5

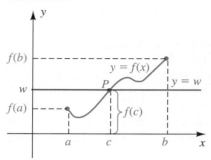

point $(a, f(a))$ to the point $(b, f(b))$, as illustrated in Figure 5, then for any number w between $f(a)$ and $f(b)$, the horizontal line $y = w$ intersects the graph in at least one point P. The x-coordinate c of P is a number such that $f(c) = w$.

A consequence of the intermediate value theorem is that if $f(a)$ and $f(b)$ have opposite signs, there is at least one number c between a and b such that $f(c) = 0$; that is, f *has a zero at* c. Thus, if the point $(a, f(a))$ lies below the x-axis and the point $(b, f(b))$ lies above the x-axis, or vice versa, the graph crosses the x-axis at least once between $x = a$ and $x = b$, as illustrated in Figure 6.

FIGURE 6

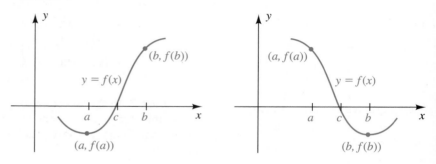

EXAMPLE 2   *Using the intermediate value theorem*

Show that $f(x) = x^5 + 2x^4 - 6x^3 + 2x - 3$ has a zero between 1 and 2.

SOLUTION   Substituting 1 and 2 for x gives us the following function values:

$$f(1) = 1 + 2 - 6 + 2 - 3 = -4$$

$$f(2) = 32 + 32 - 48 + 4 - 3 = 17$$

Since $f(1)$ and $f(2)$ have opposite signs, we see that $f(c) = 0$ for at least one real number c between 1 and 2.

Example 2 illustrates a method for locating real zeros of polynomials. By using *successive approximations*, we can approximate each zero to any degree of accuracy by locating it in smaller and smaller intervals.

If c and d are *successive* real zeros of $f(x)$—that is, there are no other zeros between c and d—then $f(x)$ *does not change sign on the interval* (c, d). Thus, if we choose any number k such that $c < k < d$ and if $f(k)$ is positive, then $f(x)$ is positive throughout (c, d). Similarly, if $f(k)$ is negative, then $f(x)$ is negative throughout (c, d). We shall call $f(k)$ a **test value** for $f(x)$ on the interval (c, d). Test values may also be used on infinite intervals of the form $(-\infty, a)$ or (a, ∞), provided that $f(x)$ has no zeros on these intervals. The use of test values in graphing is similar to the technique used for inequalities in Section 2.7.

EXAMPLE 3   *Sketching the graph of a polynomial function of degree 3*

Let $f(x) = x^3 + x^2 - 4x - 4$. Find all values of x such that $f(x) > 0$ and all x such that $f(x) < 0$, and then sketch the graph of f.

SOLUTION   The graph of f lies above the x-axis for values of x such that $f(x) > 0$, and it lies below the x-axis for all x such that $f(x) < 0$. We may factor $f(x)$ as follows:

$$
\begin{aligned}
f(x) &= x^3 + x^2 - 4x - 4 & &\text{given} \\
&= (x^3 + x^2) - (4x + 4) & &\text{group terms} \\
&= x^2(x + 1) - 4(x + 1) & &\text{factor } x^3 + x^2 \text{ and } 4x + 4 \\
&= (x^2 - 4)(x + 1) & &\text{factor out } (x + 1) \\
&= (x + 2)(x - 2)(x + 1) & &\text{factor } x^2 - 4
\end{aligned}
$$

FIGURE 7

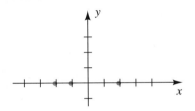

We see from the last equation that the zeros of $f(x)$ (the x-intercepts of the graph) are -2, -1, and 2. The corresponding points on the graph (see Figure 7) divide the x-axis into four parts, and we consider the open intervals

$$(-\infty, -2), \quad (-2, -1), \quad (-1, 2), \quad (2, \infty).$$

As in our work with inequalities in Section 2.7, the sign of $f(x)$ in each of these intervals can be determined by using a sign chart, as follows.

FIGURE 8

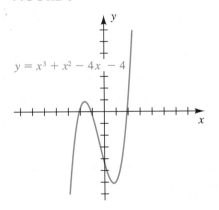

$y = x^3 + x^2 - 4x - 4$

Interval	$(-\infty, -2)$	$(-2, -1)$	$(-1, 2)$	$(2, \infty)$
Sign of $x + 2$	$-$	$+$	$+$	$+$
Sign of $x + 1$	$-$	$-$	$+$	$+$
Sign of $x - 2$	$-$	$-$	$-$	$+$
Sign of $f(x)$	$-$	$+$	$-$	$+$
Position of graph	Below x-axis	Above x-axis	Below x-axis	Above x-axis

Using the information from the sign chart leads to the sketch in Figure 8. To find the turning points of the graph, it would be necessary to use a computational device or methods developed in calculus.

The graph of every polynomial function of degree 3 has an appearance similar to that of Figure 8, or it has an inverted version of that graph if the coefficient of x^3 is negative. Sometimes, however, the graph may have only one x-intercept or the shape may be elongated, as in Figures 1 and 2.

EXAMPLE 4   Sketching the graph of a polynomial function of degree 4

Let $f(x) = x^4 - 4x^3 + 3x^2$. Find all values of x such that $f(x) > 0$ and all x such that $f(x) < 0$, and then sketch the graph of f.

SOLUTION   We shall follow the same steps used in the solution of Example 3. Thus, we begin by factoring $f(x)$:

$$
\begin{aligned}
f(x) &= x^4 - 4x^3 + 3x^2 && \text{given}\\
&= x^2(x^2 - 4x + 3) && \text{factor out } x^2\\
&= x^2(x - 1)(x - 3) && \text{factor}
\end{aligned}
$$

FIGURE 9

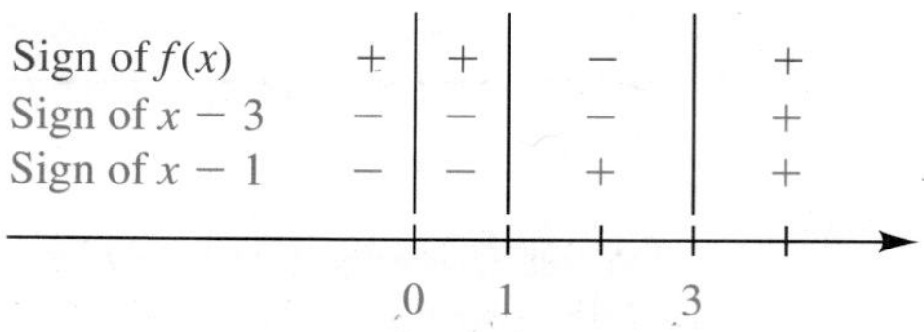

Sign of $f(x)$	$+$	$+$	$-$	$+$
Sign of $x - 3$	$-$	$-$	$-$	$+$
Sign of $x - 1$	$-$	$-$	$+$	$+$

```
                0    1        3
```

FIGURE 10

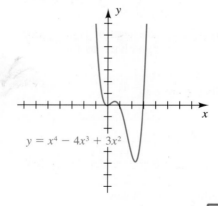

$y = x^4 - 4x^3 + 3x^2$

Next, we construct the sign diagram in Figure 9, where the vertical lines indicate the zeros 0, 1, and 3 of the factors. Since the factor x^2 is always positive if $x \ne 0$, it has no effect on the sign of the product and hence may be omitted from the diagram.

Referring to the sign of $f(x)$ in the diagram, we see that

$$f(x) > 0 \quad \text{in } (-\infty, 0) \cup (0, 1) \cup (3, \infty) \qquad \text{and} \qquad f(x) < 0 \quad \text{in } (1, 3).$$

Making use of these facts leads to the sketch in Figure 10.

In the following example we use a graphing utility to estimate coordinates of important points on a graph.

EXAMPLE 5   Estimating high and low points on a graph

(a) Estimate the real zeros of $f(x) = x^3 - 4.6x^2 + 5.72x - 0.656$ to three decimal places.

(b) Estimate the coordinates of the highest and lowest points on the graph for $0.5 \le x \le 2.5$.

SOLUTION

(a) We assign $f(x)$ to Y_1 and use a standard viewing rectangle to obtain a display similar to Figure 11(a). Since all the real roots appear to lie between 0 and 3, let us regraph, using the viewing rectangle $[-1, 3]$ by

$[-1, 3]$. This gives us a display similar to Figure 11(b), which shows that there is only one x-intercept and hence one real root. Using the tracing and zoom-in features, we estimate the zero as 0.127.

FIGURE 11
(a) $[-15, 15]$ by $[-10, 10]$ **(b)** $[-1, 3]$ by $[-1, 3]$

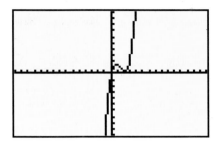 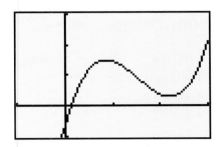

(b) We see from Figure 11(b) that if $0.5 \leq x \leq 2.5$, then $0 < y < 2$. This suggests that we use the viewing rectangle $[0.5, 2.5]$ by $[0, 2]$. Again using the tracing and zoom-in features, we estimate the highest point to be $(0.866, 1.497)$ and the lowest point to be $(2.200, 0.312)$.

In the next example we use a graphing utility to solve an inequality involving a polynomial of degree 3.

EXAMPLE 6   *Solving an inequality graphically*

Estimate the solutions of the inequality

$$6x^2 - 3x^3 < 2.$$

SOLUTION   Let us subtract 2 from both sides and consider the equivalent inequality

$$6x^2 - 3x^3 - 2 < 0.$$

FIGURE 12 $[-15, 15]$ by $[-10, 10]$

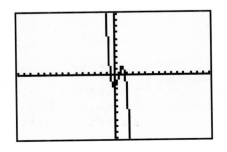

We assign $6x^2 - 3x^3 - 2$ to Y_1 and use a standard viewing rectangle to obtain a display similar to Figure 12. We see that there are three x-intercepts. If we denote them by x_1, x_2, and x_3, then the solutions to the inequality are given by

$$(x_1, x_2) \cup (x_3, \infty),$$

since these are the intervals on which Y_1 is less than 0. Using the tracing and zoom-in features for each x-intercept, we find that

$$x_1 \approx -0.515, \qquad x_2 \approx 0.722, \qquad x_3 \approx 1.793.$$

4.1 EXERCISES

Exer. 1–4: Sketch the graph of f for the indicated value of c or a.

1 $f(x) = 2x^3 + c$
  (a) $c = 3$ (b) $c = -3$

2 $f(x) = -2x^3 + c$
  (a) $c = -2$ (b) $c = 2$

3 $f(x) = ax^3 + 2$
  (a) $a = 2$ (b) $a = -\frac{1}{3}$

4 $f(x) = ax^3 - 3$
  (a) $a = -2$ (b) $a = \frac{1}{4}$

Exer. 5–10: Use the intermediate value theorem to show that f has a zero between a and b.

5 $f(x) = x^3 - 4x^2 + 3x - 2;$ $a = 3,$ $b = 4$

6 $f(x) = 2x^3 + 5x^2 - 3;$ $a = -3,$ $b = -2$

7 $f(x) = -x^4 + 3x^3 - 2x + 1;$ $a = 2,$ $b = 3$

8 $f(x) = 2x^4 + 3x - 2;$ $a = \frac{1}{2},$ $b = \frac{3}{4}$

9 $f(x) = x^5 + x^3 + x^2 + x + 1;$ $a = -\frac{1}{2},$ $b = -1$

10 $f(x) = x^5 - 3x^4 - 2x^3 + 3x^2 - 9x - 6;$ $a = 3,$ $b = 4$

Exer. 11–26: Find all values of x such that $f(x) > 0$ and all x such that $f(x) < 0$, and then sketch the graph of f.

11 $f(x) = \frac{1}{4}x^3 - 2$ 12 $f(x) = -\frac{1}{9}x^3 - 3$

13 $f(x) = -\frac{1}{16}x^4 + 1$ 14 $f(x) = x^5 + 1$

15 $f(x) = x^4 - 4x^2$ 16 $f(x) = 9x - x^3$

17 $f(x) = -x^3 + 3x^2 + 10x$

18 $f(x) = x^4 + 3x^3 - 4x^2$

19 $f(x) = \frac{1}{6}(x + 2)(x - 3)(x - 4)$

20 $f(x) = -\frac{1}{8}(x + 4)(x - 2)(x - 6)$

21 $f(x) = x^3 + 2x^2 - 4x - 8$

22 $f(x) = x^3 - 3x^2 - 9x + 27$

23 $f(x) = x^4 - 6x^2 + 8$

24 $f(x) = -x^4 + 12x^2 - 27$

25 $f(x) = x^2(x + 2)(x - 1)^2(x - 2)$

26 $f(x) = x^3(x + 1)^2(x - 2)(x - 4)$

27 Let $f(x)$ be a polynomial such that the coefficient of every odd power of x is 0. Show that f is an even function.

28 Let $f(x)$ be a polynomial such that the coefficient of every even power of x is 0. Show that f is an odd function.

29 If $f(x) = 3x^3 - kx^2 + x - 5k$, find a number k such that the graph of f contains the point $(-1, 4)$.

30 If one zero of $f(x) = x^3 - 2x^2 - 16x + 16k$ is 2, find two other zeros.

31 **A Legendre polynomial** The third-degree Legendre polynomial $P(x) = \frac{1}{2}(5x^3 - 3x)$ occurs in the solution of heat transfer problems in physics and engineering. Find all values of x such that $P(x) > 0$ and all x such that $P(x) < 0$, and sketch the graph of P.

32 **A Chebyshev polynomial** The fourth-degree Chebyshev polynomial $f(x) = 8x^4 - 8x^2 + 1$ occurs in statistical studies. Find all values of x such that $f(x) > 0$. (*Hint:* Let $z = x^2$, and use the quadratic formula.)

33 **Constructing a box** From a rectangular piece of cardboard having dimensions 20 inches × 30 inches, an open box is to be made by cutting out identical squares of area x^2 from each corner and turning up the sides (see Exercise 49 of Section 3.4).

  (a) Show that the volume of the box is given by the function $V(x) = x(20 - 2x)(30 - 2x)$.

  (b) Find all positive values of x such that $V(x) > 0$, and sketch the graph of V for $x > 0$.

34 **Constructing a crate** The frame for a shipping crate is to be constructed from 24 feet of 2 × 2 lumber (see the figure).

EXERCISE 34

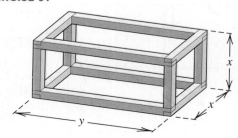

(a) If the crate is to have square ends of side x feet, express the outer volume V of the crate as a function of x (disregard the thickness of the lumber).

(b) Sketch the graph of V for $x > 0$.

35 *Determining temperatures* A meteorologist determines that the temperature T (in °F) for a certain 24-hour period in winter was given by $T = \frac{1}{20}t(t - 12)(t - 24)$ for $0 \le t \le 24$, where t is time in hours and $t = 0$ corresponds to 6 A.M.

(a) When was $T > 0$, and when was $T < 0$?

(b) Sketch the graph of T.

(c) Show that the temperature was 32 °F sometime between 12 noon and 1 P.M. (*Hint:* Use the intermediate value theorem.)

36 *Deflections of diving boards* A diver stands at the very end of a diving board before beginning a dive. The deflection d of the board at a position s feet from the stationary end is given by $d = cs^2(3L - s)$ for $0 \le s \le L$, where L is the length of the board and c is a positive constant that depends on the weight of the diver and on the physical properties of the board (see the figure). Suppose the board is 10 feet long.

(a) If the deflection at the end of the board is 1 foot, find c.

(b) Show that the deflection is $\frac{1}{2}$ foot somewhere between $s = 6.5$ and $s = 6.6$.

EXERCISE 36

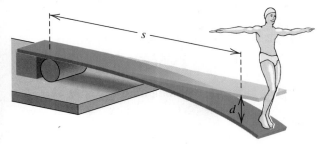

37 *Deer population* A herd of 100 deer is introduced onto a small island. At first the herd increases rapidly, but eventually food resources dwindle and the population declines. Suppose that the number $N(t)$ of deer after t years is given by $N(t) = -t^4 + 21t^2 + 100$, where $t > 0$.

(a) Determine the values of t for which $N(t) > 0$, and sketch the graph of N.

(b) Does the population become extinct? If so, when?

38 *Deer population* Refer to Exercise 37. It can be shown by means of calculus that the rate R (in deer per year) at which the deer population changes at time t is given by $R = -4t^3 + 42t$.

(a) When does the population cease to grow?

(b) Determine the positive values of t for which $R > 0$.

c **Exer. 39–42: Graph f and estimate its zeros.**

39　$f(x) = x^3 + 0.2x^2 - 2.6x + 1.1$

40　$f(x) = -x^4 + 0.1x^3 + 4x^2 - 0.5x - 3$

41　$f(x) = x^3 - 3x + 1$

42　$f(x) = 2x^3 - 4x^2 - 3x + 1$

c **Exer. 43–46: Graph f and estimate all values of x such that $f(x) > k$.**

43　$f(x) = x^3 + 5x - 2;$　　　　　　$k = 1$

44　$f(x) = x^4 - 4x^3 + 3x^2 - 8x + 5;$　$k = 3$

45　$f(x) = x^5 - 2x^2 + 2;$　　　　　$k = -2$

46　$f(x) = x^4 - 2x^3 + 10x - 26;$　$k = -1$

c **Exer. 47–48: Graph f and g on the same coordinate plane, and estimate the points of intersection.**

47　$f(x) = x^3 - 2x^2 - 1.5x + 2.8;$
　　$g(x) = -x^3 - 1.7x^2 + 2x + 2.5$

48　$f(x) = x^4 - 5x^2 + 4;$
　　$g(x) = x^4 - 3x^3 - 0.25x^2 + 3.75x$

4.2　PROPERTIES OF DIVISION

In this section we use $f(x)$, $g(x)$, and so on, to denote polynomials in x. If $g(x)$ is a factor of $f(x)$, then $f(x)$ is **divisible** by $g(x)$. For example, $x^4 - 16$ is divisible by $x^2 - 4$, by $x^2 + 4$, by $x + 2$, and by $x - 2$.

The polynomial $x^4 - 16$ is not divisible by $x^2 + 3x + 1$; however, we can use the process called **long division** to find a *quotient* and a *remainder*, as in the following illustration, where we have inserted terms with zero coefficients.

ILLUSTRATION

Long Division of Polynomials

$$
\begin{array}{r}
\overset{\text{quotient}}{\overbrace{x^2 - 3x + 8}} \\
x^2 + 3x + 1 \,\big|\, \overline{x^4 + 0x^3 + 0x^2 + 0x - 16} \\
\end{array}
$$

$$
\begin{array}{rl}
x^4 + 3x^3 + x^2 & \qquad x^2(x^2 + 3x + 1) \\
\hline
-3x^3 - x^2 & \qquad \text{subtract} \\
-3x^3 - 9x^2 - 3x & \qquad -3x(x^2 + 3x + 1) \\
\hline
8x^2 + 3x - 16 & \qquad \text{subtract} \\
8x^2 + 24x + 8 & \qquad 8(x^2 + 3x + 1) \\
\hline
\underbrace{-21x - 24}_{\text{remainder}} & \qquad \text{subtract}
\end{array}
$$

The long division process ends when we arrive at a polynomial (the remainder) that either is 0 or has smaller degree than the divisor. The result of the long division in the preceding illustration can be written

$$
\frac{x^4 - 16}{x^2 + 3x + 1} = (x^2 - 3x + 8) + \left(\frac{-21x - 24}{x^2 + 3x + 1} \right).
$$

Multiplying both sides of this equation by $x^2 + 3x + 1$, we obtain

$$
x^4 - 16 = (x^2 + 3x + 1)(x^2 - 3x + 8) + (-21x - 24).
$$

This example illustrates the following theorem, which we state without proof.

Division Algorithm
for Polynomials

> If $f(x)$ and $p(x)$ are polynomials and if $p(x) \neq 0$, then there exist unique polynomials $q(x)$ and $r(x)$ such that
>
> $$ f(x) = p(x)q(x) + r(x), $$
>
> where either $r(x) = 0$ or the degree of $r(x)$ is less than the degree of $p(x)$. The polynomial $q(x)$ is the **quotient**, and $r(x)$ is the **remainder** in the division of $f(x)$ by $p(x)$.

A useful special case of the division algorithm occurs if $f(x)$ is divided by $x - c$, where c is a real number. If $x - c$ is a factor of $f(x)$, then

$$f(x) = (x - c)q(x)$$

for some quotient $q(x)$; that is, the remainder $r(x)$ is 0. If $x - c$ is not a factor of $f(x)$, then the degree of the remainder $r(x)$ is less than the degree of $x - c$, and hence $r(x)$ must have degree 0. This means that the remainder is a nonzero number. Consequently, for every $x - c$ we have

$$f(x) = (x - c)q(x) + d,$$

where the remainder d is a real number (possibly $d = 0$). If we substitute c for x, we obtain

$$\begin{aligned} f(c) &= (c - c)q(c) + d \\ &= 0 \cdot q(c) + d \\ &= d. \end{aligned}$$

This proves the following theorem.

Remainder Theorem

> If a polynomial $f(x)$ is divided by $x - c$, then the remainder is $f(c)$.

EXAMPLE I *Using the remainder theorem*

If $f(x) = x^3 - 3x^2 + x + 5$, use the remainder theorem to find $f(2)$.

SOLUTION According to the remainder theorem, $f(2)$ is the remainder when $f(x)$ is divided by $x - 2$. By long division,

$$
\begin{array}{r}
x^2 - x - 1 \\
x - 2 \,\overline{\smash{\big)}\, x^3 - 3x^2 + x + 5} \\
\underline{x^3 - 2x^2} \qquad\qquad x^2(x-2) \\
-x^2 + x \qquad\quad \text{subtract} \\
\underline{-x^2 + 2x} \qquad -x(x-2) \\
-x + 5 \qquad \text{subtract} \\
\underline{-x + 2} \qquad (-1)(x-2) \\
3 \qquad \text{subtract}
\end{array}
$$

Hence, $f(2) = 3$. We may check this fact by direct substitution:

$$f(2) = 2^3 - 3(2)^2 + 2 + 5 = 3.$$

We shall use the remainder theorem to prove the following important result.

Factor Theorem

A polynomial $f(x)$ has a factor $x - c$ if and only if $f(c) = 0$.

PROOF   By the remainder theorem,

$$f(x) = (x - c)q(x) + f(c)$$

for some quotient $q(x)$.

If $f(c) = 0$, then $f(x) = (x - c)q(x)$; that is, $x - c$ is a factor of $f(x)$. Conversely, if $x - c$ is a factor of $f(x)$, then the remainder upon division of $f(x)$ by $x - c$ must be 0, and hence, by the remainder theorem, $f(c) = 0$. ∎

The factor theorem is useful for finding factors of polynomials, as illustrated in the next example.

EXAMPLE 2   *Using the factor theorem*

Show that $x - 2$ is a factor of $f(x) = x^3 - 4x^2 + 3x + 2$.

SOLUTION   Since $f(2) = 8 - 16 + 6 + 2 = 0$, we see from the factor theorem that $x - 2$ is a factor of $f(x)$. Another method of solution would be to divide $f(x)$ by $x - 2$ and show that the remainder is 0. The quotient in the division would be another factor of $f(x)$.

EXAMPLE 3   *Finding a polynomial with prescribed zeros*

Find a polynomial $f(x)$ of degree 3 that has zeros 2, -1, and 3.

SOLUTION   By the factor theorem, $f(x)$ has factors $x - 2$, $x + 1$, and $x - 3$. Thus,

$$f(x) = a(x - 2)(x + 1)(x - 3),$$

where any nonzero value may be assigned to a. If we let $a = 1$ and multiply, we obtain

$$f(x) = x^3 - 4x^2 + x + 6.$$

To apply the remainder theorem it is necessary to divide a polynomial $f(x)$ by $x - c$. The method of **synthetic division** may be used to simplify this work. The following guidelines state how to proceed. The method can be justified by a careful (and lengthy) comparison with the method of long division.

Guidelines for Synthetic Division of $a_n x^n + a_{n-1} x^{n-1} + \cdots + a_1 x + a_0$ **by** $x - c$

1 Begin with the following display, supplying zeros for any missing coefficients in the given polynomial.

$$
\begin{array}{c|cccccc}
c & a_n & a_{n-1} & a_{n-2} & \cdots & a_1 & a_0 \\
\hline
& a_n
\end{array}
$$

2 Multiply a_n by c, and place the product ca_n underneath a_{n-1}, as indicated by the arrow in the following display. (This arrow, and others, is used only to clarify these guidelines and will not appear in *specific* synthetic divisions.) Next find the sum $b_1 = a_{n-1} + ca_n$, and place it below the line as shown.

$$
\begin{array}{c|cccccccc}
c & a_n & a_{n-1} & a_{n-2} & \cdots & & a_1 & a_0 \\
& & ca_n & cb_1 & cb_2 & \cdots & cb_{n-2} & cb_{n-1} \\
\hline
& a_n & b_1 & b_2 & \cdots & & b_{n-2} & b_{n-1} & r
\end{array}
$$

3 Multiply b_1 by c, and place the product cb_1 underneath a_{n-2}, as indicated by the second arrow. Proceeding, we next find the sum $b_2 = a_{n-2} + cb_1$, and place it below the line as shown.

4 Continue this process, as indicated by the arrows, until the final sum $r = a_0 + cb_{n-1}$ is obtained. The numbers

$$a_n, \quad b_1, \quad b_2, \quad \ldots, \quad b_{n-2}, \quad b_{n-1}$$

are the coefficients of the quotient $q(x)$; that is,

$$q(x) = a_n x^{n-1} + b_1 x^{n-2} + \cdots + b_{n-2} x + b_{n-1},$$

and r is the remainder.

The following examples illustrate synthetic division for some special cases.

EXAMPLE 4 *Using synthetic division to find a quotient and remainder*

Use synthetic division to find the quotient and remainder if the polynomial $2x^4 + 5x^3 - 2x - 8$ is divided by $x + 3$.

SOLUTION Since the divisor is $x + 3$, the c in the expression $x - c$ is -3. Hence, the synthetic division takes this form:

$$
\begin{array}{c|ccccc}
-3 & 2 & 5 & 0 & -2 & -8 \\
& & -6 & 3 & -9 & 33 \\
\hline
& 2 & -1 & 3 & -11 & 25
\end{array}
$$

$$\underbrace{\qquad\qquad\qquad}_{\substack{\text{coefficients} \\ \text{of quotient}}} \quad \underbrace{}_{\text{remainder}}$$

(continued)

As we have indicated, the first four numbers in the third row are the coefficients of the quotient $q(x)$, and the last number is the remainder r. Thus,

$$q(x) = 2x^3 - x^2 + 3x - 11 \quad \text{and} \quad r = 25.$$

Synthetic division can be used to find values of polynomial functions, as illustrated in the next example.

EXAMPLE 5   *Using synthetic division to find values of a polynomial*

If $f(x) = 3x^5 - 38x^3 + 5x^2 - 1$, use synthetic division to find $f(4)$.

SOLUTION   By the remainder theorem, $f(4)$ is the remainder when $f(x)$ is divided by $x - 4$. Dividing synthetically, we obtain

$$
\begin{array}{r|rrrrrr}
4 & 3 & 0 & -38 & 5 & 0 & -1 \\
  &   & 12 & 48 & 40 & 180 & 720 \\
\hline
  & 3 & 12 & 10 & 45 & 180 & 719
\end{array}
$$

$$\underbrace{\qquad\qquad\qquad\qquad}_{\substack{\text{coefficients} \\ \text{of quotient}}} \quad \underbrace{\quad}_{\text{remainder}}$$

Consequently, $f(4) = 719$.

Synthetic division may be employed to help find zeros of polynomials. By the method illustrated in the preceding example, $f(c) = 0$ if and only if the remainder in the synthetic division by $x - c$ is 0.

EXAMPLE 6   *Using synthetic division to find zeros of a polynomial*

Show that -11 is a zero of the polynomial

$$f(x) = x^3 + 8x^2 - 29x + 44.$$

SOLUTION   Dividing synthetically by $x - (-11) = x + 11$ gives us

$$
\begin{array}{r|rrrr}
-11 & 1 & 8 & -29 & 44 \\
    &   & -11 & 33 & -44 \\
\hline
    & 1 & -3 & 4 & 0
\end{array}
$$

$$\underbrace{\qquad\qquad\qquad}_{\substack{\text{coefficients} \\ \text{of quotient}}} \quad \underbrace{\quad}_{\text{remainder}}$$

Thus, $f(-11) = 0$.

Example 6 shows that the number -11 is a solution of the equation $x^3 + 8x^2 - 29x + 44 = 0$. In Section 4.4 we shall use synthetic division to find rational solutions of equations.

If $f(x) = x^3 - 5x^2 + 7x - 11$ and k is a specific number, then $f(k)$ is equal to $k^3 - 5k^2 + 7k - 11$. If we use synthetic division to find $f(k)$, we obtain a form that can be used to find values of f easily with a calculator. The following synthetic division of f by $(x - k)$ gives us this form of $f(k)$ as the remainder.

$$
\begin{array}{r|llll}
k & 1 & -5 & 7 & -11 \\
  &   & k & (k-5)k & ((k-5)k+7)k \\
\hline
  & 1 & k-5 & (k-5)k+7 & \underbrace{((k-5)k+7)k-11} \\
  &   &   &   & \text{remainder}
\end{array}
$$

The remainder shows that $f(x) = ((x - 5)x + 7)x - 11$, which is called the **nested form**, or **Horner's method** of writing a polynomial. The next example illustrates the use of this form when $f(k)$ is evaluated with a calculator.

EXAMPLE 7   *Finding the nested form of a polynomial*

Let $f(x) = 3x^4 + 11x^3 - 5x^2 + 7x - 4$.

(a) Express $f(x)$ in nested form.

(b) Use the nested form to evaluate $f(-3.712)$.

SOLUTION

(a) We could obtain the nested form using synthetic division as in the preceding discussion, but it is easier to successively factor out x from the nonconstant terms of the polynomial as follows:

$$
\begin{aligned}
f(x) &= 3x^4 + 11x^3 - 5x^2 + 7x - 4 & \text{given} \\
     &= (3x^3 + 11x^2 - 5x + 7)x - 4 & \text{factor out } x \text{ from the first 4 terms} \\
     &= ((3x^2 + 11x - 5)x + 7)x - 4 & \text{factor out } x \text{ from the first 3 terms} \\
     &= (((3x + 11)x - 5)x + 7)x - 4 & \text{factor out } x \text{ from the first 2 terms}
\end{aligned}
$$

(b) To evaluate $f(-3.712)$ with a calculator, we first store -3.712 in a memory location. (We assume that a memory retrieval key such as $\boxed{\text{RCL}}$ or $\boxed{\text{MR}}$ exists on your calculator.) We next use the keystroke sequence

$$
3 \;\; \boxed{\times} \;\; \boxed{\text{RCL}} \;\; \boxed{+} \;\; 11 \;\; \boxed{=}
$$

$$
\boxed{\times} \;\; \boxed{\text{RCL}} \;\; \boxed{-} \;\; 5 \;\; \boxed{=}
$$

$$
\boxed{\times} \;\; \boxed{\text{RCL}} \;\; \boxed{+} \;\; 7 \;\; \boxed{=}
$$

$$
\boxed{\times} \;\; \boxed{\text{RCL}} \;\; \boxed{-} \;\; 4 \;\; \boxed{=}
$$

to obtain the value -91.923 (rounded to three decimal places). It should be noted that this is not the method to use if computer software or a typical graphics calculator is available. In that case, it is best to enter $f(x)$, either substitute the value for x or store the value for x, and then simply query the function value.

The following illustration shows special cases of the nested form for polynomials in which some of the coefficients are zero.

ILLUSTRATION

Nested Form of Polynomials

■ $5x^4 - 3x^2 + 4x + 1 = (((5x)x - 3)x + 4)x + 1$

■ $x^5 - x^2 = ((x(x(x)) - 1)x)x$

■ $-3x^4 + 2x^2 - 7x = (((-3x)x + 2)x - 7)x$

4.2 EXERCISES

Exer. 1–8: Find the quotient and remainder if $f(x)$ is divided by $p(x)$.

1 $f(x) = 2x^4 - x^3 - 3x^2 + 7x - 12;$ $p(x) = x^2 - 3$

2 $f(x) = 3x^4 + 2x^3 - x^2 - x - 6;$ $p(x) = x^2 + 1$

3 $f(x) = 3x^3 + 2x - 4;$ $p(x) = 2x^2 + 1$

4 $f(x) = 3x^3 - 5x^2 - 4x - 8;$ $p(x) = 2x^2 + x$

5 $f(x) = 7x + 2;$ $p(x) = 2x^2 - x - 4$

6 $f(x) = -5x^2 + 3;$ $p(x) = x^3 - 3x + 9$

7 $f(x) = 9x + 4;$ $p(x) = 2x - 5$

8 $f(x) = 7x^2 + 3x - 10;$ $p(x) = x^2 - x + 10$

Exer. 9–12: Use the remainder theorem to find $f(c)$.

9 $f(x) = 3x^3 - x^2 + 5x - 4;$ $c = 2$

10 $f(x) = 2x^3 + 4x^2 - 3x - 1;$ $c = 3$

11 $f(x) = x^4 - 6x^2 + 4x - 8;$ $c = -3$

12 $f(x) = x^4 + 3x^2 - 12;$ $c = -2$

Exer. 13–16: Use the factor theorem to show that $x - c$ is a factor of $f(x)$.

13 $f(x) = x^3 + x^2 - 2x + 12;$ $c = -3$

14 $f(x) = x^3 + x^2 - 11x + 10;$ $c = 2$

15 $f(x) = x^{12} - 4096;$ $c = -2$

16 $f(x) = x^4 - 3x^3 - 2x^2 + 5x + 6;$ $c = 2$

Exer. 17–20: Find a polynomial $f(x)$ with leading coefficient 1 and having the given degree and zeros.

17 degree 3;   zeros $-2, 0, 5$

18 degree 3;   zeros $\pm 2, 3$

19 degree 4;   zeros $-2, \pm 1, 4$

20 degree 4;   zeros $-3, 0, 1, 5$

Exer. 21–28: Use synthetic division to find the quotient and remainder if the first polynomial is divided by the second.

21 $2x^3 - 3x^2 + 4x - 5;$ $x - 2$

22 $3x^3 - 4x^2 - x + 8;$ $x + 4$

23 $x^3 - 8x - 5;$ $x + 3$

24 $5x^3 - 6x^2 + 15;$ $x - 4$

25 $3x^5 + 6x^2 + 7;$ $x + 2$

26 $-2x^4 + 10x - 3;$ $x - 3$

27 $4x^4 - 5x^2 + 1;$ $x - \frac{1}{2}$

28 $9x^3 - 6x^2 + 3x - 4;$ $x - \frac{1}{3}$

Exer. 29–34: Use synthetic division to find $f(c)$.

29 $f(x) = 2x^3 + 3x^2 - 4x + 4;$ $c = 3$

30 $f(x) = -x^3 + 4x^2 + x;$ $c = -2$

31 $f(x) = 0.3x^3 + 0.04x - 0.034;$ $c = -0.2$

32 $f(x) = 8x^5 - 3x^2 + 7;$ $c = \frac{1}{2}$

33 $f(x) = x^2 + 3x - 5;$ $c = 2 + \sqrt{3}$

34 $f(x) = x^3 - 3x^2 - 8;$ $c = 1 + \sqrt{2}$

Exer. 35–38: Use synthetic division to show that c is a zero of $f(x)$.

35 $f(x) = 3x^4 + 8x^3 - 2x^2 - 10x + 4;$ $c = -2$

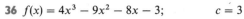

36 $f(x) = 4x^3 - 9x^2 - 8x - 3;$ $c = 3$

37 $f(x) = 4x^3 - 6x^2 + 8x - 3;$ $c = \frac{1}{2}$

38 $f(x) = 27x^4 - 9x^3 + 3x^2 + 6x + 1;$ $c = -\frac{1}{3}$

Exer. 39–40: Find all values of k such that $f(x)$ is divisible by the given linear polynomial.

39 $f(x) = kx^3 + x^2 + k^2x + 3k^2 + 11;$ $x + 2$

40 $f(x) = k^2x^3 - 4kx + 3;$ $x - 1$

Exer. 41–42: Show that $x - c$ is not a factor of $f(x)$ for any real number c.

41 $f(x) = 3x^4 + x^2 + 5$ **42** $f(x) = -x^4 - 3x^2 - 2$

43 Find the remainder if the polynomial
$$3x^{100} + 5x^{85} - 4x^{38} + 2x^{17} - 6$$
is divided by $x + 1$.

Exer. 44–46: Use the factor theorem to verify the statement.

44 $x - y$ is a factor of $x^n - y^n$ for every positive integer n.

45 $x + y$ is a factor of $x^n - y^n$ for every positive even integer n.

46 $x + y$ is a factor of $x^n + y^n$ for every positive odd integer n.

47 Let $P(x, y)$ be a first-quadrant point on $y = 6 - x$, and consider the vertical line segment PQ shown in the figure.

  (a) If PQ is rotated about the y-axis, determine the volume V of the resulting cylinder.

  (b) For what point $P(x, y)$ with $x \neq 1$ is the volume V in part (a) the same as the volume of the cylinder of radius 1 and altitude 5 shown in the figure?

EXERCISE 47

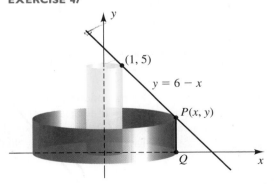

48 *Strength of a beam*   The strength of a rectangular beam is directly proportional to the product of its width and the square of the depth of a cross section (see the figure). A beam of width 1.5 feet has been cut from a cylindrical log of radius 1 foot. Find the width of a second rectangular beam of equal strength that could have been cut from the log.

EXERCISE 48

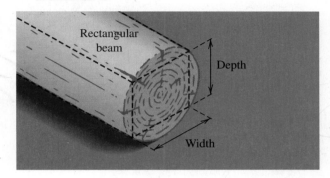

Rectangular beam

Depth

Width

49 *Parabolic arch*   An arch has the shape of the parabola $y = 4 - x^2$. A rectangle is fit under the arch by selecting a point (x, y) on the parabola (see the figure).

  (a) Express the area A of the rectangle in terms of x.

  (b) If $x = 1$, the rectangle has base 2 and height 3. Find the base of a second rectangle that has the same area.

EXERCISE 49

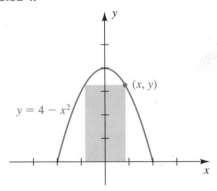

$y = 4 - x^2$

(x, y)

50 *Dimensions of a capsule*   An aspirin tablet in the shape of a right circular cylinder has height $\frac{1}{3}$ centimeter and radius $\frac{1}{2}$ centimeter. The manufacturer also wishes to market the aspirin in capsule form. The capsule is to be

$\frac{3}{2}$ centimeters long, in the shape of a right circular cylinder with hemispheres attached at both ends (see the figure).

(a) If r denotes the radius of a hemisphere, find a formula for the volume of the capsule.

(b) Find the radius of the capsule so that its volume is equal to that of the tablet.

EXERCISE 50

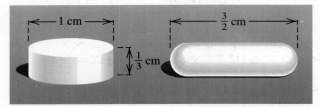

Exer. 51–52: Express $f(x)$ in nested form, and use this form to evaluate $f(a)$ without a calculator.

51 $f(x) = 4x^4 - 5x^2 + 7x - 3; \quad a = -2$

52 $f(x) = 3x^4 + 5x^3 - 6x + 11; \quad a = 2$

Exer. 53–54: Express $f(x)$ in nested form, and use this form to approximate $f(a)$.

53 $f(x) = 2x^5 + 3x^4 - 6x^3 - 4x^2 + x - 9; \quad a = 0.0325$

54 $f(x) = -3.43x^6 + 8.39x^4 - 5.75x^3 + 8.6x^2 - x + 12.1;$
$a = 0.1274$

Exer. 55–56: Approximate the remainder if the polynomial is divided by $x - 0.21$.

55 $x^8 - 7.9x^5 - 0.8x^4 + x^3 + 1.2x - 9.81$

56 $3.33x^6 - 2.5x^5 + 6.9x^3 - 4.1x^2 + 1.22x - 6.78$

c Exer. 57–58: Use the graph of f to approximate all values of k such that $f(x)$ is divisible by the given linear polynomial.

57 $f(x) = x^3 + k^3x^2 + 2kx - 2k^4; \quad x - 1.6$

58 $f(x) = k^5x^3 - 2.1x^2 + k^3x - 1.2k^2; \quad x + 0.4$

4.3 ZEROS OF POLYNOMIALS

The **zeros of a polynomial** $f(x)$ are the solutions of the equation $f(x) = 0$. Each real zero is an x-intercept of the graph of f. In applied fields, calculators and computers are usually employed to find or approximate zeros. Before using a calculator, however, it is worth knowing what type of zeros to expect. Some questions we could ask are

1. How many zeros of $f(x)$ are real?
2. How many real zeros of $f(x)$ are positive? negative?
3. How many real zeros of $f(x)$ are rational? irrational?
4. Are the real zeros of $f(x)$ large or small in value?

In this and the following section we shall discuss results that help answer some of these questions. These results form the basis of the *theory of equations*.

The factor and remainder theorems can be extended to the system of complex numbers. Thus, a complex number $c = a + bi$ is a zero of a polynomial $f(x)$ if and only if $x - c$ is a factor of $f(x)$. Except in special cases, zeros of polynomials are very difficult to find. For example, there are no obvious zeros of $f(x) = x^5 - 3x^4 + 4x^3 - 4x - 10$. Although we have no formula that can be used to find the zeros, the next theorem states that there is at *least* one zero c, and hence, by the factor theorem, $f(x)$ has a factor of the form $x - c$.

Fundamental Theorem of Algebra

If a polynomial $f(x)$ has positive degree and complex coefficients, then $f(x)$ has at least one complex zero.

The standard proof of this theorem requires results from an advanced field of mathematics called *functions of a complex variable*. A prerequisite for studying this field is a strong background in calculus. The first proof of the fundamental theorem of algebra was given by the German mathematician Carl Friedrich Gauss (1777–1855), who is considered by many to be the greatest mathematician of all time.

As a special case of the fundamental theorem, if all the coefficients of $f(x)$ are real, then $f(x)$ has at least one complex zero. If $a + bi$ is a complex zero, it may happen that $b = 0$, in which case the number a is a real zero.

The fundamental theorem of algebra enables us, at least in theory, to express every polynomial $f(x)$ of positive degree as a product of polynomials of degree 1, as in the next theorem.

Complete Factorization Theorem for Polynomials

If $f(x)$ is a polynomial of degree $n > 0$, then there exist n complex numbers $c_1, c_2, \ldots, c_n$ such that

$$f(x) = a(x - c_1)(x - c_2) \cdots (x - c_n),$$

where a is the leading coefficient of $f(x)$. Each number c_k is a zero of $f(x)$.

PROOF   If $f(x)$ has degree $n > 0$, then, by the fundamental theorem of algebra, $f(x)$ has a complex zero c_1. Hence, by the factor theorem, $f(x)$ has a factor $x - c_1$; that is,

$$f(x) = (x - c_1)f_1(x),$$

where $f_1(x)$ is a polynomial of degree $n - 1$. If $n - 1 > 0$, then, by the same argument, $f_1(x)$ has a complex zero c_2 and therefore a factor $x - c_2$. Thus,

$$f_1(x) = (x - c_2)f_2(x),$$

where $f_2(x)$ is a polynomial of degree $n - 2$. Hence,

$$f(x) = (x - c_1)(x - c_2)f_2(x).$$

Continuing this process, after n steps we arrive at a polynomial $f_n(x)$ of degree 0. Thus, $f_n(x) = a$ for some nonzero number a, and we may write

$$f(x) = a(x - c_1)(x - c_2) \cdots (x - c_n),$$

where each complex number c_k is a zero of $f(x)$. The leading coefficient of the polynomial on the right-hand side in the last equation is a, and therefore a is the leading coefficient of $f(x)$. ■

We may now prove the following.

Theorem on the Maximum Number of Zeros of a Polynomial

> A polynomial of degree $n > 0$ has at most n different complex zeros.

PROOF We will give an indirect proof; that is, we will suppose $f(x)$ has *more* than n different complex zeros and show that this supposition leads to a contradiction. Let us choose $n + 1$ of the zeros and label them $c_1, c_2, \ldots, c_n$, and c. We may use the c_k to obtain the factorization indicated in the statement of the complete factorization theorem. Substituting c for x and using the fact that $f(c) = 0$, we obtain

$$0 = a(c - c_1)(c - c_2) \cdots (c - c_n).$$

However, each factor on the right-hand side is different from zero because $c \neq c_k$ for every k. Since the product of nonzero numbers cannot equal zero, we have a contradiction. ■

EXAMPLE I *Finding a polynomial with prescribed zeros*

Find a polynomial $f(x)$ in factored form that has degree 3; zeros 2, -1, and 3; and satisfies $f(1) = 5$.

SOLUTION By the factor theorem, $f(x)$ has factors $x - 2$, $x + 1$, and $x - 3$. No other factors of degree 1 exist, since, by the factor theorem, another linear factor $x - c$ would produce a fourth zero of $f(x)$, contrary to the preceding theorem. Hence, $f(x)$ has the form

$$f(x) = a(x - 2)(x + 1)(x - 3)$$

for some number a. Since $f(1) = 5$, we see that

$$5 = a(1 - 2)(1 + 1)(1 - 3) \quad \text{let } x = 1 \text{ in } f(x)$$

$$5 = 4a \quad \text{simplify}$$

$$a = \tfrac{5}{4} \quad \text{solve for } a$$

Consequently,

$$f(x) = \tfrac{5}{4}(x - 2)(x + 1)(x - 3).$$

If we multiply the factors, we obtain the polynomial

$$f(x) = \tfrac{5}{4}x^3 - 5x^2 + \tfrac{5}{4}x + \tfrac{15}{2}.$$

The numbers $c_1, c_2, \ldots, c_n$ in the complete factorization theorem are not necessarily all different. To illustrate, $f(x) = x^3 + x^2 - 5x + 3$ has the factorization

$$f(x) = (x + 3)(x - 1)(x - 1).$$

If a factor $x - c$ occurs m times in the factorization, then c is a **zero of multiplicity m** of $f(x)$, or a **root of multiplicity m** of the equation $f(x) = 0$. In the preceding display, 1 is a zero of multiplicity 2, and -3 is a zero of multiplicity 1.

If c is a real zero of $f(x)$ of multiplicity m, then $f(x)$ has the factor $(x - c)^m$, and the graph of f has an x-intercept c. The general shape of the graph at $(c, 0)$ depends on whether m is an odd integer or an even integer. If m is odd, then $(x - c)^m$ changes sign as x increases through c, and hence the graph of f crosses the x-axis at $(c, 0)$, as indicated in the first row of the following chart. The figures in the chart do not show the complete graph of f, but only its general shape near $(c, 0)$. If m is even, then $(x - c)^m$ does not change sign at c, and the graph of f near $(c, 0)$ has the appearance of one of the two figures in the second row.

Factor of $f(x)$	General shape of the graph of f near $(c, 0)$	
$(x - c)^m$, with m odd and $m \neq 1$	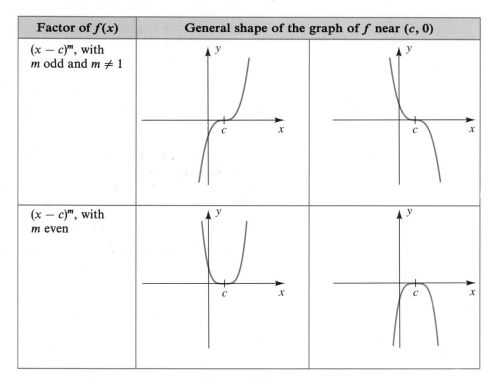	
$(x - c)^m$, with m even		

EXAMPLE 2 *Finding multiplicities of zeros*

Find the zeros of the polynomial $f(x) = \frac{1}{16}(x - 2)(x - 4)^3(x + 1)^2$, state the multiplicity of each, and then sketch the graph of f.

FIGURE 13

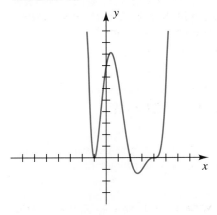

SOLUTION   We see from the factored form that $f(x)$ has three distinct zeros, 2, 4, and -1. The zero 2 has multiplicity 1, the zero 4 has multiplicity 3, and the zero -1 has multiplicity 2. Note that $f(x)$ has degree 6.

The x-intercepts of the graph of f are the real zeros -1, 2, and 4. Since the multiplicity of -1 is an even integer, the graph intersects, but does not cross, the x-axis at $(-1, 0)$. Since the multiplicities of 2 and 4 are odd, the graph crosses the x-axis at $(2, 0)$ and $(4, 0)$. The y-intercept is $f(0) = 8$. The graph is shown in Figure 13.

If $f(x) = a(x - c_1)(x - c_2) \cdots (x - c_n)$ is a polynomial of degree n, then the n complex numbers $c_1, c_2, \ldots, c_n$ are zeros of $f(x)$. Counting a zero of multiplicity m as m zeros tells us that $f(x)$ has at least n zeros (not necessarily all different). Combining this fact with the fact that $f(x)$ has at most n zeros gives us the next result.

Theorem on the Exact Number of Zeros of a Polynomial

> If $f(x)$ is a polynomial of degree $n > 0$ and if a zero of multiplicity m is counted m times, then $f(x)$ has precisely n zeros.

EXAMPLE 3   *Finding the zeros of a polynomial*

Express $f(x) = x^5 - 4x^4 + 13x^3$ as a product of linear factors, and find the five zeros of $f(x)$.

SOLUTION   We begin by factoring out x^3:

$$f(x) = x^3(x^2 - 4x + 13)$$

By the quadratic formula, the zeros of the polynomial $x^2 - 4x + 13$ are

$$\frac{-(-4) \pm \sqrt{(-4)^2 - 4(1)(13)}}{2(1)} = \frac{4 \pm \sqrt{-36}}{2} = \frac{4 \pm 6i}{2} = 2 \pm 3i.$$

Hence, by the factor theorem, $x^2 - 4x + 13$ has factors $x - (2 + 3i)$ and $x - (2 - 3i)$, and we obtain the factorization

$$f(x) = x \cdot x \cdot x \cdot (x - 2 - 3i)(x - 2 + 3i).$$

Since $x - 0$ occurs as a factor three times, the number 0 is a zero of multiplicity 3, and the five zeros of $f(x)$ are 0, 0, 0, $2 + 3i$, and $2 - 3i$.

We next show how to use *Descartes' rule of signs* to obtain information about the zeros of a polynomial $f(x)$ with real coefficients. In the statement of the rule we assume that the terms of $f(x)$ are arranged in order of decreasing powers of x and that terms with zero coefficients are deleted. We also assume that the **constant term**—that is, the term that does not contain x—is different from 0. We say there is a **variation of sign** in $f(x)$ if two consecutive coefficients have opposite signs. To illustrate, the polynomial $f(x)$ in the following illustration has three variations in sign, as indicated by the braces—one variation from $2x^5$ to $-7x^4$, a second from $-7x^4$ to $3x^2$, and a third from $6x$ to -5.

ILLUSTRATION

Variations of Sign in $f(x) = 2x^5 - 7x^4 + 3x^2 + 6x - 5$

$$f(x) = 2x^5 \overbrace{-}^{+\text{ to }-} 7x^4 \overbrace{+}^{-\text{ to }+} 3x^2 \overbrace{+}^{\substack{\text{no} \\ \text{variation}}} 6x \overbrace{-}^{+\text{ to }-} 5$$

Descartes' rule also refers to the variations of sign in $f(-x)$. Using the previous illustration, note that

$$f(-x) = 2(-x)^5 - 7(-x)^4 + 3(-x)^2 + 6(-x) - 5$$
$$= -2x^5 - 7x^4 + 3x^2 - 6x - 5.$$

Hence, as indicated in the next illustration, there are two variations of sign in $f(-x)$—one from $-7x^4$ to $3x^2$ and a second from $3x^2$ to $-6x$.

ILLUSTRATION

Variations of Sign in $f(-x)$ if $f(x) = 2x^5 - 7x^4 + 3x^2 + 6x - 5$

$$f(-x) = -2x^5 \overbrace{-}^{\substack{\text{no} \\ \text{variation}}} 7x^4 \overbrace{+}^{-\text{ to }+} 3x^2 \overbrace{-}^{+\text{ to }-} 6x \overbrace{-}^{\substack{\text{no} \\ \text{variation}}} 5$$

We may state Descartes' rule as follows.

Descartes' Rule of Signs

> Let $f(x)$ be a polynomial with real coefficients and nonzero constant term.
>
> **(1)** The number of positive real zeros of $f(x)$ either is equal to the number of variations of sign in $f(x)$ or is less than that number by an even integer.
>
> **(2)** The number of negative real zeros of $f(x)$ either is equal to the number of variations of sign in $f(-x)$ or is less than that number by an even integer.

A proof of Descartes' rule will not be given.

EXAMPLE 4   *Using Descartes' rule of signs*

Discuss the number of possible positive and negative real solutions and nonreal complex solutions of the equation

$$2x^5 - 7x^4 + 3x^2 + 6x - 5 = 0.$$

SOLUTION   The polynomial $f(x)$ on the left-hand side of the equation is the same as the one given in the two previous illustrations. Since there are three variations of sign in $f(x)$, the equation has either three positive real solutions or one positive real solution.

Since $f(-x) = -2x^5 - 7x^4 + 3x^2 - 6x - 5$ has two variations of sign, the given equation has either two negative solutions or no negative solution. The solutions that are not real numbers are complex numbers of the form $a + bi$ for real numbers a and b with $b \neq 0$. The following table summarizes the various possibilities that can occur for solutions of the equation.

Total number of solutions	5	5	5	5
Number of positive real solutions	1	1	3	3
Number of negative real solutions	0	2	0	2
Number of nonreal, complex solutions	4	2	2	0

Descartes' rule stipulates that the constant term of the polynomial $f(x)$ is different from 0. If the constant term is 0, as in the equation

$$x^4 - 3x^3 + 2x^2 - 5x = 0,$$

we factor out the lowest power of x, obtaining

$$x(x^3 - 3x^2 + 2x - 5) = 0.$$

Thus, one solution is $x = 0$, and we apply Descartes' rule to the polynomial $x^3 - 3x^2 + 2x - 5$ to determine the nature of the remaining three solutions.

EXAMPLE 5   *Using Descartes' rule of signs*

Discuss the nature of the roots of the equation $3x^5 + 4x^3 + 2x - 5 = 0$.

SOLUTION   The polynomial $f(x)$ on the left-hand side of the equation has one variation of sign (from $2x$ to -5), and hence, by part (1) of Descartes' rule, the equation has precisely one positive real root. Since

$$f(-x) = 3(-x)^5 + 4(-x)^3 + 2(-x) - 5$$
$$= -3x^5 - 4x^3 - 2x - 5,$$

$f(-x)$ has no variation of sign, and hence, by part (2) of Descartes' rule, there is no negative real root. Thus, the equation has one real root and four nonreal, complex roots.

When applying Descartes' rule, we count roots of multiplicity k as k roots. For example, given $x^2 - 2x + 1 = 0$, the polynomial $x^2 - 2x + 1$ has two variations of sign, and hence the equation has either two positive real roots or none. The factored form of the equation is $(x - 1)^2 = 0$, and hence 1 is a root of multiplicity 2.

We conclude this section with a discussion of *bounds* for the real zeros of a polynomial $f(x)$ with real coefficients. By definition, a real number b is an **upper bound** for the zeros if no zero is greater than b. A real number a is a **lower bound** for the zeros if no zero is less than a. Thus, if r is any real zero of $f(x)$, then $a \le r \le b$; that is, r is in the closed interval $[a, b]$, as illustrated in Figure 14. Note that upper and lower bounds are not unique, since any number greater than b is also an upper bound, and any number less than a is a lower bound.

FIGURE 14

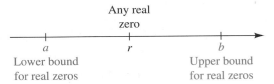

We may use synthetic division to find upper and lower bounds for the zeros of $f(x)$. Recall that if we divide $f(x)$ synthetically by $x - c$, the third row in the division process contains the coefficients of the quotient $q(x)$ together with the remainder $f(c)$. The following theorem indicates how this third row may be used to find upper and lower bounds for the real solutions.

Bounds for Real Zeros of Polynomials

Suppose that $f(x)$ is a polynomial with real coefficients and a positive leading coefficient and that $f(x)$ is divided synthetically by $x - c$.

(1) If $c > 0$ and if all numbers in the third row of the division process are either positive or zero, then c is an upper bound for the real zeros of $f(x)$.

(2) If $c < 0$ and if the numbers in the third row of the division process are alternately positive and negative (and a 0 in the third row is considered to be either positive or negative), then c is a lower bound for the real zeros of $f(x)$.

A general proof of this result can be patterned after the solution given in the next example.

EXAMPLE 6 *Finding bounds for the solutions of an equation*

Find upper and lower bounds for the real solutions of the equation $2x^3 + 5x^2 - 8x - 7 = 0$.

SOLUTION   We divide the polynomial $2x^3 + 5x^2 - 8x - 7$ synthetically by $x - 1$ and $x - 2$:

$$
\begin{array}{r|rrrr}
1 & 2 & 5 & -8 & -7 \\
  &   & 2 & 7 & -1 \\
\hline
  & 2 & 7 & -1 & -8
\end{array}
\qquad
\begin{array}{r|rrrr}
2 & 2 & 5 & -8 & -7 \\
  &   & 4 & 18 & 20 \\
\hline
  & 2 & 9 & 10 & 13
\end{array}
$$

The third row of the synthetic division by $x - 1$ contains negative numbers, and hence part (1) of the result on bounds for real zeros does not apply. However, since all numbers in the third row of the synthetic division by $x - 2$ are positive, it follows from part (1) that 2 is an upper bound for the real solutions of the equation. This fact is also evident if we express the division by $x - 2$ in the division algorithm form

$$2x^3 + 5x^2 - 8x - 7 = (x - 2)(2x^2 + 9x + 10) + 13,$$

for if $x > 2$, then the right-hand side of the equation is positive and hence is not zero. Consequently, $2x^3 + 5x^2 - 8x - 7$ is not zero if $x > 2$.

We now find the lower bound. After some trial-and-error attempts using $x - (-1)$, $x - (-2)$, and $x - (-3)$, we see that synthetic division by $x - (-4)$ gives us

$$
\begin{array}{r|rrrr}
-4 & 2 & 5 & -8 & -7 \\
   &   & -8 & 12 & -16 \\
\hline
   & 2 & -3 & 4 & -23
\end{array}
$$

Since the numbers in the third row are alternately positive and negative, it follows from part (2) of the preceding theorem that -4 is a lower bound for the real solutions. This can also be proved by expressing the division by $x + 4$ in the form

$$2x^3 + 5x^2 - 8x - 7 = (x + 4)(2x^2 - 3x + 4) - 23,$$

for if $x < -4$, then the right-hand side of this equation is negative and therefore is not zero, Hence, $2x^3 + 5x^2 - 8x - 7$ is not zero if $x < -4$.

Since lower and upper bounds for the real solutions are -4 and 2, respectively, it follows that all real solutions are in the closed interval $[-4, 2]$.

The graph of $f(x) = 2x^3 + 5x^2 - 8x - 7$ is sketched in Figure 15. The graph shows that the three zeros of f are in the intervals $[-4, -3]$, $[-1, 0]$, and $[1, 2]$, respectively.

FIGURE 15

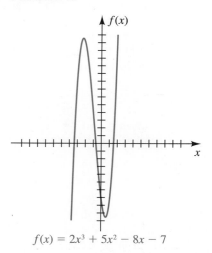

$f(x) = 2x^3 + 5x^2 - 8x - 7$

4.3 EXERCISES

Exer. 1–6: Find a polynomial $f(x)$ of degree 3 that has the indicated zeros and satisfies the given condition.

1 $-1, 2, 3$; $f(-2) = 80$

2 $-5, 2, 4$; $f(3) = -24$

3 $-4, 3, 0$; $f(2) = -36$

4 $-3, -2, 0$; $f(-4) = 16$

5 $-2i, 2i, 3;\quad f(1) = 20$

6 $-3i, 3i, 4;\quad f(-1) = 50$

7 Find a polynomial $f(x)$ of degree 4 with leading coefficient 1 such that both -4 and 3 are zeros of multiplicity 2, and sketch the graph of f.

8 Find a polynomial $f(x)$ of degree 4 with leading coefficient 1 such that both -5 and 2 are zeros of multiplicity 2, and sketch the graph of f.

9 Find a polynomial $f(x)$ of degree 6 such that 0 and 3 are both zeros of multiplicity 3 and $f(2) = -24$. Sketch the graph of f.

10 Find a polynomial $f(x)$ of degree 7 such that -2 and 2 are both zeros of multiplicity 2, 0 is a zero of multiplicity 3, and $f(-1) = 27$. Sketch the graph of f.

11 Find the third-degree polynomial function whose graph is shown in the figure.

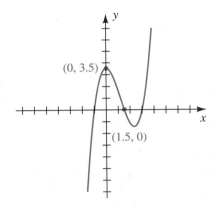

12 Find the fourth-degree polynomial function whose graph is shown in the figure.

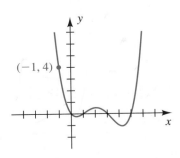

Exer. 13–14: Find the polynomial function of degree 3 whose graph is shown in the figure.

13

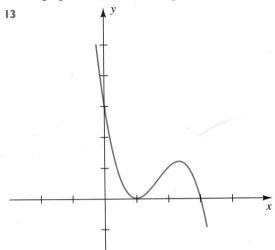

14

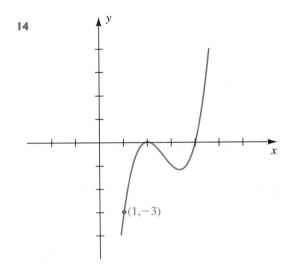

Exer. 15–22: Find the zeros of $f(x)$, and state the multiplicity of each zero.

15 $f(x) = x^2(3x + 2)(2x - 5)^3$

16 $f(x) = x(x + 1)^4(3x - 7)^2$

17 $f(x) = 4x^5 + 12x^4 + 9x^3$

18 $f(x) = (4x^2 - 5)^2$

19 $f(x) = (x^2 + x - 12)^3(x^2 - 9)^2$

20 $f(x) = (6x^2 + 7x - 5)^4(4x^2 - 1)^2$

21 $f(x) = x^4 + 7x^2 - 144$

22 $f(x) = x^4 + 21x^2 - 100$

Exer. 23–26: Show that the number is a zero of $f(x)$ of the given multiplicity, and express $f(x)$ as a product of linear factors.

23 $f(x) = x^4 + 7x^3 + 13x^2 - 3x - 18$;
-3 (multiplicity 2)

24 $f(x) = x^4 - 9x^3 + 22x^2 - 32$;   4 (multiplicity 2)

25 $f(x) = x^6 - 4x^5 + 5x^4 - 5x^2 + 4x - 1$;
1 (multiplicity 5)

26 $f(x) = x^5 + x^4 - 6x^3 - 14x^2 - 11x - 3$;
-1 (multiplicity 4)

Exer. 27–34: Use Descartes' rule of signs to determine the number of possible positive, negative, and nonreal complex solutions of the equation.

27 $4x^3 - 6x^2 + x - 3 = 0$ **28** $5x^3 - 6x - 4 = 0$

29 $4x^3 + 2x^2 + 1 = 0$ **30** $3x^3 - 4x^2 + 3x + 7 = 0$

31 $3x^4 + 2x^3 - 4x + 2 = 0$

32 $2x^4 - x^3 + x^2 - 3x + 4 = 0$

33 $x^5 + 4x^4 + 3x^3 - 4x + 2 = 0$

34 $2x^6 + 5x^5 + 2x^2 - 3x + 4 = 0$

Exer. 35–40: Find the smallest and largest integers that are upper and lower bounds, respectively, for the real solutions of the equation.

35 $x^3 - 4x^2 - 5x + 7 = 0$

36 $2x^3 - 5x^2 + 4x - 8 = 0$

37 $x^4 - x^3 - 2x^2 + 3x - 6 = 0$

38 $2x^4 - 9x^3 - 8x - 10 = 0$

39 $2x^5 - 13x^3 + 2x - 5 = 0$

40 $3x^5 + 2x^4 - x^3 - 8x^2 - 7 = 0$

41 *Using limited data*   A scientist has limited data on the temperature T (in °C) during a 24-hour period. If t denotes time in hours and $t = 0$ corresponds to midnight, find the fourth-degree polynomial that fits the information in the following table.

t (hours)	0	5	12	19	24
T (°C)	0	0	10	0	0

42 *Lagrange interpolation polynomial*   A polynomial $f(x)$ of degree 3 with zeros at c_1, c_2, and c_3 and with $f(c) = 1$ for $c_2 < c < c_3$ is a third-degree *Lagrange interpolation polynomial*. Find an explicit formula for $f(x)$ in terms of c_1, c_2, c_3, and c.

c **Exer. 43–44:** Graph f for each value of n on the same coordinate plane, and describe how the multiplicity of a zero affects the graph of f.

43 $f(x) = (x - 0.5)^n(x^2 + 1)$; $n = 1, 2, 3, 4$

44 $f(x) = (x - 1)^n(x + 1)^n$; $n = 1, 2, 3, 4$

c **Exer. 45–46:** Graph f, estimate all real zeros, and determine the multiplicity of each zero.

45 $f(x) = x^3 + 1.3x^2 - 1.2x - 1.584$

46 $f(x) = x^5 - \frac{1}{4}x^4 - \frac{19}{8}x^3 - \frac{9}{32}x^2 + \frac{405}{256}x + \frac{675}{1024}$

Exer. 47–48: Is there a polynomial of the given degree n whose graph contains the indicated points?

47 $n = 3$;
$(1.1, -49.815)$, $(2, 0)$, $(3.5, 25.245)$, $(5.2, 0)$,
$(6.4, -29.304)$, $(10.1, 0)$

48 $n = 4$;
$(1.25, 0)$, $(2, 0)$, $(2.5, 56.25)$, $(3, 128.625)$, $(6.5, 0)$,
$(9, -307.75)$, $(10, 0)$

c **49** *Greenhouse effect*   Because of the combustion of fossil fuels, the concentration of carbon dioxide in the atmosphere is increasing. Research indicates that this will result in a *greenhouse effect* that will change the average global surface temperature. Assuming a vigorous expansion of coal use, the future amount $A(t)$ of atmospheric carbon dioxide concentration can be approximated (in parts per million) by

$$A(t) = -\frac{1}{2400}t^3 + \frac{1}{20}t^2 + \frac{7}{6}t + 340,$$

where t is in years, $t = 0$ corresponds to 1980, and $0 \le t \le 60$. Use the graph of A to estimate the year when the carbon dioxide concentration will be 400.

c **50** *Greenhouse effect*   The average increase in global surface temperature due to the greenhouse effect can be approximated by

$$T(t) = \frac{21}{5,000,000}t^3 - \frac{127}{1,000,000}t^2 + \frac{1293}{50,000}t,$$

where $0 \le t \le 60$ and $t = 0$ corresponds to 1980. Use the graph of T to estimate the year when the average temperature will have risen by 1 °C.

4.4 COMPLEX AND RATIONAL ZEROS OF POLYNOMIALS

Example 3 of the preceding section illustrates an important fact about polynomials with real coefficients: The two complex zeros $2 + 3i$ and $2 - 3i$ of $x^5 - 4x^4 + 13x^3$ are conjugates of each other. The relationship is not accidental, since the following general result is true.

Theorem on Conjugate Pair Zeros of a Polynomial

> If a polynomial $f(x)$ of degree $n > 1$ has real coefficients and if $z = a + bi$ with $b \neq 0$ is a complex zero of $f(x)$, then the conjugate $\bar{z} = a - bi$ is also a zero of $f(x)$.

PROOF Let us consider the polynomial

$$f(x) = a_n x^n + a_{n-1} x^{n-1} + \cdots + a_1 x + a_0,$$

where each coefficient a_k is a real number and $a_n \neq 0$. If $f(z) = 0$, then

$$a_n z^n + a_{n-1} z^{n-1} + \cdots + a_1 z + a_0 = 0.$$

If two complex numbers are equal, then so are their conjugates. Hence, the conjugate of the left-hand side of the last equation equals the conjugate of the right-hand side; that is,

$$\overline{a_n z^n + a_{n-1} z^{n-1} + \cdots + a_1 z + a_0} = \bar{0} = 0.$$

The fact that $\bar{0} = 0$ follows from $\bar{0} = \overline{0 + 0i} = 0 - 0i = 0$.

If z and w are arbitrary complex numbers, then it can be shown that $\overline{z + w} = \bar{z} + \bar{w}$. More generally, the conjugate of *any* sum of complex numbers is the sum of the conjugates. Consequently,

$$\overline{a_n z^n} + \overline{a_{n-1} z^{n-1}} + \cdots + \overline{a_1 z} + \overline{a_0} = 0.$$

It can also be shown that $\overline{z \cdot w} = \bar{z} \cdot \bar{w}$, $\overline{z^n} = \bar{z}^n$ for every positive integer n, and $\bar{z} = z$ if and only if z is real. Thus, for every k,

$$\overline{a_k z^k} = \overline{a_k} \cdot \overline{z^k} = \overline{a_k} \cdot \bar{z}^k = a_k \bar{z}^k,$$

and therefore

$$a_n \bar{z}^n + a_{n-1} \bar{z}^{n-1} + \cdots + a_1 \bar{z} + a_0 = 0.$$

The last equation states that $f(\bar{z}) = 0$, which completes the proof. ■

EXAMPLE I   *Finding a polynomial with prescribed zeros*

Find a polynomial $f(x)$ of degree 4 that has real coefficients and zeros $2 + i$ and $-3i$.

SOLUTION   By the theorem on conjugate pair zeros, $f(x)$ must also have zeros $2 - i$ and $3i$. Applying the factor theorem, we find that $f(x)$ has the following factors:

$$x - (2 + i), \quad x - (2 - i), \quad x - (-3i), \quad x - (3i)$$

Multiplying these four factors gives us

$$\begin{aligned}
f(x) &= [x - (2 + i)][x - (2 - i)](x + 3i)(x - 3i) \\
&= (x^2 - 4x + 5)(x^2 + 9) \\
&= x^4 - 4x^3 + 14x^2 - 36x + 45.
\end{aligned}$$

If a polynomial with real coefficients is factored as in the complete factorization theorem (see page 243), some of the factors $x - c_k$ may contain a complex number c_k. However, it is always possible to obtain a factorization into polynomials with real coefficients, as stated in the next theorem.

Theorem on Expressing a Polynomial as a Product of Linear and Quadratic Factors

> Every polynomial with real coefficients and positive degree n can be expressed as a product of linear and quadratic polynomials with real coefficients such that the quadratic factors are irreducible over $\mathbb{R}$.

PROOF   Since $f(x)$ has precisely n complex zeros $c_1, c_2, \ldots, c_n$, we may write

$$f(x) = a(x - c_1)(x - c_2) \cdots (x - c_n),$$

where a is the leading coefficient of $f(x)$. Of course, some of the zeros may be real. In such cases we obtain the linear factors referred to in the statement of the theorem.

If a zero c_k is not real, then, by the theorem on conjugate pair zeros, the conjugate $\overline{c_k}$ is also a zero of $f(x)$ and hence must be one of the numbers $c_1, c_2, \ldots, c_n$. This implies that both $x - c_k$ and $x - \overline{c_k}$ appear in the factorization of $f(x)$. If those factors are multiplied, we obtain

$$(x - c_k)(x - \overline{c_k}) = x^2 - (c_k + \overline{c_k})x + c_k\overline{c_k},$$

which has *real* coefficients, since $c_k + \overline{c_k}$ and $c_k\overline{c_k}$ are real numbers (see page 89). Thus, if c_k is a complex zero, then the product $(x - c_k)(x - \overline{c_k})$ is a quadratic polynomial that is irreducible over $\mathbb{R}$. This completes the proof. ■

EXAMPLE 2   *Expressing a polynomial as a product of linear and quadratic factors*

Express $x^5 - 4x^3 + x^2 - 4$ as a product of

(a) linear and quadratic polynomials with real coefficients that are irreducible over $\mathbb{R}$

(b) linear polynomials

SOLUTION

(a) $x^5 - 4x^3 + x^2 - 4$

$$\begin{aligned}
&= (x^5 - 4x^3) + (x^2 - 4) &&\text{group terms}\\
&= x^3(x^2 - 4) + (x^2 - 4) &&\text{factor out } x^3\\
&= (x^3 + 1)(x^2 - 4) &&\text{factor out } x^2 - 4\\
&= (x + 1)(x^2 - x + 1)(x + 2)(x - 2) &&\text{factor as the sum of cubes and}\\
& &&\text{the difference of squares}
\end{aligned}$$

Using the quadratic formula, we see that the polynomial $x^2 - x + 1$ has the complex zeros

$$\frac{-(-1) \pm \sqrt{(-1)^2 - 4(1)(1)}}{2(1)} = \frac{1 \pm \sqrt{3}i}{2} = \frac{1}{2} \pm \frac{\sqrt{3}}{2}i$$

and hence is irreducible over $\mathbb{R}$. Thus, the desired factorization is

$$(x + 1)(x^2 - x + 1)(x + 2)(x - 2).$$

(b) Since the polynomial $x^2 - x + 1$ in part (a) has zeros $\frac{1}{2} \pm (\sqrt{3}/2)i$, it follows from the factor theorem that the polynomial has factors

$$x - \left(\frac{1}{2} + \frac{\sqrt{3}}{2}i\right) \quad \text{and} \quad x - \left(\frac{1}{2} - \frac{\sqrt{3}}{2}i\right).$$

Substituting in the factorization found in part (a), we obtain the following complete factorization into linear polynomials:

$$(x + 1)\left(x - \frac{1}{2} - \frac{\sqrt{3}}{2}i\right)\left(x - \frac{1}{2} + \frac{\sqrt{3}}{2}i\right)(x + 2)(x - 2)$$

We previously pointed out that it is generally very difficult to find the zeros of a polynomial of high degree. If all the coefficients are rational numbers, however, there is a method for finding the *rational* zeros, if they exist. The method is a consequence of the following result.

Theorem on Rational Zeros of a Polynomial

If the polynomial

$$f(x) = a_n x^n + a_{n-1} x^{n-1} + \cdots + a_1 x + a_0$$

has *integer* coefficients and if c/d is a rational zero of $f(x)$ such that c and d have no common prime factor, then

(1) the numerator c of the zero is a factor of the constant term a_0

(2) the denominator d of the zero is a factor of the leading coefficient a_n

PROOF   Assume that $c > 0$. (The proof for $c < 0$ is similar.) Let us show that c is a factor of a_0. The case $c = 1$ is trivial, since 1 is a factor of *any* number. Thus, suppose $c \neq 1$. In this case $c/d \neq 1$, for if $c/d = 1$, we obtain $c = d$, and since c and d have no prime factor in common, this implies that $c = d = 1$, a contradiction. Hence, in the following discussion we have $c \neq 1$ and $c \neq d$.

Since $f(c/d) = 0$,

$$a_n \frac{c^n}{d^n} + a_{n-1} \frac{c^{n-1}}{d^{n-1}} + \cdots + a_1 \frac{c}{d} + a_0 = 0.$$

We multiply by d^n and then add $-a_0 d^n$ to both sides:

$$a_n c^n + a_{n-1} c^{n-1} d + \cdots + a_1 c d^{n-1} = -a_0 d^n$$

$$c(a_n c^{n-1} + a_{n-1} c^{n-2} d + \cdots + a_1 d^{n-1}) = -a_0 d^n$$

The last equation shows that c is a factor of the integer $a_0 d^n$. Since c and d have no common factor, c is a factor of a_0. A similar argument may be used to prove that d is a factor of a_n. ■

The technique of using the theorem on rational zeros to find rational solutions of equations with integral coefficients is illustrated in the following examples.

EXAMPLE 3   *A polynomial with no rational zeros*

Show that $f(x) = x^3 - 4x - 2$ has no rational zeros.

SOLUTION   If $f(x)$ has a rational zero c/d such that c and d have no common prime factor, then, by the theorem on rational zeros, c is a factor of the constant term -2 and hence is either 2 or -2 (which we write as ± 2). The denominator d is a factor of the leading coefficient 1 and hence is ± 1. Thus, the only possibilities for c/d are

$$\frac{\pm 1}{\pm 1} \quad \text{and} \quad \frac{\pm 2}{\pm 1} \qquad \text{or, equivalently,} \qquad \pm 1 \quad \text{and} \quad \pm 2.$$

Substituting each of these numbers for x, we obtain

$$f(1) = -5, \quad f(-1) = 1, \quad f(2) = -2, \quad \text{and} \quad f(-2) = -2.$$

Since $f(\pm 1) \neq 0$ and $f(\pm 2) \neq 0$, it follows that $f(x)$ has no rational zeros.

In the solution of the following example we shall assume that a graphing utility is not available. In Example 5 we will rework the problem to demonstrate the advantage of using a graphing utility.

EXAMPLE 4   *Finding the rational solutions of an equation*

Find all rational solutions of

$$3x^4 + 14x^3 + 14x^2 - 8x - 8 = 0.$$

SOLUTION   The problem is equivalent to finding the rational zeros of the polynomial on the left-hand side of the equation. If c/d is a rational zero and c and d have no common factor, then c is a factor of the constant term -8 and d is a factor of the leading coefficient 3. All possible choices are listed in the following table.

Choices for the numerator c	$\pm 1, \pm 2, \pm 4, \pm 8$
Choices for the denominator d	$\pm 1, \pm 3$
Choices for c/d	$\pm 1, \pm 2, \pm 4, \pm 8, \pm\frac{1}{3}, \pm\frac{2}{3}, \pm\frac{4}{3}, \pm\frac{8}{3}$

We can reduce the number of choices by finding upper and lower bounds for the real solutions; however, we shall not do so here. It is necessary to determine which of the choices for c/d, if any, are zeros. We see by substitution that neither 1 nor -1 is a solution. If we divide synthetically by $x + 2$, we obtain

$$\begin{array}{r|rrrrr} -2 & 3 & 14 & 14 & -8 & -8 \\ & & -6 & -16 & 4 & 8 \\ \hline & 3 & 8 & -2 & -4 & 0 \end{array}$$

This result shows that -2 is a zero. Moreover, the synthetic division provides the coefficients of the quotient in the division of the polynomial by $x + 2$. Hence, we have the following factorization of the given polynomial:

$$(x + 2)(3x^3 + 8x^2 - 2x - 4)$$

The remaining solutions of the equation must be zeros of the second factor, so we use that polynomial to check for solutions. (Note that $\pm\frac{8}{3}$ are no longer candidates, since the numerator must be a factor of 4.) Again proceeding by trial and error, we ultimately find that synthetic division by $x + \frac{2}{3}$ gives us the following result:

$$\begin{array}{r|rrrr} -\frac{2}{3} & 3 & 8 & -2 & -4 \\ & & -2 & -4 & 4 \\ \hline & 3 & 6 & -6 & 0 \end{array}$$

Therefore, $-\frac{2}{3}$ is also a zero.

Using the coefficients of the quotient, we know that the remaining zeros are solutions of the equation $3x^2 + 6x - 6 = 0$. Dividing both sides by 3 gives us the equivalent equation $x^2 + 2x - 2 = 0$. By the quadratic formula, this equation has solutions

$$\frac{-2 \pm \sqrt{2^2 - 4(1)(-2)}}{2(1)} = \frac{-2 \pm \sqrt{12}}{2} = \frac{-2 \pm 2\sqrt{3}}{2} = -1 \pm \sqrt{3}.$$

Hence, the given polynomial has two rational roots, -2 and $-\frac{2}{3}$, and two irrational roots, $-1 + \sqrt{3} \approx 0.732$ and $-1 - \sqrt{3} \approx -2.732$.

EXAMPLE 5   Finding the rational solutions of an equation

Find all rational solutions of the equation

$$3x^4 + 14x^3 + 14x^2 - 8x - 8 = 0.$$

FIGURE 16 $[-7.5, 7.5]$ by $[-5, 5]$

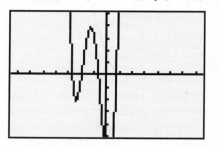

SOLUTION   Assigning the indicated polynomial to Y_1 and choosing the viewing rectangle $[-7.5, 7.5]$ by $[-5, 5]$, we obtain a sketch similar to Figure 16. The graph indicates that -2 is a solution and that there is one solution in each of the intervals $(-3, -2)$, $(-1, 0)$, and $(0, 1)$. Referring to the choices for a rational zero c/d in the preceding example, we see that the only possibilities are $-\frac{8}{3}$ in $(-3, -2)$, $-\frac{2}{3}$ or $-\frac{1}{3}$ in $(-1, 0)$, and $\frac{1}{3}$ or $\frac{2}{3}$ in $(0, 1)$. Thus, by referring to the graph, we have reduced the number of choices for zeros from sixteen, as in Example 4, to five.

If we zoom in on the graph of Y_1, we see that $-\frac{8}{3}$, $-\frac{2}{3}$, and $\frac{2}{3}$ are the most reasonable choices for zeros. As in Example 4, synthetic division can be used to determine that the only rational solutions are -2 and $-\frac{2}{3}$.

The theorem on rational zeros may be applied to equations with rational coefficients. We merely multiply both sides of the equation by the lcd of all the coefficients to obtain an equation with integral coefficients, and then we proceed as in Example 4.

EXAMPLE 6   Finding the radius of a grain silo

A grain silo has the shape of a right circular cylinder with a hemisphere attached to the top. If the total height of the structure is 30 feet, find the radius of the cylinder that results in a total volume of 1008π ft^3.

FIGURE 17

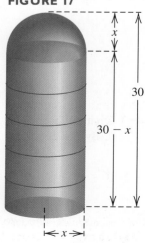

SOLUTION   Let x denote the radius of the cylinder. A sketch of the silo, labeled appropriately, is shown in Figure 17. Since the volume of the cylinder is $\pi x^2(30 - x)$ and the volume of the hemisphere is $\frac{2}{3}\pi x^3$, we must solve the following:

$$\pi x^2(30 - x) + \tfrac{2}{3}\pi x^3 = 1008\pi \qquad \text{total volume is } 1008$$

$$3x^2(30 - x) + 2x^3 = 3024 \qquad \text{multiply by } \frac{3}{\pi}$$

$$90x^2 - x^3 = 3024 \qquad \text{simplify}$$

$$x^3 - 90x^2 + 3024 = 0 \qquad \text{equivalent equation}$$

Since the leading coefficient of the polynomial on the left side of the last equation is 1, any rational root has the form $c/1 = c$, where c is a factor of 3024. If we factor 3024 into primes, we find that $3024 = 2^4 \cdot 3^3 \cdot 7$. It follows that some of the positive factors of 3024 are

$$1, \quad 2, \quad 3, \quad 4, \quad 6, \quad 8, \quad 9, \quad 12, \quad \ldots.$$

To help us decide which of these numbers to test first, let us make a rough estimate of the radius by assuming that the silo has the shape of a right circular cylinder of height 30 feet. In that case, the volume would be $\pi r^2 h = 30\pi r^2$. Since this volume should be close to 1008π, we see that

$$30r^2 = 1008, \quad \text{or} \quad r^2 = 1008/30 \approx 33.6.$$

This suggests that we use 6 in our first synthetic division, as follows:

$$\begin{array}{r|rrrr} 6 & 1 & -90 & 0 & 3024 \\ & & 6 & -504 & -3024 \\ \hline & 1 & -84 & -504 & 0 \end{array}$$

FIGURE 18 [0, 20] by [0, 4000]

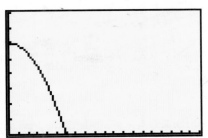

Thus, 6 is a solution of the equation $x^3 - 90x^2 + 3024 = 0$.

The remaining two solutions of the equation can be found by solving $x^2 - 84x - 504 = 0$, which we obtain from the quotient in the synthetic division. These zeros are approximately -5.62 and 89.62—neither of which satisfies the conditions of the problem. Hence, the desired radius is 6 feet.

Using a graphing utility with viewing rectangle [0, 20] by [0, 4000], we obtain the graph of $f(x) = x^3 - 90x^2 + 3024$ in Figure 18, which shows the zero $x = 6$. An extended graph would also indicate the other two zeros.

4.4 EXERCISES

Exer. 1–10: A polynomial $f(x)$ with real coefficients and leading coefficient 1 has the given zero(s) and degree. Express $f(x)$ as a product of linear and quadratic polynomials with real coefficients that are irreducible over $\mathbb{R}$.

1 $3 + 2i$;   degree 2

2 $-4 + 3i$;   degree 2

3 $2, -2 - 5i$;   degree 3

4 $-3, 1 - 7i$;   degree 3

5 $-1, 0, 3 + i$;   degree 4

6 $0, 2, -2 - i$;   degree 4

7 $4 + 3i, -2 + i$; degree 4

8 $3 + 5i, -1 - i$; degree 4

9 $0, -2i, 1 - i$; degree 5

10 $0, 3i, 4 + i$; degree 5

Exer. 11–14: Show that the equation has no rational root.

11 $x^3 + 3x^2 - 4x + 6 = 0$

12 $3x^3 - 4x^2 + 7x + 5 = 0$

13 $x^5 - 3x^3 + 4x^2 + x - 2 = 0$

14 $2x^5 + 3x^3 + 7 = 0$

Exer. 15–24: Find all solutions of the equation.

15 $x^3 - x^2 - 10x - 8 = 0$

16 $x^3 + x^2 - 14x - 24 = 0$

17 $2x^3 - 3x^2 - 17x + 30 = 0$

18 $12x^3 + 8x^2 - 3x - 2 = 0$

19 $x^4 + 3x^3 - 30x^2 - 6x + 56 = 0$

20 $3x^5 - 10x^4 - 6x^3 + 24x^2 + 11x - 6 = 0$

21 $6x^5 + 19x^4 + x^3 - 6x^2 = 0$

22 $6x^4 + 5x^3 - 17x^2 - 6x = 0$

23 $8x^3 + 18x^2 + 45x + 27 = 0$

24 $3x^3 - x^2 + 11x - 20 = 0$

25 Does there exist a polynomial of degree 3 with real coefficients that has zeros 1, -1, and i? Justify your answer.

26 The polynomial $f(x) = x^3 - ix^2 + 2ix + 2$ has the complex number i as a zero; however, the conjugate $-i$ of i is not a zero. Why doesn't this result contradict the theorem on conjugate pair zeros of a polynomial?

27 If n is an odd positive integer, prove that a polynomial of degree n with real coefficients has at least one real zero.

28 If a polynomial of the form

$$x^n + a_{n-1}x^{n-1} + \cdots + a_1x + a_0,$$

where each a_k is an integer, has a rational root r, show that r is an integer and is a factor of a_0.

29 *Constructing a box*   From a rectangular piece of cardboard having dimensions 20 inches × 30 inches, an open box is to be made by removing squares of area x^2 from each corner and turning up the sides. (See Exercise 33 of Section 4.1.)

(a) Show that there are two boxes that have a volume of 1000 in.³.

(b) Which box has the smaller surface area?

30 *Constructing a crate*   The frame for a shipping crate is to be constructed from 24 feet of 2 × 2 lumber. Assuming the crate is to have square ends of length x feet, determine the value(s) of x that result(s) in a volume of 4 ft³. (See Exercise 34 of Section 4.1.)

31 A right triangle has area 30 ft² and a hypotenuse that is 1 foot longer than one of its sides.

(a) If x denotes the length of this side, show that $2x^3 + x^2 - 3600 = 0$.

(b) Show that there is a positive root of the equation in part (a) and that this root is less than 13.

(c) Find the lengths of the sides of the triangle.

32 *Constructing a storage tank*   A storage tank for propane gas is to be constructed in the shape of a right circular cylinder of altitude 10 feet with a hemisphere attached to each end. Determine the radius x so that the resulting volume is 27π ft³. (See Example 7 of Section 3.4.)

33 *Constructing a storage shelter*   A storage shelter is to be constructed in the shape of a cube with a triangular prism forming the roof (see the figure). The length x of a side of the cube is yet to be determined.

(a) If the total height of the structure is 6 feet, show that its volume V is given by $V = x^3 + \frac{1}{2}x^2(6 - x)$.

(b) Determine x so that the volume is 80 ft³.

EXERCISE 33

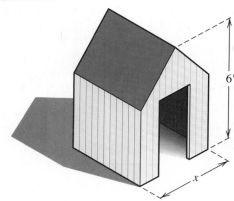

34 *Designing a tent*   A canvas camping tent is to be constructed in the shape of a pyramid with a square base. An 8-foot pole will form the center support, as illustrated in the figure. Find the length x of a side of the base so that the total canvas needed for the sides and bottom is 384 ft².

EXERCISE 34

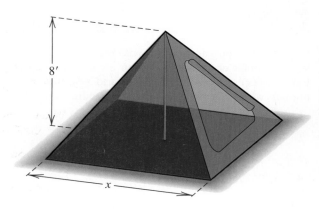

Exer. 35–36: Use a graph to determine the number of nonreal solutions of the equation.

35 $x^5 + 1.1x^4 - 3.21x^3 - 2.835x^2 + 2.7x + 0.62 = -1$

36 $x^4 - 0.4x^3 - 2.6x^2 + 1.1x + 3.5 = 2$

Exer. 37–38: Use a graph and synthetic division to find all solutions of the equation.

37 $x^4 + 1.4x^3 + 0.44x^2 - 0.56x - 0.96 = 0$

38 $x^5 + 1.1x^4 - 2.62x^3 - 4.72x^2 - 0.2x + 5.44 = 0$

c 39 *Atmospheric density* The density $D(h)$ (in kg/m³) of the earth's atmosphere at an altitude of h meters can be approximated by

$$D(h) = 1.2 - ah + bh^2 - ch^3,$$

where $a = 1.096 \times 10^{-4}$, $b = 3.42 \times 10^{-9}$, $c = 3.6 \times 10^{-14}$, and $0 \leq h \leq 30{,}000$. Use the graph of D to approximate the altitude h at which the density is 0.4.

c 40 *The earth's density* The earth's density $D(h)$ (in g/cm³) h meters underneath the surface can be approximated by

$$D(h) = 2.84 + ah + bh^2 - ch^3,$$

where $a = 1.4 \times 10^{-3}$, $b = 2.49 \times 10^{-6}$, $c = 2.19 \times 10^{-9}$, and $0 \leq h \leq 1000$. Use the graph of D to approximate the depth h at which the density of the earth is 3.7.

4.5 RATIONAL FUNCTIONS

A function f is a **rational function** if

$$f(x) = \frac{g(x)}{h(x)},$$

where $g(x)$ and $h(x)$ are polynomials. The domain of f consists of all real numbers *except* the zeros of the denominator $h(x)$.

ILLUSTRATION

Rational Functions and Their Domains

■ $f(x) = \dfrac{1}{x - 2}$; *domain:* all x except $x = 2$

■ $f(x) = \dfrac{3x}{x^2 - 9}$; *domain:* all x except $x = \pm 3$

■ $f(x) = \dfrac{x^3 - 8}{x^2 + 4}$; *domain:* all real numbers x

When sketching the graph of a rational function f, it is important to answer the following two questions.

Question 1 What can be said of the function values $f(x)$ when x is close to (but not equal to) a zero of the denominator?

Question 2 What can be said of the function values $f(x)$ when x is large positive or when x is large negative?

As we shall see, if a is a zero of the denominator, one of several situations often occurs. These are shown in Figure 19, where we have used notations from the following chart.

Notation	Terminology
$x \to a^-$	x approaches a from the left (through values *less* than a).
$x \to a^+$	x approaches a from the right (through values *greater* than a).
$f(x) \to \infty$	$f(x)$ increases without bound (can be made as large positive as desired).
$f(x) \to -\infty$	$f(x)$ decreases without bound (can be made as large negative as desired).

FIGURE 19

$f(x) \to \infty$ as $x \to a^-$

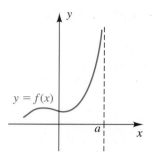

$f(x) \to \infty$ as $x \to a^+$

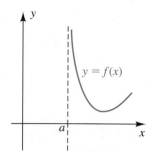

$f(x) \to -\infty$ as $x \to a^-$

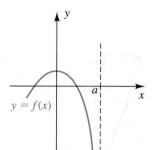

$f(x) \to -\infty$ as $x \to a^+$

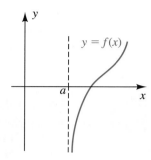

The symbols ∞ (**infinity**) and $-\infty$ (**minus infinity**) do not represent real numbers; they simply specify certain types of behavior of functions and variables.

The dashed line $x = a$ in Figure 19 is called a *vertical asymptote*, as in the following definition.

Definition of Vertical Asymptote

The line $x = a$ is a **vertical asymptote** for the graph of a function f if

$$f(x) \to \infty \qquad \text{or} \qquad f(x) \to -\infty$$

as x approaches a from either the left or the right.

Thus, the answer to Question 1 is that if a is a zero of the denominator of a rational function f, then the graph of f *may* have a vertical asymptote $x = a$. There are rational functions where this is *not* the case (see Example 7). If the numerator and denominator have no common factor, then f *must* have a vertical asymptote $x = a$.

Let us next consider Question 2. For x *large positive* or *large negative*, the graph of a rational function may look like one of those in Figure 20,

FIGURE 20

$f(x) \to c$ as $x \to \infty$

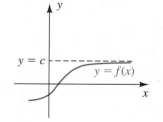

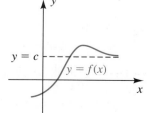

$f(x) \to c$ as $x \to -\infty$

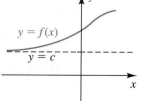

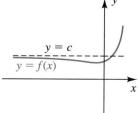

where the notation

$$f(x) \to c \quad \text{as } x \to \infty$$

is read "$f(x)$ *approaches c as x increases without bound*" and the notation

$$f(x) \to c \quad \text{as } x \to -\infty$$

is read "$f(x)$ *approaches c as x decreases without bound.*"

We call the dashed line in Figure 20 a *horizontal asymptote*, as in the next definition.

Definition of Horizontal Asymptote	The line $y = c$ is a **horizontal asymptote** for the graph of a function f if $$f(x) \to c \quad \text{as } x \to \infty \text{ or as } x \to -\infty.$$

Thus, the answer to Question 2 is that $f(x)$ *may* be very close to some number c when x is large positive or large negative; that is, the graph of f may have a horizontal asymptote $y = c$. There are rational functions where this is *not* the case (see Examples 2(c) and 5).

Note that as in the second and fourth sketches in Figure 20, the graph of f may cross a horizontal asymptote.

In the next example we find the asymptotes for the graph of a simple rational function.

EXAMPLE 1   *Sketching the graph of a rational function*

Sketch the graph of f if

$$f(x) = \frac{1}{x - 2}.$$

SOLUTION   Let us begin by considering Question 1, stated at the beginning of this section. The denominator $x - 2$ is zero at $x = 2$. If x is close to 2 and $x > 2$, then $f(x)$ is large positive, as indicated in the following table.

x	2.1	2.01	2.001	2.0001	2.00001
$\dfrac{1}{x-2}$	10	100	1000	10,000	100,000

Since we can make $1/(x - 2)$ as large as desired by taking x close to 2 (and $x > 2$), we see that

$$f(x) \to \infty \quad \text{as } x \to 2^+.$$

(continued)

If $f(x)$ is close to 2 and $x < 2$, then $f(x)$ is large negative; for example, $f(1.9999) = -10,000$ and $f(1.99999) = -100,000$. Thus,

$$f(x) \to -\infty \quad \text{as } x \to 2^-.$$

The line $x = 2$ is a vertical asymptote for the graph of f, as illustrated in Figure 21.

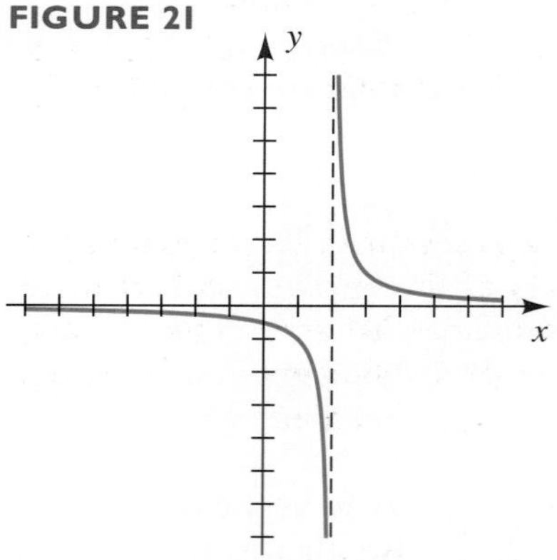

We next consider Question 2. The following table lists some approximate values for $f(x)$ when x is large and positive.

FIGURE 21

x	100	1000	10,000	100,000	1,000,000
$\dfrac{1}{x-2}$ (approx.)	0.01	0.001	0.0001	0.00001	0.000001

We may describe this behavior of $f(x)$ by writing

$$f(x) \to 0 \quad \text{as } x \to \infty.$$

Similarly, $f(x)$ is close to 0 when x is large negative; for example, $f(-100,000) = -0.00001$. Thus,

$$f(x) \to 0 \quad \text{as } x \to -\infty.$$

The line $y = 0$ (the x-axis) is a horizontal asymptote, as shown in Figure 21.

Plotting the points $(1, -1)$ and $(3, 1)$ gives us a rough sketch of the graph.

The following theorem is useful for finding the horizontal asymptote for the graph of a rational function.

Theorem on Horizontal Asymptotes

Let $f(x) = \dfrac{a_n x^n + a_{n-1} x^{n-1} + \cdots + a_1 x + a_0}{b_k x^k + b_{k-1} x^{k-1} + \cdots + b_1 x + b_0}$, where $a_n \neq 0$ and $b_k \neq 0$.

(1) If $n < k$, then the x-axis (the line $y = 0$) is a horizontal asymptote for the graph of f.

(2) If $n = k$, then the line $y = a_n/b_k$ (the ratio of leading coefficients) is a horizontal asymptote for the graph of f.

(3) If $n > k$, the graph of f has no horizontal asymptote. Instead, either $f(x) \to \infty$ or $f(x) \to -\infty$ as $x \to \infty$ or as $x \to -\infty$.

Proofs for each part of this theorem may be patterned after the solutions in the next example. Concerning part (3), if $q(x)$ is the quotient obtained by dividing the numerator by the denominator, then $f(x) \to \infty$ if $q(x) \to \infty$ or $f(x) \to -\infty$ if $q(x) \to -\infty$.

EXAMPLE 2 *Finding horizontal asymptotes*

Find the horizontal asymptote for the graph of f, if it exists.

(a) $f(x) = \dfrac{3x - 1}{x^2 - x - 6}$ **(b)** $f(x) = \dfrac{5x^2 + 1}{3x^2 - 4}$ **(c)** $f(x) = \dfrac{2x^4 - 3x^2 + 5}{x^2 + 1}$

SOLUTION

(a) The degree of the numerator $3x - 1$ is less than the degree of the denominator $x^2 - x - 6$, so, by part (1) of the theorem on horizontal asymptotes, the x-axis is a horizontal asymptote. To verify this directly, we divide the numerator and denominator of the quotient by x^2 (since 2 is the highest power on x in the denominator), obtaining

$$f(x) = \frac{\dfrac{3x - 1}{x^2}}{\dfrac{x^2 - x - 6}{x^2}} = \frac{\dfrac{3}{x} - \dfrac{1}{x^2}}{1 - \dfrac{1}{x} - \dfrac{6}{x^2}} \quad \text{for } x \neq 0.$$

If x is large positive or large negative, then $3/x$, $1/x^2$, $1/x$, and $6/x^2$ are close to 0, and hence

$$f(x) \approx \frac{0 - 0}{1 - 0 - 0} = \frac{0}{1} = 0.$$

Thus,

$$f(x) \to 0 \quad \text{as } x \to \infty \text{ or as } x \to -\infty.$$

Since $f(x)$ is the y-coordinate of a point on the graph, the last statement means that the line $y = 0$ (that is, the x-axis) is a horizontal asymptote.

(b) If $f(x) = (5x^2 + 1)/(3x^2 - 4)$, then the numerator and denominator have the same degree and the leading coefficients are 5 and 3, respectively. Hence, by part (2) of the theorem on horizontal asymptotes, the line $y = \frac{5}{3}$ is a horizontal asymptote. We can also show that $y = \frac{5}{3}$ is a horizontal asymptote by dividing the numerator and denominator of $f(x)$ by x^2, obtaining

$$f(x) = \frac{\dfrac{5x^2 + 1}{x^2}}{\dfrac{3x^2 - 4}{x^2}} = \frac{5 + \dfrac{1}{x^2}}{3 - \dfrac{4}{x^2}} \quad \text{for } x \neq 0.$$

Since $1/x^2 \to 0$ and $4/x^2 \to 0$ as $x \to \infty$ or as $x \to -\infty$, we see that

$$f(x) \to \frac{5 + 0}{3 - 0} = \frac{5}{3} \quad \text{as } x \to \infty \text{ or as } x \to -\infty.$$

(continued)

(c) The degree of the numerator $2x^4 - 3x^2 + 5$ is greater than the degree of the denominator $x^2 + 1$, so, by part (3) of the theorem on horizontal asymptotes, the graph has no horizontal asymptote. If we use long division, we obtain

$$f(x) = 2x^2 - 5 + \frac{10}{x^2 + 1}.$$

As either $x \to \infty$ or $x \to -\infty$, the quotient $2x^2 - 5$ increases without bound and $10/(x^2 + 1) \to 0$. Hence, $f(x) \to \infty$ as $x \to \infty$ or as $x \to -\infty$.

We next list some guidelines for sketching the graph of a rational function. Their use will be illustrated in Examples 3 and 4.

Guidelines for Sketching the Graph of a Rational Function

Assume that $f(x) = \dfrac{g(x)}{h(x)}$, where $g(x)$ and $h(x)$ are polynomials that have no common factor.

1 Find the x-intercepts—that is, the real zeros of the numerator $g(x)$—and plot the corresponding points on the x-axis.

2 Find the real zeros of the denominator $h(x)$. For each real zero a, sketch the vertical asymptote $x = a$ with dashes.

3 Find the y-intercept $f(0)$, if it exists, and plot the point $(0, f(0))$ on the y-axis.

4 Apply the theorem on horizontal asymptotes. If there is a horizontal asymptote $y = c$, sketch it with dashes.

5 If there is a horizontal asymptote $y = c$, determine whether it intersects the graph. The x-coordinates of the points of intersection are the solutions of the equation $f(x) = c$. Plot these points, if they exist.

6 Sketch the graph of f in each of the regions in the xy-plane determined by the vertical asymptotes in guideline 2. If necessary, use the sign of specific function values to tell whether the graph is above or below the x-axis or the horizontal asymptote. Use guideline 5 to decide whether the graph approaches the horizontal asymptote from above or below.

In the following examples our main objective is to determine the general shape of the graph, paying particular attention to how the graph approaches the asymptotes. We will plot only a few points, such as those corresponding to the x-intercepts and y-intercept or the intersection of the graph with a horizontal asymptote.

EXAMPLE 3 *Sketching the graph of a rational function*

Sketch the graph of f if

$$f(x) = \frac{x-1}{x^2 - x - 6}.$$

SOLUTION   It is useful to express both numerator and denominator in factored form. Thus, we begin by writing

$$f(x) = \frac{x-1}{x^2 - x - 6} = \frac{x-1}{(x+2)(x-3)}.$$

We next follow the guidelines.

FIGURE 22

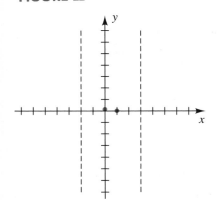

Guideline 1   To find the x-intercepts we find the zeros of the numerator. Letting $x - 1 = 0$ gives us $x = 1$, and we plot the point $(1, 0)$ on the x-axis, as shown in Figure 22.

Guideline 2   The denominator has zeros -2 and 3. Hence, the lines $x = -2$ and $x = 3$ are vertical asymptotes; we sketch them with dashes, as in Figure 22.

Guideline 3   The y-intercept is $f(0) = \frac{1}{6}$, and we plot the point $(0, \frac{1}{6})$ in Figure 22.

Guideline 4   The degree of the numerator of $f(x)$ is less than the degree of the denominator, so, by part (1) of the theorem on horizontal asymptotes, the x-axis is a horizontal asymptote.

Guideline 5   The points where the graph intersects the horizontal asymptote (the x-axis) found in guideline 4 correspond to the x-intercepts. We already plotted the point $(1, 0)$ in guideline 1.

Guideline 6   The vertical asymptotes in Figure 22 divide the xy-plane into three regions:

R_1:   the region to the left of $x = -2$

R_2:   the region between $x = -2$ and $x = 3$

R_3:   the region to the right of $x = 3$

FIGURE 23

(a)

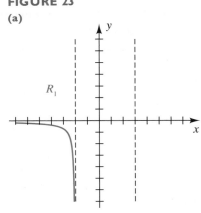

For R_1, we have $x < -2$. There are only two choices for the shape of the graph of f in R_1: as $x \to -\infty$, the graph approaches the x-axis either from above or from below. To determine which choice is correct, we will examine the *sign* of a typical function value in R_1. Choosing -10 for x, we use the factored form of $f(x)$ to find the sign of $f(-10)$ (this process is similar to the one used in Section 2.7):

$$f(-10) = \frac{(-)}{(-)(-)} = -$$

The negative value of $f(-10)$ indicates that the graph approaches the horizontal asymptote $y = 0$ from *below* as $x \to -\infty$. Moreover, as $x \to -2^-$, the graph extends *downward*; that is, $f(x) \to -\infty$. A sketch of f on R_1 is shown in Figure 23(a).

(continued)

FIGURE 23

(b)

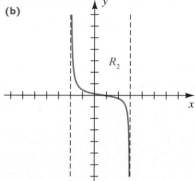

In R_2, we have $-2 < x < 3$, and the graph crosses the x-axis at $x = 1$. Since, for example, $f(0)$ is positive, it follows that the graph lies *above* the x-axis if $-2 < x < 1$. Thus, as $x \to -2^+$, the graph extends *upward*; that is, $f(x) \to \infty$. Since $f(2)$ is negative, the graph lies *below* the x-axis if $1 < x < 3$. Hence, as $x \to 3^-$, the graph extends *downward*; that is, $f(x) \to -\infty$. A sketch of f on R_2 is shown in Figure 23(b).

Finally, in R_3, $x > 3$, and the graph does not cross the x-axis. Since, for example, $f(10)$ is positive, the graph lies *above* the x-axis. It follows that $f(x) \to \infty$ as $x \to 3^-$ and that the graph approaches the horizontal asymptote $y = 0$ from *above* as $x \to \infty$. The graph of f is sketched in Figure 23(c).

(c)

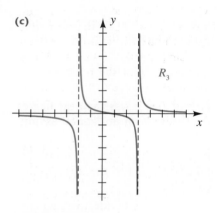

EXAMPLE 4   *Sketching the graph of a rational function*

Sketch the graph of f if

$$f(x) = \frac{x^2}{x^2 - x - 2}.$$

SOLUTION   Factoring the denominator gives us

$$f(x) = \frac{x^2}{x^2 - x - 2} = \frac{x^2}{(x + 1)(x - 2)}.$$

We again follow the guidelines.

Guideline 1 To find the x-intercepts we find the zeros of the numerator. Letting $x^2 = 0$ gives us $x = 0$, and we plot the point $(0, 0)$ on the x-axis, as shown in Figure 24.

FIGURE 24

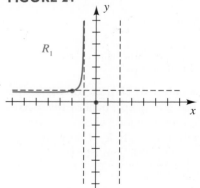

Guideline 2 The denominator has zeros -1 and 2. Hence, the lines $x = -1$ and $x = 2$ are vertical asymptotes, and we sketch them with dashes, as in Figure 24.

Guideline 3 The y-intercept is $f(0) = 0$. This gives us the same point $(0, 0)$ found in guideline 1.

Guideline 4 The numerator and denominator of $f(x)$ have the same degree, and the leading coefficients are both 1. Hence, by part (2) of the theorem on horizontal asymptotes, the line $y = \frac{1}{1} = 1$ is a horizontal asymptote. We sketch the line with dashes in Figure 24.

Guideline 5 The x-coordinates of the points where the graph intersects the horizontal asymptote $y = 1$ are solutions of the equation $f(x) = 1$. We solve this equation as follows:

$$\frac{x^2}{x^2 - x - 2} = 1 \qquad \text{let } f(x) = 1$$

$$x^2 = x^2 - x - 2 \qquad \text{multiply by } x^2 - x - 2$$

$$x = -2 \qquad \text{subtract } x^2 \text{ and add } x$$

FIGURE 25

(a)

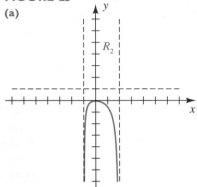

(b)

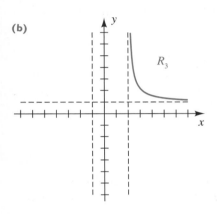

(c)

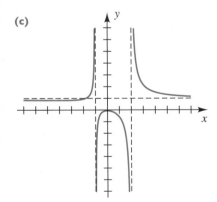

The y-coordinate of the point of intersection is $f(-2) = 1$, and we plot the point $(-2, 1)$ on the horizontal asymptote in Figure 24.

Guideline 6 The vertical asymptotes in Figure 24 divide the xy-plane into three regions:

R_1: the region to the left of $x = -1$

R_2: the region between $x = -1$ and $x = 2$

R_3: the region to the right of $x = 2$

For R_1, let us first consider the portion of the graph that corresponds to $-2 < x < -1$. From the point $(-2, 1)$ on the horizontal asymptote, the graph must extend *upward* as $x \to -1^-$ (it cannot extend downward, since there is no x-intercept between $x = -2$ and $x = -1$). As $x \to -\infty$, there will be a low point on the graph between $y = 0$ and $y = 1$, and then the graph will approach the horizontal asymptote $y = 1$ from *below*. It is difficult to see where the low point occurs in Figure 24, because the function values are very close to one another. Using calculus, it can be shown that the low point is $(-4, \frac{8}{9})$.

In R_2, we have $-1 < x < 2$, and the graph intersects the x-axis at $x = 0$. Since the function does not cross the horizontal asymptote in this region, we know that the graph extends *downward* as $x \to -1^+$ and as $x \to 2^-$, as shown in Figure 25(a).

In R_3, the graph approaches the horizontal asymptote $y = 1$ (from either above or below) as $x \to \infty$. Furthermore, the graph must extend *upward* as $x \to 2^+$, because there are no x-intercepts in R_3. This implies that as $x \to \infty$, the graph approaches the horizontal asymptote from *above*, as in Figure 25(b).

The graph of f is sketched in Figure 25(c).

EXAMPLE 5 *Sketching the graph of a rational function*

Sketch the graph of f if

$$f(x) = \frac{2x^4}{x^4 + 1}.$$

SOLUTION In this solution we shall not formally write down each guideline. Note that since $f(-x) = f(x)$, the function is even, and hence the graph is symmetric with respect to the y-axis.

The graph intersects the x-axis at $(0, 0)$. Since the denominator of $f(x)$ has no real zero, the graph has no vertical asymptote.

The numerator and denominator of $f(x)$ have the same degree. Since the leading coefficients are 2 and 1, respectively, we see from part (2) of the theorem on horizontal asymptotes that the line $y = \frac{2}{1} = 2$ is a horizontal

(continued)

FIGURE 26

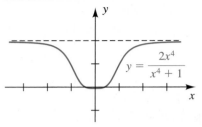

$$y = \frac{2x^4}{x^4 + 1}$$

asymptote. We represent this line with dashes in Figure 26. The graph does not cross the horizontal asymptote $y = 2$, since the equation $f(x) = 2$ has no real solution.

Plotting the points $(1, 1)$ and $(2, \frac{32}{17})$ and making use of symmetry leads to the sketch in Figure 26.

An **oblique asymptote** for a graph is a line $y = ax + b$, with $a \neq 0$, such that the graph approaches this line as $x \to \infty$ or as $x \to -\infty$. (If the graph is a line, we consider it to be its own asymptote.) If the rational function $f(x) = g(x)/h(x)$ for polynomials $g(x)$ and $h(x)$ and *if the degree of $g(x)$ is one greater than the degree of $h(x)$*, then the graph of f has an oblique asymptote. To find this oblique asymptote we may use long division to express $f(x)$ in the form

$$f(x) = \frac{g(x)}{h(x)} = (ax + b) + \frac{r(x)}{h(x)},$$

where either $r(x) = 0$ or the degree of $r(x)$ is less than the degree of $h(x)$. From part (1) of the theorem on horizontal asymptotes,

$$\frac{r(x)}{h(x)} \to 0 \quad \text{as } x \to \infty \text{ or as } x \to -\infty.$$

Consequently, $f(x)$ approaches the line $y = ax + b$ as x increases or decreases without bound; that is, $y = ax + b$ is an oblique asymptote.

EXAMPLE 6   *Finding an oblique asymptote*

Find all the asymptotes and sketch the graph of f if

$$f(x) = \frac{x^2 - 9}{2x - 4}.$$

SOLUTION   A vertical asymptote occurs if $2x - 4 = 0$ (that is, if $x = 2$).

The degree of the numerator of $f(x)$ is greater than the degree of the denominator. Hence, by part (3) of the theorem on horizontal asymptotes, there is no horizontal asymptote. Since the degree of the numerator $x^2 - 9$ is *one* greater than the degree of the denominator $2x - 4$, however, the graph has an oblique asymptote. By long division, we obtain

$$\require{enclose}
\begin{array}{r}
\frac{1}{2}x + 1 \\
2x - 4 \enclose{longdiv}{x^2 - 9} \\
\underline{x^2 - 2x } \\
2x - 9 \\
\underline{2x - 4} \\
-5
\end{array}$$

FIGURE 27

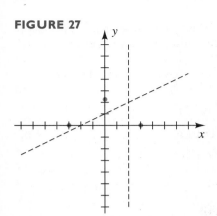

Therefore, $$\frac{x^2 - 9}{2x - 4} = \left(\frac{1}{2}x + 1\right) - \frac{5}{2x - 4}.$$

As we indicated in the discussion preceding this example, the line $y = \frac{1}{2}x + 1$ is an oblique asymptote. This line and the vertical asymptote $x = 2$ are sketched with dashes in Figure 27.

The x-intercepts of the graph are the solutions of the equation $x^2 - 9 = 0$ and hence are 3 and -3. The y-intercept is $f(0) = \frac{9}{4}$. The corresponding points are plotted in Figure 27. We may now show that the graph has the shape indicated in Figure 28.

FIGURE 28

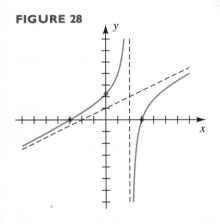

In the preceding example, the graph of f approaches the line $y = \frac{1}{2}x + 1$ *asymptotically* as $x \to \infty$ or as $x \to -\infty$. Graphs of rational functions may approach different types of curves asymptotically. For example, if

$$f(x) = \frac{x^4 - x}{x^2} = x^2 - \frac{1}{x},$$

then for large values of $|x|$, $1/x \approx 0$ and hence $f(x) \approx x^2$. Thus, the graph of f approaches the parabola $y = x^2$ asymptotically as $x \to \infty$ or as $x \to -\infty$. In general, if $f(x) = g(x)/h(x)$ and if $q(x)$ is the quotient obtained by dividing $g(x)$ by $h(x)$, then the graph of f approaches the graph of $y = q(x)$ asymptotically as $x \to \infty$ or as $x \to -\infty$.

In the guidelines for sketching the graph of a rational function we assumed that the numerator and denominator have no common prime factor. If there *is* a common factor, then there may be a *hole* in the graph—that is, a single point may be missing—as illustrated in the following example.

EXAMPLE 7   *A graph containing a hole*

Sketch the graph of f if

$$f(x) = \frac{x - 4}{(x - 2)(x - 4)}.$$

FIGURE 29

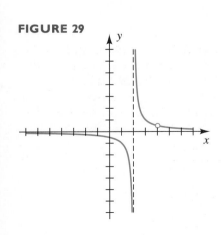

SOLUTION   If $x \neq 4$, we may cancel the factor $x - 4$; that is,

$$f(x) = \frac{1}{x - 2}, \quad \text{if } x \neq 4.$$

Thus, the graph of f is the same as that sketched in Figure 21, with one exception: There is a hole in the graph at $x = 4$. To determine the corresponding y value, we use the reduced form of the function, obtaining $f(4) = 1/(4 - 2) = \frac{1}{2}$. The point $(4, \frac{1}{2})$ is indicated by the small circle in Figure 29. It should be noted that graphics calculators or computer programs may not be designed to show this feature of a graph.

Graphs of rational functions may become increasingly complicated as the degrees of the polynomials in the numerator and denominator increase. Techniques developed in calculus must be employed for a thorough treatment of such graphs.

Formulas that represent physical quantities may determine rational functions. For example, consider Ohm's law in electrical theory, which states that $I = V/R$, where R is the resistance (in ohms) of a conductor, V is the potential difference (in volts) across the conductor, and I is the current (in amperes) that flows through the conductor. The resistance of certain alloys approaches zero as the temperature approaches absolute zero (approximately $-273\,°C$), and the alloy becomes a *superconductor* of electricity. If the voltage V is fixed, then, for such a superconductor,

$$I = \frac{V}{R} \to \infty \quad \text{as } R \to 0^+;$$

that is, as R approaches 0, the current increases without bound. Superconductors allow very large currents to be used in generating plants and motors. They also have applications in experimental high-speed ground transportation, where the strong magnetic fields produced by superconducting magnets enable trains to levitate so that there is essentially no friction between the wheels and the track. Perhaps the most important use for superconductors is in circuits for computers, because such circuits produce very little heat.

4.5 EXERCISES

Exer. 1–2: (a) Sketch the graph of f. (b) Find the domain D and range R of f. (c) Find the intervals on which f is increasing or is decreasing.

1 $f(x) = \dfrac{4}{x}$

2 $f(x) = \dfrac{1}{x^2}$

Exer. 3–24: Sketch the graph of f.

3 $f(x) = \dfrac{3}{x - 4}$

4 $f(x) = \dfrac{-3}{x + 3}$

5 $f(x) = \dfrac{-3x}{x + 2}$

6 $f(x) = \dfrac{4x}{2x - 5}$

7 $f(x) = \dfrac{x - 2}{x^2 - x - 6}$

8 $f(x) = \dfrac{x + 1}{x^2 + 2x - 3}$

9 $f(x) = \dfrac{-4}{(x - 2)^2}$

10 $f(x) = \dfrac{2}{(x + 1)^2}$

11 $f(x) = \dfrac{x - 3}{x^2 - 1}$

12 $f(x) = \dfrac{x + 4}{x^2 - 4}$

13 $f(x) = \dfrac{2x^2 - 2x - 4}{x^2 + x - 12}$

14 $f(x) = \dfrac{-3x^2 - 3x + 6}{x^2 - 9}$

15 $f(x) = \dfrac{-x^2 - x + 6}{x^2 + 3x - 4}$

16 $f(x) = \dfrac{x^2 - 3x - 4}{x^2 + x - 6}$

17 $f(x) = \dfrac{3x^2 - 3x - 36}{x^2 + x - 2}$

18 $f(x) = \dfrac{2x^2 + 4x - 48}{x^2 + 3x - 10}$

19 $f(x) = \dfrac{-2x^2 + 10x - 12}{x^2 + x}$

20 $f(x) = \dfrac{2x^2 + 8x + 6}{x^2 - 2x}$

21 $f(x) = \dfrac{x - 1}{x^3 - 4x}$

22 $f(x) = \dfrac{x^2 - 2x + 1}{x^3 - 9x}$

23 $f(x) = \dfrac{-3x^2}{x^2 + 1}$

24 $f(x) = \dfrac{x^2 - 4}{x^2 + 1}$

Exer. 25–28: Find the oblique asymptote, and sketch the graph of f.

25 $f(x) = \dfrac{x^2 - x - 6}{x + 1}$

26 $f(x) = \dfrac{2x^2 - x - 3}{x - 2}$

27 $f(x) = \dfrac{8 - x^3}{2x^2}$

28 $f(x) = \dfrac{x^3 + 1}{x^2 - 9}$

Exer. 29–34: Simplify f(x), and sketch the graph of f.

29 $f(x) = \dfrac{2x^2 + x - 6}{x^2 + 3x + 2}$

30 $f(x) = \dfrac{x^2 - x - 6}{x^2 - 2x - 3}$

31 $f(x) = \dfrac{x - 1}{1 - x^2}$

32 $f(x) = \dfrac{x + 2}{x^2 - 4}$

33 $f(x) = \dfrac{x^2 + x - 2}{x + 2}$

34 $f(x) = \dfrac{x^3 - 2x^2 - 4x + 8}{x - 2}$

35 *A container for radioactive waste* A cylindrical container for storing radioactive waste is to be constructed from lead. This container must be 6 inches thick. The volume of the outside cylinder shown in the figure is to be 16π ft^3.

(a) Express the height h of the inside cylinder as a function of the inside radius r.

EXERCISE 35

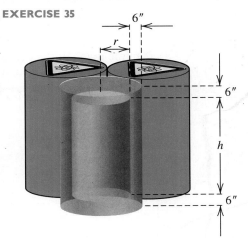

(b) Show that the inside volume $V(r)$ is given by

$$V(r) = \pi r^2 \left[\frac{16}{(r + 0.5)^2} - 1 \right].$$

(c) What values of r must be excluded in part (b)?

36 *Drug dosage* Young's rule is a formula that is used to modify adult drug dosage levels for young children. If a denotes the adult dosage (in milligrams) and if t is the age of the child (in years), then the child's dose y is given by $y = ta/(t + 12)$. Sketch the graph of this equation for $t > 0$ and $a = 100$.

37 *Salt concentration* Salt water of concentration 0.1 pound of salt per gallon flows into a large tank that initially contains 50 gallons of pure water.

(a) If the flow rate of salt water into the tank is 5 gallons per minute, find the volume $V(t)$ of water and the amount $A(t)$ of salt in the tank after t minutes.

(b) Find a formula for the salt concentration $c(t)$ (in lb/gal) after t minutes

(c) Discuss the variation of $c(t)$ as $t \to \infty$.

38 *Amount of rainfall* The total number of inches $R(t)$ of rain during a storm of length t hours can be approximated by

$$R(t) = \frac{at}{t + b},$$

where a and b are positive constants that depend on the geographical locale.

(a) Discuss the variation of $R(t)$ as $t \to \infty$.

(b) The intensity I of the rainfall (in in./hr) is defined by $I = R(t)/t$. If $a = 2$ and $b = 8$, sketch the graph of R and I on the same coordinate plane for $t > 0$.

c Exer. 39–42: Graph f.

39 $f(x) = \dfrac{20x^2 + 80x + 72}{10x^2 + 40x + 41}$

40 $f(x) = \dfrac{15x^2 - 60x + 68}{3x^2 - 12x + 13}$

41 $f(x) = \dfrac{(x - 1)^2}{(x - 0.999)^2}$

42 $f(x) = \dfrac{x^2 - 9.01}{x - 3}$

4.6 PARABOLAS

The *conic sections*, also called *conics*, can be obtained by intersecting a double-napped right circular cone with a plane. By varying the position of the plane, we obtain a *circle*, an *ellipse*, a *parabola*, or a *hyperbola*, as

FIGURE 30

(a) Circle   **(b)** Ellipse   **(c)** Parabola   **(d)** Hyperbola

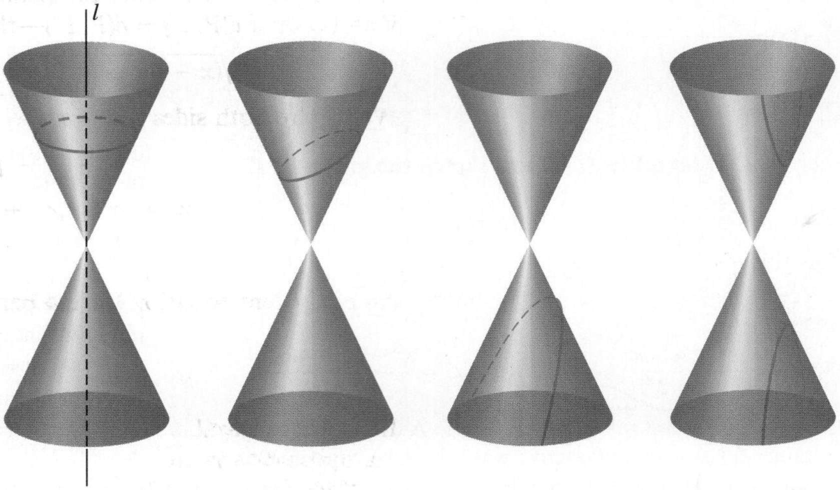

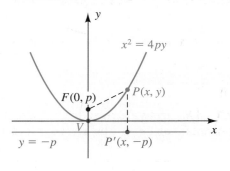

FIGURE 31

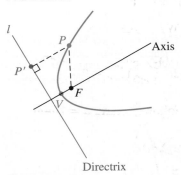

illustrated in Figure 30. *Degenerate conics* are obtained if the plane intersects the cone in only one point or along either one or two lines that lie on the cone. Conic sections were studied extensively by the ancient Greeks, who discovered properties that enable us to state their definitions in terms of points and lines, as indicated throughout our discussion.

From our work in Section 3.6, if $a \neq 0$, then the graph of the equation $y = ax^2 + bx + c$ is a parabola with a vertical axis. We shall next state a general definition of a *parabola* and derive equations for parabolas that have either a vertical axis or a horizontal axis.

Definition of a Parabola

A **parabola** is the set of all points in a plane equidistant from a fixed point F (the **focus**) and a fixed line l (the **directrix**) in the plane.

FIGURE 32

We shall assume that F is not on l, for if it were, the set of points would be a line. If P is a point in the plane and P' is the point on l determined by a line through P that is perpendicular to l (see Figure 31), then, by the preceding definition, P is on the parabola if and only if the distances $d(P, F)$ and $d(P, P')$ are equal. The **axis** of the parabola is the line through F that is perpendicular to the directrix. The **vertex** of the parabola is the point V on the axis halfway from F to l. The vertex is the point on the parabola that is closest to the directrix.

To obtain a simple equation for a parabola, place the y-axis along the axis of the parabola, with the origin at the vertex V, as shown in Figure 32. In this case, the focus F has coordinates $(0, p)$ for some real number

$p \neq 0$, and the equation of the directrix is $y = -p$. (The figure shows the case $p > 0$.) By the distance formula, a point $P(x, y)$ is on the parabola if and only if $d(P, F) = d(P, P')$—that is,

$$\sqrt{(x - 0)^2 + (y - p)^2} = \sqrt{(x - x)^2 + (y + p)^2}.$$

We square both sides and simplify:

$$x^2 + (y - p)^2 = (y + p)^2$$
$$x^2 + y^2 - 2py + p^2 = y^2 + 2py + p^2$$
$$x^2 = 4py$$

An equivalent equation for the parabola is

$$y = \frac{1}{4p} x^2.$$

If $p > 0$, the parabola opens upward, as in Figure 32. If $p < 0$, the parabola opens downward.

If we interchange the roles of x and y, we obtain

$$y^2 = 4px \qquad \text{or, equivalently,} \qquad x = \frac{1}{4p} y^2.$$

This is an equation of a parabola with vertex at the origin and focus $F(p, 0)$; it opens right if $p > 0$ (see Figure 33) or left if $p < 0$. The equation of the directrix is $x = -p$.

We may summarize our discussion as follows, letting $a = 1/(4p)$.

FIGURE 33

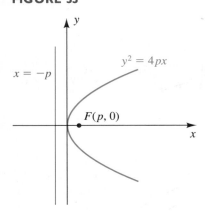

Standard Equations of a Parabola with Vertex at the Origin

For any nonzero real number a, the graph of

$$y = ax^2 \qquad \text{or} \qquad x = ay^2$$

is a parabola with vertex $V(0, 0)$ and focus $F(0, p)$ or $F(p, 0)$, respectively, where $p = 1/(4a)$. The directrix is $y = -p$ or $x = -p$, respectively.

FIGURE 34

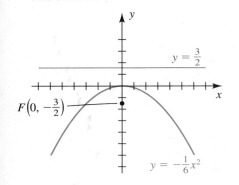

EXAMPLE I *Finding the focus and directrix of a parabola*

Find the focus and directrix of the parabola $y = -\frac{1}{6}x^2$, and sketch its graph.

SOLUTION The equation has the form $y = ax^2$ with $a = -\frac{1}{6}$. As in the preceding box,

$$p = \frac{1}{4a} = \frac{1}{4(-\frac{1}{6})} = -\frac{3}{2}.$$

Thus, the parabola opens downward and has focus $F(0, -\frac{3}{2})$, as illustrated in Figure 34. The directrix is the horizontal line $y = \frac{3}{2}$, which is a distance $\frac{3}{2}$ above V, as shown in the figure.

EXAMPLE 2   *Finding an equation of a parabola satisfying prescribed conditions*

(a) Find an equation of a parabola that has vertex at the origin, opens right, and passes through the point $P(7, -3)$.

(b) Find the focus of the parabola.

FIGURE 35

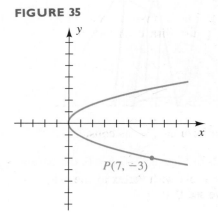

SOLUTION

(a) The parabola is sketched in Figure 35. As in the preceding box, an equation of the parabola has the form $x = ay^2$ for some number a. If $P(7, -3)$ is on the graph, then

$$7 = a(-3)^2, \quad \text{or} \quad a = \tfrac{7}{9}.$$

Hence, an equation for the parabola is $x = \tfrac{7}{9}y^2$.

(b) The focus is a distance p to the right of the vertex, where

$$p = \frac{1}{4a} = \frac{1}{4(\tfrac{7}{9})} = \frac{9}{28}.$$

Thus, the focus has coordinates $(\tfrac{9}{28}, 0)$.

To extend our discussion to a parabola with vertex *not* at the origin, we use a **translation of axes**, as illustrated in Figure 36, where the x- and y-axes are shifted to positions—denoted by x' and y'—that are parallel to their original positions. Every point P in the plane then has two different ordered-pair representations: $P(x, y)$ in the xy-system and $P(x', y')$ in the $x'y'$-system. If the origin of the new $x'y'$-system has coordinates (h, k) in the xy-plane, as illustrated in Figure 36, we see that

$$x = x' + h \quad \text{and} \quad y = y' + k.$$

FIGURE 36

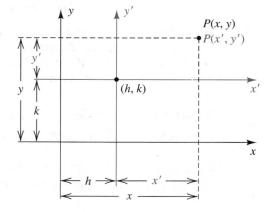

These formulas are true for all values of h and k. Equivalent formulas are

$$x' = x - h \quad \text{and} \quad y' = y - k.$$

These results are stated formally below.

Translation of Axes Formulas

> If (x, y) are the coordinates of a point P in an xy-plane and if (x', y') are the coordinates of P in an $x'y'$-plane with origin at the point (h, k) of the xy-plane, then
>
> **(1)** $x = x' + h, \quad y = y' + k$
> **(2)** $x' = x - h, \quad y' = y - k$

FIGURE 37

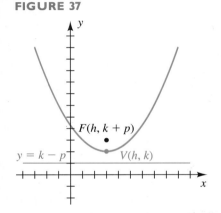

To illustrate the translation of axes formulas, let us consider

$$(x')^2 = 4py',$$

which we know is an equation of a parabola with vertex at the origin O' of the $x'y'$-plane. Using formula (2), we see that

$$(x - h)^2 = 4p(y - k)$$

is an equation of the same parabola in the xy-plane with vertex $V(h, k)$. The focus is $F(h, k + p)$, and the directrix is $y = k - p$, as shown in Figure 37 for $p > 0$.

Squaring the left side of $(x - h)^2 = 4p(y - k)$ and simplifying leads to an equation of the form

$$y = ax^2 + bx + c,$$

where a, b, and c are real numbers. Conversely, if $a \neq 0$, then the graph of $y = ax^2 + bx + c$ is a parabola with a vertical axis. As with $y = ax^2$, we can show that $a = 1/(4p)$ or, equivalently, $p = 1/(4a)$.

Similarly, if we begin with $(y')^2 = 4px'$ and use the translation of axes formulas, we obtain

$$(y - k)^2 = 4p(x - h),$$

FIGURE 38

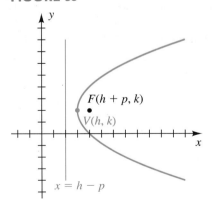

which is an equation of a parabola in the xy-plane with vertex $V(h, k)$, focus $F(h + p, k)$, and directrix $x = h - p$ and opening right if $p > 0$ (see Figure 38) or left if $p < 0$. This equation may be written in the form $x = ay^2 + by + c$, where $a = 1/(4p)$ or, equivalently, $p = 1/(4a)$.

EXAMPLE 3 *Sketching a parabola with a horizontal axis*

Discuss and sketch the graph of $2x = y^2 + 8y + 22$.

SOLUTION Since the equation can be written in the form (solve for x) $x = ay^2 + by + c$, we know that the graph is a parabola with a

(continued)

FIGURE 39

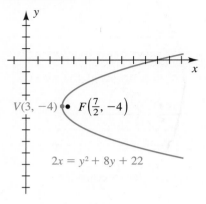

$$2x = y^2 + 8y + 22$$

horizontal axis. We first write the given equation as

$$y^2 + 8y = 2x - 22$$

and then complete the square on the left by adding 16 to both sides:

$$y^2 + 8y + 16 = 2x - 6$$

$$(y + 4)^2 = 2(x - 3)$$

As in our general discussion, $h = 3$, $k = -4$, and $4p = 2$ or, equivalently, $p = \frac{1}{2}$. This gives us the following:

The vertex $V(h, k)$ is $V(3, -4)$.

The focus is $F(h + p, k) = F(3 + \frac{1}{2}, -4)$, or $F(\frac{7}{2}, -4)$.

The directrix is $x = h - p = 3 - \frac{1}{2}$, or $x = \frac{5}{2}$.

The parabola is sketched in Figure 39.

FIGURE 40

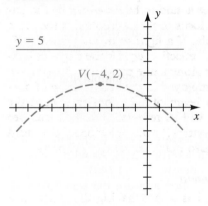

EXAMPLE 4   *Finding an equation of a parabola satisfying prescribed conditions*

A parabola has vertex $V(-4, 2)$ and directrix $y = 5$. Express the equation of the parabola in the form $y = ax^2 + bx + c$.

SOLUTION   The vertex and directrix are shown in Figure 40. The dashes indicate a possible position for the parabola. From our discussion, an equation of the parabola is

$$(x - h)^2 = 4p(y - k),$$

with $h = -4$ and $k = 2$ and with p equal to *negative* 3, since V is 3 units *below* the directrix. This gives us

$$(x + 4)^2 = -12(y - 2).$$

The last equation can be expressed in the form $y = ax^2 + bx + c$, as follows:

$$x^2 + 8x + 16 = -12y + 24$$

$$12y = -x^2 - 8x + 8$$

$$y = -\tfrac{1}{12}x^2 - \tfrac{2}{3}x + \tfrac{2}{3}$$

FIGURE 41

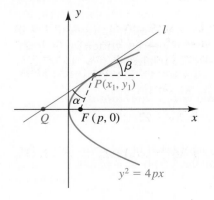

$$y^2 = 4px$$

An important property is associated with a tangent line to a parabola. (A *tangent line* to a parabola is a line that has exactly one point in common with the parabola, but does not cut through the parabola.) Suppose l is the tangent line at a point $P(x_1, y_1)$ on the graph of $y^2 = 4px$, and let F be the focus. As in Figure 41, let α denote the angle between l and the line segment FP, and let β denote the angle between l and the indicated

FIGURE 42

(a) Searchlight mirror

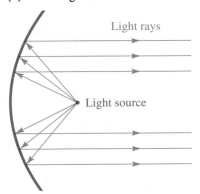

Light rays

Light source

(b) Telescope mirror

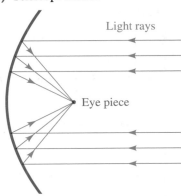

Light rays

Eye piece

FIGURE 43

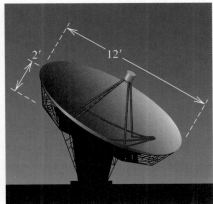

horizontal half-line with end point P. It can be shown that $\alpha = \beta$. This *reflective property* has many applications. For example, the shape of the mirror in a searchlight is obtained by revolving a parabola about its axis. The resulting three-dimensional surface is said to be *generated* by the parabola and is called a **paraboloid**. The **focus** of the paraboloid is the same as the focus of the generating parabola. If a light source is placed at F, then, by a law of physics (the angle of reflection equals the angle of incidence), a beam of light will be reflected along a line parallel to the axis (see Figure 42(a)). The same principle is employed in the construction of mirrors for telescopes or solar ovens—a beam of light coming toward the parabolic mirror, parallel to the axis, will be reflected into the focus (see Figure 42(b)). Antennas for radar systems, radio telescopes, and field microphones used at football games also make use of this property.

EXAMPLE 5 *A satellite TV antenna*

The interior of a satellite TV antenna is a dish having the shape of a (finite) paraboloid that has diameter 12 feet and is 2 feet deep, as shown in Figure 43. Find the distance from the center of the dish to the focus.

SOLUTION The generating parabola is sketched on an xy-plane in Figure 44, where we have taken the vertex of the parabola at the origin and the axis along the x-axis. An equation of the parabola is $y^2 = 4px$, where p is the required distance from the center of the dish to the focus. Since the point $(2, 6)$ is on the parabola, we obtain

$$6^2 = 4p \cdot 2, \qquad \text{or} \qquad p = \tfrac{36}{8} = 4.5 \text{ ft.}$$

FIGURE 44

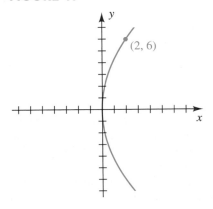

In the next example we use a graphing utility to sketch the graph of a parabola.

EXAMPLE 6   Graphing half-parabolas

Graph $x = y^2 + 2y - 4$.

SOLUTION   The graph is a parabola with a horizontal axis. We begin by solving the equivalent equation

$$y^2 + 2y - 4 - x = 0$$

for y in terms of x by using the quadratic formula with $a = 1$, $b = 2$, and $c = -4 - x$:

$$y = \frac{-2 \pm \sqrt{2^2 - 4(1)(-4 - x)}}{2(1)} \qquad \text{quadratic formula}$$

$$= \frac{-2 \pm \sqrt{20 + 4x}}{2} \qquad \text{simplify}$$

$$= -1 \pm \sqrt{x + 5} \qquad \text{simplify}$$

FIGURE 45
$[-15, 15]$ by $[-10, 10]$

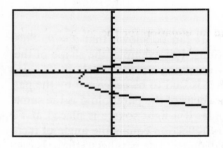

The last equation, $y = -1 \pm \sqrt{x + 5}$, represents $y = -1 + \sqrt{x + 5}$ (the top half of the parabola) and $y = -1 - \sqrt{x + 5}$ (the bottom half of the parabola). Note that $y = -1$ is the axis of the parabola.

Next, we make the assignments

$$Y_1 = \sqrt{x + 5}, \qquad Y_2 = -1 + Y_1, \qquad \text{and} \qquad Y_3 = -1 - Y_1.$$

We select the functions Y_2 and Y_3 to be graphed with a standard viewing rectangle and obtain a display similar to Figure 45.

4.6 EXERCISES

Exer. 1–12: Find the vertex, focus, and directrix of the parabola. Sketch its graph, showing the focus and the directrix.

1 $8y = x^2$

2 $20x = y^2$

3 $2y^2 = -3x$

4 $x^2 = -3y$

5 $(x + 2)^2 = -8(y - 1)$

6 $(x - 3)^2 = \frac{1}{2}(y + 1)$

7 $(y - 2)^2 = \frac{1}{4}(x - 3)$

8 $(y + 1)^2 = -12(x + 2)$

9 $y = x^2 - 4x + 2$

10 $y^2 + 14y + 4x + 45 = 0$

11 $x^2 + 20y = 10$

12 $y^2 - 4y - 2x - 4 = 0$

Exer. 13–16: Find an equation for the parabola shown in the figure.

13

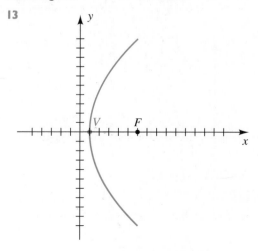

14

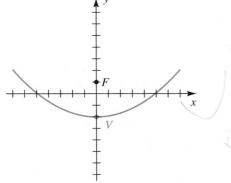

15

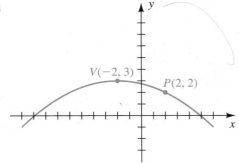

16

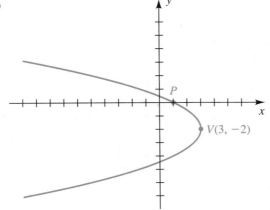

Exer. 17–28: Find an equation of the parabola that satisfies the given conditions.

17 Focus $F(2, 0)$,   directrix $x = -2$

18 Focus $F(0, -4)$,   directrix $y = 4$

19 Focus $F(6, 4)$,   directrix $y = -2$

20 Focus $F(-3, -2)$,   directrix $y = 1$

21 Vertex $V(3, -5)$,   directrix $x = 2$

22 Vertex $V(-2, 3)$,   directrix $y = 5$

23 Vertex $V(-1, 0)$,   focus $F(-4, 0)$

24 Vertex $V(1, -2)$,   focus $F(1, 0)$

25 Vertex at the origin, symmetric to the y-axis, and passing through the point $(2, -3)$

26 Vertex at the origin, symmetric to the y-axis, and passing through the point $(6, 3)$

27 Vertex $V(-3, 5)$, axis parallel to the x-axis, and passing through the point $(5, 9)$

28 Vertex $V(3, -2)$, axis parallel to the x-axis, and y-intercept 1

Exer. 29–32: Find an equation for the set of points in an xy-plane that are equidistant from the point P and the line l.

29 $P(0, 5)$; $l: y = -3$

30 $P(7, 0)$; $l: x = 1$

31 $P(-6, 3)$; $l: x = -2$

32 $P(5, -2)$; $l: y = 4$

33 *Telescope mirror* A mirror for a reflecting telescope has the shape of a (finite) paraboloid of diameter 8 inches and depth 1 inch. How far from the center of the mirror will the incoming light collect?

34 *Satellite antenna dish* A satellite antenna dish has the shape of a paraboloid that is 10 feet across at the open end and is 3 feet deep. At what distance from the center of the dish should the receiver be placed to receive the greatest intensity of sound waves?

35 *Searchlight reflector* A searchlight reflector has the shape of a paraboloid with the light source at the focus. If the reflector is 3 feet across at the opening and 1 foot deep, where is the focus?

36 *Flashlight mirror* A flashlight mirror has the shape of a paraboloid of diameter 4 inches and depth $\frac{3}{4}$ inch. Where should the bulb be placed so that the emitted light rays are parallel to the axis of the paraboloid?

37 *Sound receiving dish* A sound receiving dish used at outdoor sporting events is constructed in the shape of a paraboloid with its focus 5 inches from the vertex. Determine the width of the dish if the depth is to be 2 feet.

38 *Sound receiving dish*   Work Exercise 37 if the receiver is 9 inches from the vertex.

39 *Reflector*

   (a) The focal length of the (finite) paraboloid in the figure is the distance p between its vertex and focus. Express p in terms of r and h.

   (b) A reflector is to be constructed with a focal length of 10 feet and a depth of 5 feet. Find the radius of the reflector.

EXERCISE 39

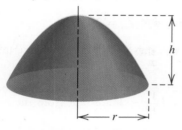

40 *Confocal parabolas*   The parabola $y^2 = 4p(x + p)$ has its focus at the origin and axis along the x-axis. By assigning different values to p, we can obtain a family of confocal parabolas, as shown in the figure. Such families occur in the study of electricity and magnetism. Show

that there are exactly two parabolas in the family that pass through a given point $P(x_1, y_1)$ if $y_1 \neq 0$.

EXERCISE 40

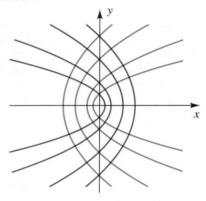

c **Exer. 41–42: Graph the equation.**

41 $x = -y^2 + 2y + 5$ **42** $x = 2y^2 + 3y - 7$

c **Exer. 43–44: Graph the parabolas on the same coordinate plane, and estimate the points of intersection.**

43 $y = x^2 - 2.1x - 1;$ $x = y^2 + 1$

44 $y = -2.1x^2 + 0.1x + 1.2;$ $x = 0.6y^2 + 1.7y - 1.1$

4.7 ELLIPSES

An ellipse may be defined as follows. (*Foci* is the plural of *focus*.)

Definition of an Ellipse

> An **ellipse** is the set of all points in a plane, the sum of whose distances from two fixed points (the **foci**) in the plane is a positive constant.

We can construct an ellipse on paper as follows: Insert two pushpins in the paper at any points F and F', and fasten the ends of a piece of string to the pins. After looping the string around a pencil and drawing it tight, as at point P in Figure 46, move the pencil, keeping the string tight. The sum of the distances $d(F, P)$ and $d(F', P)$ is the length of the string and hence is constant; thus, the pencil will trace out an ellipse with foci at F and F'. The midpoint of the segment $F'F$ is called the **center** of the ellipse. By changing the positions of F and F' while keeping the length

FIGURE 46

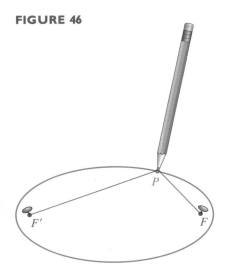

of the string fixed, we can vary the shape of the ellipse considerably. If F and F' are far apart so that $d(F, F')$ is almost the same as the length of the string, the ellipse is flat. If $d(F, F')$ is close to zero, the ellipse is almost circular. If $F = F'$, we obtain a circle with center F.

To obtain a simple equation for an ellipse, choose the x-axis as the line through the two foci F and F', with the center of the ellipse at the origin. If F has coordinates $(c, 0)$ with $c > 0$, then, as in Figure 47, F' has coordinates $(-c, 0)$. Hence, the distance between F and F' is $2c$. The constant sum of the distances of P from F and F' will be denoted by $2a$. To obtain points that are not on the x-axis, we must have $2a > 2c$—that is, $a > c$. By definition, $P(x, y)$ is on the ellipse if and only if

$$d(P, F) + d(P, F') = 2a.$$

If we use the distance formula and eliminate radicals, we obtain the following equation, where $b^2 = a^2 - c^2$:

$$\frac{x^2}{a^2} + \frac{y^2}{b^2} = 1$$

FIGURE 47

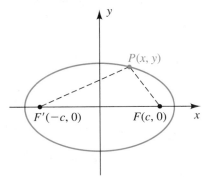

Since $c > 0$ and $b^2 = a^2 - c^2$, it follows that $a^2 > b^2$ and hence $a > b$.

We may find the x-intercepts of the ellipse by letting $y = 0$ in the equation. Doing so gives us $x^2/a^2 = 1$, or $x^2 = a^2$, and consequently the x-intercepts are a and $-a$. The corresponding points $V(a, 0)$ and $V'(-a, 0)$ on the graph are called the **vertices** of the ellipse (see Figure 48). The line segment $V'V$ is called the **major axis**. Similarly, letting $x = 0$ in the equation, we obtain $y^2/b^2 = 1$, or $y^2 = b^2$. Hence, the y-intercepts are b and $-b$. The segment between $M'(0, -b)$ and $M(0, b)$ is called the **minor axis** of the ellipse. The major axis is always longer than the minor axis, since $a > b$.

Applying tests for symmetry, we see that the ellipse is symmetric with respect to the x-axis, the y-axis, and the origin.

FIGURE 48

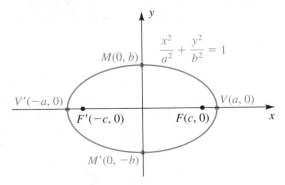

Similarly, if we take the foci on the y-axis, we obtain the equation

$$\frac{x^2}{b^2} + \frac{y^2}{a^2} = 1.$$

In this case the vertices of the ellipse are $(0, \pm a)$, and the end points of the minor axis are $(\pm b, 0)$, as shown in Figure 49.

The preceding discussion may be summarized as follows.

Standard Equations of an Ellipse with Center at the Origin

FIGURE 49

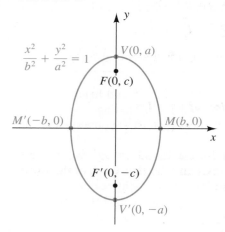

The graph of

$$\frac{x^2}{a^2} + \frac{y^2}{b^2} = 1 \qquad \text{or} \qquad \frac{x^2}{b^2} + \frac{y^2}{a^2} = 1,$$

where $a > b > 0$, is an ellipse with center at the origin. The length of the major axis is $2a$, and the length of the minor axis is $2b$. The foci are a distance c from the origin, where $c^2 = a^2 - b^2$.

We have shown that an equation of an ellipse with center at the origin and foci on a coordinate axis can always be written in the form

$$\frac{x^2}{p} + \frac{y^2}{q} = 1, \qquad \text{or} \qquad qx^2 + py^2 = pq,$$

with p and q positive and $p \neq q$. If $p > q$, the major axis is on the x-axis, and if $p < q$, the major axis is on the y-axis. It is unnecessary to memorize these facts, because in any given problem the major axis can be determined by examining the x- and y-intercepts.

EXAMPLE 1   *Sketching an ellipse with center O*

Sketch the graph of $2x^2 + 9y^2 = 18$, and find the foci.

SOLUTION   The graph is an ellipse with center at the origin and foci on a coordinate axis. To find the x-intercepts, we let $y = 0$, obtaining

$$2x^2 = 18, \qquad \text{or} \qquad x = \pm 3.$$

To find the y-intercepts, we let $x = 0$, obtaining

$$9y^2 = 18, \qquad \text{or} \qquad y = \pm\sqrt{2}.$$

This enables us to sketch the ellipse shown in Figure 50. Since $\sqrt{2} < 3$, the major axis is on the x-axis.

To find the foci, we let $a = 3$, $b = \sqrt{2}$ and calculate

$$c^2 = a^2 - b^2 = 3^2 - (\sqrt{2})^2 = 7.$$

Thus, $c = \sqrt{7}$, and the foci are $(\pm\sqrt{7}, 0)$.

FIGURE 50

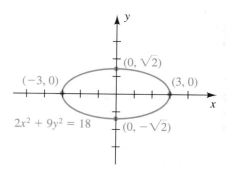

FIGURE 51

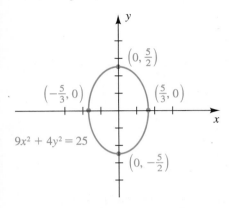

EXAMPLE 2   *Sketching an ellipse with center O*

Sketch the graph of $9x^2 + 4y^2 = 25$, and find the foci.

SOLUTION   As in Example 1, the graph is an ellipse with center at the origin and foci on a coordinate axis. To find the x-intercepts, we let $y = 0$, obtaining

$$9x^2 = 25, \quad \text{or} \quad x = \pm\tfrac{5}{3}.$$

To find the y-intercepts, we let $x = 0$, obtaining

$$4y^2 = 25, \quad \text{or} \quad y = \pm\tfrac{5}{2}.$$

This gives us the sketch in Figure 51. Since $\tfrac{5}{3} < \tfrac{5}{2}$, the major axis is on the y-axis.

To find the foci, we let $a = \tfrac{5}{2}$, $b = \tfrac{5}{3}$ and calculate

$$c^2 = a^2 - b^2 = (\tfrac{5}{2})^2 - (\tfrac{5}{3})^2 = \tfrac{125}{36}.$$

Thus, $c = \sqrt{125/36} = 5\sqrt{5}/6 \approx 1.86$, and the foci are $(0, \pm 5\sqrt{5}/6)$.

EXAMPLE 3   *Finding an equation of an ellipse*

Find an equation of the ellipse with vertices $(\pm 4, 0)$ and foci $(\pm 2, 0)$.

SOLUTION   Since the foci are on the x-axis and are equidistant from the origin, the major axis is on the x-axis and the ellipse has center $(0, 0)$. Thus, a general equation of an ellipse is

$$\frac{x^2}{a^2} + \frac{y^2}{b^2} = 1.$$

Since the vertices are $(\pm 4, 0)$, we conclude that $a = 4$. Since the foci are $(\pm 2, 0)$, we have $c = 2$. Hence,

$$b^2 = a^2 - c^2 = 4^2 - 2^2 = 12,$$

and an equation of the ellipse is

$$\frac{x^2}{16} + \frac{y^2}{12} = 1.$$

We can use the translation of axes formulas from the last section to extend our work to an ellipse with center at any point $C(h, k)$ in the xy-plane. For example, since the graph of

$$\frac{(x')^2}{a^2} + \frac{(y')^2}{b^2} = 1$$

FIGURE 52

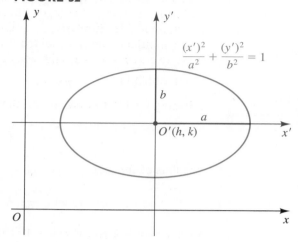

$$\frac{(x')^2}{a^2} + \frac{(y')^2}{b^2} = 1$$

is an ellipse with center at O' in an $x'y'$-plane (see Figure 52), its equation relative to the xy-coordinate system is

$$\frac{(x-h)^2}{a^2} + \frac{(y-k)^2}{b^2} = 1.$$

Squaring terms in the last equation and simplifying gives us an equation of the form

$$Ax^2 + Cy^2 + Dx + Ey + F = 0,$$

where the coefficients are real numbers and both A and C are positive. Conversely, if we start with such an equation, then by completing squares we can obtain a form that displays the center of the ellipse and the lengths of the major and minor axes. This technique is illustrated in the next example.

EXAMPLE 4   *Sketching an ellipse with center* (h, k)

Discuss and sketch the graph of the equation

$$16x^2 + 9y^2 + 64x - 18y - 71 = 0.$$

SOLUTION   We begin by grouping the terms containing x and those containing y:

$$(16x^2 + 64x) + (9y^2 - 18y) = 71$$

Next, we factor out the coefficients of x^2 and y^2, as follows:

$$16(x^2 + 4x \quad) + 9(y^2 - 2y \quad) = 71$$

We now complete the squares for the expressions within parentheses:

$$16(x^2 + 4x + 4) + 9(y^2 - 2y + 1) = 71 + 64 + 9$$

By adding 4 to the expression within the first parentheses we have added 64 to the left side of the equation, so we must compensate by adding 64 to the right side. Similarly, by adding 1 to the expression within the second parentheses we have added 9 to the left side, and consequently we must also add 9 to the right side. The last equation may be written

$$16(x + 2)^2 + 9(y - 1)^2 = 144.$$

Dividing by 144 to obtain 1 on the right side gives us

$$\frac{(x + 2)^2}{9} + \frac{(y - 1)^2}{16} = 1,$$

which is of the form

$$\frac{(x')^2}{9} + \frac{(y')^2}{16} = 1,$$

with $x' = x + 2$ and $y' = y - 1$. This corresponds to letting $h = -2$ and $k = 1$ in the translation of axes formulas.

The graph of the equation $((x')^2/9) + ((y')^2/16) = 1$ is an ellipse with center at the origin O' in the $x'y'$-plane and major axis on the y'-axis. It follows that the graph of the given equation is an ellipse with center $C(-2, 1)$ in the xy-plane and major axis on the vertical line $x = -2$. Using $a = 4$ and $b = 3$ gives us the ellipse in Figure 53.

FIGURE 53

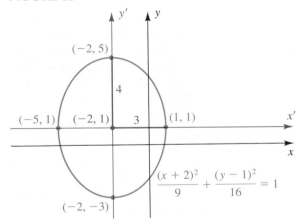

To find the foci, we first calculate

$$c^2 = a^2 - b^2 = 4^2 - 3^2 = 7.$$

The distance from the center of the ellipse to the foci is $c = \sqrt{7}$. Since the center is $(-2, 1)$, the foci are $(-2, 1 \pm \sqrt{7})$.

Graphics calculators and computer programs are often unable to plot the graph of an equation of the form

$$Ax^2 + Cy^2 + Dx + Ey + F = 0,$$

such as that considered in the last example. In most cases, we must first solve the equation for y in terms of x, and then plot the two resulting functions, as illustrated in the next example.

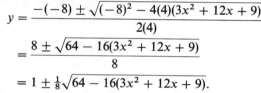

EXAMPLE 5   *Graphing half-ellipses*

Use a graphing utility to sketch the graph of

$$3x^2 + 4y^2 + 12x - 8y + 9 = 0.$$

SOLUTION   The equation may be regarded as a quadratic equation in y of the form $ay^2 + by + c = 0$ by rearranging terms as follows:

$$4y^2 - 8y + (3x^2 + 12x + 9) = 0$$

Applying the quadratic formula to the above equation with $a = 4$, $b = -8$, and $c = 3x^2 + 12x + 9$ gives us

$$y = \frac{-(-8) \pm \sqrt{(-8)^2 - 4(4)(3x^2 + 12x + 9)}}{2(4)}$$

$$= \frac{8 \pm \sqrt{64 - 16(3x^2 + 12x + 9)}}{8}$$

$$= 1 \pm \tfrac{1}{8}\sqrt{64 - 16(3x^2 + 12x + 9)}.$$

FIGURE 54 $[-6, 6]$ by $[-4, 4]$

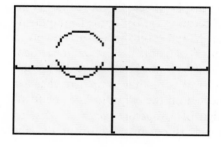

Note that we did not completely simplify the radicand, since we will be using a graphing utility.

As in Example 6 of the previous section, we now make the assignments

$$Y_1 = \tfrac{1}{8}\sqrt{64 - 16(3x^2 + 12x + 9)}, \quad Y_2 = 1 + Y_1, \quad \text{and} \quad Y_3 = 1 - Y_1.$$

Finally, we select Y_2 and Y_3 to be graphed, choose the viewing rectangle $[-6, 6]$ by $[-4, 4]$, and obtain a display similar to Figure 54.

Ellipses can be very flat or almost circular. To obtain information about the roundness of an ellipse, we sometimes use the *eccentricity*, which is defined as follows, with a, b, and c having the same meanings as before.

Definition of Eccentricity

The **eccentricity** e of an ellipse is

$$e = \frac{c}{a} = \frac{\sqrt{a^2 - b^2}}{a}.$$

Consider the ellipse $(x^2/a^2) + (y^2/b^2) = 1$. Suppose that the length $2a$ of the major axis is fixed and the length $2b$ of the minor axis is vari-

FIGURE 55

(a) Eccentricity almost 1

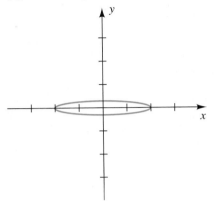

(b) Eccentricity almost 0

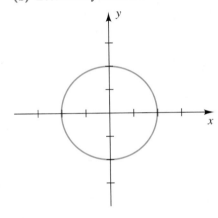

able (note that $0 < b < a$). Since b^2 is positive, $a^2 - b^2 < a^2$, and hence $\sqrt{a^2 - b^2} < a$. Dividing both sides of the last inequality by a gives us $\sqrt{a^2 - b^2}/a < 1$, or $0 < e < 1$. If b is close to 0, then $\sqrt{a^2 - b^2} \approx a$, $e \approx 1$, and the ellipse is very flat. This case is illustrated in Figure 55(a), with $a = 2$, $b = 0.3$, and $e \approx 0.99$.

If b is close to a, then $\sqrt{a^2 - b^2} \approx 0$, $e \approx 0$, and the ellipse is almost circular. This case is illustrated in Figure 55(b), with $a = 2$, $b = 1.9999$, and $e \approx 0.01$.

After many years of analyzing an enormous amount of empirical data, the German astronomer Johannes Kepler (1571–1630) formulated three laws that describe the motion of planets about the sun. Kepler's first law states that the orbit of each planet in the solar system is an ellipse with the sun at one focus. Most of these orbits are almost circular, so their corresponding eccentricities are close to 0. For Earth, $e \approx 0.017$; for Mars, $e \approx 0.093$; and for Uranus, $e \approx 0.046$. The orbits of Mercury and Pluto are less circular, with eccentricities of 0.206 and 0.249, respectively. Many comets have elliptical orbits with the sun at a focus. In this case the eccentricity e is close to 1, and the ellipse is very flat.

An ellipse has a reflective property analogous to that of the parabola discussed at the end of the previous section. To illustrate, let l denote the tangent line at a point P on an ellipse with foci F and F', as shown in Figure 56. If α is the acute angle between $F'P$ and l and if β is the acute angle between FP and l, it can be shown that $\alpha = \beta$. Thus, if a ray of light or sound emanates from one focus, it is reflected to the other focus. This property is used in the design of certain types of optical equipment.

If the ellipse with center O and foci F' and F on the x-axis is revolved about the x-axis, as illustrated in Figure 57, we obtain a three-dimensional surface called an **ellipsoid**. The upper half or lower half is a **hemi-ellipsoid**, as is the right half or left half. Sound waves or other impulses that are emitted from the focus F' will be reflected off the ellipsoid into the focus F. This property is used in the design of *whispering galleries*—that is, structures with ellipsoidal ceilings, in which a person who whispers at one

FIGURE 56

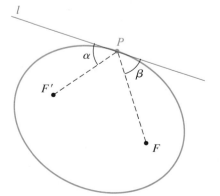

FIGURE 57

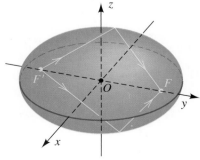

focus can be heard at the other focus. Examples of whispering galleries may be found in the Rotunda of the Capitol Building in Washington, D.C., and in the Mormon Tabernacle in Salt Lake City.

The reflective property of ellipsoids (and hemi-ellipsoids) is used in modern medicine in a device called a *lithotripter*, which disintegrates kidney stones by means of high-energy underwater shock waves. After taking extremely accurate measurements, the operator positions the patient so that the stone is at a focus. Ultra-high-frequency shock waves are then produced at the other focus, and reflected waves break up the kidney stone. Recovery time following this technique is usually 3–4 days instead of the 2–3 weeks needed when conventional methods are used. Moreover, the mortality rate is less than 0.01%, as compared to 2–3% using traditional surgery.

4.7 EXERCISES

Exer. 1–14: Find the vertices and foci of the ellipse. Sketch its graph, showing the foci.

1 $\dfrac{x^2}{9} + \dfrac{y^2}{4} = 1$

2 $\dfrac{x^2}{25} + \dfrac{y^2}{16} = 1$

3 $\dfrac{x^2}{15} + \dfrac{y^2}{16} = 1$

4 $\dfrac{x^2}{45} + \dfrac{y^2}{49} = 1$

5 $4x^2 + y^2 = 16$

6 $y^2 + 9x^2 = 9$

7 $4x^2 + 25y^2 = 1$

8 $10y^2 + x^2 = 5$

9 $\dfrac{(x-3)^2}{16} + \dfrac{(y+4)^2}{9} = 1$

10 $\dfrac{(x+2)^2}{25} + \dfrac{(y-3)^2}{4} = 1$

11 $4x^2 + 9y^2 - 32x - 36y + 64 = 0$

12 $x^2 + 2y^2 + 2x - 20y + 43 = 0$

13 $25x^2 + 4y^2 - 250x - 16y + 541 = 0$

14 $4x^2 + y^2 = 2y$

Exer. 15–18: Find an equation for the ellipse shown in the figure.

15

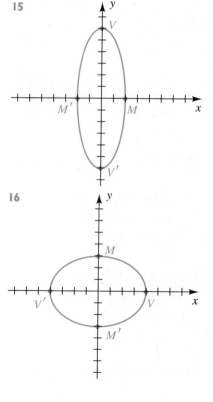

16

17

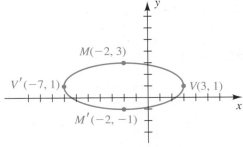

18

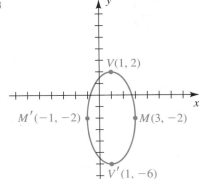

Exer. 19–30: Find an equation for the ellipse that has its center at the origin and satisfies the given conditions.

19 Vertices $V(\pm 8, 0)$, foci $F(\pm 5, 0)$

20 Vertices $V(0, \pm 7)$, foci $F(0, \pm 2)$

21 Vertices $V(0, \pm 5)$, minor axis of length 3

22 Foci $F(\pm 3, 0)$, minor axis of length 2

23 Vertices $V(0, \pm 6)$, passing through $(3, 2)$

24 Passing through $(2, 3)$ and $(6, 1)$

25 Eccentricity $\frac{3}{4}$, vertices $V(0, \pm 4)$

26 Eccentricity $\frac{1}{2}$, vertices on the x-axis, passing through $(1, 3)$

27 x-intercepts ± 2, y-intercepts $\pm \frac{1}{3}$

28 x-intercepts $\pm \frac{1}{2}$, y-intercepts ± 4

29 Horizontal major axis of length 8, minor axis of length 5

30 Vertical major axis of length 7, minor axis of length 6

Exer. 31–34: Find an equation for the set of points in an xy-plane such that the sum of the distances from F and F' is k.

31 $F(3, 0)$, $F'(-3, 0)$; $k = 10$

32 $F(12, 0)$, $F'(-12, 0)$; $k = 26$

33 $F(0, 15)$, $F'(0, -15)$; $k = 34$

34 $F(0, 8)$, $F'(0, -8)$; $k = 20$

35 *An elliptical arch* An arch of a bridge is semielliptical, with the major axis horizontal. The base of the arch is 30 feet across, and the highest part of the arch is 10 feet above the horizontal roadway, as shown in the figure. Find the height of the arch 6 feet from the center of the base.

EXERCISE 35

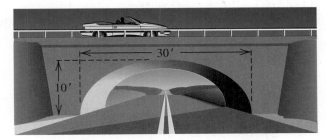

36 *An elliptical arch* A bridge is to be constructed across a river that is 200 feet wide. The arch of the bridge is to be semielliptical and must be constructed so that a ship less than 50 feet wide and 30 feet high can pass safely through the arch, as shown in the figure.

(a) Find an equation for the arch.

(b) Approximate the height of the arch in the middle of the bridge.

EXERCISE 36

37 Earth's orbit Assume that the length of the major axis of the earth's orbit is 186,000,000 miles and the eccentricity is 0.017. Approximate, to the nearest 1000 miles, the maximum and minimum distances between the earth and the sun.

38 Mercury's orbit The planet Mercury travels in an elliptical orbit that has eccentricity 0.206 and major axis of length 0.774 AU, where 1 AU (astronomical unit) is approximately 93,000,000 miles. Find the maximum and minimum distances between Mercury and the sun.

39 Elliptical reflector The basic shape of an elliptical reflector is a hemi-ellipsoid of height h and diameter k, as shown in the figure. Waves emitted from focus F will reflect off the surface into focus F'.

EXERCISE 39

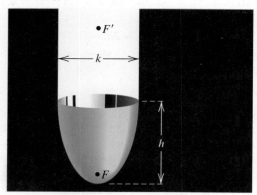

(a) Express the distances $d(V, F)$ and $d(V, F')$ in terms of h and k.

(b) An elliptical reflector of height 17 centimeters is to be constructed so that waves emitted from F are reflected to a point F' that is 32 centimeters from V. Find the diameter of the reflector and the location of F.

40 Lithotripter A lithotripter of height 15 centimeters and diameter 18 centimeters is to be constructed as shown in the figure at the top of the next column. High-energy underwater shock waves will be emitted from the focus F that is closest to the vertex V.

(a) Find the distance from V to F.

(b) How far from V (in the vertical direction) should a kidney stone be located?

EXERCISE 40

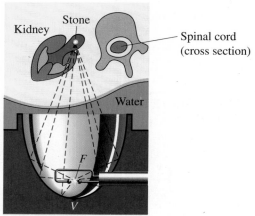

41 Whispering gallery The ceiling of a whispering gallery has the shape of the hemi-ellipsoid shown in Figure 55, with the highest point of the ceiling 15 feet above the elliptical floor and the vertices of the floor 50 feet apart. If two people are standing at the foci F' and F, how far from the vertices are their feet?

42 An ellipse has a vertex at the origin and foci $F_1(p, 0)$ and $F_2(p + 2c, 0)$, as shown in the figure. If the focus at F_1 is fixed and (x, y) is on the ellipse, show that y^2 approaches $4px$ as $c \to \infty$. (Thus, as $c \to \infty$, the ellipse takes on the shape of a parabola.)

EXERCISE 42

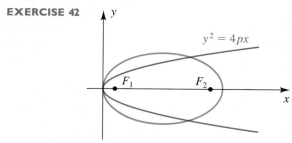

c Exer. 43–46: Graph the ellipses on the same coordinate plane, and estimate their points of intersection.

43 $\dfrac{x^2}{2.9} + \dfrac{y^2}{2.1} = 1; \qquad \dfrac{x^2}{4.3} + \dfrac{(y - 2.1)^2}{4.9} = 1$

44 $\dfrac{x^2}{3.9} + \dfrac{y^2}{2.4} = 1; \qquad \dfrac{(x + 1.9)^2}{4.1} + \dfrac{y^2}{2.5} = 1$

45 $\dfrac{(x + 0.1)^2}{1.7} + \dfrac{y^2}{0.9} = 1; \qquad \dfrac{x^2}{0.9} + \dfrac{(y - 0.25)^2}{1.8} = 1$

46 $\dfrac{x^2}{3.1} + \dfrac{(y - 0.2)^2}{2.8} = 1; \qquad \dfrac{(x + 0.23)^2}{1.8} + \dfrac{y^2}{4.2} = 1$

4.8 HYPERBOLAS

The definition of a hyperbola is similar to that of an ellipse. The only change is that instead of using the *sum* of distances from two fixed points, we use the *difference*.

Definition of a Hyperbola

A **hyperbola** is the set of all points in a plane, the difference of whose distances from two fixed points (the **foci**) in the plane is a positive constant.

FIGURE 58

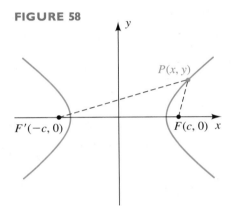

To find a simple equation for a hyperbola, we choose a coordinate system with foci at $F(c, 0)$ and $F'(-c, 0)$, as in Figure 58, and denote the (constant) distance by $2a$. The midpoint of the segment $F'F$ (the origin) is called the **center** of the hyperbola. By definition, a point $P(x, y)$ is on the hyperbola if and only if

$$\left| d(P, F) - d(P, F') \right| = 2a.$$

Using the distance formula and eliminating radicals, we obtain the following equation, where $c^2 = a^2 + b^2$:

$$\frac{x^2}{a^2} - \frac{y^2}{b^2} = 1$$

Applying tests for symmetry, we see that the hyperbola is symmetric with respect to both axes and the origin. We may find the x-intercepts of the hyperbola by letting $y = 0$ in the equation. Doing so gives us $x^2/a^2 = 1$, or $x^2 = a^2$, and consequently the x-intercepts are a and $-a$. The corresponding points $V(a, 0)$ and $V'(-a, 0)$ on the graph are called the **vertices** of the hyperbola (see Figure 59 on the following page). The line segment $V'V$ is called the **transverse axis**. The graph has no y-intercept, since the equation $-y^2/b^2 = 1$ has the *complex* solutions $y = \pm bi$. The points $W(0, b)$ and $W'(0, -b)$ are end points of the **conjugate axis** $W'W$. Although the points W and W' are not on the hyperbola, they are, as we shall see, useful for describing the graph.

Solving the equation $(x^2/a^2) - (y^2/b^2) = 1$ for y gives us

$$y = \pm \frac{b}{a}\sqrt{x^2 - a^2}.$$

If $x^2 - a^2 < 0$ or, equivalently, $-a < x < a$, then there are no points (x, y) on the graph. There *are* points $P(x, y)$ on the graph if $x \geq a$ or $x \leq -a$.

It can be shown that *the lines $y = \pm (b/a)x$ are asymptotes for the hyperbola*. These asymptotes serve as excellent guides for sketching the graph. A convenient way to sketch the asymptotes is to first plot the vertices $V(a, 0)$, $V'(-a, 0)$ and the points $W(0, b)$, $W'(0, -b)$ (see Figure 59). If vertical and horizontal lines are drawn through these end points of the transverse and conjugate axes, respectively, then the diagonals of the

FIGURE 59

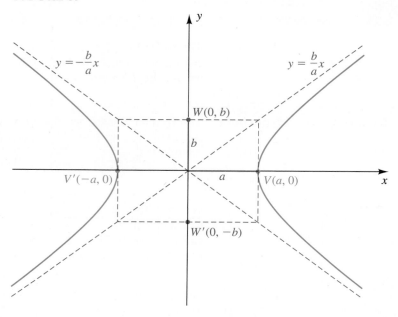

resulting rectangle have slopes b/a and $-b/a$. Hence, by extending these diagonals we obtain the asymptotes $y = \pm(b/a)x$. The hyperbola is then sketched, as in Figure 59, using the asymptotes as guides. The two parts that make up the hyperbola are called the **right branch** and the **left branch** of the hyperbola.

Similarly, if we take the foci on the y-axis, we obtain the equation

$$\frac{y^2}{a^2} - \frac{x^2}{b^2} = 1.$$

In this case the vertices of the hyperbola are $(0, \pm a)$, and the end points of the conjugate axis are $(\pm b, 0)$, as shown in Figure 60. The asymptotes are $y = \pm(a/b)x$ (*not* $y = \pm(b/a)x$, as in the previous case), and we now refer to the **upper branch** and the **lower branch** of the hyperbola.

The preceding discussion may be summarized as follows.

Standard Equations of a Hyperbola with Center at the Origin

The graph of

$$\frac{x^2}{a^2} - \frac{y^2}{b^2} = 1 \qquad \text{or} \qquad \frac{y^2}{a^2} - \frac{x^2}{b^2} = 1$$

is a hyperbola with center at the origin. The length of the transverse axis is $2a$, and the length of the conjugate axis is $2b$. The foci are a distance c from the origin, where $c^2 = a^2 + b^2$.

FIGURE 60

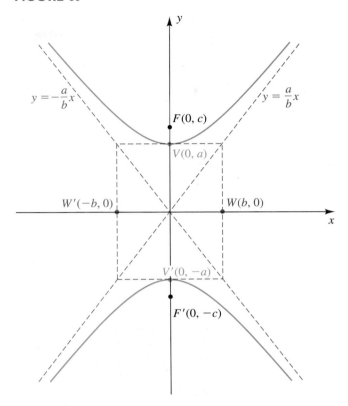

We have shown that an equation of a hyperbola with center at the origin and foci on a coordinate axis can always be written in the form

$$\frac{x^2}{p} + \frac{y^2}{q} = 1, \quad \text{or} \quad qx^2 + py^2 = pq,$$

where p and q have opposite signs. The vertices are on the x-axis if p is positive and on the y-axis if q is positive.

EXAMPLE 1 *Sketching a hyperbola with center O*

Sketch the graph of $9x^2 - 4y^2 = 36$. Find the foci and equations of the asymptotes.

SOLUTION From the remarks preceding this example, the graph is a hyperbola with center at the origin. To express the given equation in a standard form, we divide both sides by 36 and simplify, obtaining

$$\frac{x^2}{4} - \frac{y^2}{9} = 1.$$

(continued)

FIGURE 61

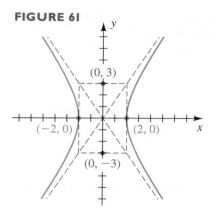

Comparing $(x^2/4) - (y^2/9) = 1$ to $(x^2/a^2) - (y^2/b^2) = 1$, we see that $a^2 = 4$ and $b^2 = 9$; that is, $a = 2$ and $b = 3$. The hyperbola has its vertices on the x-axis, since there are x-intercepts and no y-intercepts. The vertices $(\pm 2, 0)$ and the end points $(0, \pm 3)$ of the conjugate axis determine a rectangle whose diagonals (extended) give us the asymptotes. The graph of the equation is sketched in Figure 61.

To find the foci, we calculate

$$c^2 = a^2 + b^2 = 4 + 9 = 13.$$

Thus, $c = \sqrt{13}$, and the foci are $(\pm\sqrt{13}, 0)$.

The equations of the asymptotes, $y = \pm\frac{3}{2}x$, can be found by referring to the graph or to the equations $y = \pm(b/a)x$.

The preceding example indicates that for hyperbolas it is not always true that $a > b$, as is the case for ellipses. In fact, we may have $a < b$, $a > b$, or $a = b$.

EXAMPLE 2   *Sketching a hyperbola with center O*

Sketch the graph of $4y^2 - 2x^2 = 1$. Find the foci and equations of the asymptotes.

SOLUTION   To express the given equation in a standard form, we write

$$\frac{y^2}{\frac{1}{4}} - \frac{x^2}{\frac{1}{2}} = 1.$$

FIGURE 62

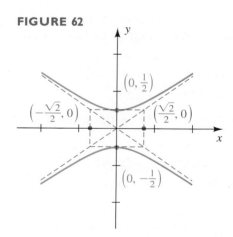

Thus,

$$a^2 = \tfrac{1}{4}, \qquad b^2 = \tfrac{1}{2}, \qquad c^2 = a^2 + b^2 = \tfrac{3}{4},$$

and consequently

$$a = \frac{1}{2}, \qquad b = \frac{1}{\sqrt{2}}, \qquad c = \frac{\sqrt{3}}{2}.$$

The hyperbola has its vertices on the y-axis, since there are y-intercepts and no x-intercepts. The vertices are $(0, \pm\frac{1}{2})$, the end points of the conjugate axes are $(\pm 1/\sqrt{2}, 0)$, and the foci are $(0, \pm\sqrt{3}/2)$. The graph is sketched in Figure 62.

To find the equations of the asymptotes, we refer to the figure or use $y = \pm(a/b)x$, obtaining $y = \pm(\sqrt{2}/2)x$.

EXAMPLE 3   *Finding an equation of a hyperbola*

A hyperbola has vertices $(\pm 3, 0)$ and passes through the point $P(5, 2)$. Find its equation, foci, and asymptotes.

FIGURE 63

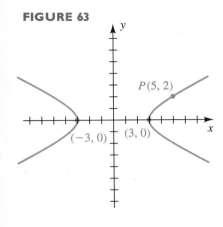

SOLUTION   We begin by sketching a hyperbola with vertices $(\pm 3, 0)$ that passes through the point $P(5, 2)$, as in Figure 63.

An equation of the hyperbola has the form

$$\frac{x^2}{3^2} - \frac{y^2}{b^2} = 1.$$

Since $P(5, 2)$ is on the hyperbola, the x- and y-coordinates satisfy this equation; that is,

$$\frac{5^2}{3^2} - \frac{2^2}{b^2} = 1.$$

Solving for b^2 gives us $b^2 = \frac{9}{4}$, and hence an equation for the hyperbola is

$$\frac{x^2}{9} - \frac{y^2}{\frac{9}{4}} = 1$$

or, equivalently,

$$x^2 - 4y^2 = 9.$$

To find the foci, we first calculate

$$c^2 = a^2 + b^2 = 9 + \frac{9}{4} = \frac{45}{4}.$$

Hence, $c = \sqrt{\frac{45}{4}} = \frac{3}{2}\sqrt{5} \approx 3.35$, and the foci are $(\pm\frac{3}{2}\sqrt{5}, 0)$.

The general equations of the asymptotes are $y = \pm(b/a)x$. Substituting $a = 3$ and $b = \frac{3}{2}$ gives us $y = \pm\frac{1}{2}x$.

As was the case for ellipses, we may use the translation of axes formulas to generalize our work. The following example illustrates this technique.

EXAMPLE 4   *Sketching a hyperbola with center (h, k)*

Discuss and sketch the graph of the equation

$$9x^2 - 4y^2 - 54x - 16y + 29 = 0.$$

SOLUTION   Using a procedure similar to the one used for ellipses in Example 4 of the previous section, we arrange our work as follows:

$(9x^2 - 54x) + (-4y^2 - 16y) = -29$	group terms
$9(x^2 - 6x \quad) - 4(y^2 + 4y \quad) = -29$	factor out 9 and -4
$9(x^2 - 6x + 9) - 4(y^2 + 4y + 4) = -29 + 81 - 16$	complete the squares
$9(x - 3)^2 - 4(y + 2)^2 = 36$	factor
$\dfrac{(x - 3)^2}{4} - \dfrac{(y + 2)^2}{9} = 1$	divide by 36

(continued)

FIGURE 64

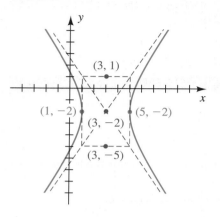

FIGURE 65

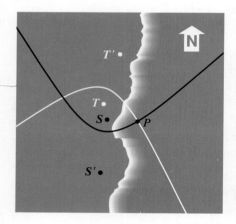

FIGURE 66

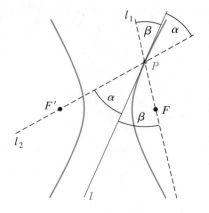

The last equation has the form

$$\frac{(x')^2}{4} - \frac{(y')^2}{9} = 1,$$

with $x' = x - 3$ and $y' = y + 2$. Thus, we translate the x- and y-axes to the new origin $C(3, -2)$. The graph is a hyperbola with vertices on the x'-axis (the line $y = -2$) and

$$a^2 = 4, \qquad b^2 = 9, \qquad c^2 = a^2 + b^2 = 13.$$

Hence,

$$a = 2, \qquad b = 3, \qquad c = \sqrt{13}.$$

As illustrated in Figure 64, the vertices are $(3 \pm 2, -2)$—that is, $(5, -2)$ and $(1, -2)$. The end points of the conjugate axis are $(3, -2 \pm 3)$—that is, $(3, 1)$ and $(3, -5)$. The foci are $(3 \pm \sqrt{13}, -2)$, and equations of the asymptotes are

$$y + 2 = \pm \tfrac{3}{2}(x - 3).$$

The results of Sections 4.6–4.8 indicate that the graph of every equation of the form

$$Ax^2 + Cy^2 + Dx + Ey + F = 0$$

is a conic, except for certain degenerate cases in which a point, one or two lines, or no graph is obtained. Although we have considered only special examples, our methods can be applied to any such equation. If A and C are equal and not 0, then the graph, when it exists, is a circle or, in exceptional cases, a point. If A and C are unequal but have the same sign, then by completing squares and properly translating axes we obtain an equation whose graph, when it exists, is an ellipse (or a point). If A and C have opposite signs, an equation of a hyperbola is obtained or possibly, in the degenerate case, two intersecting straight lines. If either A or C (but not both) is 0, the graph is a parabola or, in certain cases, a pair of parallel lines.

Hyperbolas are the basis for the navigational system LORAN (for Long Range Navigation). This system involves two pairs of radio transmitters, such as those located at T, T' and S, S' in Figure 65. Suppose that signals sent out by the transmitters at T and T' reach a radio receiver on a ship located at some point P. The difference in the times of arrival of the signals can be used to determine the difference in the distances of P from T and T'. Thus, P lies on one branch of a hyperbola with foci at T and T'. Repeating this process for the other pair of transmitters, we see that P also lies on one branch of a hyperbola with foci at S and S'. The intersection of these two branches determines the position of P.

A hyperbola has a reflective property analogous to that of the ellipse discussed at the end of the previous section. To illustrate, let l denote the tangent line at a point P on a hyperbola with foci F and F', as shown in Figure 66. If α is the acute angle between $F'P$ and l and if β is the acute angle between FP and l, it can be shown that $\alpha = \beta$. If a ray of light is directed along the line l_1 toward F, it will be reflected back at P along the line l_2 toward F'. This property is used in the design of telescopes of the Cassegrain type (see Exercise 42).

4.8 EXERCISES

Exer. 1–16: Find the vertices, the foci, and equations of the asymptotes of the hyperbola. Sketch its graph, showing the asymptotes and the foci.

1 $\dfrac{x^2}{9} - \dfrac{y^2}{4} = 1$

2 $\dfrac{y^2}{49} - \dfrac{x^2}{16} = 1$

3 $\dfrac{y^2}{9} - \dfrac{x^2}{4} = 1$

4 $\dfrac{x^2}{49} - \dfrac{y^2}{16} = 1$

5 $x^2 - \dfrac{y^2}{24} = 1$

6 $y^2 - \dfrac{x^2}{15} = 1$

7 $y^2 - 4x^2 = 16$

8 $x^2 - 2y^2 = 8$

9 $16x^2 - 36y^2 = 1$

10 $y^2 - 16x^2 = 1$

11 $\dfrac{(y + 2)^2}{9} - \dfrac{(x + 2)^2}{4} = 1$

12 $\dfrac{(x - 3)^2}{25} - \dfrac{(y - 1)^2}{4} = 1$

13 $144x^2 - 25y^2 + 864x - 100y - 2404 = 0$

14 $y^2 - 4x^2 - 12y - 16x + 16 = 0$

15 $4y^2 - x^2 + 40y - 4x + 60 = 0$

16 $25x^2 - 9y^2 + 100x - 54y + 10 = 0$

Exer. 17–20: Find an equation for the hyperbola shown in the figure.

17

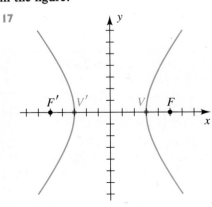

18

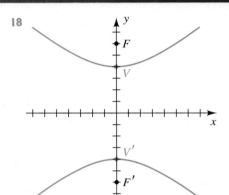

19

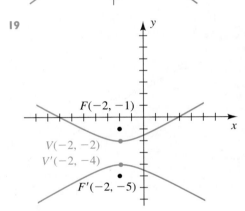

20

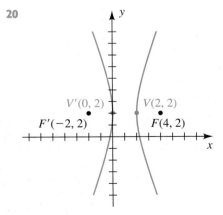

Exer. 21–32: Find an equation for the hyperbola that has its center at the origin and satisfies the given conditions.

21 Foci $F(0, \pm 4)$,   vertices $V(0, \pm 1)$

22 Foci $F(\pm 8, 0)$,   vertices $V(\pm 5, 0)$

23 Foci $F(\pm 5, 0)$,   vertices $V(\pm 3, 0)$

24 Foci $F(0, \pm 3)$,   vertices $V(0, \pm 2)$

25 Foci $F(0, \pm 5)$,   conjugate axis of length 4

26 Vertices $V(\pm 4, 0)$,   passing through $(8, 2)$

27 Vertices $V(\pm 3, 0)$,   asymptotes $y = \pm 2x$

28 Foci $F(0, \pm 10)$,   asymptotes $y = \pm \frac{1}{3}x$

29 x-intercepts ± 5,   asymptotes $y = \pm 2x$

30 y-intercepts ± 2,   asymptotes $y = \pm \frac{1}{4}x$

31 Vertical transverse axis of length 10, conjugate axis of length 14

32 Horizontal transverse axis of length 6, conjugate axis of length 2

Exer. 33–36: Find an equation for the set of points in an xy-plane such that the difference of the distances from F and F' is k.

33 $F(13, 0)$, $F'(-13, 0)$; $k = 24$

34 $F(5, 0)$, $F'(-5, 0)$; $k = 8$

35 $F(0, 10)$, $F'(0, -10)$; $k = 16$

36 $F(0, 17)$, $F'(0, -17)$; $k = 30$

37 The graphs of the equations

$$\frac{x^2}{a^2} - \frac{y^2}{b^2} = 1 \quad \text{and} \quad \frac{x^2}{a^2} - \frac{y^2}{b^2} = -1$$

are called *conjugate hyperbolas*. Sketch the graphs of both equations on the same coordinate plane, with $a = 5$ and $b = 3$, and describe the relationship between the two graphs.

38 Find an equation of the hyperbola with foci $(h \pm c, k)$ and vertices $(h \pm a, k)$, where $0 < a < c$ and $c^2 = a^2 + b^2$.

39 *Alpha particles*   In 1911 the physicist Ernest Rutherford (1871–1937) discovered that if alpha particles are shot toward the nucleus of an atom, they are eventually repulsed away from the nucleus along hyperbolic paths. The figure illustrates the path of a particle that starts toward the origin along the line $y = \frac{1}{2}x$ and comes within 3 units of the nucleus. Find an equation of the path.

EXERCISE 39

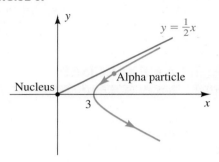

40 *Airplane maneuver*   An airplane is flying along the hyperbolic path illustrated in the figure. If an equation of the path is $2y^2 - x^2 = 8$, determine how close the airplane comes to a town located at $(3, 0)$. (*Hint:* Let S denote the square of the distance from a point (x, y) on the path to $(3, 0)$, and find the minimum value of S.)

EXERCISE 40

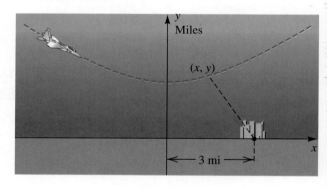

41 *Locating a ship*   A ship is traveling a course that is 100 miles from, and parallel to, a straight shoreline. The

EXERCISE 41

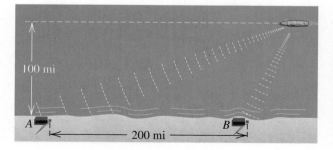

ship sends out a distress signal, which is received by Coast Guard stations A and B, located 200 miles apart, as shown in the figure. By measuring the difference in signal reception times, it is determined that the ship is 160 miles closer to B than to A. Where is the ship?

42 *Cassegrain telescope* The Cassegrain telescope design (dating to 1672) makes use of the reflective properties of both the parabola and the hyperbola. Shown in the figure is a (split) parabolic mirror, with focus at F_1 and axis along the line l, and a second hyperbolic mirror, with one focus also at F_1 and transverse axis along l. Where do incoming light waves parallel to the common axis finally collect?

EXERCISE 42

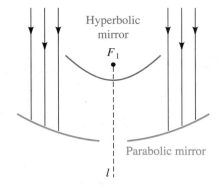

Hyperbolic mirror

F_1

Parabolic mirror

l

c **Exer. 43–44: Graph the hyperbolas on the same coordinate plane, and estimate their first-quadrant point of intersection.**

43 $\dfrac{(y-0.1)^2}{1.6} - \dfrac{(x+0.2)^2}{0.5} = 1;\quad \dfrac{(y-0.5)^2}{2.7} - \dfrac{(x-0.1)^2}{5.3} = 1$

44 $\dfrac{(x-0.1)^2}{0.12} - \dfrac{y^2}{0.1} = 1;\quad \dfrac{x^2}{0.9} - \dfrac{(y-0.3)^2}{2.1} = 1$

c **Exer. 45–46: Graph the hyperbolas on the same coordinate plane, and determine the number of points of intersection.**

45 $\dfrac{(x-0.3)^2}{1.3} - \dfrac{y^2}{2.7} = 1;\quad \dfrac{y^2}{2.8} - \dfrac{(x-0.2)^2}{1.2} = 1$

46 $\dfrac{(x+0.2)^2}{1.75} - \dfrac{(y-0.5)^2}{1.6} = 1;\quad \dfrac{(x-0.6)^2}{2.2} - \dfrac{(y+0.4)^2}{2.35} = 1$

CHAPTER 4 REVIEW EXERCISES

Exer. 1–6: Find all values of x such that $f(x) > 0$ and all x such that $f(x) < 0$, and then sketch the graph of f.

1 $f(x) = (x+2)^3$

2 $f(x) = x^6 - 32$

3 $f(x) = -\frac{1}{4}(x+2)(x-1)^2(x-3)$

4 $f(x) = 2x^2 + x^3 - x^4$

5 $f(x) = x^3 + 2x^2 - 8x$

6 $f(x) = \frac{1}{15}(x^5 - 20x^3 + 64x)$

7 If $f(x) = x^3 - 5x^2 + 7x - 9$, use the intermediate value theorem for polynomial functions to prove that there is a real number a such that $f(a) = 100$.

8 Prove that the equation $x^5 - 3x^4 - 2x^3 - x + 1 = 0$ has a solution between 0 and 1.

Exer. 9–10: Find the quotient and remainder if $f(x)$ is divided by $p(x)$.

9 $f(x) = 3x^5 - 4x^3 + x + 5;\quad p(x) = x^3 - 2x + 7$

10 $f(x) = 4x^3 - x^2 + 2x - 1;\quad p(x) = x^2$

11 If $f(x) = -4x^4 + 3x^3 - 5x^2 + 7x - 10$, use the remainder theorem to find $f(-2)$.

12 Use the factor theorem to show that $x - 3$ is a factor of $f(x) = 2x^4 - 5x^3 - 4x^2 + 9$.

Exer. 13–14: Use synthetic division to find the quotient and remainder if $f(x)$ is divided by $p(x)$.

13 $f(x) = 6x^5 - 4x^2 + 8;\quad p(x) = x + 2$

14 $f(x) = 2x^3 + 5x^2 - 2x + 1;\quad p(x) = x - \sqrt{2}$

Exer. 15–16: A polynomial $f(x)$ with real coefficients has the indicated zero(s) and degree and satisfies the given condition. Express $f(x)$ as a product of linear and quadratic polynomials with real coefficients that are irreducible over $\mathbb{R}$.

15 $-3 + 5i, -1;$ degree 3; $f(1) = 4$

16 $1 - i, 3, 0;$ degree 4; $f(2) = -1$

17 Find a polynomial $f(x)$ of degree 7 with leading coefficient 1 such that -3 is a zero of multiplicity 2 and 0 is a zero of multiplicity 5, and sketch the graph of f.

18 Show that 2 is a zero of multiplicity 3 of the polynomial $f(x) = x^5 - 4x^4 - 3x^3 + 34x^2 - 52x + 24$, and express $f(x)$ as a product of linear factors.

Exer. 19–20: Find the zeros of $f(x)$, and state the multiplicity of each zero.

19 $f(x) = (x^2 - 2x + 1)^2(x^2 + 2x - 3)$

20 $f(x) = x^6 + 2x^4 + x^2$

Exer. 21–22: (a) Use Descartes' rule of signs to determine the number of possible positive, negative, and nonreal complex solutions of the equation. **(b)** Find the smallest and largest integers that are upper and lower bounds, respectively, for the real solutions of the equation.

21 $2x^4 - 4x^3 + 2x^2 - 5x - 7 = 0$

22 $x^5 - 4x^3 + 6x^2 + x + 4 = 0$

23 Show that $7x^6 + 2x^4 + 3x^2 + 10$ has no real zero.

Exer. 24–26: Find all solutions of the equation.

24 $x^4 + 9x^3 + 31x^2 + 49x + 30 = 0$

25 $16x^3 - 20x^2 - 8x + 3 = 0$

26 $x^4 - 7x^2 + 6 = 0$

Exer. 27–36: Sketch the graph of f.

27 $f(x) = \dfrac{-2}{(x + 1)^2}$

28 $f(x) = \dfrac{1}{(x - 1)^3}$

29 $f(x) = \dfrac{3x^2}{16 - x^2}$

30 $f(x) = \dfrac{x}{(x + 5)(x^2 - 5x + 4)}$

31 $f(x) = \dfrac{-3x^2}{x^2 + 1}$

32 $f(x) = \dfrac{x^2 - 4}{x^2 + 1}$

33 $f(x) = \dfrac{3x^2 + x - 10}{x^2 + 2x}$

34 $f(x) = \dfrac{-2x^2 - 8x - 6}{x^2 - 6x + 8}$

35 $f(x) = \dfrac{x^2 + 2x - 8}{x + 3}$

36 $f(x) = \dfrac{x^4 - 16}{x^3}$

37 *Deflection of a beam*   A horizontal beam l feet long is supported at one end and unsupported at the other end (see the figure). If the beam is subjected to a uniform load and if y denotes the deflection of the beam at a position x feet from the supported end, then it can be shown that $y = cx^2(x^2 - 4lx + 6l^2)$, where c is a positive constant that depends on the weight of the load and the physical properties of the beam.

(a) If the beam is 10 feet long and the deflection at the unsupported end of the board is 2 feet, find c.

(b) Show that the deflection is 1 foot somewhere between $x = 6.1$ and $x = 6.2$.

EXERCISE 37

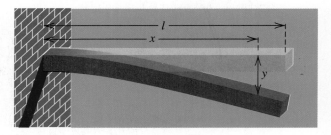

38 *Elastic cylinder*   A rectangle made of elastic material is to be made into a cylinder by joining edge AD to edge BC, as shown in the figure. A wire of fixed length l is placed along the diagonal of the rectangle to support the structure. Let x denote the height of the cylinder.

(a) Express the volume V of the cylinder in terms of x.

(b) For what positive values of x is $V > 0$?

EXERCISE 38

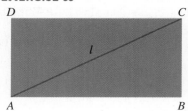 

39 *Determining temperatures*   A meteorologist determines that the temperature T (in °F) for a certain 24-hour period in winter was given by $T = \frac{1}{20}t(t - 12)(t - 24)$ for $0 \le t \le 24$, where t is time in hours and $t = 0$ corresponds to 6 A.M. At what time(s) was the temperature 32 °F?

40 *Deer propagation*   A herd of 100 deer is introduced onto a small island. Assuming the number $N(t)$ of deer after t years is given by $N(t) = -t^4 + 21t^2 + 100$ (for $t > 0$), determine when the herd size exceeds 180.

41 *Threshold response curve* In biochemistry, the general threshold-response curve is the graph of an equation

$$R = \frac{kS^n}{S^n + a^n},$$

where R is the chemical response when the level of the substance being acted on is S and a, k, and n are positive constants. An example is the removal rate R of alcohol from the bloodstream by the liver when the blood alcohol concentration is S.

(a) Find an equation of the horizontal asymptote for the graph.

(b) In the case of alcohol removal, $n = 1$ and a typical value of k is 0.22 gram per liter per minute. What is the interpretation of k in this setting?

42 *Oil spill clean-up* The cost $C(x)$ of cleaning up x percent of an oil spill that has washed ashore increases greatly as x approaches 100. Suppose that

$$C(x) = \frac{20x}{101 - x} \text{ (thousand dollars)}.$$

(a) Compare $C(100)$ to $C(90)$.

(b) Sketch the graph of C for $0 < x < 100$.

Exer. 43–58: Find the vertices and foci of the conic, and sketch its graph.

43 $y^2 = 64x$

44 $y = 8x^2 + 32x + 33$

45 $9y^2 = 144 - 16x^2$

46 $9y^2 = 144 + 16x^2$

47 $x^2 - y^2 - 4 = 0$

48 $25x^2 + 36y^2 = 1$

49 $25y = 100 - x^2$

50 $3x^2 + 4y^2 - 18x + 8y + 19 = 0$

51 $x^2 - 9y^2 + 8x + 90y - 210 = 0$

52 $x = 2y^2 + 8y + 3$

53 $4x^2 + 9y^2 + 24x - 36y + 36 = 0$

54 $4x^2 - y^2 - 40x - 8y + 88 = 0$

55 $y^2 - 8x + 8y + 32 = 0$

56 $4x^2 + y^2 - 24x + 4y + 36 = 0$

57 $x^2 - 9y^2 + 8x + 7 = 0$

58 $y^2 - 2x^2 + 6y + 8x - 3 = 0$

Exer. 59–60: Find the standard equation of a parabola that has a vertical axis and satisfies the given conditions.

59 x-intercepts -10 and -4, y-intercept 80

60 x-intercepts -11 and 3, passing through $(2, 39)$

Exer. 61–70: Find an equation for the conic that satisfies the given conditions.

61 Hyperbola, with vertices $V(0, \pm 7)$ and end points of conjugate axis $(\pm 3, 0)$

62 Parabola, with focus $F(-4, 0)$ and directrix $x = 4$

63 Parabola, with focus $F(0, -10)$ and directrix $y = 10$

64 Parabola, with vertex at the origin, symmetric to the x-axis, and passing through the point $(5, -1)$

65 Ellipse, with vertices $V(0, \pm 10)$ and foci $F(0, \pm 5)$

66 Hyperbola, with foci $F(\pm 10, 0)$ and vertices $V(\pm 5, 0)$

67 Hyperbola, with vertices $V(0, \pm 6)$ and asymptotes $y = \pm 9x$

68 Ellipse, with foci $F(\pm 2, 0)$ and passing through the point $(2, \sqrt{2})$

69 Ellipse, with eccentricity $\frac{2}{3}$ and end points of minor axis $(\pm 5, 0)$

70 Ellipse, with eccentricity $\frac{3}{4}$ and foci $F(\pm 12, 0)$

71 (a) Determine A so that the point $(2, -3)$ is on the conic $Ax^2 + 2y^2 = 4$.

(b) Is the conic an ellipse or a hyperbola?

72 If a square with sides parallel to the coordinate axes is inscribed in the ellipse $(x^2/a^2) + (y^2/b^2) = 1$, express the area A of the square in terms of a and b.

73 Find the standard equation of the circle that has its center at the focus of the parabola $y = \frac{1}{8}x^2$ and passes through the origin.

74 A point $P(x, y)$ is the same distance from $(4, 0)$ as it is from the circle $x^2 + y^2 = 4$, as illustrated in the figure. Show that the collection of all such points forms a branch of a hyperbola, and sketch its graph.

EXERCISE 74

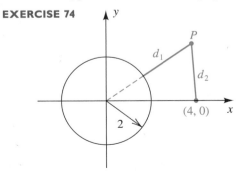

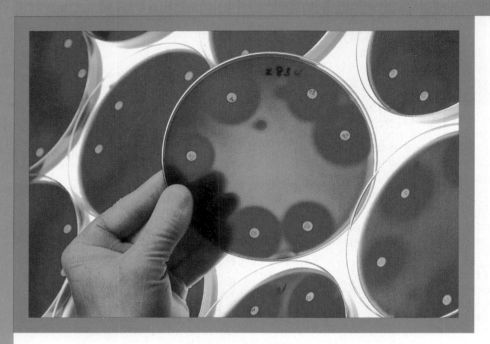

The number of bacteria in a culture often increases exponentially.

■ *Exponential and logarithmic functions are transcendental functions, since they cannot be defined in terms of only addition, subtraction, multiplication, division, and rational powers of a variable x, as is the case for the algebraic functions considered in previous chapters. Such functions are of major importance in mathematics and have applications in almost every field of human endeavor. They are especially useful in the fields of chemistry, biology, physics, and engineering, where they help describe the manner in which quantities in nature grow or decay. As we shall see in this chapter, there is a close relationship between specific exponential and logarithmic functions—they are inverse functions of each other.* ■

CHAPTER 5

EXPONENTIAL AND LOGARITHMIC FUNCTIONS

5.1   EXPONENTIAL FUNCTIONS

Let us consider the function f defined by

$$f(x) = 2^x,$$

where x is restricted to *rational* numbers. (Recall that if $x = m/n$ for integers m and n with $n > 0$, then $2^x = 2^{m/n} = (\sqrt[n]{2})^m$.) Coordinates of several points on the graph of $y = 2^x$ are listed in the following table.

FIGURE 1

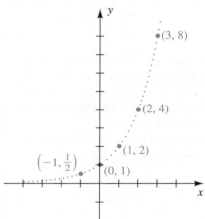

x	-10	-3	-2	-1	0	1	2	3	10
$y = 2^x$	$\frac{1}{1024}$	$\frac{1}{8}$	$\frac{1}{4}$	$\frac{1}{2}$	1	2	4	8	1024

Other values of y for x rational, such as $2^{1/3}$, $2^{-9/7}$, and $2^{5.143}$, can be approximated with a calculator. We can show algebraically that if x_1 and x_2 are rational numbers such that $x_1 < x_2$, then $2^{x_1} < 2^{x_2}$. Thus, f is an increasing function, and its graph rises. Plotting points leads to the sketch in Figure 1, where the small dots indicate that only the points with *rational* x-coordinates are on the graph. There is a *hole* in the graph whenever the x-coordinate of a point is irrational.

To extend the domain of f to all real numbers, it is necessary to define 2^x for every *irrational* exponent x. To illustrate, if we wish to define 2^π, we could use the nonterminating decimal representing 3.1415926 . . . for π and consider the following *rational* powers of 2:

$$2^3, \quad 2^{3.1}, \quad 2^{3.14}, \quad 2^{3.141}, \quad 2^{3.1415}, \quad 2^{3.14159}, \quad \ldots$$

FIGURE 2

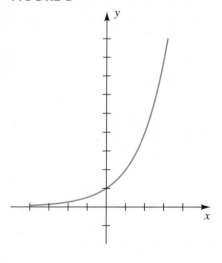

It can be shown, using calculus, that each successive power gets closer to a unique real number, denoted by 2^π. Thus,

$$2^x \to 2^\pi \quad \text{as} \quad x \to \pi, \quad \text{with } x \text{ rational.}$$

The same technique can be used for any other irrational power of 2. To sketch the graph of $y = 2^x$ with x *real*, we replace the holes in the graph in Figure 1 with points, and we obtain the graph in Figure 2. The function f defined by $f(x) = 2^x$ for every real number x is called the *exponential function with base* 2.

Let us next consider *any* base a, where a is a positive real number different from 1. As in the preceding discussion, to each real number x there corresponds exactly one positive number a^x such that the laws of exponents are true. Thus, as in the following chart, we may define a function f whose domain is $\mathbb{R}$ and range is the set of positive real numbers.

Terminology	Definition	Graph of f for $a > 1$	Graph of f for $0 < a < 1$
Exponential function f with base a	$f(x) = a^x$ for every x in $\mathbb{R}$, where $a > 0$ and $a \neq 1$		

The graphs in the chart show that if $a > 1$, then f is increasing on $\mathbb{R}$, and if $0 < a < 1$, then f is decreasing on $\mathbb{R}$. (These facts can be proved using calculus.) The graphs merely indicate the *general* appearance—the *exact* shape depends on the value of a. Note, however, that since $a^0 = 1$, the y-intercept is 1 for every a.

If $a > 1$, then as x *decreases* through negative values, the graph of f approaches the x-axis (see the third column in the chart). Thus, the x-axis is a *horizontal asymptote*. As x increases through positive values, the graph rises rapidly. This type of variation is characteristic of the **exponential law of growth**, and f is sometimes called a **growth function**.

If $0 < a < 1$, then as x *increases*, the graph of f approaches the x-axis asymptotically (see the last column of the chart). This type of variation is known as **exponential decay**.

When considering a^x we exclude the cases $a \leq 0$ and $a = 1$. Note that if $a < 0$, then a^x is not a real number for many values of x, such as $\frac{1}{2}, \frac{3}{4}$, and $\frac{11}{6}$. If $a = 0$, then $a^0 = 0^0$ is undefined. Finally, if $a = 1$, then $a^x = 1$ for every x, and the graph of $y = a^x$ is a horizontal line.

The graph of an exponential function f is either increasing throughout its domain or decreasing throughout its domain. Thus, f is one-to-one by the theorem on page 205. Combining this result with the definition of a one-to-one function (see page 204) gives us parts (1) and (2) of the following theorem.

Theorem: Exponential Functions Are One-to-One

The exponential function f given by

$$f(x) = a^x \quad \text{for } 0 < a < 1 \text{ or } a > 1$$

is one-to-one. Thus, the following equivalent conditions are satisfied for real numbers x_1 and x_2:

(1) If $x_1 \neq x_2$, then $a^{x_1} \neq a^{x_2}$.

(2) If $a^{x_1} = a^{x_2}$, then $x_1 = x_2$.

When using this theorem as a reason for a step in the solution to an example, we will state that *exponential functions are one-to-one*.

ILLUSTRATION

Exponential Functions Are One-to-One

■ If $7^{3x} = 7^{2x+5}$, then $3x = 2x + 5$, or $x = 5$.

In the following example we solve a simple *exponential equation*—that is, an equation in which the variable appears in an exponent.

EXAMPLE I   *Solving an exponential equation*

Solve the equation $3^{5x-8} = 9^{x+2}$.

SOLUTION

$3^{5x-8} = 9^{x+2}$	given
$3^{5x-8} = (3^2)^{x+2}$	express both sides with the same base
$3^{5x-8} = 3^{2x+4}$	law of exponents
$5x - 8 = 2x + 4$	exponential functions are one-to-one
$3x = 12$	subtract $2x$ and add 8
$x = 4$	divide by 3

In the next two examples we sketch the graphs of several different exponential functions.

FIGURE 3

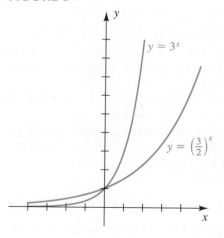

EXAMPLE 2   *Sketching graphs of exponential functions*

If $f(x) = \left(\frac{3}{2}\right)^x$ and $g(x) = 3^x$, sketch the graphs of f and g on the same coordinate plane.

SOLUTION   Since $\frac{3}{2} > 1$ and $3 > 1$, each graph *rises* as x increases. The following table displays coordinates for several points on the graphs.

x	-2	-1	0	1	2	3	4
$y = \left(\frac{3}{2}\right)^x$	$\frac{4}{9} \approx 0.4$	$\frac{2}{3} \approx 0.7$	1	$\frac{3}{2}$	$\frac{9}{4} \approx 2.3$	$\frac{27}{8} \approx 3.4$	$\frac{81}{16} \approx 5.1$
$y = 3^x$	$\frac{1}{9} \approx 0.1$	$\frac{1}{3} \approx 0.3$	1	3	9	27	81

Plotting points and being familiar with the general graph of $y = a^x$ leads to the graphs in Figure 3.

Example 2 illustrates the fact that if $1 < a < b$, then $a^x < b^x$ for positive values of x and $b^x < a^x$ for negative values of x. In particular, since $\frac{3}{2} < 2 < 3$, the graph of $y = 2^x$ in Figure 1 lies between the graphs of f and g in Figure 3.

EXAMPLE 3   *Sketching the graph of an exponential function*

Sketch the graph of the equation $y = \left(\frac{1}{2}\right)^x$.

SOLUTION   Since $0 < \frac{1}{2} < 1$, the graph *falls* as x increases. Coordinates of some points on the graph are listed in the following table.

x	-3	-2	-1	0	1	2	3
$y = \left(\frac{1}{2}\right)^x$	8	4	2	1	$\frac{1}{2}$	$\frac{1}{4}$	$\frac{1}{8}$

The graph is sketched in Figure 4. Since $\left(\frac{1}{2}\right)^x = 2^{-x}$, the graph is the same as the graph of the equation $y = 2^{-x}$. Note that the graph is a reflection through the y-axis of the graph of $y = 2^x$ in Figure 2.

Equations of the form $y = a^u$, where u is some expression in x, occur in applications. The next two examples illustrate equations of this form.

EXAMPLE 4   *Shifting graphs of exponential functions*

Sketch the graph of the equation:

(a) $y = 3^{x-2}$ **(b)** $y = 3^x - 2$

SOLUTION

(a) The graph of $y = 3^x$ was sketched in Figure 3 and is resketched in Figure 5. From the discussion of horizontal shifts in Section 3.5 we can obtain the graph of $y = 3^{x-2}$ by shifting the graph of $y = 3^x$ two units to the right, as shown in Figure 5.

The graph of $y = 3^{x-2}$ can also be obtained by plotting several points and using them as a guide to sketch an exponential-type curve.

(b) From the discussion of vertical shifts in Section 3.5 we can obtain the graph of $y = 3^x - 2$ by shifting the graph of $y = 3^x$ two units downward, as shown in Figure 6. Note that the y-intercept is -1 and the line $y = -2$ is a horizontal asymptote for the graph.

The bell-shaped graph of the function in the next example is similar to a *normal probability curve* used in statistical studies.

FIGURE 4

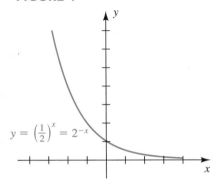

$y = \left(\frac{1}{2}\right)^x = 2^{-x}$

FIGURE 5

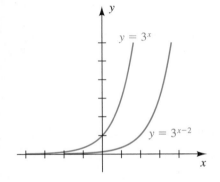

$y = 3^x$

$y = 3^{x-2}$

FIGURE 6

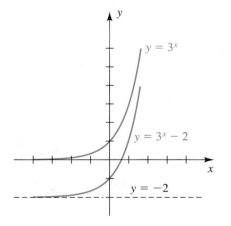

$y = 3^x$

$y = 3^x - 2$

$y = -2$

EXAMPLE 5   *Sketching a bell-shaped graph*

If $f(x) = 2^{-x^2}$, sketch the graph of f.

FIGURE 7

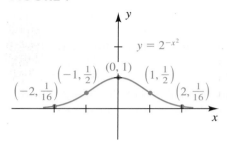

SOLUTION   If we rewrite $f(x)$ as

$$f(x) = \frac{1}{2^{(x^2)}}$$

we see that as x increases through positive values, $f(x)$ decreases rapidly, and hence the graph approaches the x-axis asymptotically. The maximum value of f is $f(0) = 1$. Since f is an even function, the graph is symmetric with respect to the y-axis. Some points on the graph are $(0, 1)$, $(1, \frac{1}{2})$, and $(2, \frac{1}{16})$. Plotting and using symmetry gives us the sketch in Figure 7.

APPLICATION

Bacterial Growth

Exponential functions may be used to describe the growth of certain populations. As an illustration, suppose it is observed experimentally that the number of bacteria in a culture doubles every day. If 1000 bacteria are present at the start, then we obtain the following table, where t is the time in days and $f(t)$ is the bacteria count at time t.

FIGURE 8

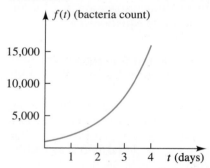

t (time in days)	0	1	2	3	4
$f(t)$ (bacteria count)	1000	2000	4000	8000	16,000

It appears that $f(t) = (1000)2^t$. With this formula we can predict the number of bacteria present at any time t. For example, at $t = 1.5 = \frac{3}{2}$,

$$f(t) = (1000)2^{3/2} \approx 2828.$$

The graph of f is sketched in Figure 8.

APPLICATION

Radioactive Decay

Certain physical quantities *decrease* exponentially. In such cases, if a is the base of the exponential function, then $0 < a < 1$. One of the most common examples of exponential decrease is the decay of a radioactive substance, or isotope. The **half-life** of an isotope is the time it takes for one-half the original amount in a given sample to decay. The half-life is the principal characteristic used to distinguish one radioactive substance from another. The polonium isotope ^{210}Po has a half-life of approximately 140 days; that is, given any amount, one-half of it will disintegrate in 140 days. If 20 milligrams of ^{210}Po is present initially, then the following table indicates the amount remaining after various intervals of time.

FIGURE 9

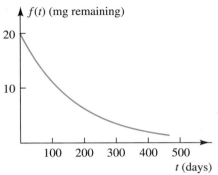

t (time in days)	0	140	280	420	560
$f(t)$ (mg remaining)	20	10	5	2.5	1.25

The sketch in Figure 9 illustrates the exponential nature of the disintegration.

Other radioactive substances have much longer half-lives. In particular, a by-product of nuclear reactors is the radioactive plutonium isotope ^{239}Pu, which has a half-life of approximately 24,000 years. It is for this reason that the disposal of radioactive waste is a major problem in modern society.

APPLICATION

Compound Interest

Compound interest provides a good illustration of exponential growth. If a sum of money P, the *principal*, is invested at a *simple* interest rate r, then the interest at the end of one interest period is the product Pr when r is expressed as a decimal (see page 64). For example, if $P = \$1000$ and the interest rate is 9% per year, then $r = 0.09$, and the interest at the end of one year is $\$1000(0.09)$, or $\$90$.

If the interest is reinvested with the principal at the end of the interest period, then the new principal is

$$P + Pr, \quad \text{or} \quad P(1 + r).$$

Note that to find the new principal we may multiply the original principal by $(1 + r)$. In the preceding example, the new principal is $\$1000(1.09)$, or $\$1090$.

After another interest period has elapsed, the new principal may be found by multiplying $P(1 + r)$ by $(1 + r)$. Thus, the principal after two interest periods is $P(1 + r)^2$. If we continue to reinvest, the principal after three periods is $P(1 + r)^3$; after four it is $P(1 + r)^4$; and, in general, the amount A accumulated after k interest periods is

$$A = P(1 + r)^k.$$

Interest accumulated by means of this formula is **compound interest**. Note that A is expressed in terms of an exponential function with base $1 + r$. The interest period may be measured in years, months, weeks, days, or any other suitable unit of time. When applying the formula for A, remember that r *is the interest rate per interest period expressed as a decimal*. For example, if the rate is stated as 6% *per year compounded monthly*, then the rate per month is $\frac{6}{12}\%$ or, equivalently, 0.5%. Thus, $r = 0.005$ and k is the number of months. If $\$100$ is invested at this rate, then the formula for A is

$$A = 100(1 + 0.005)^k = 100(1.005)^k.$$

In general, we have the following formula.

Compound Interest Formula

$$A = P\left(1 + \frac{r}{n}\right)^{nt},$$

where P = principal

   r = interest rate expressed as a decimal

   n = number of interest periods per year

   t = number of years P is invested

   A = amount after t years.

The next example illustrates a special case of the compound interest formula.

EXAMPLE 6   *Using the compound interest formula*

Suppose that $1000 is invested at an interest rate of 9% compounded monthly. Find the new amount of principal after 5 years, after 10 years, and after 15 years. Illustrate graphically the growth of the investment.

SOLUTION   Applying the compound interest formula with $r = 0.09$, $n = 12$, and $P = \$1000$, we find that the amount after t years is

$$A = 1000\left(1 + \frac{0.09}{12}\right)^{12t} = 1000(1.0075)^{12t}.$$

Substituting $t = 5$, 10, and 15 and using a calculator, we obtain the following table.

FIGURE 10
Compound interest: $A = 1000(1.0075)^{12t}$

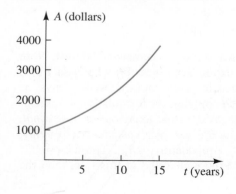

Number of years	Amount
5	$A = \$1000(1.0075)^{60} = \$1565.68$
10	$A = \$1000(1.0075)^{120} = \$2451.36$
15	$A = \$1000(1.0075)^{180} = \$3838.04$

The exponential nature of the increase is indicated by the fact that during the first five years, the growth in the investment is $565.68; during the second five-year period, the growth is $885.68; and during the last five-year period, it is $1368.68.

The sketch in Figure 10 illustrates the growth of $1000 invested over a period of 15 years.

We conclude this section with an example involving a graphing utility.

EXAMPLE 7   *Estimating amounts of a drug in the bloodstream*

If an adult takes a 100-milligram tablet of a certain prescription drug orally, the rate R at which the drug enters the bloodstream t minutes later is predicted to be

$$R = 5(0.95)^t \text{ mg/min.}$$

It can be shown using calculus that the amount A of the drug in the bloodstream at time t can be approximated by

$$A = 97.4786[1 - (0.95)^t] \text{ mg.}$$

(a) Estimate how long it takes for 50 milligrams of the drug to enter the bloodstream.

(b) Estimate the number of milligrams of the drug in the bloodstream when the drug is entering at a rate of 3 mg/min.

FIGURE 11 [0, 100] by [0, 100]

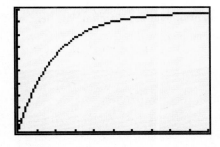

FIGURE 12 [0, 15] by [0, 10]

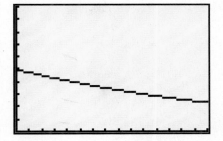

SOLUTION

(a) We wish to determine t when A is equal to 50. Since the value of A cannot exceed 97.4786, we choose the viewing rectangle to be [0, 100] by [0, 100].

We next assign $97.4786[1 - (0.95)^x]$ to Y_1 and graph Y_1, obtaining a display similar to that in Figure 11 (note that $x = t$). Using the tracing and zoom-in features, we estimate that $A = 50$ mg when $x \approx 14$ min.

(b) We wish to determine t when R is equal to 3. Let us first assign $5(0.95)^x$ to Y_2. Since the maximum value of Y_2 is 5 (at $t = 0$), we use a viewing rectangle of dimensions [0, 15] by [0, 10] and obtain a display similar to that in Figure 12. Tracing Y_2 until $y = 3$ gives us $x \approx 9.96$. Thus, after almost 10 minutes, the drug will be entering the bloodstream at a rate of 3 mg/min. (Note that the initial rate, at $t = 0$, is 5 mg/min.) Finding the value of Y_1 at $x = 10$, we see that there is almost 39 milligrams of the drug in the bloodstream after 10 minutes.

5.1 EXERCISES

Exer. 1–8: Solve the equation.

1 $7^{x+6} = 7^{3x-4}$

2 $6^{7-x} = 6^{2x+1}$

3 $3^{2x+3} = 3^{(x^2)}$

4 $9^{(x^2)} = 3^{3x+2}$

5 $2^{-100x} = (0.5)^{x-4}$

6 $(\frac{1}{2})^{6-x} = 2$

7 $4^{x-3} = 8^{4-x}$

8 $27^{x-1} = 9^{2x-3}$

9 Sketch the graph of f if $a = 2$.

   (a) $f(x) = a^x$

   (b) $f(x) = -a^x$

   (c) $f(x) = 3a^x$

   (d) $f(x) = a^{x+3}$

   (e) $f(x) = a^x + 3$

   (f) $f(x) = a^{x-3}$

   (g) $f(x) = a^x - 3$

   (h) $f(x) = a^{-x}$

   (i) $f(x) = \left(\frac{1}{a}\right)^x$

   (j) $f(x) = a^{3-x}$

10 Work Exercise 9 if $a = \frac{1}{2}$.

Exer. 11–20: Sketch the graph of f.

11 $f(x) = \left(\frac{2}{5}\right)^{-x}$

12 $f(x) = \left(\frac{2}{5}\right)^{x}$

13 $f(x) = -\left(\frac{1}{2}\right)^{x} + 4$

14 $f(x) = -3^{x} + 9$

15 $f(x) = 2^{|x|}$

16 $f(x) = 2^{-|x|}$

17 $f(x) = 3^{1-x^2}$

18 $f(x) = 2^{-(x+1)^2}$

19 $f(x) = 3^{x} + 3^{-x}$

20 $f(x) = 3^{x} - 3^{-x}$

21 *Elk population* One hundred elk, each 1 year old, are introduced into a game preserve. The number $N(t)$ alive after t years is predicted to be $N(t) = 100(0.9)^t$. Estimate the number alive after

(a) 1 year   (b) 5 years   (c) 10 years

22 *Drug dosage* A drug is eliminated from the body through urine. Suppose that for an initial dose of 10 milligrams, the amount $A(t)$ in the body t hours later is given by $A(t) = 10(0.8)^t$.

(a) Estimate the amount of the drug in the body 8 hours after the initial dose.

(b) What percentage of the drug still in the body is eliminated each hour?

23 *Bacterial growth* The number of bacteria in a certain culture increased from 600 to 1800 between 7:00 A.M. and 9:00 A.M. Assuming growth is exponential, the number $f(t)$ of bacteria t hours after 7:00 A.M. is given by $f(t) = 600(3)^{t/2}$.

(a) Estimate the number of bacteria in the culture at 8:00 A.M., 10:00 A.M., and 11:00 A.M.

(b) Sketch the graph of f for $0 \le t \le 4$.

24 *Newton's law of cooling* According to Newton's law of cooling, the rate at which an object cools is directly proportional to the difference in temperature between the object and the surrounding medium. The face of a household iron cools from $125°$ to $100°$ in 30 minutes in a room that remains at a constant temperature of $75°$. From calculus, the temperature $f(t)$ of the face after t hours of cooling is given by $f(t) = 50(2)^{-2t} + 75$.

(a) Assuming $t = 0$ corresponds to 1:00 P.M., approximate to the nearest tenth of a degree the temperature at 2:00 P.M., 3:30 P.M., and 4:00 P.M.

(b) Sketch the graph of f for $0 \le t \le 4$.

25 *Radioactive decay* The radioactive bismuth isotope ^{210}Bi has a half-life of 5 days. If there is 100 milligrams of ^{210}Bi present at $t = 0$, then the amount $f(t)$ remaining after t days is given by $f(t) = 100(2)^{-t/5}$.

(a) How much ^{210}Bi remains after 5 days? 10 days? 12.5 days?

(b) Sketch the graph of f for $0 \le t \le 30$.

26 *Light penetration in an ocean* An important problem in oceanography is to determine the amount of light that can penetrate to various ocean depths. The Beer-Lambert law asserts that the exponential function given by $I(x) = I_0 c^x$ is a model for this phenomenon (see the figure). For a certain location, $I(x) = 10(0.4)^x$ is the amount of light (in calories/cm²/sec) reaching a depth of x meters.

(a) Find the amount of light at a depth of 2 meters.

(b) Sketch the graph of I for $0 \le x \le 5$.

EXERCISE 26

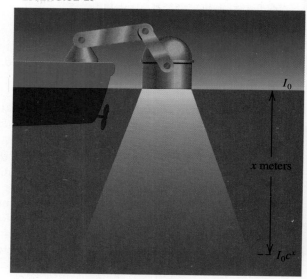

27 *Decay of radium* The half-life of radium is 1600 years. If the initial amount is q_0 milligrams, then the quantity $q(t)$ remaining after t years is given by $q(t) = q_0 2^{kt}$. Find k.

28 *Dissolving salt in water* If 10 grams of salt is added to a quantity of water, then the amount $q(t)$ that is undissolved after t minutes is given by $q(t) = 10\left(\frac{4}{5}\right)^t$. Sketch a graph that shows the value $q(t)$ at any time from $t = 0$ to $t = 10$.

29 *Compound interest* If $1000 is invested at a rate of 12% per year compounded monthly, find the principal after

(a) 1 month   (b) 2 months

(c) 6 months   (d) 1 year

30 *Compound interest* If a savings fund pays interest at a rate of 10% compounded semiannually, how much money invested now will amount to $5000 after 1 year?

31 *Automobile trade-in value* If a certain make of automobile is purchased for C dollars, its trade-in value $V(t)$ at the end of t years is given by $V(t) = 0.78C(0.85)^{t-1}$. If the original cost is $10,000, calculate, to the nearest dollar, the value after

(a) 1 year   (b) 4 years   (c) 7 years

32 *Real estate appreciation* If the value of real estate increases at a rate of 5% per year, after t years the value V of a house purchased for P dollars is $V = P(1.05)^t$. A graph for the value of a house purchased for $80,000 in 1986 is shown in the figure. Approximate the value of the house, to the nearest $1000, in the year 2000.

EXERCISE 32

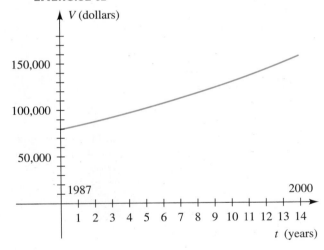

33 *Compound interest* If $1000 is invested at an interest rate of 6% per year compounded quarterly, find the principal at the end of

(a) 1 year   (b) 2 years   (c) 5 years   (d) 10 years

34 *Credit-card interest* A certain department store requires its credit-card customers to pay interest on unpaid bills at the rate of 18% per year compounded monthly. If a customer buys a television set for $500 on credit and makes no payments for one year, how much is owed at the end of the year?

35 *Depreciation* The declining balance method is an accounting method in which the amount of depreciation taken each year is a fixed percentage of the present value of the item. If y is the value of the item in a given year, the depreciation taken is ay for some depreciation rate a with $0 < a < 1$, and the new value is $(1 - a)y$.

(a) If the initial value of the item is y_0, show that the value after n years of depreciation is $(1 - a)^n y_0$.

(b) At the end of T years, the item has a salvage value of s dollars. The taxpayer wishes to choose a depreciation rate such that the value of the item after T years will equal the salvage value (see the figure). Show that $a = 1 - \sqrt[T]{s/y_0}$.

EXERCISE 35

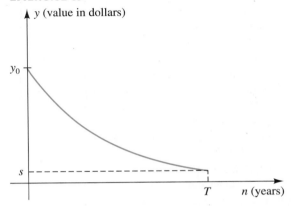

36 *Language dating* Glottochronology is a method of dating a language at a particular stage, based on the theory that over a long period of time linguistic changes take place at a fairly constant rate. Suppose that a language originally had N_0 basic words and that at time t, measured in millennia (1 millennium = 1000 years), the number $N(t)$ of basic words that remain in common use is given by $N(t) = N_0(0.805)^t$.

(a) Approximate the percentage of basic words lost every 100 years.

(b) If $N_0 = 200$, sketch the graph of N for $0 \le t \le 5$.

Exer. 37–40: Some lending institutions calculate the monthly payment M on a loan of L dollars at an interest rate r (expressed as a decimal) by using the formula

$$M = \frac{Lrk}{12(k - 1)},$$

where $k = [1 + (r/12)]^{12t}$ and t is the number of years that the loan is in effect.

37 *Home mortgage*

(a) Find the monthly payment on a 30-year $90,000 home mortgage if the interest rate is 12%.

(b) Find the total interest paid on the loan in part (a).

38 *Home mortgage*   Find the largest 25-year home mortgage that can be obtained at an interest rate of 10% if the monthly payment is to be $800.

39 *Car loan*   An automobile dealer offers customers no-down-payment 3-year loans at an interest rate of 15%. If a customer can afford to pay $220 per month, find the price of the most expensive car that can be purchased.

40 *Business loan*   The owner of a small business decides to finance a new computer by borrowing $3000 for 2 years at an interest rate of 12.5%.

(a) Find the monthly payment.

(b) Find the total interest paid on the loan.

c 41 *Trout population*   One thousand trout, each 1 year old, are introduced into a large pond. The number $N(t)$ still alive after t years is predicted to be given by $N(t) = 1000(0.9)^t$. Use the graph of N to approximate when 500 trout will be alive.

c 42 *Buying power*   An economist predicts that the buying power $B(t)$ of a dollar t years from now will be given by $B(t) = (0.95)^t$. Use the graph of B to approximate when the buying power will be half of what it is today.

c Exer. 43–44: Sketch the graph of the equation. (a) Estimate y if $x = 40$. (b) Estimate x if $y = 2$.

43 $y = (1.085)^x$ **44** $y = (1.0525)^x$

c Exer. 45–46: Use a graph to estimate the roots of the equation.

45 $1.4x^2 - 2.2^x = 1$

46 $1.21^{3x} + 1.4^{-1.1x} - 2x = 0.5$

c Exer. 47–48: Graph f on the given interval. (a) Determine whether f is one-to-one. (b) Estimate the zeros of f.

47 $f(x) = \dfrac{3.1^x - 2.5^{-x}}{2.7^x + 4.5^{-x}}$; $[-3, 3]$

48 $f(x) = \pi^{0.6x} - 1.3^{(x^{1.8})}$; $[-4, 4]$
(*Hint:* Change $x^{1.8}$ to an equivalent form that is defined for $x < 0$.)

c Exer. 49–50: Graph f on the given interval. (a) Estimate where f is increasing or is decreasing. (b) Estimate the range of f.

49 $f(x) = 0.7x^3 + 1.7^{(-1.8x)}$; $[-4, 1]$

50 $f(x) = \dfrac{3.1^{-x} - 4.1^x}{4.4^{-x} + 5.3^x}$; $[-3, 3]$

c 51 *Gompertz function*   The **Gompertz function**,

$$y = ka^{(b^x)} \quad \text{with } k > 0, 0 < a < 1, \text{ and } 0 < b < 1,$$

is sometimes used to describe the sales of a new product whose sales are initially large but then level off toward a maximum saturation level. Graph, on the same coordinate plane, the line $y = k$ and the Gompertz function with $k = 4$, $a = \frac{1}{8}$, and $b = \frac{1}{4}$. What is the significance of the constant k?

c 52 *Logistic function*   The **logistic function**,

$$y = \frac{1}{k + ab^x} \quad \text{with } k > 0, a > 0, \text{ and } 0 < b < 1,$$

is sometimes used to describe the sales of a new product that experiences slower sales initially, followed by growth toward a maximum saturation level. Graph, on the same coordinate plane, the line $y = 1/k$ and the logistic function with $k = \frac{1}{4}$, $a = \frac{1}{8}$, and $b = \frac{5}{8}$. What is the significance of the value $1/k$?

c Exer. 53–54: If monthly payments p are deposited in a savings account paying an annual interest rate r, then the amount A in the account after n years is given by

$$A = \frac{p\left(1 + \dfrac{r}{12}\right)\left[\left(1 + \dfrac{r}{12}\right)^{12n} - 1\right]}{\dfrac{r}{12}}.$$

Graph A for each value of p and r, and estimate n for $A = \$100{,}000$.

53 $p = 100$, $r = 0.05$ **54** $p = 250$, $r = 0.09$

5.2   THE NATURAL EXPONENTIAL FUNCTION

The *compound interest formula* discussed in the preceding section is

$$A = P\left(1 + \frac{r}{n}\right)^{nt},$$

where P is the principal invested, r is the interest rate (expressed as a decimal), n is the number of interest periods per year, and t is the number of years that the principal is invested. The next example illustrates what happens if the rate and total time invested are fixed, but the *interest period* is varied.

EXAMPLE 1   *Using the compound interest formula*

Suppose $1000 is invested at a compound interest rate of 9%. Find the new amount of principal after one year if the interest is compounded quarterly, monthly, weekly, daily, hourly, and each minute.

SOLUTION   If we let $P = \$1000$, $t = 1$, and $r = 0.09$ in the compound interest formula, then

$$A = \$1000\left(1 + \frac{0.09}{n}\right)^n$$

for n interest periods per year. The values of n we wish to consider are listed in the following table, where we have assumed that there are 365 days in a year and hence $(365)(24) = 8760$ hours and $(8760)(60) = 525{,}600$ minutes. (In actual business transactions an investment year is considered to be 360 days.)

Interest period	Quarter	Month	Week	Day	Hour	Minute
n	4	12	52	365	8760	525,600

Using the compound interest formula (and a calculator), we obtain the amounts given in the following table.

Interest period	Amount after one year
Quarter	$\$1000\left(1 + \dfrac{0.09}{4}\right)^4 = \$1093.08$
Month	$\$1000\left(1 + \dfrac{0.09}{12}\right)^{12} = \$1093.81$
Week	$\$1000\left(1 + \dfrac{0.09}{52}\right)^{52} = \$1094.09$
Day	$\$1000\left(1 + \dfrac{0.09}{365}\right)^{365} = \$1094.16$
Hour	$\$1000\left(1 + \dfrac{0.09}{8760}\right)^{8760} = \$1094.17$
Minute	$\$1000\left(1 + \dfrac{0.09}{525{,}600}\right)^{525{,}600} = \$1094.17$

Note that, in the preceding example, after we reach an interest period of one hour, the number of interest periods per year has no effect on the final amount. If interest had been compounded each *second*, the result would still be $1094.17, provided we truncate A to the nearest cent. (Some decimal places *beyond* the first two *do* change.) Thus, the amount approaches a fixed value as n increases. Interest is said to be **compounded continuously** if the number n of time periods per year increases without bound.

If we let $P = 1$, $r = 1$, and $t = 1$ in the compound interest formula, we obtain

$$A = \left(1 + \frac{1}{n}\right)^n.$$

The expression on the right-hand side of the equation is important in calculus. In Example 1 we considered a similar situation: as n increased, A approached a limiting value. The same phenomenon occurs for this formula, as illustrated by the following table, which was obtained using a calculator.

n	Approximation to $\left(1 + \dfrac{1}{n}\right)^n$
1	2.00000000
10	2.59374246
100	2.70481383
1000	2.71692393
10,000	2.71814593
100,000	2.71826824
1,000,000	2.71828047
10,000,000	2.71828169
100,000,000	2.71828181
1,000,000,000	2.71828183

In calculus it is shown that as n increases without bound, the value of $[1 + (1/n)]^n$ approaches a certain irrational number, denoted by e. The number e arises in the investigation of many physical phenomena. An approximation is $e \approx 2.71828$. Using the notation we developed for rational functions in Section 4.5, we denote this fact as follows.

The Number e

If n is a positive integer, then

$$\left(1 + \frac{1}{n}\right)^n \to e \approx 2.71828 \quad \text{as } n \to \infty.$$

In the following definition we use e as a base for an important exponential function.

Definition of the Natural Exponential Function

The **natural exponential function** f is defined by

$$f(x) = e^x$$

for every real number x.

The natural exponential function is one of the most useful functions in advanced mathematics and applications. Since $2 < e < 3$, the graph of $y = e^x$ lies between the graphs of $y = 2^x$ and $y = 3^x$, as shown in Figure 13. Scientific calculators have an $\boxed{e^x}$ key for approximating values of the natural exponential function.

APPLICATION

Continuously Compounded Interest

FIGURE 13

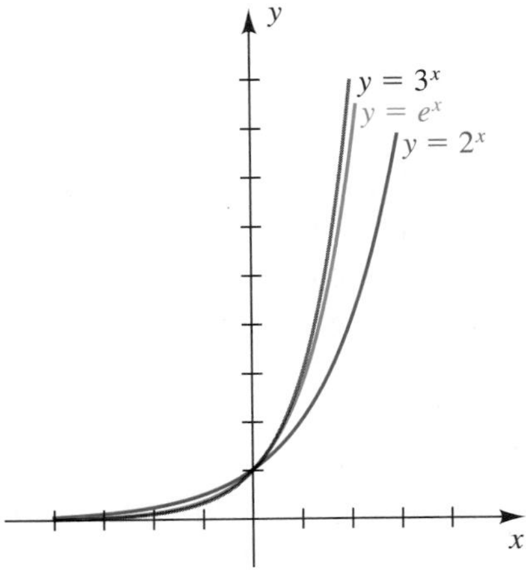

$y = 3^x$
$y = e^x$
$y = 2^x$

The compound interest formula is

$$A = P\left(1 + \frac{r}{n}\right)^{nt}.$$

If we let $k = n/r$, then $n = kr$ and $nt = krt$, and we may rewrite the formula as

$$A = P\left(1 + \frac{1}{k}\right)^{krt} = P\left[\left(1 + \frac{1}{k}\right)^k\right]^{rt}.$$

For continuously compounded interest we let n (the number of interest periods per year) increase without bound, denoted by $n \to \infty$ or, equivalently, by $k \to \infty$. Using the definition of e, we see that

$$P\left[\left(1 + \frac{1}{k}\right)^k\right]^{rt} \to P[e]^{rt} = Pe^{rt} \quad \text{as } k \to \infty.$$

This result gives us the following formula.

Continuously Compounded Interest Formula

$$A = Pe^{rt},$$

where P = principal
r = interest rate expressed as a decimal
t = number of years P is invested
A = amount after t years.

The next example illustrates a special case of this formula.

EXAMPLE 2   *Using the continuously compounded interest formula*

Suppose $20,000 is deposited in a money market account that pays interest at a rate of 8% per year compounded continuously. Determine the balance in the account after 5 years.

SOLUTION   Applying the formula for continuously compounded interest with $P = 20{,}000$, $r = 0.08$, and $t = 5$, we have

$$A = Pe^{rt} = 20{,}000e^{0.08(5)} = 20{,}000e^{0.4}.$$

Using a calculator, we find that $A = \$29{,}836.49$.

The function f in the next example is important in advanced applications of mathematics.

EXAMPLE 3   *A graph involving two exponential functions*

Sketch the graph of f if

$$f(x) = \frac{e^x + e^{-x}}{2}.$$

SOLUTION   Note that f is an even function, because

$$f(-x) = \frac{e^{-x} + e^{-(-x)}}{2} = \frac{e^{-x} + e^x}{2} = f(x).$$

Thus, the graph is symmetric with respect to the y-axis. Using a calculator or Table 2 of Appendix II, we obtain the following approximations of $f(x)$.

FIGURE 14

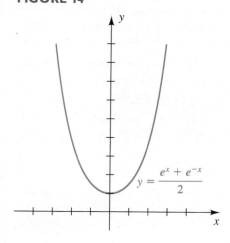

$$y = \frac{e^x + e^{-x}}{2}$$

x	0	0.5	1.0	1.5	2.0
$f(x)$ (approx.)	1	1.13	1.54	2.35	3.76

Plotting points and using symmetry with respect to the y-axis gives us the sketch in Figure 14. The graph *appears* to be a parabola; however, this is not actually the case.

APPLICATION

Flexible Cables

FIGURE 15

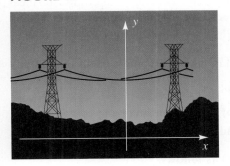

The function f of Example 3 occurs in applied mathematics and engineering, where it is called the **hyperbolic cosine function**. This function can be used to describe the shape of a uniform flexible cable or chain whose ends are supported from the same height, such as a telephone or power line (see Figure 15). If we introduce a coordinate system, as indicated in the figure, then it can be shown that an equation that corresponds to the shape of the cable is

$$y = \frac{a}{2}(e^{x/a} + e^{-x/a}),$$

where a is a real number. The graph is called a **catenary**, after the Latin word for *chain*. The function in Example 3 is the special case in which $a = 1$.

APPLICATION

Radiotherapy

Exponential functions play an important role in the field of *radiotherapy*, the treatment of tumors by radiation. The fraction of cells in a tumor that survive a treatment, called the *surviving fraction*, depends not only on the energy and nature of the radiation, but also on the depth, size, and characteristics of the tumor itself. The exposure to radiation may be thought of as a number of potentially damaging events, where at least one *hit* is required to kill a tumor cell. For instance, suppose that each cell has exactly one *target* that must be hit. If k denotes the average target size of a tumor cell and if x is the number of damaging events (the *dose*), then the surviving fraction $f(x)$ is given by

$$f(x) = e^{-kx}.$$

This is called the *one target–one hit surviving fraction*.

Suppose next that each cell has n targets and that each target must be hit once for the cell to die. In this case, the *n target–one hit surviving fraction* is given by

$$f(x) = 1 - (1 - e^{-kx})^n.$$

The graph of f may be analyzed to determine what effect increasing the dosage x will have on decreasing the surviving fraction of tumor cells. Note that $f(0) = 1$; that is, if there is no dose, then all cells survive. As an example, if $k = 1$ and $n = 2$, then

$$\begin{aligned}
f(x) &= 1 - (1 - e^{-x})^2 \\
&= 1 - (1 - 2e^{-x} + e^{-2x}) \\
&= 2e^{-x} - e^{-2x}.
\end{aligned}$$

FIGURE 16

Surviving fraction of tumor cells after a radiation treatment

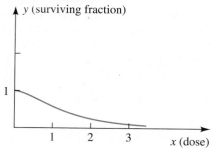

A complete analysis of the graph of f requires calculus. The graph is sketched in Figure 16. The *shoulder* on the curve near the point $(0, 1)$ represents the threshold nature of the treatment—that is, a small dose results in very little tumor cell elimination. Note that for a large x, an increase in dosage has little effect on the surviving fraction. To determine the ideal dose to administer to a patient, specialists in radiation therapy must also take into account the number of healthy cells that are killed during a treatment.

Problems of the type illustrated in the next example occur in the study of calculus.

EXAMPLE 4   *Finding zeros of a function involving exponentials*

If $f(x) = x^2(-2e^{-2x}) + 2xe^{-2x}$, find the zeros of f.

SOLUTION   We may factor $f(x)$ as follows:

$$f(x) = 2xe^{-2x} - 2x^2e^{-2x}$$
$$= 2xe^{-2x}(1 - x)$$

To find the zeros of f, we solve the equation $f(x) = 0$. Since $e^{-2x} > 0$ for every x, we see that $f(x) = 0$ if and only if $x = 0$ or $1 - x = 0$. Thus, the zeros of f are 0 and 1.

EXAMPLE 5   *Sketching a Gompertz growth curve*

In biology, the **Gompertz growth function** G given by

$$G(t) = ke^{(-Ae^{-Bt})},$$

where k, A, and B are positive constants, is used to estimate the size of certain quantities at time t. The graph of G is called a **Gompertz growth curve**. The function is always positive and increasing, and as t increases without bound, $G(t)$ levels off and approaches the value k. Graph G on the interval $[0, 5]$ for $k = 1.1$, $A = 3.2$, and $B = 1.1$, and estimate the time t at which $G(t) = 1$.

FIGURE 17   [0, 5] by [0, 2]

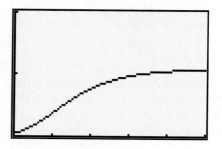

SOLUTION   We begin by assigning

$$1.1e^{(-3.2e^{-1.1t})}$$

to Y_1. Since we wish to graph G on the interval $[0, 5]$, we choose Xmin $= 0$ and Xmax $= 5$. Because $G(t)$ is always positive and does not exceed the value $k = 1.1$, we choose Ymin $= 0$ and Ymax $= 2$. Hence, the viewing rectangle dimensions are $[0, 5]$ by $[0, 2]$. Graphing G gives us a display similar to Figure 17. The end point values of the graph are approximately $(0, 0.045)$ and $(5, 1.086)$.

To determine the time when $y = G(t) = 1$, we use the tracing and zoom-in features and obtain $x = t \approx 3.194$.

5.2 EXERCISES

Exer. 1–4: Use the graph of $y = e^x$ to help sketch the graph of f.

1 (a) $f(x) = e^{-x}$ (b) $f(x) = -e^x$

2 (a) $f(x) = e^{2x}$ (b) $f(x) = 2e^x$

3 (a) $f(x) = e^{x+4}$ (b) $f(x) = e^x + 4$

4 (a) $f(x) = e^{-2x}$ (b) $f(x) = -2e^x$

Exer. 5–6: If P dollars is deposited in a savings account that pays interest at a rate of $r\%$ per year compounded continuously, find the balance after t years.

5 $P = 1000$, $r = 8\frac{1}{4}$, $t = 5$

6 $P = 100$, $r = 12\frac{1}{2}$, $t = 10$

Exer. 7–8: How much money, invested at an interest rate of $r\%$ per year compounded continuously, will amount to A dollars after t years?

7 $A = 100,000$, $r = 11$, $t = 18$

8 $A = 15,000$, $r = 9.5$, $t = 4$

Exer. 9–10: An investment of P dollars increased to A dollars in t years. If interest was compounded continuously, find the interest rate, using Table 2 of Appendix II.

9 $A = 13,464$, $P = 1000$, $t = 20$

10 $A = 890.20$, $P = 400$, $t = 16$

Exer. 11–12: Solve the equation.

11 $e^{(x^2)} = e^{7x-12}$ 12 $e^{3x} = e^{2x-1}$

Exer. 13–16: Find the zeros of f.

13 $f(x) = xe^x + e^x$

14 $f(x) = -x^2 e^{-x} + 2xe^{-x}$

15 $f(x) = x^3(4e^{4x}) + 3x^2 e^{4x}$

16 $f(x) = x^2(2e^{2x}) + 2xe^{2x} + e^{2x} + 2xe^{2x}$

Exer. 17–18: Simplify the expression.

17 $\dfrac{(e^x + e^{-x})(e^x + e^{-x}) - (e^x - e^{-x})(e^x - e^{-x})}{(e^x + e^{-x})^2}$

18 $\dfrac{(e^x - e^{-x})^2 - (e^x + e^{-x})^2}{(e^x + e^{-x})^2}$

19 *Crop growth* An exponential function W such that $W(t) = W_0 e^{kt}$ for $k > 0$ describes the first month of growth for crops such as maize, cotton, and soybeans. The function value $W(t)$ is the total weight in milligrams, W_0 is the weight on the day of emergence, and t is the time in days. If, for a species of soybean, $k = 0.2$ and $W_0 = 68$ mg, predict the weight at the end of 30 days.

20 *Crop growth* Refer to Exercise 19. It is often difficult to measure the weight W_0 of a plant when it first emerges from the soil. If, for a species of cotton, $k = 0.21$ and the weight after 10 days is 575 milligrams, estimate W_0.

21 *U.S. population growth* The 1980 population of the United States was approximately 227 million, and the population has been growing at a rate of 0.7% per year. The population $N(t)$, t years after 1980, may be approximated by $N(t) = 227e^{0.007t}$. Predict the population in the year 2000 if this growth trend continues.

22 *Population growth in India* The 1985 population estimate for India was 762 million, and the population has been growing at a rate of about 2.2% per year. The population $N(t)$, t years later, may be approximated by $N(t) = 762e^{0.022t}$. Assuming that this rapid growth rate continues, estimate the population of India in the year 2000.

23 *Longevity of halibut* In fishery science, a cohort is the collection of fish that results from one annual reproduction. It is usually assumed that the number of fish $N(t)$ still alive after t years is given by an exponential function. For Pacific halibut, $N(t) = N_0 e^{-0.2t}$, where N_0 is the initial size of the cohort. Approximate the percentage of the original number still alive after 10 years.

24 *Radioactive tracer* The radioactive tracer $^{51}\mathrm{Cr}$ can be used to locate the position of the placenta in a pregnant woman. Often the tracer must be ordered from a medical laboratory. If A_0 units (microcuries) are shipped, then because of radioactive decay, the number of units $A(t)$ present after t days is given by $A(t) = A_0 e^{-0.0249t}$

(a) If 35 units are shipped and it takes 2 days for the tracer to arrive, approximately how many units will be available for the test?

(b) If 35 units are needed for the test, approximately how many units should be shipped?

25 *Blue whale population growth* In 1978, the population of blue whales in the southern hemisphere was thought to number 5000. Since whaling has been outlawed and an abundant food supply is available, the population $N(t)$ is expected to grow exponentially according to the formula $N(t) = 5000e^{0.0036t}$, where t is in years and $t = 0$ corresponds to 1978. Predict the population in the year 2000.

26 *Halibut growth* The length (in centimeters) of many common commercial fish t years old can be approximated by a von Bertalanffy growth function of the form $f(t) = a(1 - be^{-kt})$, where a, b, and k are constants.

(a) For Pacific halibut, $a = 200$, $b = 0.956$, and $k = 0.18$. Estimate the length of a 10-year-old halibut.

(b) Use the graph of f to estimate the maximum attainable length of the Pacific halibut.

27 *Atmospheric pressure* Under certain conditions the atmospheric pressure p (in inches) at altitude h feet is given by $p = 29e^{-0.000034h}$. What is the pressure at an altitude of 40,000 feet?

28 *Polonium isotope decay* If we start with c milligrams of the polonium isotope ^{210}Po, the amount remaining after t days may be approximated by $A = ce^{-0.00495t}$. If the initial amount is 50 milligrams, approximate, to the nearest hundredth, the amount remaining after

(a) 30 days   (b) 180 days   (c) 365 days

29 *Growth of children* The Jenss model is generally regarded as the most accurate formula for predicting the height of preschool children. If y is height (in centimeters) and x is age (in years), then

$$y = 79.041 + 6.39x - e^{3.261 - 0.993x}$$

for $\frac{1}{4} \le x \le 6$. From calculus, the rate of growth R (in cm/year) is given by $R = 6.39 + 0.993e^{3.261 - 0.993x}$. Find the height and rate of growth of a typical 1-year-old child.

30 *Particle velocity* A very small spherical particle (on the order of 5 microns in diameter) is projected into still air with an initial velocity of v_0 m/sec, but its velocity decreases because of drag forces. Its velocity t seconds later is given by $v(t) = v_0 e^{-at}$ for some $a > 0$, and the distance $s(t)$ the particle travels is given by

$$s(t) = \frac{v_0}{a}(1 - e^{-at}).$$

The stopping distance is the distance traveled by the particle before it comes to rest.

(a) Express the stopping distance in terms of v_0 and a.

(b) Use the formula in part (a) to estimate the stopping distance if $v_0 = 10$ m/sec and $a = 8 \times 10^5$.

31 *Minimum wage* In 1971 the minimum wage in the United States was $1.60 per hour. Assuming that the rate of inflation increases continuously at 5% per year, find the equivalent minimum wage in the year 2000.

32 *Land value* In 1867 the United States purchased Alaska from Russia for $7,200,000. There is 586,400 square miles of land in Alaska. Assuming that the value of the land increases continuously at 3% per year and that land can be purchased at an equivalent price, determine the price of 1 acre in the year 2000. (One square mile is equivalent to 640 acres.)

Exer. 33–34: The *effective yield* (or effective annual interest rate) for an investment is the simple interest rate that would yield at the end of one year the same amount as is yielded by the compounded rate that is actually applied. Approximate, to the nearest 0.01%, the effective yield corresponding to an interest rate of r% per year compounded (a) quarterly and (b) continuously.

33 $r = 7$ **34** $r = 12$

Exer. 35–36: Sketch the graph of the equation. (a) Estimate y if $x = 40$. (b) Estimate x if $y = 2$.

35 $y = e^{0.085x}$ **36** $y = e^{0.0525x}$

Exer. 37–39: (a) Graph f using a graphing utility. (b) Sketch the graph of g by taking the reciprocals of y-coordinates in (a), *without* using a graphing utility.

37 $f(x) = \dfrac{e^x - e^{-x}}{2}$; $g(x) = \dfrac{2}{e^x - e^{-x}}$

38 $f(x) = \dfrac{e^x + e^{-x}}{2}$; $g(x) = \dfrac{2}{e^x + e^{-x}}$

39 $f(x) = \dfrac{e^x - e^{-x}}{e^x + e^{-x}}$; $g(x) = \dfrac{e^x + e^{-x}}{e^x - e^{-x}}$

40 *Probability density function* In statistics, the probability density function for the normal distribution is defined by

$$f(x) = \frac{1}{\sigma\sqrt{2\pi}} e^{-z^2/2} \quad \text{with} \quad z = \frac{x - \mu}{\sigma},$$

where μ and σ are real numbers (μ is the *mean* and σ^2 is the *variance* of the distribution). Sketch the graph of f for the case $\sigma = 1$ and $\mu = 0$.

Exer. 41–42: Graph f and g on the same coordinate plane, and estimate the solutions of the equation $f(x) = g(x)$.

41 $f(x) = e^{0.5x} - e^{-0.4x}$; $\quad g(x) = x^2 - 2$

42 $f(x) = 0.3e^x$; $\qquad\qquad g(x) = x^3 - x$

Exer. 43–44: The functions f and g can be used to approximate e^x on the interval $[0, 1]$. Graph f, g, and $y = e^x$ on the same coordinate plane, and compare the accuracy of $f(x)$ and $g(x)$ as an approximation to e^x.

43 $f(x) = x + 1$; $\qquad\qquad g(x) = 1.72x + 1$

44 $f(x) = \frac{1}{2}x^2 + x + 1$; $\quad g(x) = 0.84x^2 + 0.878x + 1$

Exer. 45–46: Graph f and estimate its zeros.

45 $f(x) = x^2 e^x - x e^{(x^2)} + 0.1$

46 $f(x) = x^3 e^x - x^2 e^{2x} + 1$

Exer. 47–48: Graph f on the interval $(0, 200]$. Find an approximate equation for the horizontal asymptote.

47 $f(x) = \left(1 + \dfrac{1}{x}\right)^x$ $\qquad$ 48 $f(x) = \left(1 + \dfrac{2}{x}\right)^x$

Exer. 49–50: Approximate the real root of the equation.

49 $e^{-x} = x$ $\qquad\qquad$ 50 $e^{3x} = 5 - 2x$

Exer. 51–52: Graph f and determine where f is increasing or is decreasing.

51 $f(x) = xe^x$ $\qquad\qquad$ 52 $f(x) = x^2 e^{-2x}$

53 *Pollution from a smokestack* The concentration C (in units/m^3) of pollution near a ground-level point that is downwind from a smokestack source of height h is sometimes given by

$$C = \frac{Q}{\pi vab} e^{-y^2/(2a^2)} \left[e^{-(z-h)^2/(2b^2)} + e^{-(z+h)^2/(2b^2)} \right],$$

where Q is the source strength (in units/sec), v is the average wind velocity (in m/sec), z is the height (in meters) above the downwind point, y is the distance from the

downwind point in the direction that is perpendicular to the wind (the cross-wind direction), and a and b are constants that depend on the downwind distance (see the figure).

(a) How does the concentration of pollution change at the ground-level, downwind position ($y = 0$ and $z = 0$) if the height of the smokestack is increased?

(b) How does the concentration of pollution change at ground level ($z = 0$) for a smokestack of fixed height h if a person moves in the cross-wind direction by increasing y?

EXERCISE 53

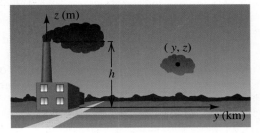

54 *Pollution concentration* Refer to Exercise 53. If the smokestack height is 100 meters and $b = 12$, use a graph to estimate the height z above the downwind point ($y = 0$) where the maximum pollution concentration occurs. (*Hint:* Let $h = 100$, $b = 12$, and graph the equation $C = e^{-(z-h)^2/(2b^2)} + e^{-(z+h)^2/(2b^2)}$.)

55 *Computer chips* For manufacturers of computer chips, it is important to consider the fraction F of chips that will fail after t years of service. This fraction can sometimes be approximated by the formula $F = 1 - e^{-ct}$, where c is a positive constant.

(a) How does the value of c affect the reliability of a chip?

(b) If $c = 0.125$, after how many years will 35% of the chips have failed?

5.3 LOGARITHMIC FUNCTIONS

In Section 5.1 we observed that the exponential function given by $f(x) = a^x$ for $0 < a < 1$ or $a > 1$ is one-to-one. Hence, f has an inverse function f^{-1} (see Section 3.8). This inverse of the exponential function with base a is called the **logarithmic function with base a** and is denoted by $\log_a$. Its values are written $\log_a (x)$ or $\log_a x$, read "*the logarithm of x with base a.*" Since, by the definition of an inverse function f^{-1},

$$y = f^{-1}(x) \quad \text{if and only if} \quad x = f(y),$$

the definition of $\log_a$ may be expressed as follows.

Definition of $\log_a$

Let a be a positive real number different from 1. The **logarithm of x with base a** is defined by

$$y = \log_a x \quad \text{if and only if} \quad x = a^y$$

for every $x > 0$ and every real number y.

Note that the two equations in the definition are equivalent. We call the first equation the **logarithmic form** and the second the **exponential form**. You should strive to become an expert in changing each form into the other. The following diagram may help you achieve this goal.

Logarithmic form   Exponential form

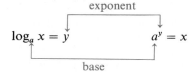

Observe that when forms are changed, *the bases of the logarithmic and exponential forms are the same.* The number y (that is, $\log_a x$) corresponds to the exponent in the exponential form. In words, $\log_a x$ *is the exponent to which the base a must be raised to obtain x.*

The following illustration contains examples of equivalent forms.

ILLUSTRATION

Equivalent Forms

Logarithmic form	Exponential form
$\log_5 u = 2$	$5^2 = u$
$\log_b 8 = 3$	$b^3 = 8$
$r = \log_p q$	$p^r = q$
$w = \log_4 (2t + 3)$	$4^w = 2t + 3$
$\log_3 x = 5 + 2z$	$3^{5+2z} = x$

The next example contains an application that involves changing from an exponential form to a logarithmic form.

EXAMPLE 1 *Changing exponential form to logarithmic form*

The number N of bacteria in a certain culture after t hours is given by $N = (1000)2^t$. Express t as a logarithmic function of N with base 2.

SOLUTION If $N = (1000)2^t$, then

$$2^t = \frac{N}{1000}.$$

Changing to logarithmic form, we obtain

$$t = \log_2 \frac{N}{1000}.$$

Some special cases of logarithms are given in the next example.

EXAMPLE 2 *Finding logarithms*

Find the number.

(a) $\log_{10} 100$ **(b)** $\log_2 \frac{1}{32}$ **(c)** $\log_9 3$ **(d)** $\log_7 1$

SOLUTION In each case we are given $\log_a x$ and must find the exponent y such that $a^y = x$. We obtain the following:

(a) $\log_{10} 100 = 2$ because $10^2 = 100$.

(b) $\log_2 \frac{1}{32} = -5$ because $2^{-5} = \frac{1}{32}$.

(c) $\log_9 3 = \frac{1}{2}$ because $9^{1/2} = 3$.

(d) $\log_7 1 = 0$ because $7^0 = 1$.

The following general properties follow from the interpretation of $\log_a x$ as an exponent.

Property of $\log_a x$	Reason	Illustration
(1) $\log_a 1 = 0$	$a^0 = 1$	$\log_3 1 = 0$
(2) $\log_a a = 1$	$a^1 = a$	$\log_{10} 10 = 1$
(3) $\log_a a^x = x$	$a^x = a^x$	$\log_2 8 = \log_2 2^3 = 3$
(4) $a^{\log_a x} = x$	see below	$5^{\log_5 7} = 7$

The reason for property (4) follows directly from the definition of $\log_a$, since

$$\text{if} \quad y = \log_a x, \quad \text{then} \quad x = a^y, \quad \text{or} \quad x = a^{\log_a x}.$$

The logarithmic function with base a is the inverse of the exponential function with base a, so the graph of $y = \log_a x$ can be obtained by reflecting the graph of $y = a^x$ through the line $y = x$ (see Section 3.8). This procedure is illustrated in Figure 18 for the case $a > 1$. Note that the x-intercept of the graph is 1, the domain is the set of positive real numbers, the range is $\mathbb{R}$, and the y-axis is a vertical asymptote. Logarithms with base $a < 1$ are seldom used, and hence we will not emphasize their graphs.

We see from Figure 18 that if $a > 1$, then $\log_a x$ is increasing on $(0, \infty)$ and hence is one-to-one by the theorem on page 205. Combining this result with parts (1) and (2) of the definition of one-to-one function on page 204 gives us the following theorem, which can also be proved if $0 < a < 1$.

Theorem: Logarithmic Functions Are One-to-One

FIGURE 18

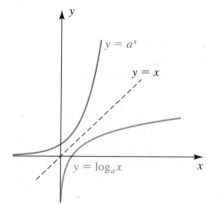

The logarithmic function with base a is one-to-one. Thus, the following equivalent conditions are satisfied for real numbers x_1 and x_2:

(1) If $x_1 \neq x_2$, then $\log_a x_1 \neq \log_a x_2$.

(2) If $\log_a x_1 = \log_a x_2$, then $x_1 = x_2$.

When using this theorem as a reason for a step in the solution to an example, we will state that *logarithmic functions are one-to-one*.

In the following example we solve a simple *logarithmic equation*—that is, an equation involving a logarithm of an expression that contains a variable.

EXAMPLE 3   *Solving a logarithmic equation*

Solve the equation $\log_3 (4x - 5) = \log_3 (2x + 1)$.

SOLUTION

$$\log_3 (4x - 5) = \log_3 (2x + 1) \qquad \text{given}$$
$$4x - 5 = 2x + 1 \qquad \text{logarithmic functions are one-to-one}$$
$$2x = 6 \qquad \text{subtract } 2x \text{ and add } 5$$
$$x = 3 \qquad \text{divide by } 2$$

✓ **Check $x = 3$**   We must check solutions of logarithmic equations to make sure that we are taking logarithms of *only positive real numbers*, because a logarithmic function is not defined for nonpositive real numbers.

$$\text{LS:} \quad \log_3 (4 \cdot 3 - 5) = \log_3 7$$
$$\text{RS:} \quad \log_3 (2 \cdot 3 + 1) = \log_3 7$$

Since $\log_3 7 = \log_3 7$ is a true statement, $x = 3$ is a solution.

In the next example we use the definition of logarithm to solve a logarithmic equation.

EXAMPLE 4   *Solving a logarithmic equation*

Solve the equation $\log_4 (5 + x) = 3$.

SOLUTION

$$\log_4 (5 + x) = 3 \qquad \text{given}$$
$$5 + x = 4^3 \qquad \text{definition of logarithm}$$
$$x = 59 \qquad \text{subtract 5}$$

✓ **Check $x = 59$**   LS: $\log_4 (5 + 59) = \log_4 64 = \log_4 4^3 = 3$

RS: 3

Since $3 = 3$ is a true statement, $x = 59$ is a solution.

We next sketch the graph of a specific logarithmic function.

EXAMPLE 5   *Sketching the graph of a logarithmic function*

Sketch the graph of f if $f(x) = \log_3 x$.

SOLUTION   We will describe three methods for sketching the graph.

Method 1   Since the functions given by $\log_3 x$ and 3^x are inverses of each other, we proceed as we did for $y = \log_a x$ in Figure 18; that is, we first sketch the graph of $y = 3^x$ and then reflect it through the line $y = x$. This gives us the sketch in Figure 19. Note that points $(-1, 3^{-1})$, $(0, 1)$, and $(1, 3)$ on the graph of $y = 3^x$ reflect into the points $(3^{-1}, -1)$, $(1, 0)$, and $(3, 1)$ on the graph of $y = \log_3 x$.

FIGURE 19

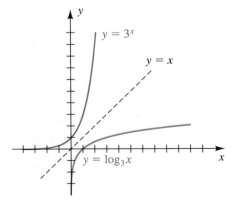

(continued)

Method 2 We can find points on the graph of $y = \log_3 x$ by letting $x = 3^k$, where k is a real number, and then applying property (3) of logarithms on page 327, as follows:

$$y = \log_3 x = \log_3 3^k = k$$

Using this formula, we obtain the points on the graph listed in the following table.

$x = 3^k$	3^{-3}	3^{-2}	3^{-1}	3^0	3^1	3^2	3^3
$y = \log_3 x = k$	-3	-2	-1	0	1	2	3

This gives us the same points obtained using the first method.

Method 3 We can sketch the graph of $y = \log_3 x$ by sketching the graph of the exponential form $x = 3^y$.

As in the following examples, we often wish to sketch the graph of $f(x) = \log_a u$, where u is some expression involving x.

EXAMPLE 6 *Sketching the graph of a logarithmic function*

Sketch the graph of f if $f(x) = \log_3 |x|$ for $x \neq 0$.

FIGURE 20

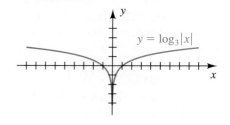

SOLUTION   The graph is symmetric with respect to the y-axis, since

$$f(-x) = \log_3 |-x| = \log_3 |x| = f(x).$$

If $x > 0$, then $|x| = x$ and the graph coincides with the graph of $y = \log_3 x$ sketched in Figure 19. Using symmetry, we reflect that part of the graph through the y-axis, obtaining the sketch in Figure 20.

EXAMPLE 7 *Reflecting the graph of a logarithmic function*

Sketch the graph of f if $f(x) = \log_3 (-x)$.

FIGURE 21

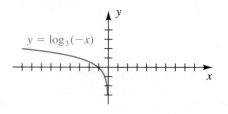

SOLUTION   The domain of f is the set of negative real numbers, since $\log_3 (-x)$ exists only if $-x > 0$ or, equivalently, $x < 0$. We can obtain the graph of f from the graph of $y = \log_3 x$ by replacing each point (x, y) in Figure 19 by $(-x, y)$. This is equivalent to reflecting the graph of $y = \log_3 x$ through the y-axis. The graph is sketched in Figure 21.

Another method is to change $y = \log_3 (-x)$ to the exponential form $3^y = -x$ and then sketch the graph of $x = -3^y$.

FIGURE 22

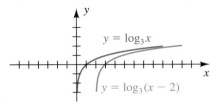

FIGURE 23

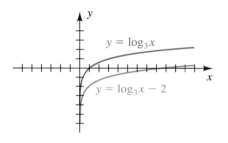

FIGURE 24

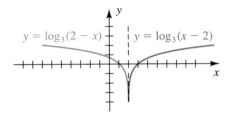

EXAMPLE 8 *Shifting graphs of logarithmic equations*

Sketch the graph of the equation:

(a) $y = \log_3 (x - 2)$ **(b)** $y = \log_3 x - 2$

SOLUTION

(a) The graph of $y = \log_3 x$ was sketched in Figure 19 and is resketched in Figure 22. From the discussion of horizontal shifts in Section 3.5, we can obtain the graph of $y = \log_3 (x - 2)$ by shifting the graph of $y = \log_3 x$ two units to the right, as shown in Figure 22.

(b) From the discussion of vertical shifts in Section 3.5, the graph of $y = \log_3 x - 2$ can be obtained by shifting the graph of $y = \log_3 x$ two units downward, as shown in Figure 23. Note that the x-intercept is given by $\log_3 x = 2$, or $x = 3^2 = 9$.

EXAMPLE 9 *Reflecting the graph of a logarithmic function*

Sketch the graph of f if $f(x) = \log_3 (2 - x)$.

SOLUTION   If we write

$$f(x) = \log_3 (2 - x) = \log_3 [-(x - 2)],$$

then, by applying the same technique used to obtain the graph of the equation $y = \log_3 (-x)$ in Example 7 (with x replaced by $x - 2$), we see that the graph of f is the reflection of the graph of $y = \log_3 (x - 2)$ through the vertical line $x = 2$. This gives us the sketch in Figure 24.

Another method is to change $y = \log_3 (2 - x)$ to the exponential form $3^y = 2 - x$ and then sketch the graph of $x = 2 - 3^y$.

Before electronic calculators were invented, logarithms with base 10 were used for complicated numerical computations involving products, quotients, and powers of real numbers. Base 10 was employed because it is well suited for numbers that are expressed in scientific form. Logarithms with base 10 are called **common logarithms**. The symbol **log x** is used as an abbreviation for $\log_{10} x$.

Definition of Common Logarithm

$$\log x = \log_{10} x \quad \text{for every } x > 0$$

Since inexpensive calculators are now available, there is no need for common logarithms as a tool for computational work. Base 10 does occur in applications, however, and hence many calculators have a ⌐LOG⌐ key, which can be used to approximate common logarithms.

The natural exponential function is given by $f(x) = e^x$ (see Section 5.2). The logarithmic function with base e is called the **natural logarithmic function**. The symbol **ln x** (read "*ell-en of x*") is an abbreviation for $\log_e x$, and we refer to it as the **natural logarithm of x**. Thus, *the natural logarithmic function and the natural exponential function are inverse functions of each other.*

Definition of Natural Logarithm

$$\ln x = \log_e x \quad \text{for every } x > 0$$

Many calculators have a key labeled $\boxed{\text{LN}}$, which can be used to approximate natural logarithms. The next illustration gives several examples of equivalent forms involving common and natural logarithms.

ILLUSTRATION

Equivalent Forms

Logarithmic form	Exponential form
$\log x = 2$	$10^2 = x$
$\log z = y + 3$	$10^{y+3} = z$
$\ln x = 2$	$e^2 = x$
$\ln z = y + 3$	$e^{y+3} = z$

To find x when given $\log x$ or $\ln x$, we may use the $\boxed{10^x}$ key or the $\boxed{e^x}$ key, respectively, on a calculator, as in the next example. If your calculator has an $\boxed{\text{INV}}$ key (for inverse), you may enter x and successively press $\boxed{\text{INV}}$ $\boxed{\text{LOG}}$ or $\boxed{\text{INV}}$ $\boxed{\text{LN}}$.

EXAMPLE 10   *Solving a logarithmic equation*

Find x if

(a) $\log x = 1.7959$ **(b)** $\ln x = 4.7$

SOLUTION

(a) Changing $\log x = 1.7959$ to its equivalent exponential form gives us

$$x = 10^{1.7959}.$$

Evaluating the last expression to three-decimal-place accuracy yields

$$x \approx 62.503.$$

(b) Changing $\ln x = 4.7$ to its equivalent exponential form gives us

$$x = e^{4.7} \approx 109.95.$$

The following chart lists common and natural logarithmic forms for some of the properties discussed earlier.

Logarithms with base a	Common logarithms	Natural logarithms
$\log_a 1 = 0$	$\log 1 = 0$	$\ln 1 = 0$
$\log_a a = 1$	$\log 10 = 1$	$\ln e = 1$
$\log_a a^x = x$	$\log 10^x = x$	$\ln e^x = x$
$a^{\log_a x} = x$	$10^{\log x} = x$	$e^{\ln x} = x$

The next three examples illustrate applications of common and natural logarithms.

EXAMPLE 11 *The Richter scale*

On the Richter scale, the magnitude R of an earthquake of intensity I is given by

$$R = \log \frac{I}{I_0},$$

where I_0 is a certain minimum intensity.

(a) If the intensity of an earthquake is $1000 I_0$, find R.

(b) Express I in terms of R and I_0.

SOLUTION

(a)

$$R = \log \frac{I}{I_0} \qquad \text{given}$$

$$= \log \frac{1000 I_0}{I_0} \qquad \text{let } I = 1000 I_0$$

$$= \log 1000 \qquad \text{cancel } I_0$$

$$= \log 10^3 \qquad 1000 = 10^3$$

$$= 3 \qquad \log 10^x = x \text{ for every } x$$

(b)

$$R = \log \frac{I}{I_0} \qquad \text{given}$$

$$\frac{I}{I_0} = 10^R \qquad \text{definition of } \log_{10}$$

$$I = I_0 \cdot 10^R \qquad \text{multiply by } I_0$$

EXAMPLE 12   *Newton's law of cooling*

Newton's law of cooling states that the rate at which an object cools is directly proportional to the difference in temperature between the object and its surrounding medium. Newton's law can be used to show that under certain conditions the temperature T (in °C) of an object at time t (in hours) is given by $T = 75e^{-2t}$. Express t as a function of T.

SOLUTION

$$T = 75e^{-2t} \qquad \text{given}$$

$$e^{-2t} = \frac{T}{75} \qquad \text{isolate the exponential expression}$$

$$-2t = \ln \frac{T}{75} \qquad \text{change to logarithmic form}$$

$$t = -\frac{1}{2} \ln \frac{T}{75} \qquad \text{divide by } -2$$

EXAMPLE 13   *The half-life of a radioactive substance*

A physicist finds that an unknown radioactive substance registers 2000 counts per minute on a Gieger counter. Ten days later the substance registers 1500 counts per minute. Using calculus, it can be shown that after t days the amount of radioactive material, and hence the number of counts per minute $N(t)$, is directly proportional to e^{ct} for some constant c. Determine the half-life of the substance.

SOLUTION   Since $N(t)$ is directly proportional to e^{ct},

$$N(t) = ke^{ct},$$

where k is a constant. Letting $t = 0$ and using $N(0) = 2000$, we obtain

$$2000 = ke^{c0} = k \cdot 1 = k.$$

Hence, the formula for $N(t)$ may be written

$$N(t) = 2000e^{ct}.$$

Since $N(10) = 1500$, we may determine c as follows:

$$1500 = 2000e^{c \cdot 10} \qquad \text{let } t = 10 \text{ in } N(t)$$

$$\tfrac{3}{4} = e^{10c} \qquad \text{divide by 2000 and simplify}$$

$$10c = \ln \tfrac{3}{4} \qquad \text{change to logarithmic form}$$

$$c = \tfrac{1}{10} \ln \tfrac{3}{4} \qquad \text{divide by 10}$$

Finally, since the half-life corresponds to the time t at which $N(t)$ is equal to 1000, we have the following:

$$1000 = 2000e^{ct} \qquad \text{let } N(t) = 1000$$

$$\tfrac{1}{2} = e^{ct} \qquad \text{divide by 2000}$$

$$ct = \ln \tfrac{1}{2} \qquad \text{change to logarithmic form}$$

$$t = \frac{1}{c}\ln\frac{1}{2} \qquad \text{divide by } c$$

$$= \frac{1}{\frac{1}{10}\ln\frac{3}{4}}\ln\frac{1}{2} \quad c = \frac{1}{10}\ln\frac{3}{4}$$

$$\approx 24 \text{ days} \qquad \text{approximate}$$

5.3 EXERCISES

Exer. 1–2: Change to logarithmic form.

1 (a) $4^3 = 64$ (b) $4^{-3} = \frac{1}{64}$ (c) $t^r = s$

(d) $3^x = 4 - t$ (e) $5^{7t} = \dfrac{a+b}{a}$ (f) $(0.7)^t = 5.3$

2 (a) $3^5 = 243$ (b) $3^{-4} = \frac{1}{81}$ (c) $c^p = d$

(d) $7^x = 100p$ (e) $3^{-2x} = \dfrac{P}{F}$ (f) $(0.9)^t = \frac{1}{2}$

Exer. 3–4: Change to exponential form.

3 (a) $\log_2 32 = 5$ (b) $\log_3 \frac{1}{243} = -5$

(c) $\log_t r = p$ (d) $\log_3 (x + 2) = 5$

(e) $\log_2 m = 3x + 4$ (f) $\log_b 512 = \frac{3}{2}$

4 (a) $\log_3 81 = 4$ (b) $\log_4 \frac{1}{256} = -4$

(c) $\log_v w = q$ (d) $\log_6 (2x - 1) = 3$

(e) $\log_4 p = 5 - x$ (f) $\log_a 343 = \frac{3}{4}$

Exer. 5–8: Solve for t using logarithms with base a.

5 $2a^{t/3} = 5$ **6** $3a^{4t} = 10$

7 $A = Ba^{Ct} + D$ **8** $L = Ma^{t/N} - P$

Exer. 9–10: Change to logarithmic form.

9 (a) $10^5 = 100,000$ (b) $10^{-3} = 0.001$

(c) $10^x = y + 1$ (d) $e^7 = p$

(e) $e^{2t} = 3 - x$

10 (a) $10^4 = 10,000$ (b) $10^{-2} = 0.01$

(c) $10^x = 38z$ (d) $e^4 = D$

(e) $e^{0.1t} = x + 2$

Exer. 11–12: Change to exponential form.

11 (a) $\log x = 50$ (b) $\log x = 20t$

(c) $\ln x = 0.1$ (d) $\ln w = 4 + 3x$

(e) $\ln (z - 2) = \frac{1}{6}$

12 (a) $\log x = -8$ (b) $\log x = y - 2$

(c) $\ln x = \frac{1}{2}$ (d) $\ln z = 7 + x$

(e) $\ln (t - 5) = 1.2$

Exer. 13–14: Find the number, if possible.

13 (a) $\log_5 1$ (b) $\log_3 3$ (c) $\log_4 (-2)$

(d) $\log_7 7^2$ (e) $3^{\log_3 8}$ (f) $\log_5 125$

(g) $\log_4 \frac{1}{16}$

14 (a) $\log_8 1$ (b) $\log_9 9$ (c) $\log_5 0$

(d) $\log_6 6^7$ (e) $5^{\log_5 4}$ (f) $\log_3 243$

(g) $\log_2 128$

Exer. 15–16: Find the number.

15 (a) $10^{\log 3}$ (b) $\log 10^5$ (c) $\log 100$

(d) $\log 0.0001$ (e) $e^{\ln 2}$ (f) $\ln e^{-3}$

(g) $e^{2 + \ln 3}$

16 (a) $10^{\log 7}$ (b) $\log 10^{-6}$ (c) $\log 100{,}000$

(d) $\log 0.001$ (e) $e^{\ln 8}$ (f) $\ln e^{2/3}$

(g) $e^{1+\ln 5}$

Exer. 17–30: Solve the equation.

17 $\log_4 x = \log_4 (8 - x)$

18 $\log_3 (x + 4) = \log_3 (1 - x)$

19 $\log_5 (x - 2) = \log_5 (3x + 7)$

20 $\log_7 (x - 5) = \log_7 (6x)$

21 $\log x^2 = \log (-3x - 2)$ 22 $\ln x^2 = \ln (12 - x)$

23 $\log_3 (x - 4) = 2$ 24 $\log_2 (x - 5) = 4$

25 $\log_9 x = \frac{3}{2}$ 26 $\log_4 x = -\frac{3}{2}$

27 $\ln x^2 = -2$ 28 $\log x^2 = -4$

29 $e^{2 \ln x} = 9$ 30 $e^{-\ln x} = 0.2$

31 Sketch the graph of f if $a = 4$:

(a) $f(x) = \log_a x$ (b) $f(x) = -\log_a x$

(c) $f(x) = 2 \log_a x$ (d) $f(x) = \log_a (x + 2)$

(e) $f(x) = (\log_a x) + 2$ (f) $f(x) = \log_a (x - 2)$

(g) $f(x) = (\log_a x) - 2$ (h) $f(x) = \log_a |x|$

(i) $f(x) = \log_a (-x)$ (j) $f(x) = \log_a (3 - x)$

(k) $f(x) = |\log_a x|$

32 Work Exercise 31 if $a = 5$.

Exer. 33–36: Sketch the graph of f.

33 $f(x) = \log x$ 34 $f(x) = \ln x$

35 $f(x) = \log_2 |x - 5|$ 36 $f(x) = \log_3 |x + 1|$

Exer. 37–44: Shown in the figure is the graph of a function f. Express $f(x)$ in terms of logarithms with base 2.

37

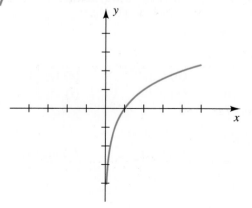

38

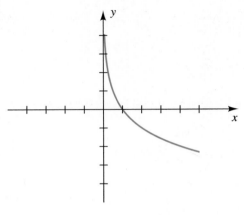

39

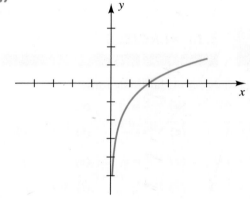

40

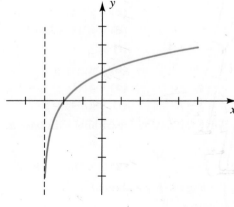

41

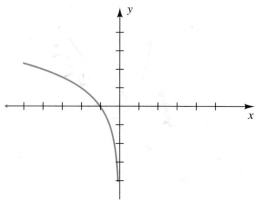

42

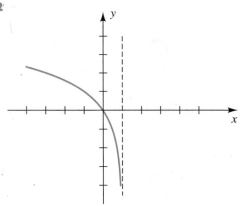

43

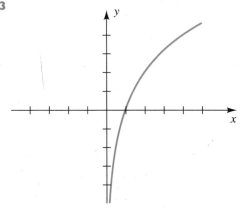

44

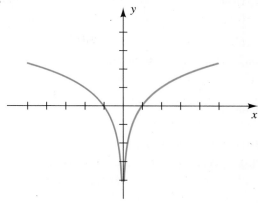

Exer. 45–46: Approximate x to three significant figures.

45 (a) $\log x = 3.6274$ (b) $\log x = 0.9469$
    (c) $\log x = -1.6253$ (d) $\ln x = 2.3$
    (e) $\ln x = 0.05$ (f) $\ln x = -1.6$

46 (a) $\log x = 1.8965$ (b) $\log x = 4.9680$
    (c) $\log x = -2.2118$ (d) $\ln x = 3.7$
    (e) $\ln x = 0.95$ (f) $\ln x = -5$

47 *Radium decay* If we start with q_0 milligrams of radium, the amount q remaining after t years is given by $q = q_0(2)^{-t/1600}$. Express t in terms of q and q_0.

48 *Bismuth isotope decay* The radioactive bismuth isotope ^{210}Bi disintegrates according to the formula $Q = k(2)^{-t/5}$, where k is a constant and t is the time in days. Express t in terms of Q and k.

49 *Electrical circuit* A schematic of a simple electrical circuit consisting of a resistor and an inductor is shown in the figure. The current I at time t is given by $I = 20e^{-Rt/L}$, where R is the resistance and L is the inductance. Solve this equation for t.

EXERCISE 49

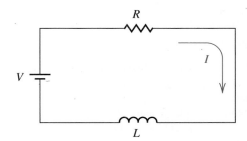

50 Electrical condenser An electrical condenser with initial charge Q_0 is allowed to discharge. After t seconds the charge Q is $Q = Q_0 e^{kt}$, where k is a constant. Solve this equation for t.

51 Richter scale Use the Richter scale formula $R = \log(I/I_0)$ to find the magnitude of an earthquake that has an intensity

(a) 100 times that of I_0

(b) 10,000 times that of I_0

(c) 100,000 times that of I_0

52 Richter scale Refer to Exercise 51. The largest recorded magnitudes of earthquakes have been between 8 and 9 on the Richter scale. Find the corresponding intensities in terms of I_0.

53 Sound intensity The loudness of a sound, as experienced by the human ear, is based on its intensity level. A formula used for finding the intensity level α (in decibels) that corresponds to a sound intensity I is $\alpha = 10 \log(I/I_0)$, where I_0 is a special value of I agreed to be the weakest sound that can be detected by the ear under certain conditions. Find α if

(a) I is 10 times as great as I_0

(b) I is 1000 times as great as I_0

(c) I is 10,000 times as great as I_0 (This is the intensity level of the average voice.)

54 Sound intensity Refer to Exercise 53. A sound intensity level of 140 decibels produces pain in the average human ear. Approximately how many times greater than I_0 must I be in order for α to reach this level?

55 U.S. population growth The population $N(t)$ (in millions) of the United States t years after 1980 may be approximated by the formula $N(t) = 227e^{0.007t}$. When will the population be twice what it was in 1980?

56 Population growth in India The population $N(t)$ (in millions) of India t years after 1985 may be approximated by the formula $N(t) = 762e^{0.022t}$. When will the population reach 1 billion?

57 Children's weight The Ehrenberg relation

$$\ln W = \ln 2.4 + (1.84)h$$

is an empirically based formula relating the height h (in meters) to the average weight W (in kilograms) for children 5 through 13 years old.

(a) Express W as a function of h that does not contain ln.

(b) Estimate the average weight of an 8-year-old child who is 1.5 meters tall.

58 Continuously compounded interest If interest is compounded continuously at the rate of 10% per year, approximate the number of years it will take an initial deposit of $6000 to grow to $25,000.

59 Air pressure The air pressure $p(h)$ (in lb/in.²) at an altitude of h feet above sea level may be approximated by the formula $p(h) = 14.7e^{-0.0000385h}$. At approximately what altitude h is the air pressure

(a) 10 lb/in.²?   **(b)** one-half its value at sea level?

60 Vapor pressure A liquid's vapor pressure P (in lb/in.²), a measure of its volatility, is related to its temperature T (in °F) by the Antoine equation

$$\log P = a + \frac{b}{c + T},$$

where a, b, and c are constants. Vapor pressure increases rapidly with an increase in temperature. Express P as a function of T.

61 Elephant growth The weight W (in kilograms) of a female African elephant at age t (in years) may be approximated by

$$W = 2600(1 - 0.51e^{-0.075t})^3.$$

(a) Approximate the weight at birth.

(b) Estimate the age of a female African elephant weighing 1800 kilograms by using (1) the accompanying graph and (2) the formula for W.

EXERCISE 61

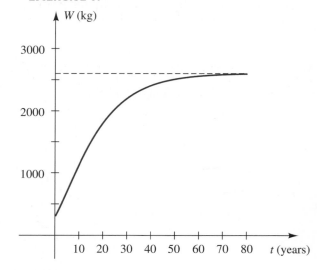

62 Coal consumption A country presently has coal reserves of 50 million tons. Last year 6.5 million tons of coal was consumed. Past years' data and population projections suggest that the rate of consumption R (in million tons/year) will increase according to the formula $R = 6.5e^{0.02t}$, and the total amount T (in million tons) of coal that will be used in t years is given by the formula $T = 325(e^{0.02t} - 1)$. If the country uses only its own resources, when will the coal reserves be depleted?

63 Urban population density An urban density model is a formula that relates the population density D (in thousands/mi^2) to the distance x (in miles) from the center of the city. The formula $D = ae^{-bx}$ for the central density a and coefficient of decay b has been found to be appropriate for many large U.S. cities. For the city of Atlanta in 1970, $a = 5.5$ and $b = 0.10$. At approximately what distance was the population density 2000 per square mile?

64 Brightness of stars Stars are classified into categories of brightness called magnitudes. The faintest stars, with light flux L_0, are assigned a magnitude 6. Brighter stars of light flux L are assigned a magnitude m by means of the formula

$$m = 6 - (2.5) \log \frac{L}{L_0}.$$

(a) Find m if $L = 10^{0.4}L_0$.

(b) Solve the formula for L in terms of m and L_0.

65 Radioactive iodine decay Radioactive iodine ^{131}I is frequently used in tracer studies involving the thyroid gland. The substance decays according to the formula $A(t) = A_0 a^{-t}$, where A_0 is the initial dose and t is the

time in days. Find a, assuming the half-life of ^{131}I is eight days.

66 Radioactive contamination Radioactive strontium ^{90}Sr has been deposited in a large field by acid rain. If sufficient amounts make their way through the food chain to humans, bone cancer can result. It has been determined that the radioactivity level in the field is 2.5 times the safe level S. ^{90}Sr decays according to the formula

$$A(t) = A_0 e^{-0.0239t},$$

where A_0 is the amount currently in the field and t is the time in years. For how many years will the field be contaminated?

67 Walking speed In a survey of 15 cities ranging in population P from 300 to 3,000,000, it was found that the average walking speed S (in ft/sec) of a pedestrian could be approximated by the equation $S = 0.05 + 0.86 \log P$.

(a) How does the population affect the average walking speed?

(b) For what population is the average walking speed 5 ft/sec?

Exer. 68–69: **(a) Show that f takes on both positive and negative values on the interval $[1, 2]$. (b) Use the method of Exercise 67 in Section 2.5, with $x_1 = 1.5$, to approximate a zero of f to two-decimal-place accuracy.**

68 $f(x) = x - 2 + \log x$

69 $f(x) = \log x - 10^{-x}$ (*Hint:* Solve for x in $\log x$.)

5.4 PROPERTIES OF LOGARITHMS

In the preceding section we observed that $\log_a x$ can be interpreted as an exponent. Thus, it seems reasonable to expect that the laws of exponents can be used to obtain corresponding laws of logarithms. This is demonstrated in the proofs of the following laws, which are fundamental for all work with logarithms.

Laws of Logarithms

If u and w denote positive real numbers, then

(1) $\log_a (uw) = \log_a u + \log_a w$

(2) $\log_a \dfrac{u}{w} = \log_a u - \log_a w$

(3) $\log_a (u^c) = c \log_a u$ for every real number c

PROOF   Throughout the proof, let
$$r = \log_a u \quad \text{and} \quad s = \log_a w.$$
Applying the definition of logarithm gives us
$$u = a^r \quad \text{and} \quad w = a^s.$$
We now proceed as follows:

(1)	$uw = a^r a^s$	$u = a^r$ and $w = a^s$
	$uw = a^{r+s}$	law of exponents
	$\log_a uw = r + s$	definition of $\log_a$
	$\log_a uw = \log_a u + \log_a w$	$r = \log_a u, s = \log_a w$
(2)	$\dfrac{u}{w} = \dfrac{a^r}{a^s}$	$u = a^r$ and $w = a^s$
	$\dfrac{u}{w} = a^{r-s}$	law of exponents
	$\log_a \dfrac{u}{w} = r - s$	definition of $\log_a$
	$\log_a \dfrac{u}{w} = \log_a u - \log_a w$	$r = \log_a u, s = \log_a w$
(3)	$u^c = (a^r)^c$	$u = a^r$
	$u^c = a^{cr}$	law of exponents
	$\log_a u^c = cr$	definition of $\log_a$
	$\log_a u^c = c \log_a u$	$r = \log_a u$ ∎

The laws of logarithms for the special cases $a = 10$ (common logs) and $a = e$ (natural logs) are written as in the following chart.

Common logarithms	Natural logarithms
$\log (uw) = \log u + \log w$	$\ln (uw) = \ln u + \ln w$
$\log \dfrac{u}{w} = \log u - \log w$	$\ln \dfrac{u}{w} = \ln u - \ln w$
$\log (u^c) = c \log u$	$\ln (u^c) = c \ln u$

As indicated by the following warnings, there are no general laws for expressing $\log_a (u + w)$ or $\log_a (u - w)$ in terms of simpler logarithms. The expressions on the right side equal $\log_a uw$ and $\log_a \dfrac{u}{w}$, respectively.

Warnings

$$\log_a (u + w) \neq \log_a u + \log_a w; \qquad \log_a (u - w) \neq \log_a u - \log_a w$$

The following examples illustrate uses of the laws of logarithms.

EXAMPLE 1   *Using laws of logarithms*

Express $\log_a \dfrac{x^3 \sqrt{y}}{z^2}$ in terms of logarithms of x, y, and z.

SOLUTION   We write $\sqrt{y}$ as $y^{1/2}$ and use laws of logarithms:

$$\log_a \frac{x^3 \sqrt{y}}{z^2} = \log_a (x^3 y^{1/2}) - \log_a z^2 \qquad \text{law (2)}$$

$$= \log_a x^3 + \log_a y^{1/2} - \log_a z^2 \qquad \text{law (1)}$$

$$= 3 \log_a x + \tfrac{1}{2} \log_a y - 2 \log_a z \qquad \text{law (3)}$$

EXAMPLE 2   *Using laws of logarithms*

Express as one logarithm:

$$\tfrac{1}{3} \log_a (x^2 - 1) - \log_a y - 4 \log_a z$$

SOLUTION   We apply the laws of logarithms as follows:

$$\tfrac{1}{3} \log_a (x^2 - 1) - \log_a y - 4 \log_a z$$

$$= \log_a (x^2 - 1)^{1/3} - \log_a y - \log_a z^4 \qquad \text{law (3)}$$

$$= \log_a \sqrt[3]{x^2 - 1} - (\log_a y + \log_a z^4) \qquad \text{algebra}$$

$$= \log_a \sqrt[3]{x^2 - 1} - \log_a (yz^4) \qquad \text{law (1)}$$

$$= \log_a \frac{\sqrt[3]{x^2 - 1}}{yz^4} \qquad \text{law (2)}$$

EXAMPLE 3   *Solving a logarithmic equation*

Solve the equation $\log_5 (2x + 3) = \log_5 11 + \log_5 3$.

SOLUTION

$$\log_5 (2x + 3) = \log_5 11 + \log_5 3 \qquad \text{given}$$

$$\log_5 (2x + 3) = \log_5 (11 \cdot 3) \qquad \text{law (1)}$$

$$\log_5 (2x + 3) = \log_5 33 \qquad \text{multiply}$$

$$2x + 3 = 33 \qquad \text{logarithmic functions are one-to-one}$$

$$2x = 30 \qquad \text{subtract 3}$$

$$x = 15 \qquad \text{divide by 2}$$

(continued)

✓ **Check** $x = 15$ LS: $\log_5 (2 \cdot 15 + 3) = \log_5 33$

RS: $\log_5 11 + \log_5 3 = \log_5 (11 \cdot 3) = \log_5 33$

Since $\log_5 33 = \log_5 33$ is a true statement, $x = 15$ is a solution.

The laws of logarithms were proved for logarithms of *positive* real numbers u and w. If we apply these laws to equations in which u and w are expressions involving a variable, then extraneous solutions may occur. Answers should therefore be substituted for the variable in u and w to determine whether these expressions are defined.

EXAMPLE 4   *Solving a logarithmic equation*

Solve the equation $\log_2 x + \log_2 (x + 2) = 3$.

SOLUTION

$$
\begin{array}{ll}
\log_2 x + \log_2 (x + 2) = 3 & \text{given} \\
\log_2 [x(x + 2)] = 3 & \text{law (1)} \\
x(x + 2) = 2^3 & \text{definition of logarithm} \\
x^2 + 2x - 8 = 0 & \text{simplify} \\
(x - 2)(x + 4) = 0 & \text{factor} \\
x - 2 = 0, \quad x + 4 = 0 & \text{set each factor equal to 0} \\
x = 2, \qquad x = -4 & \text{solve for } x
\end{array}
$$

✓ **Check** $x = 2$ LS: $\log_2 2 + \log_2 (2 + 2) = 1 + \log_2 4$

$= 1 + \log_2 2^2 = 1 + 2 = 3$

RS: 3

Since $3 = 3$ is a true statement, $x = 2$ is a solution.

✓ **Check** $x = -4$ LS: $\log_2 (-4) + \log_2 (-4 + 2)$

Since logarithms of negative numbers are undefined, $x = -4$ is not a solution.

EXAMPLE 5   *Solving a logarithmic equation*

Solve the equation $\ln (x + 6) - \ln 10 = \ln (x - 1) - \ln 2$.

SOLUTION

$$\ln (x + 6) - \ln (x - 1) = \ln 10 - \ln 2 \quad \text{rearrange terms}$$

$$\ln \left(\frac{x + 6}{x - 1} \right) = \ln \frac{10}{2} \quad \text{law (2)}$$

$$\frac{x + 6}{x - 1} = 5 \quad \text{ln is one-to-one}$$

$$x + 6 = 5x - 5 \quad \text{multiply by } x - 1$$

$$4x = 11 \quad \text{subtract } x \text{ and add 5}$$

$$x = \tfrac{11}{4} \quad \text{divide by 4}$$

  Check   Since both $\ln (x + 6)$ and $\ln (x - 1)$ are defined at $x = \tfrac{11}{4}$ (they are logarithms of positive real numbers) and since our algebraic steps are correct, it follows that $\tfrac{11}{4}$ is a solution of the given equation.

EXAMPLE 6   *Shifting the graph of a logarithmic equation*

Sketch the graph of $y = \log_3 (81x)$.

FIGURE 25

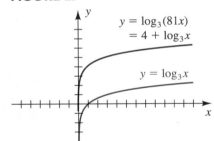

SOLUTION   We may rewrite the equation as follows:

$$y = \log_3 (81x) \quad \text{given}$$
$$= \log_3 81 + \log_3 x \quad \text{law (1)}$$
$$= \log_3 3^4 + \log_3 x \quad 81 = 3^4$$
$$= 4 + \log_3 x \quad \log_a a^x = x$$

Thus, we can obtain the graph of $y = \log_3 (81x)$ by vertically shifting the graph of $y = \log_3 x$ in Figure 19 upward four units. This gives us the sketch in Figure 25.

EXAMPLE 7   *Sketching graphs of logarithmic equations*

Sketch the graph of the equation:

(a) $y = \log_3 (x^2)$ **(b)** $y = 2 \log_3 x$

SOLUTION

(a) Since $x^2 = |x|^2$, we may rewrite the given equation as

$$y = \log_3 |x|^2.$$

Using a law of logarithms, we have

$$y = 2 \log_3 |x|.$$

(continued)

FIGURE 26
(a)

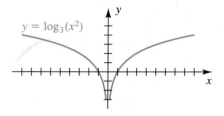

$y = \log_3(x^2)$

(b)

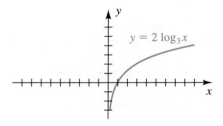

$y = 2\log_3 x$

We can obtain the graph of $y = 2\log_3 |x|$ by multiplying the y-coordinates of points on the graph of $y = \log_3 |x|$ in Figure 20 by 2. This gives us the graph in Figure 26(a).

(b) If $y = 2\log_3 x$, then x must be positive. Hence, the graph is identical to that part of the graph of $y = 2\log_3 |x|$ in Figure 26(a) that lies to the right of the y-axis. This gives us Figure 26(b).

EXAMPLE 8　*A relationship between selling price and demand*

In the study of economics, the demand D for a product is often related to its selling price p by an equation of the form

$$\log_a D = \log_a c - k\log_a p,$$

where a, c, and k are positive constants.

(a) Solve the equation for D.

(b) How does increasing or decreasing the selling price affect the demand?

SOLUTION

(a)　　　$\log_a D = \log_a c - k\log_a p$　　given

$\log_a D = \log_a c - \log_a p^k$　　law (3)

$\log_a D = \log_a \dfrac{c}{p^k}$　　law (2)

$D = \dfrac{c}{p^k}$　　$\log_a$ is one-to-one

(b) If the price p is increased, the denominator p^k in $D = c/p^k$ will also increase and hence the demand D for the product will decrease. If the price is decreased, then p^k will decrease and the demand D will increase.

5.4 EXERCISES

Exer. 1–8: Express in terms of logarithms of x, y, z, or w.

1 (a) $\log_4 (xz)$　　(b) $\log_4 (y/x)$　　(c) $\log_4 \sqrt[3]{z}$

2 (a) $\log_3 (xyz)$　　(b) $\log_3 (xz/y)$　　(c) $\log_3 \sqrt[5]{y}$

3 $\log_a \dfrac{x^3 w}{y^2 z^4}$

4 $\log_a \dfrac{y^5 w^2}{x^4 z^3}$

5 $\log \dfrac{\sqrt[3]{z}}{x\sqrt{y}}$

6 $\log \dfrac{\sqrt{y}}{x^4 \sqrt[3]{z}}$

7 $\ln \sqrt[4]{\dfrac{x^7}{y^5 z}}$

8 $\ln x \sqrt[3]{\dfrac{y^4}{z^5}}$

Exer. 9–16: Write the expression as one logarithm.

9 (a) $\log_3 x + \log_3 (5y)$　　(b) $\log_3 (2z) - \log_3 x$
(c) $5\log_3 y$

10 (a) $\log_4 (3z) + \log_4 x$　　(b) $\log_4 x - \log_4 (7y)$
(c) $\frac{1}{3}\log_4 w$

11 $2 \log_a x + \frac{1}{3} \log_a (x - 2) - 5 \log_a (2x + 3)$

12 $5 \log_a x - \frac{1}{2} \log_a (3x - 4) - 3 \log_a (5x + 1)$

13 $\log (x^3 y^2) - 2 \log x \sqrt[3]{y} - 3 \log \left(\dfrac{x}{y} \right)$

14 $2 \log \dfrac{y^3}{x} - 3 \log y + \dfrac{1}{2} \log x^4 y^2$

15 $\ln y^3 + \frac{1}{3} \ln (x^3 y^6) - 5 \ln y$

16 $2 \ln x - 4 \ln (1/y) - 3 \ln (xy)$

Exer. 17–32: Solve the equation.

17 $\log_6 (2x - 3) = \log_6 12 - \log_6 3$

18 $\log_4 (3x + 2) = \log_4 5 + \log_4 3$

19 $2 \log_3 x = 3 \log_3 5$

20 $3 \log_2 x = 2 \log_2 3$

21 $\log x - \log (x + 1) = 3 \log 4$

22 $\log (x + 2) - \log x = 2 \log 4$

23 $\ln (-4 - x) + \ln 3 = \ln (2 - x)$

24 $\ln x + \ln (x + 6) = \frac{1}{2} \ln 9$

25 $\log_2 (x + 7) + \log_2 x = 3$

26 $\log_6 (x + 5) + \log_6 x = 2$

27 $\log_3 (x + 3) + \log_3 (x + 5) = 1$

28 $\log_3 (x - 2) + \log_3 (x - 4) = 2$

29 $\log (x + 3) = 1 - \log (x - 2)$

30 $\log (57x) = 2 + \log (x - 2)$

31 $\ln x = 1 - \ln (x + 2)$

32 $\ln x = 1 + \ln (x + 1)$

Exer. 33–44: Sketch the graph of f.

33 $f(x) = \log_3 (3x)$ 34 $f(x) = \log_4 (16x)$

35 $f(x) = 3 \log_3 x$ 36 $f(x) = \frac{1}{3} \log_3 x$

37 $f(x) = \log_3 (x^2)$ 38 $f(x) = \log_2 (x^2)$

39 $f(x) = \log_2 (x^3)$ 40 $f(x) = \log_3 (x^3)$

41 $f(x) = \log_2 \sqrt{x}$ 42 $f(x) = \log_2 \sqrt[3]{x}$

43 $f(x) = \log_3 \left(\dfrac{1}{x} \right)$ 44 $f(x) = \log_2 \left(\dfrac{1}{x} \right)$

Exer. 45–48: Shown in the figure is the graph of a function f. Express $f(x)$ as one logarithm with base 2.

45

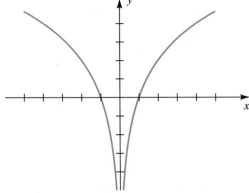

46

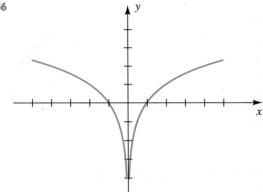

47

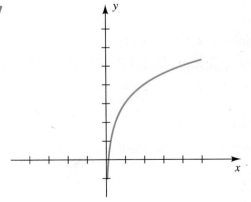

48

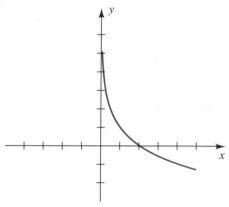

49 Pareto's law Pareto's law for capitalist countries states that the relationship between annual income x and the number y of individuals whose income exceeds x is

$$\log y = \log b - k \log x,$$

where b and k are positive constants. Solve this equation for y.

50 Price and demand If p denotes the selling price (in dollars) of a commodity and x is the corresponding demand (in number sold per day), then the relationship between p and x is sometimes given by $p = p_0 e^{-ax}$, where p_0 and a are positive constants. Express x as a function of p.

51 Wind velocity If v denotes the wind velocity (in m/sec) at a height of z meters above the ground, then under certain conditions $v = c \ln (z/z_0)$, where c is a positive constant and z_0 is the height at which the velocity is zero. Sketch the graph of this equation on a zv-plane for $c = 0.5$ and $z_0 = 0.1$ m.

52 Eliminating pollution If the pollution of Lake Erie were stopped suddenly, it has been estimated that the level y of pollutants would decrease according to the formula $y = y_0 e^{-0.3821t}$, where t is the time in years and y_0 is the pollutant level at which further pollution ceased. How many years would it take to clear 50% of the pollutants?

53 Reaction to a stimulus Let R denote the reaction of a subject to a stimulus of strength x. There are many possibilities for R and x. If the stimulus x is saltiness (in grams of salt per liter), R may be the subject's estimate of how salty the solution tasted, based on a scale from 0 to 10. One relationship between R and x is given by the Weber-Fechner formula $R(x) = a \log (x/x_0)$,

where a is a positive constant and x_0 is called the threshold stimulus.

(a) Find $R(x_0)$.

(b) Find a relationship between $R(x)$ and $R(2x)$.

54 Electron energy The energy $E(x)$ of an electron after passing through material of thickness x is given by $E(x) = E_0 e^{-x/x_0}$, where E_0 is the initial energy and x_0 is the radiation length.

(a) Express, in terms of E_0, the energy of an electron after it passes through material of thickness x_0.

(b) Express, in terms of x_0, the thickness at which the electron loses 99% of its initial energy.

55 Ozone layer One method of estimating the thickness of the ozone layer is to use the formula $\ln I_0 - \ln I = kx$, where I_0 is the intensity of a particular wavelength of light from the sun before it reaches the atmosphere, I is the intensity of the same wavelength after passing through a layer of ozone x centimeters thick, and k is the absorption constant of ozone for that wavelength. Suppose for a wavelength of 3176×10^{-8} cm with $k \approx 0.39$, I_0/I is measured as 1.12. Approximate the thickness of the ozone layer to the nearest 0.01 centimeter.

56 Ozone layer Refer to Exercise 55. Approximate the percentage decrease in the intensity of light with a wavelength of 3176×10^{-8} cm if the ozone layer is 0.24 centimeter thick.

[c] **Exer. 57–58: Graph f and g on the same coordinate plane, and estimate the solution of the inequality $f(x) \geq g(x)$.**

57 $f(x) = x^3 - 3.5x^2 + 3x$; $g(x) = \log 3x$

58 $f(x) = 3^{-0.5x}$; $g(x) = \log x$

[c] **Exer. 59–60: Use a graph to estimate the roots of the equation on the given interval.**

59 $e^{-x} - 2 \log (1 + x^2) + 0.5x = 0$; $[0, 8]$

60 $2 \log 2x - \log_3 x^2 = 0$; $(0, 3)$

[c] **Exer. 61–62: Graph f on the interval [0.2, 16]. (a) Estimate the intervals where f is increasing or is decreasing. (b) Estimate the maximum and minimum values of f on [0.2, 16].**

61 $f(x) = 2 \log 2x - 1.5x + 0.1x^2$

62 $f(x) = 1.1^{3x} + x - 1.35^x - \log x + 5$

5.5 EXPONENTIAL AND LOGARITHMIC EQUATIONS

In this section we shall consider various types of exponential and logarithmic equations and their applications.

EXAMPLE I *Solving an exponential equation*

Solve the equation $3^x = 21$.

SOLUTION

$$3^x = 21 \qquad \text{given}$$

$$\log(3^x) = \log 21 \qquad \text{take log of both sides}$$

$$x \log 3 = \log 21 \qquad \text{law (3) of logarithms}$$

$$x = \frac{\log 21}{\log 3} \qquad \text{divide by log 3}$$

We could also have used natural logarithms to obtain

$$x = \frac{\ln 21}{\ln 3}.$$

Using a calculator gives us the approximate solution $x \approx 2.77$. A partial check is to note that since $3^2 = 9$ and $3^3 = 27$, the number x such that $3^x = 21$ should lie between 2 and 3, somewhat closer to 3 than to 2.

We could also have solved the equation in Example 1 by changing the exponential form $3^x = 21$ to logarithmic form, as we did in Section 5.3, obtaining

$$x = \log_3 21.$$

This is, in fact, the solution of the equation; however, since calculators typically have keys only for log and ln, we cannot approximate $\log_3 21$ directly. The next theorem gives us a simple *change of base formula* for finding $\log_b u$ if $u > 0$ and b is *any* logarithmic base.

Theorem: Change of Base Formula

If $u > 0$ and if a and b are positive real numbers different from 1, then

$$\log_b u = \frac{\log_a u}{\log_a b}.$$

PROOF We begin with the equivalent equations

$$w = \log_b u \qquad \text{and} \qquad b^w = u$$

and proceed as follows:

$$b^w = u \qquad \text{given}$$

$$\log_a b^w = \log_a u \qquad \text{take } \log_a \text{ of both sides}$$

$$w \log_a b = \log_a u \qquad \text{law (3) of logarithms}$$

$$w = \frac{\log_a u}{\log_a b} \qquad \text{divide by } \log_a b$$

Since $w = \log_b u$, we obtain the formula. ▬

The following special case of the change of base formula is obtained by letting $u = a$ and using the fact that $\log_a a = 1$:

$$\log_b a = \frac{1}{\log_a b}$$

The change of base formula is sometimes confused with law (2) of logarithms. The following warning could be remembered by the phrase "a quotient of logs is *not* the log of the quotient."

Warning

$$\frac{\log_a u}{\log_a b} \neq \log_a \frac{u}{b}$$

The most frequently used special cases of the change of base formula are those for $a = 10$ (common logarithms) and $a = e$ (natural logarithms), as stated in the next box.

**Special Change of
Base Formulas**

$$\textbf{(1)} \quad \log_b u = \frac{\log u}{\log b} \qquad\qquad \textbf{(2)} \quad \log_b u = \frac{\ln u}{\ln b}$$

We shall next rework Example 1 using a change of base formula.

EXAMPLE 2   *Using a change of base formula*

Solve the equation $3^x = 21$.

SOLUTION   We proceed as follows:

$$3^x = 21 \qquad \text{given}$$

$$x = \log_3 21 \qquad \text{change to logarithmic form}$$

$$= \frac{\log 21}{\log 3} \qquad \text{special change of base formula (1)}$$

Another method is to use special change of base formula (2), obtaining

$$x = \frac{\ln 21}{\ln 3}.$$

Logarithms with base 2 are used in computer science. The next example indicates how to find logarithms with base 2 using change of base formulas.

EXAMPLE 3 *Approximating a logarithm with base 2*

Approximate $\log_2 5$ using

(a) common logarithms   **(b)** natural logarithms

SOLUTION   Using special change of base formulas (1) and (2), we obtain the following:

(a)
$$\log_2 5 = \frac{\log 5}{\log 2} \approx 2.322$$

(b)
$$\log_2 5 = \frac{\ln 5}{\ln 2} \approx 2.322$$

EXAMPLE 4 *Solving an exponential equation*

Solve the equation $5^{2x+1} = 6^{x-2}$.

SOLUTION   We can use either common or natural logarithms. Using base 10 gives us the following:

$5^{2x+1} = 6^{x-2}$	given
$\log (5^{2x+1}) = \log (6^{x-2})$	take log of both sides
$(2x + 1) \log 5 = (x - 2) \log 6$	law (3) of logarithms
$2x \log 5 + \log 5 = x \log 6 - 2 \log 6$	multiply
$2x \log 5 - x \log 6 = -\log 5 - 2 \log 6$	subtract log 5 and x log 6
$x(\log 5^2 - \log 6) = -(\log 5 + \log 6^2)$	factor and use law (3) of logarithms
$x = -\dfrac{\log (5 \cdot 36)}{\log \frac{25}{6}}$	solve for x and use laws of logarithms

An approximation is $x \approx -3.64$.

EXAMPLE 5 *Solving an exponential equation*

Solve the equation $\dfrac{5^x - 5^{-x}}{2} = 3$.

SOLUTION

$$\frac{5^x - 5^{-x}}{2} = 3 \qquad \text{given}$$

$$5^x - 5^{-x} = 6 \qquad \text{multiply by 2}$$

$$5^x - \frac{1}{5^x} = 6 \qquad \text{definition of negative exponent}$$

$$5^x(5^x) - \frac{1}{5^x}(5^x) = 6(5^x) \qquad \text{multiply by the lcd } 5^x$$

$$5^{2x} - 6(5^x) - 1 = 0 \qquad \text{simplify and subtract } 6(5^x)$$

We recognize this form of the equation as a quadratic in 5^x and proceed as follows:

$$(5^x)^2 - 6(5^x) - 1 = 0 \qquad \text{law of exponents}$$
$$u^2 - 6u - 1 = 0 \qquad \text{let } u = 5^x$$
$$u = \frac{6 \pm \sqrt{36 + 4}}{2} \qquad \text{quadratic formula}$$
$$5^x = 3 \pm \sqrt{10} \qquad u = 5^x$$
$$5^x = 3 + \sqrt{10} \qquad 5^x > 0, \text{ but } 3 - \sqrt{10} < 0$$
$$\log 5^x = \log(3 + \sqrt{10}) \qquad \text{take log of both sides}$$
$$x \log 5 = \log(3 + \sqrt{10}) \qquad \text{law (3) of logarithms}$$
$$x = \frac{\log(3 + \sqrt{10})}{\log 5} \qquad \text{divide by log 5}$$

Natural logarithms could have also been used to obtain

$$x = \frac{\ln(3 + \sqrt{10})}{\ln 5}.$$

An approximation is $x \approx 1.13$.

EXAMPLE 6   *Light penetration in an ocean*

The Beer-Lambert law states that the amount of light I that penetrates to a depth of x meters in an ocean is given by $I = I_0 c^x$, where $0 < c < 1$ and I_0 is the amount of light at the surface.

(a) Solve for x using common logarithms.

(b) Solve for x using natural logarithms.

(c) If $c = \frac{1}{4}$, approximate the depth at which $I = 0.01 I_0$. (This determines the zone where photosynthesis can take place.)

SOLUTION

(a)

$$I = I_0 c^x \qquad \text{given}$$

$$\frac{I}{I_0} = c^x \qquad \text{isolate the exponential expression}$$

$$x = \log_c \frac{I}{I_0} \qquad \text{change to logarithmic form}$$

$$= \frac{\log (I/I_0)}{\log c} \qquad \text{special change of base formula (1)}$$

(b) Replacing log with ln throughout part (a), we obtain

$$x = \frac{\ln (I/I_0)}{\ln c}.$$

(c) Letting $I = 0.01 I_0$ and $c = \frac{1}{4}$ in the formula for x obtained in part (a), we have

$$x = \frac{\log (0.01 I_0/I_0)}{\log \frac{1}{4}} = \frac{\log (0.01)}{\log 1 - \log 4} = \frac{\log 10^{-2}}{0 - \log 4} = \frac{-2}{-\log 4} = \frac{2}{\log 4}.$$

An approximation is $x \approx 3.32$ m.

EXAMPLE 7   *Comparing light intensities*

If a beam of light that has intensity I_0 is projected vertically downward into water, then its intensity $I(x)$ at a depth of x meters is $I(x) = I_0 e^{-1.4x}$ (see Figure 27). At what depth is the intensity one-half its value at the surface?

FIGURE 27

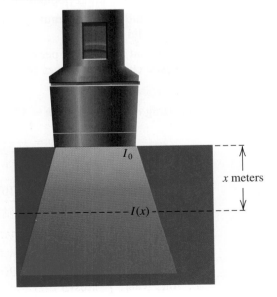

SOLUTION   At the surface, $x = 0$, and the intensity is

$$I(0) = I_0 e^0 = I_0.$$

We wish to find the value of x such that $I(x) = \frac{1}{2}I_0$. This leads to the following:

$$
\begin{array}{ll}
I(x) = \frac{1}{2}I_0 & \text{desired intensity} \\[1mm]
I_0 e^{-1.4x} = \frac{1}{2}I_0 & \text{formula for } I(x) \\[1mm]
e^{-1.4x} = \frac{1}{2} & \text{divide by } I_0 \\[1mm]
-1.4x = \ln \frac{1}{2} & \text{change to logarithmic form} \\[1mm]
x = \dfrac{\ln \frac{1}{2}}{-1.4} & \text{divide by } -1.4
\end{array}
$$

An approximation is $x \approx 0.495$ m.

EXAMPLE 8   A logistic curve

A **logistic curve** is the graph of an equation of the form

$$y = \frac{k}{1 + be^{-cx}},$$

where k, b, and c are positive constants. Such curves are useful for describing a population y that grows rapidly initially, but whose growth rate decreases after x reaches a certain value. In a famous study of the growth of protozoa by Gause, a population of Paramecium caudata was found to be described by a logistic equation with $c = 1.1244$, $k = 105$, and x the time in days.

(a) Find b if the initial population was 3 protozoa.

(b) In the study, the maximum growth rate took place at $y = 52$. At what time x did this occur?

(c) Show that after a long period of time, the population described by any logistic curve approaches the constant k.

SOLUTION

(a) Letting $c = 1.1244$ and $k = 105$ in the logistic equation, we obtain

$$y = \frac{105}{1 + be^{-1.1244x}}.$$

We now proceed as follows:

$$
\begin{array}{ll}
3 = \dfrac{105}{1 + be^0} = \dfrac{105}{1 + b} & y = 3 \text{ when } x = 0 \\[4mm]
1 + b = 35 & \text{multiply by } \dfrac{1 + b}{3} \\[4mm]
b = 34 & \text{solve for } b
\end{array}
$$

(b) Using the fact that $b = 34$ leads to the following:

$$52 = \frac{105}{1 + 34e^{-1.1244x}} \qquad \text{let } y = 52 \text{ in part (a)}$$

$$1 + 34e^{-1.1244x} = \frac{105}{52} \qquad \text{multiply by } \frac{1 + 34e^{-1.1244x}}{52}$$

$$e^{-1.1244x} = \left(\frac{105}{52} - 1\right) \cdot \frac{1}{34} = \frac{53}{1768} \qquad \text{isolate } e^{-1.1244x}$$

$$-1.1244x = \ln \frac{53}{1768} \qquad \text{change to logarithmic form}$$

$$x = \frac{\ln \frac{53}{1768}}{-1.1244} \approx 3.12 \text{ days} \qquad \text{divide by } -1.1244$$

(c) As $x \to \infty$, $e^{-cx} \to 0$. Hence,

$$y = \frac{k}{1 + be^{-cx}} \quad \to \quad \frac{k}{1 + b \cdot 0} = k.$$

In the next example we graph the equation obtained in part (a) of the preceding example.

EXAMPLE 9 *Sketching the graph of a logistic curve*

Graph the logistic curve given by

$$y = \frac{105}{1 + 34e^{-1.1244x}},$$

and estimate the value of x for $y = 52$.

SOLUTION We begin by assigning

$$\frac{105}{1 + 34e^{-1.1244x}}$$

FIGURE 28 [0, 10] by [0, 105]

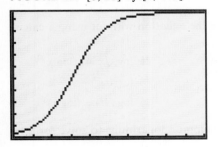

to Y_1. Since the time x is nonnegative, we choose Xmin $= 0$. We select Xmax $= 10$ in order to include the value of x found in part (b) of Example 8. By part (c), we know that the value of y cannot exceed 105. Thus, we choose Ymin $= 0$ and Ymax $= 105$ and obtain a display similar to Figure 28.

Using the tracing and zoom-in features, we see that for $y = 52$, the value of x is approximately 3.12, which agrees with the approximation found in (b) of Example 8.

The following example is a good illustration of the power of a graphing utility, since it is impossible to find the exact solution using only algebraic methods.

EXAMPLE 10 *Estimating points of intersection of logarithmic graphs*

Estimate the point of intersection of the graphs of

$$f(x) = \log_3 x \qquad \text{and} \qquad g(x) = \log_6 (x + 2).$$

SOLUTION Most graphing utilities are equipped to work with only common and natural logarithmic functions. Thus, we first use a change of base formula to rewrite f and g as

$$f(x) = \frac{\ln x}{\ln 3} \qquad \text{and} \qquad g(x) = \frac{\ln (x + 2)}{\ln 6}.$$

FIGURE 29 $[-2, 4]$ by $[-2, 2]$

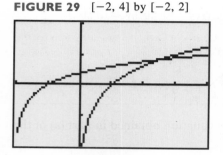

We next assign $(\ln x)/\ln 3$ and $(\ln (x + 2))/\ln 6$ to Y_1 and Y_2, respectively. After graphing Y_1 and Y_2 using a standard viewing rectangle, we see that there is a point of intersection in the first quadrant with $2 < x < 3$. Using the tracing and zoom-in features, we find that the point of intersection is approximately $(2.52, 0.84)$.

Figure 29 was obtained using viewing rectangle dimensions $[-2, 4]$ by $[-2, 2]$. There are no other points of intersection, since f increases more rapidly than g for $x > 3$.

5.5 EXERCISES

Exer. 1–4: Find the exact solution and a two-decimal-place approximation for it by using **(a)** the method of Example 1 and **(b)** the method of Example 2.

1 $5^x = 8$

2 $4^x = 3$

3 $3^{4-x} = 5$

4 $(\frac{1}{3})^x = 100$

Exer. 5–10: Evaluate using the change of base formula.

5 $\log_5 6$

6 $\log_2 20$

7 $\log_9 0.2$

8 $\log_6 \frac{1}{2}$

9 $\dfrac{\log_5 16}{\log_5 4}$

10 $\dfrac{\log_7 243}{\log_7 3}$

Exer. 11–24: Find the exact solution, using common logarithms, and a two-decimal-place approximation of each solution, when appropriate.

11 $3^{x+4} = 2^{1-3x}$

12 $4^{2x+3} = 5^{x-2}$

13 $2^{2x-3} = 5^{x-2}$

14 $3^{2-3x} = 4^{2x+1}$

15 $2^{-x} = 8$

16 $2^{-x^2} = 5$

17 $\log x = 1 - \log (x - 3)$

18 $\log (5x + 1) = 2 + \log (2x - 3)$

19 $\log (x^2 + 4) - \log (x + 2) = 2 + \log (x - 2)$

20 $\log (x - 4) - \log (3x - 10) = \log (1/x)$

21 $5^x + 125(5^{-x}) = 30$

22 $3(3^x) + 9(3^{-x}) = 28$

23 $4^x - 3(4^{-x}) = 8$

24 $2^x - 6(2^{-x}) = 6$

Exer. 25–30: Solve the equation without using a calculator or a table.

25 $\log (x^2) = (\log x)^2$

26 $\log \sqrt{x} = \sqrt{\log x}$

27 $\log (\log x) = 2$

28 $\log \sqrt{x^3 - 9} = 2$

29 $x^{\sqrt{\log x}} = 10^8$

30 $\log (x^3) = (\log x)^3$

Exer. 31–34: Use common logarithms to solve for x in terms of y.

31 $y = \dfrac{10^x + 10^{-x}}{2}$

32 $y = \dfrac{10^x - 10^{-x}}{2}$

33 $y = \dfrac{10^x - 10^{-x}}{10^x + 10^{-x}}$

34 $y = \dfrac{10^x + 10^{-x}}{10^x - 10^{-x}}$

Exer. 35–38: Use natural logarithms to solve for x in terms of y.

35 $y = \dfrac{e^x - e^{-x}}{2}$

36 $y = \dfrac{e^x + e^{-x}}{2}$

37 $y = \dfrac{e^x + e^{-x}}{e^x - e^{-x}}$

38 $y = \dfrac{e^x - e^{-x}}{e^x + e^{-x}}$

Exer. 39–40: Sketch the graph of f, and use the change of base formula to approximate the y-intercept.

39 $f(x) = \log_2 (x + 3)$

40 $f(x) = \log_3 (x + 5)$

Exer. 41–42: Sketch the graph of f, and use the change of base formula to approximate the x-intercept.

41 $f(x) = 4^x - 3$

42 $f(x) = 3^x - 6$

Exer. 43–46: Chemists use a number denoted by pH to describe quantitatively the acidity or basicity of solutions. By definition, $\text{pH} = -\log [\text{H}^+]$, where $[\text{H}^+]$ is the hydrogen ion concentration in moles per liter.

43 Approximate the pH of each substance:

   (a) vinegar: $[\text{H}^+] \approx 6.3 \times 10^{-3}$

   (b) carrots: $[\text{H}^+] \approx 1.0 \times 10^{-5}$

   (c) sea water: $[\text{H}^+] \approx 5.0 \times 10^{-9}$

44 Approximate the hydrogen ion concentration $[\text{H}^+]$ of each substance:

   (a) apples: $\text{pH} \approx 3.0$

   (b) beer: $\text{pH} \approx 4.2$

   (c) milk: $\text{pH} \approx 6.6$

45 A solution is considered basic if $[\text{H}^+] < 10^{-7}$ or acidic if $[\text{H}^+] > 10^{-7}$. Find the corresponding inequalities involving pH.

46 Many solutions have a pH between 1 and 14. Find the corresponding range of $[\text{H}^+]$.

47 *Compound interest* Use the compound interest formula to determine how long it will take for a sum of money to double if it is invested at a rate of 6% per year compounded monthly.

48 *Compound interest* Solve the compound interest formula

$$A = P\left(1 + \frac{r}{n}\right)^{nt}$$

for t by using natural logarithms.

49 *Photic zone* Refer to Example 6. The most important zone in the sea from the viewpoint of marine biology is the photic zone, in which photosynthesis takes place. The photic zone ends at the depth where about 1% of the surface light penetrates. In very clear waters in the Caribbean, 50% of the light at the surface reaches a depth of about 13 meters. Estimate the depth of the photic zone.

50 *Photic zone* In contrast to the situation described in the previous exercise, in parts of New York harbor, 50% of the surface light does not reach a depth of 10 centimeters. Estimate the depth of the photic zone.

51 Shown in the figure is a graph of $f(x) = (\ln x)/x$ for $x > 0$. The maximum value of $f(x)$ occurs at $x = e$.

   (a) The integers 2 and 4 have the unusual property that $2^4 = 4^2$. Show that if $x^y = y^x$ for positive real numbers x and y, then $(\ln x)/x = (\ln y)/y$.

   (b) Use the graph of f to explain why many pairs of real numbers satisfy the equation $x^y = y^x$.

EXERCISE 51

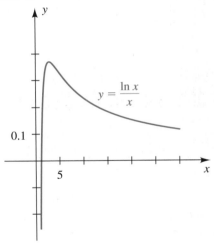

$$y = \frac{\ln x}{x}$$

52 *Language history* Refer to Exercise 36 of Section 5.1. If a language originally had N_0 basic words of which $N(t)$ are still in use, then $N(t) = N_0(0.805)^t$, where time t is measured in millennia. After how many years are one-half the basic words still in use?

53 *Drug absorption* If a 100-milligram tablet of an asthma drug is taken orally and if none of the drug is present in the body when the tablet is first taken, the total amount A in the bloodstream after t minutes is predicted to be

$$A = 100[1 - (0.9)^t] \quad \text{for } 0 \le t \le 10.$$

(a) Sketch the graph of the equation.

(b) Determine the number of minutes needed for 50 milligrams of the drug to have entered the bloodstream.

54 Drug dosage   A drug is eliminated from the body through urine. Suppose that for a dose of 10 milligrams, the amount $A(t)$ remaining in the body t hours later is given by $A(t) = 10(0.8)^t$ and that in order for the drug to be effective, at least 2 milligrams must be in the body.

(a) Determine when 2 milligrams is left in the body.

(b) What is the half-life of the drug?

55 Genetic mutation   The basic source of genetic diversity is mutation, or changes in the chemical structure of genes. If a gene mutates at a constant rate m and if other evolutionary forces are negligible, then the frequency F of the original gene after t generations is given by $F = F_0(1 - m)^t$, where F_0 is the frequency at $t = 0$.

(a) Solve the equation for t using common logarithms.

(b) If $m = 5 \times 10^{-5}$, after how many generations does $F = \frac{1}{2}F_0$?

c 56 Employee productivity   Certain learning processes may be illustrated by the graph of $f(x) = a + b(1 - e^{-cx})$, where a, b, and c are positive constants. Suppose a manufacturer estimates that a new employee can produce five items the first day on the job. As the employee becomes more proficient, the daily production increases until a certain maximum production is reached. Suppose that on the nth day on the job, the number $f(n)$ of items produced is approximated by the formula

$$f(n) = 3 + 20(1 - e^{-0.1n}).$$

(a) Estimate the number of items produced on the fifth day, the ninth day, the twenty-fourth day, and the thirtieth day.

(b) Sketch the graph of f from $n = 0$ to $n = 30$. (Graphs of this type are called *learning curves* and are used frequently in education and psychology.)

(c) What happens as n increases without bound?

57 Height of trees   The growth in height of trees is frequently described by a logistic equation. Suppose the height h (in feet) of a tree at age t (in years) is

$$h = \frac{120}{1 + 200e^{-0.2t}},$$

as illustrated by the graph in the figure.

(a) What is the height of the tree at age 10?

(b) At what age is the height 50 feet?

EXERCISE 57

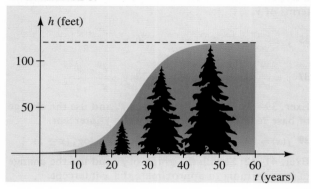

58 Employee productivity   Manufacturers sometimes use empirically based formulas to predict the time required to produce the nth item on an assembly line for an integer n. If $T(n)$ denotes the time required to assemble the nth item and T_1 denotes the time required for the first, or prototype, item, then typically $T(n) = T_1 n^{-k}$ for some positive constant k.

(a) For many airplanes, the time required to assemble the second airplane, $T(2)$, is equal to $(0.80)T_1$. Find the value of k.

(b) Express, in terms of T_1, the time required to assemble the fourth airplane.

(c) Express, in terms of $T(n)$, the time $T(2n)$ required to assemble the $(2n)$th airplane.

59 Vertical wind shear   Refer to Exercises 65–66 in Section 3.3. If v_0 is the wind speed at height h_0, and v_1 is the wind speed at height h_1, then the vertical wind shear can be described by the equation

$$\frac{v_0}{v_1} = \left(\frac{h_0}{h_1}\right)^P,$$

where P is a constant. During a one-year period in Montreal, the maximum vertical wind shear occurred when the winds at the 200-foot level were 25 mi/hr while the winds at the 35-foot level were 6 mi/hr. Find P for these conditions.

60 Vertical wind shear   Refer to Exercise 59. The average vertical wind shear is given by the equation

$$s = \frac{v_1 - v_0}{h_1 - h_0}.$$

Suppose that the velocity of the wind increases with increasing altitude and that all values for wind speeds taken at the 35-foot and 200-foot altitudes are greater than 1 mi/hr. Does increasing the value of P produce larger or smaller values of s?

Exer. 61–62: An economist suspects that the following data points lie on the graph of $y = c2^{kx}$, where c and k are constants. If the data points have three-decimal-place accuracy, is this suspicion correct?

61 $(0, 4)$, $(1, 3.249)$, $(2, 2.639)$, $(3, 2.144)$

62 $(0, -0.3)$, $(0.5, -0.345)$, $(1, -0.397)$, $(1.5, -0.551)$, $(2, -0.727)$

Exer. 63–64: It is suspected that the following data points lie on the graph of $y = c \log (kx + 10)$, where c and k are constants. If the data points have three-decimal-place accuracy, is this suspicion correct?

63 $(0, 1.5)$, $(1, 1.619)$, $(2, 1.720)$, $(3, 1.997)$

64 $(0, 0.7)$, $(1, 0.782)$, $(2, 0.847)$, $(3, 0.900)$, $(4, 0.945)$

[c] **Exer. 65–66:** Approximate the real root of the equation.

65 $x \ln x = 1$ **66** $\ln x + x = 0$

[c] **Exer. 67–68:** Graph f and g on the same coordinate plane, and estimate the solution to the inequality $f(x) > g(x)$.

67 $f(x) = 3^{-x} - 4^{0.2x}$, $g(x) = \ln (1.2) - x$

68 $f(x) = 3 \log_4 x - \log x$, $g(x) = e^x - 0.25x^4$

CHAPTER 5 REVIEW EXERCISES

Exer. 1–16: Sketch the graph of f.

1 $f(x) = 3^{x+2}$ **2** $f(x) = (\frac{3}{5})^x$

3 $f(x) = (\frac{3}{2})^{-x}$ **4** $f(x) = 3^{-2x}$

5 $f(x) = 3^{-x^2}$ **6** $f(x) = 1 - 3^{-x}$

7 $f(x) = e^{x/2}$ **8** $f(x) = \frac{1}{2}e^x$

9 $f(x) = e^{x-2}$ **10** $f(x) = e^{2-x}$

11 $f(x) = \log_6 x$ **12** $f(x) = \log_6 (36x)$

13 $f(x) = \log_4 (x^2)$ **14** $f(x) = \log_4 \sqrt[3]{x}$

15 $f(x) = \log_2 (x + 4)$ **16** $f(x) = \log_2 (4 - x)$

Exer. 17–18: Evaluate without using a calculator or table.

17 (a) $\log_2 \frac{1}{16}$ (b) $\log_\pi 1$ (c) $\ln e$
      (d) $6^{\log_6 4}$ (e) $\log 1{,}000{,}000$ (f) $10^{3 \log 2}$
      (g) $\log_4 2$

18 (a) $\log_5 \sqrt[3]{5}$ (b) $\log_5 1$ (c) $\log 10$
      (d) $e^{\ln 5}$ (e) $\log \log 10^{10}$ (f) $e^{2 \ln 5}$
      (g) $\log_{27} 3$

Exer. 19–36: Solve the equation without using a calculator or table.

19 $2^{3x-1} = \frac{1}{2}$ **20** $\log \sqrt{x} = \log (x - 6)$

21 $\log_8 (x - 5) = \frac{2}{3}$

22 $\log_4 (x + 1) = 2 + \log_4 (3x - 2)$

23 $2 \ln (x + 3) - \ln (x + 1) = 3 \ln 2$

24 $\log \sqrt[4]{x + 1} = \frac{1}{2}$ **25** $2^{5-x} = 6$

26 $3^{(x^2)} = 7$ **27** $2^{5x+3} = 3^{2x+1}$

28 $\log_3 (3x) = \log_3 x + \log_3 (4 - x)$

29 $\log_4 x = \sqrt[3]{\log_4 x}$ **30** $e^{x + \ln 4} = 3e^x$

31 $10^{2 \log x} = 5$ **32** $e^{\ln (x+1)} = 3$

33 $x^2(-2xe^{-x^2}) + 2xe^{-x^2} = 0$

34 $e^x + 2 = 8e^{-x}$

35 (a) $\log x^2 = \log (6 - x)$ (b) $2 \log x = \log (6 - x)$

36 (a) $\ln (e^x)^2 = 16$ (b) $\ln e^{(x^2)} = 16$

37 Express $\log x^4 \sqrt[3]{y^2/z}$ in terms of logarithms of x, y, and z.

38 Express $\log (x^2/y^3) + 4 \log y - 6 \log \sqrt{xy}$ as one logarithm.

Exer. 39–40: Use common logarithms to solve the equation for x in terms of y.

39 $y = \dfrac{1}{10^x + 10^{-x}}$ **40** $y = \dfrac{1}{10^x - 10^{-x}}$

Exer. 41–42: Approximate x to three significant figures.

41 (a) $x = \ln 6.6$ (b) $\log x = 1.8938$
      (c) $\ln x = -0.75$

42 (a) $x = \log 8.4$

(b) $\log x = -2.4260$

(c) $\ln x = 1.8$

43 *Bacteria growth*   The number of bacteria in a certain culture at time t (in hours) is given by $Q(t) = 2(3^t)$, where $Q(t)$ is measured in thousands.

(a) What is the number of bacteria at $t = 0$?

(b) Find the number after 10 minutes, 30 minutes, and 1 hour.

44 *Compound interest*   If $1000 is invested at a rate of 12% compounded quarterly, what is the principal after one year?

45 *Radioactive iodine decay*   Radioactive iodine ^{131}I, which is frequently used in tracer studies involving the thyroid gland, decays according to $N = N_0(0.5)^{t/8}$, where N_0 is the initial dose and t is the time in days.

(a) Sketch the graph of the equation if $N_0 = 64$.

(b) Find the half-life of ^{131}I.

46 *Trout population*   A pond is stocked with 1000 trout. Three months later, it is estimated that 600 remain. Find a formula of the form $N = N_0 a^{ct}$ that can be used to estimate the number of trout remaining after t months.

47 *Continuously compounded interest*   Ten thousand dollars is invested in a savings fund in which interest is compounded continuously at the rate of 11% per year.

(a) When will the account contain $35,000?

(b) How long does it take for money to double in the account?

48 *Electrical current*   The current $I(t)$ in a certain electrical circuit at time t is given by $I(t) = I_0 e^{-Rt/L}$, where R is the resistance, L is the inductance, and I_0 is the initial current at $t = 0$. Find the value of t, in terms of L and R, for which $I(t)$ is 1% of I_0.

49 *Sound intensity*   The sound intensity level formula is $\alpha = 10 \log (I/I_0)$.

(a) Solve for I in terms of α and I_0.

(b) Show that a one-decibel rise in the intensity level α corresponds to a 26% increase in the intensity I.

50 *Fish growth*   The length L of a fish is related to its age by means of the von Bertalanffy growth formula

$$L = a(1 - be^{-kt}),$$

where a, b, and k are positive constants that depend on the type of fish. Solve this equation for t to obtain a formula that can be used to estimate the age of a fish from a length measurement.

51 *Earthquake area in the West*   In the western United States, the area A (in mi^2) affected by an earthquake is related to the magnitude R of the quake by the formula

$$R = 2.3 \log (A + 3000) - 5.1.$$

Solve for A in terms of R.

52 *Earthquake area in the East*   Refer to Exercise 51. For the eastern United States, the area-magnitude formula has the form

$$R = 2.3 \log (A + 34,000) - 7.5.$$

If A_1 is the area affected by an earthquake of magnitude R in the West and A_2 is the area affected by a similar quake in the East, find a formula for A_1/A_2 in terms of R.

53 *Earthquake area in the Central states*   Refer to Exercise 51. For the Rocky Mountain and Central states, the area-magnitude formula has the form

$$R = 2.3 \log (A + 14,000) - 6.6.$$

If an earthquake has magnitude 4 on the Richter scale, estimate the area A of the region that will feel the quake.

54 *Atmospheric pressure*   Under certain conditions, the atmospheric pressure p at altitude h is given by the formula $p = 29e^{-0.000034h}$. Express h as a function of p.

55 *Rocket velocity*   A rocket of mass m_1 is filled with fuel of initial mass m_2. If frictional forces are disregarded, the total mass m of the rocket at time t after ignition is related to its upward velocity v by $v = -a \ln m + b$, where a and b are constants. At ignition time $t = 0$, $v = 0$ and $m = m_1 + m_2$. At burnout, $m = m_1$. Use this information to find a formula, in terms of one logarithm, for the velocity of the rocket at burnout.

56 *Earthquake frequency*   Let n be the average number of earthquakes per year that have magnitudes between R and $R + 1$ on the Richter scale. A formula that approximates the relationship between n and R is $\log n = 7.7 - (0.9)R$.

(a) Solve the equation for n in terms of R.

(b) Find n if $R = 4$, 5, and 6.

57 *Earthquake energy*   The energy E (in ergs) released during an earthquake of magnitude R may be approximated by using the formula $\log E = 11.4 + (1.5)R$.

(a) Solve for E in terms of R.

(b) Find the energy released during the famous Alaskan quake of 1964, which measured 8.4 on the Richter scale.

58 Radioactive decay A certain radioactive substance decays according to the formula $q(t) = q_0 e^{-0.0063t}$, where q_0 is the initial amount of the substance and t is the time in days. Approximate the half-life of the substance.

59 Children's growth The Count Model is a formula that can be used to predict the height of preschool children. If h is height (in centimeters) and t is age (in years), then

$$h = 70.228 + 5.104t + 9.222 \ln t$$

for $\frac{1}{4} \le t \le 6$. From calculus, the rate of growth R (in cm/year) is given by $R = 5.104 + (9.222/t)$. Predict the height and rate of growth of a typical 2-year-old.

60 Electrical circuit The current I in a certain electrical circuit at time t is given by

$$I = \frac{V}{R}(1 - e^{-Rt/L}),$$

where V is the electromotive force, R is the resistance, and L is the inductance. Solve the equation for t.

61 Carbon-14 dating The technique of carbon-14 (^{14}C) dating is used to determine the age of archeological and geological specimens. The formula $T = -8310 \ln x$ is sometimes used to predict the age T (in years) of a bone fossil, where x is the percentage (expressed as a decimal) of ^{14}C still present in the fossil.

(a) Estimate the age of a bone fossil that contains 4% of the ^{14}C found in an equal amount of carbon in present-day bone.

(b) Approximate the percentage of ^{14}C present in a fossil that is 10,000 years old.

62 Population of Kenya Based on present birth and death rates, the population of Kenya is expected to increase according to the formula $N = 20.2e^{0.041t}$, with N in millions and $t = 0$ corresponding to 1985. How many years will it take for the population to double?

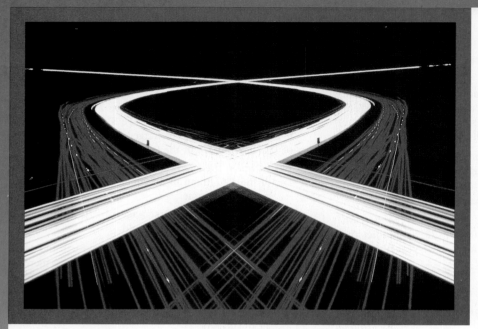

Systems of equations may be applied to finding points at which curved paths intersect.

CHAPTER 6

SYSTEMS OF EQUATIONS AND INEQUALITIES

■ Applications of mathematics sometimes require working simultaneously with more than one equation in several variables—that is, with a system of equations. In this chapter we develop methods for finding solutions that are common to all the equations in a system. Of particular importance are the techniques involving matrices, because they are well suited for computer programs and can be readily applied to systems containing any number of linear equations in any number of variables. We shall also consider systems of inequalities and linear programming—topics that are of major importance in business applications and statistics. The final four sections provide an introduction to the algebra of matrices and determinants. ■

6.1   SYSTEMS OF EQUATIONS

FIGURE 1

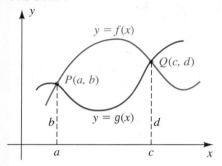

Consider the graphs of the two functions f and g, illustrated in Figure 1. In applications it is often necessary to find points such as $P(a, b)$ and $Q(c, d)$ at which the graphs intersect. Since $P(a, b)$ is on each graph, the pair (a, b) is a solution of *both* of the equations $y = f(x)$ and $y = g(x)$; that is,

$$b = f(a) \qquad \text{and} \qquad b = g(a).$$

We say that (a, b) is a solution of the **system of equations** (or simply **system**)

$$\begin{cases} y = f(x) \\ y = g(x) \end{cases}$$

where the brace is used to indicate that the equations are to be treated simultaneously. Similarly, the pair (c, d) is a solution of the system. To **solve** a system of equations means to find all the solutions.

As a special case, consider the system

$$\begin{cases} y = x^2 \\ y = 2x + 3 \end{cases}$$

FIGURE 2

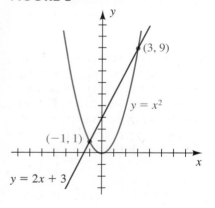

The graphs of the equations are the parabola and line sketched in Figure 2. The following table shows that the points $(-1, 1)$ and $(3, 9)$ are on both graphs.

(x, y)	$y = x^2$	$y = 2x + 3$
$(-1, 1)$	$1 = (-1)^2$, or $1 = 1$	$1 = 2(-1) + 3$, or $1 = 1$
$(3, 9)$	$9 = 3^2$, or $9 = 9$	$9 = 2(3) + 3$, or $9 = 9$

Hence, $(-1, 1)$ and $(3, 9)$ are solutions of the system.

The preceding discussion does not give us a strategy for actually finding the solutions. The next two examples illustrate how to find the solutions of the system using only algebraic methods.

EXAMPLE 1   *Solving a system of two equations*

Solve the system

$$\begin{cases} y = x^2 \\ y = 2x + 3 \end{cases}$$

SOLUTION   If (x, y) is a solution of the system, then the variable y in the equation $y = 2x + 3$ must satisfy the condition $y = x^2$. Hence, we *substitute x^2 for y in $y = 2x + 3$:*

$$x^2 = 2x + 3 \qquad \text{substitute } y = x^2 \text{ in } y = 2x + 3$$
$$x^2 - 2x - 3 = 0 \qquad \text{subtract } 2x + 3$$
$$(x + 1)(x - 3) = 0 \qquad \text{factor}$$
$$x = -1, \quad x = 3 \qquad \text{set each factor equal to 0}$$

This gives us the x-values for the solutions (x, y) of the system. To find the corresponding y-values, we may use either $y = x^2$ or $y = 2x + 3$. Using $y = x^2$, we find that

$$\text{if} \quad x = -1, \quad \text{then} \quad y = (-1)^2 = 1;$$

and

$$\text{if} \quad x = 3, \quad \text{then} \quad y = 3^2 = 9.$$

Hence, the solutions of the system are $(-1, 1)$ and $(3, 9)$.

We could also have found the solutions by substituting $y = 2x + 3$ in the *first* equation, obtaining

$$2x + 3 = x^2.$$

The remainder of the solution is the same.

Given the system in Example 1, we *could* have solved one of the equations for x in terms of y and then substituted in the other equation, obtaining an equation in y alone. Solving the latter equation would give us the y-values for the solutions of the system. The x-values could then be found using one of the given equations. In general, we may use the following guidelines, where u and v denote any two variables (*possibly x and y*). This technique is called the **method of substitution**.

Guidelines for the Method of Substitution for Two Equations in Two Variables

1 Solve one of the equations for one variable u in terms of the other variable v.

2 Substitute the expression for u found in guideline 1 in the other equation, obtaining an equation in v alone.

3 Find the solutions of the equation in v obtained in guideline 2.

4 Substitute the v-values found in guideline 3 in the equation of guideline 1 to find the corresponding u-values.

5 Check each pair (u, v) found in guideline 4 in the given system.

EXAMPLE 2   *Using the method of substitution*

Solve the following system and then sketch the graph of each equation, showing the points of intersection:

$$\begin{cases} x + y^2 = 6 \\ x + 2y = 3 \end{cases}$$

SOLUTION   We shall follow the guidelines with $u = x$ and $v = y$.

Guideline 1   Solve the second equation for x in terms of y:

$$x = 3 - 2y$$

Guideline 2   Substitute the expression for x found in guideline 1 in the first equation of the system:

$$(3 - 2y) + y^2 = 6 \quad \text{substitute } x = 3 - 2y \text{ in } x + y^2 = 6$$

$$y^2 - 2y - 3 = 0 \quad \text{simplify}$$

Guideline 3   Solve the equation in guideline 2 for y:

$$(y - 3)(y + 1) = 0 \qquad \text{factor } y^2 - 2y - 3$$

$$y - 3 = 0, \quad y + 1 = 0 \qquad \text{set each factor equal to 0}$$

$$y = 3, \qquad y = -1 \quad \text{solve for } y$$

These are the only possible y-values for the solutions of the system.

Guideline 4   Use the equation $x = 3 - 2y$ from guideline 1 to find the corresponding x-values:

$$\text{if} \quad y = 3, \quad \text{then} \quad x = 3 - 2(3) = 3 - 6 = -3$$

$$\text{if} \quad y = -1, \quad \text{then} \quad x = 3 - 2(-1) = 3 + 2 = 5$$

Thus, possible solutions are $(-3, 3)$ and $(5, -1)$.

Guideline 5   Substituting $x = -3$ and $y = 3$ in $x + y^2 = 6$, the first equation of the system, yields $-3 + 9 = 6$, a true statement. Substituting $x = -3$ and $y = 3$ in $x + 2y = 3$, the second equation of the system, yields $-3 + 6 = 3$, also a true statement. Hence, $(-3, 3)$ is a solution of the system. In a similar manner, we may check that $(5, -1)$ is also a solution.

The graphs of the two equations (a parabola and a line) are sketched in Figure 3, showing the two points of intersection.

FIGURE 3

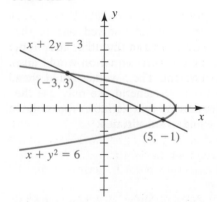

In future examples we will not list the specific guidelines that are used in finding solutions of systems.

In solving certain systems using the method of substitution, it is convenient to let u or v in the guidelines denote an *expression* involving another variable. This technique is illustrated in the next example with $u = x^2$.

EXAMPLE 3 *Using the method of substitution*

Solve the following system and then sketch the graph of each equation, showing the points of intersection.

$$\begin{cases} x^2 + y^2 = 25 \\ x^2 + y = 19 \end{cases}$$

SOLUTION We proceed as follows:

$$x^2 = 19 - y \qquad \text{solve } x^2 + y = 19 \text{ for } x^2$$

$$(19 - y) + y^2 = 25 \qquad \text{substitute } x^2 = 19 - y \text{ in } x^2 + y^2 = 25$$

$$y^2 - y - 6 = 0 \qquad \text{simplify}$$

$$(y - 3)(y + 2) = 0 \qquad \text{factor}$$

$$y - 3 = 0, \quad y + 2 = 0 \qquad \text{set each factor equal to 0}$$

$$y = 3, \qquad y = -2 \qquad \text{solve for } y$$

These are the only possible y-values for the solutions of the system. To find the corresponding x-values, we use $x^2 = 19 - y$:

$$\text{if} \quad y = 3, \quad \text{then} \quad x^2 = 19 - 3 = 16 \quad \text{and} \quad x = \pm 4$$

$$\text{if} \quad y = -2, \quad \text{then} \quad x^2 = 19 - (-2) = 21 \quad \text{and} \quad x = \pm\sqrt{21}$$

Thus, the only possible solutions of the system are

$$(4, 3), \quad (-4, 3), \quad (\sqrt{21}, -2), \quad \text{and} \quad (-\sqrt{21}, -2).$$

We can check by substitution in the given equations that all four pairs are solutions.

The graph of $x^2 + y^2 = 25$ is a circle of radius 5 with center at the origin, and the graph of $y = 19 - x^2$ is a parabola with a vertical axis. The graphs are sketched in Figure 4. The points of intersection correspond to the solutions of the system.

There are, of course, other ways to find the solutions. We could solve the first equation for x^2, $x^2 = 25 - y^2$, and then substitute in the second, obtaining $25 - y^2 + y = 19$. Another method is to solve the second equation for y, $y = 19 - x^2$, and substitute in the first.

FIGURE 4

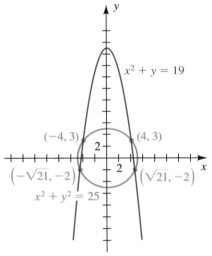

We can also consider equations in three variables x, y, and z, such as

$$x^2 y + xz + 3^y = 4z^3.$$

Such an equation has a **solution** (a, b, c) if substitution of a, b, and c, for x, y, and z, respectively, yields a true statement. We refer to (a, b, c) as an **ordered triple** of real numbers. Equivalent equations have the same solutions. A system of equations in three variables and the solutions of the system are defined as in the two-variable case. Similarly, we can consider systems of *any* number of equations in *any* number of variables.

The method of substitution can be extended to these more complicated systems. For example, given three equations in three variables, suppose that it is possible to solve one of the equations for one variable in terms of the remaining two variables. By substituting that expression in each of the other equations, we obtain a system of two equations in two variables. The solutions of the two-variable system can then be used to find the solutions of the original system.

EXAMPLE 4   *Solving a system of three equations*

Solve the system

$$\begin{cases} x - y + z = 2 \\ xyz = 0 \\ 2y + z = 1 \end{cases}$$

SOLUTION   We proceed as follows:

$$z = 1 - 2y \quad \text{solve } 2y + z = 1 \text{ for } z$$

$$\begin{cases} x - y + (1 - 2y) = 2 \\ xy(1 - 2y) = 0 \end{cases} \quad \text{substitute } z = 1 - 2y \text{ in the first two equations}$$

$$\begin{cases} x - 3y - 1 = 0 \\ xy(1 - 2y) = 0 \end{cases} \quad \text{equivalent system}$$

We now find the solutions of the last system:

$$x = 3y + 1 \quad \text{solve } x - 3y - 1 = 0 \text{ for } x$$

$$(3y + 1)y(1 - 2y) = 0 \quad \begin{array}{l} \text{substitute } x = 3y + 1 \text{ in} \\ xy(1 - 2y) = 0 \end{array}$$

$$3y + 1 = 0, \quad y = 0, \quad 1 - 2y = 0 \quad \text{set each factor equal to 0}$$

$$y = -\tfrac{1}{3}, \quad y = 0, \quad y = \tfrac{1}{2} \quad \text{solve for } y$$

These are the only possible y-values for the solutions of the system.

To obtain the corresponding x-values, we substitute for y in the equation $x = 3y + 1$, obtaining

$$x = 0, \quad x = 1, \quad \text{and} \quad x = \tfrac{5}{2}.$$

Using $z = 1 - 2y$ gives us the corresponding z-values

$$z = \tfrac{5}{3}, \quad z = 1, \quad \text{and} \quad z = 0.$$

Thus, the solutions (x, y, z) of the original system must be among the ordered triples

$$(0, -\tfrac{1}{3}, \tfrac{5}{3}), \quad (1, 0, 1), \quad \text{and} \quad (\tfrac{5}{2}, \tfrac{1}{2}, 0).$$

Checking each shows that the three ordered triples are solutions of the system.

EXAMPLE 5 *An application of a system of equations*

Is it possible to construct an aquarium with a glass top and two square ends that holds 16 ft^3 of water and requires 40 ft^2 of glass? (Disregard the thickness of the glass.)

SOLUTION We begin by sketching a typical aquarium and labeling it as in Figure 5, with x and y in feet. Referring to the figure and using formulas for volume and area, we see that

volume of the aquarium $= x^2 y$ length × width × height

square feet of glass required $= 2x^2 + 4xy$ 2 ends, 2 sides, top and bottom

FIGURE 5

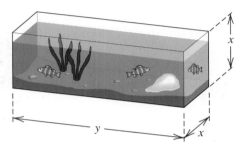

Since the volume is to be 16 ft^3 and the area of the glass required is 40 ft^2, we obtain the following system of equations:

$$\begin{cases} x^2 y = 16 \\ 2x^2 + 4xy = 40 \end{cases}$$

We find the solutions as follows:

$$y = \frac{16}{x^2} \qquad \text{solve } x^2 y = 16 \text{ for } y$$

$$2x^2 + 4x\left(\frac{16}{x^2}\right) = 40 \qquad \text{substitute } y = 16/x^2 \text{ in } 2x^2 + 4xy = 40$$

$$x^2 + \frac{32}{x} = 20 \qquad \text{cancel } x \text{ and divide by 2}$$

$$x^3 + 32 = 20x \qquad \text{multiply by } x$$

$$x^3 - 20x + 32 = 0 \qquad \text{subtract } 20x$$

We next look for rational solutions of the last equation. Dividing the polynomial $x^3 - 20x + 32$ synthetically by $x - 2$ gives us

$$\begin{array}{r|rrrr} 2 & 1 & 0 & -20 & 32 \\ & & 2 & 4 & -32 \\ \hline & 1 & 2 & -16 & 0 \end{array}$$

(continued)

Thus, one solution of $x^3 - 20x + 32 = 0$ is 2, and the remaining two solutions are zeros of the quotient $x^2 + 2x - 16$—that is, roots of the equation

$$x^2 + 2x - 16 = 0.$$

By the quadratic formula,

$$x = \frac{-2 \pm \sqrt{4 + 64}}{2} = \frac{-2 \pm 2\sqrt{17}}{2} = -1 \pm \sqrt{17}.$$

Since x is positive, we may discard $-1 - \sqrt{17}$. Hence, the only possible values of x are

$$x = 2 \qquad \text{and} \qquad x = -1 + \sqrt{17} \approx 3.12.$$

The corresponding y-values can be found by substituting for x in the equation $y = 16/x^2$. Letting $x = 2$ gives us $y = \frac{16}{4} = 4$. Using these values, we obtain the dimensions 2 feet by 2 feet by 4 feet for an aquarium.

Letting $x = -1 + \sqrt{17}$ in $y = 16/x^2$, we obtain $y = 16/(-1 + \sqrt{17})^2$, which simplifies to $y = \frac{1}{8}(9 + \sqrt{17}) \approx 1.64$. Thus, approximate dimensions for another aquarium are 3.12 feet by 3.12 feet by 1.64 feet.

6.1 EXERCISES

Exer. 1–30: Use the method of substitution to solve the system.

1. $\begin{cases} y = x^2 - 4 \\ y = 2x - 1 \end{cases}$

2. $\begin{cases} y = x^2 + 1 \\ x + y = 3 \end{cases}$

3. $\begin{cases} y^2 = 1 - x \\ x + 2y = 1 \end{cases}$

4. $\begin{cases} y^2 = x \\ x + 2y + 3 = 0 \end{cases}$

5. $\begin{cases} 2y = x^2 \\ y = 4x^3 \end{cases}$

6. $\begin{cases} x - y^3 = 1 \\ 2x = 9y^2 + 2 \end{cases}$

7. $\begin{cases} x + 2y = -1 \\ 2x - 3y = 12 \end{cases}$

8. $\begin{cases} 3x - 4y + 20 = 0 \\ 3x + 2y + 8 = 0 \end{cases}$

9. $\begin{cases} 2x - 3y = 1 \\ -6x + 9y = 4 \end{cases}$

10. $\begin{cases} 4x - 5y = 2 \\ 8x - 10y = -5 \end{cases}$

11. $\begin{cases} x + 3y = 5 \\ x^2 + y^2 = 25 \end{cases}$

12. $\begin{cases} 3x - 4y = 25 \\ x^2 + y^2 = 25 \end{cases}$

13. $\begin{cases} x^2 + y^2 = 8 \\ y - x = 4 \end{cases}$

14. $\begin{cases} x^2 + y^2 = 25 \\ 3x + 4y = -25 \end{cases}$

15. $\begin{cases} x^2 + y^2 = 9 \\ y - 3x = 2 \end{cases}$

16. $\begin{cases} x^2 + y^2 = 16 \\ y + 2x = -1 \end{cases}$

17. $\begin{cases} x^2 + y^2 = 16 \\ 2y - x = 4 \end{cases}$

18. $\begin{cases} x^2 + y^2 = 1 \\ y + 2x = -3 \end{cases}$

19. $\begin{cases} (x - 1)^2 + (y + 2)^2 = 10 \\ x + y = 1 \end{cases}$

20. $\begin{cases} xy = 2 \\ 3x - y + 5 = 0 \end{cases}$

21. $\begin{cases} y = 20/x^2 \\ y = 9 - x^2 \end{cases}$

22. $\begin{cases} x = y^2 - 4y + 5 \\ x - y = 1 \end{cases}$

23. $\begin{cases} y^2 - 4x^2 = 4 \\ 9y^2 + 16x^2 = 140 \end{cases}$

24. $\begin{cases} 25y^2 - 16x^2 = 400 \\ 9y^2 - 4x^2 = 36 \end{cases}$

25. $\begin{cases} x^2 - y^2 = 4 \\ x^2 + y^2 = 12 \end{cases}$

26. $\begin{cases} 6x^3 - y^3 = 1 \\ 3x^3 + 4y^3 = 5 \end{cases}$

27. $\begin{cases} x + 2y - z = -1 \\ 2x - y + z = 9 \\ x + 3y + 3z = 6 \end{cases}$

28. $\begin{cases} 2x - 3y - z^2 = 0 \\ x - y - z^2 = -1 \\ x^2 - xy = 0 \end{cases}$

29. $\begin{cases} x^2 + z^2 = 5 \\ 2x + y = 1 \\ y + z = 1 \end{cases}$

30. $\begin{cases} x + 2z = 1 \\ 2y - z = 4 \\ xyz = 0 \end{cases}$

Exer. 31–32: Find the points of intersection of the graphs of the equations. Sketch both graphs on the same coordinate plane, and show the points of intersection.

31 $\begin{cases} x^2 + 4y^2 = 20 \\ x + 2y = 6 \end{cases}$

32 $\begin{cases} x^2 + 4y^2 = 36 \\ x^2 + y^2 = 12 \end{cases}$

Exer. 33–34: Find the points of intersection of the graphs of the equations. Sketch both graphs on the same coordinate plane, and show the points of intersection.

33 $\begin{cases} y^2 - 4x^2 = 16 \\ y - x = 4 \end{cases}$

34 $\begin{cases} x^2 - y^2 = 4 \\ y^2 - 3x = 0 \end{cases}$

35 The perimeter of a rectangle is 40 inches, and its area is 96 in.². Find its length and width.

36 Find the values of b such that the system
$$\begin{cases} x^2 + y^2 = 4 \\ y = x + b \end{cases}$$
has

(a) one solution   (b) two solutions

(c) no solution

Interpret (a)–(c) graphically.

37 Is there a real number x such that $x = 2^{-x}$? Decide by displaying graphically the system
$$\begin{cases} y - x = 0 \\ y - 2^{-x} = 0 \end{cases}$$

38 Is there a real number x such that $x = \log x$? Decide by displaying graphically the system
$$\begin{cases} y - x = 0 \\ y - \log x = 0 \end{cases}$$

39 *Constructing tubing* Sections of cylindrical tubing are to be made from thin rectangular sheets that have an area of 200 in.² (see the figure). It is possible to construct a

EXERCISE 39

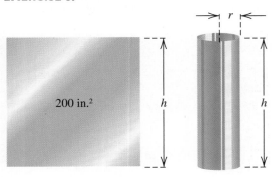

200 in.²

tube that has a volume of 200 in.³? If so, find the dimensions of the rectangular sheet.

40 Shown in the figure is the graph of $y = x^2$ and a line of slope m that passes through the point $(1, 1)$. Find the value of m such that the line intersects the graph only at $(1, 1)$, and interpret graphically.

EXERCISE 40

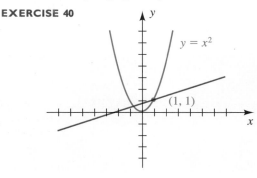

$y = x^2$

$(1, 1)$

41 *Fish population* In fishery science, spawner-recruit functions are used to predict the number of adult fish R in next year's breeding population from an estimate S of the number of fish presently spawning.

(a) For a certain species of fish, $R = aS/(S + b)$. Estimate a and b from the data in the following table.

Year	1987	1988	1989
Number spawning	40,000	60,000	72,000

(b) Predict the breeding population for the year 1990.

42 *Fish population* Refer to Exercise 41. Ricker's spawner-recruit function is given by
$$R = aSe^{-bS}$$
for positive constants a and b. This relationship predicts low recruitment from very high stocks and has been found to be appropriate for many species, such as arctic cod. Rework Exercise 41 using Ricker's spawner-recruit function.

43 *Competition for food* A *competition model* is a collection of equations that specifies how two or more species interact in competition for the food resources of an ecosystem. Let x and y denote the numbers (in hundreds) of two competing species, and suppose that the respective rates of growth R_1 and R_2 are given by
$$R_1 = 0.01x(50 - x - y),$$
$$R_2 = 0.02y(100 - y - 0.5x).$$

Determine the population levels (x, y) at which both rates of growth are zero. (Such population levels are called *stationary points*.)

44 *Fencing a region* A rancher has 2420 feet of fence to enclose a rectangular region that lies along a straight river. If no fence is used along the river (see the figure), is it possible to enclose 10 acres of land? Recall that 1 acre = 43,560 ft^2.

EXERCISE 44

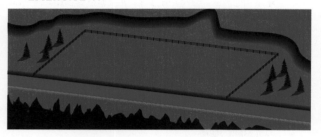

45 *Constructing an aquarium* Refer to Example 5. Is it possible to construct a small aquarium with an *open* top and two square ends that holds 2 ft^3 of water and requires 8 ft^2 of glass? If so, approximate the dimensions. (Disregard the thickness of the glass.)

46 *Isoperimetric problem* The isoperimetric problem is to prove that of all plane geometric figures with the same perimeter (isoperimetric figures), the circle has the greatest area. Show that no rectangle has both the same area and the same perimeter as any circle.

47 *Moiré pattern* A moiré pattern is formed when two geometrically regular patterns are superimposed. Shown in the figure is a pattern obtained from the family of

EXERCISE 47

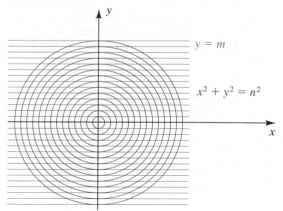

circles $x^2 + y^2 = n^2$ and the family of horizontal lines $y = m$ for integers m and n.

(a) Show that the points of intersection of the circle $x^2 + y^2 = n^2$ and the line $y = n - 1$ lie on a parabola.

(b) Work part (a) using the line $y = n - 2$.

48 *Dimensions of a pill* A spherical pill has diameter 1 centimeter. A second pill in the shape of a right circular cylinder is to be manufactured with the same volume and twice the surface area of the spherical pill.

(a) If r is the radius and h is the height of the cylindrical pill, show that $6r^2h = 1$ and $r^2 + rh = 1$. Conclude that $6r^3 - 6r + 1 = 0$.

(b) The positive solutions of $6r^3 - 6r + 1 = 0$ are approximately 0.172 and 0.903. Find the corresponding heights, and interpret these results.

49 *Hammer throw* A hammer thrower is working on his form in a small practice area. The hammer spins, generating a circle with a radius of 5 feet, and when released, it hits a tall screen that is 50 feet from the center of the throwing area. Let coordinate axes be introduced as shown in the figure (not to scale).

(a) If the hammer is released at $(-4, -3)$ and travels in the tangent direction, where will it hit the screen?

(b) If the hammer is to hit at $(0, -50)$, where on the circle should it be released?

EXERCISE 49

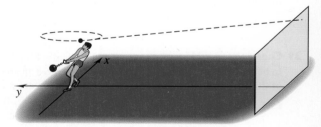

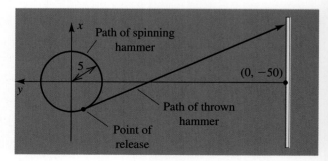

50 *Path of a tossed ball*   A person throws a ball from the edge of a hill, at an angle of 45° with the horizontal, as illustrated in the figure. The ball lands 50 feet down the

EXERCISE 50

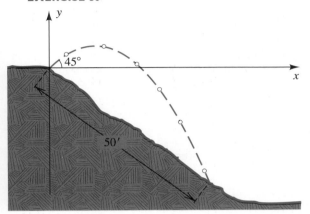

hill, which has slope $-\frac{3}{4}$. Using calculus, it can be shown that the path of the ball is given by $y = ax^2 + x + c$ for some constants a and c.

(a) Disregarding the height of the person, find an equation for the path.

(b) What is the maximum height of the ball *off the ground*?

c Exer. 51–54: Graph the two equations on the same coordinate plane and estimate the coordinates of the points of intersection.

51 $y = x^3 + x$; $x^2 + y^2 = 1$

52 $y = 3x^4 - \frac{3}{2}$; $x^2 + y^2 = 1$

53 $x^2 + (y - 1)^2 = 1$; $(x - \frac{5}{4})^2 + y^2 = 1$

54 $(x + 1)^2 + (y - 1)^2 = \frac{1}{4}$; $(x + \frac{1}{2})^2 + (y - \frac{1}{2})^2 = 1$

6.2   SYSTEMS OF LINEAR EQUATIONS IN TWO VARIABLES

An equation $ax + by = c$ (or, equivalently, $ax + by - c = 0$) with a and b not both zero is a linear equation in two variables x and y. Similarly, $ax + by + cz = d$ is a linear equation in three variables x, y, and z. We may also consider linear equations in four, five, or *any* number of variables. The most common systems of equations are those in which every equation is linear. In this section we shall consider only systems of two linear equations in two variables. Systems involving more than two variables are discussed in the next section.

Two systems of equations are equivalent if they have the same solutions. To find the solutions of a system, we may manipulate the equations until we obtain an equivalent system of simple equations for which the solutions can be found readily. Some manipulations (or *transformations*) that lead to equivalent systems are stated without proof in the next theorem.

Theorem on Equivalent Systems

Given a system of equations, an equivalent system results if

(1) two equations are interchanged.

(2) an equation is multiplied or divided by a nonzero constant.

(3) a constant multiple of one equation is added to another equation.

A *constant multiple* of an equation is obtained by multiplying *each* term of the equation by the same nonzero constant k. When applying part (3) of the theorem, we often use the phrase *add to one equation k times any other equation*. To *add* two equations means to add corresponding sides of the equations.

The next example illustrates how the theorem on equivalent systems may be used to solve a system of linear equations.

EXAMPLE I   *Using the theorem on equivalent systems*

Solve the system

$$\begin{cases} x + 3y = -1 \\ 2x - y = 5 \end{cases}$$

SOLUTION   Using the preceding theorem gives us the following equivalent systems:

$$\begin{cases} x + 3y = -1 \\ 2x - y = 5 \end{cases} \quad \text{given}$$

$$\begin{cases} x + 3y = -1 \\ 6x - 3y = 15 \end{cases} \quad \text{multiply the second equation by 3}$$

$$\begin{cases} x + 3y = -1 \\ 7x = 14 \end{cases} \quad \text{add the first equation to the second}$$

We see from the last system that $7x = 14$, and hence $x = \frac{14}{7} = 2$. To find the corresponding y-value, we substitute 2 for x in $x + 3y = -1$, obtaining $y = -1$. Thus, $(2, -1)$ is the only solution of the system.

There are many other ways to use the theorem on equivalent systems to find the solution. One such method is to proceed as follows:

$$\begin{cases} x + 3y = -1 \\ 2x - y = 5 \end{cases} \quad \text{given}$$

$$\begin{cases} -2x - 6y = 2 \\ 2x - y = 5 \end{cases} \quad \text{multiply the first equation by } -2$$

$$\begin{cases} -2x - 6y = 2 \\ -7y = 7 \end{cases} \quad \text{add the first equation to the second}$$

We see from the last system that $-7y = 7$, or $y = -1$. To find the corresponding x-value, we could substitute -1 for y in $x + 3y = -1$, obtaining $x = 2$. Hence, $(2, -1)$ is the solution.

The graphs of the two equations are lines that intersect at the point $(2, -1)$, as shown in Figure 6.

FIGURE 6

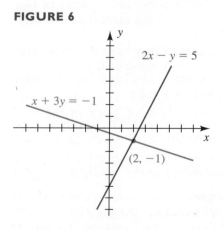

The technique used in Example 1 is called the **method of elimination**, since it involves the elimination of a variable from one of the equations.

The method of elimination usually leads to solutions in fewer steps than does the method of substitution discussed in the preceding section.

EXAMPLE 2 *A system of linear equations with an infinite number of solutions*

Solve the system

$$\begin{cases} 3x + y = 6 \\ 6x + 2y = 12 \end{cases}$$

SOLUTION Multiplying the second equation by $\frac{1}{2}$ gives us

$$\begin{cases} 3x + y = 6 \\ 3x + y = 6 \end{cases}$$

Thus, (a, b) is a solution if and only if $3a + b = 6$—that is, $b = 6 - 3a$. It follows that the solutions consist of all ordered pairs of the form $(a, 6 - 3a)$, where a is a real number. If we wish to find particular solutions, we may substitute various values for a. A few solutions are $(0, 6)$, $(1, 3)$, $(3, -3)$, $(-2, 12)$, and $(\sqrt{2}, 6 - 3\sqrt{2})$.

The graph of each equation is the same line, as shown in Figure 7.

FIGURE 7

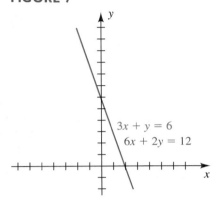

$3x + y = 6$
$6x + 2y = 12$

EXAMPLE 3 *A system of linear equations with no solutions*

Solve the system

$$\begin{cases} 3x + y = 6 \\ 6x + 2y = 20 \end{cases}$$

SOLUTION If we add to the second equation -2 times the first equation, we obtain the equivalent system

$$\begin{cases} 3x + y = 6 \\ 0 = 8 \end{cases}$$

The last equation can be written $0x + 0y = 8$, which is false for every ordered pair (x, y). Thus, the system has no solution.

The graphs of the two equations in the given system are lines that have the same slope and hence are parallel (see Figure 8). The conclusion that the system has no solution corresponds to the fact that these lines do not intersect.

FIGURE 8

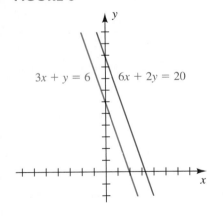

$3x + y = 6$ $6x + 2y = 20$

The preceding three examples illustrate typical outcomes of solving a system of two linear equations in two variables: there is either exactly one solution, an infinite number of solutions, or no solution. A system is **consistent** if it has at least one solution. A system with an infinite number of solutions is **dependent and consistent**. A system is **inconsistent** if it has no solution.

Since the graph of any linear equation $ax + by = c$ is a line, *exactly one* of the three cases listed in the following table holds for any system of two such equations.

Characteristics of a System of Two Linear Equations in Two Variables

Graphs	Number of solutions	Classification
Nonparallel lines	One solution	Consistent system
Identical lines	Infinite number of solutions	Dependent and consistent system
Parallel lines	No solution	Inconsistent system

In practice, there should be little difficulty determining which of the three cases occurs. The case of the unique solution will become apparent when suitable transformations are applied to the system, as illustrated in Example 1. The case of an infinite number of solutions is similar to that of Example 2, where one of the equations can be transformed into the other. The case of no solution is indicated by a contradiction, such as the statement $0 = 8$, which appeared in Example 3.

Certain applied problems can be solved by introducing systems of two linear equations, as illustrated in the next two examples.

EXAMPLE 4   *An application of a system of linear equations*

A produce company has a 100-acre farm on which it grows lettuce and cabbage. Each acre of cabbage requires 600 hours of labor, and each acre of lettuce needs 400 hours of labor. If 45,000 hours are available and if all land and labor resources are to be used, find the number of acres of each crop that should be planted.

SOLUTION   Let us introduce variables to denote the unknown quantities as follows:

$$x = \text{number of acres of cabbage}$$

$$y = \text{number of acres of lettuce}$$

Thus, the number of hours of labor required for each crop can be expressed as follows:

$$600x = \text{number of hours required for cabbage}$$

$$400y = \text{number of hours required for lettuce}$$

Using the facts that the total number of acres is 100 and the total number of hours available is 45,000 leads to the following system:

$$\begin{cases} x + y = 100 \\ 600x + 400y = 45{,}000 \end{cases}$$

We next use the method of elimination:

$$\begin{cases} x + y = 100 \\ 6x + 4y = 450 \end{cases}$$ divide the second equation by 100

$$\begin{cases} -6x - 6y = -600 \\ 6x + 4y = 450 \end{cases}$$ multiply the first equation by -6

$$\begin{cases} -6x - 6y = -600 \\ -2y = -150 \end{cases}$$ add the first equation to the second

We see from the last equation that $-2y = -150$, or $y = 75$. Substituting 75 for y in $x + y = 100$ gives us $x = 25$. Hence, the company should plant 25 acres of cabbage and 75 acres of lettuce.

EXAMPLE 5   *Finding the speed of the current in a river*

A motorboat, operating at full throttle, made a trip 4 miles upstream (against a constant current) in 15 minutes. The return trip (with the same current and at full throttle) took 12 minutes. Find the speed of the current and the equivalent speed of the boat in still water.

SOLUTION   We begin by introducing variables to denote the unknown quantities. Thus, let

$$x = \text{speed of boat (in mi/hr)}$$

$$y = \text{speed of current (in mi/hr)}.$$

We plan to use the formula $d = rt$, where d denotes the distance traveled, r the rate, and t the time. Since the current slows the boat as it travels upstream but adds to its speed as it travels downstream, we obtain

$$\text{upstream rate} = x - y \quad \text{(in mi/hr)}$$

$$\text{downstream rate} = x + y \quad \text{(in mi/hr)}.$$

The time (in hours) traveled in each direction is

$$\text{upstream time} = \tfrac{15}{60} = \tfrac{1}{4} \text{ hr}$$

$$\text{downstream time} = \tfrac{12}{60} = \tfrac{1}{5} \text{ hr}.$$

The distance is 4 miles for each trip. Substituting in $d = rt$ gives us the system

$$\begin{cases} 4 = (x - y)(\tfrac{1}{4}) \\ 4 = (x + y)(\tfrac{1}{5}) \end{cases}$$

(continued)

Applying the theorem on equivalent systems, we obtain

$$\begin{cases} x - y = 16 \\ x + y = 20 \end{cases}$$

multiply the first equation by 4 and the second by 5

$$\begin{cases} x - y = 16 \\ 2x \quad\;\; = 36 \end{cases}$$

add the first equation to the second

We see from the last equation that $2x = 36$, or $x = 18$. Substituting 18 for x in $x + y = 20$ gives us $y = 2$. Hence, the speed of the boat in still water is 18 mi/hr, and the speed of the current is 2 mi/hr.

6.2 EXERCISES

Exer. 1–20: Solve the system.

1. $\begin{cases} 2x + 3y = 2 \\ x - 2y = 8 \end{cases}$

2. $\begin{cases} 4x + 5y = 13 \\ 3x + y = -4 \end{cases}$

3. $\begin{cases} 2x + 5y = 16 \\ 3x - 7y = 24 \end{cases}$

4. $\begin{cases} 7x - 8y = 9 \\ 4x + 3y = -10 \end{cases}$

5. $\begin{cases} 3r + 4s = 3 \\ r - 2s = -4 \end{cases}$

6. $\begin{cases} 9u + 2v = 0 \\ 3u - 5v = 17 \end{cases}$

7. $\begin{cases} 5x - 6y = 4 \\ 3x + 7y = 8 \end{cases}$

8. $\begin{cases} 2x + 8y = 7 \\ 3x - 5y = 4 \end{cases}$

9. $\begin{cases} \frac{1}{3}c + \frac{1}{2}d = 5 \\ c - \frac{2}{3}d = -1 \end{cases}$

10. $\begin{cases} \frac{1}{2}t - \frac{1}{5}v = \frac{3}{2} \\ \frac{2}{3}t + \frac{1}{4}v = \frac{5}{12} \end{cases}$

11. $\begin{cases} \sqrt{3}x - \sqrt{2}y = 2\sqrt{3} \\ 2\sqrt{2}x + \sqrt{3}y = \sqrt{2} \end{cases}$

12. $\begin{cases} 0.11x - 0.03y = 0.25 \\ 0.12x + 0.05y = 0.70 \end{cases}$

13. $\begin{cases} 2x - 3y = 5 \\ -6x + 9y = 12 \end{cases}$

14. $\begin{cases} 3p - q = 7 \\ -12p + 4q = 3 \end{cases}$

15. $\begin{cases} 3m - 4n = 2 \\ -6m + 8n = -4 \end{cases}$

16. $\begin{cases} x - 5y = 2 \\ 3x - 15y = 6 \end{cases}$

17. $\begin{cases} 2y - 5x = 0 \\ 3y + 4x = 0 \end{cases}$

18. $\begin{cases} 3x + 7y = 9 \\ y = 5 \end{cases}$

19. $\begin{cases} \dfrac{2}{x} + \dfrac{3}{y} = -2 \\ \dfrac{4}{x} - \dfrac{5}{y} = 1 \end{cases}$ $\left(\textit{Hint: Let } u = \dfrac{1}{x} \text{ and } v = \dfrac{1}{y}.\right)$

20. $\begin{cases} \dfrac{3}{x-1} + \dfrac{4}{y+2} = 2 \\ \dfrac{6}{x-1} - \dfrac{7}{y+2} = -3 \end{cases}$

21. *Ticket sales* The price of admission to a high school play was $1.50 for students and $2.25 for nonstudents. If 450 tickets were sold for a total of $777.75, how many of each kind were purchased?

22. *Air travel* An airline that flies from Los Angeles to Albuquerque with a stopover in Phoenix charges a fare of $45 to Phoenix and a fare of $60 from Los Angeles to Albuquerque. A total of 185 passengers boarded the plane in Los Angeles, and fares totaled $10,500. How many passengers got off the plane in Phoenix?

23. *Crayon dimensions* A crayon 8 centimeters in length and 1 centimeter in diameter will be made from 5 cm^3 of colored wax. The crayon is to have the shape of a cylinder surmounted by a small conical tip (see the figure). Find the length x of the cylinder and the height y of the cone.

EXERCISE 23

8 cm

24 *Rowing a boat* A man rows a boat 500 feet upstream against a constant current in 10 minutes. He then rows 300 feet downstream (with the same current) in 5 minutes. Find the speed of the current and the equivalent rate at which he can row in still water.

25 *Table top dimensions* A large table for a conference room is to be constructed in the shape of a rectangle with two semicircles at the ends (see the figure). The table is to have a perimeter of 40 feet, and the area of the rectangular portion is to be twice the sum of the areas of the two ends. Find the length l and width w of the rectangular portion.

EXERCISE 25

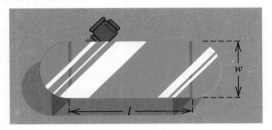

26 *Investment income* A woman has $15,000 to invest in two funds that pay simple interest at the rates of 6% and 8%. Interest on the 6% fund is tax-exempt; however, income tax must be paid on interest on the 8% fund. Being in a high tax bracket, the woman does not wish to invest the entire sum in the 8% account. Is there a way of investing the money so that she will receive $1000 in interest at the end of one year?

27 *Bobcat population* A bobcat population is classified by age into kittens (less than 1 year old) and adults (at least 1 year old). All adult females, including those born the prior year, have a litter each June, with an average litter size of 3 kittens. The springtime population of bobcats in a certain area is estimated to be 6000, and the male-female ratio is one. Estimate the number of adults and kittens in the population.

28 *Flow rates* A 300-gallon water storage tank is filled by a single inlet pipe, and two identical outlet pipes can be used to supply water to the surrounding fields (see the figure). It takes 5 hours to fill an empty tank when both outlet pipes are open. When one outlet pipe is closed, it takes 3 hours to fill the tank. Find the flow rates (in gallons per hour) in and out of the pipes.

EXERCISE 28

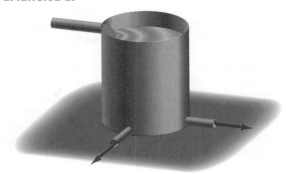

29 *Mixing a silver alloy* A silversmith has two alloys, one containing 35% silver and the other 60% silver. How much of each should be melted and combined to obtain 100 grams of an alloy containing 50% silver?

30 *Mixing nuts* A merchant wishes to mix peanuts costing $3 per pound with cashews costing $8 per pound to obtain 60 pounds of a mixture costing $5 per pound. How many pounds of each variety should be mixed?

31 *Air travel* An airplane, flying with a tail wind, travels 1200 miles in 2 hours. The return trip, against the wind, takes $2\frac{1}{2}$ hours. Find the cruising speed of the plane and the speed of the wind (assume that both rates are constant).

32 *Filling orders* A stationery company sells two types of notepads to college bookstores, the first wholesaling for 50¢ and the second for 70¢. The company receives an order for 500 notepads, together with a check for $286. If the order fails to specify the number of each type, how should the company fill the order?

33 *Acceleration* As a ball rolls down an inclined plane, its velocity $v(t)$ (in cm/sec) at time t (in seconds) is given by $v(t) = v_0 + at$ for initial velocity v_0 and acceleration a (in cm/sec²). If $v(2) = 16$ and $v(5) = 25$, find v_0 and a.

34 *Vertical projection* If an object is projected vertically upward from an altitude of s_0 feet with an initial velocity of v_0 ft/sec, then its distance $s(t)$ above the ground after t seconds is
$$s(t) = -16t^2 + v_0 t + s_0.$$
If $s(1) = 84$ and $s(2) = 116$, what are v_0 and s_0?

35 *Planning production* A small furniture company manufactures sofas and recliners. Each sofa requires 8 hours of labor and $60 in materials, while a recliner can be built for $35 in 6 hours. The company has 340 hours of labor available each week and can afford to buy $2250 worth of materials. How many recliners and sofas can be produced if all labor hours and all materials must be used?

36 Livestock diet A rancher is preparing an oat-cornmeal mixture for livestock. Each ounce of oats provides 4 grams of protein and 18 grams of carbohydrates, and an ounce of cornmeal provides 3 grams of protein and 24 grams of carbohydrates. How many ounces of each can be used to meet the nutritional goals of 200 grams of protein and 1320 grams of carbohydrates per feeding?

37 Services swap A plumber and an electrician are each doing repairs on their offices and agree to swap services. The number of hours spent on each of the projects is shown in the following table.

	Plumber's office	Electrician's office
Plumber's hours	6	4
Electrician's hours	5	6

They would prefer to call the matter even, but because of tax laws, they must charge for all work performed. They agree to select hourly wage rates so that the bill on each project will match the income that each person would ordinarily receive for a comparable job.

(a) If x and y denote the hourly wages of the plumber and electrician, respectively, show that

$$6x + 5y = 10x \quad \text{and} \quad 4x + 6y = 11y.$$

Describe the solutions to this system.

(b) If the plumber ordinarily makes $20 per hour, what should the electrician charge?

38 Find equations for the altitudes of the triangle with vertices $A(-3, 2)$, $B(5, 4)$, and $C(3, -8)$, and find the point at which the altitudes intersect.

39 Warming trend in Paris As a result of urbanization, the temperatures in Paris have increased. In 1891 the average daily minimum and maximum temperatures were 5.8 °C and 15.1 °C, respectively. Between 1891 and 1968, these average temperatures rose 0.019 °C/yr and 0.011 °C/yr, respectively. Assuming the increases were linear, find the year when the difference between the minimum and maximum temperatures was 9 °C, and determine the corresponding average maximum temperature.

40 Long distance telephone rates A telephone company charges customers a certain amount for the first minute of a long distance call and another amount for each additional minute. A customer makes two calls to the same city—a 36-minute call for $7.27 and a 13-minute call for $2.67.

(a) Determine the cost for the first minute and the cost for each additional minute.

(b) If there is a federal tax rate of 3.2% and a state tax rate of 7.2% on all long distance calls, find, to the nearest minute, the longest call to the same city whose cost will not exceed $5.00.

41 Price and demand Suppose consumers will buy 1,000,000 T-shirts if the selling price is $10, but for each $1 increase in price, they will buy 100,000 fewer T-shirts. Moreover, suppose vendors will order 2,000,000 T-shirts if the selling price is $15, and for every $1 increase in price, they will order an additional 150,000.

(a) Express the number Q of T-shirts consumers will buy if the selling price is p dollars.

(b) Express the number K of T-shirts vendors will order if the selling price is p dollars.

(c) Determine the market price—that is, the price when $Q = K$.

6.3 SYSTEMS OF LINEAR EQUATIONS IN MORE THAN TWO VARIABLES

For systems of linear equations containing more than two variables, we can use either the method of substitution explained in Section 6.1 or the method of elimination developed in Section 6.2. The method of elimination is the shorter and more straightforward technique for finding solutions. In addition, it leads to the matrix technique, discussed in this section.

EXAMPLE I   *Using the method of elimination to solve a system of linear equations*

Solve the system

$$\begin{cases} x - 2y + 3z = 4 \\ 2x + y - 4z = 3 \\ -3x + 4y - z = -2 \end{cases}$$

SOLUTION

$$\begin{cases} x - 2y + 3z = 4 \\ 5y - 10z = -5 \\ -3x + 4y - z = -2 \end{cases}$$ add -2 times the first equation to the second equation

$$\begin{cases} x - 2y + 3z = 4 \\ 5y - 10z = -5 \\ -2y + 8z = 10 \end{cases}$$ add 3 times the first equation to the third equation

$$\begin{cases} x - 2y + 3z = 4 \\ y - 2z = -1 \\ -2y + 8z = 10 \end{cases}$$ multiply the second equation by $\frac{1}{5}$

$$\begin{cases} x - 2y + 3z = 4 \\ y - 2z = -1 \\ 4z = 8 \end{cases}$$ add 2 times the second equation to the third equation

$$\begin{cases} x - 2y + 3z = 4 \\ y - 2z = -1 \\ z = 2 \end{cases}$$ multiply the third equation by $\frac{1}{4}$

The solutions of the last system are easy to find by **back substitution**. From the third equation, we see that $z = 2$. Substituting 2 for z in the second equation, $y - 2z = -1$, we get $y = 3$. Finally, we find the x-value by substituting $y = 3$ and $z = 2$ in the first equation, $x - 2y + 3z = 4$, obtaining $x = 4$. Thus, there is one solution, $(4, 3, 2)$.

Any system of three linear equations in three variables has either a *unique solution*, an *infinite number of solutions*, or *no solution*. As for two equations in two variables, the terminology used to describe these is *consistent*, *dependent and consistent*, or *inconsistent*, respectively.

If we analyze the method of solution in Example 1, we see that the symbols used for the variables are immaterial. The *coefficients* of the variables are what we must consider. Thus, if different symbols such as r, s, and t are used for the variables, we obtain the system

$$\begin{cases} r - 2s + 3t = 4 \\ 2r + s - 4t = 3 \\ -3r + 4s - t = -2 \end{cases}$$

The method of elimination can then proceed exactly as in the example. Since this is true, it is possible to simplify the process. Specifically, we introduce a scheme for keeping track of the coefficients in such a way that we do not have to write down the variables. Referring to the preceding system, we first check that variables appear in the same order in each equation and that terms not involving variables are to the right of the equal signs. We then list the numbers that are involved in the equations as follows:

$$\begin{bmatrix} 1 & -2 & 3 & 4 \\ 2 & 1 & -4 & 3 \\ -3 & 4 & -1 & -2 \end{bmatrix}$$

An array of numbers of this type is called a **matrix**. The **rows** of the matrix are the numbers that appear next to each other *horizontally*:

$$\begin{matrix} 1 & -2 & 3 & 4 & \text{first row, } R_1 \\ 2 & 1 & -4 & 3 & \text{second row, } R_2 \\ -3 & 4 & -1 & -2 & \text{third row, } R_3 \end{matrix}$$

The **columns** of the matrix are the numbers that appear next to each other *vertically*:

$$\begin{matrix} \text{first column, } C_1 & \text{second column, } C_2 & \text{third column, } C_3 & \text{fourth column, } C_4 \\ 1 & -2 & 3 & 4 \\ 2 & 1 & -4 & 3 \\ -3 & 4 & -1 & -2 \end{matrix}$$

The matrix obtained from a system of linear equations in the preceding manner is the **matrix of the system**. If we delete the last column of this matrix, the remaining array of numbers is the **coefficient matrix**. Since the matrix of the system can be obtained from the coefficient matrix by adjoining one column, we call it the **augmented coefficient matrix** or simply the **augmented matrix**. Later, when we use matrices to find the solutions of a system of linear equations, we shall introduce a vertical line segment in the augmented matrix to indicate where the equal signs would appear in the corresponding system of equations, as in the next illustration.

ILLUSTRATION

Coefficient Matrix and Augmented Matrix

$$\begin{matrix} \text{system} & \text{coefficient matrix} & \text{augmented matrix} \end{matrix}$$

$$\begin{cases} x - 2y + 3z = 4 \\ 2x + y - 4z = 3 \\ -3x + 4y - z = -2 \end{cases} \qquad \begin{bmatrix} 1 & -2 & 3 \\ 2 & 1 & -4 \\ -3 & 4 & -1 \end{bmatrix} \qquad \left[\begin{array}{ccc|c} 1 & -2 & 3 & 4 \\ 2 & 1 & -4 & 3 \\ -3 & 4 & -1 & -2 \end{array}\right]$$

Before discussing a matrix method of solving a system of linear equations, let us state a general definition of a matrix. We shall use a **double subscript notation**, denoting the number that appears in row i and column j by a_{ij}. The **row subscript** of a_{ij} is i, and the **column subscript** is j.

Definition of a Matrix

Let m and n be positive integers. An **$m \times n$ matrix** is an array of the following form, where each a_{ij} is a real number:

$$\begin{bmatrix} a_{11} & a_{12} & a_{13} & \cdots & a_{1n} \\ a_{21} & a_{22} & a_{23} & \cdots & a_{2n} \\ a_{31} & a_{32} & a_{33} & \cdots & a_{3n} \\ \vdots & \vdots & \vdots & & \vdots \\ a_{m1} & a_{m2} & a_{m3} & \cdots & a_{mn} \end{bmatrix}$$

The notation $m \times n$ in the definition is read "m by n." We often say that the matrix *is* $m \times n$ and call $m \times n$ the **size** of the matrix. It is possible to consider matrices in which the symbols a_{ij} represent complex numbers, polynomials, or other mathematical objects; however, we shall not do so in this text. The rows and columns of a matrix are defined as before. Thus, the matrix in the definition has m rows and n columns. Note that a_{23} is in row 2 and column 3 and a_{32} is in row 3 and column 2. Each a_{ij} is an **element of the matrix**. The elements $a_{ij}, a_{22}, a_{23}, \ldots$ are the **main diagonal elements**. If $m = n$, the matrix is a **square matrix of order n**.

ILLUSTRATION

$m \times n$ Matrices

$$\blacksquare \overset{2 \times 3}{\begin{bmatrix} -5 & 3 & 1 \\ 7 & 0 & -2 \end{bmatrix}} \qquad \blacksquare \overset{2 \times 2}{\begin{bmatrix} 5 & -1 \\ 2 & 3 \end{bmatrix}} \qquad \blacksquare \overset{1 \times 3}{[3 \quad 1 \quad -2]}$$

$$\blacksquare \overset{3 \times 2}{\begin{bmatrix} 2 & -1 \\ 0 & 1 \\ 8 & 3 \end{bmatrix}} \qquad \blacksquare \overset{3 \times 1}{\begin{bmatrix} -4 \\ 0 \\ 5 \end{bmatrix}}$$

To find the solutions of a system of linear equations, we begin with the augmented matrix. If a variable does not appear in an equation, we assume that the coefficient is zero. We then work with the rows of the matrix *just as though they were equations*. The only items missing are the symbols for the variables, the addition signs used between terms, and the equal signs. We simply keep in mind that the numbers in the first column are the coefficients of the first variable, the numbers in the second column are the coefficients of the second variable, and so on. The rules for transforming a matrix are formulated so that they always produce a matrix of an equivalent system of equations.

The next theorem is a restatement, in terms of matrices, of the theorem on equivalent systems in Section 6.2. In part (2) of the theorem, the terminology *a row is multiplied by a nonzero constant* means that each element in the row is multiplied by the constant. To *add* two rows of a matrix, as in part (3), we add corresponding elements in each row.

Matrix Row Transformation Theorem

Given a matrix of a system of linear equations, a matrix of an equivalent system results if

(1) two rows are interchanged.

(2) a row is multiplied by a nonzero constant.

(3) a constant multiple of one row is added to another row.

We refer to 1–3 as the **elementary row transformations** of a matrix. If a matrix is obtained from another matrix by one or more elementary row transformations, the two matrices are said to be **equivalent** or, more precisely, **row equivalent**. We shall use the symbols in the following chart to denote elementary row transformations of a matrix, where the arrow $\rightarrow$ may be read "replaces." Thus, for the transformation $k\mathrm{R}_i \rightarrow \mathrm{R}_i$, the constant multiple $k\mathrm{R}_i$ *replaces* R_i. Similarly, for $k\mathrm{R}_i + \mathrm{R}_j \rightarrow \mathrm{R}_j$, the sum $k\mathrm{R}_i + \mathrm{R}_j$ *replaces* R_j. For convenience, we shall write $(-1)\mathrm{R}_i$ as $-\mathrm{R}_i$.

Elementary Row Transformations of a Matrix

Symbol	Meaning
$\mathrm{R}_i \leftrightarrow \mathrm{R}_j$	Interchange rows R_i and R_j
$k\mathrm{R}_i \rightarrow \mathrm{R}_i$	Multiply row R_i by k
$k\mathrm{R}_i + \mathrm{R}_j \rightarrow \mathrm{R}_j$	Add $k\mathrm{R}_i$ to row R_j

We shall next rework Example 1 using matrices. You should compare the two solutions, since analogous steps are employed in each case.

EXAMPLE 2   *Using matrices to solve a system of linear equations*

Solve the system

$$\begin{cases} x - 2y + 3z = 4 \\ 2x + y - 4z = 3 \\ -3x + 4y - z = -2 \end{cases}$$

SOLUTION   We begin with the matrix of the system—that is, with the augmented matrix:

$$\begin{bmatrix} 1 & -2 & 3 & 4 \\ 2 & 1 & -4 & 3 \\ -3 & 4 & -1 & -2 \end{bmatrix}$$

We next apply elementary row transformations to obtain another (simpler) matrix of an equivalent system of equations. These transformations correspond to the manipulations used for equations in Example 1. We will place appropriate symbols between equivalent matrices.

$$\begin{bmatrix} 1 & -2 & 3 & 4 \\ 2 & 1 & -4 & 3 \\ -3 & 4 & -1 & -2 \end{bmatrix} \xrightarrow[\substack{-2R_1 + R_2 \to R_2 \\ 3R_1 + R_3 \to R_3}]{} \begin{bmatrix} 1 & -2 & 3 & 4 \\ 0 & 5 & -10 & -5 \\ 0 & -2 & 8 & 10 \end{bmatrix} \quad \begin{array}{l} \text{add } -2R_1 \text{ to } R_2 \\ \text{add } 3R_1 \text{ to } R_3 \end{array}$$

$$\xrightarrow[\frac{1}{5}R_2 \to R_2]{} \begin{bmatrix} 1 & -2 & 3 & 4 \\ 0 & 1 & -2 & -1 \\ 0 & -2 & 8 & 10 \end{bmatrix} \quad \text{multiply } R_2 \text{ by } \tfrac{1}{5}$$

$$\xrightarrow[2R_2 + R_3 \to R_3]{} \begin{bmatrix} 1 & -2 & 3 & 4 \\ 0 & 1 & -2 & -1 \\ 0 & 0 & 4 & 8 \end{bmatrix} \quad \text{add } 2R_2 \text{ to } R_3$$

$$\xrightarrow[\frac{1}{4}R_3 \to R_3]{} \begin{bmatrix} 1 & -2 & 3 & 4 \\ 0 & 1 & -2 & -1 \\ 0 & 0 & 1 & 2 \end{bmatrix} \quad \text{multiply } R_3 \text{ by } \tfrac{1}{4}$$

We use the final matrix to return to the system of equations

$$\begin{cases} x - 2y + 3z = 4 \\ y - 2z = -1 \\ z = 2 \end{cases}$$

which is equivalent to the original system. The solution $x = 4, y = 3, z = 2$ may now be found by back substitution, as in Example 1.

The final matrix in the solution of Example 2 is in **echelon form**. In general, a matrix is in echelon form if it satisfies the following conditions.

Echelon Form of a Matrix

> **(1)** The first nonzero number in each row, reading from left to right, is 1.
>
> **(2)** The column containing the first nonzero number in any row is to the left of the column containing the first nonzero number in the row below.
>
> **(3)** Rows consisting entirely of zeros may appear at the bottom of the matrix.

The following is an illustration of a 6×7 matrix in echelon form. The symbols a_{ij} represent real numbers.

ILLUSTRATION

Echelon Form

$$\begin{bmatrix} 1 & a_{12} & a_{13} & a_{14} & a_{15} & a_{16} & a_{17} \\ 0 & 1 & a_{23} & a_{24} & a_{25} & a_{26} & a_{27} \\ 0 & 0 & 0 & 1 & a_{35} & a_{36} & a_{37} \\ 0 & 0 & 0 & 0 & 0 & 1 & a_{47} \\ 0 & 0 & 0 & 0 & 0 & 0 & 0 \\ 0 & 0 & 0 & 0 & 0 & 0 & 0 \end{bmatrix}$$

The following guidelines may be used to find echelon forms.

Guidelines for Finding the Echelon Form of a Matrix

1 Locate the *first* column that contains nonzero elements, and apply elementary row transformations to get the number 1 into the first row of that column.

2 Apply elementary row transformations of the type $k\mathrm{R}_1 + \mathrm{R}_j \to \mathrm{R}_j$ for $j > 1$ to get 0 underneath the 1 obtained in guideline 1 in each of the remaining rows.

3 *Disregard the first row.* Locate the next column that contains nonzero elements, and apply elementary row transformations to get the number 1 into the *second* row of that column.

4 Apply elementary row transformations of the type $k\mathrm{R}_2 + \mathrm{R}_j \to \mathrm{R}_j$ for $j > 2$ to get 0 underneath the 1 obtained in guideline 3 in each of the remaining rows.

5 *Disregard the first and second rows.* Locate the next column that contains nonzero elements, and repeat the procedure.

6 Continue the process until the echelon form is reached.

Not all echelon forms contain rows consisting of only zeros (see Example 2).

We can use elementary row operations to transform the matrix of any system of linear equations to echelon form. The echelon form can then be used to produce a system of equations that is equivalent to the original system. The solutions of the given system may be found by back substitution. The next example illustrates this technique for a system of four linear equations.

EXAMPLE 3   *Using an echelon form to solve a system of linear equations*

Solve the system

$$\begin{cases} -2x + 3y + 4z = -1 \\ x - 2z + 2w = 1 \\ y + z - w = 0 \\ 3x + y - 2z - w = 3 \end{cases}$$

SOLUTION   We have arranged the equations so that the same variables appear in vertical columns. We begin with the augmented matrix and then obtain an echelon form as described in the guidelines.

$$\begin{bmatrix} -2 & 3 & 4 & 0 & | & -1 \\ 1 & 0 & -2 & 2 & | & 1 \\ 0 & 1 & 1 & -1 & | & 0 \\ 3 & 1 & -2 & -1 & | & 3 \end{bmatrix} \overset{R_1 \leftrightarrow R_2}{} \begin{bmatrix} 1 & 0 & -2 & 2 & | & 1 \\ -2 & 3 & 4 & 0 & | & -1 \\ 0 & 1 & 1 & -1 & | & 0 \\ 3 & 1 & -2 & -1 & | & 3 \end{bmatrix}$$

$$\begin{matrix} \\ 2R_1 + R_2 \rightarrow R_2 \\ \\ -3R_1 + R_4 \rightarrow R_4 \end{matrix} \begin{bmatrix} 1 & 0 & -2 & 2 & | & 1 \\ 0 & 3 & 0 & 4 & | & 1 \\ 0 & 1 & 1 & -1 & | & 0 \\ 0 & 1 & 4 & -7 & | & 0 \end{bmatrix}$$

$$R_2 \leftrightarrow R_3 \begin{bmatrix} 1 & 0 & -2 & 2 & | & 1 \\ 0 & 1 & 1 & -1 & | & 0 \\ 0 & 3 & 0 & 4 & | & 1 \\ 0 & 1 & 4 & -7 & | & 0 \end{bmatrix}$$

$$\begin{matrix} \\ \\ -3R_2 + R_3 \rightarrow R_3 \\ -R_2 + R_4 \rightarrow R_4 \end{matrix} \begin{bmatrix} 1 & 0 & -2 & 2 & | & 1 \\ 0 & 1 & 1 & -1 & | & 0 \\ 0 & 0 & -3 & 7 & | & 1 \\ 0 & 0 & 3 & -6 & | & 0 \end{bmatrix}$$

$$\begin{matrix} \\ \\ \\ R_3 + R_4 \rightarrow R_4 \end{matrix} \begin{bmatrix} 1 & 0 & -2 & 2 & | & 1 \\ 0 & 1 & 1 & -1 & | & 0 \\ 0 & 0 & -3 & 7 & | & 1 \\ 0 & 0 & 0 & 1 & | & 1 \end{bmatrix}$$

$$-\tfrac{1}{3}R_3 \rightarrow R_3 \begin{bmatrix} 1 & 0 & -2 & 2 & | & 1 \\ 0 & 1 & 1 & -1 & | & 0 \\ 0 & 0 & 1 & -\tfrac{7}{3} & | & -\tfrac{1}{3} \\ 0 & 0 & 0 & 1 & | & 1 \end{bmatrix}$$

(continued)

The final matrix is in echelon form and corresponds to the following system of equations:

$$\begin{cases} x & - 2z + 2w = & 1 \\ & y + \ z - \ w = & 0 \\ & z - \frac{7}{3}w = & -\frac{1}{3} \\ & w = & 1 \end{cases}$$

We now use back substitution to find the solution. From the last equation we see that $w = 1$. Substituting in the third equation, $z - \frac{7}{3}w = -\frac{1}{3}$, we get

$$z - \frac{7}{3}(1) = -\frac{1}{3}, \quad \text{or} \quad z = \frac{6}{3} = 2.$$

Substituting $w = 1$ and $z = 2$ in the second equation, $y + z - w = 0$, we obtain

$$y + 2 - 1 = 0, \quad \text{or} \quad y = -1.$$

Finally, from the first equation, $x - 2z + 2w = 1$, we have

$$x - 2(2) + 2(1) = 1, \quad \text{or} \quad x = 3.$$

Hence, the system has one solution, $x = 3$, $y = -1$, $z = 2$, and $w = 1$.

After obtaining an echelon form, it is often convenient to apply additional elementary row operations of the type $k\mathrm{R}_i + \mathrm{R}_j \to \mathrm{R}_j$ so that 0 also appears *above* the first 1 in each row. We refer to the resulting matrix as being in **reduced echelon form**. The following is an illustration of a 6×7 matrix in reduced echelon form. (Compare it with the echelon form on page 384.)

ILLUSTRATION

Reduced Echelon Form

$$\begin{bmatrix} 1 & 0 & a_{13} & 0 & a_{15} & 0 & a_{17} \\ 0 & 1 & a_{23} & 0 & a_{25} & 0 & a_{27} \\ 0 & 0 & 0 & 1 & a_{35} & 0 & a_{37} \\ 0 & 0 & 0 & 0 & 0 & 1 & a_{47} \\ 0 & 0 & 0 & 0 & 0 & 0 & 0 \\ 0 & 0 & 0 & 0 & 0 & 0 & 0 \end{bmatrix}$$

EXAMPLE 4 *Using a reduced echelon form to solve a system of linear equations*

Solve the system in Example 3 using reduced echelon form.

SOLUTION We begin with the echelon form obtained in Example 3 and apply additional row operations as follows:

$$\begin{bmatrix} 1 & 0 & -2 & 2 & | & 1 \\ 0 & 1 & 1 & -1 & | & 0 \\ 0 & 0 & 1 & -\frac{7}{3} & | & -\frac{1}{3} \\ 0 & 0 & 0 & 1 & | & 1 \end{bmatrix} \begin{matrix} -2R_4 + R_1 \to R_1 \\ R_4 + R_2 \to R_2 \\ \frac{7}{3}R_4 + R_3 \to R_3 \end{matrix} \begin{bmatrix} 1 & 0 & -2 & 0 & | & -1 \\ 0 & 1 & 1 & 0 & | & 1 \\ 0 & 0 & 1 & 0 & | & 2 \\ 0 & 0 & 0 & 1 & | & 1 \end{bmatrix}$$

$$\begin{matrix} 2R_3 + R_1 \to R_1 \\ -R_3 + R_2 \to R_2 \end{matrix} \begin{bmatrix} 1 & 0 & 0 & 0 & | & 3 \\ 0 & 1 & 0 & 0 & | & -1 \\ 0 & 0 & 1 & 0 & | & 2 \\ 0 & 0 & 0 & 1 & | & 1 \end{bmatrix}$$

The system of equations corresponding to the reduced echelon form gives us the solution *without* using back substitution:

$$x = 3, \quad y = -1, \quad z = 2, \quad w = 1$$

Sometimes it is necessary to consider systems in which the number of equations is not the same as the number of variables. The same matrix techniques are applicable, as illustrated in the next example.

EXAMPLE 5 *Solving a system of two linear equations in three variables*

Solve the system

$$\begin{cases} 2x + 3y + 4z = 1 \\ 3x + 4y + 5z = 3 \end{cases}$$

SOLUTION We shall begin with the augmented matrix and then find a reduced echelon form. There are many different ways of getting the number 1 into the first position of the first row. For example, the elementary row transformation $\frac{1}{2}R_1 \to R_1$ or $-\frac{1}{3}R_2 + R_1 \to R_1$ would accomplish this in one step. Another way, which does not involve fractions, is the following:

$$\begin{bmatrix} 2 & 3 & 4 & | & 1 \\ 3 & 4 & 5 & | & 3 \end{bmatrix} \begin{matrix} R_1 \leftrightarrow R_2 \end{matrix} \begin{bmatrix} 3 & 4 & 5 & | & 3 \\ 2 & 3 & 4 & | & 1 \end{bmatrix}$$

$$\begin{matrix} -R_2 + R_1 \to R_1 \end{matrix} \begin{bmatrix} 1 & 1 & 1 & | & 2 \\ 2 & 3 & 4 & | & 1 \end{bmatrix}$$

$$\begin{matrix} -2R_1 + R_2 \to R_2 \end{matrix} \begin{bmatrix} 1 & 1 & 1 & | & 2 \\ 0 & 1 & 2 & | & -3 \end{bmatrix}$$

$$\begin{matrix} -R_2 + R_1 \to R_1 \end{matrix} \begin{bmatrix} 1 & 0 & -1 & | & 5 \\ 0 & 1 & 2 & | & -3 \end{bmatrix}$$

(continued)

The reduced echelon form is the matrix of the system

$$\begin{cases} x & - & z = & 5 \\ & y + & 2z = & -3 \end{cases}$$

or, equivalently,

$$\begin{cases} x = & z + 5 \\ y = & -2z - 3 \end{cases}$$

There are an infinite number of solutions to this system; they can be found by assigning z any value c and then using the last two equations to express x and y in terms of c. This gives us

$$x = c + 5, \quad y = -2c - 3, \quad z = c.$$

Thus, the solutions of the system consist of all ordered triples of the form

$$(c + 5, -2c - 3, c)$$

for any real number c. The solutions may be checked by substituting $c + 5$ for x, $-2c - 3$ for y, and c for z in the two given equations.

We can obtain any number of solutions for the system by substituting specific real numbers for c. For example, if $c = 0$, we obtain $(5, -3, 0)$; if $c = 2$, we have $(7, -7, 2)$; and so on.

There are other ways to specify the general solution. For example, starting with $x = z + 5$ and $y = -2z - 3$, we could let $z = d - 5$ for any real number d. In this case,

$$x = z + 5 = (d - 5) + 5 = d$$

$$y = -2z - 3 = -2(d - 5) - 3 = -2d + 7,$$

and the solutions of the system have the form

$$(d, -2d + 7, d - 5).$$

These triples produce the same solutions as $(c + 5, -2c - 3, c)$. For example, if $d = 5$, we get $(5, -3, 0)$; if $d = 7$, we obtain $(7, -7, 2)$; and so on.

A system of linear equations is **homogeneous** if all the terms that do not contain variables—that is, the *constant terms*—are zero. A system of homogeneous equations always has the **trivial solution** obtained by substituting zero for each variable. Nontrivial solutions sometimes exist. The procedure for finding solutions is the same as that used for nonhomogeneous systems.

EXAMPLE 6   *Solving a homogeneous system of linear equations*

Solve the homogeneous system

$$\begin{cases} x - y + 4z = 0 \\ 2x + y - z = 0 \\ -x - y + 2z = 0 \end{cases}$$

SOLUTION   We begin with the augmented matrix and find a reduced echelon form:

$$\begin{bmatrix} 1 & -1 & 4 & | & 0 \\ 2 & 1 & -1 & | & 0 \\ -1 & -1 & 2 & | & 0 \end{bmatrix} \begin{matrix} \\ -2R_1 + R_2 \to R_2 \\ R_1 + R_3 \to R_3 \end{matrix} \begin{bmatrix} 1 & -1 & 4 & | & 0 \\ 0 & 3 & -9 & | & 0 \\ 0 & -2 & 6 & | & 0 \end{bmatrix}$$

$$\tfrac{1}{3}R_2 \to R_2 \begin{bmatrix} 1 & -1 & 4 & | & 0 \\ 0 & 1 & -3 & | & 0 \\ 0 & -2 & 6 & | & 0 \end{bmatrix}$$

$$\begin{matrix} R_2 + R_1 \to R_1 \\ \\ 2R_2 + R_3 \to R_3 \end{matrix} \begin{bmatrix} 1 & 0 & 1 & | & 0 \\ 0 & 1 & -3 & | & 0 \\ 0 & 0 & 0 & | & 0 \end{bmatrix}$$

The reduced echelon form corresponds to the system

$$\begin{cases} x \quad + \quad z = 0 \\ \quad y - 3z = 0 \end{cases}$$

or, equivalently,

$$\begin{cases} x = -z \\ y = 3z \end{cases}$$

Assigning any value c to z, we obtain $x = -c$ and $y = 3c$. The solutions consist of all ordered triples of the form $(-c, 3c, c)$ for any real number c.

EXAMPLE 7   *A homogeneous system with only the trivial solution*

Solve the system

$$\begin{cases} x + y + z = 0 \\ x - y + z = 0 \\ x - y - z = 0 \end{cases}$$

SOLUTION   We begin with the augmented matrix and find a reduced echelon form:

$$\begin{bmatrix} 1 & 1 & 1 & | & 0 \\ 1 & -1 & 1 & | & 0 \\ 1 & -1 & -1 & | & 0 \end{bmatrix} \begin{matrix} \\ -R_1 + R_2 \rightarrow R_2 \\ -R_1 + R_3 \rightarrow R_3 \end{matrix} \begin{bmatrix} 1 & 1 & 1 & | & 0 \\ 0 & -2 & 0 & | & 0 \\ 0 & -2 & -2 & | & 0 \end{bmatrix}$$

$$-\tfrac{1}{2}R_2 \rightarrow R_2 \begin{bmatrix} 1 & 1 & 1 & | & 0 \\ 0 & 1 & 0 & | & 0 \\ 0 & -2 & -2 & | & 0 \end{bmatrix}$$

$$\begin{matrix} -R_2 + R_1 \rightarrow R_1 \\ \\ 2R_2 + R_3 \rightarrow R_3 \end{matrix} \begin{bmatrix} 1 & 0 & 1 & | & 0 \\ 0 & 1 & 0 & | & 0 \\ 0 & 0 & -2 & | & 0 \end{bmatrix}$$

$$-\tfrac{1}{2}R_3 \rightarrow R_3 \begin{bmatrix} 1 & 0 & 1 & | & 0 \\ 0 & 1 & 0 & | & 0 \\ 0 & 0 & 1 & | & 0 \end{bmatrix}$$

$$-R_3 + R_1 \rightarrow R_1 \begin{bmatrix} 1 & 0 & 0 & | & 0 \\ 0 & 1 & 0 & | & 0 \\ 0 & 0 & 1 & | & 0 \end{bmatrix}$$

The reduced echelon form is the matrix of the system

$$x = 0, \qquad y = 0, \qquad z = 0.$$

Thus, the only solution for the given system is the trivial one, $(0, 0, 0)$.

The next two examples illustrate applied problems.

EXAMPLE 8   *Using a system of equations to determine maximum profit*

A manufacturer of electrical equipment has the following information about the weekly profit from the production and sale of a type of electric motor.

Production level x	25	50	100
Profit $P(x)$ (dollars)	5250	7500	4500

(a) Determine a, b, and c so that the graph of $P(x) = ax^2 + bx + c$ fits this information.

(b) According to the quadratic function P in part (a), how many motors should be produced each week for maximum profit? What is the maximum weekly profit?

SOLUTION

(a) We see from the table that the graph of $P(x) = ax^2 + bx + c$ contains the points (25, 5250), (50, 7500), and (100, 4500). This gives us the system of equations

$$\begin{cases} 5250 = 625a + 25b + c \\ 7500 = 2500a + 50b + c \\ 4500 = 10{,}000a + 100b + c \end{cases}$$

We can verify that the solution is $a = -2$, $b = 240$, $c = 500$.

(b) From part (a),

$$P(x) = -2x^2 + 240x + 500.$$

Since $a = -2 < 0$, the graph of the quadratic function P is a parabola that opens downward. By the formula on page 187, the x-coordinate of the vertex (the highest point on the parabola) is

$$x = \frac{-b}{2a} = \frac{-240}{2(-2)} = \frac{-240}{-4} = 60.$$

Hence, for the maximum profit, the manufacturer should produce and sell 60 motors per week. The maximum weekly profit is

$$P(60) = -2(60)^2 + 240(60) + 500 = \$7700.$$

EXAMPLE 9   *Solving a mixture problem*

A merchant wishes to mix two grades of peanuts costing $3 and $4 per pound, respectively, with cashews costing $8 per pound, to obtain 140 pounds of a mixture costing $6 per pound. If the merchant also wants the amount of cheaper-grade peanuts to be twice that of the better-grade peanuts, how many pounds of each variety should be mixed?

SOLUTION   Let us introduce three variables as follows:

$$x = \text{number of pounds of peanuts at \$3 per pound}$$

$$y = \text{number of pounds of peanuts at \$4 per pound}$$

$$z = \text{number of pounds of cashews at \$8 per pound}$$

We refer to the statement of the problem and obtain the following system:

$$\begin{cases} x + y + z = 140 \\ 3x + 4y + 8z = 6(140) \\ x = 2y \end{cases}$$

You may verify that the solution of this system is $x = 40$, $y = 20$, $z = 80$. Thus, the merchant should use 40 pounds of the $3 peanuts, 20 pounds of the $4 peanuts, and 80 pounds of cashews.

6.3 EXERCISES

Exer. 1–26: Use matrices to solve the system.

1. $\begin{cases} x - 2y - 3z = -1 \\ 2x + y + z = 6 \\ x + 3y - 2z = 13 \end{cases}$

2. $\begin{cases} x + 3y - z = -3 \\ 3x - y + 2z = 1 \\ 2x - y + z = -1 \end{cases}$

3. $\begin{cases} 5x + 2y - z = -7 \\ x - 2y + 2z = 0 \\ 3y + z = 17 \end{cases}$

4. $\begin{cases} 4x - y + 3z = 6 \\ -8x + 3y - 5z = -6 \\ 5x - 4y = -9 \end{cases}$

5. $\begin{cases} 2x + 6y - 4z = 1 \\ x + 3y - 2z = 4 \\ 2x + y - 3z = -7 \end{cases}$

6. $\begin{cases} x + 3y - 3z = -5 \\ 2x - y + z = -3 \\ -6x + 3y - 3z = 4 \end{cases}$

7. $\begin{cases} 2x - 3y + 2z = -3 \\ -3x + 2y + z = 1 \\ 4x + y - 3z = 4 \end{cases}$

8. $\begin{cases} 2x - 3y + z = 2 \\ 3x + 2y - z = -5 \\ 5x - 2y + z = 0 \end{cases}$

9. $\begin{cases} x + 3y + z = 0 \\ x + y - z = 0 \\ x - 2y - 4z = 0 \end{cases}$

10. $\begin{cases} 2x - y + z = 0 \\ x - y - 2z = 0 \\ 2x - 3y - z = 0 \end{cases}$

11. $\begin{cases} 2x + y + z = 0 \\ x - 2y - 2z = 0 \\ x + y + z = 0 \end{cases}$

12. $\begin{cases} x + y - 2z = 0 \\ x - y - 4z = 0 \\ y + z = 0 \end{cases}$

13. $\begin{cases} 3x - 2y + 5z = 7 \\ x + 4y - z = -2 \end{cases}$

14. $\begin{cases} 2x - y + 4z = 8 \\ -3x + y - 2z = 5 \end{cases}$

15. $\begin{cases} 4x - 2y + z = 5 \\ 3x + y - 4z = 0 \end{cases}$

16. $\begin{cases} 5x + 2y - z = 10 \\ y + z = -3 \end{cases}$

17. $\begin{cases} x + 2y - z - 3w = 2 \\ 3x + y - 2z - w = 6 \\ x + y + 3z - 2w = -3 \\ -2x - 2y + 3z + w = -9 \end{cases}$

18. $\begin{cases} x - 2y - 5z + w = -1 \\ 2x - y + z + w = 1 \\ 3x - 2y - 4z - 2w = 1 \\ x + y + 3z - 2w = 2 \end{cases}$

19. $\begin{cases} 2x - y - 2z + 2s - 5t = 2 \\ x + 3y - 2z + s - 2t = -5 \\ -x + 4y + 2z - 3s + 8t = -4 \\ 3x - 2y - 4z + s - 3t = -3 \\ 4x - 6y + z - 2s + t = 10 \end{cases}$

20. $\begin{cases} 3x + 2y + z + 3u + v + w = 1 \\ 2x + y - 2z + 3u - v + 4w = 6 \\ 6x + 3y + 4z - u + 2v + w = -6 \\ x + y + z + u - v - w = 8 \\ -2x - 2y + z - 3u + 2v - 3w = -10 \\ x - 3y + 2z + u + 3v + w = -1 \end{cases}$

21. $\begin{cases} 5x + 2z = 1 \\ y - 3z = 2 \\ 2x + y = 3 \end{cases}$

22. $\begin{cases} 2x - 3y = 12 \\ 3y + z = -2 \\ 5x - 3z = 3 \end{cases}$

23. $\begin{cases} 4x - 3y = 1 \\ 2x + y = -7 \\ -x + y = -1 \end{cases}$

24. $\begin{cases} 2x + 3y = -2 \\ x + y = 1 \\ x - 2y = 13 \end{cases}$

25. $\begin{cases} 2x + 3y = 5 \\ x - 3y = 4 \\ x + y = -2 \end{cases}$

26. $\begin{cases} 4x - y = 2 \\ 2x + 2y = 1 \\ 4x - 5y = 3 \end{cases}$

27. *Mixing acid solutions* Three solutions contain a certain acid. The first contains 10% acid, the second 30%, and the third 50%. A chemist wishes to use all three solutions to obtain a 50-liter mixture containing 32% acid. If the chemist wants to use twice as much of the 50% solution as of the 30% solution, how many liters of each solution should be used?

28. *Filling a pool* A swimming pool can be filled by three pipes A, B, and C. Pipe A alone can fill the pool in 8 hours. If pipes A and C are used together, the pool can be filled in 6 hours. If B and C are used together, it takes 10 hours. How long does it take to fill the pool if all three pipes are used?

29. *Production capability* A company has three machines A, B, and C that are each capable of producing a certain item. However, because of a lack of skilled operators, only two of the machines can be used simultaneously. The following table indicates production over a three-day period using various combinations of the machines.

Machines used	Hours used	Items produced
A and B	6	4500
A and C	8	3600
B and C	7	4900

How long would it take each machine, if used alone, to produce 1000 items?

30 Electrical resistance In electrical circuits, the formula $1/R = (1/R_1) + (1/R_2)$ is used to find the total resistance R if two resistors R_1 and R_2 are connected in parallel. Given three resistors A, B, and C, suppose that the total resistance is 48 ohms if A and B are connected in parallel, 80 ohms if B and C are connected in parallel, and 60 ohms if A and C are connected in parallel. Find the resistances of A, B, and C.

31 Mixing fertilizers A supplier of lawn products has three types of grass fertilizer G_1, G_2, and G_3, having nitrogen contents of 30%, 20%, and 15%, respectively. The supplier plans to mix them, obtaining 600 pounds of fertilizer with a 25% nitrogen content. The mixture is to contain 100 pounds more of type G_3 than of type G_2. How much of each type should be used?

32 Particle acceleration If a particle moves along a coordinate line with a constant acceleration a (in cm/sec^2), then at time t (in seconds) its distance $s(t)$ (in centimeters) from the origin is

$$s(t) = \tfrac{1}{2}at^2 + v_0t + s_0$$

for the velocity v_0 and distance s_0 from the origin at $t = 0$. If the distances of the particle from the origin at $t = \tfrac{1}{2}$, $t = 1$, and $t = \tfrac{3}{2}$ are 7, 11, and 17, respectively, find a, v_0, and s_0.

33 Electrical currents Shown in the figure is a schematic of an electrical circuit containing three resistors, a 6-volt battery, and a 12-volt battery. It can be shown, using Kirchhoff's laws, that the three currents I_1, I_2, and I_3 are solutions of the following system of equations:

$$\begin{cases} I_1 - I_2 + I_3 = 0 \\ R_1I_1 + R_2I_2 = 6 \\ R_2I_2 + R_3I_3 = 12 \end{cases}$$

Find the three currents if

(a) $R_1 = R_2 = R_3 = 3$ ohms

(b) $R_1 = 4$ ohms, $R_2 = 1$ ohm, and $R_3 = 4$ ohms

EXERCISE 33

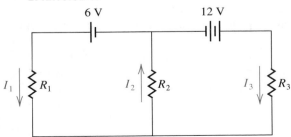

34 Bird population A stable population of 35,000 birds lives on three islands. Each year 10% of the population on island A migrates to island B, 20% of the population on island B migrates to island C, and 5% of the population on island C migrates to island A. Find the number of birds on each island if the population count on each island does not vary from year to year.

35 Blending coffees A shop specializes in preparing blends of gourmet coffees. From Columbian, Brazilian, and Kenyan coffees, the owner wishes to prepare 1-pound bags that will sell for $8.50. The cost per pound of these coffees is $10, $6, and $8, respectively. The amount of Columbian is to be three times the amount of Brazilian. Find the amount of each type of coffee in the blend.

36 Cattle population A rancher has 750 head of cattle consisting of 400 adults (aged 2 or more years), 150 yearlings, and 200 calves. The following information is known about this particular species. Each spring an adult female gives birth to a single calf, and 75% of these calves will survive the first year. The yearly survival percentages for yearlings and adults are 80% and 90%, respectively. The male-female ratio is one in all age classes. Estimate the population of each age class

(a) next spring **(b)** last spring

37 Traffic flow Shown in the figure is a system of four one-way streets leading into the center of a city. The numbers in the figure denote the average number of vehicles per hour that travel in the directions shown. A total of 300 vehicles enter the area and 300 vehicles leave the area every hour. Signals at intersections A, B, C, and D are to be timed in order to avoid congestion, and this timing will determine traffic flow rates x_1, x_2, x_3, and x_4.

(a) If the number of vehicles entering an intersection per hour must equal the number leaving the intersection per hour, describe the traffic flow rates at each intersection with a system of equations.

(b) If the signal at intersection C is timed so that x_3 is equal to 100, find x_1, x_2, and x_4.

(continued)

EXERCISE 37

(c) Make use of the system in part (a) to explain why $75 \le x_3 \le 150$.

38 If $f(x) = ax^3 + bx + c$, determine a, b, and c such that the graph of f passes through the points $P(-3, -12)$, $Q(-1, 22)$, and $R(2, 13)$.

c 39 Air pollution Between 1850 and 1985 approximately 155 billion metric tons of carbon was added to the earth's atmosphere and the climate became about $0.5\,°C$ warmer, an indication of the *greenhouse effect*. It is estimated that doubling the carbon dioxide (CO_2) in the atmosphere would result in an average global temperature increase of 4–$5\,°C$. The future amount A of CO_2 in the atmosphere in parts per million is sometimes estimated using the formula $A = a + ct + ke^{rt}$, where a, c, and k are constants, r is the percentage increase in the emission of CO_2, and t is time in years, with $t = 0$ corresponding to 1990. Suppose it is estimated that in the year 2070, A will be 800 if $r = 2.5\%$ and A will be 560 if $r = 1.5\%$. If, in 1990, $A = 340$ and $r = 1\%$, find the year in which the amount of CO_2 in the atmosphere will have doubled.

c 40 Air pollution Refer to Exercise 39. Suppose it is estimated that in the year 2030, A will be 455 if $r = 2.0\%$ and A will be 430 if $r = 1.5\%$. If, in 1990, $A = 340$ and

$r = 2.5\%$, find the year in which the amount of CO_2 in the atmosphere will have doubled.

Exer. 41–42: Find an equation of the parabola that has a vertical axis and passes through the given points.

41 $P(2, 5)$, $Q(-2, -3)$, $R(1, 6)$

42 $P(3, -1)$, $Q(1, -7)$, $R(-2, 14)$

Exer. 43–44: Find an equation of the parabola that has a horizontal axis and passes through the given points.

43 $P(-1, 1)$, $Q(11, -2)$, $R(5, -1)$

44 $P(2, 1)$, $Q(6, 2)$, $R(12, -1)$

Exer. 45–46: Find an equation of the circle of the form $x^2 + y^2 + ax + by + c = 0$ that passes through the given points.

45 $P(2, 1)$, $Q(-1, -4)$, $R(3, 0)$

46 $P(-5, 5)$, $Q(-2, -4)$, $R(2, 4)$

47 If $f(x) = ax^3 + bx^2 + cx + d$, find a, b, c, and d if the graph of f is to pass through $(-1, 2)$, $(0.5, 2)$, $(1, 3)$, and $(2, 4.5)$.

48 If $f(x) = ax^4 + bx^3 + cx^2 + dx + e$, find a, b, c, d, and e if the graph of f is to pass through $(-2, 1.5)$, $(-1, -2)$, $(1, -3)$, $(2, -3.5)$, and $(3, -4.8)$.

6.4 PARTIAL FRACTIONS

In this section we show how systems of equations can be used to help decompose rational expressions into sums of simpler expressions. This technique is useful in advanced mathematics courses.

We may verify that

$$\frac{2}{x^2 - 1} = \frac{1}{x - 1} + \frac{-1}{x + 1}$$

by adding the fractions $1/(x - 1)$ and $-1/(x + 1)$ to obtain $2/(x^2 - 1)$. The expression on the right side of this equation is called the *partial fraction decomposition* of $2/(x^2 - 1)$.

It is theoretically possible to write *any* rational expression as a sum of rational expressions whose denominators involve powers of polynomials of degree not greater than two. Specifically, if $f(x)$ and $g(x)$ are polynomials *and the degree of $f(x)$ is less than the degree of $g(x)$*, it can be proved that

$$\frac{f(x)}{g(x)} = F_1 + F_2 + \cdots + F_r$$

such that each F_k has one of the forms

$$\frac{A}{(px + q)^m} \quad \text{or} \quad \frac{Ax + B}{(ax^2 + bx + c)^n},$$

where A and B are real numbers, m and n are nonnegative integers, and the quadratic polynomial $ax^2 + bx + c$ is irreducible over $\mathbb{R}$ (that is, has no real zero). The sum $F_1 + F_2 + \cdots + F_r$ is the **partial fraction decomposition** of $f(x)/g(x)$, and each F_k is a **partial fraction**.

For the partial fraction decomposition of $f(x)/g(x)$ to be found, *it is essential that $f(x)$ have lower degree than $g(x)$.* If this is not the case, we can use long division to obtain such an expression. For example, given

$$\frac{x^3 - 6x^2 + 5x - 3}{x^2 - 1},$$

we obtain

$$\frac{x^3 - 6x^2 + 5x - 3}{x^2 - 1} = x - 6 + \frac{6x - 9}{x^2 - 1}.$$

We then find the partial fraction decomposition of $(6x - 9)/(x^2 - 1)$.

The following guidelines can be used to obtain decompositions.

Guidelines for Finding Partial Fraction Decompositions of $f(x)/g(x)$

1 If the degree of the numerator $f(x)$ is not lower than the degree of the denominator $g(x)$, use long division to obtain the proper form.

2 Factor the denominator $g(x)$ into a product of linear factors $px + q$ or irreducible quadratic factors $ax^2 + bx + c$, and collect repeated factors so that $g(x)$ is a product of *different* factors of the form $(px + q)^m$ or $(ax^2 + bx + c)^n$ for a nonnegative integer m or n.

3 Apply the following rules to the factors found in guideline 2.
Rule A: For each factor of the form $(px + q)^m$ with $m \geq 1$, the partial fraction decomposition contains a sum of m partial fractions of the form

$$\frac{A_1}{px + q} + \frac{A_2}{(px + q)^2} + \cdots + \frac{A_m}{(px + q)^m},$$

where each numerator A_k is a real number.
Rule B: For each factor of the form $(ax^2 + bx + c)^n$ with $n \geq 1$ and $ax^2 + bx + c$ irreducible, the partial fraction decomposition contains a sum of n partial fractions of the form

$$\frac{A_1 x + B_1}{ax^2 + bx + c} + \frac{A_2 x + B_2}{(ax^2 + bx + c)^2} + \cdots + \frac{A_n x + B_n}{(ax^2 + bx + c)^n},$$

where each A_k and each B_k is a real number.

4 Find the numbers A_k and B_k in guideline 3.

We shall apply the preceding guidelines in the following examples. For the sake of convenience, we will use the variables A, B, C, and so on, rather than the subscripted variables A_k and B_k given in the guidelines.

EXAMPLE I *A partial fraction decomposition in which each denominator is linear*

Find the partial fraction decomposition of

$$\frac{4x^2 + 13x - 9}{x^3 + 2x^2 - 3x}.$$

SOLUTION

Guideline 1 The degree of the numerator, 2, is less than the degree of the denominator, 3, so long division is not required.

Guideline 2 We factor the denominator:

$$x^3 + 2x^2 - 3x = x(x^2 + 2x - 3) = x(x + 3)(x - 1)$$

Guideline 3 Each factor of the denominator has the form stated in Rule A, with $m = 1$. Thus, to the factor x there corresponds a partial fraction of the form A/x. Similarly, to the factors $x + 3$ and $x - 1$ there correspond partial fractions of the form $B/(x + 3)$ and $C/(x - 1)$, respectively. The partial fraction decomposition has the form

$$\frac{4x^2 + 13x - 9}{x^3 + 2x^2 - 3x} = \frac{A}{x} + \frac{B}{x + 3} + \frac{C}{x - 1}.$$

Guideline 4 We find the values of A, B, and C in guideline 3. Multiplying both sides of the partial fraction decomposition by the lcd, $x(x + 3)(x - 1)$, gives us

$$
\begin{aligned}
4x^2 + 13x - 9 &= A(x + 3)(x - 1) + Bx(x - 1) + Cx(x + 3) \\
&= A(x^2 + 2x - 3) + B(x^2 - x) + C(x^2 + 3x) \\
&= (A + B + C)x^2 + (2A - B + 3C)x - 3A.
\end{aligned}
$$

Equating the coefficients of like powers of x on each side of the last equation, we obtain the system of equations

$$
\begin{cases}
A + B + \ C = \ \ \ 4 \\
2A - B + 3C = \ \ 13 \\
-3A \qquad\qquad = -9
\end{cases}
$$

Using the methods of the preceding section yields the solution $A = 3$, $B = -1$, and $C = 2$. Hence, the partial fraction decomposition is

$$\frac{4x^2 + 13x - 9}{x(x + 3)(x - 1)} = \frac{3}{x} + \frac{-1}{x + 3} + \frac{2}{x - 1}.$$

There is an alternative way to find A, B, and C if all factors of the denominator are linear and nonrepeated, as in this example. Instead of equating coefficients and using a system of equations, we begin with the equation

$$4x^2 + 13x - 9 = A(x + 3)(x - 1) + Bx(x - 1) + Cx(x + 3).$$

We next substitute values for x that make the factors, x, $x - 1$, and $x - 3$, zero. If we let $x = 0$ and simplify, we obtain

$$-9 = -3A, \quad \text{or} \quad A = 3.$$

Letting $x = 1$ in the equation leads to $8 = 4C$, or $C = 2$. Finally, if $x = -3$, then we have $-12 = 12B$, or $B = -1$.

EXAMPLE 2 *A partial fraction decomposition containing a repeated linear factor*

Find the partial fraction decomposition of

$$\frac{x^2 + 10x - 36}{x(x - 3)^2}.$$

SOLUTION

Guideline 1 The degree of the numerator, 2, is less than the degree of the denominator, 3, so long division is not required.

Guideline 2 The denominator, $x(x - 3)^2$, is already in factored form.

Guideline 3 By Rule A with $m = 1$, there is a partial fraction of the form A/x corresponding to the factor x. Next, applying Rule A with $m = 2$, we find that the factor $(x - 3)^2$ determines a sum of two partial fractions of the form $B/(x - 3)$ and $C/(x - 3)^2$. Thus, the partial fraction decomposition has the form

$$\frac{x^2 + 10x - 36}{x(x - 3)^2} = \frac{A}{x} + \frac{B}{x - 3} + \frac{C}{(x - 3)^2}.$$

Guideline 4 To find A, B, and C, we begin by multiplying both sides of the partial fraction decomposition in guideline 3 by the lcd, $x(x - 3)^2$:

$$\begin{aligned}
x^2 + 10x - 36 &= A(x - 3)^2 + Bx(x - 3) + Cx \\
&= A(x^2 - 6x + 9) + B(x^2 - 3x) + Cx \\
&= (A + B)x^2 + (-6A - 3B + C)x + 9A
\end{aligned}$$

We next equate the coefficients of like powers of x, obtaining the system

$$\begin{cases}
A + B & = 1 \\
-6A - 3B + C & = 10 \\
9A & = -36
\end{cases}$$

(continued)

This system of equations has the solution $A = -4$, $B = 5$, $C = 1$. The partial fraction decomposition is therefore

$$\frac{x^2 + 10x - 36}{x(x - 3)^2} = \frac{-4}{x} + \frac{5}{x - 3} + \frac{1}{(x - 3)^2}.$$

As in Example 1, we could also obtain A and C by beginning with the equation

$$x^2 + 10x - 36 = A(x - 3)^2 + Bx(x - 3) + Cx$$

and then substituting values for x that make the factors, $x - 3$ and x, zero. Thus, letting $x = 3$, we obtain $3 = 3C$, or $C = 1$. Letting $x = 0$ gives us $-36 = 9A$, or $A = -4$. The value of B may then be found by using one of the equations in the system.

EXAMPLE 3   *A partial fraction decomposition containing an irreducible quadratic factor*

Find the partial fraction decomposition of

$$\frac{4x^3 - x^2 + 15x - 29}{2x^3 - x^2 + 8x - 4}.$$

SOLUTION

Guideline 1   The degree of the numerator, 3, is *equal* to the degree of the denominator. Thus, long division is required, and we obtain

$$\frac{4x^3 - x^2 + 15x - 29}{2x^3 - x^2 + 8x - 4} = 2 + \frac{x^2 - x - 21}{2x^3 - x^2 + 8x - 4}.$$

Guideline 2   The denominator may be factored by grouping, as follows:

$$2x^3 - x^2 + 8x - 4 = x^2(2x - 1) + 4(2x - 1) = (x^2 + 4)(2x - 1)$$

Guideline 3   Applying Rule B to the irreducible quadratic factor $x^2 + 4$ in guideline 2, we see that one of the partial fractions has the form $(Ax + B)/(x^2 + 4)$. By Rule A, there is also a partial fraction $C/(2x - 1)$ corresponding to $2x - 1$. Consequently,

$$\frac{x^2 - x - 21}{2x^3 - x^2 + 8x - 4} = \frac{Ax + B}{x^2 + 4} + \frac{C}{2x - 1}.$$

Guideline 4   Multiplying both sides of the partial fraction decomposition in guideline 3 by the lcd, $(x^2 + 4)(2x - 1)$, we obtain

$$\begin{aligned}
x^2 - x - 21 &= (Ax + B)(2x - 1) + C(x^2 + 4) \\
&= 2Ax^2 - Ax + 2Bx - B + Cx^2 + 4C \\
&= (2A + C)x^2 + (-A + 2B)x - B + 4C.
\end{aligned}$$

This leads to the system

$$\begin{cases} 2A & + \; C = & 1 \\ -A + 2B & = & -1 \\ & - \; B + 4C = & -21 \end{cases}$$

This system has the solution $A = 3$, $B = 1$, and $C = -5$. Thus, the partial fraction decomposition in guideline 3 is

$$\frac{x^2 - x - 21}{2x^3 - x^2 + 8x - 4} = \frac{3x + 1}{x^2 + 4} + \frac{-5}{2x - 1},$$

and therefore the decomposition of the given expression (see guideline 1) is

$$\frac{4x^3 - x^2 + 15x - 29}{2x^3 - x^2 + 8x - 4} = 2 + \frac{3x + 1}{x^2 + 4} + \frac{-5}{2x - 1}.$$

EXAMPLE 4 *A partial fraction decomposition containing a repeated quadratic factor*

Find the partial fraction decomposition of

$$\frac{5x^3 - 3x^2 + 7x - 3}{(x^2 + 1)^2}.$$

SOLUTION

Guideline 1 The degree of the numerator, 3, is less than the degree of the denominator, 4, so long division is not required.

Guideline 2 The denominator, $(x^2 + 1)^2$, is already in the factored form.

Guideline 3 We apply Rule B, with $n = 2$, to $(x^2 + 1)^2$ to obtain the partial fraction decomposition

$$\frac{5x^3 - 3x^2 + 7x - 3}{(x^2 + 1)^2} = \frac{Ax + B}{x^2 + 1} + \frac{Cx + D}{(x^2 + 1)^2}.$$

Guideline 4 Multiplying both sides of the decomposition in guideline 3 by $(x^2 + 1)^2$ gives us

$$5x^3 - 3x^2 + 7x - 3 = (Ax + B)(x^2 + 1) + Cx + D$$
$$= Ax^3 + Bx^2 + (A + C)x + (B + D).$$

Comparing the coefficients of x^3 and x^2, we obtain $A = 5$ and $B = -3$. From the coefficients of x, we see that $A + C = 7$. Thus, $C = 7 - A = 7 - 5 = 2$. Finally, comparing the constant terms gives us $B + D = -3$ and $D = -3 - B = -3 - (-3) = 0$. Therefore, the partial fraction decomposition is

$$\frac{5x^3 - 3x^2 + 7x - 3}{(x^2 + 1)^2} = \frac{5x - 3}{x^2 + 1} + \frac{2x}{(x^2 + 1)^2}.$$

6.4 EXERCISES

Exer. 1–28: Find the partial fraction decomposition.

1. $\dfrac{8x - 1}{(x - 2)(x + 3)}$

2. $\dfrac{x - 29}{(x - 4)(x + 1)}$

3. $\dfrac{x + 34}{x^2 - 4x - 12}$

4. $\dfrac{5x - 12}{x^2 - 4x}$

5. $\dfrac{4x^2 - 15x - 1}{(x - 1)(x + 2)(x - 3)}$

6. $\dfrac{x^2 + 19x + 20}{x(x + 2)(x - 5)}$

7. $\dfrac{4x^2 - 5x - 15}{x^3 - 4x^2 - 5x}$

8. $\dfrac{37 - 11x}{(x + 1)(x^2 - 5x + 6)}$

9. $\dfrac{2x + 3}{(x - 1)^2}$

10. $\dfrac{5x^2 - 4}{x^2(x + 2)}$

11. $\dfrac{19x^2 + 50x - 25}{3x^3 - 5x^2}$

12. $\dfrac{10 - x}{x^2 + 10x + 25}$

13. $\dfrac{x^2 - 6}{(x + 2)^2(2x - 1)}$

14. $\dfrac{2x^2 + x}{(x - 1)^2(x + 1)^2}$

15. $\dfrac{3x^3 + 11x^2 + 16x + 5}{x(x + 1)^3}$

16. $\dfrac{4x^3 + 3x^2 + 5x - 2}{x^3(x + 2)}$

17. $\dfrac{x^2 + x - 6}{(x^2 + 1)(x - 1)}$

18. $\dfrac{x^2 - x - 21}{(x^2 + 4)(2x - 1)}$

19. $\dfrac{9x^2 - 3x + 8}{x^3 + 2x}$

20. $\dfrac{2x^3 + 2x^2 + 4x - 3}{x^4 + x^2}$

21. $\dfrac{4x^3 - x^2 + 4x + 2}{(x^2 + 1)^2}$

22. $\dfrac{3x^3 + 13x - 1}{(x^2 + 4)^2}$

23. $\dfrac{2x^4 - 2x^3 + 6x^2 - 5x + 1}{x^3 - x^2 + x - 1}$

24. $\dfrac{x^3}{x^3 - 3x^2 + 9x - 27}$

25. $\dfrac{3x^2 - 16}{x^2 - 4x}$

26. $\dfrac{2x^2 + 7x}{x^2 + 6x + 9}$

27. $\dfrac{4x^3 + 4x^2 - 4x + 2}{2x^2 - x - 1}$

28. $\dfrac{x^5 - 5x^4 + 7x^3 - x^2 - 4x + 12}{x^3 - 3x^2}$

6.5 SYSTEMS OF INEQUALITIES

In Chapter 2 we restricted our discussion to inequalities in one variable. We shall now consider inequalities in two variables x and y, such as those listed in the following illustration.

ILLUSTRATION

Inequalities in x and y

- $y^2 < x + 4$ ▪ $3x - 4y > 12$ ▪ $x^2 + y^2 \le 16$

A **solution** of an inequality in x and y is an ordered pair (a, b) that yields a true statement if a and b are substituted for x and y, respectively. To **solve** an inequality in x and y means to find all the solutions. The **graph** of such an inequality is the set of all points (a, b) in an xy-plane that correspond to the solutions. Two inequalities are **equivalent** if they have the same solutions.

Given an inequality in x and y, if we replace the inequality symbol with an equal sign, we obtain an equation whose graph usually separates the xy-plane into several regions. We shall consider only equations having the property that if R is one such region and if a **test point** (p, q) in R

yields a solution of the inequality, then *every* point in R yields a solution. The following guidelines may then be used to sketch the graph of the inequality.

Guidelines for Sketching the Graph of an Inequality in x and y

1 Replace the inequality symbol with an equal sign and graph the resulting equation. Use dashes if the inequality symbol is $<$ or $>$ to indicate that no point on the graph yields a solution. Use a solid line or curve for $\leq$ or $\geq$ to indicate that solutions of the equation are also solutions of the inequality.

2 If R is a region of the xy-plane determined by the graph in guideline 1 and if a test point (p, q) in R yields a solution of the inequality, then every point in R yields a solution. Shade R to indicate this fact. If (p, q) is not a solution, then *no* point in R yields a solution and R is left unshaded.

The use of these guidelines is demonstrated in the next example.

FIGURE 9

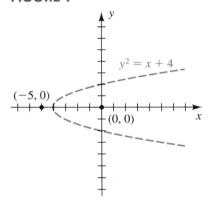

EXAMPLE I *Sketching the graph of an inequality*

Find the solutions and sketch the graph of the inequality $y^2 < x + 4$.

SOLUTION

Guideline I We replace $<$ with $=$, obtaining $y^2 = x + 4$. The graph of this equation is a parabola, symmetric with respect to the x-axis and having x-intercept -4 and y-intercepts ± 2. Since the inequality symbol is $<$, we sketch the parabola using dashes, as in Figure 9.

Guideline 2 The graph in guideline 1 separates the xy-plane into two regions, one to the *right* of the parabola and the other to the *left*. Let us choose test points $(0, 0)$ and $(-5, 0)$ in the regions (see Figure 9) and substitute for x and y in $y^2 < x + 4$ as follows:

FIGURE 10

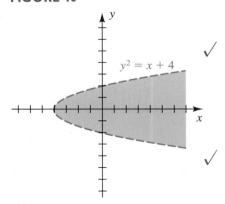

✓ *Test point* **$(0, 0)$**   LS: $0^2 = 0$

RS: $0 + 4 = 4$

Since $0 < 4$ is a *true* statement, $(0, 0)$ *is* a solution of the inequality. Hence, *all* points to the right of the parabola yield solutions, so we shade this region, as in Figure 10.

✓ *Test point* **$(-5, 0)$**   LS: $0^2 = 0$

RS: $-5 + 4 = -1$

Since $0 < -1$ is a *false* statement, $(-5, 0)$ *is not* a solution of the inequality. Hence, *no* point to the left of the parabola is a solution, and we leave that region unshaded.

FIGURE 11

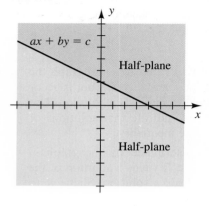

A **linear inequality** is an inequality that can be written in one of the following forms, where a, b, and c are real numbers:

$$ax + by < c, \qquad ax + by > c, \qquad ax + by \le c, \qquad ax + by \ge c$$

The line $ax + by = c$ separates the xy-plane into two **half-planes**, as illustrated in Figure 11. The solutions of the inequality consist of all points in *one* of these half-planes, where the line is included for $\le$ or $\ge$ and is not included for $<$ or $>$. *For a linear inequality, only one test point (p, q) is required,* because if (p, q) is a solution, then the half-plane with (p, q) in it contains all the solutions, whereas if (p, q) is *not* a solution, then the *other* half-plane contains the solutions.

EXAMPLE 2   *Sketching the graph of a linear inequality*

Sketch the graph of the inequality $3x - 4y > 12$.

FIGURE 12

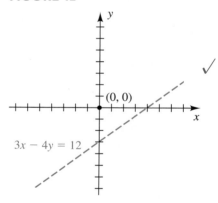

SOLUTION   Replacing $>$ with $=$ gives us the line $3x - 4y = 12$, sketched with dashes in Figure 12. This line separates the xy-plane into two half-planes, one *above* the line and the other *below* the line. It is convenient to choose the test point $(0, 0)$ above the line and substitute in $3x - 4y > 12$, as follows:

> **Test point $(0, 0)$**   LS: $3 \cdot 0 - 4 \cdot 0 = 0 - 0 = 0$
>
> RS: 12

Since $0 > 12$ is a false statement, $(0, 0)$ is not a solution. Thus, no point above the line is a solution, and the solutions of $3x - 4y > 12$ are given by the points in the half-plane *below* the line. The graph is sketched in Figure 13.

As we did with equations, we sometimes work simultaneously with several inequalities in two variables—that is, with a **system of inequalities**. The **solutions** of a system of inequalities are the solutions common to all inequalities in the system. The **graph** of a system of inequalities consists of the points corresponding to the solutions. The following examples illustrate a method for solving systems of inequalities.

FIGURE 13

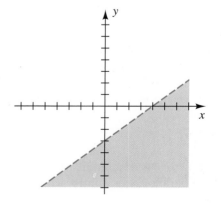

EXAMPLE 3   *Solving a system of linear inequalities*

Sketch the graph of the system

$$\begin{cases} x + y \le 4 \\ 2x - y \le 4 \end{cases}$$

SOLUTION   We replace each $\le$ with $=$ and then sketch the resulting lines, as shown in Figure 14. Using the test point $(0, 0)$, we see that the solutions of the system correspond to the points *below* (and on) the line

FIGURE 14

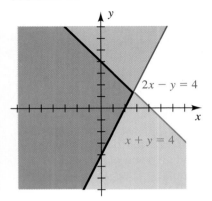

$x + y = 4$ and *above* (and on) the line $2x - y = 4$. Shading these half-planes with different colors, as in Figure 14, we have as the graph of the system the points that are in *both* regions, indicated by the purple portion of the figure.

EXAMPLE 4 *Solving a system of linear inequalities*

Sketch the graph of the system

$$\begin{cases} x + y \leq 4 \\ 2x - y \leq 4 \\ x \geq 0 \\ y \geq 0 \end{cases}$$

FIGURE 15

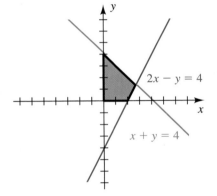

SOLUTION The first two inequalities are the same as those considered in Example 3, and hence the points on the graph of the system must lie within the purple region shown in Figure 14. In addition, the third and fourth inequalities in the system tell us that the points must lie in the first quadrant or on its boundaries. This gives us the region shown in Figure 15.

EXAMPLE 5 *Solving a system of inequalities containing absolute values*

Sketch the graph of the system

$$\begin{cases} |x| \leq 2 \\ |y| > 1 \end{cases}$$

FIGURE 16

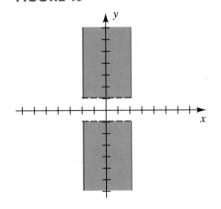

SOLUTION Using properties of absolute value, we see that (x, y) is a solution of the system if and only if *both* of the following conditions are true:

(1) $-2 \leq x \leq 2$

(2) $y < -1$ or $y > 1$

Thus, a point (x, y) on the graph of the system must lie between (or on) the vertical lines $x = \pm 2$ and also either below the horizontal line $y = -1$ or above the line $y = 1$. The graph is sketched in Figure 16.

EXAMPLE 6 *Solving a system of inequalities*

Sketch the graph of the system

$$\begin{cases} x^2 + y^2 \leq 16 \\ x + y \geq 2 \end{cases}$$

FIGURE 17

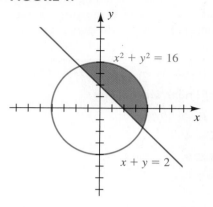

SOLUTION   The graphs of $x^2 + y^2 = 16$ and $x + y = 2$ are the circle and line shown in Figure 17. Using the test point $(0, 0)$, we see that the points that yield solutions of the system must lie inside (or on) the circle and also above (or on) the line. This gives us the region sketched in Figure 17.

EXAMPLE 7   *An application of a system of inequalities*

The manager of a baseball team wishes to buy bats and balls costing $12 and $3 each, respectively. At least five bats and ten balls are required, and the total cost is not to exceed $180. Find a system of inequalities that describes all possibilities, and sketch the graph.

SOLUTION   We begin by letting x denote the number of bats and y the number of balls. Since the cost of a bat is $12 and the cost of a ball is $3, we see that

$$12x = \text{cost of } x \text{ bats}$$

$$3y = \text{cost of } y \text{ balls.}$$

Since the total cost is not to exceed $180, we must have

$$12x + 3y \le 180$$

or, equivalently, $y \le -4x + 60.$

Since at least five bats and ten balls are required, we also have

$$x \ge 5 \qquad \text{and} \qquad y \ge 10.$$

FIGURE 18

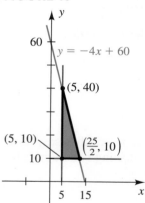

The graph of $y \le -4x + 60$ is the half-plane that lies *below* (or on) the line $y = -4x + 60$ shown in Figure 18.

The graph of $x \ge 5$ is the region to the right (or on) the vertical line $x = 5$, and the graph of $y \ge 10$ is the region above (or on) the horizontal line $y = 10$.

The graph of the system—that is, the points common to the three half-planes—is the triangular region sketched in Figure 18.

6.5 EXERCISES

Exer. 1–10: Sketch the graph of the inequality.

1 $3x - 2y < 6$

2 $4x + 3y < 12$

3 $2x + 3y \ge 2y + 1$

4 $2x - y > 3$

5 $y + 2 < x^2$

6 $y^2 - x \le 0$

7 $x^2 + 1 \le y$

8 $y - x^3 < 1$

9 $yx^2 \ge 1$

10 $x^2 + 4 \ge y$

Exer. 11–26: Sketch the graph of the system of inequalities.

11 $\begin{cases} 3x + y < 3 \\ 4 - y < 2x \end{cases}$

12 $\begin{cases} y + 2 < 2x \\ y - x > 4 \end{cases}$

13 $\begin{cases} y - x < 0 \\ 2x + 5y < 10 \end{cases}$

14 $\begin{cases} 2y - x \le 4 \\ 3y + 2x < 6 \end{cases}$

15 $\begin{cases} 3x + y \le 6 \\ y - 2x \ge 1 \\ x \ge -2 \\ y \le 4 \end{cases}$

16 $\begin{cases} 3x - 4y \ge 12 \\ x - 2y \le 2 \\ x \ge 9 \\ y \le 5 \end{cases}$

17 $\begin{cases} x + 2y \le 8 \\ 0 \le x \le 4 \\ 0 \le y \le 3 \end{cases}$

18 $\begin{cases} 2x + 3y \ge 6 \\ 0 \le x \le 5 \\ 0 \le y \le 4 \end{cases}$

19 $\begin{cases} |x| \ge 2 \\ |y| < 3 \end{cases}$

20 $\begin{cases} |x| \ge 4 \\ |y| \ge 3 \end{cases}$

21 $\begin{cases} |x + 2| \le 1 \\ |y - 3| < 5 \end{cases}$

22 $\begin{cases} |x - 2| \le 5 \\ |y - 4| > 2 \end{cases}$

23 $\begin{cases} x^2 + y^2 \le 4 \\ x + y \ge 1 \end{cases}$

24 $\begin{cases} x^2 + y^2 > 1 \\ x^2 + y^2 < 4 \end{cases}$

25 $\begin{cases} x^2 \le 1 - y \\ x \ge 1 + y \end{cases}$

26 $\begin{cases} x - y^2 < 0 \\ x + y^2 > 0 \end{cases}$

28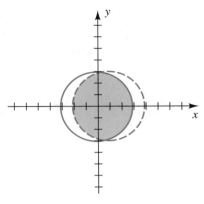

29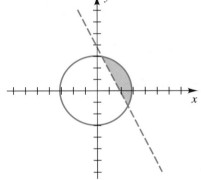

Exer. 27–34: Find a system of inequalities whose graph is shown.

27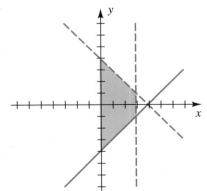

30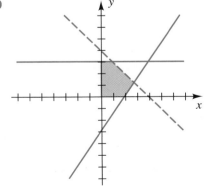

31

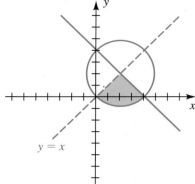

$y = x$

32

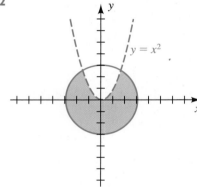

$y = x^2$

33

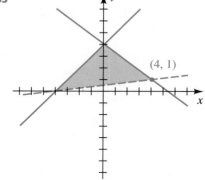

$(4, 1)$

34

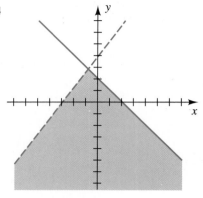

35 *Inventory levels*   A store sells two brands of television sets. Customer demand indicates that it is necessary to stock at least twice as many sets of brand A as of brand B. It is also necessary to have on hand at least 10 sets of brand B. There is room for not more than 100 sets in the store. Find and graph a system of inequalities that describes all possibilities for stocking the two brands.

36 *Ticket prices*   An auditorium contains 600 seats. For an upcoming event tickets will be priced at $8.00 for some seats and $5.00 for others. At least 225 tickets are to be priced at $5.00, and total sales of at least $3000 are desired. Find and graph a system of inequalities that describes all possibilities for pricing the two types of tickets.

37 *Investment strategy*   A woman with $15,000 to invest decides to place at least $2000 in a high-risk, high-yield investment and at least three times that amount in a low-risk, low-yield investment. Find and graph a system of inequalities that describes all possibilities for placing the money in the two investments.

38 *Inventory levels*   The manager of a college bookstore stocks two types of notebooks, the first wholesaling for 55 cents and the second for 85 cents. The maximum amount to be spent is $600, and an inventory of at least 300 of the 85-cent variety and 400 of the 55-cent variety is desired. Find and graph a system of inequalities that describes all possibilities for stocking the two types of notebooks.

39 *Dimensions of a can*   An aerosol can is to be constructed in the shape of a circular cylinder with a small cone on the top. The total height of the can is to be no more than 9 inches, and the cylinder must contain at least 75% of the total volume. In addition, the height of the conical top must be at least 1 inch. Find and graph a system of inequalities that describes all possibilities for the relationship between the height y of the cylinder and the height x of the cone.

40 *Dimensions of a window* A stained-glass window is to be constructed in the form of a rectangle surmounted by a semicircle (see the figure). The total height h of the window can be no more than 6 feet, and the area of the rectangular part must be at least twice the area of the semicircle. In addition, the diameter d of the semicircle must be at least 2 feet. Find and graph a system of inequalities that describes all possibilities for the base and height of the rectangular part.

EXERCISE 40

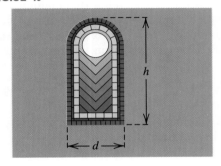

41 *Locating a power plant* A nuclear power plant will be constructed to serve the power needs of cities A and B. City B is 100 miles due east of A. The state has promised that the plant will be at least 60 miles from each city. It is not possible, however, to locate the plant south of either city because of rough terrain, and the plant must be within 100 miles of both A and B. Assuming A is at the origin, find and graph a system of inequalities that describes all possible locations for the plant.

42 *Allocating space* A man has a rectangular back yard that is 50 feet wide and 60 feet deep. He plans to construct a pool area and a patio area, as shown in the figure, where $y \geq 10$. He can spend at most $10,500 on the project. The patio area must be at least as large as the pool area. The pool area will cost $5 per square foot, and the patio will cost $3 per square foot. Find and graph a system of inequalities that describes all possibilities for the width of the patio and pool areas.

EXERCISE 42

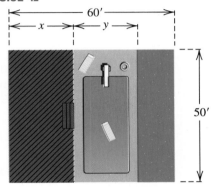

c **Exer. 43–44: Graph the inequality.**

43 $64y^3 - x^3 \leq e^{1-2x}$ **44** $e^{5y} - e^{-x} \geq x^4$

c **Exer. 45–48: Graph the system of inequalities.**

45 $\begin{cases} 5^{1-y} \geq x^4 + x^2 + 1 \\ x + 3y \geq x^{5/3} \end{cases}$ **46** $\begin{cases} x^4 + y^5 < 2^x \\ \ln(x^2 + 1) < y^3 \end{cases}$

47 $\begin{cases} x^4 - 2x < 3y \\ x + 2y < x^3 - 5 \end{cases}$ **48** $\begin{cases} e^x + x^2 \leq 2^{x+2y} \\ 2^{x+2y} \leq x^3 2^y \\ x > 0 \end{cases}$

6.6 LINEAR PROGRAMMING

If a system of inequalities contains only linear inequalities of the form

$$ax + by \leq c \quad \text{or} \quad ax + by \geq c,$$

where a, b, and c are real numbers, then the graph of the system may be a region R in the xy-plane bounded by a polygon—possibly of the type illustrated in Figure 19 on the next page (for a specific illustration, see Example 4 and Figure 15 of Section 6.5). For problems in **linear program-**

FIGURE 19

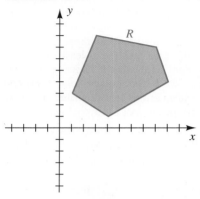

ming, we consider such systems together with an expression of the form

$$C = Ax + By + K,$$

where A, B, and K are real numbers and (x, y) is a point in R (that is, a solution of the system). Since for each (x, y) we obtain a specific value for C, we call C a *function of two variables x and y.* In linear programming, C is called an **objective function,** and the inequalities in the system are referred to as the **constraints** on C. The solutions of the system—that is, the pairs (x, y) corresponding to the points in R—are called the **feasible solutions** for the problem.

In typical business applications, the value of C may represent cost, profit, loss, or a physical resource, and the goal is to find a specific point (x, y) in R at which C takes on its maximum or minimum value. The methods of linear programming greatly simplify the task of finding this point. Specifically, it can be shown that *the maximum and minimum values of C occur at a vertex of R.* This fact is used in the next example.

EXAMPLE I   *Finding the maximum and minimum values of an objective function*

Find the maximum and minimum values of the objective function $C = 7x + 3y$ subject to the following constraints:

$$\begin{cases} x - 2y \geq -10 \\ 2x + y \leq 10 \\ x \geq 0 \\ y \geq 0 \end{cases}$$

FIGURE 20

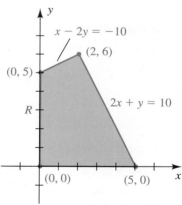

SOLUTION   The graph of the system of inequalities determined by the constraints is the region R bounded by the quadrilateral sketched in Figure 20. From the preceding discussion, the maximum and minimum values of C must occur at a vertex of R. The values at the vertices are given in the following table.

Vertex	Value of $C = 7x + 3y$
$(0, 0)$	$7(0) + 3(0) = 0$
$(0, 5)$	$7(0) + 3(5) = 15$
$(5, 0)$	$7(5) + 3(0) = 35$
$(2, 6)$	$7(2) + 3(6) = 32$

Hence, the minimum value $C = 0$ occurs if $x = 0$ and $y = 0$. The maximum value $C = 35$ occurs if $x = 5$ and $y = 0$.

FIGURE 21

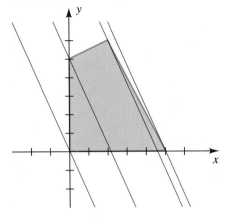

Guidelines for Solving a Linear Programming Problem

In the preceding example, we say that the maximum value of *C on R* occurs at the vertex (5, 0). To verify this fact, let us solve $C = 7x + 3y$ for *y*, obtaining

$$y = -\frac{7}{3}x + \frac{C}{2}.$$

For each *C*, the graph of this equation is a line of slope $-\frac{7}{3}$ and *y*-intercept *C*/2, as illustrated in Figure 21. To find the maximum value of *C*, we simply determine which of these lines that intersect the region has the largest *y*-intercept *C*/2. Referring to Figure 21, we see that the required line passes through (5, 0). Similarly, for the minimum value of *C*, we determine the line $y = (-7/3)x + (C/2)$ that intersects the region and has the *smallest* *y*-intercept. This is the line through (0, 0).

We shall call a problem that can be expressed in the form of Example 1 a **linear programming problem**. To solve such problems, we may use the following guidelines.

> 1 Sketch the region *R* determined by the system of constraints.
> 2 Find the vertices of *R*.
> 3 Calculate the value of the objective function *C* at each vertex of *R*.
> 4 Determine the maximum or minimum values of *C* by using the vertices of *R* in guideline 3 that yield the largest or smallest values, respectively.

In the next example we encounter a linear programming problem in which the minimum value of the objective function occurs at more than one point.

FIGURE 22

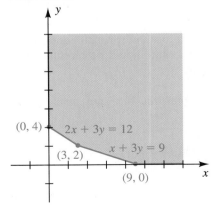

EXAMPLE 2 *Solving a linear programming problem*

Find the minimum value of the objective function $C = 2x + 6y$ subject to the following constraints:

$$\begin{cases} 2x + 3y \geq 12 \\ x + 3y \geq 9 \\ x \geq 0 \\ y \geq 0 \end{cases}$$

SOLUTION We shall follow the guidelines.

Guideline 1 The graph of the system of inequalities determined by the constraints is the unbounded region *R* sketched in Figure 22.

Guideline 2 The vertices of *R* are (0, 4), (3, 2), and (9, 0), as shown in Figure 22.

(continued)

Guideline 3 The value of C at each vertex of R is given in the following table.

Vertex	Value of $C = 2x + 6y$
(0, 4)	$2(0) + 6(4) = 24$
(3, 2)	$2(3) + 6(2) = 18$
(9, 0)	$2(9) + 6(0) = 18$

Guideline 4 The table in guideline 3 shows that the minimum value of C, 18, occurs at *two* vertices, (3, 2) and (9, 0). Moreover, if (x, y) is any point on the line segment joining these points, then (x, y) is a solution of the equation $x + 3y = 9$, and hence

$$C = 2x + 6y = 2(x + 3y) = 2(9) = 18.$$

Thus, the minimum value $C = 18$ occurs at *every* point on this line segment.

In the next two examples we consider applications of linear programming. For such problems it is necessary to use given information and data to formulate the system of constraints and the objective function. Once this has been accomplished, we may apply the guidelines as in the solution to Example 2.

EXAMPLE 3   *Maximizing profit*

A firm manufactures two products X and Y. For each product it is necessary to use three different machines A, B, and C. To manufacture one unit of product X, machine A must be used for 3 hours, machine B for 1 hour, and machine C for 1 hour. To manufacture one unit of product Y requires 2 hours on A, 2 hours on B, and 1 hour on C. The profit on product X is $500 per unit, and the profit on product Y is $350 per unit. Machine A is available for a total of 24 hours per day; however, B can be used for only 16 hours and C for 9 hours. Assuming the machines are available when needed (subject to the noted total hour restrictions), determine the number of units of each product that should be manufactured each day in order to maximize the profit.

SOLUTION   The following table summarizes the data given in the statement of the problem.

Machine	Hours required for one unit of X	Hours required for one unit of Y	Hours available
A	3	2	24
B	1	2	16
C	1	1	9

Let us introduce the following variables:

$$x = \text{number of units of X manufactured each day}$$

$$y = \text{number of units of Y manufactured each day}$$

Using the first row of the table, we note that each unit of X requires 3 hours on machine A, and hence x units require $3x$ hours. Similarly, since each unit of Y requires 2 hours on A, y units require $2y$ hours. Hence, the total number of hours per day that machine A must be used is $3x + 2y$. This, together with the fact that A can be used for at most 24 hours per day, gives us the first constraint in the following system of inequalities: $3x + 2y \le 24$. The second and third constraints are obtained by using the same type of reasoning for rows 2 and 3 of the table. The last two constraints, $x \ge 0$ and $y \ge 0$, are true because x and y cannot be negative.

FIGURE 23

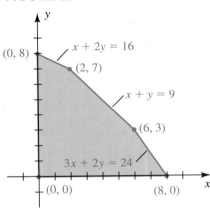

$$\begin{cases} 3x + 2y \le 24 \\ x + 2y \le 16 \\ x + y \le 9 \\ x \ge 0 \\ y \ge 0 \end{cases}$$

The graph of this system is the region R in Figure 23.

Since the production of each unit of product X yields a profit of \$500 and each unit of product Y yields a profit of \$350, the profit P obtained by producing x units of X together with y units of Y is

$$P = 500x + 350y.$$

This is the objective function for the problem. The maximum value of P must occur at one of the vertices of R in Figure 23. The values of P at these vertices are given in the following table.

Vertex	Value of $P = 500x + 350y$
(0, 0)	$500(0) + 350(0) = 0$
(0, 8)	$500(0) + 350(8) = 2800$
(8, 0)	$500(8) + 350(0) = 4000$
(2, 7)	$500(2) + 350(7) = 3450$
(6, 3)	$500(6) + 350(3) = 4050$

We see from the table that a maximum profit of \$4050 occurs for a daily production of 6 units of product X and 3 units of product Y.

Example 3 illustrates maximization of profit. The next example demonstrates how linear programming can be used to minimize the cost in a certain situation.

EXAMPLE 4   *Minimizing cost*

A distributor of compact disk players has two warehouses W_1 and W_2. There are 80 units stored at W_1 and 70 units at W_2. Two customers A and B order 35 units and 60 units, respectively. The shipping cost from each warehouse to A and B is determined according to the following table. How should the order be filled to minimize the total shipping cost?

Warehouse	Customer	Shipping cost per unit
W_1	A	$ 8
W_1	B	12
W_2	A	10
W_2	B	13

SOLUTION   Let us begin by introducing the following variables:

$$x = \text{number of units sent to A from } W_1$$

$$y = \text{number of units sent to B from } W_1$$

Since A ordered 35 units and B ordered 60 units, we must have

$$35 - x = \text{number of units sent to A from } W_2$$

$$60 - y = \text{number of units sent to B from } W_2$$

Our goal is to determine values for x and y that make the total shipping cost minimal.

The number of units shipped from W_1 cannot exceed 80, and the number shipped from W_2 cannot exceed 70. Expressing these facts in terms of inequalities gives us

$$\begin{cases} x + y \leq 80 \\ (35 - x) + (60 - y) \leq 70 \end{cases}$$

Simplifying, we obtain the first two constraints in the following system. The last two constraints are true because the largest values of x and y are 35 and 60, respectively.

$$\begin{cases} x + y \leq 80 \\ x + y \geq 25 \\ 0 \leq x \leq 35 \\ 0 \leq y \leq 60 \end{cases}$$

The graph of this system is the region R shown in Figure 24.

Let C denote the total cost (in dollars) of shipping the disk players to customers A and B. We see from the table of shipping costs that the following are true:

$$\text{cost of shipping 35 units to A} = 8x + 10(35 - x)$$

$$\text{cost of shipping 60 units to B} = 12y + 13(60 - y)$$

FIGURE 24

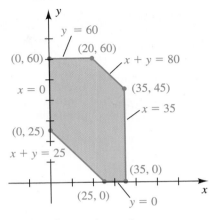

Hence, the total cost is

$$C = 8x + 10(35 - x) + 12y + 13(60 - y).$$

Simplifying gives us the following objective function:

$$C = 1130 - 2x - y$$

To determine the minimum value of C on R, we need check only the vertices shown in Figure 24, as in the following table.

Vertex	Value of $C = 1130 - 2x - y$
(0, 25)	$1130 - 2(0) - 25 = 1105$
(0, 60)	$1130 - 2(0) - 60 = 1070$
(20, 60)	$1130 - 2(20) - 60 = 1030$
(35, 45)	$1130 - 2(35) - 45 = 1015$
(35, 0)	$1130 - 2(35) - 0 = 1060$
(25, 0)	$1130 - 2(25) - 0 = 1080$

We see from the table that the minimal shipping cost $1015 occurs if $x = 35$ and $y = 45$. This means that the distributor should ship all of the disk players to A from W_1 and none from W_2. In addition, the distributor should ship 45 units to B from W_1 and 15 units to B from W_2. (Note that the *maximum* shipping cost will occur if $x = 0$ and $y = 25$—that is, if all 35 units are shipped to A from W_2 and if B receives 25 units from W_1 and 35 units from W_2.)

The examples in this section are elementary linear programming problems in two variables that can be solved by basic methods. The much more complicated problems in many variables that occur in practice may be solved by employing matrix techniques that are adapted for solution by computers.

6.6 EXERCISES

Exer. 1–2: Find the maximum and minimum values of the objective function C on the region in the figure.

1 $C = 3x + 2y + 5$

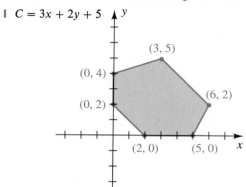

2 $C = 2x + 7y + 3$

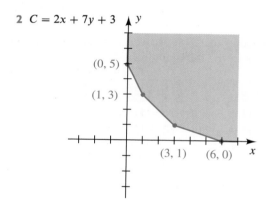

Exer. 3–4: Sketch the region R determined by the given constraints, and label its vertices. Find the maximum value of C on R.

3 $C = 3x + y;$ $x \geq 0, y \geq 0,$
$3x - 4y \geq -12,$ $3x + 2y \leq 24,$ $3x - y \leq 15$

4 $C = 4x - 2y;$
$x - 2y \geq -8,$ $7x - 2y \leq 28,$ $x + y \geq 4$

Exer. 5–6: Sketch the region R determined by the given constraints, and label its vertices. Find the minimum value of C on R.

5 $C = 3x + 6y;$ $x \geq 0, y \geq 0,$
$2x + 3y \geq 12,$ $2x + 5y \geq 16$

6 $C = 6x + y;$ $y \geq 0,$
$3x + y \geq 3,$ $x + 5y \leq 15,$ $2x + y \leq 12$

Exer. 7–8: Sketch the region R determined by the given constraints, and label its vertices. Describe the set of points for which C is a maximum on R.

7 $C = 2x + 4y;$ $x \geq 0, y \geq 0,$
$x - 2y \geq -8,$ $\frac{1}{2}x + y \leq 6,$ $3x + 2y \leq 24$

8 $C = 6x + 3y;$ $x \geq 2, y \geq 1,$
$2x + 3y \leq 19,$ $x + 0.5y \leq 6.5$

9 *Production scheduling* A manufacturer of tennis rackets makes a profit of $15 on each oversized racket and $8 on each standard racket. To meet dealer demand, daily production of standard rackets should be between 30 and 80, and production of oversized rackets should be between 10 and 30. To maintain high quality, the total number of rackets produced should not exceed 80 per day. How many of each type should be manufactured daily to maximize the profit?

10 *Production scheduling* A manufacturer of CB radios makes a profit of $25 on a deluxe model and $30 on a standard model. The company wishes to produce at least 80 deluxe models and at least 100 standard models per day. To maintain high quality, the daily production should not exceed 200 radios. How many of each type should be produced daily in order to maximize the profit?

11 *Minimizing cost* Two substances S and T each contain two types of ingredients I and G. One pound of S contains 2 ounces of I and 4 ounces of G. One pound of T contains 2 ounces of I and 6 ounces of G. A manufacturer plans to combine quantities of the two substances to obtain a mixture that contains at least 9 ounces of I and 20 ounces of G. If the cost of S is $3.00 per pound and the cost of T is $4.00 per pound, how much of each substance should be used to keep the cost to a minimum?

12 *Maximizing gross profit* A stationery company makes two types of notebooks: a deluxe notebook with subject dividers, which sells for $1.25, and a regular notebook, which sells for $0.90. The production cost is $1.00 for each deluxe notebook and $0.75 for each regular notebook. The company has the facilities to manufacture between 2000 and 3000 deluxe and between 3000 and 6000 regular, but not more than 7000 altogether. How many notebooks of each type should be manufactured to maximize the difference between the selling prices and the production cost?

13 *Minimizing shipping costs*   Refer to Example 4 of this section. If the shipping costs are $12 per unit from W_1 to A, $10 per unit from W_2 to A, $16 per unit from W_1 to B, and $12 per unit from W_2 to B, determine how the order should be filled to minimize shipping cost.

14 *Minimizing cost*   A coffee company purchases mixed lots of coffee beans and then grades them into premium, regular, and unusable beans. The company needs at least 280 tons of premium-grade and 200 tons of regular-grade coffee beans. The company can purchase ungraded coffee from two suppliers in any amount desired. Samples from the two suppliers contain the following percentages of premium, regular, and unusable beans:

Supplier	Premium	Regular	Unusable
A	20%	50%	30%
B	40%	20%	40%

If supplier A charges $125 per ton and B charges $200 per ton, how much should the company purchase from each supplier to fulfill its needs at minimum cost?

15 *Planning crop acreage*   A farmer, in the business of growing fodder for livestock, has 90 acres available for planting alfalfa and corn. The cost of seed per acre is $4 for alfalfa and $6 for corn. The total cost of labor will amount to $20 per acre for alfalfa and $10 per acre for corn. The expected income from alfalfa is $110 per acre, and from corn, $150 per acre. If the farmer does not wish to spend more than $480 for seed and $1400 for labor, how many acres of each crop should be planted to obtain the maximum profit?

16 *Machinery scheduling*   A small firm manufactures bookshelves and desks for microcomputers. For each product it is necessary to use a table saw and a power router. To manufacture each bookshelf, the saw must be used for $\frac{1}{2}$ hour and the router for 1 hour. A desk requires the use of each machine for 2 hours. The profits are $20 per bookshelf and $50 per desk. If the saw can be used 8 hours per day and the router for 12 hours per day, how many bookshelves and desks should be manufactured each day to maximize the profit?

17 *Minimizing a mixture's cost*   Three substances X, Y, and Z each contain four ingredients A, B, C, and D. The percentage of each ingredient and the cost in cents per ounce of each substance are given in the following table.

Substance	Ingredients				Cost per ounce
	A	B	C	D	
X	20%	10%	25%	45%	25¢
Y	20%	40%	15%	25%	35¢
Z	10%	20%	25%	45%	50¢

If the cost is to be minimal, how many ounces of each substance should be combined to obtain a mixture of 20 ounces containing at least 14% A, 16% B, and 20% C? What combination would make the cost greatest?

18 *Maximizing profit*   A man plans to operate a stand at a one-day fair at which he will sell bags of peanuts and bags of candy. He has $400 available to purchase his stock, which will cost 40¢ per bag of peanuts and 80¢ per bag of candy. He intends to sell the peanuts at $1.00 and the candy at $1.60 per bag. His stand can accommodate up to 500 bags of peanuts and 400 bags of candy. From past experience he knows that he will sell no more than a total of 700 bags. Find the number of bags of each that he should have available in order to maximize his profit. What is the maximum profit?

19 *Maximizing passenger capacity*   A small community wishes to purchase used vans and small buses for its public transportation system. The community can spend no more than $100,000 for the vehicles and no more than $500 per month for maintenance. The vans sell for $10,000 each and average $100 per month in maintenance costs. The corresponding cost estimates for each bus are $20,000 and $75 per month. If each van can carry 15 passengers and each bus can accommodate 25 riders, determine the number of vans and buses that should be purchased to maximize the passenger capacity of the system.

20 *Minimizing fuel cost*   Refer to Exercise 19. The monthly fuel cost (based on 5000 miles of service) is $550 for each van and $850 for each bus. Find the number of vans and buses that should be purchased to minimize the monthly fuel costs if the passenger capacity of the system must be at least 75.

21 *Stocking a fish farm*   A fish farmer will purchase no more than 5000 young trout and bass from the hatchery and will feed them a special diet for the next year. The cost of food per fish will be $0.50 for trout and $0.75 for bass, and the total cost is not to exceed $3000. At the end of the year, a typical trout will weigh 3 pounds, and a bass will weigh 4 pounds. How many fish of each type should be stocked in the pond in order to maximize the total number of pounds of fish at the end of the year?

22 *Dietary planning*   A hospital dietician wishes to prepare a corn-squash vegetable dish that will provide at least 3 grams of protein and cost no more than 36 cents per serving. An ounce of creamed corn provides $\frac{1}{2}$ gram of protein and costs 4 cents. An ounce of squash supplies $\frac{1}{4}$ gram of protein and costs 3 cents. For taste, there must be at least 2 ounces of corn and at least as much squash as corn. It is important to keep the total number of ounces in a serving as small as possible. Find the combination of corn and squash that will minimize the amount of ingredients used per serving.

23 *Planning storage units*   A contractor has a large building that she wishes to convert into a series of rental storage spaces. She will construct basic 8 ft × 10 ft units and deluxe 12 ft × 10 ft units that contain extra shelves and a clothes closet. Market considerations dictate that there be at least twice as many basic units as deluxe units

and that the basic units rent for \$40 per month and the deluxe units for \$75 per month. At most 7200 ft² is available for the storage spaces, and no more than \$30,000 can be spent on construction. If each basic unit will cost \$300 to make and each deluxe unit will cost \$600, how many units of each type should be constructed to maximize monthly revenue?

24 *A moose's diet*   A moose feeding primarily on tree leaves and aquatic plants is capable of digesting no more than 33 kilograms of these foods daily. Although the aquatic plants are lower in energy content, the animal must eat at least 17 kilograms to satisfy its sodium requirement. A kilogram of leaves provides four times as much energy as a kilogram of aquatic plants. Find the combination of foods that maximizes the daily energy intake.

6.7   THE ALGEBRA OF MATRICES

Matrices were introduced in Section 6.3 as an aid to finding solutions of systems of equations. In this section we discuss some of their properties. These properties are important in advanced fields of mathematics and in applications.

In the following definition, the symbol (a_{ij}) denotes an $m \times n$ matrix A of the type displayed in the definition on page 381. We use similar notations for the matrices B and C.

Definition of Equality and Addition of Matrices

> Let $A = (a_{ij})$, $B = (b_{ij})$, and $C = (c_{ij})$ be $m \times n$ matrices.
>
> **(1)** $A = B$ if and only if $a_{ij} = b_{ij}$ for every i and j.
> **(2)** $C = A + B$ if and only if $c_{ij} = a_{ij} + b_{ij}$ for every i and j.

Note that two matrices are equal if and only if they have the same size and corresponding elements are equal.

ILLUSTRATION

Equality of Matrices

$$\begin{bmatrix} 1 & 0 & 5 \\ \sqrt[3]{8} & 3^2 & -2 \end{bmatrix} = \begin{bmatrix} (-1)^2 & 0 & \sqrt{25} \\ 2 & 9 & -2 \end{bmatrix}$$

Using the parentheses notation for matrices, we may write the definition of addition of two $m \times n$ matrices as

$$(a_{ij}) + (b_{ij}) = (a_{ij} + b_{ij}).$$

Thus, to add two matrices, we add the elements in corresponding positions in each matrix. *Two matrices can be added only if they have the same size.*

ILLUSTRATION

Addition of Matrices

- $\begin{bmatrix} 4 & -5 \\ 0 & 4 \\ -6 & 1 \end{bmatrix} + \begin{bmatrix} 3 & 2 \\ 7 & -4 \\ -2 & 1 \end{bmatrix} = \begin{bmatrix} 7 & -3 \\ 7 & 0 \\ -8 & 2 \end{bmatrix}$

- $\begin{bmatrix} 2 & 3 \\ -4 & 1 \end{bmatrix} + \begin{bmatrix} -2 & -3 \\ 4 & -1 \end{bmatrix} = \begin{bmatrix} 0 & 0 \\ 0 & 0 \end{bmatrix}$

- $\begin{bmatrix} 1 & 3 & -2 \\ 0 & -5 & 4 \end{bmatrix} + \begin{bmatrix} 0 & 0 & 0 \\ 0 & 0 & 0 \end{bmatrix} = \begin{bmatrix} 1 & 3 & -2 \\ 0 & -5 & 4 \end{bmatrix}$

The **$m \times n$ zero matrix**, denoted by O, is the matrix with m rows and n columns in which every elements is 0.

ILLUSTRATION

Zero Matrices

- $\begin{bmatrix} 0 & 0 \\ 0 & 0 \end{bmatrix}$ - $\begin{bmatrix} 0 & 0 \\ 0 & 0 \\ 0 & 0 \end{bmatrix}$ - $\begin{bmatrix} 0 & 0 & 0 & 0 \\ 0 & 0 & 0 & 0 \end{bmatrix}$

The **additive inverse** $-A$ of the matrix $A = (a_{ij})$ is the matrix $(-a_{ij})$ obtained by changing the sign of each nonzero element of A.

ILLUSTRATION

Additive Inverse

- $-\begin{bmatrix} 2 & -3 & 4 \\ -1 & 0 & 5 \end{bmatrix} = \begin{bmatrix} -2 & 3 & -4 \\ 1 & 0 & -5 \end{bmatrix}$

The proof of the next theorem follows from the definition of addition of matrices.

Theorem

> If A, B, and C are $m \times n$ matrices and if O is the $m \times n$ zero matrix, then
>
> **(1)** $A + B = B + A$
>
> **(2)** $A + (B + C) = (A + B) + C$
>
> **(3)** $A + O = A$
>
> **(4)** $A + (-A) = O$

Subtraction of two $m \times n$ matrices is defined by

$$A - B = A + (-B).$$

Using the parentheses notation, we have

$$(a_{ij}) - (b_{ij}) = (a_{ij}) + (-b_{ij})$$
$$= (a_{ij} - b_{ij}).$$

Thus, to subtract two matrices, we subtract the elements in corresponding positions.

ILLUSTRATION

Subtraction of Matrices

$$\begin{bmatrix} 4 & -5 \\ 0 & 4 \\ -6 & 1 \end{bmatrix} - \begin{bmatrix} 3 & 2 \\ 7 & -4 \\ -2 & 1 \end{bmatrix} = \begin{bmatrix} 1 & -7 \\ -7 & 8 \\ -4 & 0 \end{bmatrix}$$

Definition

The **product** of a real number c and an $m \times n$ matrix $A = (a_{ij})$ is

$$cA = (ca_{ij}).$$

Note that to find cA, we multiply each element of A by c.

ILLUSTRATION

Product of a Real Number and a Matrix

$$3\begin{bmatrix} 4 & -1 \\ 2 & 3 \end{bmatrix} = \begin{bmatrix} 12 & -3 \\ 6 & 9 \end{bmatrix}$$

We can prove the following.

Theorem

If A and B are $m \times n$ matrices and if c and d are real numbers, then

(1) $c(A + B) = cA + cB$

(2) $(c + d)A = cA + dA$

(3) $(cd)A = c(dA)$

The next definition, of the product AB of two matrices, may seem unusual, but it has many uses in mathematics and applications. For multiplication, unlike addition, A and B may have different sizes; however, *the number of columns of A must be the same as the number of rows of B.* Thus, if A is $m \times n$, then B must be $n \times p$ for some p. As we shall see, the size of AB is then $m \times p$. If $C = AB$, then a method for finding the element c_{ij} in row i and column j of C is given in the following guidelines.

Guidelines for Finding c_{ij} in the Product $C = AB$ If A Is $m \times n$ and B Is $n \times p$

I Single out the ith row R_i of A and the jth column C_j of B:

$$\begin{bmatrix} a_{11} & a_{12} & \cdots & a_{1n} \\ \vdots & \vdots & & \vdots \\ a_{i1} & a_{i2} & \cdots & a_{in} \\ \vdots & \vdots & & \vdots \\ a_{m1} & a_{m2} & \cdots & a_{mn} \end{bmatrix} \begin{bmatrix} b_{11} & \cdots & b_{1j} & \cdots & b_{1p} \\ b_{21} & \cdots & b_{2j} & \cdots & b_{2p} \\ \vdots & & \vdots & & \vdots \\ b_{n1} & \cdots & b_{nj} & \cdots & b_{np} \end{bmatrix}$$

2 *Simultaneously* move to the right along R_i and down C_j, multiplying pairs of elements, to obtain

$$a_{i1}b_{1j}, a_{i2}b_{2j}, a_{i3}b_{3j}, \ldots, a_{in}b_{nj}.$$

3 Add the products of the pairs in guideline 2 to obtain c_{ij}.

$$c_{ij} = a_{i1}b_{1j} + a_{i2}b_{2j} + a_{i3}b_{3j} + \cdots + a_{in}b_{nj}$$

Using the guidelines, we see that the element c_{11} in the first row and the first column of AB is

$$c_{11} = a_{11}b_{11} + a_{12}b_{21} + a_{13}b_{31} + \cdots + a_{1n}b_{n1}.$$

The element c_{mp} in the last row and the last column of AB is

$$c_{mp} = a_{m1}b_{1p} + a_{m2}b_{2p} + a_{m3}b_{3p} + \cdots + a_{mn}b_{np}.$$

The preceding discussion is summarized in the next definition.

Definition of the Product of Two Matrices

Let $A = (a_{ij})$ be an $m \times n$ matrix and let $B = (b_{ij})$ be an $n \times p$ matrix. The **product** AB is the $m \times p$ matrix $C = (c_{ij})$ such that

$$c_{ij} = a_{i1}b_{1j} + a_{i2}b_{2j} + a_{i3}b_{3j} + \cdots + a_{in}b_{nj}$$

for $i = 1, 2, \ldots, m$ and $j = 1, 2, \ldots, p$.

The following diagram may help you remember the relationship between sizes of matrices when working with a product AB.

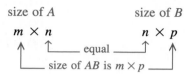

The next illustration contains some special cases.

ILLUSTRATION

Sizes of Matrices in Products

Size of A	Size of B	Size of AB
2×3	3×5	2×5
4×2	2×3	4×3
3×1	1×3	3×3
1×3	3×1	1×1
5×3	3×5	5×5
5×3	5×3	AB is not defined

In the following example we find the product of two specific matrices.

EXAMPLE I   *Finding the product of two matrices*

Find the product AB if

$$A = \begin{bmatrix} 1 & 2 & -3 \\ 4 & 0 & -2 \end{bmatrix} \quad \text{and} \quad B = \begin{bmatrix} 5 & -4 & 2 & 0 \\ -1 & 6 & 3 & 1 \\ 7 & 0 & 5 & 8 \end{bmatrix}.$$

SOLUTION   The matrix A is 2×3, and the matrix B is 3×4. Hence, the product $C = AB$ is defined and is 2×4. We next use the guidelines to find the elements $c_{11}, c_{12}, \ldots$ of the product. For instance, to find the element c_{23} we single out the second row R_2 of A and the third column C_3 of B, as illustrated below, and then use guidelines 2 and 3 to obtain

$$c_{23} = 4 \cdot 2 + 0 \cdot 3 + (-2) \cdot 5 = -2.$$

$$\begin{bmatrix} 1 & 2 & -3 \\ 4 & 0 & -2 \end{bmatrix} \begin{bmatrix} 5 & -4 & 2 & 0 \\ -1 & 6 & 3 & 1 \\ 7 & 0 & 5 & 8 \end{bmatrix} = \begin{bmatrix} & & & \\ & & -2 & \end{bmatrix}$$

Similarly, to find the element c_{12} in row 1 and column 2 of the product, we proceed as follows:

$$c_{12} = 1 \cdot (-4) + 2 \cdot 6 + (-3) \cdot 0 = 8$$

$$\begin{bmatrix} 1 & 2 & -3 \\ 4 & 0 & -2 \end{bmatrix} \begin{bmatrix} 5 & -4 & 2 & 0 \\ -1 & 6 & 3 & 1 \\ 7 & 0 & 5 & 8 \end{bmatrix} = \begin{bmatrix} & 8 & & \\ & -2 & & \end{bmatrix}$$

The remaining elements of the product are calculated as follows, where we have indicated the row of A and the column of B that are used when guideline 1 is applied.

Row of A	Column of B	Element of C
R_1	C_1	$c_{11} = 1 \cdot 5 + 2 \cdot (-1) + (-3) \cdot 7 = -18$
R_1	C_3	$c_{13} = 1 \cdot 2 + 2 \cdot 3 + (-3) \cdot 5 = -7$
R_1	C_4	$c_{14} = 1 \cdot 0 + 2 \cdot 1 + (-3) \cdot 8 = -22$
R_2	C_1	$c_{21} = 4 \cdot 5 + 0 \cdot (-1) + (-2) \cdot 7 = 6$
R_2	C_2	$c_{22} = 4 \cdot (-4) + 0 \cdot 6 + (-2) \cdot 0 = -16$
R_2	C_4	$c_{24} = 4 \cdot 0 + 0 \cdot 1 + (-2) \cdot 8 = -16$

Hence,

$$AB = \begin{bmatrix} 1 & 2 & -3 \\ 4 & 0 & -2 \end{bmatrix} \begin{bmatrix} 5 & -4 & 2 & 0 \\ -1 & 6 & 3 & 1 \\ 7 & 0 & 5 & 8 \end{bmatrix}$$

$$= \begin{bmatrix} -18 & 8 & -7 & -22 \\ 6 & -16 & -2 & -16 \end{bmatrix}.$$

A matrix is a **row matrix** if it has only one row. A **column matrix** has only one column. The following illustration contains some products involving row and column matrices. You should check each entry in the products.

ILLUSTRATION

Products Involving Row and Column Matrices

$$\begin{bmatrix} -2 & 4 \\ 0 & -1 \\ 5 & 3 \end{bmatrix} \begin{bmatrix} -2 \\ 1 \end{bmatrix} = \begin{bmatrix} 8 \\ -1 \\ -7 \end{bmatrix} \qquad \begin{bmatrix} 3 & -1 & 2 \end{bmatrix} \begin{bmatrix} -2 & 4 \\ 0 & -1 \\ 5 & 3 \end{bmatrix} = \begin{bmatrix} 4 & 19 \end{bmatrix}$$

$$\begin{bmatrix} -2 \\ 3 \end{bmatrix} \begin{bmatrix} 1 & 5 \end{bmatrix} = \begin{bmatrix} -2 & -10 \\ 3 & 15 \end{bmatrix} \qquad \begin{bmatrix} 1 & 5 \end{bmatrix} \begin{bmatrix} -2 \\ 3 \end{bmatrix} = \begin{bmatrix} 13 \end{bmatrix}$$

The product operation for matrices is not commutative. For example, if A is 2×3 and B is 3×4, then AB may be found, since the number of columns of A is the same as the number of rows of B. However, BA is undefined, since the number of columns of B is different from the number of rows of A. Even if AB and BA are both defined, it is often true that these products are different. This is illustrated in the next example, along with the fact that the product of two nonzero matrices may equal a zero matrix.

EXAMPLE 2   *Matrix multiplication is not commutative*

If $A = \begin{bmatrix} 2 & 2 \\ -1 & -1 \end{bmatrix}$ and $B = \begin{bmatrix} 1 & 2 \\ 1 & 2 \end{bmatrix}$, show that $AB \neq BA$.

SOLUTION   Using the definition of the product of two matrices, we obtain the following:

$$AB = \begin{bmatrix} 2 & 2 \\ -1 & -1 \end{bmatrix} \begin{bmatrix} 1 & 2 \\ 1 & 2 \end{bmatrix} = \begin{bmatrix} 4 & 8 \\ -2 & -4 \end{bmatrix}$$

$$BA = \begin{bmatrix} 1 & 2 \\ 1 & 2 \end{bmatrix} \begin{bmatrix} 2 & 2 \\ -1 & -1 \end{bmatrix} = \begin{bmatrix} 0 & 0 \\ 0 & 0 \end{bmatrix}$$

Hence, $AB \neq BA$. Note that the last equality shows that *the product of two nonzero matrices can equal a zero matrix.*

Although matrix multiplication is not commutative, it is associative. Thus, if A is $m \times n$, B is $n \times p$, and C is $p \times q$, then

$$A(BC) = (AB)C.$$

The distributive properties also hold if the matrices involved have the proper number of rows and columns. If A_1 and A_2 are $m \times n$ matrices and if B_1 and B_2 are $n \times p$ matrices, then

$$A_1(B_1 + B_2) = A_1B_1 + A_1B_2$$
$$(A_1 + A_2)B_1 = A_1B_1 + A_2B_1.$$

As a special case, if all matrices are square, of order n, then both the associative and distributive properties are true.

6.7 EXERCISES

Exer. 1–8: Find, if possible, $A + B$, $A - B$, $2A$, and $-3B$.

1 $A = \begin{bmatrix} 5 & -2 \\ 1 & 3 \end{bmatrix}$, $B = \begin{bmatrix} 4 & 1 \\ -3 & 2 \end{bmatrix}$

2 $A = \begin{bmatrix} 3 & 0 \\ -1 & 2 \end{bmatrix}$, $B = \begin{bmatrix} 3 & -4 \\ 1 & 1 \end{bmatrix}$

$3 \quad A = \begin{bmatrix} 6 & -1 \\ 2 & 0 \\ -3 & 4 \end{bmatrix}, \qquad B = \begin{bmatrix} 3 & 1 \\ -1 & 5 \\ 6 & 0 \end{bmatrix}$

$4 \quad A = \begin{bmatrix} 0 & -2 & 7 \\ 5 & 4 & -3 \end{bmatrix}, \qquad B = \begin{bmatrix} 8 & 4 & 0 \\ 0 & 1 & 4 \end{bmatrix}$

$5 \quad A = \begin{bmatrix} 4 & -3 & 2 \end{bmatrix}, \qquad B = \begin{bmatrix} 7 & 0 & -5 \end{bmatrix}$

$6 \quad A = \begin{bmatrix} 7 \\ -16 \end{bmatrix}, \qquad B = \begin{bmatrix} -11 \\ 9 \end{bmatrix}$

$7 \quad A = \begin{bmatrix} 3 & -2 & 2 \\ 0 & 1 & -4 \\ -3 & 2 & -1 \end{bmatrix}, \qquad B = \begin{bmatrix} 4 & 0 \\ 2 & -1 \\ -1 & 3 \end{bmatrix}$

$8 \quad A = \begin{bmatrix} 2 & 1 \end{bmatrix}, \qquad B = \begin{bmatrix} 3 & -1 & 5 \end{bmatrix}$

Exer. 9–20: Find, if possible, AB and BA.

$9 \quad A = \begin{bmatrix} 2 & 6 \\ 3 & -4 \end{bmatrix}, \qquad B = \begin{bmatrix} 5 & -2 \\ 1 & 7 \end{bmatrix}$

$10 \quad A = \begin{bmatrix} 4 & -2 \\ -2 & 1 \end{bmatrix}, \qquad B = \begin{bmatrix} 2 & 1 \\ 4 & 2 \end{bmatrix}$

$11 \quad A = \begin{bmatrix} 3 & 0 & -1 \\ 0 & 4 & 2 \\ 5 & -3 & 1 \end{bmatrix}, \qquad B = \begin{bmatrix} 1 & -5 & 0 \\ 4 & 1 & -2 \\ 0 & -1 & 3 \end{bmatrix}$

$12 \quad A = \begin{bmatrix} 5 & 0 & 0 \\ 0 & -3 & 0 \\ 0 & 0 & 2 \end{bmatrix}, \qquad B = \begin{bmatrix} 3 & 0 & 0 \\ 0 & 4 & 0 \\ 0 & 0 & -2 \end{bmatrix}$

$13 \quad A = \begin{bmatrix} 4 & -3 & 1 \\ -5 & 2 & 2 \end{bmatrix}, \qquad B = \begin{bmatrix} 2 & 1 \\ 0 & 1 \\ -4 & 7 \end{bmatrix}$

$14 \quad A = \begin{bmatrix} 2 & 1 & -1 & 0 \\ 3 & -2 & 0 & 5 \\ -2 & 1 & 4 & 2 \end{bmatrix}, \qquad B = \begin{bmatrix} 5 & -3 & 1 \\ 1 & 2 & 0 \\ -1 & 0 & 4 \\ 0 & -2 & 3 \end{bmatrix}$

$15 \quad A = \begin{bmatrix} 1 & 2 & 3 \\ 4 & 5 & 6 \\ 7 & 8 & 9 \end{bmatrix}, \qquad B = \begin{bmatrix} 1 & 0 & 0 \\ 0 & 1 & 0 \\ 0 & 0 & 1 \end{bmatrix}$

$16 \quad A = \begin{bmatrix} 1 & 2 & 3 \\ 2 & 3 & 1 \\ 3 & 1 & 2 \end{bmatrix}, \qquad B = \begin{bmatrix} 2 & 0 & 0 \\ 0 & 2 & 0 \\ 0 & 0 & 2 \end{bmatrix}$

$17 \quad A = \begin{bmatrix} -3 & 7 & 2 \end{bmatrix}, \qquad B = \begin{bmatrix} 1 \\ 4 \\ -5 \end{bmatrix}$

$18 \quad A = \begin{bmatrix} 4 & 8 \end{bmatrix}, \qquad B = \begin{bmatrix} -3 \\ 2 \end{bmatrix}$

$19 \quad A = \begin{bmatrix} 2 & 0 & 1 \\ -1 & 2 & 0 \end{bmatrix}, \qquad B = \begin{bmatrix} 1 & -1 & 2 \\ 3 & 1 & 0 \\ 0 & 2 & 1 \end{bmatrix}$

$20 \quad A = \begin{bmatrix} 3 & -1 & 4 \end{bmatrix}, \qquad B = \begin{bmatrix} -2 \\ 5 \end{bmatrix}$

Exer. 21–24: Find AB.

$21 \quad A = \begin{bmatrix} 4 & -2 \\ 0 & 3 \\ -7 & 5 \end{bmatrix}, \qquad B = \begin{bmatrix} 3 \\ 4 \end{bmatrix}$

$22 \quad A = \begin{bmatrix} 4 \\ -3 \\ 2 \end{bmatrix}, \qquad B = \begin{bmatrix} 5 & 1 \end{bmatrix}$

$23 \quad A = \begin{bmatrix} 2 & 1 & 0 & -3 \\ -7 & 0 & -2 & 4 \end{bmatrix}, \qquad B = \begin{bmatrix} 4 & -2 & 0 \\ 1 & 1 & -2 \\ 0 & 0 & 5 \\ -3 & -1 & 0 \end{bmatrix}$

$24 \quad A = \begin{bmatrix} 1 & 2 & -3 \\ 4 & -5 & 6 \end{bmatrix}, \qquad B = \begin{bmatrix} 1 & -1 & 0 & 2 \\ -2 & 3 & 1 & 0 \\ 0 & 4 & 0 & -3 \end{bmatrix}$

Exer. 25–28: Let

$$A = \begin{bmatrix} 1 & 2 \\ 0 & -3 \end{bmatrix}, \qquad B = \begin{bmatrix} 2 & -1 \\ 3 & 1 \end{bmatrix}, \qquad C = \begin{bmatrix} 3 & 1 \\ -2 & 0 \end{bmatrix}.$$

Verify the statement.

$25 \quad (A + B)(A - B) \neq A^2 - B^2,$
where $A^2 = AA$ and $B^2 = BB$.

$26 \quad (A + B)(A + B) \neq A^2 + 2AB + B^2$

$27 \quad A(B + C) = AB + AC$

$28 \quad A(BC) = (AB)C$

Exer. 29–32: Verify the identity for

$$A = \begin{bmatrix} a & b \\ c & d \end{bmatrix}, \qquad B = \begin{bmatrix} p & q \\ r & s \end{bmatrix}, \qquad C = \begin{bmatrix} w & x \\ y & z \end{bmatrix}$$

and real numbers m and n.

$29 \quad m(A + B) = mA + mB$

$30 \quad (m + n)A = mA + nA$

$31 \quad A(B + C) = AB + AC$

$32 \quad A(BC) = (AB)C$

6.8 THE INVERSE OF A MATRIX

Throughout this section we shall restrict our discussion to square matrices. The symbol I_n will denote the square matrix of order n that has 1 in each position on the main diagonal and 0 elsewhere. We call I_n the **identity matrix of order n.**

ILLUSTRATION

Identity Matrices

■ $I_2 = \begin{bmatrix} 1 & 0 \\ 0 & 1 \end{bmatrix}$ ■ $I_3 = \begin{bmatrix} 1 & 0 & 0 \\ 0 & 1 & 0 \\ 0 & 0 & 1 \end{bmatrix}$

We can show that if A is any square matrix of order n, then

$$AI_n = A = I_nA.$$

ILLUSTRATION

$AI_3 = A = I_3A$

■ $\begin{bmatrix} a_{11} & a_{12} & a_{13} \\ a_{21} & a_{22} & a_{23} \\ a_{31} & a_{32} & a_{33} \end{bmatrix} \begin{bmatrix} 1 & 0 & 0 \\ 0 & 1 & 0 \\ 0 & 0 & 1 \end{bmatrix} = \begin{bmatrix} a_{11} & a_{12} & a_{13} \\ a_{21} & a_{22} & a_{23} \\ a_{31} & a_{32} & a_{33} \end{bmatrix}$

$= \begin{bmatrix} 1 & 0 & 0 \\ 0 & 1 & 0 \\ 0 & 0 & 1 \end{bmatrix} \begin{bmatrix} a_{11} & a_{12} & a_{13} \\ a_{21} & a_{22} & a_{23} \\ a_{31} & a_{32} & a_{33} \end{bmatrix}$

Sometimes an $n \times n$ matrix A has an **inverse**—a matrix B such that $AB = I_n = BA$. If A has an inverse, we denote it by A^{-1} and write

$$AA^{-1} = I_n = A^{-1}A.$$

The symbol A^{-1} is read "A inverse." For matrices, the symbol $1/A$ does not represent the inverse A^{-1}.

If a square matrix A has an inverse, we can calculate A^{-1} using elementary row operations. If $A = (a_{ij})$ is $n \times n$, we begin with the $n \times 2n$ matrix

$$\begin{bmatrix} a_{11} & a_{12} & \cdots & a_{1n} & 1 & 0 & \cdots & 0 \\ a_{21} & a_{22} & \cdots & a_{2n} & 0 & 1 & \cdots & 0 \\ \vdots & \vdots & & \vdots & \vdots & \vdots & & \vdots \\ a_{n1} & a_{n2} & \cdots & a_{nn} & 0 & 0 & \cdots & 1 \end{bmatrix}$$

in which the $n \times n$ identity matrix I_n appears to the right of the vertical rule. We next apply a succession of elementary row transformations, as we did in Section 6.3 to find reduced echelon forms, until we arrive at a matrix of the form

$$\begin{bmatrix} 1 & 0 & \cdots & 0 & b_{11} & b_{12} & \cdots & b_{1n} \\ 0 & 1 & \cdots & 0 & b_{21} & b_{22} & \cdots & b_{2n} \\ \vdots & \vdots & & \vdots & \vdots & \vdots & & \vdots \\ 0 & 0 & \cdots & 1 & b_{n1} & b_{n2} & \cdots & b_{nn} \end{bmatrix}$$

in which the identity matrix I_n appears to the left of the vertical rule. It can be shown that the $n \times n$ matrix (b_{ij}) is the inverse A^{-1}.

EXAMPLE I   *Finding the inverse of a 2×2 matrix*

Find A^{-1} if $A = \begin{bmatrix} 3 & 5 \\ 1 & 4 \end{bmatrix}$.

SOLUTION   We begin with the matrix

$$\begin{bmatrix} 3 & 5 & 1 & 0 \\ 1 & 4 & 0 & 1 \end{bmatrix}.$$

Next we perform elementary row transformations until the identity matrix I_2 appears on the left of the vertical rule, as follows:

$$\begin{bmatrix} 3 & 5 & 1 & 0 \\ 1 & 4 & 0 & 1 \end{bmatrix} \mathbf{R_1 \leftrightarrow R_2} \begin{bmatrix} 1 & 4 & 0 & 1 \\ 3 & 5 & 1 & 0 \end{bmatrix}$$

$$-3\mathbf{R_1} + \mathbf{R_2} \to \mathbf{R_2} \begin{bmatrix} 1 & 4 & 0 & 1 \\ 0 & -7 & 1 & -3 \end{bmatrix}$$

$$-\tfrac{1}{7}\mathbf{R_2} \to \mathbf{R_2} \begin{bmatrix} 1 & 4 & 0 & 1 \\ 0 & 1 & -\tfrac{1}{7} & \tfrac{3}{7} \end{bmatrix}$$

$$-4\mathbf{R_2} + \mathbf{R_1} \to \mathbf{R_1} \begin{bmatrix} 1 & 0 & \tfrac{4}{7} & -\tfrac{5}{7} \\ 0 & 1 & -\tfrac{1}{7} & \tfrac{3}{7} \end{bmatrix}$$

By the previous discussion,

$$A^{-1} = \begin{bmatrix} \tfrac{4}{7} & -\tfrac{5}{7} \\ -\tfrac{1}{7} & \tfrac{3}{7} \end{bmatrix} = \tfrac{1}{7} \begin{bmatrix} 4 & -5 \\ -1 & 3 \end{bmatrix}.$$

Let us verify that $AA^{-1} = I_2 = A^{-1}A$:

$$\begin{bmatrix} 3 & 5 \\ 1 & 4 \end{bmatrix} \begin{bmatrix} \tfrac{4}{7} & -\tfrac{5}{7} \\ -\tfrac{1}{7} & \tfrac{3}{7} \end{bmatrix} = \begin{bmatrix} 1 & 0 \\ 0 & 1 \end{bmatrix} = \begin{bmatrix} \tfrac{4}{7} & -\tfrac{5}{7} \\ -\tfrac{1}{7} & \tfrac{3}{7} \end{bmatrix} \begin{bmatrix} 3 & 5 \\ 1 & 4 \end{bmatrix}$$

EXAMPLE 2   *Finding the inverse of a 3 × 3 matrix*

Find A^{-1} if $A = \begin{bmatrix} -1 & 3 & 1 \\ 2 & 5 & 0 \\ 3 & 1 & -2 \end{bmatrix}$.

SOLUTION

$\begin{bmatrix} -1 & 3 & 1 & | & 1 & 0 & 0 \\ 2 & 5 & 0 & | & 0 & 1 & 0 \\ 3 & 1 & -2 & | & 0 & 0 & 1 \end{bmatrix}$ $\xrightarrow{-R_1 \to R_1}$ $\begin{bmatrix} 1 & -3 & -1 & | & -1 & 0 & 0 \\ 2 & 5 & 0 & | & 0 & 1 & 0 \\ 3 & 1 & -2 & | & 0 & 0 & 1 \end{bmatrix}$

$\xrightarrow[\substack{-2R_1 + R_2 \to R_2 \\ -3R_1 + R_3 \to R_3}]{}$ $\begin{bmatrix} 1 & -3 & -1 & | & -1 & 0 & 0 \\ 0 & 11 & 2 & | & 2 & 1 & 0 \\ 0 & 10 & 1 & | & 3 & 0 & 1 \end{bmatrix}$

$\xrightarrow{-R_3 + R_2 \to R_2}$ $\begin{bmatrix} 1 & -3 & -1 & | & -1 & 0 & 0 \\ 0 & 1 & 1 & | & -1 & 1 & -1 \\ 0 & 10 & 1 & | & 3 & 0 & 1 \end{bmatrix}$

$\xrightarrow[\substack{3R_2 + R_1 \to R_1 \\ -10R_2 + R_3 \to R_3}]{}$ $\begin{bmatrix} 1 & 0 & 2 & | & -4 & 3 & -3 \\ 0 & 1 & 1 & | & -1 & 1 & -1 \\ 0 & 0 & -9 & | & 13 & -10 & 11 \end{bmatrix}$

$\xrightarrow{-\frac{1}{9}R_3 \to R_3}$ $\begin{bmatrix} 1 & 0 & 2 & | & -4 & 3 & -3 \\ 0 & 1 & 1 & | & -1 & 1 & -1 \\ 0 & 0 & 1 & | & -\frac{13}{9} & \frac{10}{9} & -\frac{11}{9} \end{bmatrix}$

$\xrightarrow[\substack{-2R_3 + R_1 \to R_1 \\ -R_3 + R_2 \to R_2}]{}$ $\begin{bmatrix} 1 & 0 & 0 & | & -\frac{10}{9} & \frac{7}{9} & -\frac{5}{9} \\ 0 & 1 & 0 & | & \frac{4}{9} & -\frac{1}{9} & \frac{2}{9} \\ 0 & 0 & 1 & | & -\frac{13}{9} & \frac{10}{9} & -\frac{11}{9} \end{bmatrix}$

Consequently,

$$A^{-1} = \begin{bmatrix} -\frac{10}{9} & \frac{7}{9} & -\frac{5}{9} \\ \frac{4}{9} & -\frac{1}{9} & \frac{2}{9} \\ -\frac{13}{9} & \frac{10}{9} & -\frac{11}{9} \end{bmatrix} = \frac{1}{9}\begin{bmatrix} -10 & 7 & -5 \\ 4 & -1 & 2 \\ -13 & 10 & -11 \end{bmatrix}.$$

You may verify that $AA^{-1} = I_3 = A^{-1}A$.

If the procedure used in Examples 1 and 2 does not lead to an identity matrix to the left of the vertical rule, then the matrix A has no inverse.

We may apply inverses of matrices to solutions of systems of linear equations. Consider the case of two linear equations in two unknowns:

$$\begin{cases} a_{11}x + a_{12}y = k_1 \\ a_{21}x + a_{22}y = k_2 \end{cases}$$

This system can be expressed in terms of matrices as

$$\begin{bmatrix} a_{11}x + a_{12}y \\ a_{21}x + a_{22}y \end{bmatrix} = \begin{bmatrix} k_1 \\ k_2 \end{bmatrix}.$$

If we let

$$A = \begin{bmatrix} a_{11} & a_{12} \\ a_{21} & a_{22} \end{bmatrix}, \quad X = \begin{bmatrix} x \\ y \end{bmatrix}, \quad \text{and} \quad B = \begin{bmatrix} k_1 \\ k_2 \end{bmatrix},$$

then a *matrix form* for the system is

$$AX = B.$$

If A^{-1} exists, then multiplying both sides of the last equation by A^{-1} gives us $A^{-1}AX = A^{-1}B$. Since $A^{-1}A = I_2$ and $I_2X = X$, this leads to

$$X = A^{-1}B,$$

from which the solution (x, y) may be found. This technique may be extended to systems of n linear equations in n unknowns.

EXAMPLE 3 *Solving a system of linear equations using an inverse of a matrix*

Solve the system of equations:

$$\begin{cases} -x + 3y + z = 1 \\ 2x + 5y = 3 \\ 3x + y - 2z = -2 \end{cases}$$

SOLUTION If we let

$$A = \begin{bmatrix} -1 & 3 & 1 \\ 2 & 5 & 0 \\ 3 & 1 & -2 \end{bmatrix}, \quad X = \begin{bmatrix} x \\ y \\ z \end{bmatrix}, \quad \text{and} \quad B = \begin{bmatrix} 1 \\ 3 \\ -2 \end{bmatrix},$$

then $AX = B$. This implies that $X = A^{-1}B$. The matrix A^{-1} was found in Example 2. Hence,

$$\begin{bmatrix} x \\ y \\ z \end{bmatrix} = \frac{1}{9}\begin{bmatrix} -10 & 7 & -5 \\ 4 & -1 & 2 \\ -13 & 10 & -11 \end{bmatrix}\begin{bmatrix} 1 \\ 3 \\ -2 \end{bmatrix} = \frac{1}{9}\begin{bmatrix} 21 \\ -3 \\ 39 \end{bmatrix} = \begin{bmatrix} \frac{7}{3} \\ -\frac{1}{3} \\ \frac{13}{3} \end{bmatrix}.$$

Thus, $x = \frac{7}{3}$, $y = -\frac{1}{3}$, $z = \frac{13}{3}$, and the ordered triple $(\frac{7}{3}, -\frac{1}{3}, \frac{13}{3})$ is the solution of the given system.

The method of solution in Example 3 is beneficial only if A^{-1} is known (or can be easily computed) or if many systems with the same coefficient matrix are to be considered. The preferred technique for solving an arbitrary system of linear equations is the matrix method discussed in Section 6.3.

There are other, more important uses for the inverse of a matrix; however, they occur in more advanced fields of mathematics and in applications of such fields.

6.8 EXERCISES

Exer. 1–10: Find the inverse of the matrix, if it exists.

1 $\begin{bmatrix} 2 & -4 \\ 1 & 3 \end{bmatrix}$ **2** $\begin{bmatrix} 3 & 2 \\ 4 & 5 \end{bmatrix}$

3 $\begin{bmatrix} 2 & 4 \\ 4 & 8 \end{bmatrix}$ **4** $\begin{bmatrix} 3 & -1 \\ 6 & -2 \end{bmatrix}$

5 $\begin{bmatrix} 3 & -1 & 0 \\ 2 & 2 & 0 \\ 0 & 0 & 4 \end{bmatrix}$ **6** $\begin{bmatrix} 3 & 0 & 2 \\ 0 & 1 & 0 \\ -4 & 0 & 2 \end{bmatrix}$

7 $\begin{bmatrix} -2 & 2 & 3 \\ 1 & -1 & 0 \\ 0 & 1 & 4 \end{bmatrix}$ **8** $\begin{bmatrix} 1 & 2 & 3 \\ -2 & 1 & 0 \\ 3 & -1 & 1 \end{bmatrix}$

9 $\begin{bmatrix} 2 & 0 & 0 \\ 0 & 4 & 0 \\ 0 & 0 & 6 \end{bmatrix}$ **10** $\begin{bmatrix} 1 & 1 & 1 \\ 2 & 2 & 2 \\ 3 & 3 & 3 \end{bmatrix}$

11 State conditions on a and b that guarantee that the matrix $\begin{bmatrix} a & 0 \\ 0 & b \end{bmatrix}$ has an inverse, and find a formula for the inverse, if it exists.

12 If $abc \neq 0$, find the inverse of $\begin{bmatrix} a & 0 & 0 \\ 0 & b & 0 \\ 0 & 0 & c \end{bmatrix}$.

13 If $A = \begin{bmatrix} a_{11} & a_{12} & a_{13} \\ a_{21} & a_{22} & a_{23} \\ a_{31} & a_{32} & a_{33} \end{bmatrix}$, show that $AI_3 = A = I_3 A$.

14 Show that $AI_4 = A = I_4 A$ for every square matrix A of order 4.

Exer. 15–18: Solve the system using the method in Example 3. (Refer to Exercises 1–2 and 7–8.)

15 $\begin{cases} 2x - 4y = c \\ x + 3y = d \end{cases}$

(a) $\begin{bmatrix} c \\ d \end{bmatrix} = \begin{bmatrix} 3 \\ 1 \end{bmatrix}$ (b) $\begin{bmatrix} c \\ d \end{bmatrix} = \begin{bmatrix} -2 \\ 5 \end{bmatrix}$

16 $\begin{cases} 3x + 2y = c \\ 4x + 5y = d \end{cases}$

(a) $\begin{bmatrix} c \\ d \end{bmatrix} = \begin{bmatrix} -1 \\ 1 \end{bmatrix}$ (b) $\begin{bmatrix} c \\ d \end{bmatrix} = \begin{bmatrix} 4 \\ 3 \end{bmatrix}$

17 $\begin{cases} -2x + 2y + 3z = c \\ x - y = d \\ y + 4z = e \end{cases}$

(a) $\begin{bmatrix} c \\ d \\ e \end{bmatrix} = \begin{bmatrix} 1 \\ 3 \\ -2 \end{bmatrix}$ (b) $\begin{bmatrix} c \\ d \\ e \end{bmatrix} = \begin{bmatrix} -1 \\ 0 \\ 4 \end{bmatrix}$

18 $\begin{cases} x + 2y + 3z = c \\ -2x + y = d \\ 3x - y + z = e \end{cases}$

(a) $\begin{bmatrix} c \\ d \\ e \end{bmatrix} = \begin{bmatrix} -1 \\ 4 \\ 2 \end{bmatrix}$ (b) $\begin{bmatrix} c \\ d \\ e \end{bmatrix} = \begin{bmatrix} -3 \\ -2 \\ 1 \end{bmatrix}$

Exer. 19–20: (a) Express the system in the matrix form $AX = B$. (b) Approximate A^{-1} using four-decimal-place accuracy for its elements. (c) Use $X = A^{-1}B$ to approximate the solution of the system to four-decimal-place accuracy.

19 $\begin{cases} 3.1x + 6.7y - 8.7z = 1.5 \\ 4.1x - 5.1y + 0.2z = 2.1 \\ 0.6x + 1.1y - 7.4z = 3.9 \end{cases}$

20 $\begin{cases} 5.6x + 8.4y - 7.2z + 4.2w = 8.1 \\ 8.4x + 9.2y - 6.1z - 6.2w = 5.3 \\ -7.2x - 6.1y + 9.2z + 4.5w = 0.4 \\ 4.2x - 6.2y - 4.5z + 5.8w = 2.7 \end{cases}$

6.9 DETERMINANTS

Throughout this section and the next we will assume that all matrices under discussion are *square* matrices. Associated with each square matrix A is a number called the **determinant of A**, denoted by $|A|$. This notation should not be confused with the symbol for the absolute value of a real number. To avoid any misunderstanding, the expression det A is sometimes used in place of $|A|$. We shall define $|A|$ by beginning with the case in which A has order 1 and then increasing the order one at a time. As we shall see in Section 6.10, these definitions arise in a natural way when systems of linear equations are solved.

If A is a square matrix of order 1, then A has only one element. Thus, $A = [a_{11}]$ and we define $|A| = a_{11}$. If A is a square matrix of order 2, then

$$A = \begin{bmatrix} a_{11} & a_{12} \\ a_{21} & a_{22} \end{bmatrix}$$

and the determinant of A is defined by

$$|A| = a_{11}a_{22} - a_{21}a_{12}.$$

Another notation for $|A|$ is obtained by replacing the brackets used for A with vertical bars, as follows.

Determinant of a 2 × 2 Matrix A

$$|A| = \begin{vmatrix} a_{11} & a_{12} \\ a_{21} & a_{22} \end{vmatrix} = a_{11}a_{22} - a_{21}a_{12}$$

EXAMPLE 1 *Finding the determinant of a 2 × 2 matrix*

Find $|A|$ if $A = \begin{bmatrix} 2 & -1 \\ 4 & -3 \end{bmatrix}$.

SOLUTION By definition,

$$|A| = \begin{vmatrix} 2 & -1 \\ 4 & -3 \end{vmatrix} = (2)(-3) - (4)(-1) = -6 + 4 = -2.$$

For square matrices of order $n > 1$, it is convenient to introduce the following terminology.

Definition of Minors and Cofactors

Let $A = (a_{ij})$ be a square matrix of order $n > 1$.

(1) The **minor** M_{ij} of the element a_{ij} is the determinant of the matrix of order $n - 1$ obtained by deleting row i and column j.

(2) The **cofactor** A_{ij} of the element a_{ij} is $A_{ij} = (-1)^{i+j}M_{ij}$.

To determine the minor of an element, we delete the row and column in which the element appears and then find the determinant of the resulting square matrix. This process is demonstrated in the following illustration, where deletions of rows and columns in a 3×3 matrix are indicated with horizontal and vertical line segments, respectively.

ILLUSTRATION

Minors

Matrix	**Minor**
$\begin{bmatrix} a_{11} & a_{12} & a_{13} \\ a_{21} & a_{22} & a_{23} \\ a_{31} & a_{32} & a_{33} \end{bmatrix}$	$M_{11} = \begin{vmatrix} a_{22} & a_{23} \\ a_{32} & a_{33} \end{vmatrix} = a_{22}a_{33} - a_{32}a_{23}$
$\begin{bmatrix} a_{11} & a_{12} & a_{13} \\ a_{21} & a_{22} & a_{23} \\ a_{31} & a_{32} & a_{33} \end{bmatrix}$	$M_{12} = \begin{vmatrix} a_{21} & a_{23} \\ a_{31} & a_{33} \end{vmatrix} = a_{21}a_{33} - a_{31}a_{23}$
$\begin{bmatrix} a_{11} & a_{12} & a_{13} \\ a_{21} & a_{22} & a_{23} \\ a_{31} & a_{32} & a_{33} \end{bmatrix}$	$M_{13} = \begin{vmatrix} a_{21} & a_{22} \\ a_{31} & a_{32} \end{vmatrix} = a_{21}a_{32} - a_{31}a_{22}$
$\begin{bmatrix} a_{11} & a_{12} & a_{13} \\ a_{21} & a_{22} & a_{23} \\ a_{31} & a_{32} & a_{33} \end{bmatrix}$	$M_{23} = \begin{vmatrix} a_{11} & a_{12} \\ a_{31} & a_{32} \end{vmatrix} = a_{11}a_{32} - a_{31}a_{12}$

For the matrix in the preceding illustration, there are five other minors—$M_{21}, M_{22}, M_{31}, M_{32},$ and M_{33}—that can be obtained in similar fashion.

To obtain the cofactor of a_{ij} of a square matrix $A = (a_{ij})$, we find the minor and multiply it by 1 or -1, depending on whether the sum of i and j is even or odd, respectively. Another way to remember the sign $(-1)^{i+j}$ associated with the cofactor A_{ij} is to consider the following checkerboard style of plus and minus signs:

$$\begin{bmatrix} + & - & + & - & \cdots \\ - & + & - & + & \cdots \\ + & - & + & - & \cdots \\ - & + & - & + & \cdots \\ \vdots & \vdots & \vdots & \vdots & \end{bmatrix}$$

EXAMPLE 2   *Finding minors and cofactors*

If $A = \begin{bmatrix} 1 & -3 & 3 \\ 4 & 2 & 0 \\ -2 & -7 & 5 \end{bmatrix}$, find $M_{11}, M_{21}, M_{22}, A_{11}, A_{21},$ and A_{22}.

SOLUTION Deleting appropriate rows and columns of A, we obtain

$$M_{11} = \begin{vmatrix} 2 & 0 \\ -7 & 5 \end{vmatrix} = (2)(5) - (-7)(0) = 10$$

$$M_{21} = \begin{vmatrix} -3 & 3 \\ -7 & 5 \end{vmatrix} = (-3)(5) - (-7)(3) = 6$$

$$M_{22} = \begin{vmatrix} 1 & 3 \\ -2 & 5 \end{vmatrix} = (1)(5) - (-2)(3) = 11.$$

To obtain the cofactors, we prefix the corresponding minors with the proper signs. Thus, using the definition of cofactor, we have

$$A_{11} = (-1)^{1+1}M_{11} = (1)(10) = 10$$

$$A_{21} = (-1)^{2+1}M_{21} = (-1)(6) = -6$$

$$A_{22} = (-1)^{2+2}M_{22} = (1)(11) = 11.$$

We can also use the checkerboard style of plus and minus signs to determine the proper signs.

The determinant $|A|$ of a square matrix of order 3 is defined as follows.

Definition of the Determinant of a 3 × 3 Matrix A

$$|A| = \begin{vmatrix} a_{11} & a_{12} & a_{13} \\ a_{21} & a_{22} & a_{23} \\ a_{31} & a_{32} & a_{33} \end{vmatrix} = a_{11}A_{11} + a_{12}A_{12} + a_{13}A_{13}$$

Since $A_{11} = (-1)^{1+1}M_{11} = M_{11}$, $A_{12} = (-1)^{1+2}M_{12} = -M_{12}$, and $A_{13} = (-1)^{1+3}M_{13} = M_{13}$, the preceding definition may also be written

$$|A| = a_{11}M_{11} - a_{12}M_{12} + a_{13}M_{13}.$$

If we express M_{11}, M_{12}, and M_{13} using elements of A and rearrange terms, we obtain the following formula for $|A|$:

$$|A| = a_{11}a_{22}a_{33} - a_{11}a_{23}a_{32} - a_{12}a_{21}a_{33} + a_{12}a_{23}a_{31} + a_{13}a_{21}a_{32} - a_{13}a_{22}a_{31}$$

The definition of $|A|$ for a square matrix A of order 3 displays a pattern of multiplying each element in row 1 by its cofactor and then adding to find $|A|$. This process is referred to as *expanding $|A|$ by the first row*. By actually carrying out the computations, we can show that $|A|$ *can be expanded in similar fashion by using any row or column.* As an illustration, the expansion by the second column is

$$|A| = a_{12}A_{12} + a_{22}A_{22} + a_{32}A_{32} = a_{12}\left(-\begin{vmatrix} a_{21} & a_{23} \\ a_{31} & a_{33} \end{vmatrix}\right) + a_{22}\left(+\begin{vmatrix} a_{11} & a_{13} \\ a_{31} & a_{33} \end{vmatrix}\right) + a_{32}\left(-\begin{vmatrix} a_{11} & a_{13} \\ a_{21} & a_{23} \end{vmatrix}\right).$$

Applying the definition to the determinants in parentheses, multiplying as indicated, and rearranging the terms in the sum, we could arrive at the formula for $|A|$ in terms of the elements of A. Similarly, the expansion by the third row is

$$|A| = a_{31}A_{31} + a_{32}A_{32} + a_{33}A_{33}.$$

Once again we can show that this result agrees with previous expansions.

EXAMPLE 3   *Finding the determinant of a* 3 × 3 *matrix*

Find $|A|$ if $A = \begin{bmatrix} -1 & 3 & 1 \\ 2 & 5 & 0 \\ 3 & 1 & -2 \end{bmatrix}$.

SOLUTION   Since the second row contains a zero, we shall expand $|A|$ by that row, because then we need to evaluate only two cofactors. Thus,

$$|A| = (2)A_{21} + (5)A_{22} + (0)A_{23}.$$

Using the definition of cofactors, we have

$$A_{21} = (-1)^{2+1}M_{21} = -\begin{vmatrix} 3 & 1 \\ 1 & -2 \end{vmatrix} = -[(3)(-2) - (1)(1)] = 7$$

$$A_{22} = (-1)^{2+2}M_{22} = \begin{vmatrix} -1 & 1 \\ 3 & -2 \end{vmatrix} = [(-1)(-2) - (3)(1)] = -1.$$

Consequently,

$$|A| = (2)(7) + (5)(-1) + (0)A_{23} = 14 - 5 + 0 = 9.$$

The following definition of the determinant of a matrix of arbitrary order n is patterned after that used for the determinant of a matrix of order 3.

Definition of the Determinant of an $n \times n$ Matrix A

The **determinant** $|A|$ **of a matrix A of order n** is the cofactor expansion by the first row:

$$|A| = a_{11}A_{11} + a_{12}A_{12} + \cdots + a_{1n}A_{1n}$$

In terms of minors,

$$|A| = a_{11}M_{11} - a_{12}M_{12} + \cdots + a_{1n}(-1)^{1+n}M_{1n}.$$

The number $|A|$ may be found by using *any* row or column, as stated in the following theorem.

Expansion Theorem for Determinants

> If A is a square matrix of order $n > 1$, then the determinant $|A|$ may be found by multiplying the elements of any row (or column) by their respective cofactors and adding the resulting products.

The proof of this theorem may be found in texts on linear algebra. The theorem is useful if many zeros appear in a row or column, as illustrated in the following example.

EXAMPLE 4 *Finding the determinant of a* 4 × 4 *matrix*

Find $|A|$ if $A = \begin{bmatrix} 1 & 0 & 2 & 5 \\ -2 & 1 & 5 & 0 \\ 0 & 0 & -3 & 0 \\ 0 & -1 & 0 & 3 \end{bmatrix}$.

SOLUTION Note that all but one of the elements in the third row are zero. Hence, if we expand $|A|$ by the third row, there will be at most one nonzero term. Specifically,

$$|A| = (0)A_{31} + (0)A_{32} + (-3)A_{33} + (0)A_{34} = -3A_{33}$$

with

$$A_{33} = (-1)^{3+3}\begin{vmatrix} 1 & 0 & 5 \\ -2 & 1 & 0 \\ 0 & -1 & 3 \end{vmatrix}.$$

We expand A_{33} by column 1:

$$A_{33} = (1)\begin{vmatrix} 1 & 0 \\ -1 & 3 \end{vmatrix} + (-2)\left(-\begin{vmatrix} 0 & 5 \\ -1 & 3 \end{vmatrix}\right) + (0)\begin{vmatrix} 0 & 5 \\ 1 & 0 \end{vmatrix}$$
$$= 3 + 10 + 0 = 13$$

Thus, $$|A| = -3A_{33} = (-3)(13) = -39.$$

In general, if all but one element in some row (or column) of A are zero and if the determinant $|A|$ is expanded by that row (or column), then all terms drop out except the product of that element with its cofactor. If *all* elements in a row (or column) are zero, we have the following.

Theorem

> If every element of a row (or column) of a square matrix A is zero, then $|A| = 0$.

PROOF If every element in a row (or column) of a matrix A is zero, the expansion by that row (or column) is zero. ■

6.9 EXERCISES

Exer. 1–4: Find all the minors and cofactors of the elements in the matrix.

$$1 \begin{bmatrix} 7 & -1 \\ 5 & 0 \end{bmatrix} \qquad 2 \begin{bmatrix} -6 & 4 \\ 3 & 2 \end{bmatrix}$$

$$3 \begin{bmatrix} 2 & 4 & -1 \\ 0 & 3 & 2 \\ -5 & 7 & 0 \end{bmatrix} \qquad 4 \begin{bmatrix} 5 & -2 & 1 \\ 4 & 7 & 0 \\ -3 & 4 & -1 \end{bmatrix}$$

Exer. 5–8: Find the determinants of the matrices in the given exercises.

5 Exercise 1   6 Exercise 2

7 Exercise 3   8 Exercise 4

Exer. 9–20: Find the determinant of the matrix.

$$9 \begin{bmatrix} -5 & 4 \\ -3 & 2 \end{bmatrix} \qquad 10 \begin{bmatrix} 6 & 4 \\ -3 & 2 \end{bmatrix}$$

$$11 \begin{bmatrix} a & -a \\ b & -b \end{bmatrix} \qquad 12 \begin{bmatrix} c & d \\ -d & c \end{bmatrix}$$

$$13 \begin{bmatrix} 3 & 1 & -2 \\ 4 & 2 & 5 \\ -6 & 3 & -1 \end{bmatrix} \qquad 14 \begin{bmatrix} 2 & -5 & 1 \\ -3 & 1 & 6 \\ 4 & -2 & 3 \end{bmatrix}$$

$$15 \begin{bmatrix} -5 & 4 & 1 \\ 3 & -2 & 7 \\ 2 & 0 & 6 \end{bmatrix} \qquad 16 \begin{bmatrix} 2 & 7 & -3 \\ 1 & 0 & 4 \\ 4 & -1 & -2 \end{bmatrix}$$

$$17 \begin{bmatrix} 3 & -1 & 2 & 0 \\ 4 & 0 & -3 & 5 \\ 0 & 6 & 0 & 0 \\ 1 & 3 & -4 & 2 \end{bmatrix} \qquad 18 \begin{bmatrix} 2 & 5 & 1 & 0 \\ -4 & 0 & -3 & 0 \\ 3 & -2 & 1 & 6 \\ -1 & 4 & 2 & 0 \end{bmatrix}$$

$$19 \begin{bmatrix} 0 & b & 0 & 0 \\ 0 & 0 & c & 0 \\ a & 0 & 0 & 0 \\ 0 & 0 & 0 & d \end{bmatrix} \qquad 20 \begin{bmatrix} a & u & v & w \\ 0 & b & x & y \\ 0 & 0 & c & z \\ 0 & 0 & 0 & d \end{bmatrix}$$

Exer. 21–28: Verify the identity by expanding each determinant.

$$21 \begin{vmatrix} a & b \\ c & d \end{vmatrix} = - \begin{vmatrix} c & d \\ a & b \end{vmatrix} \qquad 22 \begin{vmatrix} a & b \\ c & d \end{vmatrix} = - \begin{vmatrix} b & a \\ d & c \end{vmatrix}$$

$$23 \begin{vmatrix} a & kb \\ c & kd \end{vmatrix} = k \begin{vmatrix} a & b \\ c & d \end{vmatrix} \qquad 24 \begin{vmatrix} a & b \\ kc & kd \end{vmatrix} = k \begin{vmatrix} a & b \\ c & d \end{vmatrix}$$

$$25 \begin{vmatrix} a & b \\ c & d \end{vmatrix} = \begin{vmatrix} a & b \\ ka + c & kb + d \end{vmatrix}$$

$$26 \begin{vmatrix} a & b \\ c & d \end{vmatrix} = \begin{vmatrix} a & ka + b \\ c & kc + d \end{vmatrix}$$

$$27 \begin{vmatrix} a & b \\ c & d \end{vmatrix} + \begin{vmatrix} a & e \\ c & f \end{vmatrix} = \begin{vmatrix} a & b + e \\ c & d + f \end{vmatrix}$$

$$28 \begin{vmatrix} a & b \\ c & d \end{vmatrix} + \begin{vmatrix} a & b \\ e & f \end{vmatrix} = \begin{vmatrix} a & b \\ c + e & d + f \end{vmatrix}$$

29 Let $A = (a_{ij})$ be a square matrix of order n such that $a_{ij} = 0$ if $i < j$. Show that $|A| = a_{11} a_{22} \cdots a_{nn}$.

30 If $A = (a_{ij})$ is any 2×2 matrix such that $|A| \neq 0$, show that A has an inverse, and find a general formula for A^{-1}.

Exer. 31–34: Let $I = I_2$ be the identity matrix of order 2, and let $f(x) = |A - xI|$. Find (a) the polynomial $f(x)$ and (b) the zeros of $f(x)$. (In the study of matrices, $f(x)$ is the *characteristic polynomial of A*, and the zeros of $f(x)$ are the *characteristic values* (*eigenvalues*) of A.)

$$31 \; A = \begin{bmatrix} 1 & 2 \\ 3 & 2 \end{bmatrix} \qquad 32 \; A = \begin{bmatrix} 3 & 1 \\ 2 & 2 \end{bmatrix}$$

$$33 \; A = \begin{bmatrix} -3 & -2 \\ 2 & 2 \end{bmatrix} \qquad 34 \; A = \begin{bmatrix} 2 & -4 \\ -3 & 5 \end{bmatrix}$$

Exer. 35–38: Let $I = I_3$ and let $f(x) = |A - xI|$. Find (a) the polynomial $f(x)$ and (b) the zeros of $f(x)$.

$$35 \; A = \begin{bmatrix} 1 & 0 & 0 \\ 1 & 0 & -2 \\ -1 & 1 & -3 \end{bmatrix} \qquad 36 \; A = \begin{bmatrix} 2 & 1 & 0 \\ -1 & 0 & 0 \\ 1 & 3 & 2 \end{bmatrix}$$

$$37 \; A = \begin{bmatrix} 0 & 2 & -2 \\ -1 & 3 & 1 \\ -3 & 3 & 1 \end{bmatrix} \qquad 38 \; A = \begin{bmatrix} 3 & 2 & 2 \\ 1 & 0 & 2 \\ -1 & -1 & 0 \end{bmatrix}$$

Exer. 39–42: Express the determinant in the form $ai + bj + ck$ for real numbers a, b, and c.

$$39 \begin{vmatrix} i & j & k \\ 2 & -1 & 6 \\ -3 & 5 & 1 \end{vmatrix} \qquad 40 \begin{vmatrix} i & j & k \\ 1 & -2 & 3 \\ 2 & 1 & -4 \end{vmatrix}$$

$$41 \quad \begin{vmatrix} i & j & k \\ 5 & -6 & -1 \\ 3 & 0 & 1 \end{vmatrix}$$

$$42 \quad \begin{vmatrix} i & j & k \\ 4 & -6 & 2 \\ -2 & 3 & -1 \end{vmatrix}$$

c Exer. 43–44: Let $I = I_3$ and $f(x) = |A - xI|$. **(a)** Find the polynomial $f(x)$. **(b)** Graph f, and estimate the characteristic values of A.

$$43 \quad A = \begin{bmatrix} 1 & 0 & 1 \\ 0 & 2 & 1 \\ 1 & 1 & -2 \end{bmatrix}$$

$$44 \quad A = \begin{bmatrix} 3 & -1 & -1 \\ -1 & 1 & 0 \\ -1 & 0 & -2 \end{bmatrix}$$

6.10 PROPERTIES OF DETERMINANTS

Evaluating a determinant by using the expansion theorem stated in Section 6.9 is inefficient for matrices of high order. For example, if a determinant of a matrix of order 10 is expanded by any row, a sum of 10 terms is obtained, and each term contains the determinant of a matrix of order 9, which is a cofactor of the original matrix. If any of the latter determinants are expanded by a row (or column), a sum of 9 terms is obtained, each containing the determinant of a matrix of order 8. Hence, at this stage there are 90 determinants of matrices of order 8 to evaluate. The process could be continued until only determinants of matrices of order 2 remain. You may verify that there are 1,814,400 such matrices of order 2! Unless many elements of the original matrix are zero, it is an enormous task to carry out all of the computations.

In this section we discuss rules that simplify the process of evaluating determinants. The main use for these rules is to introduce zeros into the determinant. They may also be used to change the determinant to **echelon form**—that is, to a form in which the elements below the main diagonal elements are all zero (see Section 6.3). The transformations on rows stated in the next theorem are the same as the elementary row transformations of a matrix introduced in Section 6.3. However, for determinants we may also employ similar transformations on columns.

Theorem on Row and Column Transformations of a Determinant

Let A be a square matrix of order n.

(1) If a matrix B is obtained from A by interchanging two rows (or columns), then $|B| = -|A|$.

(2) If B is obtained from A by multiplying every element of one row (or column) of A by a real number k, then $|B| = k|A|$.

(3) If B is obtained from A by adding k times any row (or column) of A to another row (or column) for a real number k, then $|B| = |A|$.

When using the theorem, we refer to the rows (or columns) of the *determinant* in the obvious way. For example, property (3) may be phrased:

Adding k times any row (or column) to another row (or column) of a determinant does not affect the value of the determinant.

Row transformations of determinants will be specified by means of the symbols $R_i \leftrightarrow R_j$, $kR_i \to R_i$, and $kR_i + R_j \to R_j$, which were introduced in Section 6.3. Analogous symbols are used for column transformations. For example, $kC_i + C_j \to C_j$ means: Add k times the ith column to the jth column. The following are illustrations of the preceding theorem, with the reason for each equality stated at the right.

ILLUSTRATION

Transformations of Determinants

$$\begin{vmatrix} 2 & 0 & 1 \\ 6 & 4 & 3 \\ 0 & 3 & 5 \end{vmatrix} = -\begin{vmatrix} 6 & 4 & 3 \\ 2 & 0 & 1 \\ 0 & 3 & 5 \end{vmatrix} \qquad R_1 \leftrightarrow R_2$$

$$\begin{vmatrix} 2 & 0 & 1 \\ 6 & 4 & 3 \\ 0 & 3 & 5 \end{vmatrix} = -\begin{vmatrix} 1 & 0 & 2 \\ 3 & 4 & 6 \\ 5 & 3 & 0 \end{vmatrix} \qquad C_1 \leftrightarrow C_3$$

$$\begin{vmatrix} 1 & -3 & 4 \\ 2 & -1 & 0 \\ 3 & 1 & 6 \end{vmatrix} = \begin{vmatrix} 1 & -3 & 4 \\ 0 & 5 & -8 \\ 3 & 1 & 6 \end{vmatrix} \qquad -2R_1 + R_2 \to R_2$$

$$\begin{vmatrix} 1 & -3 & 4 \\ 2 & -1 & 0 \\ 3 & 1 & 6 \end{vmatrix} = \begin{vmatrix} -5 & -3 & 4 \\ 0 & -1 & 0 \\ 5 & 1 & 6 \end{vmatrix} \qquad 2C_2 + C_1 \to C_1$$

Theorem

> If two rows (or columns) of a square matrix A are identical, then $|A| = 0$.

PROOF If B is the matrix obtained from A by interchanging the two identical rows (or columns), then B and A are the same and, consequently, $|B| = |A|$. However, by property (1) of the theorem on row and column transformations of a determinant, $|B| = -|A|$, and hence $-|A| = |A|$. Thus, $2|A| = 0$, and therefore $|A| = 0$. ∎

EXAMPLE 1   *Using row and column transformations*

Find $|A|$ if $A = \begin{bmatrix} 2 & 3 & 0 & 4 \\ 0 & 5 & -1 & 6 \\ 1 & 0 & -2 & 3 \\ -3 & 2 & 0 & -5 \end{bmatrix}$.

SOLUTION We plan to use property (3) of the theorem on row and column transformations to introduce many zeros in some row or column. It is convenient to work with an element of the matrix that equals 1, since this enables us to avoid the use of fractions. If 1 is not an element of the original matrix, it is always possible to introduce the number 1 by using property (2) or (3) of the theorem. In this example, 1 appears in row 3, and we proceed as follows, with the reason for each equality stated at the right.

$$
\begin{vmatrix} 2 & 3 & 0 & 4 \\ 0 & 5 & -1 & 6 \\ 1 & 0 & -2 & 3 \\ -3 & 2 & 0 & -5 \end{vmatrix} = \begin{vmatrix} 0 & 3 & 4 & -2 \\ 0 & 5 & -1 & 6 \\ 1 & 0 & -2 & 3 \\ 0 & 2 & -6 & 4 \end{vmatrix}
\qquad
\begin{aligned} &-2R_3 + R_1 \to R_1 \\ \\ \\ &3R_3 + R_4 \to R_4 \end{aligned}
$$

$$
= (1) \begin{vmatrix} 3 & 4 & -2 \\ 5 & -1 & 6 \\ 2 & -6 & 4 \end{vmatrix}
\qquad \text{expand by the first column}
$$

$$
= \begin{vmatrix} 23 & 4 & 22 \\ 0 & -1 & 0 \\ -28 & -6 & -32 \end{vmatrix}
\qquad
\begin{aligned} &5C_2 + C_1 \to C_1 \\ \\ &6C_2 + C_3 \to C_3 \end{aligned}
$$

$$
= (-1) \begin{vmatrix} 23 & 22 \\ -28 & -32 \end{vmatrix}
\qquad \text{expand by the second row}
$$

$$
= (-1)[(23)(-32) - (-28)(22)] \qquad \text{definition of determinant}
$$

$$
= 120
$$

Part (2) of the theorem on row and column transformations is useful for finding factors of determinants. To illustrate, for a determinant of a matrix of order 3, we have the following:

$$
\begin{vmatrix} a_{11} & a_{12} & a_{13} \\ ka_{21} & ka_{22} & ka_{23} \\ a_{31} & a_{32} & a_{33} \end{vmatrix} = k \begin{vmatrix} a_{11} & a_{12} & a_{13} \\ a_{21} & a_{22} & a_{23} \\ a_{31} & a_{32} & a_{33} \end{vmatrix}
$$

Similar formulas hold if k is a common factor of the elements of any other row or column. When referring to this manipulation, we often use the phrase k *is a common factor of the row* (or *column*).

EXAMPLE 2 *Removing common factors from rows of a determinant*

Find $|A|$ if $A = \begin{bmatrix} 14 & -6 & 4 \\ 4 & -5 & 12 \\ -21 & 9 & -6 \end{bmatrix}$.

SOLUTION

$$|A| = 2 \begin{vmatrix} 7 & -3 & 2 \\ 4 & -5 & 12 \\ -21 & 9 & -6 \end{vmatrix} \qquad \text{2 is a common factor of row 1}$$

$$= (2)(-3) \begin{vmatrix} 7 & -3 & 2 \\ 4 & -5 & 12 \\ 7 & -3 & 2 \end{vmatrix} \qquad -3 \text{ is a common factor of row 3}$$

$$= 0 \qquad \text{two rows are identical}$$

EXAMPLE 3   *Removing a common factor from a column*

Without expanding, show that $a - b$ is a factor of $\begin{vmatrix} 1 & 1 & 1 \\ a & b & c \\ a^2 & b^2 & c^2 \end{vmatrix}$.

SOLUTION

$$\begin{vmatrix} 1 & 1 & 1 \\ a & b & c \\ a^2 & b^2 & c^2 \end{vmatrix} = \begin{vmatrix} 0 & 1 & 1 \\ a-b & b & c \\ a^2-b^2 & b^2 & c^2 \end{vmatrix} \qquad -\mathbf{C}_2 + \mathbf{C}_1 \to \mathbf{C}_1$$

$$= (a-b) \begin{vmatrix} 0 & 1 & 1 \\ 1 & b & c \\ a+b & b^2 & c^2 \end{vmatrix} \qquad \begin{array}{l} a-b \text{ is a common factor} \\ \text{of column 1} \end{array}$$

Determinants arise in the study of solutions of systems of linear equations. To illustrate, let us consider two linear equations in two variables x and y:

$$\begin{cases} a_{11}x + a_{12}y = k_1 \\ a_{21}x + a_{22}y = k_2 \end{cases}$$

where at least one nonzero coefficient appears in each equation. We may assume that $a_{11} \neq 0$, for otherwise $a_{12} \neq 0$, and we could then regard y as the first variable instead of x. We shall use elementary row transformations to obtain the matrix of an equivalent system with $a_{21} = 0$ as follows:

$$\begin{bmatrix} a_{11} & a_{12} & \bigg| & k_1 \\ a_{21} & a_{22} & \bigg| & k_2 \end{bmatrix} \underset{-\frac{a_{21}}{a_{11}}\mathbf{R_1} + \mathbf{R_2} \to \mathbf{R_2}}{} \begin{bmatrix} a_{11} & a_{12} & \bigg| & k_1 \\ 0 & a_{22} - \left(\dfrac{a_{21}a_{12}}{a_{11}}\right) & \bigg| & k_2 - \left(\dfrac{a_{21}k_1}{a_{11}}\right) \end{bmatrix}$$

$$\underset{a_{11}\mathbf{R_2} \to \mathbf{R_2}}{} \begin{bmatrix} a_{11} & a_{12} & \bigg| & k_1 \\ 0 & (a_{11}a_{22} - a_{21}a_{12}) & \bigg| & (a_{11}k_2 - a_{21}k_1) \end{bmatrix}$$

Thus, the given system is equivalent to

$$\begin{cases} a_{11}x + a_{12}y = k_1 \\ (a_{11}a_{22} - a_{21}a_{12})y = a_{11}k_2 - a_{21}k_1 \end{cases}$$

which may also be written

$$\begin{cases} a_{11}x + a_{12}y = k_1 \\ \begin{vmatrix} a_{11} & a_{12} \\ a_{21} & a_{22} \end{vmatrix} y = \begin{vmatrix} a_{11} & k_1 \\ a_{21} & k_2 \end{vmatrix} \end{cases}$$

If $\begin{vmatrix} a_{11} & a_{12} \\ a_{21} & a_{22} \end{vmatrix} \neq 0$, we can solve the second equation for y, obtaining

$$y = \frac{\begin{vmatrix} a_{11} & k_1 \\ a_{21} & k_2 \end{vmatrix}}{\begin{vmatrix} a_{11} & a_{12} \\ a_{21} & a_{22} \end{vmatrix}}.$$

The corresponding value for x may be found by substituting for y in the first equation, which leads to

$$x = \frac{\begin{vmatrix} k_1 & a_{12} \\ k_2 & a_{22} \end{vmatrix}}{\begin{vmatrix} a_{11} & a_{12} \\ a_{21} & a_{22} \end{vmatrix}}.$$

This proves that *if the determinant of the coefficient matrix of a system of two linear equations in two variables is not zero, then the system has a unique solution.* The last two formulas for x and y as quotients of determinants constitute **Cramer's rule** for two variables.

There is an easy way to remember Cramer's rule. Let

$$D = \begin{bmatrix} a_{11} & a_{12} \\ a_{21} & a_{22} \end{bmatrix}$$

be the coefficient matrix of the system, and let D_x denote the matrix obtained from D by replacing the coefficients a_{11}, a_{21} of x by the numbers k_1, k_2, respectively. Similarly, let D_y denote the matrix obtained from D by replacing the coefficients a_{12}, a_{22} of y by the numbers k_1, k_2, respectively. Thus,

$$D_x = \begin{bmatrix} k_1 & a_{12} \\ k_2 & a_{22} \end{bmatrix}, \qquad D_y = \begin{bmatrix} a_{11} & k_1 \\ a_{21} & k_2 \end{bmatrix}.$$

If $|D| \neq 0$, the solution (x, y) is given by the following formulas.

Cramer's Rule for Two Variables

$$x = \frac{|D_x|}{|D|}, \qquad y = \frac{|D_y|}{|D|}$$

EXAMPLE 4 *Using Cramer's rule to solve a system of two linear equations*

Use Cramer's rule to solve the system

$$\begin{cases} 2x - 3y = -4 \\ 5x + 7y = 1 \end{cases}$$

SOLUTION   The determinant of the coefficient matrix is

$$|D| = \begin{vmatrix} 2 & -3 \\ 5 & 7 \end{vmatrix} = 29.$$

Using the notation introduced previously, we have

$$|D_x| = \begin{vmatrix} -4 & -3 \\ 1 & 7 \end{vmatrix} = -25, \qquad |D_y| = \begin{vmatrix} 2 & -4 \\ 5 & 1 \end{vmatrix} = 22.$$

Hence, $x = \dfrac{|D_x|}{|D|} = \dfrac{-25}{29}, \qquad y = \dfrac{|D_y|}{|D|} = \dfrac{22}{29}.$

Thus, the system has the unique solution $(-\frac{25}{29}, \frac{22}{29})$.

Cramer's rule can be extended to systems of n linear equations in n variables $x_1, x_2, \ldots, x_n$, where each equation has the form

$$a_1 x_1 + a_2 x_2 + \cdots + a_n x_n = k.$$

To solve such a system, let D denote the coefficient matrix and let D_{x_i} denote the matrix obtained by replacing the coefficients of x_i in D by the numbers $k_1, \ldots, k_n$ that appear in the column to the right of the equal signs in the system. If $|D| \neq 0$, then the system has the following unique solution.

Cramer's Rule (General Form)

$$x_1 = \frac{|D_{x_1}|}{|D|}, \qquad x_2 = \frac{|D_{x_2}|}{|D|}, \qquad \ldots, \qquad x_n = \frac{|D_{x_n}|}{|D|}$$

EXAMPLE 5   *Using Cramer's rule to solve a system of three linear equations*

Use Cramer's rule to solve the system

$$\begin{cases} x & - 2z = 3 \\ & -y + 3z = 1 \\ 2x & + 5z = 0 \end{cases}$$

SOLUTION   We shall merely list the various determinants. You should check the results.

$$|D| = \begin{vmatrix} 1 & 0 & -2 \\ 0 & -1 & 3 \\ 2 & 0 & 5 \end{vmatrix} = -9, \qquad |D_x| = \begin{vmatrix} 3 & 0 & -2 \\ 1 & -1 & 3 \\ 0 & 0 & 5 \end{vmatrix} = -15$$

$$|D_y| = \begin{vmatrix} 1 & 3 & -2 \\ 0 & 1 & 3 \\ 2 & 0 & 5 \end{vmatrix} = 27, \qquad |D_z| = \begin{vmatrix} 1 & 0 & 3 \\ 0 & -1 & 1 \\ 2 & 0 & 0 \end{vmatrix} = 6$$

By Cramer's rule, the solution is

$$x = \frac{|D_x|}{|D|} = \frac{-15}{-9} = \frac{5}{3}$$

$$y = \frac{|D_y|}{|D|} = \frac{27}{-9} = -3$$

$$z = \frac{|D_z|}{|D|} = \frac{6}{-9} = -\frac{2}{3}.$$

Cramer's rule is an inefficient method to apply if the system has a large number of equations, since many determinants of matrices of high order must be evaluated. Note also that Cramer's rule cannot be used directly if $|D| = 0$ or if the number of equations is not the same as the number of variables. For numerical calculations, the matrix method is superior to Cramer's rule; however, the Cramer's rule formulation is theoretically useful.

6.10 EXERCISES

Exer. 1–14: Without expanding, explain why the statement is true.

1 $\begin{vmatrix} 1 & 0 & 1 \\ 0 & 1 & 1 \\ 1 & 1 & 0 \end{vmatrix} = - \begin{vmatrix} 1 & 0 & 1 \\ 1 & 1 & 0 \\ 0 & 1 & 1 \end{vmatrix}$

2 $\begin{vmatrix} 1 & 0 & 1 \\ 0 & 1 & 1 \\ 1 & 1 & 0 \end{vmatrix} = - \begin{vmatrix} 1 & 1 & 0 \\ 0 & 1 & 1 \\ 1 & 0 & 1 \end{vmatrix}$

3 $\begin{vmatrix} 1 & 0 & 1 \\ 2 & 1 & 0 \\ 1 & 1 & 2 \end{vmatrix} = \begin{vmatrix} 1 & 0 & 1 \\ 2 & 1 & 0 \\ 0 & 1 & 1 \end{vmatrix}$

4 $\begin{vmatrix} 1 & 1 & 2 \\ 1 & 0 & 1 \\ 2 & 1 & 1 \end{vmatrix} = \begin{vmatrix} 0 & 1 & 1 \\ 1 & 0 & 1 \\ 2 & 1 & 1 \end{vmatrix}$

5 $\begin{vmatrix} 2 & 4 & 2 \\ 1 & 2 & 4 \\ 2 & 6 & 4 \end{vmatrix} = 4 \begin{vmatrix} 1 & 2 & 1 \\ 1 & 2 & 4 \\ 1 & 3 & 2 \end{vmatrix}$

6 $\begin{vmatrix} 2 & 1 & 6 \\ 4 & 3 & 3 \\ 2 & 1 & 3 \end{vmatrix} = 6 \begin{vmatrix} 1 & 1 & 2 \\ 2 & 3 & 1 \\ 1 & 1 & 1 \end{vmatrix}$

7 $\begin{vmatrix} 1 & -1 & 2 \\ 1 & 2 & -1 \\ 1 & -1 & 2 \end{vmatrix} = 0$

8 $\begin{vmatrix} 1 & -1 & 1 \\ 0 & 1 & 0 \\ -1 & 1 & -1 \end{vmatrix} = 0$

9 $\begin{vmatrix} 1 & 5 \\ -3 & 2 \end{vmatrix} = - \begin{vmatrix} 1 & 5 \\ 3 & -2 \end{vmatrix}$

10 $\begin{vmatrix} 2 & -2 \\ 1 & 1 \end{vmatrix} = - \begin{vmatrix} -2 & 2 \\ 1 & 1 \end{vmatrix}$

11 $\begin{vmatrix} 0 & 0 & 1 \\ 1 & 0 & 0 \\ 0 & 0 & 2 \end{vmatrix} = 0$

12 $\begin{vmatrix} 1 & 0 & 1 \\ 0 & 0 & 0 \\ 1 & 1 & 0 \end{vmatrix} = 0$

13 $\begin{vmatrix} 1 & -1 & -2 \\ -1 & 2 & 1 \\ 0 & 1 & 1 \end{vmatrix} = \begin{vmatrix} 1 & -1 & 0 \\ -1 & 2 & -1 \\ 0 & 1 & 1 \end{vmatrix}$

14 $\begin{vmatrix} a & 0 & 0 \\ 0 & b & 0 \\ 0 & 0 & c \end{vmatrix} = - \begin{vmatrix} 0 & 0 & a \\ 0 & b & 0 \\ c & 0 & 0 \end{vmatrix}$

Exer. 15–24: Find the determinant of the matrix after introducing zeros, as in Example 1.

15 $\begin{bmatrix} 3 & 1 & 0 \\ -2 & 0 & 1 \\ 1 & 3 & -1 \end{bmatrix}$

16 $\begin{bmatrix} -3 & 0 & 4 \\ 1 & 2 & 0 \\ 4 & 1 & -1 \end{bmatrix}$

17 $\begin{bmatrix} 5 & 4 & 3 \\ -3 & 2 & 1 \\ 0 & 7 & -2 \end{bmatrix}$

18 $\begin{bmatrix} 0 & 2 & -6 \\ 5 & 1 & -3 \\ 6 & -2 & 5 \end{bmatrix}$

19 $\begin{bmatrix} 2 & 2 & -3 \\ 3 & 6 & 9 \\ -2 & 5 & 4 \end{bmatrix}$ 20 $\begin{bmatrix} 3 & 8 & 5 \\ 5 & 3 & -6 \\ 2 & 4 & -2 \end{bmatrix}$

21 $\begin{bmatrix} 3 & 1 & -2 & 2 \\ 2 & 0 & 1 & 4 \\ 0 & 1 & 3 & 5 \\ -1 & 2 & 0 & -3 \end{bmatrix}$

22 $\begin{bmatrix} 3 & 2 & 0 & 4 \\ -2 & 0 & 5 & 0 \\ 4 & -3 & 1 & 6 \\ 2 & -1 & 2 & 0 \end{bmatrix}$

23 $\begin{bmatrix} 2 & -2 & 0 & 0 & -3 \\ 3 & 0 & 3 & 2 & -1 \\ 0 & 1 & -2 & 0 & 2 \\ -1 & 2 & 0 & 3 & 0 \\ 0 & 4 & 1 & 0 & 0 \end{bmatrix}$

24 $\begin{bmatrix} 2 & 0 & -1 & 0 & 2 \\ 1 & 3 & 0 & 0 & 1 \\ 0 & 4 & 3 & 0 & -1 \\ -1 & 2 & 0 & -2 & 0 \\ 0 & 1 & 5 & 0 & -4 \end{bmatrix}$

25 Show that
$$\begin{vmatrix} 1 & 1 & 1 \\ a & b & c \\ a^2 & b^2 & c^2 \end{vmatrix} = (a-b)(b-c)(c-a).$$
(*Hint:* See Example 3.)

26 Show that
$$\begin{vmatrix} 1 & 1 & 1 \\ a & b & c \\ a^3 & b^3 & c^3 \end{vmatrix} = (a-b)(b-c)(c-a)(a+b+c).$$

27 If
$$A = \begin{bmatrix} a_{11} & a_{12} & a_{13} & a_{14} \\ 0 & a_{22} & a_{23} & a_{24} \\ 0 & 0 & a_{33} & a_{34} \\ 0 & 0 & 0 & a_{44} \end{bmatrix},$$
show that $|A| = a_{11}a_{22}a_{33}a_{44}$.

28 If
$$A = \begin{bmatrix} a & b & 0 & 0 \\ c & d & 0 & 0 \\ 0 & 0 & e & f \\ 0 & 0 & g & h \end{bmatrix},$$
show that
$$|A| = \begin{vmatrix} a & b \\ c & d \end{vmatrix}\begin{vmatrix} e & f \\ g & h \end{vmatrix}.$$

29 If $A = (a_{ij})$ and $B = (b_{ij})$ are arbitrary square matrices of order 2, show that $|AB| = |A||B|$.

30 If $A = (a_{ij})$ is a square matrix of order n and k is any real number, show that $|kA| = k^n|A|$. (*Hint:* Use property (2) of the theorem on row and column transformations of a determinant.)

31 Use properties of determinants to show that the following is an equation of a line through the points (x_1, y_1) and (x_2, y_2):
$$\begin{vmatrix} x & y & 1 \\ x_1 & y_1 & 1 \\ x_2 & y_2 & 1 \end{vmatrix} = 0$$

32 Use properties of determinants to show that the following is an equation of a circle through three noncollinear points (x_1, y_1), (x_2, y_2), and (x_3, y_3):
$$\begin{vmatrix} x^2 + y^2 & x & y & 1 \\ x_1^2 + y_1^2 & x_1 & y_1 & 1 \\ x_2^2 + y_2^2 & x_2 & y_2 & 1 \\ x_3^2 + y_3^2 & x_3 & y_3 & 1 \end{vmatrix} = 0$$

Exer. 33–42: Use Cramer's rule, whenever applicable, to solve the system.

33 $\begin{cases} 2x + 3y = 2 \\ x - 2y = 8 \end{cases}$ 34 $\begin{cases} 4x + 5y = 13 \\ 3x + y = -4 \end{cases}$

35 $\begin{cases} 2x + 5y = 16 \\ 3x - 7y = 24 \end{cases}$ 36 $\begin{cases} 7x - 8y = 9 \\ 4x + 3y = -10 \end{cases}$

37 $\begin{cases} 2x - 3y = 5 \\ -6x + 9y = 12 \end{cases}$ 38 $\begin{cases} 3p - q = 7 \\ -12p + 4q = 3 \end{cases}$

39 $\begin{cases} x - 2y - 3z = -1 \\ 2x + y + z = 6 \\ x + 3y - 2z = 13 \end{cases}$ 40 $\begin{cases} x + 3y - z = -3 \\ 3x - y + 2z = 1 \\ 2x - y + z = -1 \end{cases}$

41 $\begin{cases} 5x + 2y - z = -7 \\ x - 2y + 2z = 0 \\ 3y + z = 17 \end{cases}$ 42 $\begin{cases} 4x - y + 3z = 6 \\ -8x + 3y - 5z = -6 \\ 5x - 4y = -9 \end{cases}$

CHAPTER 6 REVIEW EXERCISES

Exer. 1–16: Solve the system.

1 $\begin{cases} 2x - 3y = 4 \\ 5x + 4y = 1 \end{cases}$

2 $\begin{cases} x - 3y = 4 \\ -2x + 6y = 2 \end{cases}$

3 $\begin{cases} y + 4 = x^2 \\ 2x + y = -1 \end{cases}$

4 $\begin{cases} x^2 + y^2 = 25 \\ x - y = 7 \end{cases}$

5 $\begin{cases} 9x^2 + 16y^2 = 140 \\ x^2 - 4y^2 = 4 \end{cases}$

6 $\begin{cases} 2x = y^2 + 3z \\ x = y^2 + z - 1 \\ x^2 = xz \end{cases}$

7 $\begin{cases} \dfrac{1}{x} + \dfrac{3}{y} = 7 \\ \dfrac{4}{x} - \dfrac{2}{y} = 1 \end{cases}$

8 $\begin{cases} 2^x + 3^{y+1} = 10 \\ 2^{x+1} - 3^y = 5 \end{cases}$

9 $\begin{cases} 3x + y - 2z = -1 \\ 2x - 3y + z = 4 \\ 4x + 5y - z = -2 \end{cases}$

10 $\begin{cases} x + 3y = 0 \\ y - 5z = 3 \\ 2x + z = -1 \end{cases}$

11 $\begin{cases} 4x - 3y - z = 0 \\ x - y - z = 0 \\ 3x - y + 3z = 0 \end{cases}$

12 $\begin{cases} 2x + y - z = 0 \\ x - 2y + z = 0 \\ 3x + 3y + 2z = 0 \end{cases}$

13 $\begin{cases} 4x + 2y - z = 1 \\ 3x + 2y + 4z = 2 \end{cases}$

14 $\begin{cases} 2x + y = 6 \\ x - 3y = 17 \\ 3x + 2y = 7 \end{cases}$

15 $\begin{cases} \dfrac{4}{x} + \dfrac{1}{y} + \dfrac{2}{z} = 4 \\ \dfrac{2}{x} + \dfrac{3}{y} - \dfrac{1}{z} = 1 \\ \dfrac{1}{x} + \dfrac{1}{y} + \dfrac{1}{z} = 4 \end{cases}$

16 $\begin{cases} 2x - y + 3z - w = -3 \\ 3x + 2y - z + w = 13 \\ x - 3y + z - 2w = -4 \\ -x + y + 4z + 3w = 0 \end{cases}$

Exer. 17–20: Find the partial fraction decomposition.

17 $\dfrac{4x^2 + 54x + 134}{(x + 3)(x^2 + 4x - 5)}$

18 $\dfrac{2x^2 + 7x + 9}{x^2 + 2x + 1}$

19 $\dfrac{x^2 + 14x - 13}{x^3 + 5x^2 + 4x + 20}$

20 $\dfrac{x^3 + 2x^2 + 2x + 16}{x^4 + 7x^2 + 10}$

Exer. 21–24: Sketch the graph of the system.

21 $\begin{cases} x^2 + y^2 < 16 \\ y - x^2 > 0 \end{cases}$

22 $\begin{cases} y - x \le 0 \\ y + x \ge 2 \\ x \le 5 \end{cases}$

23 $\begin{cases} x - 2y \le 2 \\ y - 3x \le 4 \\ 2x + y \le 4 \end{cases}$

24 $\begin{cases} x^2 - y < 0 \\ y - 2x < 5 \\ xy < 0 \end{cases}$

Exer. 25–34: Express as a single matrix.

25 $\begin{bmatrix} 2 & -1 & 0 \\ 3 & 0 & -2 \end{bmatrix} \begin{bmatrix} 2 & -1 & 3 \\ 0 & 3 & 0 \\ 1 & 4 & 2 \end{bmatrix}$

26 $\begin{bmatrix} 4 & 2 \\ 5 & -3 \end{bmatrix} \begin{bmatrix} 3 \\ 7 \end{bmatrix}$

27 $\begin{bmatrix} 2 & 0 \\ 1 & 4 \\ -2 & 3 \end{bmatrix} \begin{bmatrix} 0 & 2 & -3 \\ 4 & 5 & 1 \end{bmatrix}$

28 $\begin{bmatrix} 0 & -2 & 3 \\ 4 & 1 & 2 \end{bmatrix} \begin{bmatrix} 2 & 0 \\ 3 & 8 \\ 2 & -7 \end{bmatrix}$

29 $2 \begin{bmatrix} 0 & -1 & -4 \\ 3 & 2 & 1 \end{bmatrix} - 3 \begin{bmatrix} 4 & -2 & 1 \\ 0 & 5 & -1 \end{bmatrix}$

30 $\begin{bmatrix} 1 & 3 \\ 2 & 4 \end{bmatrix} \begin{bmatrix} a & 0 \\ 0 & a \end{bmatrix}$

31 $\begin{bmatrix} a & 0 \\ 0 & b \end{bmatrix} \begin{bmatrix} 1 & 3 \\ 2 & 4 \end{bmatrix}$

32 $\begin{bmatrix} 3 & 2 \\ 0 & 0 \end{bmatrix} \begin{bmatrix} -2 & 0 \\ 3 & 0 \end{bmatrix}$

33 $\begin{bmatrix} 1 & 2 \\ 3 & 4 \end{bmatrix} \left\{ \begin{bmatrix} 2 & -4 \\ 3 & 7 \end{bmatrix} + \begin{bmatrix} 1 & 5 \\ -2 & -3 \end{bmatrix} \right\}$

34 $\begin{bmatrix} 3 & 2 & 5 \\ -3 & 4 & 7 \\ 6 & 5 & 1 \end{bmatrix} \begin{bmatrix} 3 & 2 & 5 \\ -3 & 4 & 7 \\ 6 & 5 & 1 \end{bmatrix}^{-1}$

Exer. 35–38: Find the inverse of the matrix.

35 $\begin{bmatrix} 5 & -4 \\ -3 & 2 \end{bmatrix}$

36 $\begin{bmatrix} 2 & -1 & 0 \\ 1 & 4 & 2 \\ 3 & -2 & 1 \end{bmatrix}$

37 $\begin{bmatrix} 1 & 0 & 0 \\ 0 & 4 & 7 \\ 0 & 1 & 2 \end{bmatrix}$ **38** $\begin{bmatrix} 2 & 0 & 5 \\ 0 & 3 & -1 \\ 3 & 4 & 0 \end{bmatrix}$

39 Use the result of Exercise 35 to solve the system
$$\begin{cases} 5x - 4y = 30 \\ -3x + 2y = -16 \end{cases}$$

40 Use the result of Exercise 36 to solve the system
$$\begin{cases} 2x - y = -5 \\ x + 4y + 2z = 15 \\ 3x - 2y + z = -7 \end{cases}$$

Exer. 41–50: Find the determinant of the matrix.

41 $\begin{bmatrix} -6 \end{bmatrix}$ **42** $\begin{bmatrix} 3 & 4 \\ -6 & -5 \end{bmatrix}$

43 $\begin{bmatrix} 3 & -4 \\ 6 & 8 \end{bmatrix}$ **44** $\begin{bmatrix} 0 & 4 & -3 \\ 2 & 0 & 4 \\ -5 & 1 & 0 \end{bmatrix}$

45 $\begin{bmatrix} 2 & -3 & 5 \\ -4 & 1 & 3 \\ 3 & 2 & -1 \end{bmatrix}$ **46** $\begin{bmatrix} 3 & 1 & -2 \\ -5 & 2 & -4 \\ 7 & 3 & -6 \end{bmatrix}$

47 $\begin{bmatrix} 5 & 0 & 0 & 0 \\ 6 & -3 & 0 & 0 \\ 1 & 4 & -4 & 0 \\ 7 & 2 & 3 & 2 \end{bmatrix}$

48 $\begin{bmatrix} 1 & 2 & 0 & 3 & 1 \\ -2 & -1 & 4 & 1 & 2 \\ 3 & 0 & -1 & 0 & -1 \\ 2 & -3 & 2 & -4 & 2 \\ -1 & 1 & 0 & 1 & 3 \end{bmatrix}$

49 $\begin{bmatrix} 2 & 0 & 1 & 0 & -1 \\ 0 & 1 & 0 & 1 & 2 \\ 2 & -2 & 1 & -2 & 0 \\ 0 & 0 & -2 & 0 & 1 \\ 1 & -1 & 0 & -1 & 0 \end{bmatrix}$

50 $\begin{bmatrix} 1 & 2 & 0 & 0 & 0 \\ 3 & 4 & 0 & 0 & 0 \\ 0 & 0 & 1 & 2 & 3 \\ 0 & 0 & 2 & -1 & 1 \\ 0 & 0 & 1 & 3 & -1 \end{bmatrix}$

Exer. 51–52: Solve the equation $|A - xI| = 0$.

51 $A = \begin{bmatrix} 2 & 3 \\ 1 & -4 \end{bmatrix}$, $I = I_2$

52 $A = \begin{bmatrix} 2 & -1 & 3 \\ 0 & 4 & 0 \\ 1 & 0 & -2 \end{bmatrix}$, $I = I_3$

Exer. 53–54: Without expanding, explain why the statement is true.

53 $\begin{vmatrix} 2 & 4 & -6 \\ 1 & 4 & 3 \\ 2 & 2 & 0 \end{vmatrix} = 12 \begin{vmatrix} 1 & 1 & -1 \\ 1 & 2 & 1 \\ 2 & 1 & 0 \end{vmatrix}$

54 $\begin{vmatrix} a & b & c \\ d & e & f \\ g & h & k \end{vmatrix} = \begin{vmatrix} d & e & f \\ g & h & k \\ a & b & c \end{vmatrix}$

55 Find the determinant of the $n \times n$ matrix (a_{ij}) with $a_{ij} = 0$ for $i \neq j$.

56 Without expanding, show that
$$\begin{vmatrix} 1 & a & b+c \\ 1 & b & a+c \\ 1 & c & a+b \end{vmatrix} = 0.$$

Exer. 57–58: Use Cramer's rule to solve the system.

57 $\begin{cases} 5x - 6y = 4 \\ 3x + 7y = 8 \end{cases}$ **58** $\begin{cases} 2x - 3y + 2z = -3 \\ -3x + 2y + z = 1 \\ 4x + y - 3z = 4 \end{cases}$

59 *Watering a field* A rotating sprinkler head with a range of 50 feet is to be placed in the center of a rectangular field (see the figure). If the area of the field is 4000 ft^2 and the water is to just reach the corners, find the dimensions of the field.

EXERCISE 59

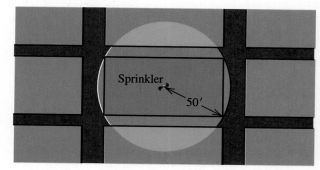

Sprinkler

50′

60 Find equations of the two lines that are tangent to the circle $x^2 + y^2 = 1$ and pass through the point $(0, 3)$. (*Hint:* Let $y = mx + 3$ and determine conditions on m that will ensure that the system has only one solution.)

61 *Payroll accounting*   An accountant must pay taxes and payroll bonuses to employees from the company's profits of $50,000. The total tax is 40% of the amount left after bonuses are paid, and the total paid in bonuses is 10% of the amount left after taxes. Find the total tax and the total bonus amount.

62 *Track dimensions*   A circular track is to have a 10-foot-wide running lane around the outside (see the figure). The inside distance around the track is to be 90% of the outside distance. Find the dimensions of the track.

EXERCISE 62

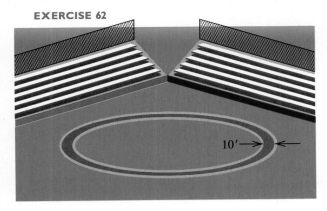

63 *Flow rates*   Three inlet pipes A, B, and C can be used to fill a 1000-ft^3 water storage tank. When all three pipes are in operation, the tank can be filled in 10 hours. When only A and B are used, the time increases to 20 hours. With pipes A and C, the tank can be filled in 12.5 hours. Find the individual flow rates (in ft^3/hr) for each of the three pipes.

64 *Warehouse shipping charges*   To fill an order for 150 office desks, a furniture distributor must ship the desks from two warehouses. The shipping cost per desk is $24 from the western warehouse and $35 from the eastern warehouse. If the total shipping charge is $4205, how many desks are shipped from each location?

65 *Express mail rates*   An express-mail company charges $15 for overnight delivery of a letter, provided the di-

mensions of the standard envelope satisfy the following three conditions: (a) the length, the larger of the two dimensions, must be at most 12 inches; (b) the width must be at most 8 inches; (c) the width must be at least one-half the length. Find and graph a system of inequalities that describes all the possibilities for dimensions of a standard envelope.

66 *Activities of a deer*   A deer spends the day in three basic activities: rest, searching for food, and grazing. At least 6 hours each day must be spent resting, and the number of hours spent searching for food will be at least two times the number of hours spent grazing. Using x as the number of hours spent searching for food and y as the number of hours spent grazing, find and graph the system of inequalities that describes the possible divisions of the day.

67 *Production scheduling*   A company manufactures a power lawn mower and a power edger. These two products are of such high quality that the company can sell all the products it makes, but production capability is limited in the areas of machining, welding, and assembly. Each week, the company has 600 hours available for machining, 300 hours for welding, and 550 hours for assembly. The number of hours required for the production of a single item is shown in the following table.

Product	Machining	Welding	Assembly
Lawn mower	6	2	5
Edger	4	3	5

The profits from the sale of a mower and an edger are $100 and $80, respectively. How many mowers and edgers should be produced each week in order to maximize the profit?

68 *Maximizing investment income*   A retired couple wishes to invest $150,000, diversifying the investment in three areas: a high-risk stock that has an expected annual rate of return (or interest) of 15%; a low-risk stock that has an expected annual return of 10%; and government-issued bonds that pay annual interest of 8% and involve no risk. To protect the value of the investment, the couple wishes to place at least twice as much in the low-risk stock as in the high-risk stock and use the remainder to buy bonds. How should the money be invested to maximize the expected annual return?

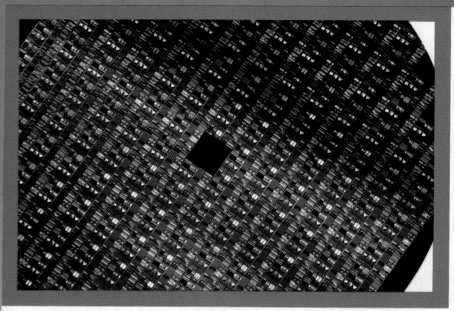

Sequences occur in high density computer chips and in many other aspects of science and industry.

Sequences and summation notation, discussed in the first section, are very important in advanced mathematics and applications. Of special interest are arithmetic and geometric sequences, considered in Sections 7.2 and 7.3. We then discuss the method of mathematical induction, a process that is often used to prove that each statement in an infinite sequence of statements is true. As an application, we use it to prove the binomial theorem in Section 7.5. The last part of the chapter deals with counting processes that occur frequently in mathematics and everyday life. These include the concepts of permutations, combinations, and probability.

CHAPTER 7

SEQUENCES, SERIES, AND PROBABILITY

7.1 INFINITE SEQUENCES AND SUMMATION NOTATION

An arbitrary *infinite sequence* may be denoted as follows.

Infinite Sequence Notation

$$a_1, a_2, a_3, \ldots, a_n, \ldots$$

For convenience, we often refer to infinite sequences as *sequences*. We may regard an infinite sequence as a collection of real numbers that is in one-to-one correspondence with the positive integers. Each number a_k is a **term** of the sequence. The sequence is *ordered* in the sense that there is a **first term** a_1, a **second term** a_2, a **forty-fifth term** a_{45}, and, if n denotes an arbitrary positive integer, an **nth term** a_n. Infinite sequences are often defined by stating a formula for the nth term.

Infinite sequences occur frequently in mathematics. For example, the sequence

$$0.6, 0.66, 0.666, 0.6666, 0.66666, \ldots$$

may be used to represent the rational number $\frac{2}{3}$. In this case the nth term gets closer and closer to $\frac{2}{3}$ as n increases.

We may regard an infinite sequence as a function. Recall from Section 3.4 that a function f is a correspondence that assigns to each number x in the domain D exactly one number $f(x)$ in the range R. If we restrict the domain to the positive integers $1, 2, 3, \ldots$, we obtain an infinite sequence, as in the following definition.

Definition of Infinite Sequence

An **infinite sequence** is a function whose domain is the set of positive integers.

In our work, the range of an infinite sequence will be a set of real numbers.

If a function f is an infinite sequence, then to each positive integer n there corresponds a real number $f(n)$. These numbers in the range of f may be represented by writing

$$f(1), f(2), f(3), \ldots, f(n), \ldots.$$

To obtain the subscript form of a sequence, as shown at the beginning of this section, we let $a_n = f(n)$ for every positive integer n.

From the definition of equality of functions we see that a sequence

$$a_1, a_2, a_3, \ldots, a_n, \ldots$$

is **equal** to a sequence

$$b_1, b_2, b_3, \ldots, b_n, \ldots$$

if and only if $a_k = b_k$ for every positive integer k.

Another notation for a sequence with nth term a_n is $\{a_n\}$. For example, the sequence $\{2^n\}$ has nth term $a_n = 2^n$. Using sequence notation, we write this sequence as follows:

$$2^1, 2^2, 2^3, \ldots, 2^n, \ldots$$

By definition, the sequence $\{2^n\}$ is the function f with $f(n) = 2^n$ for every positive integer n.

EXAMPLE I   *Finding terms of a sequence*

List the first four terms and the tenth term of each sequence:

(a) $\left\{\dfrac{n}{n+1}\right\}$ (b) $\{2 + (0.1)^n\}$ (c) $\left\{(-1)^{n+1}\dfrac{n^2}{3n-1}\right\}$ (d) $\{4\}$

SOLUTION   To find the first four terms, we substitute, successively, $n = 1, 2, 3,$ and 4 in the formula for a_n. The tenth term is found by substituting 10 for n. Doing this and simplifying gives us the following:

Sequence	nth term a_n	First four terms	Tenth term
(a) $\left\{\dfrac{n}{n+1}\right\}$	$\dfrac{n}{n+1}$	$\dfrac{1}{2}, \dfrac{2}{3}, \dfrac{3}{4}, \dfrac{4}{5}$	$\dfrac{10}{11}$
(b) $\{2 + (0.1)^n\}$	$2 + (0.1)^n$	$2.1, 2.01, 2.001, 2.0001$	2.0000000001
(c) $\left\{(-1)^{n+1}\dfrac{n^2}{3n-1}\right\}$	$(-1)^{n+1}\dfrac{n^2}{3n-1}$	$\dfrac{1}{2}, -\dfrac{4}{5}, \dfrac{9}{8}, -\dfrac{16}{11}$	$-\dfrac{100}{29}$
(d) $\{4\}$	4	$4, 4, 4, 4$	4

For some sequences we state the first term a_1, together with a rule for obtaining any term a_{k+1} from the preceding term a_k whenever $k \geq 1$. We call such a statement a **recursive definition**, and the sequence is said to be defined **recursively**.

EXAMPLE 2   *Finding terms of a recursively defined sequence*

Find the first four terms and the nth term of the infinite sequence defined recursively as follows:

$$a_1 = 3, \qquad a_{k+1} = 2a_k \quad \text{for } k \geq 1$$

SOLUTION   The first four terms are

$$a_1 = 3$$

$$a_2 = 2a_1 = 2 \cdot 3 = 6$$

$$a_3 = 2a_2 = 2 \cdot 2 \cdot 3 = 2^2 \cdot 3 = 12$$

$$a_4 = 2a_3 = 2 \cdot 2 \cdot 2 \cdot 3 = 2^3 \cdot 3 = 24.$$

We have written the terms as products to gain some insight into the nature of the nth term. Continuing, we obtain $a_5 = 2^4 \cdot 3$, $a_6 = 2^5 \cdot 3$, and, in general,

$$a_n = 2^{n-1} \cdot 3$$

for every positive integer n.

If only the first few terms of an infinite sequence are known, then it is impossible to predict additional terms. For example, if we were given $3, 6, 9, \ldots$ and asked to find the fourth term, we could not proceed without further information. The infinite sequence with nth term

$$a_n = 3n + (1 - n)^3(2 - n)^2(3 - n)$$

has for its first four terms 3, 6, 9, and 120. It is possible to describe sequences in which the first three terms are 3, 6, and 9 and the fourth term is *any* given number. This shows that when we work with an infinite sequence it is essential to have either specific information about the nth term or a general scheme for obtaining each term from the preceding one.

We sometimes need to find the sum of many terms of an infinite sequence. To express such sums easily, we use **summation notation**. Given an infinite sequence

$$a_1, a_2, a_3, \ldots, a_n, \ldots,$$

the symbol $\sum_{k=1}^{m} a_k$ represents the sum of the first m terms, as follows.

Summation Notation

$$\sum_{k=1}^{m} a_k = a_1 + a_2 + a_3 + \cdots + a_m$$

The Greek capital letter sigma, $\sum$, indicates a sum, and the symbol a_k represents the kth term. The letter k is the **index of summation**, or the **summation variable**, and the numbers 1 and m indicate the smallest and largest values of the summation variable.

EXAMPLE 3   *Evaluating a sum*

Find the sum $\sum\limits_{k=1}^{4} k^2(k-3)$.

SOLUTION   In this case, $a_k = k^2(k-3)$. To find the sum, we merely substitute, in succession, the integers 1, 2, 3, and 4 for k and add the resulting terms:

$$\sum_{k=1}^{4} k^2(k-3) = 1^2(1-3) + 2^2(2-3) + 3^2(3-3) + 4^2(4-3)$$

$$= (-2) + (-4) + 0 + 16 = 10$$

The letter we use for the summation variable is immaterial. To illustrate, if j is the summation variable, then

$$\sum_{j=1}^{m} a_j = a_1 + a_2 + a_3 + \cdots + a_m,$$

which is the same sum as $\sum_{k=1}^{m} a_k$. We may also use other symbols. For example, the sum in Example 3 can be written

$$\sum_{j=1}^{4} j^2(j-3).$$

If n is a positive integer, then the sum of the first n terms of an infinite sequence will be denoted by S_n. For example, given $a_1, a_2, a_3, \ldots, a_n, \ldots,$

$$S_1 = a_1$$
$$S_2 = a_1 + a_2$$
$$S_3 = a_1 + a_2 + a_3$$
$$S_4 = a_1 + a_2 + a_3 + a_4$$

and, in general,

$$S_n = \sum_{k=1}^{n} a_k = a_1 + a_2 + \cdots + a_n.$$

Note that we can also write

$$S_1 = a_1$$
$$S_2 = S_1 + a_2$$
$$S_3 = S_2 + a_3$$
$$S_4 = S_3 + a_4$$

and, for every $n > 1$,

$$S_n = S_{n-1} + a_n.$$

The real number S_n is called the **nth partial sum** of the infinite sequence $a_1, a_2, a_3, \ldots, a_n, \ldots$, and the sequence

$$S_1, S_2, S_3, \ldots, S_n, \ldots$$

is a **sequence of partial sums**. Sequences of partial sums are important in the study of *infinite series*, a topic in calculus. We shall discuss some special types of infinite series in Section 7.3.

EXAMPLE 4   *Finding the terms of a sequence of partial sums*

Find the first four terms and the nth term of the sequence of partial sums associated with the sequence $1, 2, 3, \ldots, n, \ldots$ of positive integers.

SOLUTION   If we let $a_n = n$, then the first four terms of the sequence of partial sums are

$$S_1 = a_1 = 1$$
$$S_2 = S_1 + a_2 = 1 + 2 = 3$$
$$S_3 = S_2 + a_3 = 3 + 3 = 6$$
$$S_4 = S_3 + a_4 = 6 + 4 = 10.$$

The nth partial sum S_n (that is, the sum of $1, 2, 3, \ldots, n$) can be written in either of the following forms:

$$S_n = 1 + 2 \quad\quad + 3 \quad\quad + \cdots + (n - 2) + (n - 1) + n$$
$$S_n = n + (n - 1) + (n - 2) + \cdots + 3 \quad\quad + 2 \quad\quad + 1$$

Adding corresponding terms on each side of these equations gives us

$$2S_n = \underbrace{(n + 1) + (n + 1) + (n + 1) + \cdots + (n + 1) + (n + 1) + (n + 1)}_{n \text{ times}}.$$

Since the expression $(n + 1)$ appears n times on the right side of the last equation, we see that

$$2S_n = n(n + 1) \quad \text{or, equivalently,} \quad S_n = \frac{n(n + 1)}{2}.$$

If a_k is the same for every positive integer k, say $a_k = c$ for a real number c, then

$$\sum_{k=1}^{n} a_k = a_1 + a_2 + a_3 + \cdots + a_n$$
$$= c + c + c + \cdots + c = nc.$$

We have proved property (1) of the following theorem.

Theorem on the Sum of a Constant

$$\textbf{(1)} \ \sum_{k=1}^{n} c = nc \qquad\qquad \textbf{(2)} \ \sum_{k=m}^{n} c = (n - m + 1)c$$

To prove (2), we may write

$$\sum_{k=m}^{n} c = \sum_{k=1}^{n} c - \sum_{k=1}^{m-1} c \qquad \text{subtract the first } m - 1 \text{ terms from the sum of } n \text{ terms}$$

$$= nc - (m - 1)c \qquad \text{use (1) of the theorem for each sum}$$

$$= (n - m + 1)c \qquad \text{factor out } c$$

ILLUSTRATION

Sum of a Constant

■ $\displaystyle\sum_{k=1}^{4} 7 = 7 + 7 + 7 + 7 = 4 \cdot 7 = 28$

■ $\displaystyle\sum_{k=1}^{10} \pi = 10\pi$

■ $\displaystyle\sum_{k=3}^{8} 9 = (8 - 3 + 1)(9) = 6(9) = 54$

■ $\displaystyle\sum_{k=10}^{20} 5 = (20 - 10 + 1)(5) = 11(5) = 55$

As shown in property (2) of the preceding theorem, the domain of the summation variable does not have to begin at 1. For example,

$$\sum_{k=4}^{8} a_k = a_4 + a_5 + a_6 + a_7 + a_8.$$

As another variation, if the first term of an infinite sequence is a_0, as in

$$a_0, a_1, a_2, \ldots, a_n, \ldots,$$

then we may consider sums of the form

$$\sum_{k=0}^{n} a_k = a_0 + a_1 + a_2 + \cdots + a_n,$$

which is the sum of the first $n + 1$ terms of the sequence.

EXAMPLE 5 *Evaluating a sum*

Find the sum $\displaystyle\sum_{k=0}^{3} \frac{2^k}{k + 1}$.

SOLUTION

$$\sum_{k=0}^{3} \frac{2^k}{k+1} = \frac{2^0}{0+1} + \frac{2^1}{1+1} + \frac{2^2}{2+1} + \frac{2^3}{3+1}$$

$$= 1 + 1 + \tfrac{4}{3} + 2 = \tfrac{16}{3}$$

Summation notation can be used to denote polynomials. Thus, if

$$f(x) = a_0 + a_1x + a_2x^2 + \cdots + a_nx^n,$$

then

$$f(x) = \sum_{k=0}^{n} a_k x^k.$$

The following theorem concerning sums has many uses.

Theorem on Sums

If $a_1, a_2, \ldots, a_n, \ldots$ and $b_1, b_2, \ldots, b_n, \ldots$ are infinite sequences, then for every positive integer n,

(1) $\displaystyle\sum_{k=1}^{n} (a_k + b_k) = \sum_{k=1}^{n} a_k + \sum_{k=1}^{n} b_k$

(2) $\displaystyle\sum_{k=1}^{n} (a_k - b_k) = \sum_{k=1}^{n} a_k - \sum_{k=1}^{n} b_k$

(3) $\displaystyle\sum_{k=1}^{n} ca_k = c\left(\sum_{k=1}^{n} a_k\right)$ for every real number c

PROOF   To prove formula (1), we first write

$$\sum_{k=1}^{n} (a_k + b_k) = (a_1 + b_1) + (a_2 + b_2) + (a_3 + b_3) + \cdots + (a_n + b_n).$$

Using commutative and associative properties many times, we may rearrange the terms on the right-hand side to produce

$$\sum_{k=1}^{n} (a_k + b_k) = (a_1 + a_2 + a_3 + \cdots + a_n) + (b_1 + b_2 + b_3 + \cdots + b_n).$$

Expressing the right-hand side in summation notation gives us formula (1).

For formula (3), we have

$$\sum_{k=1}^{n} (ca_k) = ca_1 + ca_2 + ca_3 + \cdots + ca_n$$

$$= c(a_1 + a_2 + a_3 + \cdots + a_n)$$

$$= c\left(\sum_{k=1}^{n} a_k\right).$$

The proof of (2) is left as an exercise. ■

7.1 EXERCISES

Exer. 1–16: Find the first four terms and the eighth term of the sequence.

1 $\{12 - 3n\}$

2 $\left\{\dfrac{3}{5n - 2}\right\}$

3 $\left\{\dfrac{3n - 2}{n^2 + 1}\right\}$

4 $\left\{10 + \dfrac{1}{n}\right\}$

5 $\{9\}$

6 $\{\sqrt{2}\}$

7 $\{2 + (-0.1)^n\}$

8 $\{4 + (0.1)^n\}$

9 $\left\{(-1)^{n-1} \dfrac{n + 7}{2n}\right\}$

10 $\left\{(-1)^n \dfrac{6 - 2n}{\sqrt{n + 1}}\right\}$

11 $\{1 + (-1)^{n+1}\}$

12 $\{(-1)^{n+1} + (0.1)^{n-1}\}$

13 $\left\{\dfrac{2^n}{n^2 + 2}\right\}$

14 $\{(n - 1)(n - 2)(n - 3)\}$

15 a_n is the number of decimal places in $(0.1)^n$.

16 a_n is the number of positive integers less than n^3.

Exer. 17–24: Find the first five terms of the recursively defined infinite sequence.

17 $a_1 = 2, \qquad a_{k+1} = 3a_k - 5$

18 $a_1 = 5, \qquad a_{k+1} = 7 - 2a_k$

19 $a_1 = -3, \quad a_{k+1} = a_k^2$

20 $a_1 = 128, \quad a_{k+1} = \frac{1}{4}a_k$

21 $a_1 = 5, \qquad a_{k+1} = ka_k$

22 $a_1 = 3, \qquad a_{k+1} = 1/a_k$

23 $a_1 = 2, \qquad a_{k+1} = (a_k)^k$

24 $a_1 = 2, \qquad a_{k+1} = (a_k)^{1/k}$

Exer. 25–28: Find the first four terms of the sequence of partial sums for the given sequence.

25 $\{3 + \frac{1}{2}n\}$

26 $\{1/n^2\}$

27 $\{(-1)^n n^{-1/2}\}$

28 $\{(-1)^n 1/2^n\}$

Exer. 29–42: Find the sum.

29 $\displaystyle\sum_{k=1}^{5} (2k - 7)$

30 $\displaystyle\sum_{k=1}^{6} (10 - 3k)$

31 $\displaystyle\sum_{k=1}^{4} (k^2 - 5)$

32 $\displaystyle\sum_{k=1}^{10} [1 + (-1)^k]$

33 $\displaystyle\sum_{k=0}^{5} k(k - 2)$

34 $\displaystyle\sum_{k=0}^{4} (k - 1)(k - 3)$

35 $\displaystyle\sum_{k=3}^{6} \dfrac{k - 5}{k - 1}$

36 $\displaystyle\sum_{k=1}^{6} \dfrac{3}{k + 1}$

37 $\displaystyle\sum_{k=1}^{5} (-3)^{k-1}$

38 $\displaystyle\sum_{k=0}^{4} 3(2^k)$

39 $\displaystyle\sum_{k=1}^{100} 100$

40 $\displaystyle\sum_{k=1}^{1000} 5$

41 $\displaystyle\sum_{k=253}^{571} \frac{1}{3}$

42 $\displaystyle\sum_{k=137}^{428} 2.1$

43 Prove formula (2) of the theorem on sums.

44 Extend formula (1) of the theorem on sums to

$$\sum_{k=1}^{n} (a_k + b_k + c_k).$$

45 Terms of the sequence defined recursively by $a_1 = 5$, $a_{k+1} = \sqrt{a_k}$ may be approximated with a calculator by entering 5 and pressing the $\boxed{\sqrt{x}}$ key repeatedly. Describe what happens to the terms of the sequence as k increases.

46 *Bode's sequence*   Bode's sequence, defined by

$$a_1 = 0.4, \qquad a_k = 0.1(3 \cdot 2^{k-2} + 4) \quad \text{for } k \geq 2,$$

can be used to approximate distances of planets from the sun. These distances are measured in astronomical units, with 1 AU = 93,000,000 mi. For example, the third term corresponds to earth and the fifth term to the minor planet Ceres. Approximate the first five terms of the sequence.

47 A test question lists the first four terms of a sequence as 2, 4, 6, and 8 and asks for the fifth term. Show that the fifth term can be any real number a by finding the nth term of a sequence that has for its first five terms 2, 4, 6, 8, and a.

48 *Growth of bacteria*   The number of bacteria in a certain culture is initially 500, and the culture doubles in size every day.

(a) Find the number of bacteria present after one day, two days, and three days.

(b) Find a formula for the number of bacteria present after n days.

49 The Fibonacci sequence The Fibonacci sequence is defined recursively by

$$a_1 = 1, \qquad a_2 = 1, \qquad a_{k+1} = a_k + a_{k-1} \quad \text{for } k \geq 2.$$

(a) Find the first ten terms of the sequence.

(b) The terms of the sequence $r_k = a_{k+1}/a_k$ give progressively better approximations to τ, the golden ratio. Approximate the first ten terms of this sequence.

50 The Fibonacci sequence The Fibonacci sequence can be defined by the formula

$$a_n = \frac{1}{\sqrt{5}}\left(\frac{1 + \sqrt{5}}{2}\right)^n - \frac{1}{\sqrt{5}}\left(\frac{1 - \sqrt{5}}{2}\right)^n.$$

Find the first eight terms and show that they agree with those found using the definition in Exercise 49.

c **Exer. 51–53: For the given nth term $a_n = f(n)$ of a sequence, use the graph of $y = f(x)$ on the interval [1, 100] to verify that as n increases without bound, a_n approaches some real number c.**

51 $a_n = \left(1 + \dfrac{1}{n} + \dfrac{1}{2n^2}\right)^n$

52 $a_n = n^{1/n}$

53 $a_n = \left(\dfrac{1}{n}\right)^{1/n}$

7.2 ARITHMETIC SEQUENCES

In this section and the next we consider two special types of sequences: arithmetic and geometric. The first type may be defined as follows.

Definition of Arithmetic Sequence

A sequence $a_1, a_2, \ldots, a_n, \ldots$ is an **arithmetic sequence** if there is a real number d such that for every positive integer k,

$$a_{k+1} = a_k + d.$$

The number $d = a_{k+1} - a_k$ is called the **common difference** of the sequence.

Note that the common difference d is the difference of *any* two successive terms of an arithmetic sequence.

EXAMPLE I   *Showing that a sequence is arithmetic*

Show that the sequence

$$1, 4, 7, 10, \ldots, 3n - 2, \ldots$$

is arithmetic, and find the common difference.

SOLUTION   If $a_n = 3n - 2$, then for every positive integer k,

$$a_{k+1} - a_k = [3(k + 1) - 2] - (3k - 2)$$
$$= 3k + 3 - 2 - 3k + 2 = 3.$$

Hence, the given sequence is arithmetic with common difference 3.

Given an arithmetic sequence, we know that

$$a_{k+1} = a_k + d$$

for every positive integer k. This gives us a recursive formula for obtaining successive terms. Beginning with any real number a_1, we can obtain an arithmetic sequence with common difference d simply by adding d to a_1, then to $a_1 + d$, and so on, obtaining

$$a_1, a_1 + d, a_1 + 2d, a_1 + 3d, a_1 + 4d, \ldots.$$

The nth term a_n of this sequence is given by the next formula.

The nth Term of an Arithmetic Sequence

$$a_n = a_1 + (n-1)d$$

EXAMPLE 2 *Finding a specific term of an arithmetic sequence*

The first three terms of an arithmetic sequence are 20, 16.5, and 13. Find the fifteenth term.

SOLUTION The common difference is

$$16.5 - 20 = -3.5.$$

Substituting $a_1 = 20$, $d = -3.5$, and $n = 15$ in $a_n = a_1 + (n-1)d$ gives

$$a_{15} = 20 + (15 - 1)(-3.5) = 20 - 49 = -29.$$

EXAMPLE 3 *Finding a specific term of an arithmetic sequence*

If the fourth term of an arithmetic sequence is 5 and the ninth term is 20, find the sixth term.

SOLUTION We are given $a_4 = 5$ and $a_9 = 20$ and wish to find a_6. The following are equivalent systems of equations in the variables a_1 and d:

$$\begin{cases} a_4 = a_1 + (4-1)d & \text{let } n = 4 \text{ in } a_n = a_1 + (n-1)d \\ a_9 = a_1 + (9-1)d & \text{let } n = 9 \text{ in } a_n = a_1 + (n-1)d \end{cases}$$

$$\begin{cases} 5 = a_1 + 3d & a_4 = 5 \\ 20 = a_1 + 8d & a_9 = 20 \end{cases}$$

Subtracting the first equation of the system from the second gives us $15 = 5d$, or $d = 3$. Thus, $a_1 = 5 - 3d = -4$. Hence,

$$a_6 = a_1 + (6-1)d \quad \text{let } n = 6 \text{ in } a_n = a_1 + (n-1)d$$
$$= (-4) + (5)(3) \quad a_1 = -4 \text{ and } d = 3$$
$$= 11. \quad \text{simplify}$$

The next theorem contains a formula for the nth partial sum S_n of an arithmetic sequence.

Theorem: Formulas for S_n

If $a_1, a_2, \ldots, a_n, \ldots$ is an arithmetic sequence with common difference d, then the nth partial sum S_n is given by either

$$S_n = \frac{n}{2}[2a_1 + (n-1)d] \qquad \text{or} \qquad S_n = \frac{n}{2}(a_1 + a_n).$$

PROOF   We may write

$$S_n = a_1 + a_2 + a_3 + \cdots + a_n$$
$$= a_1 + (a_1 + d) + (a_1 + 2d) + \cdots + [a_1 + (n-1)d].$$

Employing commutative and associative properties many times, we obtain

$$S_n = (a_1 + a_1 + a_1 + \cdots + a_1) + [d + 2d + \cdots + (n-1)d],$$

with a_1 appearing n times within the first pair of parentheses. Thus,

$$S_n = na_1 + d[1 + 2 + \cdots + (n-1)].$$

The expression within brackets is the sum of the first $n-1$ positive integers. Using the formula $S_n = n(n+1)/2$ from Example 4 of Section 7.1, but with $n-1$ in place of n, we have

$$1 + 2 + \cdots + (n-1) = \frac{(n-1)n}{2}.$$

Substituting in the last equation for S_n and factoring gives

$$S_n = na_1 + d\frac{n(n-1)}{2}$$
$$= \frac{n}{2}[2a_1 + (n-1)d].$$

Since $a_n = a_1 + (n-1)d$, the last equation is equivalent to

$$S_n = \frac{n}{2}(a_1 + a_n). \quad \blacksquare$$

EXAMPLE 4   *Finding a sum of even integers*

Find the sum of all the even integers from 2 through 100.

SOLUTION   This problem is equivalent to finding the sum of the first 50 terms of the arithmetic sequence

$$2, 4, 6, \ldots, 2n, \ldots.$$

Substituting $n = 50$, $a_1 = 2$, and $a_{50} = 100$ in $S_n = (n/2)(a_1 + a_n)$ produces

$$S_{50} = \tfrac{50}{2}(2 + 100) = 2550.$$

Alternatively, we may use $S_n = \frac{n}{2}[2a_1 + (n-1)d]$ with $d = 2$:

$$S_{50} = \frac{50}{2}[2(2) + (50-1)2] = 25[4 + 98] = 2550$$

The **arithmetic mean** of two numbers a and b is defined as $(a+b)/2$. This is the **average** of a and b. Note that

$$a, \frac{a+b}{2}, b$$

is a (finite) arithmetic sequence. This concept may be generalized as follows: If $c_1, c_2, \ldots, c_k$ are real numbers such that

$$a, c_1, c_2, \ldots, c_k, b$$

is a (finite) arithmetic sequence, then $c_1, c_2, \ldots, c_k$ are k **arithmetic means** between the numbers a and b. The process of determining these numbers is referred to as *inserting k arithmetic means between a and b.*

EXAMPLE 5 *Inserting arithmetic means*

Insert three arithmetic means between 2 and 9.

SOLUTION We wish to find three real numbers c_1, c_2, and c_3 such that the following is a (finite) arithmetic sequence:

$$2, c_1, c_2, c_3, 9$$

We may regard this sequence as an arithmetic sequence with first term $a_1 = 2$ and fifth term $a_5 = 9$. To find the common difference d, we proceed as follows:

$$a_5 = a_1 + (5-1)d \qquad \text{let } n = 5 \text{ in } a_n = a_1 + (n-1)d$$
$$9 = 2 + 4d \qquad a_5 = 9 \text{ and } a_1 = 2$$
$$d = \tfrac{7}{4} \qquad \text{solve for } d$$

Thus, the arithmetic means are

$$c_1 = a_1 + d = 2 + \tfrac{7}{4} = \tfrac{15}{4}$$
$$c_2 = c_1 + d = \tfrac{15}{4} + \tfrac{7}{4} = \tfrac{22}{4} = \tfrac{11}{2}$$
$$c_3 = c_2 + d = \tfrac{22}{4} + \tfrac{7}{4} = \tfrac{29}{4}.$$

EXAMPLE 6 *An application of an arithmetic sequence*

A carpenter wishes to construct a ladder with nine rungs whose lengths decrease uniformly from 24 inches at the base to 18 inches at the top. Determine the lengths of the seven intermediate rungs.

SOLUTION   The ladder is sketched in Figure 1. The lengths of the rungs are to form an arithmetic sequence $a_1, a_2, \ldots, a_9$ with $a_1 = 18$ and $a_9 = 24$. Hence, we need to insert seven arithmetic means between 18 and 24. Using $a_n = a_1 + (n - 1)d$ with $n = 9$, $a_1 = 18$, and $a_9 = 24$ gives us

$$24 = 18 + 8d, \qquad \text{or} \qquad 8d = 6.$$

FIGURE 1

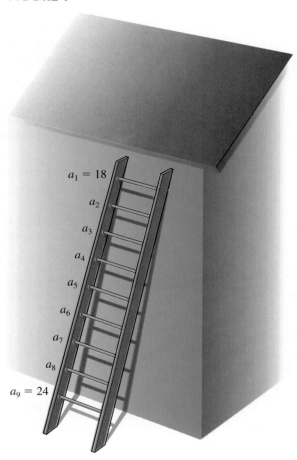

Hence, $d = \frac{6}{8} = 0.75$ and the intermediate rungs have lengths (in inches)

$$18.75, \ 19.5, \ 20.25, \ 21, \ 21.75, \ 22.5, \ 23.25.$$

It is sometimes necessary to express a sum in terms of summation notation, as illustrated in the next example.

EXAMPLE 7   *Expressing a sum in summation notation*

Express in terms of summation notation:

$$\frac{1}{4} + \frac{2}{9} + \frac{3}{14} + \frac{4}{19} + \frac{5}{24} + \frac{6}{29}$$

SOLUTION   The six terms of the sum do not form an arithmetic sequence; however, the numerators and denominators of the fractions, *considered separately*, are each arithmetic. Specifically, we have the following:

Numerators: 1, 2, 3, 4, 5, 6   common difference 1

Denominators: 4, 9, 14, 19, 24, 29   common difference 5

Using the formula for the nth term of an arithmetic sequence twice, we obtain the following nth term for each sequence:

$$a_n = a_1 + (n-1)d = 1 + (n-1)1 = n$$

$$a_n = a_1 + (n-1)d = 4 + (n-1)5 = 5n - 1$$

Hence, the nth term of the given sum is $n/(5n-1)$, and we may write

$$\frac{1}{4} + \frac{2}{9} + \frac{3}{14} + \frac{4}{19} + \frac{5}{24} + \frac{6}{29} = \sum_{n=1}^{6} \frac{n}{5n-1}.$$

7.2 EXERCISES

Exer. 1–2: Show that the given sequence is arithmetic, and find the common difference.

1 $-6, -2, 2, \ldots, 4n - 10, \ldots$

2 $53, 48, 43, \ldots, 58 - 5n, \ldots$

Exer. 3–10: Find the fifth term, the tenth term, and the nth term of the arithmetic sequence.

3 $2, 6, 10, 14, \ldots$ 4 $16, 13, 10, 7, \ldots$

5 $3, 2.7, 2.4, 2.1, \ldots$

6 $-6, -4.5, -3, -1.5, \ldots$

7 $-7, -3.9, -0.8, 2.3, \ldots$

8 $x - 8, x - 3, x + 2, x + 7, \ldots$

9 $\ln 3, \ln 9, \ln 27, \ln 81, \ldots$

10 $\log 1000, \log 100, \log 10, \log 1, \ldots$

Exer. 11–12: Find the common difference for the arithmetic sequence with the specified terms.

11 $a_2 = 21, \quad a_6 = -11$

12 $a_4 = 14, \quad a_{11} = 35$

Exer. 13–18: Find the specified term of the arithmetic sequence that has the two given terms.

13 $a_{12};$ $a_1 = 9.1,$ $a_2 = 7.5$

14 $a_{11};$ $a_1 = 2 + \sqrt{2},$ $a_2 = 3$

15 $a_1;$ $a_6 = 2.7,$ $a_7 = 5.2$

16 $a_1;$ $a_8 = 47,$ $a_9 = 53$

17 $a_{15};$ $a_3 = 7,$ $a_{20} = 43$

18 $a_{10};$ $a_2 = 1,$ $a_{18} = 49$

Exer. 19–22: Find the sum S_n of the arithmetic sequence that satisfies the stated conditions.

19 $a_1 = 40,$ $d = -3,$ $n = 30$

20 $a_1 = 5,$ $d = 0.1,$ $n = 40$

21 $a_1 = -9,$ $a_{10} = 15,$ $n = 10$

22 $a_7 = \frac{7}{3},$ $d = -\frac{2}{3},$ $n = 15$

Exer. 23–26: Find the sum.

23 $\displaystyle\sum_{k=1}^{20} (3k - 5)$ **24** $\displaystyle\sum_{k=1}^{12} (7 - 4k)$

25 $\displaystyle\sum_{k=1}^{18} (\tfrac{1}{2}k + 7)$ **26** $\displaystyle\sum_{k=1}^{10} (\tfrac{1}{4}k + 3)$

Exer. 27–32: Express the sum in terms of summation notation. (Answers are not unique.)

27 $1 + 3 + 5 + 7$ **28** $2 + 4 + 6 + 8 + 10$

29 $1 + 3 + 5 + \cdots + 73$

30 $2 + 4 + 6 + \cdots + 150$

31 $\frac{3}{7} + \frac{6}{11} + \frac{9}{15} + \frac{12}{19} + \frac{15}{23} + \frac{18}{27}$

32 $\frac{5}{13} + \frac{10}{11} + \frac{15}{9} + \frac{20}{7}$

Exer. 33–34: Find the number of terms in the arithmetic sequence with the given conditions.

33 $a_1 = -2$, $d = \frac{1}{4}$, $S = 21$

34 $a_6 = -3$, $d = 0.2$, $S = -33$

35 Insert five arithmetic means between 2 and 10.

36 Insert three arithmetic means between 3 and −5.

37 (a) Find the number of integers between 32 and 395 that are divisible by 6.

    (b) Find their sum.

38 (a) Find the number of negative integers greater than − 500 that are divisible by 33.

    (b) Find their sum.

39 *Log pile* A pile of logs has 24 logs in the bottom layer, 23 in the second layer, 22 in the third, and so on. The top layer contains 10 logs. Find the total number of logs in the pile.

40 *Stadium seating* The first ten rows of seating in a certain section of a stadium have 30 seats, 32 seats, 34 seats, and so on. The eleventh through the twentieth rows each contain 50 seats. Find the total number of seats in the section.

41 *Constructing a grain bin* A grain bin is to be constructed in the shape of a frustum of a cone (see the figure). The bin is to be 10 feet tall with 11 metal rings positioned uniformly around it, from the 4-foot opening at the bottom to the 24-foot opening at the top. Find the total length of metal needed to make the rings.

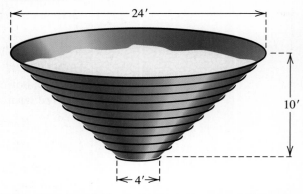

24′

10′

4′

42 *Coasting downhill* A bicycle rider coasts downhill, traveling 4 feet the first second. In each succeeding second, the rider travels 5 feet farther than in the preceding second. If the rider reaches the bottom of the hill in 11 seconds, find the total distance traveled.

43 *Prize money* A contest will have five cash prizes totaling $5000, and there will be a $100 difference between successive prizes. Find the first prize.

44 *Sales bonuses* A company is to distribute $46,000 in bonuses to its top ten salespeople. The tenth salesperson on the list will receive $1000, and the difference in bonus money between successively ranked salespeople is to be constant. Find the bonus for each salesperson.

45 *Distance an object falls* Assuming air resistance is negligible, a small object that is dropped from a hot air balloon falls 16 feet during the first second, 48 feet during the second second, 80 feet during the third second, 112 feet during the fourth second, and so on. Find an expression for the distance the object falls in n seconds.

46 If f is a linear function, show that the sequence with nth term $a_n = f(n)$ is an arithmetic sequence.

47 *Genetic sequence* The sequence defined recursively by $x_{k+1} = x_k/(1 + x_k)$ occurs in genetics in the study of the elimination of a deficient gene from a population. Show that the sequence whose nth term is $1/x_n$ is arithmetic.

48 *Dimensions of a maze* Find the total length of the broken-line curve in the figure on the next page if the width of the maze formed by the curve is 16 inches and all halls in the maze have width 1 inch. What is the length if the width of the maze is 32 inches?

EXERCISE 48

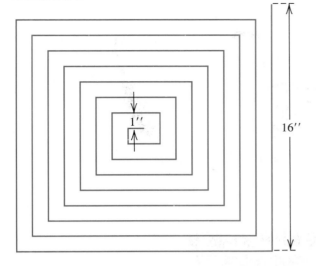

16″

1″

Exer. 49–50: Depreciation methods are sometimes used by businesses and individuals to estimate the value of an asset over a life span of n years. In the sum-of-years'-digits method, for each year $k = 1, 2, 3, \ldots, n$, the value of an asset is decreased by the fraction $A_k = \dfrac{n - k + 1}{T_n}$ of its initial cost, where $T_n = 1 + 2 + 3 + \cdots + n$.

49 (a) If $n = 8$, find $A_1, A_2, A_3, \ldots, A_8$.

   (b) Show that the sequence in (a) is arithmetic, and find S_8.

   (c) If the initial value of an asset is \$1000, how much has been depreciated after 4 years?

50 (a) If n is any positive integer, find $A_1, A_2, A_3, \ldots, A_n$.

   (b) Show that the sequence in (a) is arithmetic, and find S_n.

7.3 GEOMETRIC SEQUENCES

The second special type of sequence that we will discuss—the geometric sequence—occurs frequently in applications.

Definition of Geometric Sequence

A sequence $a_1, a_2, \ldots, a_n, \ldots$ is a **geometric sequence** if $a_1 \neq 0$ and if there is a real number $r \neq 0$ such that for every positive integer k,

$$a_{k+1} = a_k r.$$

The number $r = a_{k+1}/a_k$ is called the **common ratio** of the sequence.

Note that the common ratio $r = a_{k+1}/a_k$ is the ratio of *any* two successive terms of a geometric sequence.

The formula $a_{k+1} = a_k r$ provides a recursive method for obtaining terms of a geometric sequence. Beginning with any nonzero real number a_1, we multiply by the number r successively, obtaining

$$a_1, a_1 r, a_1 r^2, a_1 r^3, \ldots.$$

The nth term a_n of this sequence is given by the next formula.

Formula for the nth Term of a Geometric Sequence

$$a_n = a_1 r^{n-1}$$

EXAMPLE 1   *Finding terms of a geometric sequence*

A geometric sequence has first term 3 and common ratio $-\frac{1}{2}$. Find the first five terms and the tenth term.

SOLUTION   If we let $a_1 = 3$ and $r = -\frac{1}{2}$, then the first five terms are

$$3, \ -\tfrac{3}{2}, \ \tfrac{3}{4}, \ -\tfrac{3}{8}, \ \tfrac{3}{16}.$$

Using the formula $a_n = a_1 r^{n-1}$ with $n = 10$, we find that the tenth term is

$$a_{10} = 3(-\tfrac{1}{2})^9 = -\tfrac{3}{512}.$$

EXAMPLE 2   *Finding a specific term of a geometric sequence*

The third term of a geometric sequence is 5, and the sixth term is -40. Find the eighth term.

SOLUTION   We are given $a_3 = 5$ and $a_6 = -40$ and wish to find a_8. The following are equivalent systems of equations in the variables a_1 and r:

$$\begin{cases} a_3 = a_1 r^{3-1} & \text{let } n = 3 \text{ in } a_n = a_1 r^{n-1} \\ a_6 = a_1 r^{6-1} & \text{let } n = 6 \text{ in } a_n = a_1 r^{n-1} \end{cases}$$

$$\begin{cases} 5 = a_1 r^2 & a_3 = 5 \\ -40 = a_1 r^5 & a_6 = -40 \end{cases}$$

Solving the first equation of the system for a_1 gives us $a_1 = 5/r^2$. Substituting this expression in the second equation yields

$$-40 = \frac{5}{r^2} \cdot r^5.$$

Simplifying, we get $r^3 = -8$, and hence $r = -2$. We next use $a_1 = 5/r^2$ to obtain

$$a_1 = \frac{5}{(-2)^2} = \frac{5}{4}.$$

Finally, using $a_n = a_1 r^{n-1}$ with $n = 8$ gives

$$a_8 = a_1 r^7 = (\tfrac{5}{4})(-2)^7 = -160.$$

The next theorem contains a formula for the nth partial sum S_n of a geometric sequence.

Theorem: Formula for S_n

> The nth partial sum S_n of a geometric sequence with first term a_1 and common ratio $r \neq 1$ is
>
> $$S_n = a_1 \frac{1 - r^n}{1 - r}.$$

PROOF   By definition, the nth partial sum S_n is

$$S_n = a_1 + a_1 r + a_1 r^2 + \cdots + a_1 r^{n-2} + a_1 r^{n-1}.$$

Multiplying both sides of this equation by r, we obtain

$$r S_n = a_1 r + a_1 r^2 + a_1 r^3 + \cdots + a_1 r^{n-1} + a_1 r^n.$$

If we subtract this equation from the equation for S_n, all but two terms on the right-hand side cancel and we obtain the following:

$$S_n - r S_n = a_1 - a_1 r^n \quad \text{subtract}$$

$$(1 - r) S_n = a_1 (1 - r^n) \quad \text{factor}$$

$$S_n = \frac{a_1 (1 - r^n)}{1 - r} \quad \text{divide by } 1 - r \quad \blacksquare$$

EXAMPLE 3   *Finding a sum of terms of a geometric sequence*

If $1, 0.3, 0.09, 0.027, \ldots$ is a geometric sequence, find the sum of the first five terms.

SOLUTION   If we let $a_1 = 1$, $r = 0.3$, and $n = 5$ in the formula for S_n stated in the preceding theorem, we obtain

$$S_5 = (1) \frac{1 - (0.3)^5}{1 - 0.3} = 1.4251.$$

EXAMPLE 4   *The rapid growth of terms of a geometric sequence*

A man wishes to save money by setting aside 1 cent the first day, 2 cents the second day, 4 cents the third day, and so on.

(a) If he continues to double the amount set aside each day, how much must he set aside on the fifteenth day?

(b) Assuming he does not run out of money, what is the total amount saved at the end of 30 days?

SOLUTION

(a) The amount (in cents) set aside on successive days forms a geometric sequence

$$1, 2, 4, 8, \ldots$$

with first term 1 and common ratio 2. We find the amount to be set aside on the fifteenth day by using $a_n = a_1 r^{n-1}$ with $a_1 = 1$ and $n = 15$:

$$a_{15} = a_1 r^{14} = 1 \cdot 2^{14} = 16{,}384$$

Thus, \$163.84 should be set aside on the fifteenth day.

(b) To find the total amount saved after 30 days, we use the formula for S_n with $n = 30$, obtaining (in cents)

$$S_{30} = (1)\frac{1 - 2^{30}}{1 - 2} = 1{,}073{,}741{,}823.$$

Thus, the total amount saved is \$10,737,418.23.

In a manner analogous to our work with arithmetic sequences, if a and b are positive real numbers, then we call a positive number c the **geometric mean** of a and b if a, c, b is a geometric sequence. If the common ratio is r, then

$$r = \frac{c}{a} = \frac{b}{c}, \quad \text{or} \quad c^2 = ab.$$

Taking the square root of both sides of the last equation, we see that *the geometric mean of the positive numbers a and b is $\sqrt{ab}$.* As a generalization, k positive real numbers $c_1, c_2, \ldots, c_k$ are k **geometric means** between a and b if $a, c_1, c_2, \ldots, c_k, b$ is a geometric sequence. The process of determining these numbers is referred to as *inserting k geometric means between a and b.*

ILLUSTRATION

Geometric Means

Numbers	Geometric mean
20, 45	$\sqrt{20 \cdot 45} = \sqrt{900} = 30$
3, 4	$\sqrt{3 \cdot 4} = 2\sqrt{3}$

Given the geometric series with first term a_1 and common ratio $r \neq 1$, we may write the formula for S_n of the preceding theorem in the form

$$S_n = \frac{a_1}{1 - r} - \frac{a_1}{1 - r} r^n.$$

If $|r| < 1$, then r^n *approaches 0 as n increases* without bound. Thus, S_n approaches $a_1/(1 - r)$ as n increases without bound. Using the notation we developed for rational functions in Section 4.5, we have

$$S_n \rightarrow \frac{a_1}{1-r} \quad \text{as } n \rightarrow \infty.$$

The number $a_1/(1-r)$ is called the *sum S* of the **infinite geometric series**

$$a_1 + a_1 r + a_1 r^2 + \cdots + a_1 r^{n-1} + \cdots.$$

This gives us the next result.

Theorem on the Sum of an Infinite Geometric Series

If $|r| < 1$, then the infinite geometric series

$$a_1 + a_1 r + a_1 r^2 + \cdots + a_1 r^{n-1} + \cdots$$

has the sum $S = \dfrac{a_1}{1-r}$.

The preceding theorem implies that if we add more and more terms of the indicated infinite geometric series, the sums get closer and closer to $a_1/(1-r)$. The next example illustrates how the theorem can be used to show that every real number represented by a repeating decimal is rational.

EXAMPLE 5 *Expressing an infinite repeating decimal as a rational number*

Find the rational number that corresponds to the infinite repeating decimal $5.4\overline{27}$, where the block of digits underneath the bar is repeated indefinitely.

SOLUTION From the decimal expression $5.4272727\ldots$, we obtain the infinite series

$$5.4 + 0.027 + 0.00027 + 0.0000027 + \cdots.$$

The part of the expression after the first term is

$$0.027 + 0.00027 + 0.0000027 + \cdots,$$

which has the form given in the theorem on the sum of an infinite geometric series, with $a_1 = 0.027$ and $r = 0.01$. Hence, the sum S of this infinite geometric series is

$$S = \frac{0.027}{1 - 0.01} = \frac{0.027}{0.99} = \frac{27}{990} = \frac{3}{110}.$$

Thus, it appears that the desired number is $5.4 + \frac{3}{110}$, or $\frac{597}{110}$. A check by division shows that $\frac{597}{110}$ does correspond to the given repeating decimal.

In general, given any infinite sequence $a_1, a_2, \ldots, a_n, \ldots$, the expression

$$a_1 + a_2 + \cdots + a_n + \cdots$$

is called an **infinite series** or simply a **series**. We denote this series by

$$\sum_{n=1}^{\infty} a_n.$$

Each number a_k is a **term** of the series, and a_n is the **nth term**. Since only *finite* sums may be added algebraically, it is necessary to define what is meant by an *infinite sum*. Consider the sequence of partial sums

$$S_1, S_2, \ldots, S_n, \ldots.$$

If there is a number S such that $S_n \to S$ as $n \to \infty$, then, as in our discussion of infinite geometric series, S is the **sum** of the infinite series and we write

$$S = a_1 + a_2 + \cdots + a_n + \cdots.$$

In Example 5 we found that the infinite repeating decimal $5.4272727\ldots$ corresponds to the rational number $\frac{597}{110}$. Since $\frac{597}{110}$ is the sum of an infinite series determined by the decimal, we may write

$$\tfrac{597}{110} = 5.4 + 0.027 + 0.00027 + 0.0000027 + \cdots.$$

If the terms of an infinite sequence are alternately positive and negative, as in the expression

$$a_1 + (-a_2) + a_3 + (-a_4) + \cdots + [(-1)^{n+1}a_n] + \cdots$$

for positive real numbers a_k, then the expression is an **alternating infinite series**, and we write it in the form

$$a_1 - a_2 + a_3 - a_4 + \cdots + (-1)^{n+1}a_n + \cdots.$$

The most common types of alternating infinite series are infinite geometric series in which the common ratio r is negative.

EXAMPLE 6 *Finding the sum of an infinite geometric series*

Find the sum S of the alternating infinite geometric series

$$\sum_{n=1}^{\infty} 3(-\tfrac{2}{3})^{n-1} = 3 - 2 + \tfrac{4}{3} - \tfrac{8}{9} + \cdots + 3(-\tfrac{2}{3})^{n-1} + \cdots.$$

SOLUTION   Using the formula for S in the theorem on the sum of an infinite geometric series, with $a_1 = 3$ and $r = -\tfrac{2}{3}$, we obtain

$$S = \frac{a_1}{1 - r} = \frac{3}{1 - (-\tfrac{2}{3})} = \frac{3}{\tfrac{5}{3}} = \frac{9}{5}.$$

EXAMPLE 7   *An application of an infinite geometric series*

A rubber ball is dropped from a height of 10 meters. Suppose it rebounds one-half the distance after each fall, as illustrated by the arrows in Figure 2. Find the total distance the ball travels before coming to rest.

SOLUTION   If the ball always rebounds half the distance it falls, then, theoretically, it *never* comes to rest. However, the sum of the distances it travels downward and the sum of the distances it travels on the rebounds form two finite geometric series:

Downward series: $10 + 5 + 2.5 + 1.25 + 0.625 + \cdots$

Upward series: $5 + 2.5 + 1.25 + 0.625 + \cdots$

We assume that the total distance S the ball travels can be found by adding the sums of these infinite series. This gives us

$$S = 10 + 2[5 + 2.5 + 1.25 + 0.625 + \cdots]$$
$$= 10 + 2[5 + 5(\tfrac{1}{2}) + 5(\tfrac{1}{2})^2 + 5(\tfrac{1}{2})^3 + \cdots].$$

Using the formula $S = a_1/(1 - r)$ with $a_1 = 5$ and $r = \tfrac{1}{2}$, we obtain

$$S = 10 + 2\left[\frac{5}{1 - \tfrac{1}{2}}\right] = 10 + 2(10) = 30 \text{ m}.$$

FIGURE 2

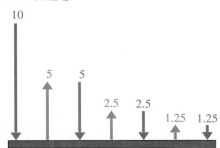

7.3 EXERCISES

Exer. 1–2: Show that the given sequence is geometric, and find the common ratio.

1 $5, -\tfrac{5}{4}, \tfrac{5}{16}, \ldots, 5(-\tfrac{1}{4})^{n-1}, \ldots$

2 $\tfrac{1}{7}, \tfrac{3}{7}, \tfrac{9}{7}, \ldots, \tfrac{1}{7}(3)^{n-1}, \ldots$

Exer. 3–14: Find the fifth term, the eighth term, and the nth term of the geometric sequence.

3 $8, 4, 2, 1, \ldots$

4 $4, 1.2, 0.36, 0.108, \ldots$

5 $300, -30, 3, -0.3, \ldots$

6 $1, -\sqrt{3}, 3, -3\sqrt{3}, \ldots$

7 $5, 25, 125, 625, \ldots$

8 $2, 6, 18, 54, \ldots$

9 $4, -6, 9, -13.5, \ldots$

10 $162, -54, 18, -6, \ldots$

11 $1, -x^2, x^4, -x^6, \ldots$

12 $1, -\dfrac{x}{3}, \dfrac{x^2}{9}, -\dfrac{x^3}{27}, \ldots$

13 $2, 2^{x+1}, 2^{2x+1}, 2^{3x+1}, \ldots$

14 $10, 10^{2x-1}, 10^{4x-3}, 10^{6x-5}, \ldots$

Exer. 15–16: Find all possible values of r for a geometric sequence with the two given terms.

15 $a_4 = 3, \quad a_6 = 9$

16 $a_3 = 4, \quad a_7 = \tfrac{1}{4}$

17 Find the sixth term of the geometric sequence whose first two terms are 4 and 6.

18 Find the seventh term of the geometric sequence whose second and third terms are 2 and $-\sqrt{2}$.

19 Given a geometric sequence such that $a_4 = 4$ and $a_7 = 12$, find r and a_{10}.

20 Given a geometric sequence such that $a_2 = 3$ and $a_5 = -81$, find r and a_9.

Exer. 21–24: Find the sum.

21 $\displaystyle\sum_{k=1}^{10} 3^k$

22 $\displaystyle\sum_{k=1}^{9} (-\sqrt{5})^k$

23 $\displaystyle\sum_{k=0}^{9} (-\tfrac{1}{2})^{k+1}$

24 $\displaystyle\sum_{k=1}^{7} (3^{-k})$

Exer. 25–28: Express the sum in terms of summation notation. (Answers are not unique.)

25 $2 + 4 + 8 + 16 + 32 + 64 + 128$

26 $2 - 4 + 8 - 16 + 32 - 64$

27 $\frac{1}{4} - \frac{1}{12} + \frac{1}{36} - \frac{1}{108}$ 28 $3 + \frac{3}{5} + \frac{3}{25} + \frac{3}{125} + \frac{3}{625}$

Exer. 29–36: Find the sum of the infinite geometric series, if it exists.

29 $1 - \frac{1}{2} + \frac{1}{4} - \frac{1}{8} + \cdots$ 30 $2 + \frac{2}{3} + \frac{2}{9} + \frac{2}{27} + \cdots$

31 $1.5 + 0.015 + 0.00015 + \cdots$

32 $1 - 0.1 + 0.01 - 0.001 + \cdots$

33 $\sqrt{2} - 2 + \sqrt{8} - 4 + \cdots$ 34 $1 + \frac{3}{2} + \frac{9}{4} + \frac{27}{8} + \cdots$

35 $256 + 192 + 144 + 108 + \cdots$

36 $250 - 100 + 40 - 16 + \cdots$

Exer. 37–44: Find the rational number represented by the repeating decimal.

37 $0.\overline{23}$ 38 $0.0\overline{71}$

39 $2.4\overline{17}$ 40 $10.\overline{5}$

41 $5.\overline{146}$ 42 $3.2\overline{394}$

43 $1.6\overline{124}$ 44 $123.61\overline{83}$

45 Find the geometric mean of 12 and 48.

46 Find the geometric mean of 20 and 25.

47 Insert two geometric means between 4 and 500.

48 Insert three geometric means between 2 and 512.

49 *Using a vacuum pump* A vacuum pump removes one-half of the air in a container with each stroke. After 10 strokes, what percentage of the original amount of air remains in the container?

50 *Calculating depreciation* The yearly depreciation of a certain machine is 25% of its value at the beginning of the year. If the original cost of the machine is $20,000, what is its value after 6 years?

51 *Growth of bacteria* A certain culture initially contains 10,000 bacteria and increases by 20% every hour.

  (a) Find a formula for the number $N(t)$ of bacteria present after t hours.

  (b) How many bacteria are in the culture at the end of 10 hours?

52 *Interest on savings* An amount of money P is deposited in a savings account that pays interest at a rate of r percent per year compounded quarterly; the principal

and accumulated interest are left in the account. Find a formula for the total amount in the account after n years.

53 *Rebounding ball* A rubber ball is dropped from a height of 60 feet. If it rebounds approximately two-thirds the distance after each fall, use an infinite geometric series to approximate the total distance the ball travels before coming to rest.

54 *Motion of a pendulum* The bob of a pendulum swings through an arc 24 centimeters long on its first swing. If each successive swing is approximately five-sixths the length of the preceding swing, use an infinite geometric series to approximate the total distance the bob travels before coming to rest.

55 *Multiplier effect* A manufacturing company that has just located in a small community will pay two million dollars per year in salaries. It has been estimated that 60% of these salaries will be spent in the local area, and 60% of the money spent will again change hands within the community. This process, called the multiplier effect, will be repeated ad infinitum. Find the total amount of local spending that will be generated by company salaries.

56 *Pest eradication* In a pest eradication program, N sterilized male flies are released into the general population each day. It is estimated that 90% of these flies will survive a given day.

  (a) Show that the number of sterilized flies in the population n days after the program has begun is

  $$N + (0.9)N + (0.9)^2 N + \cdots + (0.9)^{n-1} N.$$

  (b) If the *long-range* goal of the program is to keep 20,000 sterilized males in the population, how many flies should be released each day?

57 *Drug dosage* A certain drug has a half-life of about 2 hours in the bloodstream. The drug is formulated to be administered in doses of D milligrams every 4 hours, but D is yet to be determined.

  (a) Show that the number of milligrams of drug in the bloodstream after the nth dose has been administered is

  $$D + \tfrac{1}{4}D + \cdots + (\tfrac{1}{4})^{n-1} D,$$

  and that this sum is approximately $\frac{4}{3}D$ for large values of n.

  (b) A level of more than 500 milligrams of the drug in the bloodstream is considered to be dangerous. Find the largest possible dose that can be given repeatedly over a long period of time.

58 *Genealogy* Shown in the figure is a family tree displaying the current generation (you) and 3 prior generations, with a total of 12 grandparents. If you were to trace your family history back 10 generations, how many grandparents would you find?

EXERCISE 58

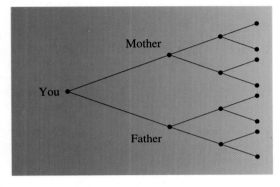

59 The first figure shows some terms of a sequence of squares $S_1, S_2, \ldots, S_k, \ldots$. Let a_k, A_k, and P_k denote the side, area, and perimeter, respectively, of the square S_k. The square S_{k+1} is constructed from S_k by connecting four points on S_k, with each point a distance of $\frac{1}{4}a_k$ from a vertex, as shown in the second figure.

(a) Find the relationship between a_{k+1} and a_k.

(b) Find a_n, A_n, and P_n.

(c) Calculate $\sum_{n=1}^{\infty} P_n$.

EXERCISE 59

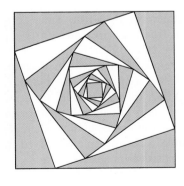

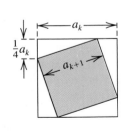

60 The figure shows several terms of a sequence consisting of alternating circles and squares. Each circle is inscribed in a square, and each square (excluding the largest) is inscribed in a circle. Let S_n denote the area of the nth square and C_n the area of the nth circle.

(a) Find the relationships between S_n and C_n and between C_n and S_{n+1}.

(b) What portion of the largest square is shaded in the figure?

EXERCISE 60

61 *Compound interest* If a deposit of $100 is made on the first day of each month into an account that pays 6% interest compounded monthly, determine the amount in the account after 18 years.

62 *Compound interest* Refer to Exercise 61. Show that if the monthly deposit is P dollars and the rate is r% compounded monthly, then the amount A in the account after n months is given by

$$A = P\left(\frac{12}{r} + 1\right)\left[\left(1 + \frac{r}{12}\right)^n - 1\right].$$

63 *Compound interest* Use Exercise 62 to find A if the monthly deposit $P = \$100$, $r = 8\%$, and $n = 60$.

64 *Compound interest* Refer to Exercise 62. If $r = 10\%$, approximately how many years are required to accumulate $100,000 if the monthly deposit P is

(a) $100   (b) $200

Exer. 65–66: The *double-declining balance method* is a method of depreciation in which, for each year $k = 1, 2, 3, \ldots, n$, the value of an asset is decreased by the fraction $A_k = \frac{2}{n}\left(1 - \frac{2}{n}\right)^{k-1}$ of its initial cost.

65 (a) If $n = 5$, find $A_1, A_2, \ldots, A_5$.

(b) Show that the sequence in (a) is geometric, and find S_5.

(c) If the initial value of an asset is $25,000, how much of its value has been depreciated after 2 years?

66 (a) If n is any positive integer, find $A_1, A_2, \ldots, A_n$.

(b) Show that the sequence in (a) is geometric, and find S_n.

7.4   MATHEMATICAL INDUCTION

If n is a positive integer and we let P_n denote the mathematical statement $(xy)^n = x^n y^n$, we obtain the following *infinite sequence of statements*:

$$\text{Statement } P_1: \quad (xy)^1 = x^1 y^1$$

$$\text{Statement } P_2: \quad (xy)^2 = x^2 y^2$$

$$\text{Statement } P_3: \quad (xy)^3 = x^3 y^3$$

$$\vdots \qquad\qquad \vdots$$

$$\text{Statement } P_n: \quad (xy)^n = x^n y^n$$

$$\vdots \qquad\qquad \vdots$$

It is easy to show that P_1, P_2, and P_3 are *true* statements. However, it is impossible to check the validity of P_n for *every* positive integer n. Showing that P_n is true for every n requires the following principle.

Principle of Mathematical Induction

> If with each positive integer n there is associated a statement P_n, then all the statements P_n are true, provided the following two conditions are satisfied:
>
> **(1)** P_1 is true.
>
> **(2)** Whenever k is a positive integer such that P_k is true, then P_{k+1} is also true.

To help understand this principle, consider an infinite sequence of statements labeled

$$P_1, P_2, P_3, \ldots, P_n, \ldots$$

that satisfy conditions (1) and (2). By (1), statement P_1 is true. Since condition (2) holds, whenever a statement P_k is true the *next* statement P_{k+1} is also true. Hence, since P_1 is true, P_2 is also true, by (2). However, if P_2 is true, then, by (2), we see that the next statement P_3 is true. Once again, if P_3 is true, then, by (2), P_4 is also true. If we continue in this manner, we can argue that if n is any *particular* integer, then P_n is true, since we can use condition (2) one step at a time, eventually reaching P_n. Although this type of reasoning does not actually *prove* the principle of mathematical induction, it certainly makes it plausible. The principle is proved in advanced algebra using postulates for the positive integers.

When applying the principle of mathematical induction, we always follow two steps.

Steps in Applying the Principle of Mathematical Induction

> **1** Show that P_1 is true.
>
> **2** *Assume* that P_k is true and then prove that P_{k+1} is true.

Step 2 often causes confusion. Note that we do not *prove* that P_k is true (except for $k = 1$). Instead, we show that *if* P_k happens to be true, then the statement P_{k+1} is also true. We refer to the assumption that P_k is true as the **induction hypothesis**.

EXAMPLE 1   *Using the principle of mathematical induction*

Use mathematical induction to prove that for every positive integer n, the sum of the first n positive integers is

$$\frac{n(n + 1)}{2}.$$

SOLUTION   If n is any positive integer, let P_n denote the statement

$$1 + 2 + 3 + \cdots + n = \frac{n(n + 1)}{2}.$$

The following are some special cases of P_n:

If $n = 1$, then P_1 is

$$1 = \frac{1(1 + 1)}{2}; \quad \text{that is,} \quad 1 = 1.$$

If $n = 2$, then P_2 is

$$1 + 2 = \frac{2(2 + 1)}{2}; \quad \text{that is,} \quad 3 = 3.$$

If $n = 3$, then P_3 is

$$1 + 2 + 3 = \frac{3(3 + 1)}{2}; \quad \text{that is,} \quad 6 = 6.$$

If $n = 4$, then P_4 is

$$1 + 2 + 3 + 4 = \frac{4(4 + 1)}{2}; \quad \text{that is,} \quad 10 = 10.$$

Although it is instructive to check the validity of P_n for several values of n as we have done, it is unnecessary to do so. We need only apply the two-step process outlined prior to this example. Thus, we proceed as follows:

Step 1   If we substitute $n = 1$ in P_n, then the left-hand side contains only the number 1 and the right-hand side is $\dfrac{1(1 + 1)}{2}$, which also equals 1. Hence, P_1 is true.

Step 2   Assume that P_k is true. Thus, the induction hypothesis is

$$1 + 2 + 3 + \cdots + k = \frac{k(k + 1)}{2}.$$

(continued)

Our goal is to show that P_{k+1} is true—that is,

$$1 + 2 + 3 + \cdots + (k + 1) = \frac{(k + 1)[(k + 1) + 1]}{2}.$$

We may prove that the last formula is true by rewriting the left side and using the induction hypothesis as follows:

$1 + 2 + 3 + \cdots + (k + 1)$

$= (1 + 2 + 3 + \cdots + k) + (k + 1)$ group the first k terms

$= \dfrac{k(k + 1)}{2} + (k + 1)$ induction hypothesis

$= \dfrac{k(k + 1) + 2(k + 1)}{2}$ add

$= \dfrac{(k + 1)(k + 2)}{2}$ factor out $k + 1$

$= \dfrac{(k + 1)[(k + 1) + 1]}{2}$ change form of $k + 2$

This shows that P_{k+1} is true, and therefore, the proof by mathematical induction is complete.

EXAMPLE 2 *Using the principle of mathematical induction*

Prove that for each positive integer n,

$$1^2 + 3^2 + \cdots + (2n - 1)^2 = \frac{n(2n - 1)(2n + 1)}{3}.$$

SOLUTION   For each positive integer n, let P_n denote the given statement. Note that this is a formula for the sum of the squares of the first n odd positive integers. We again follow the two-step procedure.

Step 1 Substituting 1 for n in P_n, we obtain

$$1^2 = \frac{(1)(2 - 1)(2 + 1)}{3} = \frac{3}{3} = 1.$$

This shows that P_1 is true.

Step 2 Assume that P_k is true. Thus, the induction hypothesis is

$$1^2 + 3^2 + \cdots + (2k - 1)^2 = \frac{k(2k - 1)(2k + 1)}{3}.$$

We wish to show that P_{k+1} is true—that is,

$$1^2 + 3^2 + \cdots + [2(k + 1) - 1]^2 = \frac{(k + 1)[2(k + 1) - 1][2(k + 1) + 1]}{3}.$$

This equation simplifies to

$$1^2 + 3^2 + \cdots + (2k + 1)^2 = \frac{(k + 1)(2k + 1)(2k + 3)}{3}.$$

Remember that the next to last term on the left-hand side of the equation is $(2k - 1)^2$. In a manner similar to that used in the solution of Example 1, we may prove the formula for P_{k+1} by rewriting the left side and using the induction hypothesis as follows:

$$
\begin{aligned}
1^2 &+ 3^2 + \cdots + (2k + 1)^2 \\
&= [1^2 + 3^2 + \cdots + (2k - 1)^2] + (2k + 1)^2 \quad &\text{group the first } k \text{ terms} \\
&= \frac{k(2k - 1)(2k + 1)}{3} + (2k + 1)^2 \quad &\text{induction hypothesis} \\
&= \frac{k(2k - 1)(2k + 1) + 3(2k + 1)^2}{3} \quad &\text{add} \\
&= \frac{(2k + 1)[k(2k - 1) + 3(2k + 1)]}{3} \quad &\text{factor out } 2k + 1 \\
&= \frac{(2k + 1)(2k^2 + 5k + 3)}{3} \quad &\text{simplify} \\
&= \frac{(k + 1)(2k + 1)(2k + 3)}{3} \quad &\text{factor and change order}
\end{aligned}
$$

This shows that P_{k+1} is true, and hence P_n is true for every n.

EXAMPLE 3   *Using the principle of mathematical induction*

Prove that 2 is a factor of $n^2 + 5n$ for every positive integer n.

SOLUTION   For each positive integer n, let P_n denote the following statement:

$$2 \text{ is a factor of } n^2 + 5n$$

We shall follow the two-step procedure.

Step 1   If $n = 1$, then

$$n^2 + 5n = 1^2 + 5 \cdot 1 = 6 = 2 \cdot 3.$$

Thus, 2 is a factor of $n^2 + 5n$ for $n = 1$; that is, P_1 is true.

Step 2   Assume that P_k is true. Thus, the induction hypothesis is

$$2 \text{ is a factor of } k^2 + 5k$$

or, equivalently, $\qquad\qquad k^2 + 5k = 2p$

for some integer p.

(continued)

We wish to show that P_{k+1} is true—that is,

$$2 \text{ is a factor of } (k+1)^2 + 5(k+1).$$

We may do this as follows:

$$
\begin{aligned}
(k+1)^2 &+ 5(k+1) \\
&= k^2 + 2k + 1 + 5k + 5 \qquad \text{multiply} \\
&= (k^2 + 5k) + (2k + 6) \qquad \text{rearrange terms} \\
&= 2p + 2(k+3) \qquad \text{induction hypothesis and factor } 2k+6 \\
&= 2(p + k + 3) \qquad \text{factor out 2}
\end{aligned}
$$

Since 2 is a factor of the last expression, P_{k+1} is true, and hence P_n is true for every n.

Let j be a positive integer, and suppose that with each integer $n \geq j$ there is associated a statement P_n. For example, if $j = 6$, then the statements are numbered $P_6, P_7, P_8, \ldots$. The principle of mathematical induction may be extended to cover this situation. To prove that the statements P_n are true for $n \geq j$, we use the following two steps, in the same manner as we did for $n \geq 1$.

Steps in Applying the Extended Principle of Mathematical Induction for P_k, $k \geq j$

> **1** Show that P_j is true.
>
> **2** Assume that P_k is true with $k \geq j$ and prove that P_{k+1} is true.

EXAMPLE 4 *Using the extended principle of mathematical induction*

Let a be a nonzero real number such that $a > -1$. Prove that

$$(1 + a)^n > 1 + na$$

for every integer $n \geq 2$.

SOLUTION   For each positive integer n, let P_n denote the inequality $(1 + a)^n > 1 + na$. Note that P_1 is *false*, since $(1 + a)^1 = 1 + (1)(a)$. However, we can show that P_n is true for $n \geq 2$ by using the extended principle with $j = 2$.

Step 1 We first note that $(1 + a)^2 = 1 + 2a + a^2$. Since $a \neq 0$, we have $a^2 > 0$, and therefore $1 + 2a + a^2 > 1 + 2a$, or $(1 + a)^2 > 1 + 2a$, Hence, P_2 is true.

Step 2 Assume that P_k is true. Thus, the induction hypothesis is

$$(1 + a)^k > 1 + ka.$$

We wish to show that P_{k+1} is true—that is,

$$(1 + a)^{k+1} > 1 + (k + 1)a.$$

To prove the last inequality, we first observe the following:

$$(1 + a)^{k+1} = (1 + a)^k(1 + a) \quad \text{law of exponents}$$
$$> (1 + ka)(1 + a) \quad \text{induction hypothesis and } 1 + a > 0$$

We next note that

$$(1 + ka)(1 + a) = 1 + ka + a + ka^2 \quad \text{multiply}$$
$$= 1 + (ka + a) + ka^2 \quad \text{group terms}$$
$$= 1 + (k + 1)a + ka^2 \quad \text{factor out } a$$
$$> 1 + (k + 1)a. \quad \text{since } ka^2 > 0$$

The last two inequalities give us

$$(1 + a)^{k+1} > 1 + (k + 1)a.$$

Thus, P_{k+1} is true, and the proof by mathematical induction is complete.

7.4 EXERCISES

Exer. 1–26: Prove that the statement is true for every positive integer n.

1 $2 + 4 + 6 + \cdots + 2n = n(n + 1)$

2 $1 + 4 + 7 + \cdots + (3n - 2) = \dfrac{n(3n - 1)}{2}$

3 $1 + 3 + 5 + \cdots + (2n - 1) = n^2$

4 $3 + 9 + 15 + \cdots + (6n - 3) = 3n^2$

5 $2 + 7 + 12 + \cdots + (5n - 3) = \frac{1}{2}n(5n - 1)$

6 $2 + 6 + 18 + \cdots + 2 \cdot 3^{n-1} = 3^n - 1$

7 $1 + 2 \cdot 2 + 3 \cdot 2^2 + \cdots + n \cdot 2^{n-1} = 1 + (n - 1) \cdot 2^n$

8 $(-1)^1 + (-1)^2 + (-1)^3 + \cdots + (-1)^n = \dfrac{(-1)^n - 1}{2}$

9 $1^2 + 2^2 + 3^2 + \cdots + n^2 = \dfrac{n(n + 1)(2n + 1)}{6}$

10 $1^3 + 2^3 + 3^3 + \cdots + n^3 = \left[\dfrac{n(n + 1)}{2}\right]^2$

11 $\dfrac{1}{1 \cdot 2} + \dfrac{1}{2 \cdot 3} + \dfrac{1}{3 \cdot 4} + \cdots + \dfrac{1}{n(n + 1)} = \dfrac{n}{n + 1}$

12 $\dfrac{1}{1 \cdot 2 \cdot 3} + \dfrac{1}{2 \cdot 3 \cdot 4} + \dfrac{1}{3 \cdot 4 \cdot 5} + \cdots + \dfrac{1}{n(n + 1)(n + 2)}$

$= \dfrac{n(n + 3)}{4(n + 1)(n + 2)}$

13 $3 + 3^2 + 3^3 + \cdots + 3^n = \frac{3}{2}(3^n - 1)$

14 $1^3 + 3^3 + 5^3 + \cdots + (2n - 1)^3 = n^2(2n^2 - 1)$

15 $n < 2^n$

16 $1 + 2n \leq 3^n$

17 $1 + 2 + 3 + \cdots + n < \frac{1}{8}(2n + 1)^2$

18 If $0 < a < b$, then $\left(\dfrac{a}{b}\right)^{n+1} < \left(\dfrac{a}{b}\right)^n$.

19 3 is a factor of $n^3 - n + 3$.

20 2 is a factor of $n^2 + n$.

21 4 is a factor of $5^n - 1$.

22 9 is a factor of $10^{n+1} + 3 \cdot 10^n + 5$.

23 If a is greater than 1, then $a^n > 1$.

24 If $r \neq 1$, then $a + ar + ar^2 + \cdots + ar^{n-1} = \dfrac{a(1 - r^n)}{1 - r}$.

25 $a - b$ is a factor of $a^n - b^n$.
   (*Hint:* $a^{k+1} - b^{k+1} = a^k(a - b) + (a^k - b^k)b$.)

26 $a + b$ is a factor of $a^{2n-1} + b^{2n-1}$.

Exer. 27–32: Find the smallest positive integer j for which the statement is true. Use the extended principle of mathematical induction to prove that the formula is true for every integer greater than j.

27 $n + 12 \leq n^2$ **28** $n^2 + 18 \leq n^3$

29 $5 + \log_2 n \leq n$ **30** $n^2 \leq 2^n$

31 $2n + 2 \leq 2^n$ **32** $n \log_2 n + 20 \leq n^2$

Exer. 33–36: Express the sum in terms of n.

33 $\displaystyle\sum_{k=1}^{n} (k^2 + 3k + 5)$

$\left(\begin{array}{l} \textit{Hint:} \text{ Use the theorem on sums to write the sum as} \\[2mm] \qquad \displaystyle\sum_{k=1}^{n} k^2 + 3 \sum_{k=1}^{n} k + \sum_{k=1}^{n} 5. \\[2mm] \text{Next employ Exercise 9, Example 4 of Section 7.1, and} \\[2mm] \text{the formula for } \displaystyle\sum_{k=1}^{n} c. \end{array}\right.$

34 $\displaystyle\sum_{k=1}^{n} (3k^2 - 2k + 1)$ **35** $\displaystyle\sum_{k=1}^{n} (2k - 3)^2$

36 $\displaystyle\sum_{k=1}^{n} (k^3 + 2k^2 - k + 4)$ (*Hint:* Use Exercise 10.)

Exer. 37–38: (a) Evaluate the given formula for the stated values of n, and solve the resulting system of equations for a, b, c, and d. (This method can sometimes be used to obtain formulas for sums.) (b) Compare the result in part (a) with the indicated exercise, and explain why this method does not prove that the formula is true for every n.

37 $1^2 + 2^2 + 3^2 + \cdots + n^2 = an^3 + bn^2 + cn$;
     $n = 1, 2, 3$ (Exercise 9)

38 $1^3 + 2^3 + 3^3 + \cdots + n^3 = an^4 + bn^3 + cn^2 + dn$;
     $n = 1, 2, 3, 4$ (Exercise 10)

Exer. 39–40: (a) Use the method of Exercises 37–38 to find a formula for the sum. (b) Verify that the formula found in part (a) is true for every n.

39 $1^4 + 2^4 + 3^4 + \cdots + n^4$

40 $2^3 + 4^3 + 6^3 + \cdots + (2n)^3$

7.5   THE BINOMIAL THEOREM

A **binomial** is a sum $a + b$, where a and b represent numbers. If n is a positive integer, then a general formula for *expanding* $(a + b)^n$ (that is, for expressing it as a sum) is given by the **binomial theorem**. In this section we shall use mathematical induction to establish this general formula. The following special cases can be obtained by multiplication:

$$(a + b)^2 = a^2 + 2ab + b^2$$

$$(a + b)^3 = a^3 + 3a^2b + 3ab^2 + b^3$$

$$(a + b)^4 = a^4 + 4a^3b + 6a^2b^2 + 4ab^3 + b^4$$

$$(a + b)^5 = a^5 + 5a^4b + 10a^3b^2 + 10a^2b^3 + 5ab^4 + b^5$$

These expansions of $(a + b)^n$ for $n = 2, 3, 4,$ and 5 have the following properties:

(1) There are $n + 1$ terms, the first being a^n and the last b^n.

(2) As we proceed from any term to the next, the power of a decreases by 1 and the power of b increases by 1. For each term, the sum of the exponents of a and b is n.

(3) Each term has the form $(c)a^{n-k}b^k$, where the coefficient c is a real number and $k = 0, 1, 2, \ldots, n$.

(4) The following formula is true for each of the first n terms of the expansion:

$$\frac{(\text{coefficient of term}) \cdot (\text{exponent of } a)}{\text{number of term}} = \text{coefficient of next term}$$

The following table illustrates property (4) for the expansion of $(a + b)^5$.

Term	Number of term	Coefficient of term	Exponent of a	Coefficient of next term
a^5	1	1	5	$\frac{1 \cdot 5}{1} = 5$
$5a^4b$	2	5	4	$\frac{5 \cdot 4}{2} = 10$
$10a^3b^2$	3	10	3	$\frac{10 \cdot 3}{3} = 10$
$10a^2b^3$	4	10	2	$\frac{10 \cdot 2}{4} = 5$
$5ab^4$	5	5	1	$\frac{5 \cdot 1}{5} = 1$

Let us next consider $(a + b)^n$ for an arbitrary positive integer n. The first term is a^n, which has coefficient 1. If we assume that property (4) is true, we obtain the successive coefficients listed in the next table.

Term	Number of term	Coefficient of term	Exponent of a	Coefficient of next term
a^n	1	1	n	$\frac{1 \cdot n}{1} = n$
$\frac{n}{1} a^{n-1}b$	2	$\frac{n}{1}$	$n - 1$	$\frac{n(n-1)}{2 \cdot 1}$
$\frac{n(n-1)}{2 \cdot 1} a^{n-2}b^2$	3	$\frac{n(n-1)}{2 \cdot 1}$	$n - 2$	$\frac{n(n-1)(n-2)}{3 \cdot 2 \cdot 1}$
$\frac{n(n-1)(n-2)}{3 \cdot 2 \cdot 1} a^{n-3}b^3$	4	$\frac{n(n-1)(n-2)}{3 \cdot 2 \cdot 1}$	$n - 3$	$\frac{n(n-1)(n-2)(n-3)}{4 \cdot 3 \cdot 2 \cdot 1}$

The pattern that appears in the fifth column leads to the following formula for the coefficient of the general term.

Coefficient of the $(k + 1)$st Term in the Expansion of $(a + b)^n$

$$\frac{n \cdot (n - 1) \cdot (n - 2) \cdot (n - 3) \cdot \cdots \cdot (n - k + 1)}{k \cdot (k - 1) \cdot \cdots \cdot 3 \cdot 2 \cdot 1}, \quad k = 1, 2, \ldots, n$$

The $(k + 1)$st coefficient can be written in a compact form by using **factorial notation**. If n is any nonnegative integer, the symbol $n!$ (*n factorial*) is defined as follows.

Definition of $n!$	**(1)** $n! = n(n - 1)(n - 2) \cdot \cdots \cdot 1$ if $n > 0$ **(2)** $0! = 1$

Thus, if $n > 0$, then $n!$ is the product of the first n positive integers. The definition $0! = 1$ is used so that certain formulas involving factorials are true for all *nonnegative* integers.

ILLUSTRATION

n Factorial

- $1! = 1$
- $2! = 2 \cdot 1 = 2$
- $3! = 3 \cdot 2 \cdot 1 = 6$
- $4! = 4 \cdot 3 \cdot 2 \cdot 1 = 24$

- $5! = 5 \cdot 4 \cdot 3 \cdot 2 \cdot 1 = 120$
- $6! = 6 \cdot 5 \cdot 4 \cdot 3 \cdot 2 \cdot 1 = 720$
- $7! = 7 \cdot 6 \cdot 5 \cdot 4 \cdot 3 \cdot 2 \cdot 1 = 5040$
- $8! = 8 \cdot 7 \cdot 6 \cdot 5 \cdot 4 \cdot 3 \cdot 2 \cdot 1 = 40{,}320$

Notice the rapid growth of $n!$ as n increases. We can show with the aid of a calculator that

$$20! \approx 2.4 \times 10^{18} \quad \text{and} \quad 50! \approx 3.0 \times 10^{64}.$$

We sometimes wish to simplify quotients where both the numerator and the denominator contain factorials, as shown in the next illustration.

ILLUSTRATION

Simplifying Quotients of Factorials

- $\dfrac{7!}{5!} = \dfrac{7 \cdot 6 \cdot 5!}{5!} = 7 \cdot 6 = 42$

- $\dfrac{10!}{6!} = \dfrac{10 \cdot 9 \cdot 8 \cdot 7 \cdot 6!}{6!} = 10 \cdot 9 \cdot 8 \cdot 7 = 5040$

As in the preceding illustration, if n and k are positive integers and $k < n$, then

$$\frac{n!}{(n - k)!} = \frac{n \cdot (n - 1) \cdot (n - 2) \cdot \cdots \cdot (n - k + 1) \cdot [(n - k)!]}{(n - k)!}$$

$$= n \cdot (n - 1) \cdot (n - 2) \cdot \cdots \cdot (n - k + 1),$$

which is the numerator of the coefficient of the $(k + 1)$st term of $(a + b)^n$. Dividing by the denominator $k!$ gives us the following alternative form for the $(k + 1)$st coefficient:

$$\frac{n \cdot (n - 1) \cdot (n - 2) \cdot \cdots \cdot (n - k + 1)}{k!} = \frac{n!}{k!(n - k)!}$$

These numbers are called **binomial coefficients** and are often denoted by either the symbol $\binom{n}{k}$ or the symbol $C(n, k)$. Thus, we have the following.

Coefficient of the $(k + 1)$st Term in the Expansion of $(a + b)^n$ (Alternative Form)

$$\binom{n}{k} = C(n, k) = \frac{n!}{k!(n - k)!}, \quad k = 0, 1, 2, \ldots, n$$

The symbols $\binom{n}{k}$ and $C(n, k)$ are sometimes read "*n choose k*."

EXAMPLE 1   *Evaluating* $\binom{n}{k}$

Find $\binom{5}{0}$, $\binom{5}{1}$, $\binom{5}{2}$, $\binom{5}{3}$, $\binom{5}{4}$, and $\binom{5}{5}$.

SOLUTION   These six numbers are the coefficients in the expansion of $(a + b)^5$, which we tabulated earlier in this section. By definition,

$$\binom{5}{0} = \frac{5!}{0!(5 - 0)!} = \frac{5!}{1 \cdot 5!} = 1$$

$$\binom{5}{1} = \frac{5!}{1!(5 - 1)!} = \frac{5!}{1 \cdot 4!} = \frac{5 \cdot 4!}{4!} = 5$$

$$\binom{5}{2} = \frac{5!}{2!(5 - 2)!} = \frac{5!}{2!3!} = \frac{5 \cdot 4 \cdot 3!}{2 \cdot 3!} = \frac{20}{2} = 10$$

$$\binom{5}{3} = \frac{5!}{3!(5 - 3)!} = \frac{5!}{3!2!} = \frac{5 \cdot 4 \cdot 3!}{3! \cdot 2} = \frac{20}{2} = 10$$

$$\binom{5}{4} = \frac{5!}{4!(5 - 4)!} = \frac{5!}{4!1!} = \frac{5 \cdot 4!}{4!} = 5$$

$$\binom{5}{5} = \frac{5!}{5!(5 - 5)!} = \frac{5!}{5!0!} = \frac{1}{0!} = \frac{1}{1} = 1.$$

The binomial theorem may be stated as follows.

The Binomial Theorem

$$(a + b)^n = a^n + \binom{n}{1}a^{n-1}b + \binom{n}{2}a^{n-2}b^2 + \cdots + \binom{n}{k}a^{n-k}b^k$$

$$+ \cdots + \binom{n}{n-1}ab^{n-1} + b^n$$

Using summation notation, we may write the binomial theorem

$$(a + b)^n = \sum_{k=0}^{n} \binom{n}{k}a^{n-k}b^k.$$

An alternative statement of the binomial theorem is as follows. (A proof is given at the end of this section.)

**The Binomial Theorem
(Alternative Form)**

$$(a + b)^n = a^n + na^{n-1}b + \frac{n(n-1)}{2!}a^{n-2}b^2 + \cdots$$

$$+ \frac{n(n-1)(n-2)\cdots(n-k+1)}{k!}a^{n-k}b^k$$

$$+ \cdots + nab^{n-1} + b^n$$

The following examples may be solved either by using the general formulas for the binomial theorem or by repeated use of property (4) stated at the beginning of this section.

EXAMPLE 2   Finding a binomial expansion
Find the binomial expansion of $(2x + 3y^2)^4$.

SOLUTION   We use the binomial theorem with $a = 2x$, $b = 3y^2$, and $n = 4$:

$$(2x + 3y^2)^4 = (2x)^4 + \binom{4}{1}(2x)^3(3y^2) + \binom{4}{2}(2x)^2(3y^2)^2$$

$$+ \binom{4}{3}(2x)(3y^2)^3 + (3y^2)^4$$

This expression can be written

$$(2x + 3y^2)^4 = 16x^4 + 4(8x^3)(3y^2) + 6(4x^2)(9y^4) + 4(2x)(27y^6) + 81y^8$$

$$= 16x^4 + 96x^3y^2 + 216x^2y^4 + 216xy^6 + 81y^8.$$

The next example illustrates that if either a or b is negative, then the terms of the expansion are alternately positive and negative.

EXAMPLE 3 *Finding a binomial expansion*

Expand $\left(\dfrac{1}{x} - 2\sqrt{x}\right)^5$.

SOLUTION The binomial coefficients for $(a + b)^5$ were calculated in Example 1. Thus, if we let $a = 1/x$, $b = -2\sqrt{x}$, and $n = 5$ in the binomial theorem, we obtain

$$\left(\frac{1}{x} - 2\sqrt{x}\right)^5 = \left(\frac{1}{x}\right)^5 + 5\left(\frac{1}{x}\right)^4(-2\sqrt{x}) + 10\left(\frac{1}{x}\right)^3(-2\sqrt{x})^2$$
$$+ 10\left(\frac{1}{x}\right)^2(-2\sqrt{x})^3 + 5\left(\frac{1}{x}\right)(-2\sqrt{x})^4 + (-2\sqrt{x})^5,$$

which can be written as

$$\left(\frac{1}{x} - 2\sqrt{x}\right)^5 = \frac{1}{x^5} - \frac{10}{x^{7/2}} + \frac{40}{x^2} - \frac{80}{x^{1/2}} + 80x - 32x^{5/2}.$$

To find a specific term in the expansion of $(a + b)^n$, it is convenient to first find the exponent k that is to be assigned to b. Notice that, by the binomial theorem, *the exponent of b is always one less than the number of the term.* Once k is found, we know that the exponent of a is $n - k$ and the coefficient is $\dbinom{n}{k}$.

EXAMPLE 4 *Finding a specific term of a binomial expansion*

Find the fifth term in the expansion of $(x^3 + \sqrt{y})^{13}$.

SOLUTION Let $a = x^3$ and $b = \sqrt{y}$. The exponent of b in the fifth term is 4, and hence the exponent of a is 9. From the discussion of the preceding paragraph we obtain

$$\binom{13}{4}(x^3)^9(\sqrt{y})^4 = \frac{13!}{4!(13 - 4)!}x^{27}y^2 = \frac{13 \cdot 12 \cdot 11 \cdot 10}{4!}x^{27}y^2$$
$$= 715x^{27}y^2.$$

EXAMPLE 5 *Finding a specific term of a binomial expansion*

Find the term involving q^{10} in the binomial expansion of $(\tfrac{1}{3}p + q^2)^{12}$.

SOLUTION From the statement of the binomial theorem with $a = \tfrac{1}{3}p$, $b = q^2$, and $n = 12$, each term in the expansion has the form

$$\binom{n}{k}a^{n-k}b^k = \binom{12}{k}(\tfrac{1}{3}p)^{12-k}(q^2)^k.$$

(continued)

Since $(q^2)^k = q^{2k}$, we must let $k = 5$ to obtain the term involving q^{10}. Doing so gives us

$$\binom{12}{5}(\tfrac{1}{3}p)^{12-5}(q^2)^5 = \frac{12!}{5!(12-5)!}\,(\tfrac{1}{3})^7 p^7 q^{10} = \frac{88}{243}\,p^7 q^{10}.$$

There is an interesting triangular array of numbers called **Pascal's triangle**, which can be used to obtain binomial coefficients. The numbers are arranged as follows.

$$
\begin{array}{ccccccccccccc}
&&&&&& 1 &&&&&& \\
&&&&& 1 && 1 &&&&& \\
&&&& 1 && 2 && 1 &&&& \\
&&& 1 && 3 && 3 && 1 &&& \\
&& 1 && 4 && 6 && 4 && 1 && \\
& 1 && 5 && 10 && 10 && 5 && 1 & \\
1 && 6 && 15 && 20 && 15 && 6 && 1 \\
\end{array}
$$

The numbers in the second row are the coefficients in the expansion of $(a + b)^1$; those in the third row are the coefficients determined by $(a + b)^2$; those in the fourth row are obtained from $(a + b)^3$; and so on. Each number in the array that is different from 1 can be found by adding the two numbers in the previous row that appear above and immediately to the left and right of the number, as illustrated in the solution of the next example.

EXAMPLE 6   *Using Pascal's triangle*

Find the eighth row of Pascal's triangle, and use it to expand $(a + b)^7$.

SOLUTION   Let us rewrite the seventh row and then use the process described above. In the following display the arrows indicate which two numbers in row seven are added to obtain the numbers in row eight.

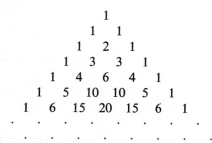

The eighth row gives us the coefficients in the expansion of $(a + b)^7$:

$$(a + b)^7 = a^7 + 7a^6b + 21a^5b^2 + 35a^4b^3$$
$$+ 35a^3b^4 + 21a^2b^5 + 7ab^6 + b^7$$

Pascal's triangle is useful for expanding small powers of $a + b$; however, for expanding large powers or finding a specific term, as in Examples 4 and 5, the general formula given by the binomial theorem is more useful.

We shall conclude this section by giving a proof of the binomial theorem using mathematical induction.

PROOF OF THE BINOMIAL THEOREM   For each positive integer n, let P_n denote the statement given in the alternative form of the binomial theorem.

Step 1 If $n = 1$, the statement reduces to $(a + b)^1 = a^1 + b^1$. Consequently, P_1 is true.

Step 2 Assume that P_k is true. Thus, the induction hypothesis is

$$(a + b)^k = a^k + ka^{k-1}b + \frac{k(k-1)}{2!} a^{k-2}b^2 + \cdots$$

$$+ \frac{k(k-1)(k-2)\cdots(k-r+2)}{(r-1)!} a^{k-r+1}b^{r-1}$$

$$+ \frac{k(k-1)(k-2)\cdots(k-r+1)}{r!} a^{k-r}b^r$$

$$+ \cdots + kab^{k-1} + b^k.$$

We have shown both the rth and the $(r + 1)$st term in the expansion.

To prove that P_{k+1} is true, we first write

$$(a + b)^{k+1} = (a + b)^k(a + b).$$

Using the induction hypothesis to substitute for $(a + b)^k$ and then multiplying that expression by $a + b$, we obtain

$$(a + b)^{k+1} = \left[a^{k+1} + ka^k b + \frac{k(k-1)}{2!} a^{k-1}b^2 + \cdots \right.$$

$$\left. + \frac{k(k-1)\cdots(k-r+1)}{r!} a^{k-r+1}b^r + \cdots + ab^k \right]$$

$$+ \left[a^k b + ka^{k-1}b^2 + \cdots + \frac{k(k-1)\cdots(k-r+2)}{(r-1)!} a^{k-r+1}b^r \right.$$

$$\left. + \cdots + kab^k + b^{k+1} \right],$$

where the terms in the first pair of brackets result from multiplying the right side of the induction hypothesis by a and the terms in the second pair of brackets result from multiplying by b. We next rearrange and combine terms:

$$(a + b)^{k+1} = a^{k+1} + (k + 1)a^k b + \left[\frac{k(k-1)}{2!} + k \right]a^{k-1}b^2 + \cdots$$

$$+ \left[\frac{k(k-1)\cdots(k-r+1)}{r!} \right.$$

$$\left. + \frac{k(k-1)\cdots(k-r+2)}{(r-1)!} \right]a^{k-r+1}b^r$$

$$+ \cdots + (1 + k)ab^k + b^{k+1}$$

If the coefficients are simplified, we obtain statement P_n with $k + 1$ substituted for n. Thus, P_{k+1} is true, and therefore P_n holds for every positive integer n, which completes the proof. ■

7.5 EXERCISES

Exer. 1–12: Evaluate the expression.

1 $2!6!$ 2 $3!4!$

3 $7!0!$ 4 $5!0!$

5 $\dfrac{8!}{5!}$ 6 $\dfrac{6!}{3!}$

7 $\dbinom{5}{5}$ 8 $\dbinom{7}{0}$

9 $\dbinom{7}{5}$ 10 $\dbinom{8}{4}$

11 $\dbinom{13}{4}$ 12 $\dbinom{52}{2}$

Exer. 13–14: Rewrite as an expression that does not contain factorials.

13 $\dfrac{(2n + 2)!}{(2n)!}$ 14 $\dfrac{(3n + 1)!}{(3n - 1)!}$

Exer. 15–28: Use the binomial theorem to expand and simplify.

15 $(4x - y)^3$ 16 $(x^2 + 2y)^3$

17 $(a + b)^6$ 18 $(a + b)^4$

19 $(a - b)^7$ 20 $(a - b)^5$

21 $(3x - 5y)^4$ 22 $(2t - s)^5$

23 $(\tfrac{1}{3}x + y^2)^5$ 24 $(\tfrac{1}{2}c + d^3)^4$

25 $\left(\dfrac{1}{x^2} + 3x\right)^6$ 26 $\left(\dfrac{1}{x^3} - 2x\right)^5$

27 $\left(\sqrt{x} - \dfrac{1}{\sqrt{x}}\right)^5$ 28 $\left(\sqrt{x} + \dfrac{1}{\sqrt{x}}\right)^5$

Exer. 29–44: Without expanding completely, find the indicated term(s) in the expansion of the expression.

29 $(3c^{2/5} + c^{4/5})^{25}$; first three terms

30 $(x^3 + 5x^{-2})^{20}$; first three terms

31 $(4b^{-1} - 3b)^{15}$;   last three terms

32 $(s - 2t^3)^{12}$;   last three terms

33 $\left(\dfrac{3}{c} + \dfrac{c^2}{4}\right)^7$;   sixth term

34 $(3a^2 - \sqrt{b})^9$;   fifth term

35 $(\tfrac{1}{3}u + 4v)^8$;   seventh term

36 $(3x^2 - y^3)^{10}$;   fourth term

37 $(x^{1/2} + y^{1/2})^8$;   middle term

38 $(rs^2 + t)^7$;   two middle terms

39 $(2y + x^2)^8$;   term that contains x^{10}

40 $(x^2 - 2y^3)^5$;   term that contains y^6

41 $(3b^3 - 2a^2)^4$;   term that contains b^9

42 $(\sqrt{c} + \sqrt{d})^8$;   term that contains c^2

43 $\left(3x - \dfrac{1}{4x}\right)^6$;   term that does not contain x

44 $(xy - 3y^{-3})^8$;   term that does not contain y

45 Approximate $(1.2)^{10}$ by using the first three terms in the expansion of $(1 + 0.2)^{10}$, and compare your answer with that obtained using a calculator.

46 Approximate $(0.9)^4$ by using the first three terms in the expansion of $(1 - 0.1)^4$, and compare your answer with that obtained using a calculator.

Exer. 47–48: Simplify the expression using the binomial theorem.

47 $\dfrac{(x + h)^4 - x^4}{h}$ 48 $\dfrac{(x + h)^5 - x^5}{h}$

49 Show that $\dbinom{n}{1} = \dbinom{n}{n - 1}$ for $n \geq 1$.

50 Show that $\dbinom{n}{0} = \dbinom{n}{n}$ for $n \geq 0$.

7.6 PERMUTATIONS

FIGURE 3

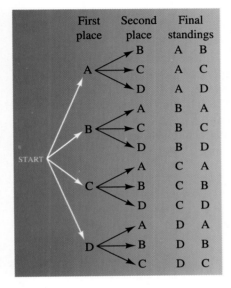

Suppose that four teams are involved in a tournament in which first, second, third, and fourth places will be determined. For identification purposes, we label the teams A, B, C, and D. Let us find the number of different ways that first and second place can be decided. It is convenient to use a **tree diagram** as in Figure 3. After the word START, the four possibilities for first place are listed. From each of these an arrow points to a possible second-place finisher. The final standings list the possible outcomes, from left to right. They are found by following the different paths (*branches* of the tree) that lead from the word START to the second-place team. The total number of outcomes is 12, which is the product of the number of choices (4) for first place and the number of choices (3) for second place (after first has been determined).

Let us now find the total number of ways that first, second, third, and fourth positions can be filled. To sketch a tree diagram, we may begin by drawing arrows from the word START to each possible first-place finisher A, B, C, or D. Next we draw arrows from those to possible second-place finishers, as was done in Figure 3. From each second-place position we then draw arrows indicating the possible third-place positions. Finally, we draw arrows to the fourth-place team. If we consider only the case in which team A finishes in first place, we have the diagram shown in Figure 4.

FIGURE 4

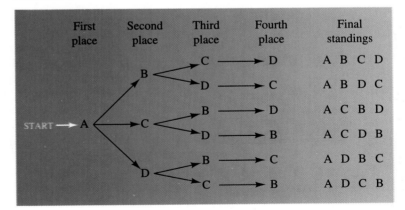

Note that there are six possible final standings in which team A occupies first place. In a complete tree diagram there would also be three other branches of this type corresponding to first-place finishes for B, C, and D. A complete diagram would display the following 24 possibilities for the final standings:

$$ABCD, ABDC, ACBD, ACDB, ADBC, ADCB,$$
$$BACD, BADC, BCAD, BCDA, BDAC, BDCA,$$
$$CABD, CADB, CBAD, CBDA, CDAB, CDBA,$$
$$DABC, DACB, DBAC, DBCA, DCAB, DCBA.$$

Note that the number of possibilities (24) is the product of the number of ways (4) that first place may occur, the number of ways (3) that second place may occur (after first place has been determined), the number of possible outcomes (2) for third place (after the first two places have been decided), and the number of ways (1) that fourth place can occur (after the first three places have been taken).

The preceding discussion illustrates the following general rule, which we accept as a basic axiom of counting.

Fundamental Counting Principle

> Let $E_1, E_2, \ldots, E_k$ be a sequence of k events. If, for each i, the event E_i can occur in m_i ways, then the total number of ways all the events may take place is the product $m_1 m_2 \cdots m_k$.

Returning to our first illustration, we let E_1 represent the determination of the first-place team, so that $m_1 = 4$. If E_2 denotes the determination of the second-place team, then $m_2 = 3$. Hence, the number of outcomes for the sequence E_1, E_2 is $4 \cdot 3 = 12$, which is the same as that found by means of the tree diagram. If we proceed to E_3, the determination of the third-place team, then $m_3 = 2$, and hence $m_1 m_2 m_3 = 24$. Finally, if E_1, E_2, and E_3 have occurred, there is only one possible outcome for E_4. Thus, $m_4 = 1$ and $m_1 m_2 m_3 m_4 = 24$.

Instead of teams, let us now regard a, b, c, and d merely as symbols and consider the various *orderings*, or *arrangements*, that may be assigned to these symbols, taking them either two at a time, three at a time, or four at a time. By abstracting in this way we may apply our methods to other similar situations. The arrangements we have discussed are **arrangements without repetitions**, since a symbol may not be used more than once in an arrangement. In Example 1 we shall consider arrangements in which repetitions *are* allowed.

Previously we defined ordered pairs and ordered triples. Similarly, an **ordered 4-tuple** is a set containing four elements x_1, x_2, x_3, x_4 in which an ordering has been specified, so that one of the elements may be referred to as the *first element*, another as the *second element*, and so on. The symbol (x_1, x_2, x_3, x_4) is used for the ordered 4-tuple having first element x_1, second element x_2, third element x_3, and fourth element x_4. In general, for any positive integer r, we speak of the **ordered r-tuple**

$$(x_1, x_2, \ldots, x_r)$$

as a set of elements in which x_1 is designated as the first element, x_2 as the second element, and so on.

EXAMPLE 1   *Determining the number of r-tuples*

Using only the letters $a, b, c,$ and d, determine how many of the following can be obtained:

(a) ordered triples   **(b)** ordered 4-tuples   **(c)** ordered r-tuples

SOLUTION

(a) We must determine the number of symbols of the form (x_1, x_2, x_3) that can be obtained using only the letters $a, b, c,$ and d. This is not the same as listing first, second, and third place as in our previous illustration, since we have not ruled out the possibility of repetitions. For example, (a, b, a), (a, a, b), and (a, a, a) are different ordered triples. If, for $i = 1, 2, 3$, we let E_i represent the determination of x_i in the ordered triple (x_1, x_2, x_3), then, since repetitions are allowed, there are four possibilities— a, b, c and d—for each of E_1, E_2, and E_3. Hence, by the fundamental counting principle, the total number of ordered triples is $4 \cdot 4 \cdot 4$, or 64.

(b) The number of possible ordered 4-tuples (x_1, x_2, x_3, x_4) is $4 \cdot 4 \cdot 4 \cdot 4$, or 256.

(c) The number of ordered r-tuples is the product $4 \cdot 4 \cdot 4 \cdots \cdot 4$, with 4 appearing as a factor r times. That product equals 4^r.

EXAMPLE 2   *Choosing class officers*

A class consists of 60 girls and 40 boys. In how many ways can a president, vice-president, treasurer, and secretary be chosen if the treasurer must be a girl, the secretary must be a boy, and a student may not hold more than one office?

SOLUTION   If an event is specialized in some way (for example, the treasurer *must* be a girl), then that event should be performed before any nonspecialized events. Thus, we let E_1 represent the choice of treasurer and E_2 the choice of secretary. Next we let E_3 and E_4 denote the choices for president and vice-president, respectively. As in the fundamental counting principle, we let m_i denote the number of different ways E_i can occur for $i = 1, 2, 3,$ and 4. It follows that $m_1 = 60$, $m_2 = 40$, $m_3 = 98$, and $m_4 = 97$. By the fundamental counting principle, the total number of possibilities is

$$60 \cdot 40 \cdot 98 \cdot 97 = 22{,}814{,}400.$$

When working with sets, we are usually not concerned about the order or arrangement of the elements. In the remainder of this section, however, the arrangement of the elements will be our main concern.

Definition of Permutation

Let S be a set of n elements and let $1 \leq r \leq n$. A **permutation** of r elements of S is an arrangement, without repetitions, of r elements.

We also use the phrase **permutation of n elements taken r at a time**. The symbol $P(n, r)$ will denote the number of different permutations of r elements that can be obtained from a set containing n elements. As a special case, $P(n, n)$ denotes the number of arrangements of n elements of S—that is, the number of ways of arranging *all* the elements of S.

In our first discussion involving the four teams A, B, C, and D, we had $P(4, 2) = 12$, since there were 12 different ways of arranging the four teams in groups of two. We also showed that the number of ways to arrange all the elements A, B, C, and D is 24. In permutation notation we would write this result as $P(4, 4) = 24$.

The next theorem gives us a general formula for $P(n, r)$.

Theorem on Permutations

Let S be a set of n elements and let $1 \leq r \leq n$. The number of different permutations of r elements of S is

$$P(n, r) = n(n - 1)(n - 2) \cdots (n - r + 1).$$

PROOF   The problem of determining $P(n, r)$ is equivalent to determining the number of different r-tuples $(x_1, x_2, \ldots, x_r)$ such that each x_i is an element of S and no element of S appears twice in the same r-tuple. We may find this number by means of the fundamental counting principle. For each $i = 1, 2, \ldots, r$, let E_i represent the determination of the element x_i and let m_i be the number of different ways of choosing x_i. We wish to apply the sequence $E_1, E_2, \ldots, E_r$. We have n possible choices for x_1, and consequently $m_1 = n$. Since repetitions are not allowed, we have $n - 1$ choices for x_2, so $m_2 = n - 1$. Continuing in this manner, we successively obtain $m_3 = n - 2$, $m_4 = n - 3$, and ultimately $m_r = n - (r - 1)$ or, equivalently, $m_r = n - r + 1$. Hence, using the fundamental counting principle, we obtain the formula for $P(n, r)$. ∎

Note that *the formula for $P(n, r)$ in the previous theorem contains exactly r factors on the right-hand side,* as shown in the following illustration.

ILLUSTRATION

Number of Different Permutations

■ $P(n, 1) = n$ ■ $P(n, 2) = n(n - 1)$

■ $P(n, 3) = n(n - 1)(n - 2)$ ■ $P(n, 4) = n(n - 1)(n - 2)(n - 3)$

EXAMPLE 3 *Evaluating $P(n, r)$*

Find $P(5, 2)$, $P(6, 4)$, and $P(5, 5)$.

SOLUTION We use the formula for $P(n, r)$ in the preceding theorem:

$$5 - 2 + 1 = 4, \text{ so } P(5, 2) = 5 \cdot 4 = 20$$

$$6 - 4 + 1 = 3, \text{ so } P(6, 4) = 6 \cdot 5 \cdot 4 \cdot 3 = 360$$

$$5 - 5 + 1 = 1, \text{ so } P(5, 5) = 5 \cdot 4 \cdot 3 \cdot 2 \cdot 1 = 120$$

EXAMPLE 4 *Arranging the batting order for a baseball team*

A baseball team consists of nine players. Find the number of ways of arranging the first four positions in the batting order if the pitcher is excluded.

SOLUTION We wish to find the number of permutations of 8 objects taken 4 at a time. Using the formula for $P(n, r)$ with $n = 8$ and $r = 4$, we have $n - r + 1 = 5$ and it follows that

$$P(8, 4) = 8 \cdot 7 \cdot 6 \cdot 5 = 1680.$$

The next result gives us a form for $P(n, r)$ that involves the factorial symbol.

Factorial Form for $P(n, r)$

If n is a positive integer and $1 \leq r \leq n$, then

$$P(n, r) = \frac{n!}{(n - r)!}.$$

PROOF If we let $r = n$ in the formula for $P(n, r)$ in the theorem on permutations, we obtain the number of different arrangements of *all* the elements of a set consisting of n elements. In this case,

$$n - r + 1 = n - n + 1 = 1$$

and $$P(n, n) = n(n - 1)(n - 2) \cdots 3 \cdot 2 \cdot 1 = n!.$$

Consequently, $P(n, n)$ is the product of the first n positive integers. This result is also given by the factorial form, for if $r = n$, then

$$P(n, n) = \frac{n!}{(n - n)!} = \frac{n!}{0!} = \frac{n!}{1} = n!.$$

If $1 \leq r < n$, then

$$\frac{n!}{(n - r)!} = \frac{n(n - 1)(n - 2) \cdots (n - r + 1) \cdot [(n - r)!]}{(n - r)!}$$

$$= n(n - 1)(n - 2) \cdots (n - r + 1).$$

This agrees with the formula for $P(n, r)$ in the theorem on permutations.

EXAMPLE 5   *Evaluating $P(n, r)$ using factorials*

Use the factorial form for $P(n, r)$ to find $P(5, 2)$, $P(6, 4)$, and $P(5, 5)$.

SOLUTION

$$P(5, 2) = \frac{5!}{(5 - 2)!} = \frac{5!}{3!} = \frac{5 \cdot 4 \cdot 3!}{3!} = 5 \cdot 4 = 20$$

$$P(6, 4) = \frac{6!}{(6 - 4)!} = \frac{6!}{2!} = \frac{6 \cdot 5 \cdot 4 \cdot 3 \cdot 2!}{2!} = 6 \cdot 5 \cdot 4 \cdot 3 = 360$$

$$P(5, 5) = \frac{5!}{(5 - 5)!} = \frac{5!}{0!} = \frac{5!}{1} = 5 \cdot 4 \cdot 3 \cdot 2 \cdot 1 = 120$$

7.6 EXERCISES

Exer. 1–8: Find the number.

1 $P(7, 3)$

2 $P(8, 5)$

3 $P(9, 6)$

4 $P(5, 3)$

5 $P(5, 5)$

6 $P(4, 4)$

7 $P(6, 1)$

8 $P(5, 1)$

9 How many three-digit numbers can be formed from the digits 1, 2, 3, 4, and 5 if repetitions

   (a) are not allowed?   (b) are allowed?

10 Work Exercise 9 for four-digit numbers.

11 How many numbers can be formed from the digits 1, 2, 3, and 4 if repetitions are not allowed? (*Note:* 42 and 231 are examples of such numbers.)

12 Determine the number of positive integers less than 10,000 that can be formed from the digits 1, 2, 3, and 4 if repetitions are allowed.

13 *Basketball standings*   If eight basketball teams are in a tournament, find the number of different ways that first, second, and third place can be decided assuming ties are not allowed.

14 *Basketball standings*   Work Exercise 13 for 12 teams.

15 *Wardrobe mix'n'match*   A girl has four skirts and six blouses. How many different skirt-blouse combinations can she wear?

16 *Wardrobe mix'n'match*   Refer to Exercise 15. If the girl also has three sweaters, how many different skirt-blouse-sweater combinations can she wear?

17 *License plate numbers* In a certain state, automobile license plates start with one letter of the alphabet, followed by five digits (0, 1, 2, . . . , 9). Find how many different license plates are possible if

(a) the first digit following the letter cannot be 0

(b) the first letter cannot be O or I and the first digit cannot be 0

18 *Tossing dice* Two dice are tossed, one after the other. In how many different ways can they fall? List the number of different ways the sum of the dots can equal

(a) 3   (b) 5   (c) 7   (d) 9   (e) 11

19 *Seating arrangement* A row of six seats in a classroom is to be filled by selecting individuals from a group of ten students.

(a) In how many different ways can the seats be occupied?

(b) If there are six boys and four girls in the group and if boys and girls are to be alternated, find the number of different seating arrangements.

20 *Scheduling courses* A student in a certain college may take mathematics at 8, 10, 11, or 2 o'clock; English at 9, 10, 1, or 2; and history at 8, 11, 2, or 3. Find the number of different ways in which the student can schedule the three courses.

21 *True-or-false test* In how many different ways can a test consisting of ten true-or-false questions be completed?

22 *Multiple-choice test* A test consists of six multiple-choice questions, and there are five choices for each question. In how many different ways can the test be completed?

23 *Seating arrangement* In how many different ways can eight people be seated in a row?

24 *Book arrangement* In how many different ways can ten books be arranged on a shelf?

25 *Semaphore* With six different flags, how many different signals can be sent by placing three flags, one above the other, on a flag pole?

26 *Selecting books* In how many different ways can five books be selected from a 12-volume set of books?

27 *Radio call letters* How many four-letter radio station call letters can be formed if the first letter must be K or W and repetitions

(a) are not allowed?   (b) are allowed?

28 *Fraternity designations* There are 24 letters in the Greek alphabet. How many fraternities may be specified by choosing three Greek letters if repetitions

(a) are not allowed?

(b) are allowed?

29 *Phone numbers* How many seven-digit phone numbers can be formed from the digits 0, 1, 2, 3, . . . , 9 if the first digit may not be 0?

30 *Baseball batting order* After selecting nine players for a baseball game, the manager of the team arranges the batting order so that the pitcher bats last and the best hitter bats third. In how many different ways can the remainder of the batting order be arranged?

31 *ATM access code* A customer remembers that 2, 4, 7, and 9 are the digits of a four-digit access code for an automatic bank-teller machine. Unfortunately, the customer has forgotten the order of the digits. Find the largest possible number of trials necessary to obtain the correct code.

32 *ATM access code* Work Exercise 31 if the digits are 2, 4, and 7 and one of these digits is repeated in the four-digit code.

33 *Selecting theater seats* Three married couples have purchased tickets for a play. Spouses are to be seated next to each other, and the six seats are in a row. In how many ways can the six people be seated?

34 *Horserace results* Ten horses are entered in a race. If the possibility of a tie for any place is ignored, in how many ways can the first-, second-, and third-place winners be determined?

35 The commutative and associative laws of addition guarantee that the sum of integers 1 through 10 is independent of the order in which the numbers are added. In how many different ways can these integers be summed?

36 *Shuffling cards*

(a) In how many ways can a standard deck of 52 cards be shuffled?

(b) In how many ways can the cards be shuffled so that the four aces appear on the top of the deck?

37 *Numerical palindromes* A palindrome is an integer, such as 45654, that reads the same backward and forward.

(a) How many five-digit palindromes are there?

(b) How many n-digit palindromes are there?

38 *Color arrangements*   Each of the six squares shown in the figure is to be filled with any one of ten possible colors. How many ways are there of coloring the strip shown in the figure so that no two adjacent squares have the same color?

EXERCISE 38

c   **39**   This exercise requires a graphing utility that can graph $x!$.

(a) Graph $y = \dfrac{x! e^x}{x^x \sqrt{2\pi x}}$ on $(0, 20]$, and estimate the horizontal asymptote.

(b) Use the graph in part (a) to find an approximation for $n!$ if n is a large positive integer.

40 (a) What happens if a calculator is used to find $P(150, 50)$? Explain.

(b) Approximate r if $P(150, 50) = 10^r$ by using the following formula from advanced mathematics:

$$\log n! \approx \frac{n \ln n - n}{\ln 10}$$

7.7   DISTINGUISHABLE PERMUTATIONS AND COMBINATIONS

Certain problems involve finding different arrangements of objects, some of which are indistinguishable. For example, suppose we are given five disks of the same size, of which three are black, one is white, and one is red. Let us find the number of ways they can be arranged in a row so that different color arrangements are obtained. If the disks were all different colors, then the number of arrangements would be 5!, or 120. However, since some of the disks have the same appearance, we cannot obtain 120 different arrangements. To clarify this point, let us write

B   B   B   W   R

for the arrangement having black disks in the first three positions in the row, the white disk in the fourth position, and the red disk in the fifth position. The first three disks can be arranged in 3!, or 6, different ways, but these arrangements cannot be distinguished from one another because the first three disks look alike. We say that those 3! permutations are **non-distinguishable**. Similarly, given any other arrangement, say

B   R   B   W   B,

there are 3! different ways of arranging the three black disks, but again each such arrangement is nondistinguishable from the others. Let us call two arrangements of objects **distinguishable permutations** if one arrangement cannot be obtained from the other by rearranging like objects. Thus, B B B W R and B R B W B are distinguishable permutations of the five disks. Let k denote the number of distinguishable permutations. Since to each such arrangement there correspond 3! *nondistinguishable* permutations, we must have $3!k = 5!$, the number of permutations of five *different* objects. Hence, $k = 5!/3! = 5 \cdot 4 = 20$. By the same type of reasoning we can obtain the following extension of this discussion.

Theorem

> If r objects in a collection of n objects are alike and if the remaining objects are different from each other and from the r objects, then the number of distinguishable permutations of the n objects is
>
> $$\frac{n!}{r!}$$

We can generalize this theorem to the case in which there are several subcollections of nondistinguishable objects. For example, consider eight disks, of which four are black, three are white, and one is red. In this case, with each arrangement, such as

$$\text{B \quad W \quad B \quad W \quad B \quad W \quad B \quad R,}$$

there are 4! arrangements of the black disks and 3! arrangements of the white disks that have no effect on the color arrangement. Hence, 4!3! possible arrangements of the disks will not produce distinguishable permutations. If we let k denote the number of *distinguishable* permutations, then $4!3!k = 8!$, since 8! is the number of permutations we would obtain if the disks were all different. Thus, the number of distinguishable permutations is

$$k = \frac{8!}{4!3!} = \frac{8 \cdot 7 \cdot 6 \cdot 5}{3!} \cdot \frac{4!}{4!} = 280.$$

The following general result can be proved.

Theorem on Distinguishable Permutations

> If, in a collection of n objects, n_1 are alike, n_2 are alike of another kind, . . . , n_k are alike of a further kind, and
>
> $$n = n_1 + n_2 + \cdots + n_k,$$
>
> then the number of distinguishable permutations of the n objects is
>
> $$\frac{n!}{n_1!n_2! \cdots n_k!}.$$

EXAMPLE 1   *Finding the number of distinguishable permutations*

Find the number of distinguishable permutations of the letters in the word *Mississippi*.

SOLUTION   In this example we are given a collection of eleven objects in which four are of one kind (the letter *s*), four are of another kind (*i*), two are of a third kind (*p*), and one is of a fourth kind (*M*). Hence, by the preceding theorem, the number of distinguishable permutations is

$$\frac{11!}{4!4!2!1!} = 34{,}650.$$

When we work with permutations our concern is with the orderings or arrangements of elements. Let us now ignore the order or arrangement of elements and consider the following question: Given a set containing *n* distinct elements, in how many ways can a subset of *r* elements be chosen with $r \le n$? Before answering, let us state a definition.

Definition of Combination

Let S be a set of n elements and let $1 \le r \le n$. A **combination** of r elements of S is a subset of S that contains r distinct elements.

If S contains n elements, we also use the phrase **combination of *n* elements taken *r* at a time**. The symbol $C(n, r)$ will denote the number of combinations of r elements that can be obtained from a set of n elements.

Theorem on Combinations

The number of combinations of r elements that can be obtained from a set of n elements is

$$C(n, r) = \frac{n!}{(n - r)!r!}, \quad 1 \le r \le n.$$

PROOF   If S contains n elements, then, to find $C(n, r)$, we must find the total number of subsets of the form

$$\{x_1, x_2, \ldots, x_r\}$$

such that the x_i are *different* elements of S. Since the elements $x_1, x_2, \ldots, x_r$ can be arranged in $r!$ different ways, each such subset produces $r!$ different r-tuples. Thus, the total number of different r-tuples is $r!C(n, r)$. However, in the previous section we found that the total number of r-tuples is

$$P(n, r) = \frac{n!}{(n - r)!}.$$

Hence, $$r!C(n, r) = \frac{n!}{(n - r)!}$$

Dividing both sides of the last equation by $r!$ gives us the formula for $C(n, r)$. ∎

The formula for $C(n, r)$ is identical to the formula for the binomial coefficient $\binom{n}{r}$ in Section 7.5.

EXAMPLE 2   *Choosing a baseball squad*

A little league baseball squad has six outfielders, seven infielders, five pitchers, and two catchers. Each outfielder can play any of the three outfield positions, and each infielder can play any of the four infield positions. In how many ways can a team of nine players be chosen?

SOLUTION   The number of ways of choosing three outfielders from the six candidates is

$$C(6, 3) = \frac{6!}{(6-3)!3!} = \frac{6!}{3!3!} = \frac{6 \cdot 5 \cdot 4 \cdot 3!}{3 \cdot 2 \cdot 1 \cdot 3!} = 20.$$

The number of ways of choosing the four infielders is

$$C(7, 4) = \frac{7!}{(7-4)!4!} = \frac{7 \cdot 6 \cdot 5 \cdot 4!}{3!4!} = \frac{7 \cdot 6 \cdot 5}{3 \cdot 2 \cdot 1} = 35.$$

There are five ways of choosing a pitcher and two choices for the catcher. It follows from the fundamental counting principle that the total number of ways to choose a team is

$$20 \cdot 35 \cdot 5 \cdot 2 = 7000.$$

Note that if $r = n$, the formula for $C(n, r)$ becomes

$$C(n, n) = \frac{n!}{(n-n)!n!} = \frac{n!}{0!n!} = 1.$$

It is convenient to assign a meaning to $C(n, r)$ if $r = 0$. If the formula is to be true in this case, then we must have

$$C(n, 0) = \frac{n!}{n!0!} = 1.$$

Hence, we *define* $C(n, 0) = 1$, which is the same as $C(n, n)$. Finally, for consistency, we also *define* $C(0, 0) = 1$. Thus, $C(n, r)$ has meaning for all nonnegative integers n and r with $r \le n$.

EXAMPLE 3   *Finding the number of subsets of a set*

Let S be a set of n elements. Find the number of distinct subsets of S.

SOLUTION   Let r be any nonnegative integer such that $r \leq n$. From our previous work, the number of subsets of S that consist of r elements is $C(n, r)$, or $\binom{n}{r}$. Hence, to find the total number of subsets, we find the sum

$$\binom{n}{0} + \binom{n}{1} + \binom{n}{2} + \binom{n}{3} + \cdots + \binom{n}{n}.$$

This is precisely the binomial expansion of $(1 + 1)^n$. Thus, there are 2^n subsets of a set of n elements. In particular, a set of 3 elements has 2^3, or 8, different subsets. A set of 4 elements has 2^4, or 16, subsets. A set of 10 elements has 2^{10}, or 1024, subsets.

7.7 EXERCISES

Exer. 1–8: Find the number.

1 $C(7, 3)$ 2 $C(8, 4)$

3 $C(9, 8)$ 4 $C(6, 2)$

5 $C(n, n - 1)$ 6 $C(n, 1)$

7 $C(7, 0)$ 8 $C(5, 5)$

Exer. 9–10: Find the number of possible color arrangements for the 12 given disks, arranged in a row.

9 5 black, 3 red, 2 white, 2 green

10 3 black, 3 red, 3 white, 3 green

11 Find the number of distinguishable permutations of the letters in the word *bookkeeper*.

12 Find the number of distinguishable permutations of the letters in the word *moon*. List all the permutations.

13 **Choosing basketball teams** Ten people wish to play in a basketball game. In how many different ways can two teams of five players each be formed?

14 **Selecting test questions** A student may answer any six of ten questions on an examination.

   (a) In how many ways can six questions be selected?

   (b) How many selections are possible if the first two questions must be answered?

Exer. 15–16: Consider any eight points such that no three are collinear.

15 How many lines are determined?

16 How many triangles are determined?

17 **Book arrangement** A student has five mathematics books, four history books, and eight fiction books. In how many different ways can they be arranged on a shelf if books in the same category are kept next to one another?

18 **Selecting a basketball team** A basketball squad consists of twelve players.

   (a) Disregarding positions, in how many ways can a team of five be selected?

   (b) If the center of a team must be selected from two specific individuals on the squad and the other four members of the team from the remaining ten players, find the number of different teams possible.

19 **Selecting a football team** A football squad consists of three centers, ten linemen who can play either guard or tackle, three quarterbacks, six halfbacks, four ends, and four fullbacks. A team must have one center, two guards, two tackles, two ends, two halfbacks, a quarterback, and a fullback. In how many different ways can a team be selected from the squad?

20 **Arranging keys on a ring** In how many different ways can seven keys be arranged on a key ring if the keys can slide completely around the ring?

21 **Committee selection** A committee of 3 men and 2 women is to be chosen from a group of 12 men and 8 women. Determine the number of different ways of selecting the committee.

22 *Birth order* Let the letters G and B denote a girl birth and a boy birth, respectively. For a family of three boys and three girls, one possible birth order is G G G B B B. How many birth orders are possible for these six children?

Exer. 23–24: Shown in the figure is a street map and a possible path from point *A* to point *B*. How many other possible paths are there from *A* to *B* if moves are restricted to the right or up? (*Hint:* If R denotes a move one unit right and U denotes a move one unit up, then the path in Exercise 23 can be specified by R U U R R R U R.)

23

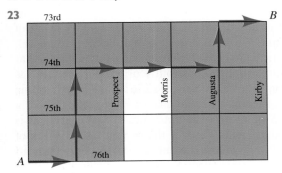

24

25 *Lotto selections* To win a state lottery game, a player must correctly select six numbers from the numbers 1 through 49.

(a) Find the total number of selections possible.

(b) Work part (a) if a player selects only even numbers.

26 *Office assignments* A mathematics department has ten faculty members but only nine offices, so one office must be shared by two individuals. In how many different ways can the offices be assigned?

27 *Tennis tournament* In a round-robin tennis tournament, every player meets every other player exactly once. How many players can participate in a tournament of 45 matches?

28 *True-or-false test* A true-or-false test has 20 questions.

(a) In how many different ways can the test be completed?

(b) In how many different ways can a student answer 10 questions correctly?

29 *Basketball championship series* The winner of the seven-game NBA championship series is the team that wins four games. In how many different ways can the series be extended to seven games?

30 A geometric design is determined by joining every pair of vertices of an octagon (see the figure).

(a) How many triangles in the design have their three vertices on the octagon?

(b) How many quadrilaterals in the design have their four vertices on the octagon?

EXERCISE 30

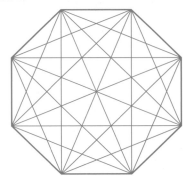

31 *Ice cream selections* An ice cream parlor stocks 31 different flavors and advertises that it serves almost 4500 different triple scoop cones, with each scoop being a different flavor. How was this number obtained?

32 *Choices of hamburger condiments* A fast food restaurant advertises that it offers any combination of 8 condiments on a hamburger, thus giving a customer 256 choices. How was this number obtained?

Exer. 33–34: (a) Calculate the sum S_n for $n = 1, 2, 3, \ldots, 10$, where if $n < r$, then $\binom{n}{r} = 0$. (b) Predict a general formula for S_n.

33 $\binom{n}{1} + \binom{n}{3} + \binom{n}{5} + \binom{n}{7} + \cdots$

34 $(1)\binom{n}{1} - (2)\binom{n}{2} + (3)\binom{n}{3} - (4)\binom{n}{4} + (5)\binom{n}{5} - \cdots$

7.8 PROBABILITY

If two dice are tossed, what are the chances of rolling a 7? If a person is dealt five cards from a standard deck of 52 playing cards, what is the likelihood of obtaining three aces? In the seventeenth century, similar questions about games of chance led to the study of *probability*. Since that time, the theory of probability has grown extensively. It is now used to predict outcomes of a large variety of situations that arise in the natural and social sciences.

Any chance process, such as flipping a coin, rolling a die, being dealt a card from a deck, determining if a manufactured item is defective, or finding the blood pressure of an individual, is an **experiment**. A result of an experiment is an **outcome**. We shall restrict our discussion to experiments for which outcomes are **equally likely**. This means, for example, that if a coin is flipped, we assume that the possibility of obtaining a head is the same as that of obtaining a tail. Similarly, if a die is tossed, we assume that the die is *fair*—that is, there is an equal chance of obtaining either a 1, 2, 3, 4, 5, or 6. The set S of all possible outcomes of an experiment is the **sample space** of the experiment. Thus, if the experiment consists of flipping a coin and we let H or T denote the outcome of obtaining a head or tail, respectively, then the sample space S may be denoted by

$$S = \{H, T\}.$$

If a fair die is tossed as an experiment, then the set S of all possible outcomes (the sample space) is

$$S = \{1, 2, 3, 4, 5, 6\}.$$

The following definition expresses, in mathematical terms, the notion of obtaining *particular* outcomes of an experiment.

Definition of Event

Let S be the sample space of an experiment. An **event** associated with the experiment is any subset E of S.

Let us consider the experiment of tossing a single die, so that the sample space is $S = \{1, 2, 3, 4, 5, 6\}$. If $E = \{4\}$, then the event E associated with the experiment consists of the outcome of obtaining a 4 on the toss. Different events may be associated with the same experiment. For example, if we let $E = \{1, 3, 5\}$, then this event consists of obtaining an odd number on a toss of the die.

As another illustration, suppose the experiment consists of flipping two coins, one after the other. If we let HH denote the outcome in which two heads appear, HT that of a head on the first coin and a tail on the second, and so on, then the sample space S of the experiment may be denoted by

$$S = \{HH, HT, TH, TT\}.$$

If we let $E = \{HT, TH\},$

then the event E consists of the appearance of a head on one of the coins and a tail on the other.

Next we shall define what is meant by the *probability* of an event. Throughout our discussion we will assume that the sample space S of an experiment contains only a finite number of elements. If E is an event, *the symbols $n(E)$ and $n(S)$ will denote the number of elements in E and S, respectively.*

Definition of the Probability of an Event

Let S be the sample space of an experiment and E an event. The **probability** $P(E)$ of E is given by

$$P(E) = \frac{n(E)}{n(S)}.$$

Since E is a subset of S, we see that

$$0 \leq n(E) \leq n(S).$$

Dividing by $n(S)$, we obtain

$$0 \leq P(E) \leq 1.$$

Note that $P(E) = 1$ if $E = S$, and $P(E) = 0$ if E contains no elements.

The next example provides several illustrations of the preceding definition if E contains exactly one element.

EXAMPLE I *Finding the probability of an event*

(a) If a coin is flipped, find the probability that a head will turn up.

(b) If a fair die is tossed, find the probability of obtaining a 4.

(c) If two coins are flipped, find the probability that both coins turn up heads.

SOLUTION For each experiment we shall list sets S and E and then use the definition of probability to find $P(E)$.

(a) $S = \{H, T\}, \quad E = \{H\}, \quad P(E) = \frac{n(E)}{n(S)} = \frac{1}{2}$

(b) $S = \{1, 2, 3, 4, 5, 6\}, \quad E = \{4\}, \quad P(E) = \frac{n(E)}{n(S)} = \frac{1}{6}$

(c) $S = \{HH, HT, TH, TT\}, \quad E = \{HH\}, \quad P(E) = \frac{n(E)}{n(S)} = \frac{1}{4}$

In part (a) of Example 1 we found that the probability of obtaining a head on a flip of a coin is $\frac{1}{2}$. We take this to mean that if a coin is flipped many times, the number of times that a head turns up should be approximately one-half the total number of flips. Thus, for 100 flips, a head should turn up approximately 50 times. It is unlikely that this number will be *exactly* 50. A probability of $\frac{1}{2}$ implies that if we let the number of flips increase, then the number of times a head turns up *approaches* one-half the total number of flips. Similar remarks can be made for parts (b) and (c) of Example 1.

In the next two examples we consider experiments in which an event contains more than one element.

EXAMPLE 2   *Finding probabilities when two dice are tossed*

If two dice are tossed, what is the probability of rolling a sum of

(a) 7?   **(b)** 9?

SOLUTION   Let us refer to one die as *the first die* and the other as *the second die*. We shall use ordered pairs to represent outcomes as follows: (2, 4) denotes the outcome of obtaining a 2 on the first die and a 4 on the second; (5, 3) represents a 5 on the first die and a 3 on the second; and so on. Since there are six different possibilities for the first number of the ordered pair and, with each of these, six possibilities for the second number, the total number of ordered pairs is 36. Hence, if S is the sample space, then $n(S) = 36$.

(a) The event E corresponding to rolling a sum of 7 is given by

$$E = \{(1, 6), (2, 5), (3, 4), (4, 3), (5, 2), (6, 1)\},$$

and consequently $P(E) = \dfrac{n(E)}{n(S)} = \dfrac{6}{36} = \dfrac{1}{6}.$

(b) If E is the event corresponding to rolling a sum of 9, then

$$E = \{(3, 6), (4, 5), (5, 4), (6, 3)\}$$

and $P(E) = \dfrac{n(E)}{n(S)} = \dfrac{4}{36} = \dfrac{1}{9}.$

In the next example (and in the exercises), when it is stated that one or more cards are drawn from a deck, we mean that each card is removed from a standard 52-card deck and is *not* replaced before the next card is drawn.

EXAMPLE 3   *Finding the probability of drawing a certain hand of cards*

Suppose five cards are drawn from a standard deck of 52 playing cards. Find the probability that all five cards are hearts.

SOLUTION The sample space S of the experiment is the set of all possible five-card hands that can be formed from the 52 cards in the deck. It follows from our work in the preceding section that $n(S) = C(52, 5)$.

Since there are 13 cards in the heart suit, the number of different ways of obtaining a hand that contains five hearts is $C(13, 5)$. Hence, if E represents this event, then

$$P(E) = \frac{n(E)}{n(S)} = \frac{C(13, 5)}{C(52, 5)} = \frac{\dfrac{13!}{5!8!}}{\dfrac{52!}{5!47!}}.$$

We may show that

$$P(E) = \frac{1287}{2,598,960} \approx 0.0005.$$

Thus,

$$P(E) \approx \frac{1}{2000}.$$

This result implies that if the experiment is performed many times, a five-card heart hand should be drawn approximately once every 2000 times.

Suppose S is the sample space of an experiment and E_1 and E_2 are two events associated with the experiment. If E_1 and E_2 have no elements in common, they are called *disjoint sets*, and we write $E_1 \cap E_2 = \varnothing$ (the *empty set*). In this case, if one event occurs, the other cannot occur; they are **mutually exclusive events**. Thus, if $E = E_1 \cup E_2$, then

$$n(E) = n(E_1 \cup E_2) = n(E_1) + n(E_2).$$

Hence,

$$P(E) = \frac{n(E_1) + n(E_2)}{n(S)} = \frac{n(E_1)}{n(S)} + \frac{n(E_2)}{n(S)},$$

or

$$P(E) = P(E_1) + P(E_2).$$

The probability of E is therefore the sum of the probabilities of E_1 and E_2. We have proved the following.

Theorem on Mutually Exclusive Events

If E_1 and E_2 are mutually exclusive events and $E = E_1 \cup E_2$, then

$$P(E) = P(E_1 \cup E_2) = P(E_1) + P(E_2).$$

The preceding theorem can be extended to any number of events E_1, $E_2, \ldots, E_k$ that are mutually exclusive in the sense that if $i \neq j$, then

$E_i \cap E_j = \varnothing$. The conclusion of the theorem is then

$$P(E) = P(E_1 \cup E_2 \cup \cdots \cup E_k) = P(E_1) + P(E_2) + \cdots + P(E_k).$$

EXAMPLE 4   *Finding probabilities when two dice are tossed*

If two dice are tossed, find the probability of rolling a sum of either 7 or 9.

SOLUTION   Let E_1 denote the event of rolling 7 and E_2 that of rolling 9. Since E_1 and E_2 cannot occur simultaneously, they are mutually exclusive events. We wish to find the probability of the event $E = E_1 \cup E_2$. From Example 2 we know that $P(E_1) = \frac{1}{6}$ and $P(E_2) = \frac{1}{9}$. Hence, by the last theorem,

$$P(E) = P(E_1) + P(E_2)$$
$$= \frac{1}{6} + \frac{1}{9} = \frac{5}{18} = 0.2\overline{7}.$$

If E_1 and E_2 are events that possibly have elements in common, then the following can be proved.

Theorem on the Probability of the Occurrence of Either of Two Events

> If E_1 and E_2 are any two events, then
> $$P(E_1 \cup E_2) = P(E_1) + P(E_2) - P(E_1 \cap E_2).$$

Note that if E_1 and E_2 are mutually exclusive, then $E_1 \cap E_2 = \varnothing$ and $P(E_1 \cap E_2) = 0$. Hence, the last theorem includes, as a special case, the theorem on mutually exclusive events.

EXAMPLE 5   *Finding the probability of selecting a certain card from a deck*

If a single card is selected from a 52-card deck, find the probability that the card is either a jack or a spade.

SOLUTION   Let E_1 denote the event that the card is a jack and E_2 the event that it is a spade. The events E_1 and E_2 are *not* mutually exclusive, since there is one card—the jack of spades—in both events, and hence $P(E_1 \cap E_2) = \frac{1}{52}$. By the preceding theorem, the probability that the card is either a jack or a spade is

$$P(E_1 \cup E_2) = P(E_1) + P(E_2) - P(E_1 \cap E_2)$$
$$= \tfrac{4}{52} + \tfrac{13}{52} - \tfrac{1}{52} = \tfrac{16}{52} = \tfrac{4}{13} \approx 0.31.$$

In solving probability problems, it is often helpful to categorize the outcomes of a sample space S into an event E and the set E' of elements of S that are not in E. We call E' the **complement** of E. Note that

$$E \cup E' = S \quad \text{and} \quad n(E) + n(E') = n(S).$$

Dividing both sides of the last equation by $n(S)$ gives us

$$\frac{n(E)}{n(S)} + \frac{n(E')}{n(S)} = 1.$$

Hence,

$$P(E) + P(E') = 1, \quad \text{or} \quad P(E) = 1 - P(E').$$

We shall use the last formula in the next example.

EXAMPLE 6 *Finding the probability of drawing a certain hand of cards*

If 13 cards are drawn from a 52-card deck, what is the probability that at least 2 of the cards are hearts?

SOLUTION If $P(k)$ denotes the probability of getting k hearts, then the probability of getting at least two hearts is

$$P(2) + P(3) + P(4) + \cdots + P(13).$$

Since the only remaining probabilities are $P(0)$ and $P(1)$, the desired probability is equal to

$$1 - [P(0) + P(1)].$$

To calculate $P(k)$ for any k, we may regard the deck as being split into two groups: hearts and non-hearts. For $P(0)$ we note that of the 13 hearts in the deck, we get none; and of the 39 non-hearts, we get 13. Since the number of ways to choose 13 cards from a 52-card deck is $C(52, 13)$, we see that

$$P(0) = \frac{n(0)}{n(S)} = \frac{C(13, 0) \cdot C(39, 13)}{C(52, 13)} \approx 0.0128.$$

The probability $P(1)$ corresponds to getting 1 of the hearts and 12 of the 39 non-hearts. Thus,

$$P(1) = \frac{n(1)}{n(S)} = \frac{C(13, 1) \cdot C(39, 12)}{C(52, 13)} \approx 0.0801.$$

Hence, the desired probability is

$$1 - [P(0) + P(1)] \approx 1 - [0.0128 + 0.0801] = 0.9071.$$

Two events E_1 and E_2 are said to be **independent events** if the occurrence of one does not influence the occurrence of the other. The following can be proved.

Theorem on Independent Events

> If E_1 and E_2 are independent events, then
>
> $$P(E_1 \cap E_2) = P(E_1) \cdot P(E_2).$$

In words, the theorem states that if E_1 and E_2 are independent events, the probability that *both* E_1 and E_2 occur simultaneously is the product of their probabilities. Note that if two events E_1 and E_2 are mutually exclusive, then $P(E_1 \cap E_2) = 0$ and they cannot be independent. (We assume that both E_1 and E_2 are not empty.)

EXAMPLE 7   *An application of probability to an electrical system*

An electrical system has open-close switches s_1, s_2, and s_3, as shown in Figure 5. The switches operate independently of one another, and current will flow from A to B either if s_1 is closed or if *both* s_2 and s_3 are closed.

(a) If S_k denotes the event that s_k is closed, where $k = 1, 2, 3$, express, in terms of $P(S_1)$, $P(S_2)$, and $P(S_3)$, the probability p that current will flow from A to B.

(b) Find p if $P(S_k) = \frac{1}{2}$ for each k.

FIGURE 5

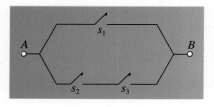

SOLUTION

(a) The probability p that either S_1 or *both* S_2 and S_3 occur is

$$p = P(S_1 \cup (S_2 \cap S_3)).$$

Using the theorem on the probability of the occurrence of either of two events S_1 or $S_2 \cap S_3$, we obtain

$$p = P(S_1) + P(S_2 \cap S_3) - P(S_1 \cap (S_2 \cap S_3)).$$

Applying the theorem on independent events twice gives us

$$p = P(S_1) + P(S_2) \cdot P(S_3) - P(S_1) \cdot P(S_2 \cap S_3).$$

Finally, using the theorem on independent events one more time, we see that

$$p = P(S_1) + P(S_2) \cdot P(S_3) - P(S_1) \cdot P(S_2) \cdot P(S_3).$$

(b) If $P(S_k) = \frac{1}{2}$ for each k, then from part (a), the probability that current will flow from A to B is

$$p = \tfrac{1}{2} + \tfrac{1}{2} \cdot \tfrac{1}{2} - \tfrac{1}{2} \cdot \tfrac{1}{2} \cdot \tfrac{1}{2} = \tfrac{5}{8} = 0.625.$$

EXAMPLE 8   *A continuation of Example 7*

Refer to Example 7. If the probability that s_k is closed is the same for each k, determine $P(S_k)$ such that $p = 0.99$.

FIGURE 6   [0.8, 1] by [−0.01, 0.01]

SOLUTION   Since the probability $P(S_k)$ is the same for each k, we let $P(S_k) = x$ for $k = 1, 2, 3$. Substituting in the formula for p obtained in part (a) of Example 7, we obtain

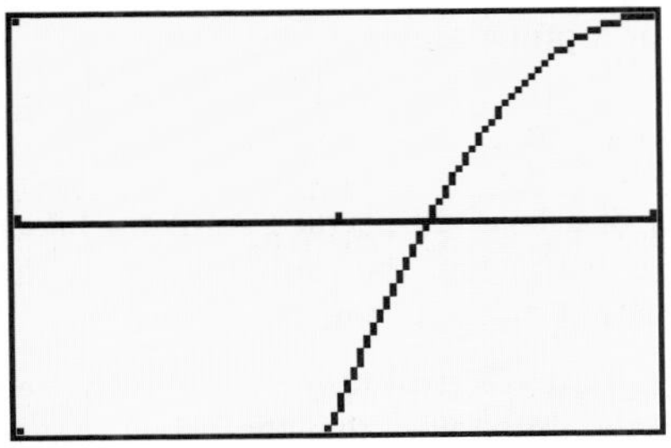

$$p = x + x \cdot x - x \cdot x \cdot x = -x^3 + x^2 + x.$$

Letting $p = 0.99$ gives us the equation

$$-x^3 + x^2 + x = 0.99.$$

Graphing $y = -x^3 + x^2 + x - 0.99$ using a standard viewing rectangle, we see that there are three x-intercepts. The desired probability must lie between $x = 0$ and $x = 1$ and should be fairly close to 1. Using the viewing rectangle dimensions [0.8, 1] by [−0.01, 0.01], we obtain a sketch similar to Figure 6. Using tracing and zoom-in features gives us $x \approx 0.93$. Hence, $P(S_k) \approx 0.93$.

Note that the probability that an individual switch is closed is *less than* the probability that current will flow through the system.

In this section we have merely introduced several basic concepts about probability. The interested person is referred to entire books and courses devoted to this branch of mathematics.

7.8 EXERCISES

1 Each suit in a 52-card deck is made up of an ace (A), nine numbered cards (2, 3, . . . , 10), and three face cards (J, Q, K). An experiment consists of drawing a single card from a 52-card deck followed by rolling a single die.

(a) Describe the sample space S of the experiment, and find $n(S)$.

(b) Let E_1 be the event consisting of the outcomes in which a numbered card is drawn and the number of dots on the die is the same as the number on the card. Find $n(E_1)$, $n(E_1')$, and $P(E_1)$.

(c) Let E_2 be the event in which the card drawn is a face card, and let E_3 be the event in which the number of dots on the die is even. Are E_2 and E_3 mutually exclusive? Are they independent? Find $P(E_2)$, $P(E_3)$, $P(E_2 \cap E_3)$, and $P(E_2 \cup E_3)$.

(d) Are E_1 and E_2 mutually exclusive? Are they independent? Find $P(E_1 \cap E_2)$ and $P(E_1 \cup E_2)$.

2 An experiment consists of selecting a letter from the alphabet and one of the digits 0, 1, . . . , 9.

(a) Describe the sample space S of the experiment, and find $n(S)$.

(b) Suppose the letters of the alphabet are assigned numbers as follows: A = 1, B = 2, . . . , Z = 26. Let E_1 be the event in which the units' digit of the number assigned to the letter of the alphabet is the same as the digit selected. Find $n(E_1)$, $n(E_1')$, and $P(E_1)$.

(c) Let E_2 be the event that the letter is one of the five vowels and E_3 the event that the digit is a prime number. Are E_2 and E_3 mutually exclusive? Are they independent? Find $P(E_2)$, $P(E_3)$, $P(E_2 \cap E_3)$, and $P(E_2 \cup E_3)$.

(d) Let E_4 be the event that the numerical value of the letter is even. Are E_2 and E_4 mutually exclusive? Are they independent? Find $P(E_2 \cap E_4)$ and $P(E_2 \cup E_4)$.

Exer. 3–4: A single card is drawn from a 52-card deck. Find the probability that the card is as specified.

3 (a) a king

(b) a king or a queen

(c) a king, a queen, or a jack

4 (a) a heart

(b) a heart or a diamond

(c) a heart, a diamond, or a club

Exer. 5–6: A single die is tossed. Find the probability that the die is as specified.

5 (a) a 4   (b) a 6   (c) a 4 or a 6

6 (a) an even number

(b) a number divisible by 5

(c) an even number or a number divisible by 5

Exer. 7–8: An urn contains five red balls, six green balls, and four white balls. If a single ball is drawn, find the probability that the ball is as specified.

7 (a) red   (b) green   (c) red or white

8 (a) white   (b) green or white   (c) not green

Exer. 9–10: Two dice are tossed. Find the probability that the sum is as specified.

9 (a) 11   (b) 8   (c) 11 or 8

10 (a) greater than 9   (b) an odd number

Exer. 11–12: Three dice are tossed. Find the probability of the specified event.

11 A sum of 5

12 A 6 turns up on exactly one die

13 If three coins are flipped, find the probability that exactly two heads turn up.

14 If four coins are flipped, find the probability of obtaining two heads and two tails.

Exer. 15–20: Suppose five cards are drawn from a 52-card deck. Find the probability of obtaining the indicated cards.

15 Four of a kind (such as four aces or four kings)

16 Three aces and two kings

17 Four diamonds and one spade

18 Five face cards

19 A flush (five cards, all of the same suit)

20 A royal flush (an ace, king, queen, jack, and 10 of the same suit)

21 If a single die is tossed, find the probability of obtaining an odd number or a prime number.

22 A single card is drawn from a 52-card deck. Find the probability that the card is either red or a face card.

23 If the probability of a baseball player's getting a hit in one time at bat is 0.326, find the probability that the player gets no hits in 4 times at bat.

24 If the probability of a basketball player's making a free throw is 0.9, find the probability that the player makes at least 1 of 2 free throws.

Exer. 25–26: The outcomes 1, 2, . . . , 6 of an experiment and their probabilities are listed in the table.

Outcome	1	2	3	4	5	6
Probability	0.25	0.10	0.15	0.20	0.25	0.05

For the indicated events, find (a) $P(E_2)$, (b) $P(E_1 \cap E_2)$, (c) $P(E_1 \cup E_2)$, and (d) $P(E_2 \cup E_3')$.

25 $E_1 = \{1, 2\};$ $E_2 = \{2, 3, 4\};$ $E_3 = \{4, 6\}$

26 $E_1 = \{1, 2, 3, 6\};$ $E_2 = \{3, 4\};$ $E_3 = \{4, 5, 6\}$

Exer. 27–28: A box contains 10 red chips, 20 blue chips, and 30 green chips. If 5 chips are drawn from the box, find the probability of drawing the indicated chips.

27 (a) all blue

(b) at least 1 green

(c) at most 1 red

28 (a) exactly 4 green

(b) at least 2 red

(c) at most 2 blue

29 *True-or-false test*   A true-or-false test consists of eight questions. If a student guesses the answer for each question, find the probability that

(a) eight answers are correct

(b) seven answers are correct and one is incorrect

(c) six answers are correct and two are incorrect

(d) at least six answers are correct

30 *Committee selection*   A 6-member committee is to be chosen by drawing names of individuals from a hat. If the hat contains the names of 8 men and 14 women, find the probability that the committee will consist of 3 men and 3 women.

Exer. 31–32: Five cards are drawn from a 52-card deck. Find the probability of the specified event.

31 Obtaining at least one ace

32 Obtaining at least one heart

33 *Tossing dice*   If two dice are tossed, find the probability that the sum is greater than 5.

34 *Tossing dice*   If three dice are tossed, find the probability that the sum is less than 16.

35 *Family make up*   Assuming that girl-boy births are equally probable, find the probability that a family with five children has

(a) all boys   (b) at least one girl

36 *Slot machine*   A standard slot machine contains three reels, and each reel contains 20 symbols. If the first reel has five bells, the middle reel four bells, and the last reel two bells, find the probability of obtaining three bells in a row.

37 *ESP experiment*   In a simple experiment designed to test ESP, four cards (jack, queen, king, and ace) are shuffled and then placed face down on a table. The subject then attempts to identify each of the four cards, giving a different name to each of the cards. If the individual is guessing, find the probability of correctly identifying

(a) all four cards   (b) exactly two of the four cards

38 *Tossing dice*   Three dice are tossed.

(a) Find the probability that all dice show the same number of dots.

(b) Find the probability that the numbers of dots on the dice are all different.

(c) Work parts (a) and (b) for *n* dice.

39 *Trick dice*   For a normal die, the sum of the dots on opposite faces is 7. Shown in the figure is a pair of trick dice in which the *same* number of dots appears on opposite faces. Find the probability of rolling a sum of

(a) 7   (b) 8

EXERCISE 39

40 *Carnival game*   In a common carnival game, three balls are rolled down an incline into slots numbered 1 through 9, as shown in the figure. Because the slots are so narrow, players have no control over where the balls collect. A prize is given if the sum of the three numbers is less than 7. Find the probability of winning a prize.

EXERCISE 40

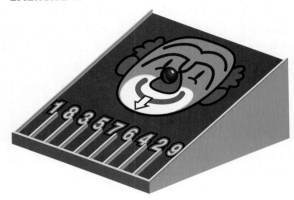

41 *Probability demonstration*   Shown in the figure is a small version of a probability demonstration device. A small ball is dropped into the top of the maze and tumbles to the bottom. Each time the ball strikes an obstacle, there is a 50% chance that the ball will move to the left. Find the probability that the ball ends up in the slot

(a) on the far left   (b) in the middle

EXERCISE 41

42 Roulette In the American version of roulette, a ball is spun around a wheel and has an equal chance of landing in any one of 38 slots numbered 0, 00, 1, 2, . . . , 36. Shown in the figure is a standard betting layout for roulette, where the color of the oval corresponds to the color of the slot on the wheel. Find the probability that the ball lands

(a) in a black slot

(b) in a black slot twice in succession

EXERCISE 42

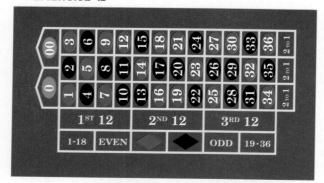

43 Selecting lottery numbers In one version of a popular lottery game, a player selects six of the numbers from 1 to 54. The agency in charge of the lottery also selects six numbers. What is the probability that the player will match the six numbers if two 50-cent tickets are purchased? (This jackpot is worth at least $2 million in prize money and grows according to the number of tickets sold.)

44 Lottery Refer to Exercise 43. The player can win about $1000 for matching five of the six numbers and about $40 for matching four of the six numbers. Find the probability that the player will win some amount of prize money on the purchase of one ticket.

45 Playing craps In the game of *craps*, there are two ways a player can win a *pass line* bet. The player wins immediately if two dice are rolled and their sum is 7 or 11. If their sum is 4, 5, 6, 8, 9, or 10, the player can still win a pass line bet if this same number (called the *point*) is rolled again before a 7 is rolled. Find the probability that the player wins

(a) a pass line bet on the first roll

(b) a pass line bet with a 4 on the first roll

(c) on any pass line bet

46 Crapless craps Refer to Exercise 45. In the game of craps, a player loses a pass line bet if a sum of 2, 3, or 12 is obtained on the first roll (referred to as "craps"). In another version of the game, called *crapless craps*, the player does not lose by rolling craps and does not win by rolling an 11 on the first roll. Instead, the player wins if the first roll is a 7 or if the point (2–12, excluding 7) is repeated before a 7 is rolled. Find the probability that the player wins on a pass line bet in crapless craps.

c **Exer. 47–48: Refer to Examples 7 and 8. (a) Find p for the electrical system shown in the figure if $P(S_k) = 0.9$ for each k. (b) Use a graph to estimate $P(S_k)$ if p = 0.99.**

47

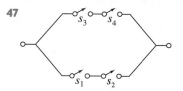

48

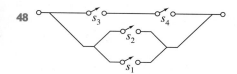

49 Birthday probability

(a) Show that the probability p that n people all have different birthdays is given by

$$p = \frac{365!}{365^n(365 - n)!}.$$

(b) If a room contains 32 people, approximate the probability that two or more people have the *same* birthday. (First approximate $\ln p$ by using the following formula from advanced mathematics:

$$\ln n! \approx n \ln n - n.)$$

c 50 Birthday probability Refer to Exercise 49. Find the smallest number of people in a room such that the probability that everyone has a different birthday is less than $\frac{1}{2}$.

CHAPTER 7 REVIEW EXERCISES

Exer. 1–4: Find the first four terms and the seventh term of the sequence that has the given nth term.

1 $\left\{\dfrac{5n}{3 - 2n^2}\right\}$

2 $\{(-1)^{n+1} - (0.1)^n\}$

3 $\{1 + (-\tfrac{1}{2})^{n-1}\}$

4 $\left\{\dfrac{2^n}{(n+1)(n+2)(n+3)}\right\}$

Exer. 5–8: Find the first five terms of the recursively defined infinite sequence.

5 $a_1 = 10, \quad a_{k+1} = 1 + (1/a_k)$

6 $a_1 = 2, \quad a_{k+1} = a_k!$

7 $a_1 = 9, \quad a_{k+1} = \sqrt{a_k}$

8 $a_1 = 1, \quad a_{k+1} = (1 + a_k)^{-1}$

Exer. 9–12: Evaluate.

9 $\displaystyle\sum_{k=1}^{5} (k^2 + 4)$

10 $\displaystyle\sum_{k=2}^{6} \dfrac{2k - 8}{k - 1}$

11 $\displaystyle\sum_{k=7}^{100} 10$

12 $\displaystyle\sum_{k=1}^{4} (2^k - 10)$

Exer. 13–24: Express the sum in terms of summation notation. (Answers are not unique.)

13 $3 + 6 + 9 + 12 + 15$

14 $4 + 2 + 1 + \tfrac{1}{2} + \tfrac{1}{4} + \tfrac{1}{8}$

15 $\dfrac{1}{1 \cdot 2} + \dfrac{1}{2 \cdot 3} + \dfrac{1}{3 \cdot 4} + \cdots + \dfrac{1}{99 \cdot 100}$

16 $\dfrac{1}{1 \cdot 2 \cdot 3} + \dfrac{1}{2 \cdot 3 \cdot 4} + \dfrac{1}{3 \cdot 4 \cdot 5} + \cdots + \dfrac{1}{98 \cdot 99 \cdot 100}$

17 $\tfrac{1}{2} + \tfrac{2}{5} + \tfrac{3}{8} + \tfrac{4}{11}$

18 $\tfrac{1}{4} + \tfrac{2}{9} + \tfrac{3}{14} + \tfrac{4}{19}$

19 $100 - 95 + 90 - 85 + 80$

20 $1 - \tfrac{1}{2} + \tfrac{1}{3} - \tfrac{1}{4} + \tfrac{1}{5} - \tfrac{1}{6} + \tfrac{1}{7}$

21 $a_0 + a_1 x^4 + a_2 x^8 + \cdots + a_{25} x^{100}$

22 $a_0 + a_1 x^3 + a_2 x^6 + \cdots + a_{20} x^{60}$

23 $1 - \dfrac{x^2}{2} + \dfrac{x^4}{4} - \dfrac{x^6}{6} + \cdots + (-1)^n \dfrac{x^{2n}}{2n}$

24 $1 + x + \dfrac{x^2}{2} + \dfrac{x^3}{3} + \cdots + \dfrac{x^n}{n}$

25 Find the tenth term and the sum of the first ten terms of the arithmetic sequence whose first two terms are $4 + \sqrt{3}$ and 3.

26 Find the sum of the first eight terms of the arithmetic sequence in which the fourth term is 9 and the common difference is -5.

27 The fifth and thirteenth terms of an arithmetic sequence are 5 and 77, respectively. Find the first term and the tenth term.

28 Insert four arithmetic means between 20 and -10.

29 Find the tenth term of the geometric sequence whose first two terms are $\tfrac{1}{8}$ and $\tfrac{1}{4}$.

30 If a geometric sequence has 3 and -0.3 as its third and fourth terms, find the eighth term.

31 Find the geometric mean of 4 and 8.

32 In a certain geometric sequence, the eighth term is 100 and the common ratio is $-\tfrac{3}{2}$. Find the first term.

33 Given an arithmetic sequence such that $S_{12} = 402$ and $a_{12} = 50$, find a_1 and d.

34 Given a geometric sequence such that $a_5 = \tfrac{1}{16}$ and $r = \tfrac{3}{2}$, find a_1 and S_5.

Exer. 35–38: Evaluate.

35 $\displaystyle\sum_{k=1}^{15} (5k - 2)$

36 $\displaystyle\sum_{k=1}^{10} (6 - \tfrac{1}{2}k)$

37 $\displaystyle\sum_{k=1}^{10} (2^k - \tfrac{1}{2})$

38 $\displaystyle\sum_{k=1}^{8} (\tfrac{1}{2} - 2^k)$

39 Find the sum of the infinite geometric series
$$1 - \tfrac{2}{5} + \tfrac{4}{25} - \tfrac{8}{125} + \cdots.$$

40 Find the rational number whose decimal representation is $6.\overline{274}$.

Exer. 41–45: Prove that the statement is true for every positive integer n.

41 $2 + 5 + 8 + \cdots + (3n - 1) = \dfrac{n(3n + 1)}{2}$

42 $2^2 + 4^2 + 6^2 + \cdots + (2n)^2 = \dfrac{2n(2n + 1)(n + 1)}{3}$

43 $\dfrac{1}{1 \cdot 3} + \dfrac{1}{3 \cdot 5} + \dfrac{1}{5 \cdot 7} + \cdots + \dfrac{1}{(2n - 1)(2n + 1)} = \dfrac{n}{2n + 1}$

44 $1 \cdot 2 + 2 \cdot 3 + 3 \cdot 4 + \cdots + n(n + 1) = \dfrac{n(n + 1)(n + 2)}{3}$

45 3 is a factor of $n^3 + 2n$.

46 Prove that $n^2 + 3 < 2^n$ for every positive integer $n \geq 5$.

Exer. 47–48: Find the smallest positive integer j for which the statement is true. Use the extended principle of mathematical induction to prove that the formula is true for every integer greater than j.

47 $2^n \leq n!$

48 $10^n \leq n^n$

Exer. 49–50: Use the binomial theorem to expand and simplify the expression.

49 $(x^2 - 3y)^6$

50 $(2a + b^3)^4$

Exer. 51–54: Without expanding completely, find the indicated term(s) in the expansion of the expression.

51 $(a^{2/5} + 2a^{-3/5})^{20}$;   first three terms

52 $(b^3 - \tfrac{1}{2}c^2)^9$;   sixth term

53 $(4a^2 - b)^7$;   term that contains a^{10}

54 $(2c^3 + 5c^{-2})^{10}$;   term that does not contain c

55 *Building blocks*   Ten-foot lengths of 2×2 lumber are to be cut into five pieces to form children's building blocks; the lengths of the five blocks are to form an arithmetic sequence.

(a) Show that the difference d in lengths must be less than 1 foot.

(b) If the smallest block is to have a length of 6 inches, find the lengths of the other four pieces.

56 *Constructing a ladder*   A ladder is to be constructed with 16 rungs whose lengths decrease uniformly from 20 inches at the base to 16 inches at the top. Find the total length of material needed for the rungs.

57 Shown in the first figure is a broken-line curve obtained by taking two adjacent sides of a square, each of length s_n, decreasing the length of the side by a factor f with $0 < f < 1$, and forming two sides of a smaller square of length $s_{n+1} = f \cdot s_n$. The process is then repeated ad infinitum. If $s_1 = 1$ in the second figure, express the length of the resulting (infinite) broken-line curve in terms of f.

EXERCISE 57

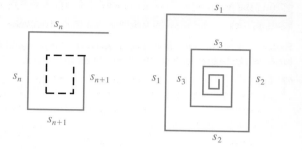

58 *Rebounding ball*   When a ball is dropped from a height of h feet, it reaches the ground in $\sqrt{h}/4$ seconds. The ball rebounds to a height of d feet in $\sqrt{d}/4$ seconds. If a rubber ball is dropped from a height of 10 feet and rebounds to three-fourths of its height after each fall, how many seconds elapse before the ball comes to rest?

59 *Selecting cards*

(a) In how many ways can 13 cards be selected from a deck of 52 cards?

(b) In how many ways can 13 cards be selected to obtain five spades, three hearts, three clubs, and two diamonds?

60 How many four-digit numbers can be formed from the digits 1, 2, 3, 4, 5, and 6 if repetitions

(a) are not allowed?   (b) are allowed?

61 *Selecting test questions*

(a) If a student must answer 8 of 12 questions on an examination, how many different selections of questions are possible?

(b) How many selections are possible if the first three questions must be answered?

62 *Color arrangements*   If six black, five red, four white, and two green disks are to be arranged in a row, what is the number of possible color arrangements?

63 *Coin toss*   Find the probability that the coins will match if

(a) two boys each toss a coin

(b) three boys each toss a coin

64 *Dealing cards*   If 4 cards are dealt from a 52-card deck, find the probability that

(a) all four cards will be the same color

(b) the cards dealt will alternate red-black-red-black

65 *Raffle probabilities* If 1000 tickets are sold for a raffle, find the probability of winning if an individual purchases

(a) 1 ticket   (b) 10 tickets   (c) 50 tickets

66 *Coin toss* If four coins are flipped, find the probability of obtaining one head and three tails.

67 *True-or-false quiz* A quiz consists of six true-or-false questions; at least four correct answers are required for a passing grade. If a student guesses at each answer, what is the probability of

(a) passing?   (b) failing?

68 *Die and card probabilities* If a single die is tossed and then a card is drawn from a 52-card deck, what is the probability of obtaining

(a) a 6 on the die and the king of hearts?

(b) a 6 on the die or the king of hearts?

69 *Population demographics* In a city of 5000 people, 1000 are over 60 years old and 2000 are female. It is known that 40% of the females are over 60. What is the probability that a randomly chosen individual from the city is either female or over 60?

70 *Backgammon moves* In the game of backgammon, players are allowed to move their counters the same number of spaces as the sum of the dots on two dice. However, if a double is rolled (that is, both dice show the same number of dots), then players may move their counters twice the sum of the dots. What is the probability that a player will be able to move his or her counters at least 10 spaces on a given roll?

APPENDIX

USING LOGARITHMIC TABLES

In this appendix we discuss how to use logarithmic tables.

If x is any positive real number and we write $x = c \cdot 10^k$ for $1 \leq c < 10$ and an integer k, then we may apply laws of logarithms, obtaining

$$\log x = \log c + \log 10^k$$

$$\log x = \log c + k.$$

From the last equation we see that to find $\log x$ for any positive real number x it is sufficient to know the logarithms of numbers between 1 and 10. The number $\log c$, for $1 \leq c < 10$, is the **mantissa**, and the integer k is the **characteristic** of $\log x$.

If $1 \leq c < 10$, then, since $\log x$ increases as x increases,

$$\log 1 \leq \log c < \log 10$$

or, equivalently, $\qquad\qquad 0 \leq \log c < 1.$

Hence, the mantissa of a logarithm is a number between 0 and 1.

In numerical problems it is usually necessary to approximate logarithms. For example, $\log 2 = 0.3010299957 \ldots$, where the decimal is nonrepeating and nonterminating. We often round off such logarithms to four decimal places and write $\log 2 \approx 0.3010$. If a number between 0 and 1 is written as a finite decimal, it is sometimes referred to as a **decimal fraction**. Thus, the equation $\log x = \log c + k$ implies that if x is any positive real number, then $\log x$ *may be approximated by the sum of a positive decimal fraction* (*the mantissa*) *and an integer* k (*the characteristic*). We shall refer to this representation as the **standard form** for $\log x$.

Common logarithms of many numbers between 1 and 10 have been calculated. Table 1 of Appendix II contains four-decimal-place approximations for logarithms of numbers between 1.00 and 9.99 at intervals of 0.01. This table can be used to find the common logarithm of any three-digit number of four-decimal-place accuracy. The use of Table 1 is illustrated in the following examples.

EXAMPLE 1

Approximate the logarithm:

(a) log 43.6   **(b)** log 43,600   **(c)** log 0.0436

SOLUTION

(a) Since $43.6 = (4.36)10^1$, the characteristic of log 43.6 is 1. Referring to Table 1, we find that the mantissa of log 4.36 may be approximated by the decimal fraction 0.6395. Hence, as in the preceding discussion,

$$\log 43.6 \approx 0.6395 + 1$$

$$\log 43.6 \approx 1.6395.$$

(b) Since $43,600 = (4.36)10^4$, the mantissa is the same as in part (a); however, the characteristic is 4. Consequently,

$$\log 43,600 \approx 0.6395 + 4$$

$$\log 43,600 \approx 4.6395.$$

(c) If we write $0.0436 = (4.36)10^{-2}$, then

$$\log 0.0436 = \log 4.36 + (-2).$$

Hence, the standard form is

$$\log 0.0436 \approx 0.6395 + (-2).$$

If we subtract 2 from 0.6395, we obtain

$$\log 0.0436 \approx -1.3605.$$

Note that this is not standard form, since $-1.3605 = -0.3605 + (-1)$, a number in which the decimal part is *negative*.

In Example 1, after obtaining $\log 0.0436 \approx 0.6395 + (-2)$, a common error is to write the answer as -2.6395. This is incorrect, since

$$-2.6395 = -0.6395 + (-2),$$

which is not the same as $0.6395 + (-2)$.

If a logarithm has a negative characteristic, we usually either leave it in standard form or rewrite the logarithm, keeping the decimal part positive. To illustrate the second technique, let us add and subtract 8 on the right side of the equation as follows:

$$\log 0.0436 \approx 0.6395 + (-2)$$

$$\log 0.0436 \approx 0.6395 + (8 - 8) + (-2)$$

$$\log 0.0436 \approx 8.6395 - 10$$

We could also write

$$\log 0.0436 \approx 18.6395 - 20$$

$$\log 0.0436 \approx 28.6395 - 30$$

$$\log 0.0436 \approx 43.6395 - 45$$

and so on, as long as the sum of the positive integer to the left of the decimal and the negative integer to the right of the decimal equals the characteristic of the logarithm.

EXAMPLE 2

Approximate the logarithm:

(a) $\log (0.00652)^2$ **(b)** $\log (0.00652)^{-2}$ **(c)** $\log (0.00652)^{1/2}$

SOLUTION

(a) By (3) of the laws of logarithms (see Section 5.4),

$$\log (0.00652)^2 = 2 \log 0.00652.$$

Since $0.00652 = (6.52)10^{-3}$,

$$\log 0.00652 = \log 6.52 + (-3).$$

Referring to Table 1, we see that $\log 6.52 \approx 0.8142$, and therefore

$$\log 0.00652 \approx 0.8142 + (-3).$$

Hence, $\log (0.00652)^2 = 2 \log 0.00652$

$$\log (0.00652)^2 \approx 2[0.8142 + (-3)]$$

$$\log (0.00652)^2 \approx 1.6284 + (-6)$$

$$\log (0.00652)^2 \approx 0.6284 + (-5).$$

The last number is the standard form for the logarithm.

(b) Again we use law (3) and the value for $\log 0.00652$ found in part (a):

$$\log (0.00652)^{-2} = -2 \log 0.00652$$

$$\log (0.00652)^{-2} \approx -2[0.8142 + (-3)]$$

$$\log (0.00652)^{-2} \approx -1.6284 + 6$$

It is important to note that -1.6284 means $-0.6284 + (-1)$, and consequently the decimal part is negative. To obtain the standard form, we may write

$$-1.6284 + 6 = 6.0000 - 1.6284$$

$$= 4.3716.$$

This shows that the mantissa is 0.3716 and the characteristic is 4.

(c) By law (3),

$$\log (0.00652)^{1/2} = \tfrac{1}{2} \log 0.00652$$

$$\log (0.00652)^{1/2} \approx \tfrac{1}{2}[0.8142 + (-3)].$$

If we multiply by $\tfrac{1}{2}$, the standard form is not obtained, since neither number in the resulting sum is the characteristic. To avoid this, we may adjust the expression within brackets by adding and subtracting a suitable number. If we use 1 in this way, we obtain

$$\log (0.00652)^{1/2} \approx \tfrac{1}{2}[1.8142 + (-4)]$$

$$\log (0.00652)^{1/2} \approx 0.9071 + (-2).$$

We could also have added and subtracted any number other than 1. For example,

$$\tfrac{1}{2}[0.8142 + (-3)] = \tfrac{1}{2}[17.8142 + (-20)]$$

$$= 8.9071 + (-10).$$

If $\log x$ is given, we can use Table 1 to find an approximation to x, as illustrated in the following example.

EXAMPLE 3

Find a decimal approximation to x:

(a) $\log x = 1.7959$ **(b)** $\log x = -3.5918$

SOLUTION

(a) The mantissa 0.7959 determines the sequence of digits in x, and the characteristic determines the position of the decimal point. Referring to the *body* of Table 1, we see that the mantissa 0.7959 is the logarithm of 6.25. Since the characteristic is 1, we know that x lies between 10 and 100. Consequently, $x \approx 62.5$.

(b) To find x from Table 1, we must express $\log x$ in standard form. To change $\log x = -3.5918$ to standard form, we may add and subtract 4, obtaining

$$\log x = (4 - 3.5918) - 4$$

$$= 0.4082 - 4.$$

Referring to Table 1, we see that the mantissa 0.4082 is the logarithm of 2.56. Since the characteristic of $\log x$ is -4, it follows that $x \approx 0.000256$.

If we use a calculator with a $\boxed{\text{LOG}}$ key to determine common logarithms, then the standard form for $\log x$ is obtained only if $x \geq 1$. For example, to find log 43.6 on a typical calculator, we enter 43.6 and press $\boxed{\text{LOG}}$, obtaining the standard form

$$1.6394865$$

If we find log 0.0436 in similar fashion, then the following number appears on the display panel:

$$-1.3605135$$

This is not the standard form for the logarithm, since the decimal part is negative (compare with Example 1(c)). To find the standard form we could add 2 to the logarithm (using a calculator) and then subtract 2 as follows:

$$\log 0.0436 \approx -1.3605135$$

$$\log 0.0436 \approx (-1.3605135 + 2) - 2$$

$$\log 0.0436 \approx 0.6394865 - 2$$

$$\log 0.0436 \approx 0.6394865 + (-2)$$

The only common logarithms that can be found *directly* from Table 1 are logarithms of numbers that contain at most three nonzero digits. If *four* nonzero digits are involved, then it is possible to obtain an approximation by using the method of linear interpolation described next. The term **linear interpolation** is used because, as we shall see, the method is based upon approximating portions of the graph of $y = \log x$ by line segments.

To illustrate the process of linear interpolation, and at the same time give some justification for it, let us consider the specific example log 12.64. Since the logarithmic function with base 10 is increasing, this number lies between log 12.60 $\approx$ 1.1004 and log 12.70 $\approx$ 1.1038. Examining the graph of $y = \log x$, we have the situation shown in Figure 1, where we have distorted the units on the x- and y-axes and also the portion of the graph shown. (A more accurate drawing would indicate that the graph of $y = \log x$ is much closer to the line segment joining $P(12.60, 1.1004)$ to $Q(12.70, 1.1038)$ than is shown in the figure.) Since log 12.64 is the y-coordinate of the point on the graph having x-coordinate 12.64, it can be approximated by the y-coordinate of the point with x-coordinate 12.64 on the *line segment PQ*. Referring to Figure 1, we see that the latter y-coordinate is $1.1004 + d$. The number d can be approximated by using similar triangles. Referring to Figure 2, where the graph of $y = \log x$ has been deleted, we may form the following proportion:

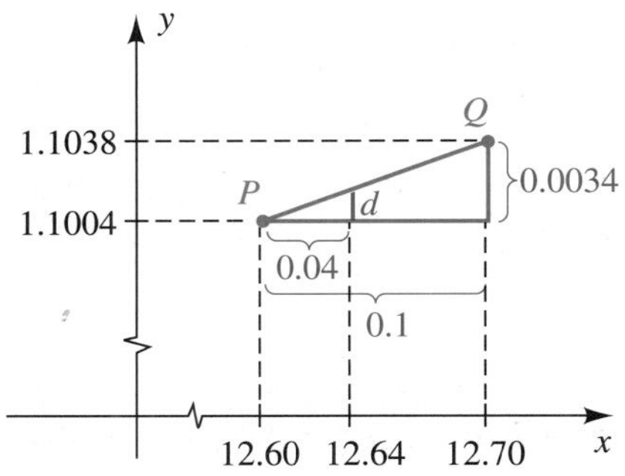

$$\frac{d}{0.0034} = \frac{0.04}{0.1}$$

Hence,

$$d = \frac{(0.04)(0.0034)}{0.1} = 0.00136.$$

FIGURE I

FIGURE 2

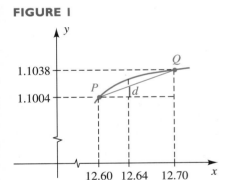

When using this technique, we always round off decimals to the same number of places as appear in the body of the table. Consequently, $d \approx 0.0014$ and

$$\log 12.64 \approx 1.1004 + 0.0014$$

$$\log 12.64 \approx 1.1018.$$

Hereafter we shall not sketch a graph when interpolating. Instead we shall use the scheme illustrated in the next example.

EXAMPLE 4

Approximate $\log 572.6$.

SOLUTION It is convenient to arrange our work as follows:

$$1.0 \left\{ 0.6 \left\{ \begin{array}{l} \log 572.0 \approx 2.7574 \\ \log 572.6 = ? \\ \log 573.0 \approx 2.7582 \end{array} \right\} d \right\} 0.0008$$

We have indicated differences next to the braces. This leads to the following:

$$\frac{d}{0.0008} = \frac{0.6}{1.0} = \frac{6}{10}$$

$$d = \tfrac{6}{10}(0.0008) = 0.00048 \approx 0.0005$$

Hence, $\log 572.6 \approx 2.7574 + 0.0005$

$$\log 572.6 \approx 2.7579.$$

Another way of working this type of problem is to reason that since 572.6 is $\frac{6}{10}$ of the way from 572.0 to 573.0, $\log 572.6$ is (approximately) $\frac{6}{10}$ of the way from 2.7574 to 2.7582. Hence,

$$\log 572.6 \approx 2.7574 + \tfrac{6}{10}(0.0008)$$

$$\log 572.6 \approx 2.7574 + 0.0005$$

$$\log 572.6 \approx 2.7579.$$

EXAMPLE 5

Approximate $\log 0.003678$.

SOLUTION We begin by arranging our work as in the solution of Example 1. Thus,

$$10 \left\{ 8 \left\{ \begin{array}{l} \log 0.003670 \approx 0.5647 + (-3) \\ \log 0.003678 = ? \\ \log 0.003680 \approx 0.5658 + (-3) \end{array} \right\} d \right\} 0.0011$$

Since we are only interested in ratios, we have used the numbers 8 and 10 on the left side because their ratio is the same as the ratio of 0.000008 to 0.000010. This leads to the following:

$$\frac{d}{0.0011} = \frac{8}{10} = 0.8$$

$$d = (0.0011)(0.8) = 0.00088 \approx 0.0009$$

Hence,

$$\log 0.003678 \approx [0.5647 + (-3)] + 0.0009$$

$$\log 0.003678 = 0.5656 + (-3).$$

If a number x is written in the form $x = c \cdot 10^k$ with $1 \leq c < 10$, then before using Table 1 to find $\log x$ by interpolation, we should round c off to three decimal places. Another way of saying this is that x should be rounded off to four **significant figures**. Some examples will help to clarify the procedure. If $x = 36.4635$, we round off to 36.46 before approximating $\log x$. The number 684,279 should be rounded off to 684,300. For a decimal such as 0.096202, we use 0.09620. The reason for doing this is that Table 1 does not guarantee more than four-digit accuracy, since the mantissas that appear in it are approximations. This means that if *more* than four-digit accuracy is required in a problem, then Table 1 cannot be used. If, in more extensive tables, the logarithm of a number containing n digits can be found directly, then interpolation is allowed for numbers involving $n + 1$ digits, and numbers should be rounded off accordingly.

The method of interpolation can also be used to find x when we are given $\log x$. If we use Table 1, then x may be found to four significant figures. In this case we are given the *y-coordinate* of a point on the graph of $y = \log x$ and are asked to find the *x-coordinate*. A geometric argument similar to the one given earlier can be used to justify the procedure illustrated in the next example.

EXAMPLE 6

Find x to four significant figures if $\log x = 1.7949$.

SOLUTION   The mantissa 0.7949 does not appear in Table 1, but it can be isolated between adjacent entries for the mantissas corresponding to 6.230 and 6.240. We shall arrange our work as follows:

$$0.1 \left\{ r \begin{cases} \log 62.30 \approx 1.7945 \\ \log x \quad\ \approx 1.7949 \end{cases} 0.0004 \\ \log 62.40 \approx 1.7952 \end{cases} 0.0007 $$

This leads to the proportion

$$\frac{r}{0.1} = \frac{0.0004}{0.0007} = \frac{4}{7}$$

$$r = (0.1)(\tfrac{4}{7}) \approx 0.06.$$

Hence, $x \approx 62.30 + 0.06$

$$x \approx 62.36.$$

A.I EXERCISES

Exer. 1–16: Use Table 1 and laws of logarithms to approximate the common logarithms of the numbers.

1 347;   0.00347;   3.47   2 86.2;   8620;   0.862

3 0.54;   540;   540,000   4 208;   2.08;   20,800

5 60.2;   0.0000602;   602   6 5;   0.5;   0.0005

7 $(44.9)^2$; $(44.9)^{1/2}$; $(44.9)^{-2}$

8 $(1810)^4$; $(1810)^{40}$; $(1810)^{1/4}$

9 $(0.943)^3$; $(0.943)^{-3}$; $(9.943)^{1/3}$

10 $(0.017)^{10}$; $10^{0.017}$; $10^{1.43}$

11 $(638)(17.3)$ 12 $\dfrac{(2.73)(78.5)}{621}$

13 $\dfrac{(47.4)^3}{(29.5)^2}$ 14 $\dfrac{(897)^4}{\sqrt{17.8}}$

15 $\sqrt[3]{20.6}(371)^3$ 16 $\dfrac{(0.0048)^{10}}{\sqrt{0.29}}$

Exer. 17–30: Use Table 1 to find a decimal approximation to x.

17 $\log x = 3.6274$ 18 $\log x = 1.8965$

19 $\log x = 0.9469$ 20 $\log x = 0.5729$

21 $\log x = 5.2095$ 22 $\log x = 6.7300 - 10$

23 $\log x = 9.7348 - 10$ 24 $\log x = 7.6739 - 10$

25 $\log x = 8.8306 - 10$ 26 $\log x = 4.9680$

27 $\log x = 2.2765$ 28 $\log x = 3.0043$

29 $\log x = -1.6253$ 30 $\log x = -2.2118$

Exer. 31–50: Use interpolation in Table 1 to approximate the common logarithm of the number.

31 25.48   32 421.6

33 5363   34 0.3817

35 0.001259   36 69,450

37 123,400   38 0.02129

39 0.7786   40 1.203

41 384.7   42 54.44

43 0.9462   44 7259

45 66,590   46 0.001428

47 0.04321   48 400,100

49 3.003   50 9.786

Exer. 51–70: Use interpolation in Table 1 to approximate x.

51 $\log x = 1.4437$ 52 $\log x = 3.7455$

53 $\log x = 4.6931$ 54 $\log x = 0.5883$

55 $\log x = 9.1664 - 10$ 56 $\log x = 8.3902 - 10$

57 $\log x = 3.8153 - 6$ 58 $\log x = 5.9306 - 9$

59 $\log x = 2.3705$ 60 $\log x = 4.2867$

61 $\log x = 0.1358$ 62 $\log x = 0.0194$

63 $\log x = 8.9752 - 10$ 64 $\log x = 2.4979 - 5$

65 $\log x = 5.0409$ 66 $\log x = 1.3796$

67 $\log x = -2.8712$ 68 $\log x = -1.8164$

69 $\log x = -0.6123$ 70 $\log x = -3.1426$

Table I: Common Logarithms

N	0	1	2	3	4	5	6	7	8	9
1.0	.0000	.0043	.0086	.0128	.0170	.0212	.0253	.0294	.0334	.0374
1.1	.0414	.0453	.0492	.0531	.0569	.0607	.0645	.0682	.0719	.0755
1.2	.0792	.0828	.0864	.0899	.0934	.0969	.1004	.1038	.1072	.1106
1.3	.1139	.1173	.1206	.1239	.1271	.1303	.1335	.1367	.1399	.1430
1.4	.1461	.1492	.1523	.1553	.1584	.1614	.1644	.1673	.1703	.1732
1.5	.1761	.1790	.1818	.1847	.1875	.1903	.1931	.1959	.1987	.2014
1.6	.2041	.2068	.2095	.2122	.2148	.2175	.2201	.2227	.2253	.2279
1.7	.2304	.2330	.2355	.2380	.2405	.2430	.2455	.2480	.2504	.2529
1.8	.2553	.2577	.2601	.2625	.2648	.2672	.2695	.2718	.2742	.2765
1.9	.2788	.2810	.2833	.2856	.2878	.2900	.2923	.2945	.2967	.2989
2.0	.3010	.3032	.3054	.3075	.3096	.3118	.3139	.3160	.3181	.3201
2.1	.3222	.3243	.3263	.3284	.3304	.3324	.3345	.3365	.3385	.3404
2.2	.3424	.3444	.3464	.3483	.3502	.3522	.3541	.3560	.3579	.3598
2.3	.3617	.3636	.3655	.3674	.3692	.3711	.3729	.3747	.3766	.3784
2.4	.3802	.3820	.3838	.3856	.3874	.3892	.3909	.3927	.3945	.3962
2.5	.3979	.3997	.4014	.4031	.4048	.4065	.4082	.4099	.4116	.4133
2.6	.4150	.4166	.4183	.4200	.4216	.4232	.4249	.4265	.4281	.4298
2.7	.4314	.4330	.4346	.4362	.4378	.4393	.4409	.4425	.4440	.4456
2.8	.4472	.4487	.4502	.4518	.4533	.4548	.4564	.4579	.4594	.4609
2.9	.4624	.4639	.4654	.4669	.4683	.4698	.4713	.4728	.4742	.4757
3.0	.4771	.4786	.4800	.4814	.4829	.4843	.4857	.4871	.4886	.4900
3.1	.4914	.4928	.4942	.4955	.4969	.4983	.4997	.5011	.5024	.5038
3.2	.5051	.5065	.5079	.5092	.5105	.5119	.5132	.5145	.5159	.5172
3.3	.5185	.5198	.5211	.5224	.5237	.5250	.5263	.5276	.5289	.5302
3.4	.5315	.5328	.5340	.5353	.5366	.5378	.5391	.5403	.5416	.5428
3.5	.5441	.5453	.5465	.5478	.5490	.5502	.5514	.5527	.5539	.5551
3.6	.5563	.5575	.5587	.5599	.5611	.5623	.5635	.5647	.5658	.5670
3.7	.5682	.5694	.5705	.5717	.5729	.5740	.5752	.5763	.5775	.5786
3.8	.5798	.5809	.5821	.5832	.5843	.5855	.5866	.5877	.5888	.5899
3.9	.5911	.5922	.5933	.5944	.5955	.5966	.5977	.5988	.5999	.6010
4.0	.6021	.6031	.6042	.6053	.6064	.6075	.6085	.6096	.6107	.6117
4.1	.6128	.6138	.6149	.6160	.6170	.6180	.6191	.6201	.6212	.6222
4.2	.6232	.6243	.6253	.6263	.6274	.6284	.6294	.6304	.6314	.6325
4.3	.6335	.6345	.6355	.6365	.6375	.6385	.6395	.6405	.6415	.6425
4.4	.6435	.6444	.6454	.6464	.6474	.6484	.6493	.6503	.6513	.6522
4.5	.6532	.6542	.6551	.6561	.6571	.6580	.6590	.6599	.6609	.6618
4.6	.6628	.6637	.6646	.6656	.6665	.6675	.6684	.6693	.6702	.6712
4.7	.6721	.6730	.6739	.6749	.6758	.6767	.6776	.6785	.6794	.6803
4.8	.6812	.6821	.6830	.6839	.6848	.6857	.6866	.6875	.6884	.6893
4.9	.6902	.6911	.6920	.6928	.6937	.6946	.6955	.6964	.6972	.6981
5.0	.6990	.6998	.7007	.7016	.7024	.7033	.7042	.7050	.7059	.7067
5.1	.7076	.7084	.7093	.7101	.7110	.7118	.7126	.7135	.7143	.7152
5.2	.7160	.7168	.7177	.7185	.7193	.7202	.7210	.7218	.7226	.7235
5.3	.7243	.7251	.7259	.7267	.7275	.7284	.7292	.7300	.7308	.7316
5.4	.7324	.7332	.7340	.7348	.7356	.7364	.7372	.7380	.7388	.7396

N	0	1	2	3	4	5	6	7	8	9
5.5	.7404	.7412	.7419	.7427	.7435	.7443	.7451	.7459	.7466	.7474
5.6	.7482	.7490	.7497	.7505	.7513	.7520	.7528	.7536	.7543	.7551
5.7	.7559	.7566	.7574	.7582	.7589	.7597	.7604	.7612	.7619	.7627
5.8	.7634	.7642	.7649	.7657	.7664	.7672	.7679	.7686	.7694	.7701
5.9	.7709	.7716	.7723	.7731	.7738	.7745	.7752	.7760	.7767	.7774
6.0	.7782	.7789	.7796	.7803	.7810	.7818	.7825	.7832	.7839	.7846
6.1	.7853	.7860	.7868	.7875	.7882	.7889	.7896	.7903	.7910	.7917
6.2	.7924	.7931	.7938	.7945	.7952	.7959	.7966	.7973	.7980	.7987
6.3	.7993	.8000	.8007	.8014	.8021	.8028	.8035	.8041	.8048	.8055
6.4	.8062	.8069	.8075	.8082	.8089	.8096	.8102	.8109	.8116	.8122
6.5	.8129	.8136	.8142	.8149	.8156	.8162	.8169	.8176	.8182	.8189
6.6	.8195	.8202	.8209	.8215	.8222	.8228	.8235	.8241	.8248	.8254
6.7	.8261	.8267	.8274	.8280	.8287	.8293	.8299	.8306	.8312	.8319
6.8	.8325	.8331	.8338	.8344	.8351	.8357	.8363	.8370	.8376	.8382
6.9	.8388	.8395	.8401	.8407	.8414	.8420	.8426	.8432	.8439	.8445
7.0	.8451	.8457	.8463	.8470	.8476	.8482	.8488	.8494	.8500	.8506
7.1	.8513	.8519	.8525	.8531	.8537	.8543	.8549	.8555	.8561	.8567
7.2	.8573	.8579	.8585	.8591	.8597	.8603	.8609	.8615	.8621	.8627
7.3	.8633	.8639	.8645	.8651	.8657	.8663	.8669	.8675	.8681	.8686
7.4	.8692	.8698	.8704	.8710	.8716	.8722	.8727	.8733	.8739	.8745
7.5	.8751	.8756	.8762	.8768	.8774	.8779	.8785	.8791	.8797	.8802
7.6	.8808	.8814	.8820	.8825	.8831	.8837	.8842	.8848	.8854	.8859
7.7	.8865	.8871	.8876	.8882	.8887	.8893	.8899	.8904	.8910	.8915
7.8	.8921	.8927	.8932	.8938	.8943	.8949	.8954	.8960	.8965	.8971
7.9	.8976	.8982	.8987	.8993	.8998	.9004	.9009	.9015	.9020	.9025
8.0	.9031	.9036	.9042	.9047	.9053	.9058	.9063	.9069	.9074	.9079
8.1	.9085	.9090	.9096	.9101	.9106	.9112	.9117	.9122	.9128	.9133
8.2	.9138	.9143	.9149	.9154	.9159	.9165	.9170	.9175	.9180	.9186
8.3	.9191	.9196	.9201	.9206	.9212	.9217	.9222	.9227	.9232	.9238
8.4	.9243	.9248	.9253	.9258	.9263	.9269	.9274	.9279	.9284	.9289
8.5	.9294	.9299	.9304	.9309	.9315	.9320	.9325	.9330	.9335	.9340
8.6	.9345	.9350	.9355	.9360	.9365	.9370	.9375	.9380	.9385	.9390
8.7	.9395	.9400	.9405	.9410	.9415	.9420	.9425	.9430	.9435	.9440
8.8	.9445	.9450	.9455	.9460	.9465	.9469	.9474	.9479	.9484	.9489
8.9	.9494	.9499	.9504	.9509	.9513	.9518	.9523	.9528	.9533	.9538
9.0	.9542	.9547	.9552	.9557	.9562	.9566	.9571	.9576	.9581	.9586
9.1	.9590	.9595	.9600	.9605	.9609	.9614	.9619	.9624	.9628	.9633
9.2	.9638	.9643	.9647	.9652	.9657	.9661	.9666	.9671	.9675	.9680
9.3	.9685	.9689	.9694	.9699	.9703	.9708	.9713	.9717	.9722	.9727
9.4	.9731	.9736	.9741	.9745	.9750	.9754	.9759	.9763	.9768	.9773
9.5	.9777	.9782	.9786	.9791	.9795	.9800	.9805	.9809	.9814	.9818
9.6	.9823	.9827	.9832	.9836	.9841	.9845	.9850	.9854	.9859	.9863
9.7	.9868	.9872	.9877	.9881	.9886	.9890	.9894	.9899	.9903	.9908
9.8	.9912	.9917	.9921	.9926	.9930	.9934	.9939	.9943	.9948	.9952
9.9	.9956	.9961	.9965	.9969	.9974	.9978	.9983	.9987	.9991	.9996

Table 3: Natural Logarithms

n	0.0	0.1	0.2	0.3	0.4	0.5	0.6	0.7	0.8	0.9
0*		7.697	8.391	8.796	9.084	9.307	9.489	9.643	9.777	9.895
1	0.000	0.095	0.182	0.262	0.336	0.405	0.470	0.531	0.588	0.642
2	0.693	0.742	0.788	0.833	0.875	0.916	0.956	0.993	1.030	1.065
3	1.099	1.131	1.163	1.194	1.224	1.253	1.281	1.308	1.335	1.361
4	1.386	1.411	1.435	1.459	1.482	1.504	1.526	1.548	1.569	1.589
5	1.609	1.629	1.649	1.668	1.686	1.705	1.723	1.740	1.758	1.775
6	1.792	1.808	1.825	1.841	1.856	1.872	1.887	1.902	1.917	1.932
7	1.946	1.960	1.974	1.988	2.001	2.015	2.028	2.041	2.054	2.067
8	2.079	2.092	2.104	2.116	2.128	2.140	2.152	2.163	2.175	2.186
9	2.197	2.208	2.219	2.230	2.241	2.251	2.262	2.272	2.282	2.293
10	2.303	2.313	2.322	2.332	2.342	2.351	2.361	2.370	2.380	2.389

*Subtract 10 if $n < 1$; for example, $\ln 0.3 \approx 8.796 - 10 = -1.204$.

Table 2: Natural Exponential Function Values

x	e^x	e^{-x}	x	e^x	e^{-x}
0.00	1.0000	1.0000	2.50	12.182	0.0821
0.05	1.0513	0.9512	2.60	13.464	0.0743
0.10	1.1052	0.9048	2.70	14.880	0.0672
0.15	1.1618	0.8607	2.80	16.445	0.0608
0.20	1.2214	0.8187	2.90	18.174	0.0550
0.25	1.2840	0.7788	3.00	20.086	0.0498
0.30	1.3499	0.7408	3.10	22.198	0.0450
0.35	1.4191	0.7047	3.20	24.533	0.0408
0.40	1.4918	0.6703	3.30	27.113	0.0369
0.45	1.5683	0.6376	3.40	29.964	0.0334
0.50	1.6487	0.6065	3.50	33.115	0.0302
0.55	1.7333	0.5769	3.60	36.598	0.0273
0.60	1.8221	0.5488	3.70	40.447	0.0247
0.65	1.9155	0.5220	3.80	44.701	0.0224
0.70	2.0138	0.4966	3.90	49.402	0.0202
0.75	2.1170	0.4724	4.00	54.598	0.0183
0.80	2.2255	0.4493	4.10	60.340	0.0166
0.85	2.3396	0.4274	4.20	66.686	0.0150
0.90	2.4596	0.4066	4.30	73.700	0.0136
0.95	2.5857	0.3867	4.40	81.451	0.0123
1.00	2.7183	0.3679	4.50	99.017	0.0111
1.10	3.0042	0.3329	4.60	99.484	0.0101
1.20	3.3201	0.3012	4.70	109.95	0.0091
1.30	3.6693	0.2725	4.80	121.51	0.0082
1.40	4.0552	0.2466	4.90	134.29	0.0074
1.50	4.4817	0.2231	5.00	148.41	0.0067
1.60	4.9530	0.2019	6.00	403.43	0.0025
1.70	5.4739	0.1827	7.00	1096.6	0.0009
1.80	6.0496	0.1653	8.00	2981.0	0.0003
1.90	6.6859	0.1496	9.00	8103.1	0.0001
2.00	7.3891	0.1353	10.00	22026.0	0.00005
2.10	8.1662	0.1225			
2.20	9.0250	0.1108			
2.30	9.9742	0.1003			
2.40	11.0232	0.0907			

ANSWERS TO

SELECTED

EXERCISES

A *Student's Solutions Manual* to accompany this textbook is available from your college bookstore. The guide contains detailed solutions to approximately one-third of the exercises, as well as strategies for solving other exercises in the text.

CHAPTER 1

EXERCISES 1.1

1 (a) Negative   (b) Positive   (c) Negative
   (d) Positive
3 (a) $<$ (b) $>$ (c) $=$
5 (a) $>$ (b) $>$ (c) $>$
7 (a) $x < 0$ (b) $y \geq 0$ (c) $q \leq \pi$ (d) $2 < d < 4$
   (e) $t \geq 5$ (f) $-z \leq 3$ (g) $\dfrac{p}{q} \leq 7$ (h) $\dfrac{1}{w} \geq 9$
   (i) $|x| > 7$
9 (a) 5   (b) 3   (c) 11
11 (a) -15 (b) -3 (c) 11
13 (a) $4 - \pi$ (b) $4 - \pi$ (c) $1.5 - \sqrt{2}$
15 (a) 4   (b) 12   (c) 12   (d) 8
17 (a) 10   (b) 9   (c) 9   (d) 19
19 $|7 - x| < 5$ **21** $|-3 - x| \geq 8$ **23** $|x - 4| \leq 3$
25 $-x - 3$ **27** $2 - x$ **29** $b - a$ **31** $x^2 + 4$
33 $\neq$ **35** $=$ **37** $\neq$ **39** $=$
41 Construct a right triangle with sides of lengths $\sqrt{2}$ and 1. The hypotenuse will have length $\sqrt{3}$. Next construct a right triangle with sides of lengths $\sqrt{3}$ and $\sqrt{2}$. The hypotenuse will have length $\sqrt{5}$.
43 The large rectangle has area $a(b + c)$. The sum of the areas of the two small rectangles is $ab + ac$.

45 (a) 4.27×10^5 (b) 9.8×10^{-8} (c) 8.1×10^8
47 (a) 830,000   (b) 0.000 000 000 002 9
   (c) 563,000,000
49 1.7×10^{-24} **51** 5.87×10^{12} **53** 1.678×10^{-24} g
55 4.1472×10^6 frames   **57** (a) 125   (b) 21

EXERCISES 1.2

1 $\dfrac{16}{81}$ **3** $\dfrac{9}{8}$ **5** $\dfrac{-47}{3}$ **7** $\dfrac{1}{8}$ **9** $\dfrac{1}{25}$ **11** $8x^9$
13 $\dfrac{6}{x}$ **15** $-2a^{14}$ **17** $\dfrac{9}{2}$ **19** $\dfrac{12u^{11}}{v^2}$ **21** $\dfrac{4}{xy}$
23 $\dfrac{9y^6}{x^8}$ **25** $\dfrac{81}{64}y^6$ **27** $\dfrac{s^6}{4r^8}$ **29** $\dfrac{20y}{x^3}$ **31** $9x^{10}y^{14}$
33 $8a^2$ **35** $24x^{3/2}$ **37** $\dfrac{1}{9a^4}$ **39** $\dfrac{8}{x^{1/2}}$ **41** $4x^2y^4$
43 $\dfrac{3}{x^3y^2}$ **45** 1   **47** $x^{3/4}$ **49** $(a + b)^{2/3}$
51 $(x^2 + y^2)^{1/2}$ **53** (a) $4x\sqrt{x}$ (b) $8x\sqrt{x}$
55 (a) $8 - \sqrt[3]{y}$ (b) $\sqrt[3]{8 - y}$ **57** 9
59 $-2\sqrt[5]{2}$ **61** $\dfrac{1}{2}\sqrt[3]{4}$ **63** $\dfrac{3y^3}{x^2}$ **65** $\dfrac{2a^2}{b}$
67 $\dfrac{1}{2y^2}\sqrt{6xy}$ **69** $\dfrac{xy}{3}\sqrt[3]{6y}$ **71** $\dfrac{x}{3}\sqrt[4]{15x^2y^3}$
73 $\dfrac{1}{2}\sqrt[5]{20x^4y^2}$ **75** $\dfrac{3x^5}{y^2}$ **77** $\dfrac{2x}{y^2}\sqrt[5]{x^2y^4}$
79 $-3tv^2$ **81** $|x^3|y^2$ **83** $|x(y - 1)|\sqrt[4]{x(y - 1)^2}$
85 $\neq$; $(a^r)^2 = a^{2r} \neq a^{(r^2)}$ **87** $\neq$; $(ab)^{xy} = a^{xy}b^{xy} \neq a^x b^y$
89 $=$; $\sqrt[n]{\dfrac{1}{c}} = \left(\dfrac{1}{c}\right)^{1/n} = \dfrac{1^{1/n}}{c^{1/n}} = \dfrac{1}{\sqrt[n]{c}}$ **91** \$232,825.78
93 2.82 m   **95** The 120-kg lifter

EXERCISES 1.3

1 $12x^3 - 13x + 1$ **3** $x^3 - 2x^2 + 4$

5 $6x^2 + x - 35$ **7** $15x^2 + 31xy + 14y^2$

9 $6u^2 - 13u - 12$ **11** $6x^3 + 37x^2 + 30x - 25$

13 $3t^4 + 5t^3 - 15t^2 + 9t - 10$

15 $2x^6 + 2x^5 - 2x^4 + 8x^3 + 10x^2 - 10x - 10$

17 $4y^2 - 5x$ **19** $3v^2 - 2u^2 + uv^2$ **21** $4x^2 - 9y^2$

23 $x^4 - 4y^2$ **25** $x^4 + 5x^2 - 36$

27 $9x^2 + 12xy + 4y^2$ **29** $x^4 - 6x^2y^2 + 9y^4$

31 $x^4 - 8x^2 + 16$ **33** $x - y$ **35** $x - y$

37 $x^3 - 6x^2y + 12xy^2 - 8y^3$

39 $8x^3 + 36x^2y + 54xy^2 + 27y^3$

41 $a^2 + b^2 + c^2 + 2ab - 2ac - 2bc$

43 $4x^2 + y^2 + 9z^2 + 4xy - 12xz - 6yz$ **45** $s(r + 4t)$

47 $3a^2b(b - 2)$ **49** $3x^2y^2(y - 3x)$

51 $5x^3y^2(3y^3 - 5x + 2x^3y^2)$ **53** $(8x + 3)(x - 7)$

55 Irreducible   **57** $(3x - 4)(2x + 5)$

59 $(3x - 5)(4x - 3)$ **61** $(2x - 5)^2$ **63** $(5z + 3)^2$

65 $(5x + 2y)(9x + 4y)$ **67** $(6r + 5t)(6r - 5t)$

69 $(z^2 + 8w)(z^2 - 8w)$ **71** $x^2(x + 2)(x - 2)$

73 Irreducible   **75** $3(5x + 4y)(5x - 4y)$

77 $(4x + 3)(16x^2 - 12x + 9)$

79 $(4x - y^2)(16x^2 + 4xy^2 + y^4)$

81 $(7x + y^3)(49x^2 - 7xy^3 + y^6)$ **83** $(2x + y)(a - 3b)$

85 $3(x + 3)(x - 3)(x + 1)$

87 $(x - 1)(x + 2)(x^2 + x + 1)$ **89** $(a^2 + b^2)(a - b)$

91 $(a + b)(a - b)(a^2 - ab + b^2)(a^2 + ab + b^2)$

93 $(x + 2 + 3y)(x + 2 - 3y)$ **95** $(y + 4 + x)(y + 4 - x)$

97 $(y + 2)(y^2 - 2y + 4)(y - 1)(y^2 + y + 1)$

99 $(x^8 + 1)(x^4 + 1)(x^2 + 1)(x + 1)(x - 1)$

101 Area of I is $(x - y)x$, area of II is $(x - y)y$, and
$A = x^2 - y^2 = (x - y)x + (x - y)y = (x - y)(x + y)$.

EXERCISES 1.4

1 $\dfrac{22}{75}$ **3** $\dfrac{7}{120}$ **5** $\dfrac{x + 3}{x - 4}$ **7** $\dfrac{y + 5}{y^2 + 5y + 25}$

9 $\dfrac{4 - r}{r^2}$ **11** $\dfrac{x}{x - 1}$ **13** $\dfrac{a}{(a^2 + 4)(5a + 2)}$ **15** $\dfrac{-3}{x + 2}$

17 $\dfrac{6s - 7}{(3s + 1)^2}$ **19** $\dfrac{5x^2 + 2}{x^3}$ **21** $\dfrac{4(2t + 5)}{t + 2}$

23 $\dfrac{2(2x + 3)}{3x - 4}$ **25** $\dfrac{2x - 1}{x}$ **27** $\dfrac{p^2 + 2p + 4}{p - 3}$

29 $\dfrac{11u^2 + 18u + 5}{u(3u + 1)}$ **31** $-\dfrac{x + 5}{(x + 2)^2}$ **33** $a + b$

35 $\dfrac{x^2 + xy + y^2}{x + y}$ **37** $\dfrac{2x^2 + 7x + 15}{x^2 + 10x + 7}$ **39** $2x + h - 3$

41 $-\dfrac{3x^2 + 3xh + h^2}{x^3(x + h)^3}$ **43** $\dfrac{-12}{(3x + 3h - 1)(3x - 1)}$

45 $\dfrac{t + 10\sqrt{t} + 25}{t - 25}$ **47** $(9x + 4y)(3\sqrt{x} + 2\sqrt{y})$

49 $\dfrac{\sqrt[3]{a^2} + \sqrt[3]{ab} + \sqrt[3]{b^2}}{a - b}$ **51** $\dfrac{1}{(a + b)(\sqrt{a} + \sqrt{b})}$

53 $\dfrac{2}{\sqrt{2(x + h) + 1} + \sqrt{2x + 1}}$ **55** $\dfrac{-1}{\sqrt{1 - x - h} + \sqrt{1 - x}}$

57 $4x^{4/3} - x^{1/3} + 5x^{-2/3}$ **59** $x^{-1} + 4x^{-3} + 4x^{-5}$

61 $\dfrac{1 + x^5}{x^3}$ **63** $\dfrac{1 - x^2}{x^{1/2}}$ **65** $(3x + 2)^3(36x^2 - 37x + 6)$

67 $\dfrac{(2x + 1)^2(8x^2 + x - 24)}{(x^2 - 4)^{1/2}}$ **69** $\dfrac{(3x + 1)^5(39x - 89)}{(2x - 5)^{1/2}}$

71 $\dfrac{27x^2 - 24x + 2}{(6x + 1)^4}$ **73** $\dfrac{x^2 + 12}{(x^2 + 4)^{4/3}}$ **75** $\dfrac{6(3 - 2x)}{(4x^2 + 9)^{3/2}}$

CHAPTER 1 REVIEW EXERCISES

1 (a) $-\dfrac{5}{12}$ (b) $\dfrac{39}{20}$ (c) $-\dfrac{13}{56}$ (d) $\dfrac{5}{8}$

2 (a) $<$ (b) $>$ (c) $>$

3 (a) $x < 0$ (b) $\dfrac{1}{3} < a < \dfrac{1}{2}$ (c) $|x| \leq 4$

4 (a) 7 (b) -1 (c) $\dfrac{1}{6}$ **5** (a) 5 (b) 5 (c) 7

6 (a) No   (b) No   (c) Yes

7 (a) 9.37×10^{10} (b) 4.02×10^{-6}

8 (a) $68,000,000$ (b) $0.000\,73$

9 $-x - 3$ **10** $-(x - 2)(x - 3)$ **11** $\dfrac{-71}{9}$ **12** $\dfrac{1}{8}$

13 $18a^5b^5$ **14** $\dfrac{3y}{r^2}$ **15** $\dfrac{xy^5}{9}$ **16** $\dfrac{b^3}{a^8}$ **17** $-\dfrac{p^8}{2q}$

18 $c^{1/3}$ **19** $\dfrac{x^3z}{y^{10}}$ **20** $\dfrac{16x^2}{z^4y^6}$ **21** $\dfrac{b^6}{a^2}$ **22** $\dfrac{27u^2v^{27}}{16w^{20}}$

23 $s + r$ **24** $u + v$ **25** s **26** $\dfrac{y - x^2}{x^2y}$ **27** $\dfrac{x^8}{y^2}$

28 $2xyz\sqrt[3]{x^2z}$ **29** $\dfrac{1}{2}\sqrt[3]{2}$ **30** $\dfrac{ab}{c}\sqrt{bc}$ **31** $2x^2y\sqrt[3]{x}$

32 $2ab\sqrt{ac}$ **33** $\dfrac{1 - \sqrt{t}}{t}$ **34** c^2d^4 **35** $\dfrac{2x}{y^2}$

36 $a + 2b$ **37** $\dfrac{1}{2\pi}\sqrt[3]{4\pi}$ **38** $\dfrac{1}{3y}\sqrt[3]{3x^2y^2}$

39 $\dfrac{1 - 2\sqrt{x} + x}{1 - x}$ **40** $\dfrac{\sqrt{a} - \sqrt{a - 2}}{2}$

41 $(9x + y)(3\sqrt{x} - \sqrt{y})$ **42** $\dfrac{x + 6\sqrt{x} + 9}{9 - x}$

43 $x^4 + x^3 - x^2 + x - 2$ **44** $3z^4 - 4z^3 - 3z^2 + 4z + 1$

45 $-x^2 + 18x + 7$ **46** $8x^3 + 2x^2 - 43x + 35$

47 $3y^5 - 2y^4 - 8y^3 + 10y^2 - 3y - 12$

48 $15x^3 - 53x^2 - 102x - 40$ **49** $a^4 - b^4$

50 $3p^2q - 2q^2 + \dfrac{5}{3}p$ **51** $6a^2 + 11ab - 35b^2$

52 $16r^4 - 24r^2s + 9s^2$ **53** $169a^4 - 16b^2$

54 $a^6 - 2a^5 + a^4$ **55** $8a^3 + 12a^2b + 6ab^2 + b^3$

56 $c^6 - 3c^4d^2 + 3c^2d^4 - d^6$ **57** $81x^4 - 72x^2y^2 + 16y^4$

58 $a^2 + b^2 + c^2 + d^2 + 2(ab + ac + ad + bc + bd + cd)$

59 $10w(6x + 7)$ **60** $2r^2s^3(r + 2s)(r - 2s)$

61 $(14x + 9)(2x - 1)$ **62** $(4a^2 + 3b^2)^2$

63 $(y - 4z)(2w + 3x)$ **64** $(2c^2 + 3)(c - 6)$

65 $8(x + 2y)(x^2 - 2xy + 4y^2)$

66 $u^3v(v - u)(v^2 + uv + u^2)$

67 $(p^4 + q^4)(p^2 + q^2)(p + q)(p - q)$ **68** $x^2(x - 4)^2$

69 $(w^2 + 1)(w^4 - w^2 + 1)$ **70** $3(x + 2)$

71 Irreducible   **72** $(x - 7 + 7y)(x - 7 - 7y)$

73 $(x - 2)(x + 2)^2(x^2 - 2x + 4)$ **74** $4x^2(x^2 + 3x + 5)$

75 $\dfrac{3x - 5}{2x + 1}$ **76** $\dfrac{r^2 + rt + t^2}{r + t}$ **77** $\dfrac{3x + 2}{x(x - 2)}$

78 $\dfrac{27}{(4x - 5)(10x + 1)}$ **79** $\dfrac{5x^2 - 6x - 20}{x(x + 2)^2}$ **80** $\dfrac{x^3 + 1}{x^2 + 1}$

81 $\dfrac{-2x^2 - x - 3}{x(x + 1)(x + 3)}$ **82** $\dfrac{ab}{a + b}$ **83** $x + 5$ **84** $\dfrac{1}{x + 3}$

85 $(x^2 + 1)^{1/2}(x + 5)^3(7x^2 + 15x + 4)$

86 $\dfrac{2(5x^2 + x + 4)}{(6x + 1)^{2/3}(4 - x^2)^2}$ **87** 2.236068   **88** 4.242641

89 2.75×10^{13} cells

90 Between 2.94×10^9 and 3.78×10^9 beats

91 0.54 m^2 **92** 0.13 dyne-cm

CHAPTER 2

EXERCISES 2.1

1 $\dfrac{5}{3}$ **3** 1   **5** $\dfrac{26}{7}$ **7** $\dfrac{35}{17}$ **9** $\dfrac{23}{18}$ **11** $-\dfrac{1}{40}$

13 $\dfrac{49}{4}$ **15** $\dfrac{4}{3}$ **17** $-\dfrac{24}{29}$ **19** $\dfrac{7}{31}$ **21** $-\dfrac{3}{61}$

23 $\dfrac{29}{4}$ **25** $\dfrac{31}{18}$ **27** No solution

29 All real numbers except $\dfrac{1}{2}$ **31** $\dfrac{5}{9}$ **33** $-\dfrac{2}{3}$

35 No solution   **37** 0

39 All real numbers except ± 2 **41** No solution

43 No solution

45 $(4x - 3)^2 - 16x^2 = (16x^2 - 24x + 9) - 16x^2 = 9 - 24x$

47 $\dfrac{x^2 - 9}{x + 3} = \dfrac{(x + 3)(x - 3)}{x + 3} = x - 3$

49 $\dfrac{3x^2 + 8}{x} = \dfrac{3x^2}{x} + \dfrac{8}{x} = \dfrac{8}{x} + 3x$ **51** $-\dfrac{19}{3}$

53 (a) Yes

    (b) No, 5 is not a solution of the first equation.

55 Choose any a and b such that $b = -\dfrac{5}{3}a$.

57 $x + 1 = x + 2$ **59** $P = \dfrac{I}{rt}$ **61** $h = \dfrac{2A}{b}$

63 $m = \dfrac{Fd^2}{gM}$ **65** $w = \dfrac{P - 2l}{2}$ **67** $b_1 = \dfrac{2A - hb_2}{h}$

69 $q = \dfrac{p(1 - S)}{S(1 - p)}$ **71** $q = \dfrac{fp}{p - f}$

EXERCISES 2.2

1 88   **3** \$820   **5** 180 mo (or 15 yr)

7 Not possible   **9** 200 children

11 $\dfrac{14}{3}$ oz of 30% glucose solution and $\dfrac{7}{3}$ oz of water

13 194.6 g of British sterling silver and 5.4 g of copper

15 (a) After 64 sec   **(b)** 96 m and 128 m, respectively

17 6 mi/hr   **19 (a)** $\dfrac{5}{9}$ mi/hr   **(b)** $2\dfrac{2}{9}$ mi   **21** 1237.5 ft

23 (a) 4050 ft^2 **(b)** 2592 ft^2 **(c)** 3600 ft^2

25 $\dfrac{19}{2} - \dfrac{3\pi}{8} \approx 8.32$ ft   **27** 55 ft   **29** 36 min

31 36 min   **33** After an additional 50 games

EXERCISES 2.3

1 $-\dfrac{3}{2}, \dfrac{4}{3}$ **3** $-\dfrac{6}{5}, \dfrac{2}{3}$ **5** $-\dfrac{9}{2}, \dfrac{3}{4}$ **7** $-\dfrac{2}{3}, \dfrac{1}{5}$

9 $-\dfrac{5}{2}$ **11** $-\dfrac{1}{2}$ **13** $-\dfrac{34}{5}$

15 (a) No, -4 is not a solution of $x = 4$.   **(b)** Yes

17 ± 13 **19** $\pm \dfrac{3}{5}$ **21** $3 \pm \sqrt{17}$ **23** $-2 \pm \dfrac{1}{2}\sqrt{11}$

25 (a) $\dfrac{81}{4}$ **(b)** 16   **(c)** ± 12 **(d)** ± 7

27 $-3 \pm \sqrt{2}$ **29** $\dfrac{3}{2} \pm \sqrt{5}$ **31** $-\dfrac{1}{2}, \dfrac{2}{3}$

33 $-2 \pm \sqrt{2}$ **35** $\dfrac{3}{4} \pm \dfrac{1}{4}\sqrt{41}$ **37** $\dfrac{4}{3} \pm \dfrac{1}{3}\sqrt{22}$

39 $\dfrac{5}{2} \pm \dfrac{1}{2}\sqrt{15}$ **41** $\dfrac{9}{2}$ **43** No real solutions

45 (a) $x = \dfrac{y \pm \sqrt{2y^2 - 1}}{2}$ **(b)** $y = -2x \pm \sqrt{8x^2 + 1}$

47 $v = \sqrt{\dfrac{2K}{m}}$ **49** $r = \dfrac{-\pi h + \sqrt{\pi^2 h^2 + 2\pi A}}{2\pi}$

51 $r = r_0\sqrt{1 - (V/V_{max})}$ **53** $\sqrt{150/\pi} \approx 6.9$ cm

55 (a) After 1 sec and after 3 sec   (b) After 4 sec

57 (a) 4320 m   (b) 96.86°C   **59** 2 ft

61 12 ft by 12 ft

63 $3 + \dfrac{1}{2}\sqrt{14} \approx 4.9$ mi or $3 - \dfrac{1}{2}\sqrt{14} \approx 1.1$ mi

65 (a) $d = 100\sqrt{20t^2 + 4t + 1}$ (b) 3:30 P.M.

67 14 in. by 27 in.   **69** 7 mi/hr   **71** 300 pairs

73 2 ft   **75** 15.89 sec

77 (a) $0; -4,500,000$ (b) 2.13×10^{-7}

EXERCISES 2.4

1 $2 + 4i$ **3** $18 - 3i$ **5** $41 - 11i$ **7** $17 - i$

9 $21 - 20i$ **11** $-24 - 7i$ **13** 25 **15** $-i$

17 i **19** $\dfrac{3}{10} - \dfrac{3}{5}i$ **21** $\dfrac{1}{2} - i$ **23** $\dfrac{34}{53} + \dfrac{40}{53}i$

25 $\dfrac{2}{5} + \dfrac{4}{5}i$ **27** $-142 - 65i$ **29** $-2 - 14i$

31 $-\dfrac{44}{113} + \dfrac{95}{113}i$ **33** $\dfrac{21}{2}i$ **35** $x = 4, y = -16$

37 $x = 1, y = 3$ **39** $3 \pm 2i$ **41** $-2 \pm 3i$

43 $\dfrac{5}{2} + \dfrac{1}{2}\sqrt{55}i$ **45** $-\dfrac{1}{8} \pm \dfrac{1}{8}\sqrt{47}i$ **47** $-5, \dfrac{5}{2} \pm \dfrac{5}{2}\sqrt{3}i$

49 $\pm 4, \pm 4i$ **51** $\pm 2i, \pm \dfrac{3}{2}i$ **53** $0, -\dfrac{3}{2} \pm \dfrac{1}{2}\sqrt{7}i$

55 If $w = c + di$, then $\overline{z + w} = \overline{(a + bi) + (c + di)}$
$= \overline{(a + c) + (b + d)i} = (a + c) - (b + d)i$
$= (a - bi) + (c - di) = \bar{z} + \bar{w}$.

57 $\overline{z \cdot w} = \overline{(a + bi) \cdot (c + di)} = \overline{(ac - bd) + (ad + bc)i}$
$= (ac - bd) - (ad + bc)i = ac - adi - bd - bci$
$= a(c - di) - bi(c - di) = (a - bi) \cdot (c - di) = \bar{z} \cdot \bar{w}$

59 If $\bar{z} = z$, then $a - bi = a + bi$ and hence $-bi = bi$, or
$2bi = 0$. Thus, $b = 0$ and $z = a$ is real. Conversely,
if z is real, then $b = 0$ and hence $\bar{z} = \overline{a + 0i} = a - 0i =$
$a + 0i = z$.

EXERCISES 2.5

1 $-15, 7$ **3** $-\dfrac{2}{3}, 2$ **5** No solution   **7** $\pm\dfrac{2}{3}, 2$

9 $\pm\dfrac{1}{2}\sqrt{6}, -\dfrac{5}{2}, 0$ **11** $0, 25$ **13** $-\dfrac{57}{5}$ **15** $\dfrac{9}{5}$

17 $\pm\dfrac{1}{2}\sqrt{62}$ **19** 6 **21** 6 **23** $5, 7$ **25** -3

27 -1 **29** $-\dfrac{5}{4}$ **31** 3 **33** $0, 4$ **35** $\pm 3, \pm 4$

37 $\pm\dfrac{1}{10}\sqrt{70 \pm 10\sqrt{29}}$ **39** $\pm 2, \pm 3$ **41** $\dfrac{8}{27}, -8$

43 $\dfrac{25}{4}, \dfrac{16}{9}$ **45** $-\dfrac{4}{3}, -\dfrac{2}{3}$ **47** $\dfrac{5}{2}$ **49** $0, 4096$

51 (a) 8 (b) ± 8 (c) No real solutions   (d) 625
(e) No real solutions

53 $l = \dfrac{gT^2}{4\pi^2}$ **55** $h = \dfrac{1}{\pi r}\sqrt{S^2 - \pi^2 r^4}$ **57** 9.16 ft/sec

59 $4.00   **61** $2\sqrt[3]{\dfrac{432}{\pi}} \approx 10.3$ cm   **63** 53.4%

65 There are two possible routes corresponding to
$x \approx 0.6743$ mi and $x \approx 2.2887$ mi.

67 2.4493

EXERCISES 2.6

1 (a) $-2 < 2$ (b) $-11 < -7$ (c) $-\dfrac{7}{3} < -1$

    (d) $1 < \dfrac{7}{3}$

3 $(-\infty, -2)$ **5** $[4, \infty)$

7 $(-2, 4]$ **9** $[3, 7]$

11 $[-2, 5)$

13 $-5 < x \le 8$ **15** $-4 \le x \le -1$ **17** $x \ge 4$

19 $x < -5$ **21** $\left(\dfrac{16}{3}, \infty\right)$ **23** $\left(-\infty, -\dfrac{4}{3}\right]$

25 $(12, \infty)$ **27** $[-6, \infty)$ **29** $(1, 6)$ **31** $[9, 19)$

33 $\left(-\dfrac{26}{3}, \dfrac{16}{3}\right]$ **35** $(6, 12]$ **37** $\left(-\infty, \dfrac{8}{53}\right)$

39 $\left(-\infty, \dfrac{4}{5}\right)$ **41** $\left(-\dfrac{2}{3}, \infty\right)$ **43** $\left(\dfrac{4}{3}, \infty\right)$

45 All real numbers except 1   **47** $(-3, 3)$

49 $(-\infty, -5] \cup [5, \infty)$ **51** $(-3.01, -2.99)$

53 $(-\infty, -2.1] \cup [-1.9, \infty)$ **55** $\left(-\dfrac{9}{2}, -\dfrac{1}{2}\right)$

57 $\left[\dfrac{3}{5}, \dfrac{9}{5}\right]$ **59** $(-\infty, \infty)$ **61** $(-\infty, 3) \cup (3, \infty)$

63 $\left(-\infty, -\dfrac{8}{3}\right] \cup [4, \infty)$ **65** $\left(-\infty, \dfrac{7}{4}\right) \cup \left(\dfrac{13}{4}, \infty\right)$

67 $(-2, 1) \cup (3, 6)$

69 (a) $-8, -2$ **(b)** $-8 < x < -2$
(c) $(-\infty, -8) \cup (-2, \infty)$

71 $|w - 148| \le 2$ **73** $5 < |T_1 - T_2| < 10$

75 $86 \le F \le 104$ **77** $R \ge 11$ **79** $4 \le p < 6$

81 $6\dfrac{2}{3}$ yr

EXERCISES 2.7

1 $\left(-\dfrac{1}{3}, \dfrac{1}{2}\right)$ **3** $[-2, 1] \cup [4, \infty)$ **5** $(-2, 3)$

7 $(-\infty, -2) \cup (4, \infty)$ **9** $\left(-\infty, -\dfrac{5}{2}\right] \cup [1, \infty)$

11 $(2, 4)$ **13** $(-4, 4)$ **15** $\left(-\dfrac{3}{5}, \dfrac{3}{5}\right)$

17 $(-\infty, 0] \cup \left[\dfrac{9}{16}, \infty\right)$ **19** $(-\infty, -2] \cup [2, \infty)$

21 $\{-2\} \cup [2, \infty)$ **23** $(-\infty, -2) \cup (-2, -1) \cup \{0\}$
25 $(-2, 0) \cup (0, 1]$ **27** $(-2, 2] \cup (5, \infty)$

29 $(-\infty, -3) \cup (0, 3)$ **31** $\left(\dfrac{3}{2}, \dfrac{7}{3}\right)$

33 $(-\infty, -1) \cup \left(2, \dfrac{7}{2}\right]$ **35** $\left(-1, \dfrac{2}{3}\right) \cup [4, \infty)$

37 $\left(1, \dfrac{5}{3}\right) \cup [2, 5]$ **39** $(-1, 0) \cup (1, \infty)$

41 $[0, 2] \cup [3, 5]$ **43** $\dfrac{1}{2}$ sec   **45** $0 \le v < 30$

47 $0 < S < 4000$ **49** height $> 25{,}600$ km

CHAPTER 2 REVIEW EXERCISES

1 $-\dfrac{5}{6}$ **2** 5 **3** -32 **4** No solution

5 Every $x > 0$ **6** $-4, \dfrac{3}{2}$ **7** $-\dfrac{2}{3} \pm \dfrac{1}{3}\sqrt{19}$

8 $\dfrac{5}{2} \pm \dfrac{1}{2}\sqrt{29}$ **9** $\dfrac{1}{2} \pm \dfrac{1}{2}\sqrt{21}$ **10** $\pm\dfrac{5}{2}, \pm\sqrt{2}$

11 $-27, 125$ **12** $\pm\dfrac{1}{2}\sqrt{7}, -\dfrac{2}{5}$ **13** $\dfrac{1}{5} \pm \dfrac{1}{5}\sqrt{14}i$

14 $-\dfrac{1}{6} \pm \dfrac{1}{6}\sqrt{71}i$ **15** $\pm\dfrac{1}{2}\sqrt{14}i, \pm\dfrac{2}{3}\sqrt{3}i$

16 $\pm\dfrac{1}{2}\sqrt{6 \pm 2\sqrt{5}}$ **17** $-\dfrac{3}{2}, 2$ **18** $-5, 4$ **19** $\dfrac{1}{4}, \dfrac{1}{9}$

20 $\dfrac{13}{4}$ **21** 2 **22** $-3, 1$ **23** 5 **24** ± 8

25 $2 \pm \sqrt{3}$ **26** $-5 \pm \sqrt{13}i$ **27** 3 **28** $\left(\dfrac{2}{3}, \infty\right)$

29 $\left(-\dfrac{11}{4}, \dfrac{9}{4}\right)$ **30** $\left[\dfrac{13}{23}, \infty\right)$ **31** $\left(-\infty, -\dfrac{3}{10}\right)$

32 $\left(-7, \dfrac{7}{2}\right)$ **33** $(-\infty, 1) \cup (5, \infty)$ **34** $[0, 6]$

35 $\left(-\infty, \dfrac{11}{3}\right] \cup [7, \infty)$ **36** $(2, 4) \cup (8, 10)$

37 $\left(-\infty, -\dfrac{3}{2}\right) \cup \left(\dfrac{2}{5}, \infty\right)$ **38** $[-2, 5]$

39 $(-\infty, -2) \cup \{0\} \cup [3, \infty)$ **40** $(-3, -1) \cup (-1, 2]$

41 $\left(-\infty, -\dfrac{3}{2}\right) \cup (2, 9)$ **42** $(-\infty, -5) \cup [-1, 5)$

43 $(1, \infty)$ **44** $(0, 1) \cup (2, 3)$ **45** $r = \sqrt[3]{\dfrac{3V}{4\pi}}$

46 $R = \sqrt[4]{\dfrac{8FVL}{\pi P}}$ **47** $h = R \pm \dfrac{1}{2}\sqrt{4R^2 - c^2}$

48 $r = \dfrac{-\pi hR + \sqrt{12\pi hV - 3\pi^2 h^2 R^2}}{2\pi h}$ **49** $15 + 2i$

50 $-28 + 6i$ **51** $-55 + 48i$ **52** $\dfrac{9}{85} + \dfrac{2}{85}i$

53 $-\dfrac{9}{53} - \dfrac{48}{53}i$ **54** $-2 - 5i$ **55** $R_2 = \dfrac{10}{3}$ ohms

56 11.055% **57** 60.3 g

58 6 oz of vegetables and 4 oz of meat

59 315.8 g of ethyl alcohol and 84.2 g of water

60 80 gal of 20% solution and 40 gal of 50% solution

61 75 mi   **62** 2   **63** 64 mi/hr   **64** $\dfrac{640}{11} \approx 58.2$ mi/hr

65 1 hr 40 min   **66** 165 mi   **67** $10 - 5\sqrt{3} \approx 1.34$ mi

68 $3\sqrt{5} - 6 \approx 0.71$ micron

69 (a) $d = \sqrt{2900t^2 - 200t + 4}$

(b) $t = \dfrac{5 + 2\sqrt{19{,}603}}{145} \approx 1.97$, or approximately
11:58 A.M.

70 There are two arrangements: 40 ft $\times$ 25 ft and
50 ft $\times$ 20 ft.

71 (a) $2\sqrt{2}$ ft   **(b)** 2 ft   **72** 12 ft by 48 ft

73 10 ft by 4 ft   **74** After $7\dfrac{2}{3}$ yr   **75** $4 \le p \le 8$

76 Over \$100,000   **77** $T > 279.57$ K

78 $\dfrac{\pi}{5}\sqrt{10} \le T \le \dfrac{2\pi}{7}\sqrt{5}$

79 $v < \dfrac{626.4}{\sqrt{6472}} \approx 7.786$ km/sec

80 $20 \le w \le 25$ **81** 36 to 38 trees/acre
82 $320 to $340

CHAPTER 3

EXERCISES 3.1

1

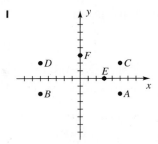

3 The line bisecting quadrants I and III

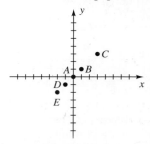

5 $A(3, 3), B(-3, 3), C(-3, -3), D(3, -3), E(3, 0), F(0, 3)$
7 (a) The line parallel to the y-axis that intersects the
   x-axis at $(-2, 0)$
   (b) The line parallel to the x-axis that intersects the
   y-axis at $(0, 3)$
   (c) All points to the right of and on the y-axis
   (d) All points in quadrants I and III
   (e) All points below the x-axis
   (f) All points on the y-axis
9 (a) $\sqrt{29}$ **(b)** $\left(5, -\dfrac{1}{2}\right)$

11 (a) $\sqrt{13}$ **(b)** $\left(-\dfrac{7}{2}, -1\right)$ **13 (a)** 4   **(b)** $(5, -3)$
15 $d(A, C)^2 = d(A, B)^2 + d(B, C)^2$; area $= 28$
17 $d(A, B) = d(B, C) = d(C, D) = d(D, A)$ and
   $d(A, C)^2 = d(A, B)^2 + d(B, C)^2$
19 $(13, -28)$ **21** $d(A, C) = d(B, C) = \sqrt{145}$
23 $5x + 2y = 3$

25 $\sqrt{x^2 + y^2} = 5$; a circle of radius 5 with center at the origin
27 $(0, 3 + \sqrt{11}), (0, 3 - \sqrt{11})$ **29** $(-2, -1)$
31 $a < \dfrac{2}{5}$ or $a > 4$
33 Let M be the midpoint of the hypotenuse. Show that
   $d(A, M) = d(B, M) = d(O, M) = \dfrac{1}{2}\sqrt{a^2 + b^2}$.

EXERCISES 3.2

1   **3**

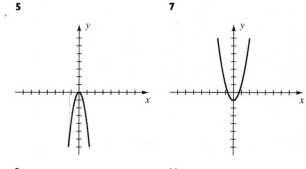

5   **7**

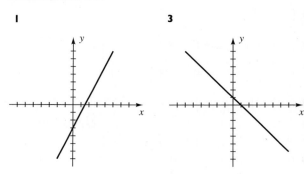

9   **11**

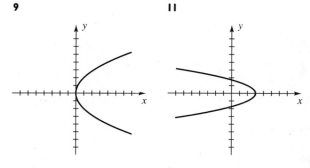

13

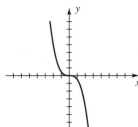

15

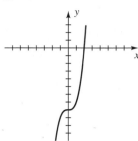

17

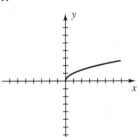

19

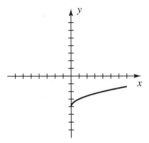

21 (a) 5, 7   **(b)** 9, 11   **(c)** 13

23

25

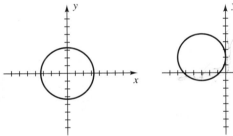

27

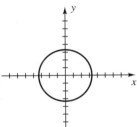

29

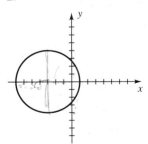

31

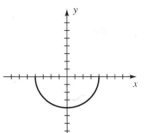

33

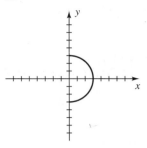

35 $(x - 2)^2 + (y + 3)^2 = 25$ **37** $\left(x - \dfrac{1}{4}\right)^2 + y^2 = 5$

39 $(x + 4)^2 + (y - 6)^2 = 41$ **41** $(x + 3)^2 + (y - 6)^2 = 9$

43 $(x + 4)^2 + (y - 4)^2 = 16$

45 $(x - 1)^2 + (y - 2)^2 = 34$ **47** $C(2, -3); r = 7$

49 $C(0, -2); r = 11$ **51** $C(3, -1); r = \dfrac{1}{2}\sqrt{70}$

53 $C(-2, 1); r = 0$ (a point)

55 Not a circle, since r^2 cannot equal -2

57 $y = \sqrt{36 - x^2}; y = -\sqrt{36 - x^2}; x = \sqrt{36 - y^2};$
$x = -\sqrt{36 - y^2}$

59 $y = -1 + \sqrt{49 - (x - 2)^2}; y = -1 - \sqrt{49 - (x - 2)^2};$
$x = 2 + \sqrt{49 - (y + 1)^2}; x = 2 - \sqrt{49 - (y + 1)^2}$

61 (a) Inside   **(b)** On   **(c)** Outside

63 (a) 2   **(b)** $3 \pm \sqrt{5}$ **65** $(x + 2)^2 + (y - 3)^2 = 25$

67 $[-15, -3) \cup (2, 15]$ **69** $(-1, 0) \cup (0, 1)$

71

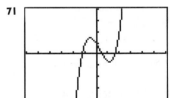

$-1.2, 0.5, 1.6$

$[-6, 6]$ by $[-4, 4]$

73

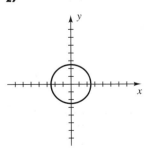

$(0.6, 0.8), (-0.6, -0.8)$

$[-6, 6]$ by $[-4, 4]$

75

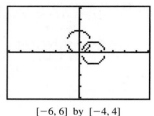

[−6, 6] by [−4, 4]

(0.999, 0.968),
(0.251, 0.032)

17

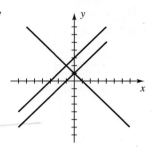

EXERCISES 3.3

1 $m = -\dfrac{3}{4}$

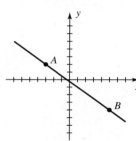

3 $m = 0$

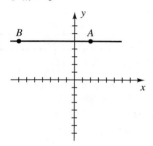

5 m is undefined

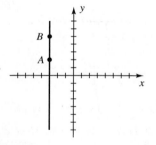

7 The slopes of opposite sides are equal.
9 The slopes of opposite sides are equal, and the
slopes of two adjacent sides are negative reciprocals.
11 $(-12, 0)$
13

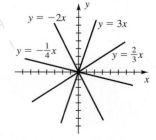

15

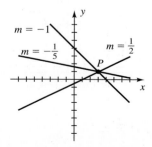

19 (a) $x = 5$　**(b)** $y = -2$　**21** $4x + y = 17$
23 $3x + y = 12$　**25** $11x + 7y = 9$　**27** $5x - 2y = 18$
29 $5x + 2y = 29$　**31** $y = \dfrac{3}{4}x - 3$　**33** $y = -\dfrac{1}{3}x + \dfrac{11}{3}$
35 $5x - 7y = -15$　**37** $y = -x$
39 $m = -\dfrac{2}{3}, b = 5$　　　**41** $m = \dfrac{4}{3}, b = -3$

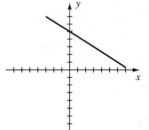 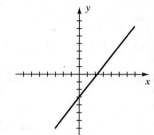

43 (a) $y = 3$　**(b)** $y = -\dfrac{1}{2}x$　**(c)** $y = -\dfrac{3}{2}x + 1$
(d) $y + 2 = -(x - 3)$
45 $\dfrac{x}{3/2} + \dfrac{y}{-3} = 1$　**47** $(x - 3)^2 + (y + 2)^2 = 49$
49 Approximately 23 weeks
51 (a) 25.2 tons　**(b)** As large as 3.4 tons
53 (a) $y = \dfrac{5}{14}x$　**(b)** 58
55 (a) $W = \dfrac{20}{3}t + 10$　**(d)**
(b) 50 lb
(c) 9 yrs

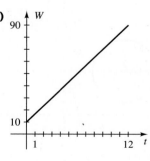

57 $H = -\dfrac{8}{3}T + \dfrac{7520}{3}$

59 (a) $T = 0.032t + 13.5$ (b) $16.22\,°C$
61 (a) $E = 0.55R + 3600$ (b) $P = 0.45R - 3600$
(c) $8000
63 (a) Yes: the creature at $x = 3$ (b) No
65 34.95 mi/hr
67 $(-0.8, -0.6)$, $(4.8, -3.4)$, $(2, 5)$; right isosceles triangle
69 $a = 0.321$; $b = -0.9425$

EXERCISES 3.4

1 $-6, -4, -24$ **3** $-12, -22, -36$
5 (a) $5a - 2$ (b) $-5a - 2$ (c) $-5a + 2$
(d) $5a + 5h - 2$ (e) $5a + 5h - 4$ (f) 5
7 (a) $a^2 - a + 3$ (b) $a^2 + a + 3$ (c) $-a^2 + a - 3$
(d) $a^2 + 2ah + h^2 - a - h + 3$
(e) $a^2 + h^2 - a - h + 6$ (f) $2a + h - 1$
9 (a) $\dfrac{4}{a^2}$ (b) $\dfrac{1}{4a^2}$ (c) $4a$ (d) $2a$

11 (a) $\dfrac{2a}{a^2 + 1}$ (b) $\dfrac{a^2 + 1}{2a}$ (c) $\dfrac{2\sqrt{a}}{a + 1}$ (d) $\dfrac{\sqrt{2a^3 + 2a}}{a^2 + 1}$

13 (a) $[-3, 4]$ (b) $[-2, 2]$ (c) 0 (d) $-1, \dfrac{1}{2}, 2$

(e) $\left(-1, \dfrac{1}{2}\right) \cup (2, 4]$

15 $\left[-\dfrac{7}{2}, \infty\right)$ **17** $[-3, 3]$
19 All real numbers except $-2, 0,$ and 2

21 $\left[\dfrac{3}{2}, 4\right) \cup (4, \infty)$ **23** $(2, \infty)$ **25** $[-2, 2]$

27 (a)

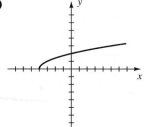

(b) $D = (-\infty, \infty)$,
$R = (-\infty, \infty)$
(c) Increasing on
$(-\infty, \infty)$

29 (a)

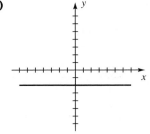

(b) $D = (-\infty, \infty)$,
$R = (-\infty, 4]$
(c) Increasing on
$(-\infty, 0]$,
decreasing on
$[0, \infty)$

31 (a)

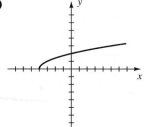

(b) $D = [-4, \infty)$,
$R = [0, \infty)$
(c) Increasing on
$[-4, \infty)$

33 (a)

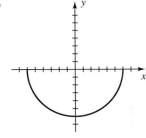

(b) $D = (-\infty, \infty)$,
$R = \{-2\}$
(c) Constant on
$(-\infty, \infty)$

35 (a)

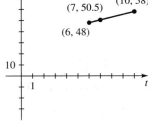

(b) $D = [-6, 6]$,
$R = [-6, 0]$
(c) Decreasing on
$[-6, 0]$,
increasing on
$[0, 6]$

37 $f(x) = \dfrac{1}{6}x + \dfrac{3}{2}$ **39** Yes   **41** No   **43** Yes
45 No   **47** No   **49** $V = 4x(15 - x)(10 - x)$
51 (a) $y = \dfrac{500}{x}$ (b) $C = 300x + \dfrac{100,000}{x} - 600$
53 (a) $y = 2.5t + 33$

(b)

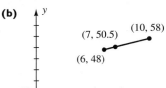

The yearly increase
in height

(c) 58 in.

55 $d = 2\sqrt{t^2 + 2500}$

57 (a) $y = \sqrt{h^2 + 2hr}$ **(b)** 1280.6 mi

59 $d = \sqrt{90{,}400 + x^2}$

61 (a)

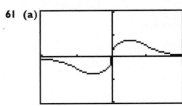

[-2, 2] by [-2, 2]

(b) $[-0.75, 0.75]$
(c) Decreasing on $[-2, -0.55]$ and on $[0.55, 2]$, increasing on $[-0.55, 0.55]$

63 (a)

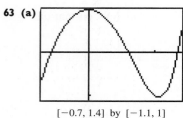

[-0.7, 1.4] by [-1.1, 1]

(b) $[-1.03, 1]$
(c) Increasing on $[-0.7, 0]$ and on $[1.06, 1.4]$, decreasing on $[0, 1.06]$

65 (a) 8   **(b)** ± 8 **(c)** No real solutions   **(d)** 625
(e) No real solutions

EXERCISES 3.5

1 Odd   **3** Even
5 Neither   **7** Even   **9** Odd

11

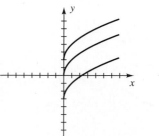

13

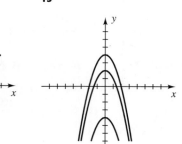

15

17

19

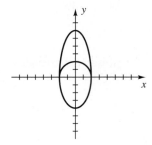

21

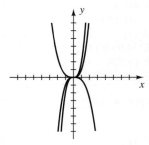

23

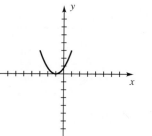

25 (a)

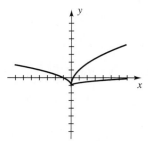

(b)

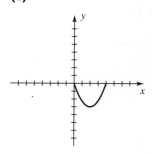

(c)

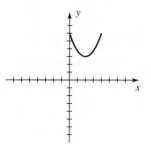

(d)

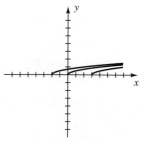

(e)

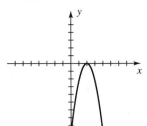

(f)

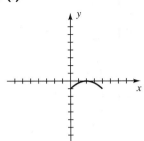

35

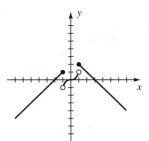

(g)

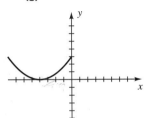

(h)

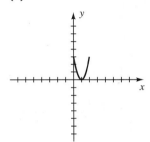

37 (a)

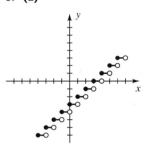

(b)

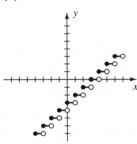

(i)

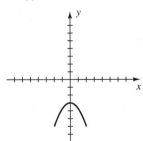

(j)

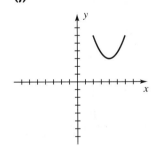

(c)

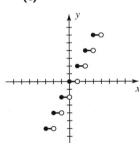

(d)

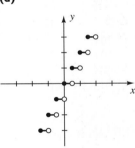

27 (a) $y = f(x + 9) + 1$ **(b)** $y = -f(x)$
(c) $y = -f(x + 7) - 1$

29 (a) $y = f(x + 4)$ **(b)** $y = f(x) + 1$ **(c)** $y = f(-x)$

31

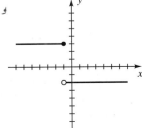

33

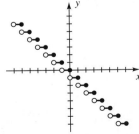

(e)

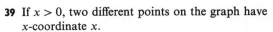

39 If $x > 0$, two different points on the graph have x-coordinate x.

41 Odd

43

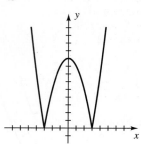

45

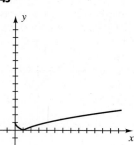

47

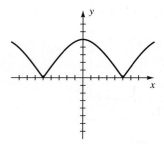

49 $T(x) = \begin{cases} 0.15x & \text{if } x \le 20{,}000 \\ 0.20x - 1000 & \text{if } x > 20{,}000 \end{cases}$

51 $\begin{cases} 1.20x & \text{if } 0 \le x \le 10{,}000 \\ 1.50x - 3000 & \text{if } 10{,}000 < x \le 15{,}000 \\ 1.80x - 7500 & \text{if } x > 15{,}000 \end{cases}$

53 $(-3.12, 22.00)$

55 $(-\infty, -3) \cup (-3, 1.87) \cup (4.13, \infty)$

EXERCISES 3.6

1 $y = a(x + 3)^2 + 1$ **3** $y = ax^2 - 3$

5 $f(x) = -(x + 2)^2 - 4$ **7** $f(x) = 2(x - 3)^2 + 4$

9 $f(x) = -3(x + 1)^2 - 2$ **11** $f(x) = -\dfrac{3}{4}(x - 6)^2 - 7$

13 **(a)** Min: $f(2) = -4$ **(c)**
(b) 0, 4

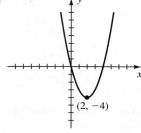

15 **(a)** Max: $f\left(\dfrac{11}{24}\right) = \dfrac{841}{48}$ **(c)**

(b) $-\dfrac{3}{4}, \dfrac{5}{3}$

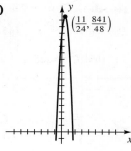

17 **(a)** Min: $f\left(-\dfrac{4}{3}\right) = 0$ **19** **(a)** Min: $f(-2) = 5$
(b) None
(b) $-\dfrac{4}{3}$ **(c)**

(c)

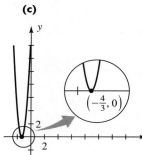

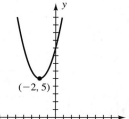

21 **(a)** Max: $f(5) = 7$ **(b)** $5 \pm \dfrac{1}{2}\sqrt{14} \approx 6.87, 3.13$

(c)

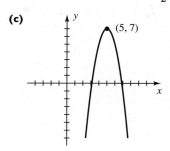

23 $y = \dfrac{1}{8}(x - 4)^2 - 1$ **25** $y = -\dfrac{4}{9}(x + 2)^2 + 4$

27 $y = 3(x - 0)^2 - 2$ **29** $y = -\dfrac{5}{9}(x - 3)^2 + 5$

31 $y = -\dfrac{1}{4}(x - 1)^2 + 4$ **33** 24.72 km **35** 10.5 lb

37 (a) 424 ft　　**(b)** 100 ft　　**39** 20 and 20

41 (a) $y = 250 - \dfrac{3}{4}x$　　**(b)** $A = x\left(250 - \dfrac{3}{4}x\right)$

　　(c) $166\dfrac{2}{3}$ ft by 125 ft

43 $y = -\dfrac{4}{27}\left(x - \dfrac{9}{2}\right)^2 + 3$

45 (a) $y = \dfrac{1}{500}x^2 + 10$　　**(b)** 282 ft　　**47** 2 ft

49 500 pairs

51 (a) $R(x) = 500x(30 - x)$

　　(b) \$15

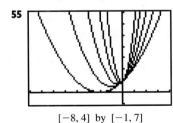

53

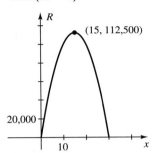

(−0.57, 0.64),
(0.02, −0.27),
(0.81, −0.41)

$[-3, 3]$ by $[-2, 2]$

55

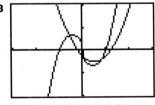

Smaller values of a result in a wider parabola; larger values of a result in a narrower parabola.

$[-8, 4]$ by $[-1, 7]$

EXERCISES 3.7

1 (a) 15　　**(b)** −3　　**(c)** 54　　**(d)** $\dfrac{2}{3}$

3 (a) $3x^2 + 1$; $3 - x^2$; $2x^4 + 3x^2 - 2$; $\dfrac{x^2 + 2}{2x^2 - 1}$

　　(b) $\mathbb{R}$　　**(c)** All real numbers except $\pm\dfrac{1}{2}\sqrt{2}$

5 (a) $2\sqrt{x + 5}$; 0; $x + 5$; 1　　**(b)** $[-5, \infty)$
　　(c) $(-5, \infty)$

7 (a) $\dfrac{3x^2 + 6x}{(x - 4)(x + 5)}$; $\dfrac{x^2 + 14x}{(x - 4)(x + 5)}$; $\dfrac{2x^2}{(x - 4)(x + 5)}$; $\dfrac{2(x + 5)}{x - 4}$

　　(b) All real numbers except −5 and 4
　　(c) All real numbers except −5, 0, and 4

9 (a) $-2x^2 - 1$　　**(b)** $-4x^2 + 4x - 1$　　**(c)** $4x - 3$
　　(d) $-x^4$

11 (a) $6x + 9$　　**(b)** $6x - 8$　　**(c)** −3　　**(d)** 10

13 (a) $75x^2 + 4$　　**(b)** $15x^2 + 20$　　**(c)** 304　　**(d)** 155

15 (a) $8x^2 - 2x - 5$　　**(b)** $4x^2 + 6x - 9$　　**(c)** 31
　　(d) 45

17 (a) $8x^3 - 20x$　　**(b)** $128x^3 - 20x$　　**(c)** −24
　　(d) 3396

19 (a) 7　　**(b)** −7　　**(c)** 7　　**(d)** −7

21 (a) $x + 2 - 3\sqrt{x + 2}$; $[-2, \infty)$
　　(b) $\sqrt{x^2 - 3x + 2}$; $(-\infty, 1] \cup [2, \infty)$

23 (a) $3x - 4$; $[0, \infty)$
　　(b) $\sqrt{3x^2 - 12}$; $(-\infty, -2] \cup [2, \infty)$

25 (a) $\sqrt{\sqrt{x + 5} - 2}$; $[-1, \infty)$
　　(b) $\sqrt{\sqrt{x - 2} + 5}$; $[2, \infty)$

27 (a) $\sqrt{3 - \sqrt{x^2 - 16}}$; $[-5, -4] \cup [4, 5]$
　　(b) $\sqrt{-x - 13}$; $(-\infty, -13]$

29 (a) x; $\mathbb{R}$　　**(b)** x; $\mathbb{R}$

31 (a) $\dfrac{1}{x^6}$; all nonzero real numbers

　　(b) $\dfrac{1}{x^6}$; all nonzero real numbers

33 (a) $\dfrac{1}{5 - x}$; all real numbers except 4 and 5

　　(b) $\dfrac{-2x + 5}{-3x + 7}$; all real numbers except 2 and $\dfrac{7}{3}$

35 $-3 \pm \sqrt{2}$　　**37 (a)** 5　　**(b)** 6　　**(c)** 6　　**(d)** 5

39 $20\sqrt{x^2 + 1}$　　**41** $A = 36\pi t^2$　　**43** $r = 9\sqrt[3]{t}$

45 $h = 5\sqrt{t^2 + 8t}$　　**47** $d = \sqrt{90,400 + (500 + 150t)^2}$

Exer. 49–56: Answers are not unique.

49 $u = x^2 + 3x$, $y = u^{1/3}$　　**51** $u = x - 3$, $y = u^{-4}$
53 $u = x^4 - 2x^2 + 5$, $y = u^5$

55 $u = \sqrt{x + 4}$, $y = \dfrac{u - 2}{u + 2}$　　**57** 5×10^{-13}

EXERCISES 3.8

1 Yes　　**3** No　　**5** Yes　　**7** No　　**9** No
11 Yes

Exer. 13–16: Show that $f(g(x)) = x = g(f(x))$.

13

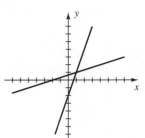

15

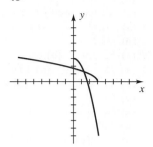

17 $f^{-1}(x) = \dfrac{x-5}{3}$ **19** $f^{-1}(x) = \dfrac{2x+1}{3x}$

21 $f^{-1}(x) = \dfrac{5x+2}{2x-3}$ **23** $f^{-1}(x) = -\sqrt{\dfrac{2-x}{3}}$

25 $f^{-1}(x) = \sqrt[3]{\dfrac{x+5}{2}}$ **27** $f^{-1}(x) = 3 - x^2,\ x \ge 0$

29 $f^{-1}(x) = (x-1)^3$ **31** $f^{-1}(x) = x$

33 (a) Since f is one-to-one, an inverse exists;

$f^{-1}(x) = \dfrac{x-b}{a}$

(b) No; not one-to-one

35 (a)

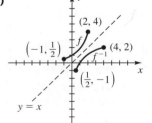

(b) $D = [-1, 2]$;

$R = \left[\dfrac{1}{2}, 4\right]$

(c) $D_1 = \left[\dfrac{1}{2}, 4\right]$;

$R_1 = [-1, 2]$

37 (a)

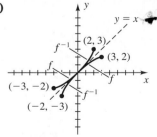

(b) $D = [-3, 3]$;

$R = [-2, 2]$

(c) $D_1 = [-2, 2]$;

$R_1 = [-3, 3]$

39 (c) The graph of f is symmetric about the line $y = x$.
Thus, $f(x) = f^{-1}(x)$.

41 Yes

43

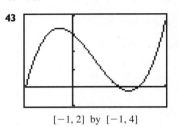

$[-1, 2]$ by $[-1, 4]$

(a) $[-0.27, 1.22]$
(b) $[-0.20, 3.31]$;
$[-0.27, 1.22]$

45 $x = \dfrac{0.4996 + \sqrt{(-0.4996)^2 - 4(0.0833)(3.5491 - D)}}{2(0.0833)}$

EXERCISES 3.9

1 $u = kv;\ k = \dfrac{2}{5}$ **3** $r = k\dfrac{s}{t};\ k = -14$

5 $y = k\dfrac{x^2}{z^3};\ k = 27$ **7** $z = kx^2y^3;\ k = -\dfrac{2}{49}$

9 $y = k\dfrac{x}{z^2};\ k = 36$ **11** $y = k\dfrac{\sqrt{x}}{z^3};\ k = \dfrac{40}{3}$

13 (a) $P = kd$ **(b)** 59 **(c)** 295 lb/ft^2

15 (a) $R = k\dfrac{l}{d^2}$ **(b)** $\dfrac{1}{40,000}$ **(c)** $\dfrac{50}{9}$ ohms

17 (a) $P = k\sqrt{l}$ **(b)** $\dfrac{3}{4}\sqrt{2}$ **(c)** $\dfrac{3}{2}\sqrt{3}$ sec

19 (a) $T = kd^{3/2}$ **(b)** $\dfrac{365}{(93)^{3/2}}$ **(c)** 223.2 days

21 (a) $V = k\sqrt{L}$ **(b)** $\dfrac{7}{2}\sqrt{2}$ **(c)** 60.6 mi/hr

23 (a) $W = kh^3$ **(b)** $\dfrac{25}{27}$ **(c)** 154 lb

25 (a) $F = kPr^4$ **(b)** About 2.05 times as hard

27 Increases 250% **29** $y = 1.2x$ **31** $y = -\dfrac{10.1}{x^2}$

CHAPTER 3 REVIEW EXERCISES

1 The points in quadrants II and IV
2 $d(A, B)^2 + d(A, C)^2 = d(B, C)^2$; area $= 10$
3 (a) $\sqrt{265}$ **(b)** $\left(-\dfrac{13}{2}, 1\right)$ **(c)** $(-11, -23)$

4 $(0, 1), (0, 11)$　　**5** $-2 < a < 1$

6 $(x - 7)^2 + (y + 4)^2 = 149$

7 $(x - 3)^2 + (y + 2)^2 = 169$　　**8** $x = -2 - \sqrt{9 - y^2}$

9 $-\dfrac{11}{19}$　　**10** The slope of AD and BC is $\dfrac{2}{3}$.

11 **(a)** $18x + 6y = 7$　　**(b)** $2x - 6y = 3$

12 $y = -\dfrac{8}{3}x + 8$　　**13** $(x + 5)^2 + (y + 1)^2 = 81$

14 $x + y = -3$　　**15** $5x - y = 23$　　**16** $2x - 3y = 5$

17 $C(0, 6);\ r = \sqrt{5}$　　**18** $C(-3, 2);\ r = \dfrac{1}{2}\sqrt{13}$

19 **(a)** $\dfrac{1}{2}$　　**(b)** $-\dfrac{1}{\sqrt{2}}$　　**(c)** 0　　**(d)** $-\dfrac{x}{\sqrt{3 - x}}$

　　(e) $-\dfrac{x}{\sqrt{x + 3}}$　　**(f)** $\dfrac{x^2}{\sqrt{x^2 + 3}}$　　**(g)** $\dfrac{x^2}{x + 3}$

20 **(a)** $\left[\dfrac{4}{3}, \infty\right);\ [0, \infty)$

　　(b) All real numbers except -3; $(0, \infty)$

21 $-2a - h + 1$　　**22** $-\dfrac{1}{(a + h + 2)(a + 2)}$

23 $f(x) = \dfrac{5}{2}x - \dfrac{1}{2}$

24 **(a)** Odd　　**(b)** Neither　　**(c)** Even

25

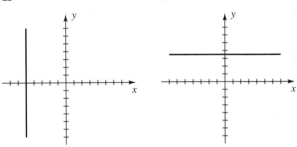

26

27

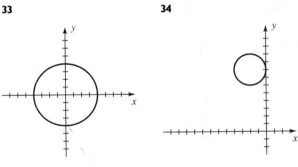

28

29

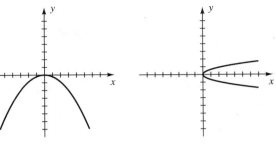

30

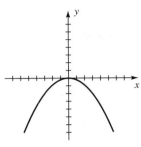

31

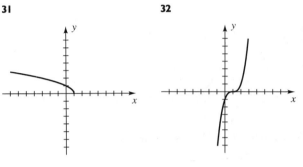

32

33

34

35

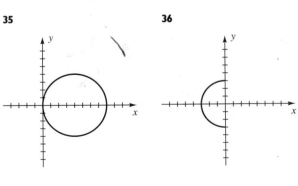

36

37

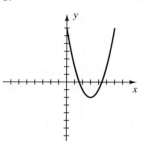

38

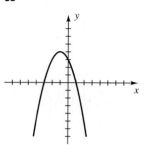

42 (a)

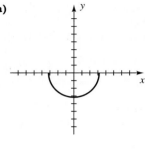

(b) $D = [-\sqrt{10}, \sqrt{10}]$;
$R = [-\sqrt{10}, 0]$
(c) Decreasing on
$[-\sqrt{10}, 0]$,
increasing on
$[0, \sqrt{10}]$

39 (a)

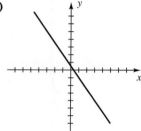

(b) $D = \mathbb{R}$; $R = \mathbb{R}$
(c) Decreasing on
$(-\infty, \infty)$

43 (a)

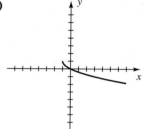

(b) $D = [-1, \infty)$;
$R = (-\infty, 1]$
(c) Decreasing on
$[-1, \infty)$

40 (a)

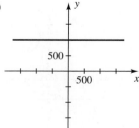

(b) $D = \mathbb{R}$;
$R = \{1000\}$
(c) Constant on
$(-\infty, \infty)$

44 (a)

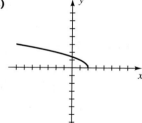

(b) $D = (-\infty, 2]$;
$R = [0, \infty)$
(c) Decreasing on
$(-\infty, 2]$

41 (a)

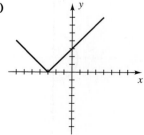

(b) $D = \mathbb{R}$;
$R = [0, \infty)$
(c) Decreasing on
$(-\infty, -3]$,
increasing on
$[-3, \infty)$

45 (a)

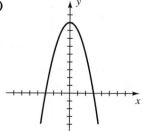

(b) $D = \mathbb{R}$;
$R = (-\infty, 9]$
(c) Increasing on
$(-\infty, 0]$,
decreasing on
$[0, \infty)$

46 (a)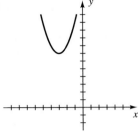

(b) $D = \mathbb{R}$; $R = [7, \infty)$
(c) Decreasing on $(-\infty, -3]$, increasing on $[-3, \infty)$

(c)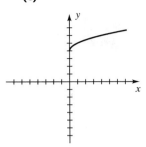

(d)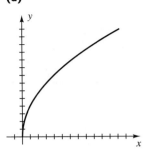

47 (a)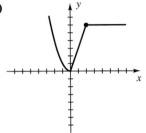

(b) $D = \mathbb{R}$; $R = [0, \infty)$
(c) Decreasing on $(-\infty, 0]$, increasing on $[0, 2]$, constant on $[2, \infty)$

(e)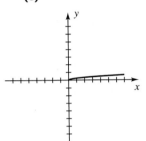

(f)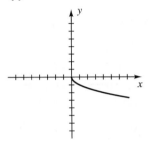

48 (a)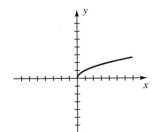

(b) $D = \mathbb{R}$; $R = \{\ldots, -3, -1, 1, 3, \ldots\}$
(c) Constant on $[n, n + 1)$, where n is any integer

50 (a) **(b)**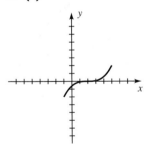

(c) 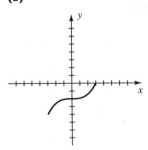 **(d)**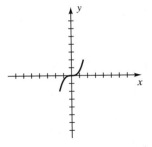

49 (a) 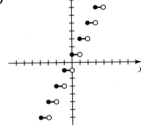 **(b)**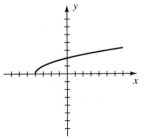

(e)

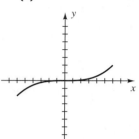

(f)

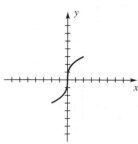

51 $2x - 5y = 10$ **52** $(x + 2)^2 + (y - 1)^2 = 25$

53 $y = \dfrac{1}{2}(x - 2)^2 - 4$ **54** $y = -|x - 2| - 1$

55 Min: $f(-3) = 4$ **56** Max: $f(5) = -7$

57 $f(x) = -2(x - 3)^2 + 4$ **58** $y = \dfrac{3}{2}(x - 3)^2 - 2$

59 (a) $[0, 2]$ **(b)** $(0, 2]$ **60 (a)** -1 **(b)** $\sqrt{13}$

61 (a) $18x^2 + 9x - 1$ **(b)** $6x^2 - 15x + 5$

62 (a) $\sqrt{\dfrac{3 + 2x^2}{x^2}}$ **(b)** $\dfrac{1}{3x + 2}$

63 (a) $\sqrt{28 - x}$; $[3, 28]$ **(b)** $\sqrt{\sqrt{25 - x^2} - 3}$; $[-4, 4]$

64 (a) $\dfrac{1}{x + 3}$; all real numbers except -3 and 0

     (b) $\dfrac{6x + 4}{x}$; all real numbers except $-\dfrac{2}{3}$ and 0

65 $u = x^2 - 5x,\ y = \sqrt[3]{u}$ **66** Yes

67 (a) $f^{-1}(x) = \dfrac{10 - x}{15}$ **68 (a)** $f^{-1}(x) = -\sqrt{\dfrac{9 - x}{2}}$

(b)

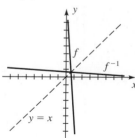

(b)

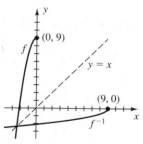

69 27   **70 (a)** 259 ft   **(b)** 2002

71 (a) $V = 6000t + 89{,}000$ **(b)** $2\dfrac{1}{3}$

72 (a) $F = \dfrac{9}{5}C + 32$ **(b)** $1.8\,°F$

73 (a) $C_1(x) = \dfrac{1}{16}x$ **(b)** $C_2(x) = \dfrac{5}{88}x + 50$ **(c)** 8800

74 (a) $y = -\dfrac{4}{5}x + 20$ **(b)** $V = 4x\left(-\dfrac{4}{5}x + 20\right)$

75 $C = \dfrac{3\pi(r^3 + 16)}{10r}$

76 (a) $V = 10t$

     (b) $V = 200h^2$ for $0 \le h \le 6$; $V = 7200 + 3200(h - 6)$ for $6 < h \le 9$

     (c) $h = \sqrt{\dfrac{t}{20}}$ for $0 \le t \le 720$; $h = 6 + \dfrac{t - 720}{320}$ for $720 < t \le 1680$

77 (a) $r = \dfrac{1}{2}x$ **(b)** $y = \dfrac{5}{4\pi} - \dfrac{1}{48}x^3$

78 (a) $y = \dfrac{bh}{a - b}$ **(b)** $V = \dfrac{1}{3}\pi h(a^2 + ab + b^2)$

     (c) $\dfrac{200}{7\pi} \approx 9.1$ ft

79 $\dfrac{18}{13}$ hr after 1:00 P.M., or about 2:23 P.M.

80 Radius of semicircle is $\dfrac{1}{8\pi}$ mi; length of rectangle is $\dfrac{1}{8}$ mi.

81 (a) 1 sec   **(b)** 4 ft   **(c)** On the moon, 6 sec and 24 ft

82 (a) $(87.5, 17.5)$ **(b)** 30.625 units   **83** 375

84 10,125 watts

CHAPTER 4

EXERCISES 4.1

1 (a)

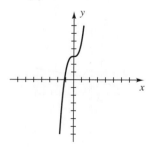

(b)

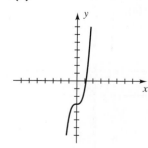

3 (a)

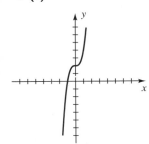

(b)

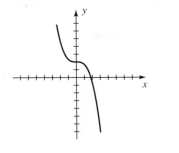

5 $f(3) = -2 < 0,\ f(4) = 10 > 0$

7 $f(2) = 5 > 0,\ f(3) = -5 < 0$

9 $f\left(-\dfrac{1}{2}\right) = \dfrac{19}{32} > 0,\ f(-1) = -1 < 0$

11 $f(x) > 0$ if $x > 2$,
$f(x) < 0$ if $x < 2$

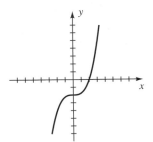

13 $f(x) > 0$ if $|x| < 2$,
$f(x) < 0$ if $|x| > 2$

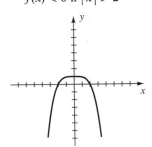

15 $f(x) > 0$ if $|x| > 2$,
$f(x) < 0$ if $0 < |x| < 2$

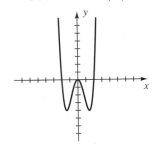

17 $f(x) > 0$ if $x < -2$ or
$0 < x < 5$, $f(x) < 0$ if
$-2 < x < 0$ or $x > 5$

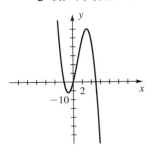

19 $f(x) > 0$ if $-2 < x < 3$
or $x > 4$, $f(x) < 0$ if
$x < -2$ or $3 < x < 4$

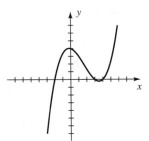

21 $f(x) > 0$ if $x > 2$,
$f(x) < 0$ if $x < -2$
or $|x| < 2$

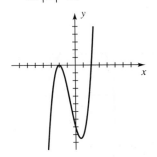

23 $f(x) > 0$ if $|x| > 2$ or
$|x| < \sqrt{2}$, $f(x) < 0$ if
$\sqrt{2} < |x| < 2$

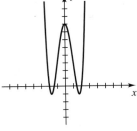

25 $f(x) > 0$ if $|x| > 2$,
$f(x) < 0$ if $|x| < 2$,
$x \neq 0,\ x \neq 1$

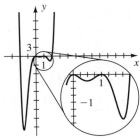
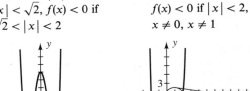

27 If n is even, then $(-x)^n = x^n$ and hence $f(-x) = f(x)$.
Thus, f is an even function.

29 $-\dfrac{4}{3}$

31 $P(x) > 0$ on $\left(-\dfrac{1}{5}\sqrt{15}, 0\right)$ and $\left(\dfrac{1}{5}\sqrt{15}, \infty\right)$;
$P(x) < 0$ on $\left(-\infty, -\dfrac{1}{5}\sqrt{15}\right)$ and $\left(0, \dfrac{1}{5}\sqrt{15}\right)$

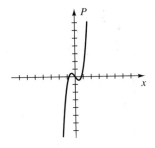

33 (b) $V(x) > 0$ on $(0, 10)$ and $(15, \infty)$; allowable values for x are in $(0, 10)$.

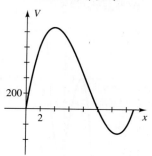

35 (a) $T > 0$ for $0 < t < 12$; $T < 0$ for $12 < t < 24$

(b)

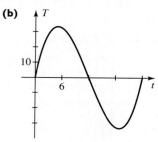

(c) $T(6) = 32.4 > 32$, $T(7) = 29.75 < 32$

37 (a) $N(t) > 0$ for $0 < t < 5$

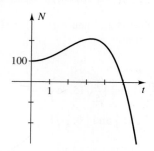

(b) The population becomes extinct after 5 years.

39

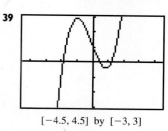

$[-4.5, 4.5]$ by $[-3, 3]$

$-1.89, 0.49, 1.20$

41

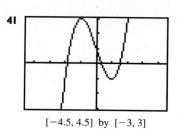

$[-4.5, 4.5]$ by $[-3, 3]$

$-1.88, 0.35,$ and 1.53

43

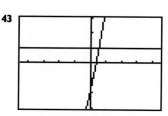

$[-4.5, 4.5]$ by $[-3, 3]$

$(0.56, \infty)$

45

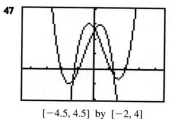

$[-4.5, 4.5]$ by $[-3, 3]$

$(-1.10, \infty)$

47

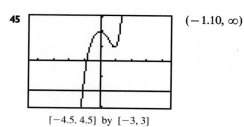

$[-4.5, 4.5]$ by $[-2, 4]$

$(-1.29, -0.77),$
$(0.085, 2.66),$
$(1.36, -0.42)$

EXERCISES 4.2

1 $2x^2 - x + 3; 4x - 3$ **3** $\frac{3}{2}x; \frac{1}{2}x - 4$ **5** $0; 7x + 2$

7 $\frac{9}{2}; \frac{53}{2}$ **9** 26 **11** 7 **13** $f(-3) = 0$

15 $f(-2) = 0$ **17** $x^3 - 3x^2 - 10x$

19 $x^4 - 2x^3 - 9x^2 + 2x + 8$ **21** $2x^2 + x + 6$; 7

23 $x^2 - 3x + 1$; -8

25 $3x^4 - 6x^3 + 12x^2 - 18x + 36$; -65

27 $4x^3 + 2x^2 - 4x - 2$; 0   **29** 73   **31** -0.0444

33 $8 + 7\sqrt{3}$ **35** $f(-2) = 0$ **37** $f\left(\dfrac{1}{2}\right) = 0$

39 3, 5   **41** $f(c) > 0$ **43** -14

45 If $f(x) = x^n - y^n$ and n is even, then $f(-y) = 0$.

47 (a) $V = \pi x^2(6 - x)$ (b) $\left(\dfrac{1}{2}(5 + \sqrt{45}), \dfrac{1}{2}(7 - \sqrt{45})\right)$

49 (a) $A = 8x - 2x^3$ (b) $\sqrt{13} - 1 \approx 2.61$ **51** 27

53 $f(x) = ((((2x + 3)x - 6)x - 4)x + 1)x - 9$; -8.9719

55 -9.55 **57** -0.75, 1.96

EXERCISES 4.3

1 $-4x^3 + 16x^2 - 4x - 24$ **3** $3x^3 + 3x^2 - 36x$

5 $-2x^3 + 6x^2 - 8x + 24$

7 $x^4 + 2x^3 - 23x^2 - 24x + 144$

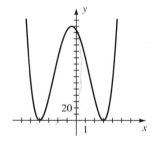

9 $3x^6 - 27x^5 + 81x^4 - 81x^3$

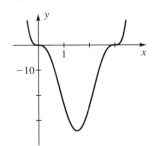

11 $f(x) = \dfrac{7}{9}(x + 1)\left(x - \dfrac{3}{2}\right)(x - 3)$

13 $f(x) = -1(x - 1)^2(x - 3)$

15 $-\dfrac{2}{3}$ (multiplicity 1); 0 (multiplicity 2); $\dfrac{5}{2}$ (multiplicity 3)

17 $-\dfrac{3}{2}$ (multiplicity 2); 0 (multiplicity 3)

19 -4 (multiplicity 3); -3 (multiplicity 2);
  3 (multiplicity 5)

21 $\pm 4i$, ± 3 (each of multiplicity 1)

23 $f(x) = (x + 3)^2(x + 2)(x - 1)$

25 $f(x) = (x - 1)^5(x + 1)$

Exer. 27–34: The types of possible solutions are listed in the order positive, negative, nonreal complex.

27 3, 0, 0 or 1, 0, 2   **29** 0, 1, 2

31 2, 2, 0; 2, 0, 2; 0, 2, 2; 0, 0, 4

33 2, 3, 0; 2, 1, 2; 0, 3, 2; 0, 1, 4

35 Upper 5, lower -2 **37** Upper 2, lower -2

39 Upper 3, lower -3

41 $f(t) = \dfrac{5}{3528}t(t - 5)(t - 19)(t - 24)$

43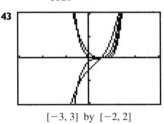

As the multiplicity increases, the graph becomes more horizontal at $(0.5, 0)$.

$[-3, 3]$ by $[-2, 2]$

45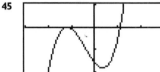

-1.2 (multiplicity 2); 1.1 (multiplicity 1)

$[-3, 3]$ by $[-3, 1]$

47 Yes: $1.5(x - 2)(x - 5.2)(x - 10.1)$

49 2007 (when t $\approx$ 27.1)

EXERCISES 4.4

1 $x^2 - 6x + 13$ **3** $(x - 2)(x^2 + 4x + 29)$

5 $x(x + 1)(x^2 - 6x + 10)$

7 $(x^2 - 8x + 25)(x^2 + 4x + 5)$

9 $x(x^2 + 4)(x^2 - 2x + 2)$

Exer. 11–14: Show that none of the possible rational roots listed satisfy the equation.

11 $\pm 1, \pm 2, \pm 3, \pm 6$　　**13** $\pm 1, \pm 2$　　**15** $-2, -1, 4$

17 $-3, 2, \dfrac{5}{2}$　　**19** $-7, \pm\sqrt{2}, 4$

21 $-3, -\dfrac{2}{3}, 0$ (multiplicity 2), $\dfrac{1}{2}$　　**23** $-\dfrac{3}{4}, -\dfrac{3}{4} \pm \dfrac{3}{4}\sqrt{7}\,i$

25 No. If i is a root, then $-i$ is also a root. Hence, the polynomial would have factors $x - 1, x + 1, x - i, x + i$ and therefore would be of degree greater than 3.

27 Since n is odd and nonreal complex zeros occur in conjugate pairs for polynomials with real coefficients, there must be at least one real zero.

29 **(a)** The two boxes correspond to $x = 5$ and
　　　$x = 5(2 - \sqrt{2})$.
　　(b) The box corresponding to $x = 5$

31 **(c)** In feet: 5, 12, and 13　　**33** **(b)** 4 ft

35 None　　**37** $-1.2, 0.8, -\dfrac{1}{2} \pm \dfrac{\sqrt{3}}{2}\,i$　　**39** 10,200 m

7
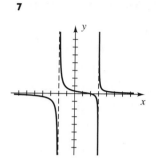

9
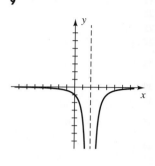

11
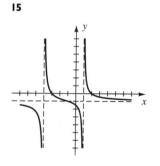

13
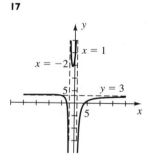

EXERCISES 4.5

1 (a)
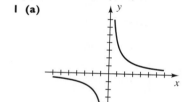

(b) D = all nonzero real numbers; $R = D$
(c) Decreasing on $(-\infty, 0)$ and on $(0, \infty)$

15
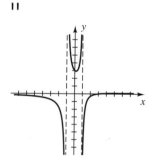

17
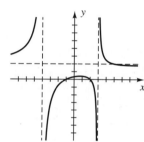

3
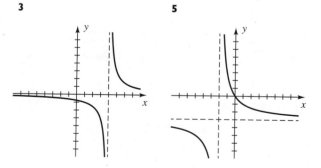

5

19
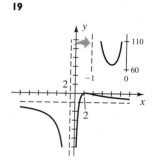

21
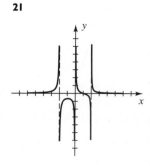

23

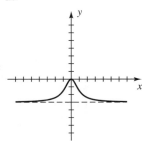

25 $y = x - 2$

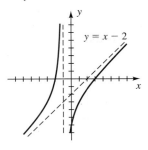

37 (a) $V(t) = 50 + 5t$, $A(t) = 0.5t$ **(b)** $\dfrac{t}{10t + 100}$

(c) As $t \to \infty$, $c(t) \to 0.1$ lb of salt per gal.

39

41

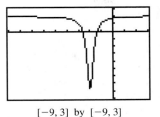
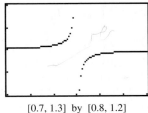

[−9, 3] by [−9, 3]   [0.7, 1.3] by [0.8, 1.2]

27 $y = -\dfrac{1}{2}x$

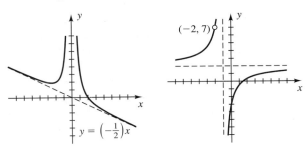

29 $f(x) = \dfrac{2x - 3}{x + 1}$ for $x \neq -2$

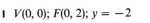

EXERCISES 4.6

1 $V(0, 0)$; $F(0, 2)$; $y = -2$ **3** $V(0, 0)$; $F\left(-\dfrac{3}{8}, 0\right)$; $x = \dfrac{3}{8}$

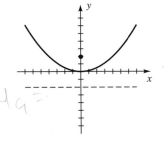

31 $f(x) = \dfrac{-1}{x + 1}$ for $x \neq 1$ **33** $f(x) = x - 1$ for $x \neq -2$

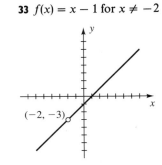

5 $V(-2, 1)$; $F(-2, -1)$;   **7** $V(3, 2)$; $F\left(\dfrac{49}{16}, 2\right)$; $x = \dfrac{47}{16}$
    $y = 3$

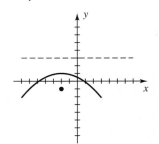

35 (a) $h = \dfrac{16}{(r + 0.5)^2} - 1$ **(b)** $V(r) = \pi r^2 h$

(c) Exclude $r \leq 0$ and $r \geq 3.5$.

9 $V(2, -2)$; $F\left(2, -\frac{7}{4}\right)$;

$y = -\frac{9}{4}$

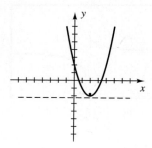

11 $V\left(0, \frac{1}{2}\right)$; $F\left(0, -\frac{9}{2}\right)$;

$y = \frac{11}{2}$

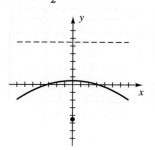

1 $V(\pm 3, 0)$; $F(\pm\sqrt{5}, 0)$

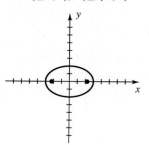

3 $V(0, \pm 4)$; $F(0, \pm 1)$

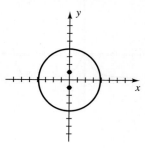

13 $y^2 = 20(x - 1)$　　**15** $(x + 2)^2 = -16(y - 3)$
17 $y^2 = 8x$　　**19** $(x - 6)^2 = 12(y - 1)$
21 $(y + 5)^2 = 4(x - 3)$　　**23** $y^2 = -12(x + 1)$
25 $3x^2 = -4y$　　**27** $(y - 5)^2 = 2(x + 3)$
29 $x^2 = 16(y - 1)$　　**31** $(y - 3)^2 = -8(x + 4)$

33 4 in.　　**35** $\frac{9}{16}$ ft from the center of the paraboloid

37 $2\sqrt{480} \approx 43.82$ in.

39 (a) $p = \frac{r^2}{4h}$　　(b) $10\sqrt{2}$ ft

5 $V(0, \pm 4)$; $F(0, \pm 2\sqrt{3})$

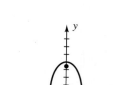

7 $V\left(\pm\frac{1}{2}, 0\right)$;

$F\left(\pm\frac{1}{10}\sqrt{21}, 0\right)$

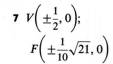

41

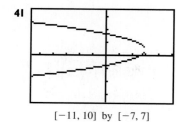

$[-11, 10]$ by $[-7, 7]$

43

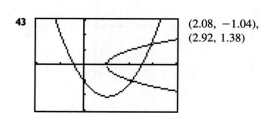

$(2.08, -1.04)$,
$(2.92, 1.38)$

$[-2, 4]$ by $[-3, 3]$

9 $V(3 \pm 4, -4)$;
　　$F(3 \pm \sqrt{7}, -4)$

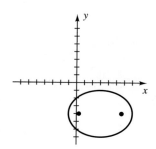

11 $V(4 \pm 3, 2)$;
　　$F(4 \pm \sqrt{5}, 2)$

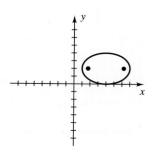

13 $V(5, 2 \pm 5)$; $F(5, 2 \pm \sqrt{21})$

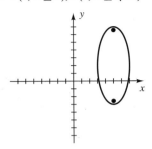

15 $\dfrac{x^2}{4} + \dfrac{y^2}{36} = 1$ **17** $\dfrac{(x+2)^2}{25} + \dfrac{(y-1)^2}{4} = 1$

19 $\dfrac{x^2}{64} + \dfrac{y^2}{39} = 1$ **21** $\dfrac{4x^2}{9} + \dfrac{y^2}{25} = 1$ **23** $\dfrac{8x^2}{81} + \dfrac{y^2}{36} = 1$

25 $\dfrac{x^2}{7} + \dfrac{y^2}{16} = 1$ **27** $\dfrac{x^2}{4} + 9y^2 = 1$ **29** $\dfrac{x^2}{16} + \dfrac{4y^2}{25} = 1$

31 $\dfrac{x^2}{25} + \dfrac{y^2}{16} = 1$ **33** $\dfrac{x^2}{64} + \dfrac{y^2}{289} = 1$ **35** $\sqrt{84} \approx 9.2$ ft

37 94,581,000 mi; 91,419,000 mi

39 (a) $d = h - \sqrt{h^2 - \frac{1}{4}k^2}$; $d' = h + \sqrt{h^2 - \frac{1}{4}k^2}$

(b) 16 cm; 2 cm from V

41 5 ft

43
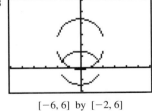
$[-6, 6]$ by $[-2, 6]$

$(\pm 1.540, 0.618)$

45
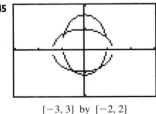
$[-3, 3]$ by $[-2, 2]$

$(-0.88, 0.76)$,
$(-0.48, -0.91)$,
$(0.58, -0.81)$,
$(0.92, 0.59)$

EXERCISES 4.8

1 $V(\pm 3, 0)$; $F(\pm\sqrt{13}, 0)$; $y = \pm\dfrac{2}{3}x$

3 $V(0, \pm 3)$; $F(0, \pm\sqrt{13})$; $y = \pm\dfrac{3}{2}x$

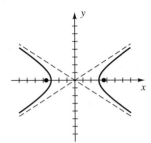
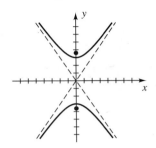

5 $V(\pm 1, 0)$; $F(\pm 5, 0)$; $y = \pm\sqrt{24}\,x$

7 $V(0, \pm 4)$; $F(0, \pm 2\sqrt{5})$; $y = \pm 2x$

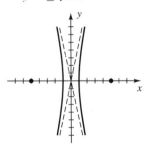
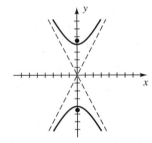

9 $V\left(\pm\dfrac{1}{4}, 0\right)$; $F\left(\pm\dfrac{1}{12}\sqrt{13}, 0\right)$; $y = \pm\dfrac{2}{3}x$

11 $V(-2, -2 \pm 3)$; $F(-2, -2 \pm \sqrt{13})$; $y + 2 = \pm\dfrac{3}{2}(x + 2)$

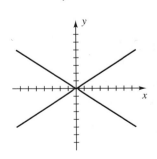
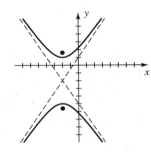

13 $V(-3 \pm 5, -2)$;
$F(-3 \pm 13, -2)$;
$(y + 2) = \pm\dfrac{12}{5}(x + 3)$

15 $V(-2, -5 \pm 3)$;
$F(-2, -5 \pm 3\sqrt{5})$;
$y + 5 = \pm\dfrac{1}{2}(x + 2)$

45

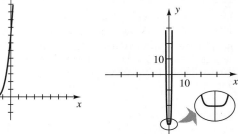

$[-15, 15]$ by $[-10, 10]$

None

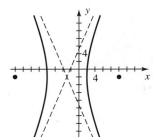

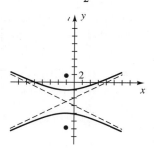

17 $\dfrac{x^2}{9} - \dfrac{y^2}{16} = 1$ **19** $(y + 3)^2 - \dfrac{(x + 2)^2}{3} = 1$

21 $y^2 - \dfrac{x^2}{15} = 1$ **23** $\dfrac{x^2}{9} - \dfrac{y^2}{16} = 1$ **25** $\dfrac{y^2}{21} - \dfrac{x^2}{4} = 1$

27 $\dfrac{x^2}{9} - \dfrac{y^2}{36} = 1$ **29** $\dfrac{x^2}{25} - \dfrac{y^2}{100} = 1$ **31** $\dfrac{y^2}{25} - \dfrac{x^2}{49} = 1$

33 $\dfrac{x^2}{144} - \dfrac{y^2}{25} = 1$ **35** $\dfrac{y^2}{64} - \dfrac{x^2}{36} = 1$

37

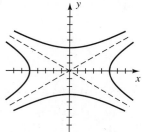

The graphs have the same asymptotes.

CHAPTER 4 REVIEW EXERCISES

1 $f(x) > 0$ if $x > -2$,
$f(x) < 0$ if $x < -2$

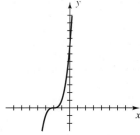

2 $f(x) > 0$ if $x < -\sqrt[6]{32}$
or $x > \sqrt[6]{32}$, $f(x) < 0$ if
$-\sqrt[6]{32} < x < \sqrt[6]{32}$

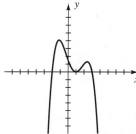

39 $x = \sqrt{9 + 4y^2}$

41 If a coordinate system having origin at the midpoint of AB is introduced, then the ship's coordinates are $\left(\dfrac{80}{3}\sqrt{34}, 100\right) \approx (155.5, 100)$.

43

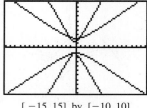

$[-15, 15]$ by $[-10, 10]$

$(0.741, 2.206)$

3 $f(x) > 0$ if $-2 < x < 1$
or $1 < x < 3$, $f(x) < 0$
if $x < -2$ or $x > 3$

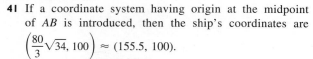

4 $f(x) > 0$ if $-1 < x < 0$
or $0 < x < 2$, $f(x) < 0$
if $x < -1$ or $x > 2$

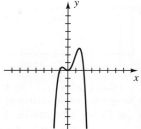

5 $f(x) > 0$ if $-4 < x < 0$
or $x > 2$, $f(x) < 0$ if
$x < -4$ or $0 < x < 2$

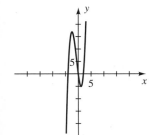

6 $f(x) > 0$ if $-4 < x < -2$,
$0 < x < 2$, or $x > 4$,
$f(x) < 0$ if $x < -4$,
$-2 < x < 0$, or
$2 < x < 4$

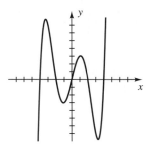

7 $f(0) = -9 < 100$ and $f(10) = 561 > 100$. By the
intermediate value theorem for polynomial functions,
f takes on every value between -9 and 561. Hence,
there is at least one real number a in $[0, 10]$ such that
$f(a) = 100$.

8 Let $f(x) = x^5 - 3x^4 - 2x^3 - x + 1$. $f(0) = 1 > 0$ and
$f(1) = -4 < 0$. By the intermediate value theorem
for polynomial functions, f takes on every value
between -4 and 1. Hence, there is at least one real
number a in $[0, 1]$ such that $f(a) = 0$.

9 $3x^2 + 2$; $-21x^2 + 5x - 9$ **10** $4x - 1$; $2x - 1$

11 -132 **12** $f(3) = 0$

13 $6x^4 - 12x^3 + 24x^2 - 52x + 104$; -200

14 $2x^2 + (5 + 2\sqrt{2})x + (2 + 5\sqrt{2})$; $11 + 2\sqrt{2}$

15 $\frac{2}{41}(x^2 + 6x + 34)(x + 1)$ **16** $\frac{1}{4}x(x^2 - 2x + 2)(x - 3)$

17 $x^7 + 6x^6 + 9x^5$

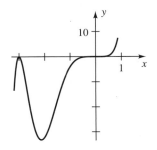

18 $(x - 2)^3(x + 3)(x - 1)$
19 1 (multiplicity 5); -3 (multiplicity 1)
20 0, $\pm i$ (all have multiplicity 2)
21 **(a)** Either 3 positive and 1 negative or 1 positive,
1 negative, and 2 nonreal complex
(b) Upper bound 3; lower bound -1

22 **(a)** Either 2 positive and 3 negative; 2 positive,
1 negative, and 2 nonreal complex; 3 negative and
2 nonreal complex; or 1 negative and 4 nonreal
complex
(b) Upper bound 2; lower bound -3

23 Since there are only even powers,
$7x^6 + 2x^4 + 3x^2 + 10 \geq 10$ for every real number x.

24 $-3, -2, -2 \pm i$ **25** $-\frac{1}{2}, \frac{1}{4}, \frac{3}{2}$ **26** $\pm\sqrt{6}, \pm 1$

27

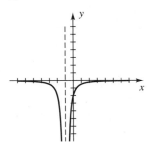

28

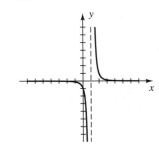

29

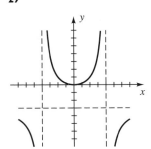

30

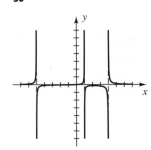

31

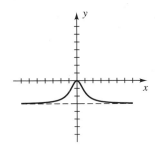

32

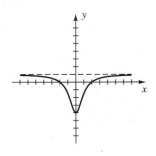

33

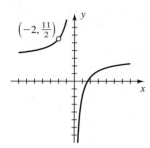

34

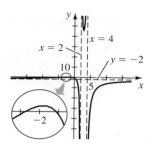

43 $V(0, 0)$; $F(16, 0)$

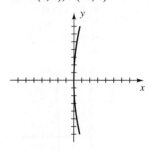

44 $V(-2, 1)$; $F\left(-2, \dfrac{33}{32}\right)$

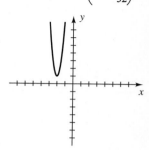

35

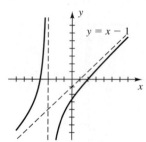

36

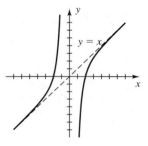

45 $V(0, \pm 4)$; $F(0, \pm\sqrt{7})$

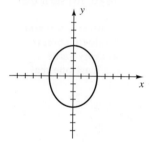

46 $V(0, \pm 4)$; $F(0, \pm 5)$

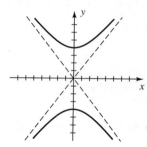

37 (a) $\dfrac{1}{15,000}$

(b) $y \approx 0.9754 < 1$ if $x = 6.1$, and $y \approx 1.0006 > 1$ if $x = 6.2$

38 (a) $V = \dfrac{1}{4\pi}x(l^2 - x^2)$

(b) If $x > 0$, $V > 0$ when $0 < x < l$.

39 $t = 4$ (10:00 A.M.) and $t = 16 - 4\sqrt{6} \approx 6.2020$ (12:12 P.M.)

40 $\sqrt{5} < t < 4$

41 (a) $R = k$

(b) k is the maximum rate at which the liver can remove alcohol from the bloodstream.

42 (a) $C(100) = \$2,000,000.00$ and $C(90) \approx \$163,636.36$

(b)

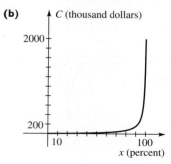

47 $V(\pm 2, 0)$; $F(\pm 2\sqrt{2}, 0)$

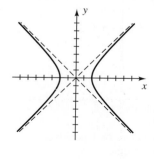

48 $V\left(\pm\dfrac{1}{5}, 0\right)$;

$F\left(\pm\dfrac{1}{30}\sqrt{11}, 0\right)$

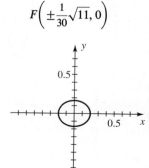

49 $V(0, 4)$; $F\left(0, -\dfrac{9}{4}\right)$

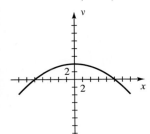

50 $V(3 \pm 2, -1)$; $F(3 \pm 1, -1)$

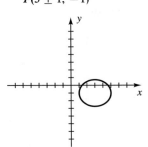

55 $V(2, -4)$; $F(4, -4)$

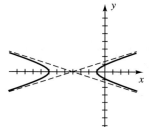

56 $V(3, -2 \pm 2)$; $F(3, -2 \pm \sqrt{3})$

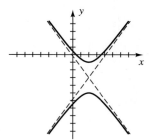

57 $V(-4 \pm 3, 0)$; $F(-4 \pm \sqrt{10}, 0)$

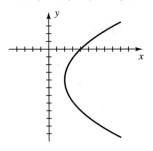

58 $V(2, -3 \pm 2)$; $F(2, -3 \pm \sqrt{6})$

51 $V(-4 \pm 1, 5)$; $F\left(-4 \pm \dfrac{1}{3}\sqrt{10}, 5\right)$

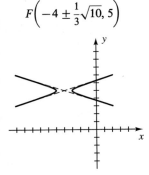

52 $V(-5, -2)$; $F\left(-\dfrac{39}{8}, -2\right)$

59 $y = 2(x + 7)^2 - 18$ **60** $y = -3(x + 4)^2 + 147$

61 $\dfrac{y^2}{49} - \dfrac{x^2}{9} = 1$ **62** $y^2 = -16x$ **63** $x^2 = -40y$

64 $x = 5y^2$ **65** $\dfrac{x^2}{75} + \dfrac{y^2}{100} = 1$ **66** $\dfrac{x^2}{25} - \dfrac{y^2}{75} = 1$

67 $\dfrac{y^2}{36} - \dfrac{x^2}{4/9} = 1$ **68** $\dfrac{x^2}{8} + \dfrac{y^2}{4} = 1$ **69** $\dfrac{x^2}{25} + \dfrac{y^2}{45} = 1$

70 $\dfrac{x^2}{256} + \dfrac{y^2}{112} = 1$ **71** **(a)** $-\dfrac{7}{2}$ **(b)** Hyperbola

72 $A = \dfrac{4a^2b^2}{a^2 + b^2}$ **73** $x^2 + (y - 2)^2 = 4$

74 $\dfrac{(x - 2)^2}{1} - \dfrac{y^2}{3} = 1,\ x \ge 3$ or $x = 2 + \sqrt{1 + \dfrac{y^2}{3}}$

53 $V(-3 \pm 3, 2)$; $F(-3 \pm \sqrt{5}, 2)$

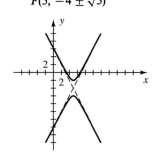

54 $V(5, -4 \pm 2)$; $F(5, -4 \pm \sqrt{5})$

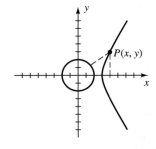

CHAPTER 5

EXERCISES 5.1

1 5 **3** $-1, 3$ **5** $-\dfrac{4}{99}$ **7** $\dfrac{18}{5}$

9 (a)

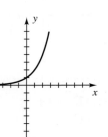

(b)

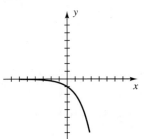

(c)

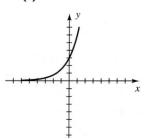

(d)

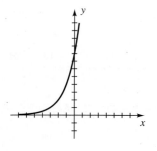

(e)

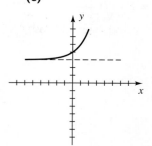

(f)

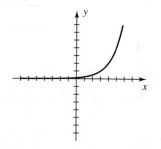

(g)

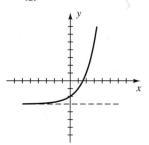

(h)

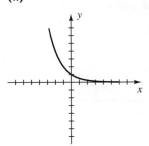

(i)

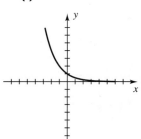

(j)

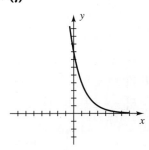

11

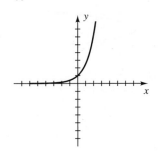

13

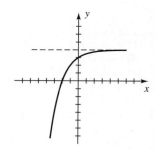

15

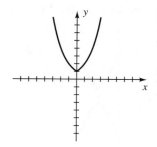

17

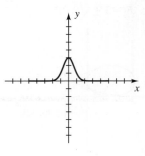

19

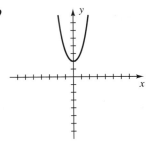

45

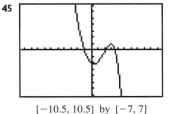

$[-10.5, 10.5]$ by $[-7, 7]$

$-1.02, 2.14, 3.62$

21 (a) 90 **(b)** 59 **(c)** 35

23 (a) 1039; 3118; 5400 **(b)**

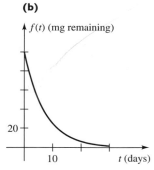

25 (a) 50 mg; 25 mg;

$$\frac{25}{2}\sqrt{2} \approx 17.7 \text{ mg}$$

(b)

47

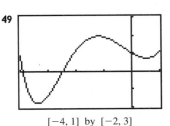

$[-3, 3]$ by $[-2, 2]$

(a) Not one-to-one
(b) 0

27 $-\dfrac{1}{1600}$

29 (a) \$1010.00 **(b)** \$1020.10 **(c)** \$1061.52
(d) \$1126.83

31 (a) \$7800 **(b)** \$4790 **(c)** \$2942

33 (a) \$1061.36 **(b)** \$1126.49 **(c)** \$1346.86
(d) \$1814.02

35 (a) Examine the pattern formed by the value y in the year n.
(b) Solve $s = (1 - a)^T y_0$ for a.

37 (a) \$925.75 **(b)** \$243,270 **39** \$6346.40

41 6.58 yr

49

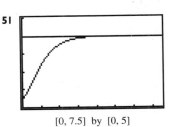

$[-4, 1]$ by $[-2, 3]$

(a) Increasing: $[-3.37, -1.19] \cup [0.52, 1]$;
decreasing: $[-4, -3.37] \cup [-1.19, 0.52]$
(b) $[-1.79, 1.94]$

43

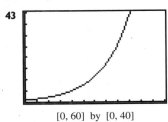

$[0, 60]$ by $[0, 40]$

(a) 26.13 **(b)** 8.50

51

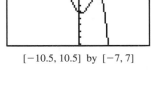

$[0, 7.5]$ by $[0, 5]$

The maximum number of sales approaches k.

53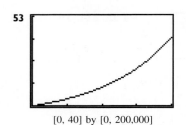

After approximately 32.8 yr

[0, 40] by [0, 200,000]

35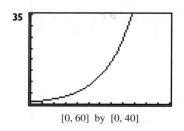

[0, 60] by [0, 40]

(a) 29.96   **(b)** 8.15

EXERCISES 5.2

1 (a)   **(b)**

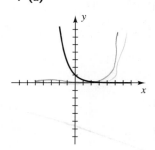

3 (a)   **(b)**

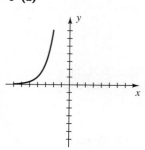

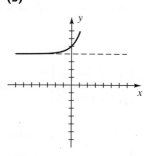

37 (a)   **(b)**

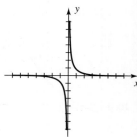

[−7.5, 7.5] by [−5, 5]

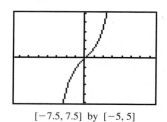

39 (a)   **(b)**

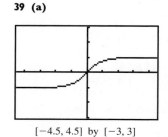

[−4.5, 4.5] by [−3, 3]

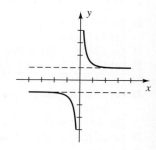

41   −1.04, 2.11

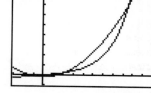

[−3, 11] by [−10, 80]

5 $1510.59   **7** $13,806.92   **9** 13%   **11** 3, 4

13 −1   **15** $-\frac{3}{4}, 0$ **17** $\dfrac{4}{(e^x + e^{-x})^2}$ **19** 27.43 g

21 261.1 million   **23** 13.5%   **25** 5412   **27** 7.44 in.

29 75.77 cm; 15.98 cm/yr   **31** $6.82 per hr

33 (a) 7.19%   **(b)** 7.25%

43

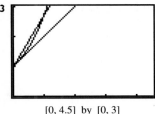

[0, 4.5] by [0, 3]

$f(x)$ is closer to e^x if $x \approx 0$; $g(x)$ is closer to e^x if $x \approx 1$.

45

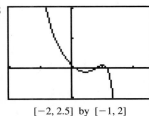

[−2, 2.5] by [−1, 2]

0.11, 0.79, 1.13

47

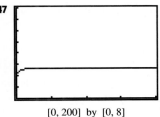

[0, 200] by [0, 8]

$y \approx 2.71 \approx e$

49 0.567

51

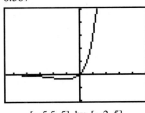

[−5.5, 5] by [−2, 5]

Increasing on $[-1, \infty)$; decreasing on $(-\infty, -1]$

53 (a) As h increases, C decreases.
(b) As y increases, C decreases.

55 (a) Larger values for c indicate a less reliable chip.

(b) About 3.45 yr

EXERCISES 5.3

1 (a) $\log_4 64 = 3$ (b) $\log_4 \frac{1}{64} = -3$ (c) $\log_t s = r$

(d) $\log_3 (4 - t) = x$ (e) $\log_5 \frac{a+b}{a} = 7t$

(f) $\log_{0.7} (5.3) = t$

3 (a) $2^5 = 32$ (b) $3^{-5} = \frac{1}{243}$ (c) $t^p = r$

(d) $3^5 = (x + 2)$ (e) $2^{3x+4} = m$ (f) $b^{3/2} = 512$

5 $t = 3 \log_a \dfrac{5}{2}$ **7** $t = \dfrac{1}{C} \log_a \left(\dfrac{A - D}{B} \right)$

9 (a) $\log 100{,}000 = 5$ (b) $\log 0.001 = -3$
(c) $\log (y + 1) = x$ (d) $\ln p = 7$
(e) $\ln (3 - x) = 2t$

11 (a) $10^{50} = x$ (b) $10^{20t} = x$ (c) $e^{0.1} = x$
(d) $e^{4+3x} = w$ (e) $e^{1/6} = z - 2$

13 (a) 0   (b) 1   (c) Not possible   (d) 2   (e) 8
(f) 3   (g) −2

15 (a) 3   (b) 5   (c) 2   (d) −4   (e) 2
(f) −3   (g) $3e^2$

17 4   **19** No solution   **21** −1, −2   **23** 13

25 27   **27** $\pm \dfrac{1}{e}$ **29** 3

31 (a)   (b)

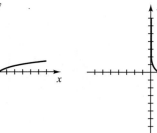

(c)   (d)

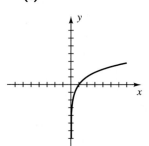

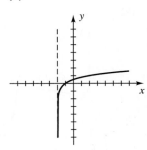

(e)

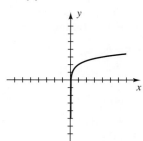

(f)

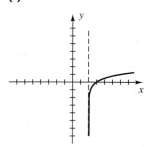

33

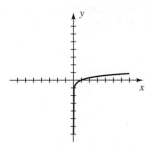

35

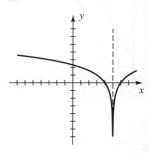

(g)

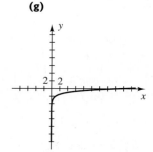

(h)

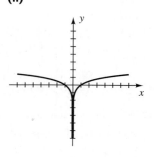

37 $f(x) = \log_2 x$ **39** $f(x) = \log_2 x - 1$
41 $f(x) = \log_2 (-x)$ **43** $f(x) = 2 \log_2 x$
45 (a) 4240   **(b)** 8.85   **(c)** 0.0237   **(d)** 9.97
     (e) 1.05   **(f)** 0.202
47 $t = -1600 \log_2 \left(\dfrac{q}{q_0} \right)$ **49** $t = -\dfrac{L}{R} \ln \left(\dfrac{I}{20} \right)$
51 (a) 2   **(b)** 4   **(c)** 5
53 (a) 10   **(b)** 30   **(c)** 40   **55** In the year 2079
57 (a) $W = 2.4e^{1.84h}$ **(b)** 37.92 kg
59 (a) 10,007 ft   **(b)** 18,004 ft
61 (a) 305.9 kg   **(b)** (1) 20 yr (2) 19.8 yr
63 10.1 mi   **65** $2^{1/8} \approx 1.09$
67 (a) Pedestrians have faster average walking speeds in
       large cities.
     (b) 570,000
69 (a) $f(1) = -0.1 < 0$, $f(2) \approx 0.29 > 0$ **(b)** 1.17

(i)

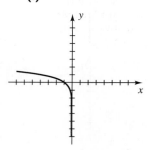

(j)

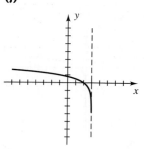

(k)

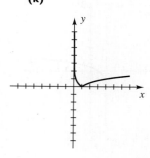

EXERCISES 5.4

1 (a) $\log_4 x + \log_4 z$ **(b)** $\log_4 y - \log_4 x$
   (c) $\dfrac{1}{3} \log_4 z$
3 $3 \log_a x + \log_a w - 2 \log_a y - 4 \log_a z$
5 $\dfrac{1}{3} \log z - \log x - \dfrac{1}{2} \log y$ **7** $\dfrac{7}{4} \ln x - \dfrac{5}{4} \ln y - \dfrac{1}{4} \ln z$
9 (a) $\log_3 (5xy)$ **(b)** $\log_3 \dfrac{2z}{x}$ **(c)** $\log_3 y^5$
11 $\log_a \dfrac{x^2 \sqrt[3]{x - 2}}{(2x + 3)^5}$ **13** $\log \dfrac{y^{13/3}}{x^2}$ **15** $\ln x$ **17** $\dfrac{7}{2}$
19 $5\sqrt{5}$ **21** No solution   **23** -7 **25** 1
27 -2 **29** $\dfrac{-1 + \sqrt{65}}{2}$ **31** $-1 + \sqrt{1 + e}$

33

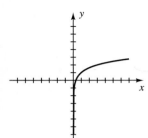

35

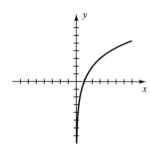

37

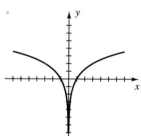

39

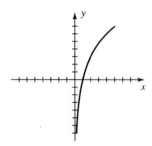

41

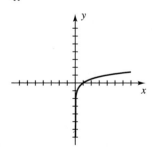

43

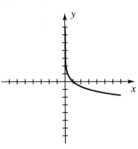

45 $f(x) = \log_2 x^2$ **47** $f(x) = \log_2 (8x)$ **49** $y = \dfrac{b}{x^k}$

51

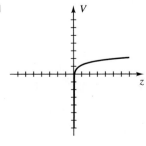

53 (a) 0   **(b)** $R(2x) = R(x) + a \log 2$ **55** 0.29 cm

57

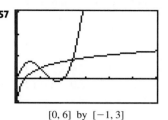

$(0, 1.01] \cup [2.4, \infty)$

[0, 6] by [−1, 3]

59

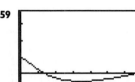

1.41, 6.59

[0, 8] by [−1.67, 3.67]

61

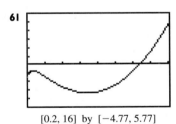

(a) Increasing on [0.2, 0.63] and [6.87, 16]; decreasing on [0.63, 6.87]
(b) 4.61; −3.31

[0.2, 16] by [−4.77, 5.77]

EXERCISES 5.5

1 $\dfrac{\log 8}{\log 5} \approx 1.29$ **3** $4 - \dfrac{\log 5}{\log 3} \approx 2.54$ **5** 1.1133

7 −0.7325   **9** 2   **11** $\dfrac{\log (2/81)}{\log 24} \approx -1.16$

13 $\dfrac{\log (8/25)}{\log (4/5)} \approx 5.11$ **15** −3   **17** 5

19 $\dfrac{2}{3}\sqrt{\dfrac{101}{11}} \approx 2.02$ **21** 1, 2   **23** $\dfrac{\log (4 + \sqrt{19})}{\log 4} \approx 1.53$

25 1 or 100   **27** 10^{100} **29** 10,000

31 $x = \log (y \pm \sqrt{y^2 - 1})$ **33** $x = \dfrac{1}{2} \log \left(\dfrac{1 + y}{1 - y} \right)$

35 $x = \ln (y + \sqrt{y^2 + 1})$ **37** $x = \dfrac{1}{2} \ln \left(\dfrac{y + 1}{y - 1} \right)$

39 y-intercept $= \log_2 3$
     ≈ 1.5850

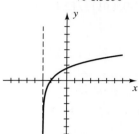

41 x-intercept $= \log_4 3$
     ≈ 0.7925

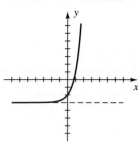

CHAPTER 5 REVIEW EXERCISES

1

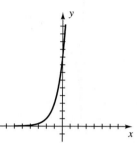

2

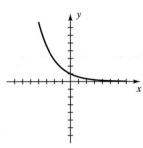

43 (a) 2.2   **(b)** 5   **(c)** 8.3
45 Basic if pH > 7, acidic if pH < 7
47 11.58 yr $\approx$ 11 yr 7 mo   **49** 86.4 m
51 (a) *Hint:* Take the natural logarithm of both sides first.

    (b) Note that $f(e) = \dfrac{1}{e}$. Any horizontal line $y = k$,

    with $0 < k < \dfrac{1}{e}$, will intersect the graph at points

    $\left(x_1, \dfrac{\ln x_1}{x_1}\right)$ and $\left(x_2, \dfrac{\ln x_2}{x_2}\right)$, where $1 < x_1 < e$ and

    $x_2 > e$.

3

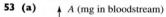

4

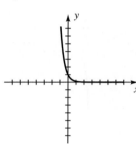

53 (a)

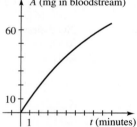

    (b) 6.58 min

5

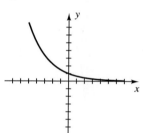

6

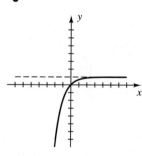

55 (a) $t = \dfrac{\log (F/F_0)}{\log (1 - m)}$ **(b)** After 13,863 generations

57 (a) 4.28 ft   **(b)** 24.8 yr   **59** $\dfrac{\ln (25/6)}{\ln (200/35)} \approx 0.82$

61 The suspicion is correct.
63 The suspicion is incorrect.   **65** 1.763

67

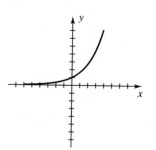

$[-5, 10]$ by $[-8, 2]$

$(-\infty, -0.32) \cup$
$(1.52, 6.84)$

7

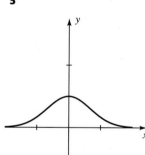

8

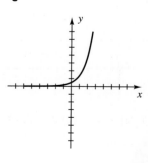

9

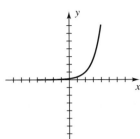

10

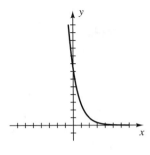

11

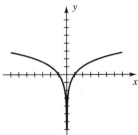

12

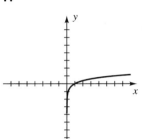

13

14

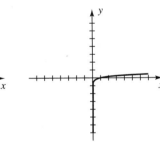

15

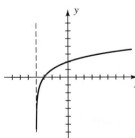

16

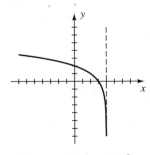

17 (a) -4 (b) 0 (c) 1 (d) 4 (e) 6 (f) 8
(g) $\dfrac{1}{2}$

18 (a) $\dfrac{1}{3}$ (b) 0 (c) 1 (d) 5 (e) 1 (f) 25
(g) $\dfrac{1}{3}$

19 0 **20** 9 **21** 9 **22** $\dfrac{33}{47}$ **23** 1 **24** 99

25 $5 - \dfrac{\log 6}{\log 2}$ **26** $\pm\sqrt{\dfrac{\log 7}{\log 3}}$ **27** $\dfrac{\log (3/8)}{\log (32/9)}$ **28** 1

29 $\dfrac{1}{4}$, 1, 4 **30** No solution   **31** $\sqrt{5}$ **32** 2

33 0, ±1 **34** $\ln 2$ **35** (a) -3, 2 (b) 2

36 (a) 8 (b) ±4 **37** $4\log x + \dfrac{2}{3}\log y - \dfrac{1}{3}\log z$

38 $-\log (xy^2)$ **39** $x = \log\left(\dfrac{1 \pm \sqrt{1 - 4y^2}}{2y}\right)$

40 If $y < 0$, then $x = \log\left(\dfrac{1 - \sqrt{1 + 4y^2}}{2y}\right)$. If $y > 0$, then
$x = \log\left(\dfrac{1 + \sqrt{1 + 4y^2}}{2y}\right)$.

41 (a) 1.89 (b) 78.3 (c) 0.472
42 (a) 0.924 (b) 0.00375 (c) 6.05
43 (a) 2000
(b) $2000(3^{1/6}) \approx 2401$; $2000(3^{1/2}) \approx 3464$; 6000
44 $1125.51

45 (a)

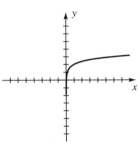

(b) 8 days

46 $N = 1000\left(\dfrac{3}{5}\right)^{t/3}$ **47** (a) After 11.39 yr   (b) 6.30 yr

48 $t = (\ln 100)\dfrac{L}{R} \approx 4.6\dfrac{L}{R}$

49 (a) $I = I_0 10^{\alpha/10}$
(b) Examine $I(\alpha + 1)$, where $I(\alpha)$ is the intensity corresponding to α decibels.

50 $t = -\dfrac{1}{k}\ln\left(\dfrac{a - L}{ab}\right)$ **51** $A = 10^{(R + 5.1)/2.3} - 3000$

52 $\dfrac{A_1}{A_2} = \dfrac{10^{(R + 5.1)/2.3} - 3000}{10^{(R + 7.5)/2.3} - 34{,}000}$ **53** $26{,}615.9$ mi^2

54 $h = \dfrac{\ln (29/p)}{0.000034}$ **55** $v = a\ln\left(\dfrac{m_1 + m_2}{m_1}\right)$

56 (a) $n = 10^{7.7 - 0.9R}$ (b) $12{,}589$; 1585; 200

57 (a) $E = 10^{11.4+1.5R}$　**(b)** 10^{24} ergs　**58** 110 days

59 86.8 cm; 9.715 cm/yr　**60** $t = -\dfrac{L}{R}\ln\left(\dfrac{V-RI}{V}\right)$

61 (a) 26,749 yr　**(b)** 30%　**62** 16.91 yr

CHAPTER 6

EXERCISES 6.1

1 $(3, 5), (-1, -3)$　**3** $(1, 0), (-3, 2)$　**5** $(0, 0), \left(\dfrac{1}{8}, \dfrac{1}{128}\right)$

7 $(3, -2)$　**9** No solution　**11** $(-4, 3), (5, 0)$

13 $(-2, 2)$

15 $\left(-\dfrac{3}{5} + \dfrac{1}{10}\sqrt{86}, \dfrac{1}{5} + \dfrac{3}{10}\sqrt{86}\right), \left(-\dfrac{3}{5} - \dfrac{1}{10}\sqrt{86}, \dfrac{1}{5} - \dfrac{3}{10}\sqrt{86}\right)$

17 $(-4, 0), \left(\dfrac{12}{5}, \dfrac{16}{5}\right)$　**19** $(0, 1), (4, -3)$

21 $(\pm 2, 5), (\pm\sqrt{5}, 4)$　**23** $(\sqrt{2}, \pm 2\sqrt{3}), (-\sqrt{2}, \pm 2\sqrt{3})$

25 $(2\sqrt{2}, \pm 2), (-2\sqrt{2}, \pm 2)$　**27** $(3, -1, 2)$

29 $(1, -1, 2), (-1, 3, -2)$

31 $(2, 2), (4, 1)$　**33** $(0, 4), \left(\dfrac{8}{3}, \dfrac{20}{3}\right)$

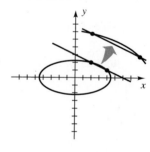

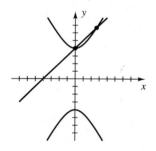

35 12 in. × 8 in.　　**37** Yes; a solution occurs between 0 and 1.

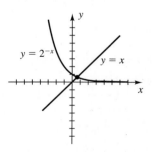

39 Yes; $r = 2$ in., $h = \dfrac{50}{\pi}$ in.

41 (a) $a = 120,000, b = 40,000$　**(b)** 77,143

43 $(0, 0), (0, 100), (50, 0)$; the fourth solution $(-100, 150)$ is not meaningful.

45 Yes; 1 ft × 1 ft × 2 ft or

$$\dfrac{\sqrt{13}-1}{2}\text{ ft} \times \dfrac{\sqrt{13}-1}{2}\text{ ft} \times \dfrac{8}{(\sqrt{13}-1)^2}\text{ ft}$$

$$\approx 1.30\text{ ft} \times 1.30\text{ ft} \times 1.18\text{ ft}$$

47 The points are on the parabola **(a)** $y = \dfrac{1}{2}x^2 - \dfrac{1}{2}$ and

(b) $y = \dfrac{1}{4}x^2 - 1.$

49 (a) $(31.25, -50)$

(b) $\left(-\dfrac{3}{2}\sqrt{11}, -\dfrac{1}{2}\right) \approx (-4.975, -0.5)$

51 $(0.6, 0.8), (-0.6, -0.8)$　**53** $(0.999, 0.968),$
$(0.251, 0.032)$

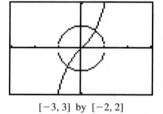

$[-3, 3]$ by $[-2, 2]$

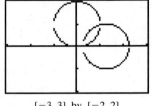

$[-3, 3]$ by $[-2, 2]$

EXERCISES 6.2

1 $(4, -2)$　**3** $(8, 0)$　**5** $\left(-1, \dfrac{3}{2}\right)$　**7** $\left(\dfrac{76}{53}, \dfrac{28}{53}\right)$

9 $\left(\dfrac{51}{13}, \dfrac{96}{13}\right)$　**11** $\left(\dfrac{8}{7}, -\dfrac{3}{7}\sqrt{6}\right)$　**13** No solution

15 All ordered pairs (m, n) such that $3m - 4n = 2$

17 $(0, 0)$　**19** $\left(-\dfrac{22}{7}, -\dfrac{11}{5}\right)$

21 313 students, 137 nonstudents

23 $x = \left(\dfrac{30}{\pi}\right) - 4 \approx 5.55$ cm, $y = 12 - \left(\dfrac{30}{\pi}\right) \approx 2.45$ cm

25 $l = 10$ ft, $w = \dfrac{20}{\pi}$ ft　**27** 2400 adults, 3600 kittens

29 40 g of 35% alloy, 60 g of 60% alloy

31 540 mi/hr, 60 mi/hr　**33** $v_0 = 10, a = 3$

35 20 sofas, 30 recliners

37 (a) $\left(c, \dfrac{4}{5}c\right)$ for an arbitrary $c > 0$　**(b)** $16 per hour

39 1928; 15.5 °C

41 (a) $Q = -100,000p + 2,000,000$
(b) $K = 150,000p - 250,000$　**(c)** $9.00

EXERCISES 6.3

1 $(2, 3, -1)$ **3** $(-2, 4, 5)$ **5** No solution

7 $\left(\dfrac{2}{3}, \dfrac{31}{21}, \dfrac{1}{21}\right)$

Exer. 9–16: There are other forms for the answers; c is any real number.

9 $(2c, -c, c)$ **11** $(0, -c, c)$

13 $\left(\dfrac{12}{7} - \dfrac{9}{7}c, \dfrac{4}{7}c - \dfrac{13}{14}, c\right)$ **15** $\left(\dfrac{7}{10}c + \dfrac{1}{2}, \dfrac{19}{10}c - \dfrac{3}{2}, c\right)$

17 $(1, 3, -1, 2)$ **19** $(2, -1, 3, 4, 1)$ **21** $\left(\dfrac{1}{11}, \dfrac{31}{11}, \dfrac{3}{11}\right)$

23 $(-2, -3)$ **25** No solution
27 17 of 10%, 11 of 30%, 22 of 50%
29 4 hr for A, 2 hr for B, 5 hr for C
31 380 lb of G_1, 60 lb of G_2, 160 lb of G_3

33 (a) $I_1 = 0, I_2 = 2, I_3 = 2$ (b) $I_1 = \dfrac{3}{4}, I_2 = 3, I_3 = \dfrac{9}{4}$

35 $\dfrac{3}{8}$ lb Columbian, $\dfrac{1}{8}$ lb Brazilian, $\dfrac{1}{2}$ lb Kenyan

37 (a) A: $x_1 + x_4 = 75$, B: $x_1 + x_2 = 150$,
    C: $x_2 + x_3 = 225$, D: $x_3 + x_4 = 150$
  (b) $x_1 = 25, x_2 = 125, x_4 = 50$
  (c) $x_3 = 150 - x_4 \le 150$;
    $x_3 = 225 - x_2 = 225 - (150 - x_1) = 75 + x_1 \ge 75$
39 2134   **41** $y = -x^2 + 2x + 5$
43 $x = y^2 - 3y + 1$ **45** $x^2 + y^2 - x + 3y - 6 = 0$
47 $a = -\dfrac{4}{9}, b = \dfrac{11}{9}, c = \dfrac{17}{18}, d = \dfrac{23}{18}$

EXERCISES 6.4

1 $\dfrac{3}{x-2} + \dfrac{5}{x+3}$ **3** $\dfrac{5}{x-6} - \dfrac{4}{x+2}$

5 $\dfrac{2}{x-1} + \dfrac{3}{x+2} - \dfrac{1}{x-3}$ **7** $\dfrac{3}{x} + \dfrac{2}{x-5} - \dfrac{1}{x+1}$

9 $\dfrac{2}{x-1} + \dfrac{5}{(x-1)^2}$ **11** $-\dfrac{7}{x} + \dfrac{5}{x^2} + \dfrac{40}{3x-5}$

13 $\dfrac{24/25}{x+2} + \dfrac{2/5}{(x+2)^2} - \dfrac{23/25}{2x-1}$ **15** $\dfrac{5}{x} - \dfrac{2}{x+1} + \dfrac{3}{(x+1)^3}$

17 $-\dfrac{2}{x-1} + \dfrac{3x+4}{x^2+1}$ **19** $\dfrac{4}{x} + \dfrac{5x-3}{x^2+2}$

21 $\dfrac{4x-1}{x^2+1} + \dfrac{3}{(x^2+1)^2}$ **23** $2x + \dfrac{1}{x-1} + \dfrac{3x}{x^2+1}$

25 $3 + \dfrac{4}{x} + \dfrac{8}{x-4}$ **27** $2x + 3 + \dfrac{2}{x-1} - \dfrac{3}{2x+1}$

EXERCISES 6.5

1
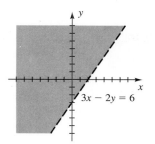

3
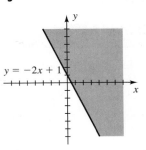

5
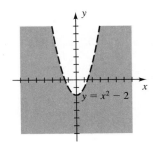

7
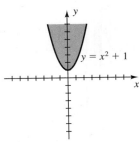

9
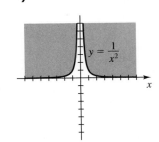

11
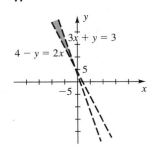

13
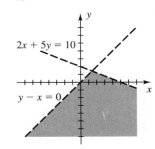

15
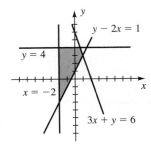

17

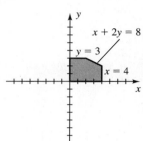

19

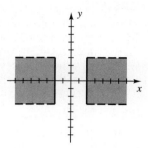

37 If x and y denote the amount placed in the high-risk and low-risk investment, respectively, then a system is $x \geq 2000$, $y \geq 3x$, $x + y \leq 15{,}000$.

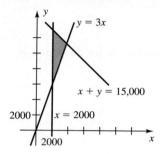

21

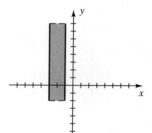

23

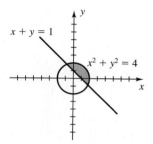

39 $x + y \leq 9$, $y \geq x$, $x \geq 1$

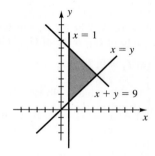

25

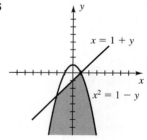

41 If the plant is located at (x, y), then a system is $(60)^2 \leq x^2 + y^2 \leq (100)^2$, $(60)^2 \leq (x - 100)^2 + y^2 \leq (100)^2$, $y \geq 0$.

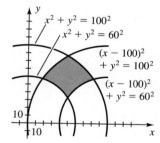

27 $0 \leq x < 3$, $y < -x + 4$, $y \geq x - 4$

29 $x^2 + y^2 \leq 9$, $y > -2x + 4$

31 $y < x$, $y \leq -x + 4$, $(x - 2)^2 + (y - 2)^2 \leq 8$

33 $y > \frac{1}{8}x + \frac{1}{2}$, $y \leq x + 4$, $y \leq -\frac{3}{4}x + 4$

35 If x and y denote the number of sets of brand A and brand B, respectively, then a system is $x \geq 20$, $y \geq 10$, $x \geq 2y$, $x + y \leq 100$.

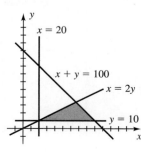

43

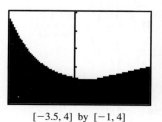

$[-3.5, 4]$ by $[-1, 4]$

45

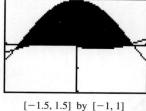

$[-1.5, 1.5]$ by $[-1, 1]$

47 There is no solution.

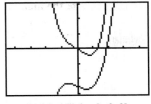

$[-4.5, 4.5]$ by $[-3, 3]$

EXERCISES 6.6

1 Maximum of 27 at $(6, 2)$; minimum of 9 at $(0, 2)$

3 Maximum of 21 at $(6, 3)$

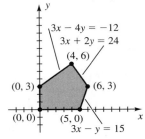

5 Minimum of 21 at $(3, 2)$

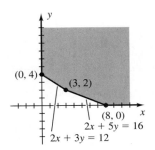

7 C has the maximum value 24 for any point on the line segment from $(2, 5)$ to $(6, 3)$.

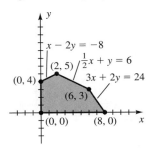

9 50 standard and 30 oversized

11 3.5 lb of S and 1 lb of T

13 Send 25 from W_1 to A and 0 from W_1 to B. Send 10 from W_2 to A and 60 from W_2 to B.

15 None of alfalfa and 80 acres of corn

17 Minimum cost: 16 oz X, 4 oz Y, 0 oz Z; maximum cost: 0 oz X, 8 oz Y, 12 oz Z

19 2 vans and 4 buses   **21** 3000 trout and 2000 bass

23 60 small units and 20 deluxe units

EXERCISES 6.7

1 $\begin{bmatrix} 9 & -1 \\ -2 & 5 \end{bmatrix}, \begin{bmatrix} 1 & -3 \\ 4 & 1 \end{bmatrix}, \begin{bmatrix} 10 & -4 \\ 2 & 6 \end{bmatrix}, \begin{bmatrix} -12 & -3 \\ 9 & -6 \end{bmatrix}$

3 $\begin{bmatrix} 9 & 0 \\ 1 & 5 \\ 3 & 4 \end{bmatrix}, \begin{bmatrix} 3 & -2 \\ 3 & -5 \\ -9 & 4 \end{bmatrix}, \begin{bmatrix} 12 & -2 \\ 4 & 0 \\ -6 & 8 \end{bmatrix}, \begin{bmatrix} -9 & -3 \\ 3 & -15 \\ -18 & 0 \end{bmatrix}$

5 $\begin{bmatrix} 11 & -3 & -3 \end{bmatrix}, \begin{bmatrix} -3 & -3 & 7 \end{bmatrix},$ $\begin{bmatrix} 8 & -6 & 4 \end{bmatrix}, \begin{bmatrix} -21 & 0 & 15 \end{bmatrix}$

7 Not possible, not possible, $\begin{bmatrix} 6 & -4 & 4 \\ 0 & 2 & -8 \\ -6 & 4 & -2 \end{bmatrix}, \begin{bmatrix} -12 & 0 \\ -6 & 3 \\ 3 & -9 \end{bmatrix}$

9 $\begin{bmatrix} 16 & 38 \\ 11 & -34 \end{bmatrix}, \begin{bmatrix} 4 & 38 \\ 23 & -22 \end{bmatrix}$

11 $\begin{bmatrix} 3 & -14 & -3 \\ 16 & 2 & -2 \\ -7 & -29 & 9 \end{bmatrix} \begin{bmatrix} 3 & -20 & -11 \\ 2 & 10 & -4 \\ 15 & -13 & 1 \end{bmatrix}$

13 $\begin{bmatrix} 4 & 8 \\ -18 & 11 \end{bmatrix}, \begin{bmatrix} 3 & -4 & 4 \\ -5 & 2 & 2 \\ -51 & 26 & 10 \end{bmatrix}$

15 $\begin{bmatrix} 1 & 2 & 3 \\ 4 & 5 & 6 \\ 7 & 8 & 9 \end{bmatrix}, \begin{bmatrix} 1 & 2 & 3 \\ 4 & 5 & 6 \\ 7 & 8 & 9 \end{bmatrix}$

17 $\begin{bmatrix} 15 \end{bmatrix}, \begin{bmatrix} -3 & 7 & 2 \\ -12 & 28 & 8 \\ 15 & -35 & -10 \end{bmatrix}$

19 $\begin{bmatrix} 2 & 0 & 5 \\ 5 & 3 & -2 \end{bmatrix}$, not possible   **21** $\begin{bmatrix} 4 \\ 12 \\ -1 \end{bmatrix}$

23 $\begin{bmatrix} 18 & 0 & -2 \\ -40 & 10 & -10 \end{bmatrix}$

EXERCISES 6.8

1 $\frac{1}{10}\begin{bmatrix} 3 & 4 \\ -1 & 2 \end{bmatrix}$ **3** Does not exist   **5** $\frac{1}{8}\begin{bmatrix} 2 & 1 & 0 \\ -2 & 3 & 0 \\ 0 & 0 & 2 \end{bmatrix}$

7 $\dfrac{1}{3}\begin{bmatrix} -4 & -5 & 3 \\ -4 & -8 & 3 \\ 1 & 2 & 0 \end{bmatrix}$ **9** $\begin{bmatrix} \dfrac{1}{2} & 0 & 0 \\ 0 & \dfrac{1}{4} & 0 \\ 0 & 0 & \dfrac{1}{6} \end{bmatrix}$

11 $ab \neq 0;\ \begin{bmatrix} \dfrac{1}{a} & 0 \\ 0 & \dfrac{1}{b} \end{bmatrix}$ **15 (a)** $\left(\dfrac{13}{10}, -\dfrac{1}{10}\right)$ **(b)** $\left(\dfrac{7}{5}, \dfrac{6}{5}\right)$

17 (a) $\left(-\dfrac{25}{3}, -\dfrac{34}{3}, \dfrac{7}{3}\right)$ **(b)** $\left(\dfrac{16}{3}, \dfrac{16}{3}, -\dfrac{1}{3}\right)$

19 (a) $\begin{bmatrix} 3.1 & 6.7 & -8.7 \\ 4.1 & -5.1 & 0.2 \\ 0.6 & 1.1 & -7.4 \end{bmatrix}\begin{bmatrix} x \\ y \\ z \end{bmatrix} = \begin{bmatrix} 1.5 \\ 2.1 \\ 3.9 \end{bmatrix}$

(b) $\begin{bmatrix} 0.1474 & 0.1572 & -0.1691 \\ 0.1197 & -0.0696 & -0.1426 \\ 0.0297 & 0.0024 & -0.1700 \end{bmatrix}$

(c) $x \approx -0.1081,\ y \approx -0.5227,\ z \approx -0.6135$

EXERCISES 6.9

1 $M_{11} = 0 = A_{11};\quad M_{12} = 5;\quad A_{12} = -5;$
$M_{21} = -1;\qquad A_{21} = 1;\quad M_{22} = 7 = A_{22}$
3 $M_{11} = -14 = A_{11};\quad M_{12} = 10;\quad A_{12} = -10;$
$M_{13} = 15 = A_{13};\quad M_{21} = 7;\quad A_{21} = -7;$
$M_{22} = -5 = A_{22};\quad M_{23} = 34;\quad A_{23} = -34;$
$M_{31} = 11 = A_{31};\quad M_{32} = 4;\quad A_{32} = -4;$
$M_{33} = 6 = A_{33}$
5 5 **7** -83 **9** 2 **11** 0
13 -125 **15** 48 **17** -216 **19** $abcd$
31 (a) $x^2 - 3x - 4$ **(b)** $-1, 4$
33 (a) $x^2 + x - 2$ **(b)** $-2, 1$
35 (a) $-x^3 - 2x^2 + x + 2$ **(b)** $-2, -1, 1$
37 (a) $-x^3 + 4x^2 + 4x - 16$ **(b)** $-2, 2, 4$
39 $-31i - 20j + 7k$ **41** $-6i - 8j + 18k$
43 (a) $-x^3 + x^2 + 6x - 7$ **(b)** $-2.51, 1.22, 2.29$

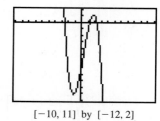

$[-10, 11]$ by $[-12, 2]$

EXERCISES 6.10

1 $R_2 \leftrightarrow R_3$ **3** $-R_1 + R_3 \to R_3$
5 2 is a common factor of R_1 and R_3
7 R_1 and R_3 are identical
9 -1 is a common factor of R_2
11 Every number in C_2 is 0   **13** $2C_1 + C_3 \to C_3$
15 -10 **17** -142 **19** -183 **21** 44 **23** 359
33 $(4, -2)$ **35** $(8, 0)$
37 $|D| = 0$ so Cramer's rule cannot be used.
39 $(2, 3, -1)$ **41** $(-2, 4, 5)$

CHAPTER 6 REVIEW EXERCISES

1 $\left(\dfrac{19}{23}, -\dfrac{18}{23}\right)$ **2** No solution   **3** $(-3, 5), (1, -3)$
4 $(4, -3), (3, -4)$ **5** $(2\sqrt{3}, \pm\sqrt{2}), (-2\sqrt{3}, \pm\sqrt{2})$
6 $(-1, \pm 1, -1), \left(0, \pm\dfrac{1}{2}\sqrt{6}, -\dfrac{1}{2}\right)$ **7** $\left(\dfrac{14}{17}, \dfrac{14}{27}\right)$
8 $\left(\log_2 \dfrac{25}{7}, \log_3 \dfrac{15}{7}\right)$ **9** $\left(\dfrac{6}{11}, -\dfrac{7}{11}, 1\right)$
10 $\left(-\dfrac{6}{29}, \dfrac{2}{29}, -\dfrac{17}{29}\right)$
11 $(-2c, -3c, c)$ for any real number c **12** $(0, 0, 0)$
13 $\left(5c - 1, -\dfrac{19}{2}c + \dfrac{5}{2}, c\right)$ for any real number c
14 $(5, -4)$ **15** $\left(-1, \dfrac{1}{2}, \dfrac{1}{3}\right)$ **16** $(3, -1, -2, 4)$
17 $\dfrac{8}{x-1} - \dfrac{3}{x+5} - \dfrac{1}{x+3}$ **18** $2 + \dfrac{3}{x+1} + \dfrac{4}{(x+1)^2}$
19 $-\dfrac{2}{x+5} + \dfrac{3x-1}{x^2+4}$ **20** $\dfrac{4}{x^2+2} + \dfrac{x-2}{x^2+5}$
21   **22**

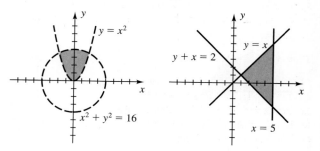

23

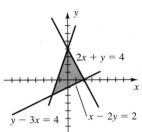

24

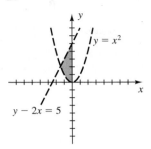

25 $\begin{bmatrix} 4 & -5 & 6 \\ 4 & -11 & 5 \end{bmatrix}$ **26** $\begin{bmatrix} 26 \\ -6 \end{bmatrix}$

27 $\begin{bmatrix} 0 & 4 & -6 \\ 16 & 22 & 1 \\ 12 & 11 & 9 \end{bmatrix}$ **28** $\begin{bmatrix} 0 & -37 \\ 15 & -6 \end{bmatrix}$

29 $\begin{bmatrix} -12 & 4 & -11 \\ 6 & -11 & 5 \end{bmatrix}$ **30** $\begin{bmatrix} a & 3a \\ 2a & 4a \end{bmatrix}$

31 $\begin{bmatrix} a & 3a \\ 2b & 4b \end{bmatrix}$ **32** $\begin{bmatrix} 0 & 0 \\ 0 & 0 \end{bmatrix}$

33 $\begin{bmatrix} 5 & 9 \\ 13 & 19 \end{bmatrix}$ **34** $\begin{bmatrix} 1 & 0 & 0 \\ 0 & 1 & 0 \\ 0 & 0 & 1 \end{bmatrix}$

35 $-\dfrac{1}{2}\begin{bmatrix} 2 & 4 \\ 3 & 5 \end{bmatrix}$ **36** $\dfrac{1}{11}\begin{bmatrix} 8 & 1 & -2 \\ 5 & 2 & -4 \\ -14 & 1 & 9 \end{bmatrix}$

37 $\begin{bmatrix} 1 & 0 & 0 \\ 0 & 2 & -7 \\ 0 & -1 & 4 \end{bmatrix}$ **38** $\dfrac{1}{37}\begin{bmatrix} -4 & -20 & 15 \\ 3 & 15 & -2 \\ 9 & 8 & -6 \end{bmatrix}$

39 $(2, -5)$ **40** $(-1, 3, 2)$ **41** -6 **42** 9

43 48 **44** -86 **45** -84 **46** 0 **47** 120

48 -76 **49** 0 **50** -50 **51** $-1 \pm 2\sqrt{3}$

52 $4, \pm\sqrt{7}$

53 2 is a common factor of R_1, 2 is a common factor of C_2, and 3 is a common factor of C_3.

54 Interchange R_1 with R_2 and then R_2 with R_3 to obtain the determinant on the right. The effect is to multiply by -1 twice.

55 $a_{11}a_{22}a_{33} \cdots a_{nn}$ **57** $\left(\dfrac{76}{53}, \dfrac{28}{53}\right)$ **58** $\left(\dfrac{2}{3}, \dfrac{31}{21}, \dfrac{1}{21}\right)$

59 $40\sqrt{5}$ ft $\times 20\sqrt{5}$ ft   **60** $y = \pm 2\sqrt{2}x + 3$

61 Tax = \$18,750; bonus = \$3,125

62 Inside radius = 90 ft, outside radius = 100 ft

63 In ft³/hr: A, 30; B, 20; C, 50

64 Western 95, eastern 55

65 If x and y denote the length and width, respectively, then a system is $x \le 12$, $y \le 8$, $y \ge \dfrac{1}{2}x$.

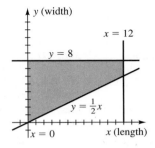

66 $x + y \le 18$, $x \ge 2y$, $x \ge 0$, $y \ge 0$

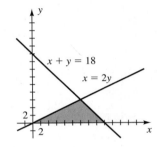

67 80 mowers and 30 edgers

68 High-risk \$50,000; low-risk \$100,000; bonds \$0

CHAPTER 7

EXERCISES 7.1

1 $9, 6, 3, 0; -12$ **3** $\dfrac{1}{2}, \dfrac{4}{5}, \dfrac{7}{10}, \dfrac{10}{17}; \dfrac{22}{65}$

5 $9, 9, 9, 9; 9$ **7** $1.9, 2.01, 1.999, 2.0001; 2.00000001$

9 $4, -\dfrac{9}{4}, \dfrac{5}{3}, -\dfrac{11}{8}; -\dfrac{15}{16}$ **11** $2, 0, 2, 0; 0$

13 $\dfrac{2}{3}, \dfrac{2}{3}, \dfrac{8}{11}, \dfrac{8}{9}; \dfrac{128}{33}$ **15** $1, 2, 3, 4; 8$

17 $2, 1, -2, -11, -38$ **19** $-3, 3^2, 3^4, 3^8, 3^{16}$

21 $5, 5, 10, 30, 120$ **23** $2, 2, 4, 4^3, 4^{12}$

25 $\dfrac{7}{2}, \dfrac{15}{2}, 12, 17$

27 $-1, -1 + \dfrac{1}{\sqrt{2}}, -1 + \dfrac{1}{\sqrt{2}} - \dfrac{1}{\sqrt{3}}, -\dfrac{1}{2} + \dfrac{1}{\sqrt{2}} - \dfrac{1}{\sqrt{3}}$

29 -5 **31** 10 **33** 25 **35** $-\dfrac{17}{15}$ **37** 61

39 $10,000$ **41** $\dfrac{319}{3}$

43 $\displaystyle\sum_{k=1}^{n} (a_k - b_k)$

$$= (a_1 - b_1) + (a_2 - b_2) + \cdots + (a_n - b_n)$$
$$= (a_1 + a_2 + \cdots + a_n) + (-b_1 - b_2 - \cdots - b_n)$$
$$= (a_1 + a_2 + \cdots + a_n) - (b_1 + b_2 + \cdots + b_n)$$
$$= \sum_{k=1}^{n} a_k - \sum_{k=1}^{n} b_k$$

45 As k increases, the terms approach 1.

47 $a_n = 2n + \dfrac{1}{24}(n-1)(n-2)(n-3)(n-4)(a-10)$

(The answer is not unique.)

49 (a) 1, 1, 2, 3, 5, 8, 13, 21, 34, 55

(b) 1, 2, 1.5, 1.$\overline{6}$, 1.6, 1.625, 1.6153846, 1.6190476, 1.6176471, 1.6181818

51 The graph has a horizontal asymptote $y \approx 2.718 \approx e$. Hence, a_n approaches e.

53 The graph has a horizontal asymptote $y \approx 1$. Hence, a_n approaches 1.

EXERCISES 7.2

1 Show that $a_{k+1} - a_k = 4$.   **3** 18; 38; $4n - 2$

5 1.8; 0.3; $3.3 - 0.3n$ **7** 5.4; 20.9; $(3.1)n - (10.1)$

9 $\ln 3^5$; $\ln 3^{10}$; $\ln 3^n$ **11** -8 **13** -8.5

15 -9.8 **17** $\dfrac{551}{17}$ **19** -105 **21** 30   **23** 530

25 $\dfrac{423}{2}$ **27** $\displaystyle\sum_{n=1}^{4}(2n-1)$ **29** $\displaystyle\sum_{n=1}^{37}(2n-1)$

31 $\displaystyle\sum_{n=1}^{6}\dfrac{3n}{4n+3}$ **33** 24   **35** $\dfrac{10}{3}, \dfrac{14}{3}, 6, \dfrac{22}{3}, \dfrac{26}{3}$

37 (a) 60   (b) 12,780   **39** 255   **41** 154π ft

43 $1200   **45** $16n^2$

47 Show that the $(n+1)$st term is 1 greater than the nth term.

49 (a) $\dfrac{8}{36}, \dfrac{7}{36}, \dfrac{6}{36}, \ldots, \dfrac{1}{36}$ (b) $d = -\dfrac{1}{36}$; 1   (c) $722.22

EXERCISES 7.3

1 Show that $\dfrac{a_{k+1}}{a_k} = -\dfrac{1}{4}$.   **3** $\dfrac{1}{2}; \dfrac{1}{16}; 8\left(\dfrac{1}{2}\right)^{n-1} = 2^{4-n}$

5 0.03; -0.00003; $300(-0.1)^{n-1}$ **7** 3125; 390,625; 5^n

9 20.25; -68.34375; $4(-1.5)^{n-1}$

11 x^8; $-x^{14}$; $(-1)^{n-1}x^{2n-2}$ **13** 2^{4x+1}; 2^{7x+1}; $2^{(n-1)x+1}$

15 $\pm\sqrt{3}$ **17** $\dfrac{243}{8}$ **19** $\sqrt[3]{3}$; 36   **21** 88,572

23 $-\dfrac{1023}{3072}$ **25** $\displaystyle\sum_{n=1}^{7} 2^n$ **27** $\displaystyle\sum_{n=1}^{4}(-1)^{n+1}\dfrac{1}{4}\left(\dfrac{1}{3}\right)^{n-1}$

29 $\dfrac{2}{3}$ **31** $\dfrac{50}{33}$

33 Since $|r| = \sqrt{2} > 1$, the sum does not exist.

35 1024   **37** $\dfrac{23}{99}$ **39** $\dfrac{2393}{990}$ **41** $\dfrac{5141}{999}$ **43** $\dfrac{16,123}{9999}$

45 24   **47** 4, 20, 100, 500   **49** $\dfrac{25}{256}\% \approx 0.1\%$

51 (a) $N(t) = 10,000(1.2)^t$ (b) 61,917   **53** 300 ft

55 $3,000,000   **57** (b) 375 mg

59 (a) $a_{k+1} = \dfrac{1}{4}\sqrt{10}\,a_k$

(b) $a_n = \left(\dfrac{1}{4}\sqrt{10}\right)^{n-1} a_1$, $A_n = \left(\dfrac{5}{8}\right)^{n-1} A_1$,

$P_n = \left(\dfrac{1}{4}\sqrt{10}\right)^{n-1} P_1$

(c) $\dfrac{16a_1}{4 - \sqrt{10}}$

61 $38,929.00   **63** $7396.67

65 (a) $\dfrac{2}{5}, \dfrac{6}{25}, \dfrac{18}{125}, \dfrac{54}{625}, \dfrac{162}{3125}$

(b) $r = \dfrac{3}{5}$; $\dfrac{2882}{3125} = 0.92224$ (c) $16,000

EXERCISES 7.4

Exer. 1–32: A typical proof is given for Exercise 1, 5, 9, ..., 29.

1 (1) P_1 is true, since $2(1) = 1(1 + 1) = 2$.

(2) Assume P_k is true:

$2 + 4 + 6 + \cdots + 2k = k(k + 1)$. Hence,

$2 + 4 + 6 + \cdots + 2k + 2(k + 1)$

$$= k(k + 1) + 2(k + 1)$$
$$= (k + 1)(k + 2)$$
$$= (k + 1)(k + 1 + 1).$$

Thus, P_{k+1} is true, and the proof is complete.

5 (1) P_1 is true, since $5(1) - 3 = \dfrac{1}{2}(1)[5(1) - 1] = 2$.

(2) Assume P_k is true:

$2 + 7 + 12 + \cdots + (5k - 3) = \dfrac{1}{2}k(5k - 1)$. Hence,

$2 + 7 + 12 + \cdots + (5k - 3) + 5(k + 1) - 3$

$$= \dfrac{1}{2}k(5k - 1) + 5(k + 1) - 3$$
$$= \dfrac{5}{2}k^2 + \dfrac{9}{2}k + 2$$
$$= \dfrac{1}{2}(5k^2 + 9k + 4)$$
$$= \dfrac{1}{2}(k + 1)(5k + 4)$$
$$= \dfrac{1}{2}(k + 1)[5(k + 1) - 1].$$

Thus, P_{k+1} is true, and the proof is complete.

9 (1) P_1 is true, since $(1)^1 = \dfrac{1(1+1)[2(1)+1]}{6} = 1$.

(2) Assume P_k is true:

$$1^2 + 2^2 + 3^2 + \cdots + k^2 = \frac{k(k+1)(2k+1)}{6}. \text{ Hence,}$$

$$1^2 + 2^2 + 3^2 + \cdots + k^2 + (k+1)^2$$
$$= \frac{k(k+1)(2k+1)}{6} + (k+1)^2$$
$$= (k+1)\left[\frac{k(2k+1)}{6} + \frac{6(k+1)}{6}\right]$$
$$= \frac{(k+1)(2k^2 + 7k + 6)}{6}$$
$$= \frac{(k+1)(k+2)(2k+3)}{6}.$$

Thus, P_{k+1} is true, and the proof is complete.

13 (1) P_1 is true, since $3^1 = \dfrac{3}{2}(3^1 - 1) = 3$.

(2) Assume P_k is true:

$$3 + 3^2 + 3^3 + \cdots + 3^k = \frac{3}{2}(3^k - 1). \text{ Hence,}$$

$$3 + 3^2 + 3^3 + \cdots + 3^k + 3^{k+1} = \frac{3}{2}(3^k - 1) + 3^{k+1}$$
$$= \frac{3}{2} \cdot 3^k - \frac{3}{2} + 3 \cdot 3^k$$
$$= \frac{9}{2} \cdot 3^k - \frac{3}{2}$$
$$= \frac{3}{2}(3 \cdot 3^k - 1)$$
$$= \frac{3}{2}(3^{k+1} - 1).$$

Thus, P_{k+1} is true, and the proof is complete.

17 (1) P_1 is true, since $1 < \dfrac{1}{8}[2(1)+1]^2 = \dfrac{9}{8}$.

(2) Assume P_k is true:

$$1 + 2 + 3 + \cdots + k < \frac{1}{8}(2k+1)^2. \text{ Hence,}$$

$$1 + 2 + 3 + \cdots + k + (k+1) < \frac{1}{8}(2k+1)^2 + (k+1)$$
$$= \frac{1}{2}k^2 + \frac{3}{2}k + \frac{9}{8}$$
$$= \frac{1}{8}(4k^2 + 12k + 9)$$
$$= \frac{1}{8}(2k+3)^2$$
$$= \frac{1}{8}[2(k+1)+1]^2.$$

Thus, P_{k+1} is true, and the proof is complete.

21 (1) For $n = 1$, $5^n - 1 = 4$ and 4 is a factor of 4.

(2) Assume 4 is a factor of $5^k - 1$. The $(k+1)$st term is
$$5^{k+1} - 1 = 5 \cdot 5^k - 1$$
$$= 5 \cdot 5^k - 5 + 4$$
$$= 5(5^k - 1) + 4.$$
By the induction hypothesis, 4 is a factor of $5^k - 1$ and 4 is a factor of 4, so 4 is a factor of the $(k+1)$st term. Thus, P_{k+1} is true, and the proof is complete.

25 (1) For $n = 1$, $a - b$ is a factor of $a^1 - b^1$.

(2) Assume $a - b$ is a factor of $a^k - b^k$. Following the hint for the $(k+1)$st term,
$$a^{k+1} - b^{k+1} = a^k \cdot a - b \cdot a^k + b \cdot a^k - b^k \cdot b$$
$$= a^k(a - b) + (a^k - b^k)b.$$
Since $(a - b)$ is a factor of $a^k(a - b)$ and since by the induction hypothesis $a - b$ is a factor of $(a^k - b^k)$, it follows that $a - b$ is a factor of the $(k+1)$st term. Thus, P_{k+1} is true, and the proof is complete.

29 (1) P_8 is true, since $5 + \log_2 8 \le 8$.

(2) Assume P_k is true: $5 + \log_2 k \le k$. Hence,
$$5 + \log_2 (k+1) < 5 + \log_2 (k + k)$$
$$= 5 + \log_2 2k$$
$$= 5 + \log_2 2 + \log_2 k$$
$$= (5 + \log_2 k) + 1$$
$$\le k + 1.$$

Thus, P_{k+1} is true, and the proof is complete.

33 $\dfrac{n^3 + 6n^2 + 20n}{3}$　　**35** $\dfrac{4n^3 - 12n^2 + 11n}{3}$

37 (a) $a + b + c = 1$, $8a + 4b + 2c = 5$,
$$27a + 9b + 3c = 14; \ a = \frac{1}{3}, \ b = \frac{1}{2}, \ c = \frac{1}{6}$$

(b) The method used in part (a) shows that the formula is true for only $n = 1, 2, 3$.

39 (a) $\dfrac{1}{5}n^5 + \dfrac{1}{2}n^4 + \dfrac{1}{3}n^3 - \dfrac{1}{30}n$

(b) Use mathematical induction.

EXERCISES 7.5

1 1440　　**3** 5040　　**5** 336　　**7** 1　　**9** 21

11 715　　**13** $(2n+2)(2n+1)$

15 $64x^3 - 48x^2y + 12xy^2 - y^3$

17 $a^6 + 6a^5b + 15a^4b^2 + 20a^3b^3 + 15a^2b^4 + 6ab^5 + b^6$

19 $a^7 - 7a^6b + 21a^5b^2 - 35a^4b^3 + 35a^3b^4 - 21a^2b^5$
$$+ 7ab^6 - b^7$$

21 $81x^4 - 540x^3y + 1350x^2y^2 - 1500xy^3 + 625y^4$

23 $\dfrac{1}{243}x^5 + \dfrac{5}{81}x^4y^2 + \dfrac{10}{27}x^3y^4 + \dfrac{10}{9}x^2y^6 + \dfrac{5}{3}xy^8 + y^{10}$

25 $x^{-12} + 18x^{-9} + 135x^{-6} + 540x^{-3} + 1215$
$$+ 1458x^3 + 729x^6$$

27 $x^{5/2} - 5x^{3/2} + 10x^{1/2} - 10x^{-1/2} + 5x^{-3/2} - x^{-5/2}$

29 $3^{25}c^{10} + 25 \cdot 3^{24}c^{52/5} + 300 \cdot 3^{23}c^{54/5}$

31 $-1680 \cdot 3^{13}b^{11} + 60 \cdot 3^{14}b^{13} - 3^{15}b^{15}$ **33** $\dfrac{189}{1024}c^8$

35 $\dfrac{114{,}688}{9}u^2v^6$ **37** $70x^2y^2$ **39** $448y^3x^{10}$

41 $-216b^9a^2$ **43** $-\dfrac{135}{16}$ **45** 4.8, 6.19

47 $4x^3 + 6x^2h + 4xh^2 + h^3$

49 $\dbinom{n}{1} = \dfrac{n!}{(n-1)!1!} = n$ and

$\dbinom{n}{n-1} = \dfrac{n!}{[n-(n-1)]!(n-1)!}$

$\qquad\qquad = \dfrac{n!}{1!(n-1)!} = n$

EXERCISES 7.6

1 210   **3** 60,480   **5** 120   **7** 6
9 (a) 60   **(b)** 125   **11** 64   **13** $P(8, 3) = 336$
15 24   **17 (a)** 2,340,000   **(b)** 2,160,000
19 (a) 151,200   **(b)** 5760   **21** 1024
23 $P(8, 8) = 40{,}320$ **25** $P(6, 3) = 120$
27 (a) 27,600   **(b)** 35,152   **29** 9,000,000
31 $P(4, 4) = 24$ **33** $3! \cdot 2^3 = 48$
35 $P(10, 10) = 3{,}628{,}800$
37 (a) 900
     (b) If n is even, $9 \cdot 10^{(n/2)-1}$; if n is odd, $9 \cdot 10^{(n-1)/2}$

39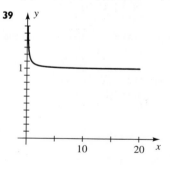

     (a) $y = 1$

     (b) $n! \approx \dfrac{n^n\sqrt{2\pi n}}{e^n}$

EXERCISES 7.7

1 35   **3** 9   **5** n **7** 1   **9** $\dfrac{12!}{5!3!2!2!} = 166{,}320$

11 $\dfrac{10!}{3!2!2!1!1!1!} = 151{,}200$ **13** $C(10, 5) = 252$

15 $C(8, 2) = 28$ **17** $(5! \cdot 4! \cdot 8!) \cdot 3! = 696{,}729{,}600$
19 $3 \cdot C(10, 2) \cdot C(8, 2) \cdot C(4, 2) \cdot C(6, 2) \cdot 3 \cdot 4 = 4{,}082{,}400$
21 $C(12, 3) \cdot C(8, 2) = 6160$ **23** $C(8, 3) = 56$
25 (a) $C(49, 6) = 13{,}983{,}816$ **(b)** $C(24, 6) = 134{,}596$

27 $C(n, 2) = 45$ and hence $n = 10$ **29** $C(6, 3) = 20$
31 By finding $C(31, 3) = 4495$
33 (a) 1, 2, 4, 8, 16, 32, 64, 128, 256, 512
     (b) $S_n = 2^{n-1}$

EXERCISES 7.8

1 (a) A representative outcome is (nine of clubs, 3); 312

     (b) 20; 292; $\dfrac{20}{312}$ **(c)** No; yes; $\dfrac{72}{312}; \dfrac{156}{312}; \dfrac{36}{312}; \dfrac{192}{312}$

     (d) Yes; no; 0; $\dfrac{92}{312}$

3 (a) $\dfrac{4}{52}$ **(b)** $\dfrac{8}{52}$ **(c)** $\dfrac{12}{52}$ **5 (a)** $\dfrac{1}{6}$ **(b)** $\dfrac{1}{6}$ **(c)** $\dfrac{2}{6}$

7 (a) $\dfrac{5}{15}$ **(b)** $\dfrac{6}{15}$ **(c)** $\dfrac{9}{15}$

9 (a) $\dfrac{2}{36}$ **(b)** $\dfrac{5}{36}$ **(c)** $\dfrac{7}{36}$ **11** $\dfrac{6}{216}$ **13** $\dfrac{3}{8}$

15 $\dfrac{48 \cdot 13}{C(52, 5)} \approx 0.00024$ **17** $\dfrac{C(13, 4) \cdot C(13, 1)}{C(52, 5)} \approx 0.00358$

19 $\dfrac{C(13, 5) \cdot 4}{C(52, 5)} \approx 0.00198$ **21** $\dfrac{4}{6}$ **23** $(0.674)^4 \approx 0.2064$

25 (a) 0.45   **(b)** 0.10   **(c)** 0.70   **(d)** 0.95

27 (a) $\dfrac{C(20, 5) \cdot C(40, 0)}{C(60, 5)} \approx 0.0028$

     (b) $1 - \dfrac{C(30, 0) \cdot C(30, 5)}{C(60, 5)} \approx 0.9739$

     (c) $\dfrac{C(10, 0) \cdot C(50, 5)}{C(60, 5)} + \dfrac{C(10, 1) \cdot C(50, 4)}{C(60, 5)} \approx 0.8096$

29 (a) $\dfrac{C(8, 8)}{2^8} \approx 0.00391$ **(b)** $\dfrac{C(8, 7)}{2^8} = 0.03125$

     (c) $\dfrac{C(8, 6)}{2^8} = 0.109375$

     (d) $\dfrac{C(8, 6) + C(8, 7) + C(8, 8)}{2^8} \approx 0.14453$

31 $1 - \dfrac{C(48, 5)}{C(52, 5)} \approx 0.34116$ **33** $1 - \dfrac{10}{36} = \dfrac{26}{36}$

35 (a) $\dfrac{1}{32}$ **(b)** $1 - \dfrac{1}{32} = \dfrac{31}{32}$

37 (a) $\dfrac{C(4, 4)}{4!} = \dfrac{1}{24}$ **(b)** $\dfrac{C(4, 2)}{4!} = \dfrac{1}{4}$

39 (a) 0   **(b)** $\dfrac{1}{9}$ **41 (a)** $\dfrac{1}{16}$ **(b)** $\dfrac{C(4, 2)}{2^4} = \dfrac{6}{16}$

43 $\dfrac{2}{25{,}827{,}165}$ (about 1 chance in 13 million)

45 (a) $\dfrac{8}{36}$ **(b)** $\dfrac{1}{36}$ **(c)** $\dfrac{244}{495} \approx 0.4929$

47 (a) 0.9639   **(b)** 0.95   **49 (b)** 0.76

CHAPTER 7 REVIEW EXERCISES

1 $5, -2, -1, -\dfrac{20}{29}; -\dfrac{7}{19}$

2 $0.9, -1.01, 0.999, -1.0001; 0.9999999$

3 $2, \dfrac{1}{2}, \dfrac{5}{4}, \dfrac{7}{8}; \dfrac{65}{64}$ **4** $\dfrac{1}{12}, \dfrac{1}{15}, \dfrac{1}{15}, \dfrac{8}{105}; \dfrac{8}{45}$

5 $10, \dfrac{11}{10}, \dfrac{21}{11}, \dfrac{32}{21}, \dfrac{53}{32}$ **6** $2, 2, 2, 2, 2$

7 $9, 3, \sqrt{3}, \sqrt[4]{3}, \sqrt[8]{3}$ **8** $1, \dfrac{1}{2}, \dfrac{2}{3}, \dfrac{3}{5}, \dfrac{5}{8}$

9 75 **10** $-\dfrac{37}{10}$ **11** 940 **12** -10 **13** $\displaystyle\sum_{n=1}^{5} 3n$

14 $\displaystyle\sum_{n=1}^{6} 2^{3-n}$ **15** $\displaystyle\sum_{n=1}^{99} \dfrac{1}{n(n+1)}$ **16** $\displaystyle\sum_{n=1}^{98} \dfrac{1}{n(n+1)(n+2)}$

17 $\displaystyle\sum_{n=1}^{4} \dfrac{n}{3n-1}$ **18** $\displaystyle\sum_{n=1}^{4} \dfrac{n}{5n-1}$

19 $\displaystyle\sum_{n=1}^{5} (-1)^{n+1}(105-5n)$ **20** $\displaystyle\sum_{n=1}^{7} (-1)^{n-1}\dfrac{1}{n}$

21 $\displaystyle\sum_{n=0}^{25} a_n x^{4n}$ **22** $\displaystyle\sum_{n=0}^{20} a_n x^{3n}$ **23** $1 + \displaystyle\sum_{k=1}^{n} (-1)^k \dfrac{x^{2k}}{2k}$

24 $1 + \displaystyle\sum_{k=1}^{n} \dfrac{x^k}{k}$ **25** $-5 - 8\sqrt{3}; -5 - 35\sqrt{3}$ **26** 52

27 $-31; 50$ **28** $20, 14, 8, 2, -4, -10$

29 64 **30** -0.00003 **31** $4\sqrt{2}$ **32** $-\dfrac{12{,}800}{2187}$

33 $17; 3$ **34** $\dfrac{1}{81}; \dfrac{211}{1296}$ **35** 570 **36** 32.5

37 2041 **38** -506 **39** $\dfrac{5}{7}$ **40** $\dfrac{6268}{999}$

41 (1) P_1 is true, since $3(1) - 1 = \dfrac{1[3(1)+1]}{2} = 2$.

(2) Assume P_k is true:
$$2 + 5 + 8 + \cdots + (3k-1) = \dfrac{k(3k+1)}{2}. \text{ Hence,}$$
$$2 + 5 + 8 + \cdots + (3k-1) + 3(k+1) - 1$$
$$= \dfrac{k(3k+1)}{2} + 3(k+1) - 1$$
$$= \dfrac{3k^2 + k + 6k + 4}{2}$$
$$= \dfrac{3k^2 + 7k + 4}{2}$$
$$= \dfrac{(k+1)(3k+4)}{2}$$
$$= \dfrac{(k+1)[3(k+1)+1]}{2}.$$

Thus, P_{k+1} is true, and the proof is complete.

42 (1) P_1 is true, since $[2(1)]^2 = \dfrac{[2(1)][2(1)+1][1+1]}{3} = 4$.

(2) Assume P_k is true:
$$2^2 + 4^2 + 6^2 + \cdots + (2k)^2 = \dfrac{(2k)(2k+1)(k+1)}{3}.$$
Hence,
$$2^2 + 4^2 + 6^2 + \cdots + (2k)^2 + [2(k+1)]^2$$
$$= \dfrac{(2k)(2k+1)(k+1)}{3} + [2(k+1)]^2$$
$$= (k+1)\left(\dfrac{4k^2 + 2k}{3} + \dfrac{12(k+1)}{3}\right)$$
$$= \dfrac{(k+1)(4k^2 + 14k + 12)}{3}$$
$$= \dfrac{2(k+1)(2k+3)(k+2)}{3}.$$

Thus, P_{k+1} is true, and the proof is complete.

43 (1) P_1 is true, since $\dfrac{1}{[2(1)-1][2(1)+1]} = \dfrac{1}{2(1)+1} = \dfrac{1}{3}$.

(2) Assume P_k is true:
$$\dfrac{1}{1 \cdot 3} + \dfrac{1}{3 \cdot 5} + \dfrac{1}{5 \cdot 7} + \cdots + \dfrac{1}{(2k-1)(2k+1)} = \dfrac{k}{2k+1}.$$
Hence,
$$\dfrac{1}{1 \cdot 3} + \dfrac{1}{3 \cdot 5} + \dfrac{1}{5 \cdot 7} + \cdots + \dfrac{1}{(2k-1)(2k+1)}$$
$$+ \dfrac{1}{(2k+1)(2k+3)} = \dfrac{k}{2k+1} + \dfrac{1}{(2k+1)(2k+3)}$$
$$= \dfrac{k(2k+3) + 1}{(2k+1)(2k+3)}$$
$$= \dfrac{2k^2 + 3k + 1}{(2k+1)(2k+3)}$$
$$= \dfrac{(2k+1)(k+1)}{(2k+1)(2k+3)}$$
$$= \dfrac{k+1}{2(k+1)+1}.$$

Thus, P_{k+1} is true, and the proof is complete.

44 (1) P_1 is true, since $1(1+1) = \dfrac{(1)(1+1)(1+2)}{3} = 2$.

(2) Assume P_k is true:
$$1 \cdot 2 + 2 \cdot 3 + 3 \cdot 4 + \cdots + k(k+1)$$
$$= \dfrac{k(k+1)(k+2)}{3}.$$

Hence,
$$1 \cdot 2 + 2 \cdot 3 + 3 \cdot 4 + \cdots + k(k+1) + (k+1)(k+2)$$
$$= \dfrac{k(k+1)(k+2)}{3} + (k+1)(k+2)$$
$$= (k+1)(k+2)\left(\dfrac{k}{3} + 1\right)$$
$$= \dfrac{(k+1)(k+2)(k+3)}{3}.$$

Thus, P_{k+1} is true, and the proof is complete.

45 (1) For $n = 1$, $n^3 + 2n = 3$ and 3 is a factor of 3.

(2) Assume 3 is a factor of $k^3 + 2k$. The $(k + 1)$st term is
$$(k + 1)^3 + 2(k + 1) = k^3 + 3k^2 + 5k + 3$$
$$= (k^3 + 2k) + (3k^2 + 3k + 3)$$
$$= (k^3 + 2k) + 3(k^2 + k + 1).$$
By the induction hypothesis, 3 is a factor of $k^3 + 2k$ and 3 is a factor of $3(k^2 + k + 1)$, so 3 is a factor of the $(k + 1)$st term. Thus, P_{k+1} is true, and the proof is complete.

46 (1) P_5 is true, since $5^2 + 3 < 2^5$.
(2) Assume P_k is true: $k^2 + 3 < 2^k$. Hence,
$$(k + 1)^2 + 3 = k^2 + 2k + 4$$
$$= (k^2 + 3) + (k + 1)$$
$$< 2^k + (k + 1)$$
$$< 2^k + 2^k$$
$$= 2 \cdot 2^k = 2^{k+1}$$
Thus, P_{k+1} is true, and the proof is complete.

47 (1) P_4 is true, since $2^4 \le 4!$.
(2) Assume P_k is true: $2^k \le k!$. Hence,
$$2^{k+1} = 2 \cdot 2^k \le 2 \cdot k! < (k + 1) \cdot k! = (k + 1)!.$$
Thus, P_{k+1} is true, and the proof is complete.

48 (1) P_{10} is true, since $10^{10} \le 10^{10}$.
(2) Assume P_k is true: $10^k \le k^k$. Hence,
$$10^{k+1} = 10 \cdot 10^k \le 10 \cdot k^k < (k + 1) \cdot k^k$$
$$< (k + 1) \cdot (k + 1)^k = (k + 1)^{k+1}.$$
Thus, P_{k+1} is true, and the proof is complete.

49 $x^{12} - 18x^{10}y + 135x^8y^2 - 540x^6y^3 + 1215x^4y^4 \\ \qquad\qquad - 1458x^2y^5 + 729y^6$

50 $16a^4 + 32a^3b^3 + 24a^2b^6 + 8ab^9 + b^{12}$

51 $a^8 + 40a^7 + 760a^6$ **52** $-\dfrac{63}{16}b^{12}c^{10}$

53 $21,504a^{10}b^2$ **54** $52,500,000$

55 (a) $d = 1 - \dfrac{1}{2}a_1$ (b) In ft: $1\dfrac{1}{4}, 2, 2\dfrac{3}{4}, 3\dfrac{1}{2}$

56 24 ft   **57** $\dfrac{2}{1 - f}$ **58** $\dfrac{7}{4}\sqrt{10} + \sqrt{30} \approx 11.01$

59 (a) $P(52, 13) \approx 3.954 \times 10^{21}$
     (b) $P(13, 5) \cdot P(13, 3) \cdot P(13, 3) \cdot P(13, 2) \approx 7.094 \times 10^{13}$

60 (a) $P(6, 4) = 360$ (b) $6^4 = 1296$

61 (a) $C(12, 8) = 495$ (b) $C(9, 5) = 126$

62 $\dfrac{17!}{6!5!4!2!} = 85,765,680$ **63** (a) $\dfrac{2}{4}$ (b) $\dfrac{2}{8}$

64 (a) $\dfrac{P(26, 4) \cdot 2}{P(52, 4)} \approx 0.1104$ (b) $\dfrac{26^2 \cdot 25^2}{P(52, 4)} \approx 0.0650$

65 (a) $\dfrac{1}{1000}$ (b) $\dfrac{10}{1000}$ (c) $\dfrac{50}{1000}$ **66** $\dfrac{C(4, 1)}{2^4} = \dfrac{4}{16}$

67 (a) $\dfrac{C(6, 4) + C(6, 5) + C(6, 6)}{2^6} = \dfrac{22}{64}$ (b) $1 - \dfrac{22}{64} = \dfrac{42}{64}$

68 (a) $\dfrac{1}{312}$ (b) $\dfrac{57}{312}$ **69** 0.44 **70** $\dfrac{8}{36}$

INDEX

QUICK REFERENCE CARD

THE SWOKOWSKI · COLE SERIES

◆ ALGEBRA AND TRIGONOMETRY WITH ANALYTIC GEOMETRY, Eighth Edition

▲ FUNDAMENTALS OF TRIGONOMETRY, Eighth Edition

■ FUNDAMENTALS OF COLLEGE ALGEBRA, Eighth Edition

● FUNDAMENTALS OF ALGEBRA AND TRIGONOMETRY, Eighth Edition

PWS-KENT Publishing Company

FORMULAS FROM TRIGONOMETRY

FUNDAMENTAL IDENTITIES

$$\csc t = \frac{1}{\sin t}$$

$$\sec t = \frac{1}{\cos t}$$

$$\cot t = \frac{1}{\tan t}$$

$$\tan t = \frac{\sin t}{\cos t}$$

$$\cot t = \frac{\cos t}{\sin t}$$

$$\sin^2 t + \cos^2 t = 1$$

$$1 + \tan^2 t = \sec^2 t$$

$$1 + \cot^2 t = \csc^2 t$$

FORMULAS FOR NEGATIVES

$$\sin(-t) = -\sin t$$

$$\cos(-t) = \cos t$$

$$\tan(-t) = -\tan t$$

$$\cot(-t) = -\cot t$$

$$\sec(-t) = \sec t$$

$$\csc(-t) = -\csc t$$

ADDITION FORMULAS

$$\sin(u + v) = \sin u \cos v + \cos u \sin v$$

$$\cos(u + v) = \cos u \cos v - \sin u \sin v$$

$$\tan(u + v) = \frac{\tan u + \tan v}{1 - \tan u \tan v}$$

SUBTRACTION FORMULAS

$$\sin(u - v) = \sin u \cos v - \cos u \sin v$$

$$\cos(u - v) = \cos u \cos v + \sin u \sin v$$

$$\tan(u - v) = \frac{\tan u - \tan v}{1 + \tan u \tan v}$$

HALF-ANGLE FORMULAS

$$\sin \frac{u}{2} = \pm\sqrt{\frac{1 - \cos u}{2}}$$

$$\cos \frac{u}{2} = \pm\sqrt{\frac{1 + \cos u}{2}}$$

$$\tan \frac{u}{2} = \frac{1 - \cos u}{\sin u} = \frac{\sin u}{1 + \cos u}$$

DOUBLE-ANGLE FORMULAS

$$\sin 2u = 2 \sin u \cos u$$

$$\cos 2u = \cos^2 u - \sin^2 u$$
$$= 1 - 2\sin^2 u$$
$$= 2\cos^2 u - 1$$

$$\tan 2u = \frac{2 \tan u}{1 - \tan^2 u}$$

FORMULAS FROM TRIGONOMETRY

TRIGONOMETRIC FUNCTIONS

OF REAL NUMBERS

$$\sin t = y \qquad \csc t = \frac{1}{y}$$

$$\cos t = x \qquad \sec t = \frac{1}{x}$$

$$\tan t = \frac{y}{x} \qquad \cot t = \frac{x}{y}$$

OF ACUTE ANGLES

$$\sin \theta = \frac{\text{opp}}{\text{hyp}} \qquad \csc \theta = \frac{\text{hyp}}{\text{opp}}$$

$$\cos \theta = \frac{\text{adj}}{\text{hyp}} \qquad \sec \theta = \frac{\text{hyp}}{\text{adj}}$$

$$\tan \theta = \frac{\text{opp}}{\text{adj}} \qquad \cot \theta = \frac{\text{adj}}{\text{opp}}$$

OBLIQUE TRIANGLE

LAW OF SINES

$$\frac{\sin \alpha}{a} = \frac{\sin \beta}{b} = \frac{\sin \gamma}{c}$$

LAW OF COSINES

$$a^2 = b^2 + c^2 - 2bc \cos \alpha$$
$$b^2 = a^2 + c^2 - 2ac \cos \beta$$
$$c^2 = a^2 + b^2 - 2ab \cos \gamma$$

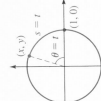

$y = \sin t, \quad 0 \le t \le 2\pi$

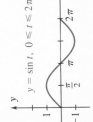

$y = \cos t, \quad 0 \le t \le 2\pi$

$y = \tan t$

area A circumference (or perimeter) C
volume V curved surface area S
altitude h radius r

RIGHT TRIANGLE

Pythagorean theorem:
$$c^2 = a^2 + b^2$$

TRIANGLE

$$A = \frac{1}{2}bh \qquad C = a + b + c$$

CIRCLE

$$A = \pi r^2 \qquad C = 2\pi r$$

SPHERE

$$V = \frac{4}{3}\pi r^3 \qquad S = 4\pi r^2$$

RIGHT CIRCULAR CYLINDER

$$V = \pi r^2 h \qquad S = 2\pi r h$$

RIGHT CIRCULAR CONE

$$V = \frac{1}{3}\pi r^2 h \qquad S = \pi r \sqrt{r^2 + h^2}$$

QUADRATIC FORMULA

If $a \neq 0$, the roots of $ax^2 + bx + c = 0$ are

$$x = \frac{-b \pm \sqrt{b^2 - 4ac}}{2a}$$

SPECIAL FACTORING FORMULAS

$$x^2 - y^2 = (x + y)(x - y)$$
$$x^2 + 2xy + y^2 = (x + y)^2$$
$$x^2 - 2xy + y^2 = (x - y)^2$$
$$x^3 - y^3 = (x - y)(x^2 + xy + y^2)$$
$$x^3 + y^3 = (x + y)(x^2 - xy + y^2)$$

EXPONENTIALS AND LOGARITHMS

$$y = \log_a x \quad \text{means} \quad a^y = x$$
$$\log_a xy = \log_a x + \log_a y$$
$$\log_a \frac{x}{y} = \log_a x - \log_a y$$
$$\log_a x^r = r \log_a x$$
$$a^{\log_a x} = x$$
$$\log_a a^x = x$$
$$\log_a 1 = 0$$
$$\log_a a = 1$$
$$\log x = \log_{10} x$$
$$\ln x = \log_e x$$
$$\log_b u = \frac{\log_a u}{\log_a b}$$

EXPONENTS AND RADICALS

$$a^m a^n = a^{m+n} \qquad a^{m/n} = \sqrt[n]{a^m} \qquad a^{1/n} = \sqrt[n]{a}$$
$$(a^m)^n = a^{mn} \qquad a^{m/n} = \sqrt[n]{a^m} = (\sqrt[n]{a})^m$$
$$(ab)^n = a^n b^n \qquad \sqrt[n]{ab} = \sqrt[n]{a}\,\sqrt[n]{b}$$
$$\left(\frac{a}{b}\right)^n = \frac{a^n}{b^n} \qquad \sqrt[n]{\frac{a}{b}} = \frac{\sqrt[n]{a}}{\sqrt[n]{b}}$$
$$\frac{a^m}{a^n} = a^{m-n} \qquad \sqrt[m]{\sqrt[n]{a}} = \sqrt[mn]{a}$$
$$a^{-n} = \frac{1}{a^n}$$

POINT-SLOPE FORM OF A LINE

$$y - y_1 = m(x - x_1)$$
m is the slope

SLOPE-INTERCEPT FORM OF A LINE

$$y = mx + b \quad m \text{ is the slope}$$

CIRCLE

$$(x - h)^2 + (y - k)^2 = r^2$$

PARABOLA

$$x^2 = 4py$$

ELLIPSE

$$\frac{x^2}{a^2} + \frac{y^2}{b^2} = 1 \quad \text{with} \quad a^2 = b^2 + c^2$$

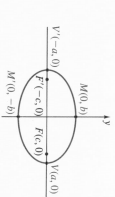

HYPERBOLA

$$\frac{x^2}{a^2} - \frac{y^2}{b^2} = 1 \quad \text{with} \quad c^2 = a^2 + b^2$$

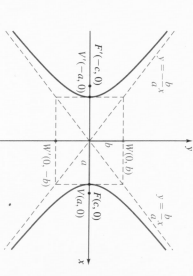

ANALYTIC GEOMETRY

DISTANCE FORMULA

$$d(P_1, P_2) = \sqrt{(x_2 - x_1)^2 + (y_2 - y_1)^2}$$

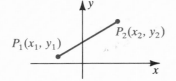

SLOPE m OF A LINE

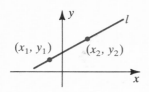

$$m = \frac{y_2 - y_1}{x_2 - x_1}$$

POINT-SLOPE FORM OF A LINE

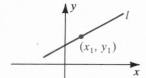

$$y - y_1 = m(x - x_1)$$

SLOPE-INTERCEPT FORM OF A LINE

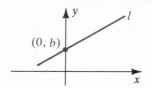

$$y = mx + b$$

INTERCEPT FORM OF A LINE

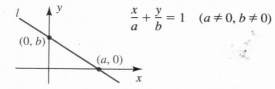

$$\frac{x}{a} + \frac{y}{b} = 1 \quad (a \neq 0, b \neq 0)$$

EQUATION OF A CIRCLE

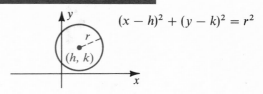

$$(x - h)^2 + (y - k)^2 = r^2$$

GRAPH OF A QUADRATIC FUNCTION

$y = ax^2, a > 0$

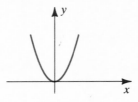

$y = ax^2 + bx + c, a > 0$

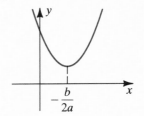

CONSTANTS

$\pi \approx 3.14159$

$e \approx 2.71828$

CONVERSIONS

1 centimeter $\approx$ 0.3937 inch

1 meter $\approx$ 3.2808 feet

1 kilometer $\approx$ 0.6214 mile

1 gram $\approx$ 0.0353 ounce

1 kilogram $\approx$ 2.2046 pounds

1 liter $\approx$ 0.2642 gallon

1 milliliter $\approx$ 0.0381 fluid ounce

1 joule $\approx$ 0.7376 foot-pound

1 newton $\approx$ 0.2248 pound

1 lumen $\approx$ 0.0015 watt

1 acre = 43,560 square feet